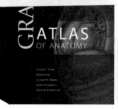

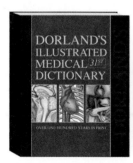

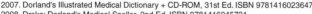

Richard L. Drake, PhD
Director of Anatomy
Professor of Surgery
Cleveland Clinic Lerner College of Medicine
Cleveland, Ohio, USA

A. Wayne Vogl, PhD
Professor of Anatomy and Cell Biology
Department of Cellular and Physiological Sciences
Faculty of Medicine
University of British Columbia
Vancouver, British Columbia, Canada

Adam W. M. Mitchell, MBBS, FRCS, FRCR
Joint Head of Graduate Entry Anatomy
Imperial College
University of London
Consultant Radiologist
Department of Imaging
Charing Cross Hospital
London, UK

DORLAND'S

GRAY'S

Pocket Atlas of Anatomy

DORLAND'S
GRAY'S

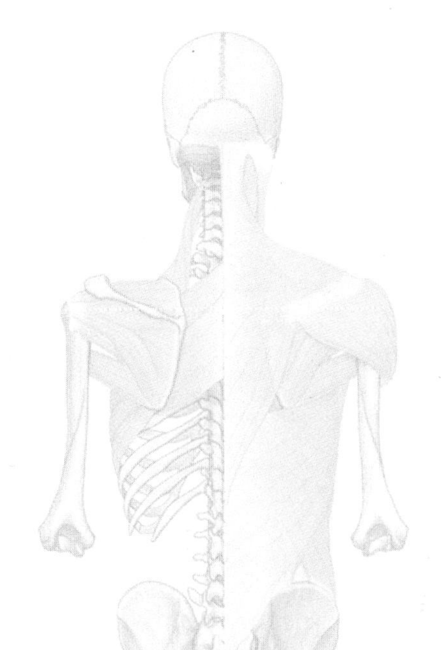

Pocket Atlas of Anatomy

Richard L. Drake
A. Wayne Vogl
Adam W. M. Mitchell

With illustrations from
Gray's Anatomy for Students
Sobotta Atlas of Human Anatomy, 14th edition
Gray's Anatomy, 39th edition
Gray's Atlas of Anatomy

Definitions based on
Dorland's Illustrated Medical Dictionary, 31st edition

CHURCHILL
LIVINGSTONE
ELSEVIER

1600 John F. Kennedy Blvd.
Ste 1800
Philadelphia, PA 19103-2899

DORLAND'S/GRAY'S POCKET ATLAS OF ANATOMY ISBN: 978-0-443-06761-7

Notice

Neither the Publisher nor the Editors assume any responsibility for any loss or injury and/or damage to persons or property arising out of or related to any use of the material contained in this book. It is the responsibility of the treating practitioner, relying on independent expertise and knowledge of the patient, to determine the best treatment and method of application for the patient.

The Publisher

Library of Congress Cataloging-in-Publication Data
Drake, Richard L., Ph.D.
Dorland's / Gray's pocket atlas of anatomy / Richard L. Drake, A. Wayne Vogl, Adam W. M. Mitchell. — 1st ed.
p. ; cm.
ISBN 978–0–443–06761–7
1. Anatomy—Atlases. 2. Anatomy—Dictionaries. 3. Anatomy—Handbooks, manuals, etc. I. Vogl, Wayne. II. Mitchell, Adam W. M. III. Dorland, W. A. Newman (William Alexander Newman), 1864–1956. Dorland's illustrated medical dictionary. IV. Title. V. Title: Pocket atlas of anatomy.
[DNLM: 1. Anatomy—Atlases. 2. Anatomy—Dictionary—English. 3. Anatomy —Handbooks. QS 39 D762d 2008]

QM25.D73 2009
611.0022'2—dc22 2007031028

Acquisitions Editor: William Schmitt
Developmental Editor: Christine Abshire
Publishing Services Manager: Linda Van Pelt
Design Direction: Steve Stave

Printed in Canada
Last digit is the print number: 9 8 7 6 5 4 3 2 1

To my wife, Cheryl, for her continued support.
—R. L. D.

To my family, colleagues and students.
—A. W. V.

To Cathy, Elsa and Max.
—A. W. M. M.

Preface

This dictionary/atlas is designed to be used as a companion to the other educational materials in the *Gray's Anatomy for Students* family and as a quick reference both for the definitions of anatomical structures and for the locations of these structures in the body. The book consists of short definitions of anatomical structures, extracted from *Dorland's Medical Dictionary*, linked to full-color illustrations that indicate the positions of these structures in the body. The terms used for anatomical structures are from *Terminologia Anatomica* (Thieme, 1998). Illustrations used in the dictionary/atlas are from *Gray's Anatomy for Students*, *Gray's Atlas of Anatomy*, *Gray's Anatomy*, *Sobotta Atlas of Human Anatomy*, and *Dorland's Illustrated Medical Dictionary*. We organized the book regionally because this is how we anticipate that users will generally search for terms and locations. Within each region, terms and structures are grouped according to which sub-region or organ in which they are found. For example, within Head and Neck, terms and structures associated with the larynx are grouped together. This organization allows the best utilization of diagrams and therefore enables the maximum number of terms to be included with the least number of pages.

Acknowledgments

First, we would like to thank the individuals at Elsevier who have participated in all aspects of the development of *Dorland's/Gray's Pocket Atlas of Anatomy*. Included in this group are Bill Schmitt, Doug Anderson, Patti Novak, Michelle Elliott, Christine Abshire, Rebecca Gruliow, Linda Van Pelt, Pamela Poll, and Karen Giacomucci. Without their help and guidance this project would not have been possible.

We would also like to thank Richard Tibbitts and Paul Richardson, whose artworks form the basis of this, as well as the other educational materials in the *Gray's Anatomy for Students* family.

RICHARD L. DRAKE, PHD

A. WAYNE VOGL, PHD

ADAM W. M. MITCHELL, MBBS, FRCS, FRCR

Figure Credits

All figures not listed below are from Drake R, Vogl AW, and Mitchell AWM: *Gray's Anatomy for Students.* Philadelphia: Churchill Livingstone, 2005.

Dorland's Illustrated Medical Dictionary, 30th ed. Philadelphia: Saunders, 2003.
A, page 3; B, page 5; C, page 5; D, page 5; B, page 7; A, page 9; B, page 9; D, page 9; D, page 11; E, page 11; A, page 17; A, page 19; B, page 19; A, page 23; B, page 25; A, page 27; B, page 27; A, page 29; B, page 29; C, page 29; D, page 29; A, page 31; B, page 31; A, page 33; C, page 35; B, page 157; B, page 169; C, page 193; F, page 327; C, page 425; D, page 487; E, page 487; A, page 489; B, page 489

Dorland's Illustrated Medical Dictionary, 31st ed. Philadelphia: Saunders, 2007.
B, page 9; B, page 35; A, page 37; B, page 87; B, page 137

Drake RA, et al: Gray's Atlas of Anatomy. Philadelphia: Churchill Livingstone, 2008.
B, page 367; B, page 407

Putz R, Pabst R (eds.): Sobotta Atlas of Human Anatomy, 14th ed., Volumes 1 and 2. Munich: Urban & Fischer, 2006.
C, page 11; B, page 15; F, page 41; G, page 41; B, page 43; C, page 43; B, page 45; C, page 45; D, page 45; A, page 51; E, page 53; E, page 75; D, page 97; C, page 119; D, page 125; E, page 125; A, page 127; A, page 159; A, page 173; B, page 173; C, page 173; D, page 173; E, page 173; E, page 181; A, page 195; B, page 195; B, page 199; A, page 201; E, page 205; A, page 207; B, page 207; C, page 207; E, page 209; A, page 221; B, page 221; C, page 221; C, page 223; A, page 225; B, page 225; C, page 225; B, page 229; C, page 229; A, page 243; C, page 243; D, page 243; G, page 243; D, page 247; E, page 247; E, page 249; G, page 249; A, page 251; B, page 251; C, page 253; D, page 253; E, page 253; F, page 253; B, page 255; C, page 257; C, page 259; F, page 261; E, page 263; B, page 265; F, page 267; G, page 267; A, page 269; B, page 269; F, page 269; C, page 271; F, page 275; F, page 271; E, page 273; F, page 273; B, page 275; A, page 277; F, page 277; C, page 281; C, page 283; B, page 285; A, page 289; A, page 291; B, page 297; D, page 301; A, page 303; B, page 303; C, page 303; D, page 317; C, page 321; E, page 325; D, page 327; A, page 329; D, page 333; C, page 335; A, page 335; B, page 335; D, page 335; A, page 337; B, page 341; B, page 353; B, page 365; E, page 365; B, page 369; A, page 371; F, page 373; A, page 387; A, page 447; B, page 447; C, page 447; D, page 447; F, page 461; D, page 497

Standring S, et al: Gray's Anatomy, 39th ed. Philadelphia: Churchill Livingstone, 2005.
A, page 13; A, page 15; B, page 23; A, page 43; A, page 55; B, page 55; B, page 71; C, page 71; C, page 73; D, page 73; A, page 75; B, page 75; B, page 129; C, page 129; D, page 129; E, page 133; B, page 133; D, page 133; A, page 135; C, page 143; A, page 157; C, page 169; C, page 195; A, page 229; B, page 237; B, page 239; A, page 259; E, page 261; D, page 281; A, page 319; A, page 323; A, page 331; B, page 331; C, page 387

A, page 223: ***From Eardley I, Sethia K: Erectile Dysfunction. London: Mosby, 1998.***

Contents

GENERAL

ANATOMY

1 ***aditus*** the entrance or approach to an organ or part.

2 ***agger*** an eminence or projection.

3 ***ala*** a winglike structure or process; called also *wing.*

4 **alveolus** a small saclike structure, especially in the jaws or lungs. A

5 ***ampulla*** a flasklike dilatation of a tubular structure.

6 **angle** a triangular area or the angle of a particular structure or part of the body. <angulus> B

7 ***ansa*** a loop or looplike structure.

8 ***antrum*** a cavity or chamber, such as one within a bone or organ.

9 ***anulus*** ring: a circular or ringlike structure; written also *annulus.*

10 ***aperture*** an opening in the body. <apertura>

11 ***apex*** the superior aspect of a body, organ, or part, or the pointed extremity of a conical structure such as the heart or lung; called also *tip.*

12 ***apical*** pertaining to or located at an apex. <apicalis>

13 ***apophysis*** any outgrowth or swelling, especially a bony outgrowth that has never been entirely separated from the bone of which it forms a part, such as a process, tubercle, or tuberosity.

14 ***appendix*** a supplementary, accessory, or dependent part attached to a main structure; called also *appendage.* Frequently used alone to refer to the *vermiform appendix* of the colon.

15 ***aqueduct*** a passage or channel in a body structure or organ, especially a channel for the conduction of fluid. <aqueductus>

16 ***arch*** any structure having a curved or bowlike outline. <arcus>

17 ***area*** a specific surface or a region with a given function.

18 ***atrium*** a chamber; used to designate a chamber affording entrance to another structure or organ. Usually used alone to designate an atrium of the heart.

19 ***axial*** a general term denoting relationship to an axis or location near the long axis or central part of the body. <axialis>

20 ***basal*** denoting relationship to or location near a base. <basalis>

21 ***base*** the lowest or fundamental part of a structure or organ, or the part opposite to or distinguished from the apex. <basis>

22 ***basilar*** a general term denoting relationship to a base or location at a base. <basilaris>

23 ***bifurcation*** the site where a single structure divides into two, as in blood vessels or teeth. <bifurcatio>

24 ***bladder*** a membranous sac or receptacle, especially the urinary bladder or gallbladder. <vesica>

25 ***branch*** a smaller structure given off by a larger one, or into which the larger structure, such as a blood vessel or nerve, divides. <ramus>

26 ***bulb*** a rounded mass or enlargement. <bulbus>

27 ***bursa*** a sac or saclike cavity filled with a viscid fluid and situated at places in the tissues at which friction would otherwise develop.

28 ***calyx*** a cup-shaped organ or cavity; also spelled *calix.*

29 ***canal*** a relatively narrow tubular passage or channel. <canalis>

30 ***canaliculus*** an extremely narrow tubular passage or channel; any of various small channels.

31 ***capsule*** a cartilaginous, fatty, fibrous, or membranous structure enveloping another structure, organ, or part. <capsula>

32 ***cardiovascular system*** the heart and blood vessels, by which blood is pumped and circulated through the body. <systema cardiovasculare>

33 ***cartilage*** a mass of specialized fibrous connective tissue at a particular site in the body. <cartilago>

34 ***caruncle*** a small fleshy eminence, which may be normal or abnormal. <caruncula>

35 ***cauda*** a tail or taillike appendage; a structure resembling such an appendage.

36 ***cave*** a type of cavity. <cavum>

37 ***cavern*** a type of cavity. <caverna>

38 ***cavity*** a hollow space or depression within the body. Called also *cavum.* <cavitas>

39 ***cellula*** a small, more or less enclosed space.

40 ***center*** the middle point of a body. <centrum>

41 ***central*** situated at or pertaining to a center; not peripheral. <centralis>

42 ***cervix*** a constricted portion of a body part or organ.

43 ***chamber*** an enclosed space or ventricle. <camera>

44 ***chiasm*** a decussation or X-shaped crossing, such as of nerves. <chiasma>

45 ***chorda*** any cord or sinew.

46 ***circle*** a ringlike arrangement, usually of arteries or veins. <circulus>

47 ***cistern*** a closed space serving as a reservoir for lymph or other body fluid, especially one of the enlarged subarachnoid spaces containing cerebrospinal fluid. <cisterna>

48 ***column*** a pillarlike structure or part. <columna>

49 ***compartment*** a small enclosure within a larger space. <compartimentum>

50 ***connexus*** a connecting structure; written also *conexus.*

51 ***conus*** cone: a structure resembling a cone in shape.

A

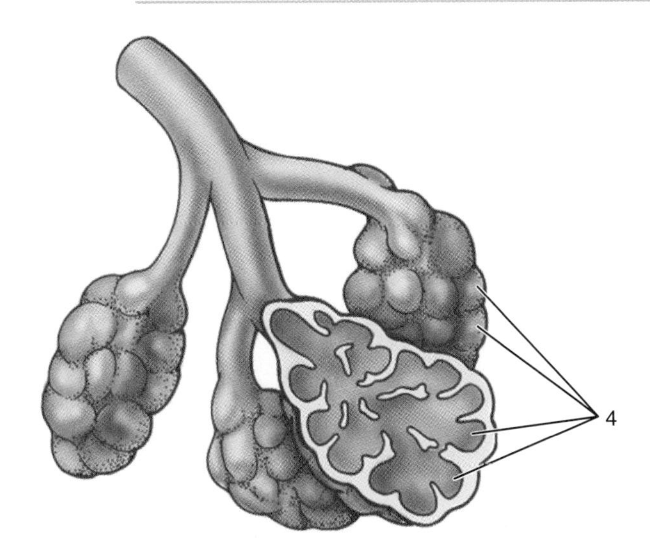

B

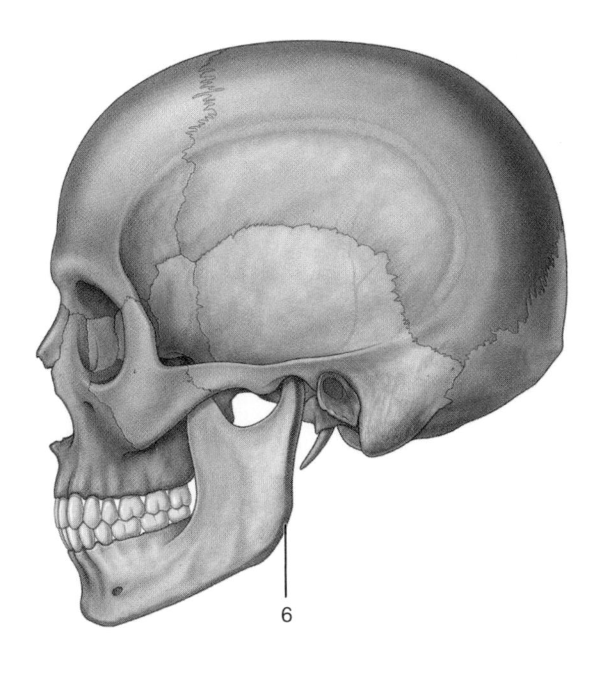

1 **cornu** a structure resembling a horn in shape. A

2 *coronal* denoting something situated in the direction of the coronal suture. <coronalis>

3 **cortex** the outer layer of an organ or other body structure, as distinguished from the internal substance. D

4 *crista* a projection or projecting structure, or ridge, especially one surmounting a bone or its border; called also *crest* and *ridge.*

5 *crypt* a blind pit or tube opening on a free surface. <crypta>

6 *curvature* a nonangular deviation from a straight course in a line or surface. <curvatura>

7 *decussation* the intercrossing of fellow parts or structures in the form of an X. <decussatio>

8 *deep* a structure situated farther than another from the surface of the body. <profundus>

9 *digestive system* the organs associated with the ingestion and digestion of food, including the mouth and associated structures, pharynx, and components of the digestive tube, as well as the associated organs and glands. <systema digestorium>

10 *digit* a finger or a toe. <digitus>

11 **duct** a passage with well-defined walls, especially such a channel for the passage of excretions or secretions. <ductus> B

12 *ductule* a minute duct; applied especially to branches of ducts nearest to the alveoli of a gland, or the smallest beginnings of the duct system of an organ. <ductulus>

13 *eminence* a prominence or projection, especially one on the surface of a bone. <eminentia>

14 **endocrine glands** ductless organs that secrete specific substances (hormones) that are released directly into the circulatory system and influence metabolism and other body processes. The endocrine glands include the hypothalamus, pituitary, thyroid, parathyroid, and adrenal glands, the pancreatic islets, the pineal body, and the gonads. <glandulae endocrinae> C

15 *excavation* a hollowed-out space, pouch, or cavity. <excavatio>

16 *external* a structure or an aspect farther from the center of a part or cavity.

17 *extremity* the distal or terminal portion of an elongated or pointed structure. <extremitas>

18 *fascia* a sheet or band of fibrous tissue such as lies deep to the skin or forms an investment for muscles and various other organs of the body.

19 *fasciculus* a small bundle of nerve, muscle, or tendon fibers.

20 **female genital system** the internal and external reproductive organs in the female. <systema genitale femininum> D

21 *fiber* an elongated, threadlike structure. <fibra>

22 *filum* a threadlike structure or part.

23 *fimbria* a fringe, border, or edge.

24 *fissure* a cleft or groove, especially a deep fold in the cerebral cortex that involves its entire thickness. <fissura>

25 *flexure* a bending, such as a bent portion of a structure or organ. <flexura>

26 *fold* a ridge or fold, as of peritoneum or other membrane. <plica>

27 *folium* a leaflike structure, especially one of the leaflike subdivisions of the cerebellar cortex.

28 *follicle* a very small excretory or secretory sac or gland. <folliculus>

A

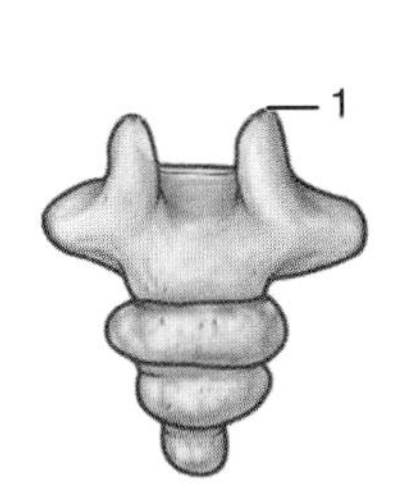

B

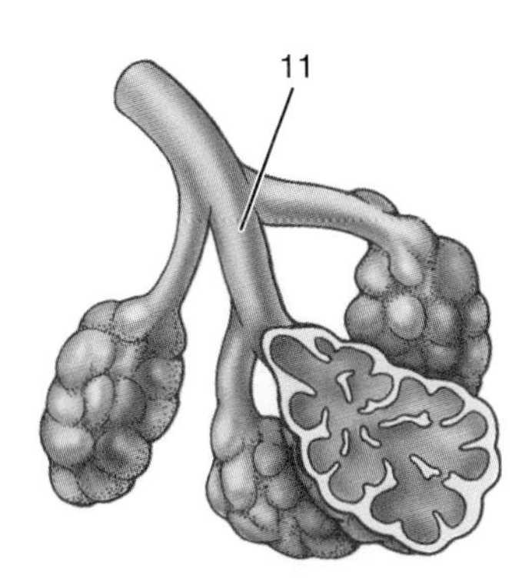

C

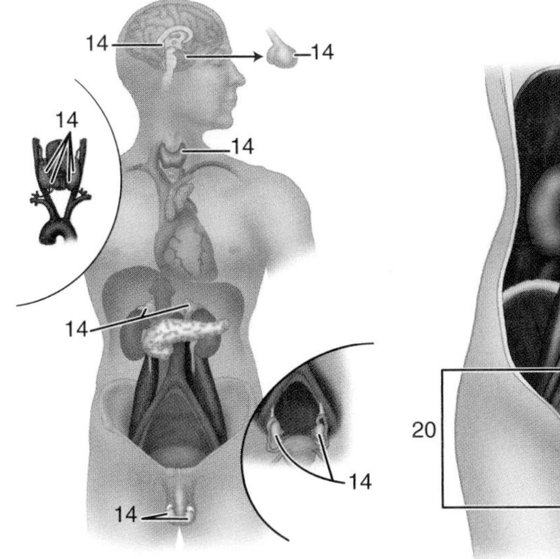

D

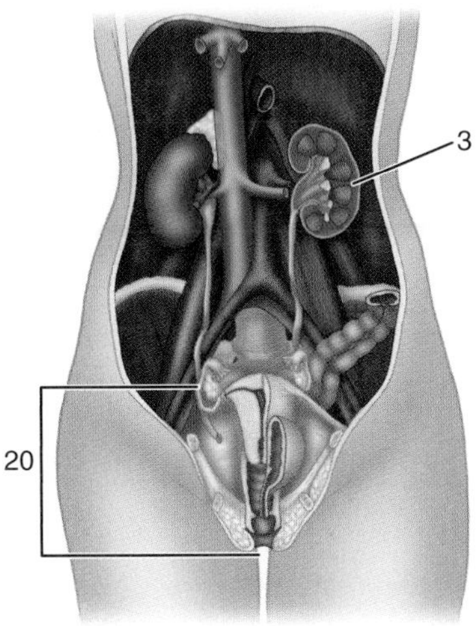

1 **foramen** a natural opening or passage, especially one into or through a bone. A

2 ***forceps*** an organ or part shaped like the surgical instrument, particularly the terminal fibers of the corpus callosum.

3 ***formation*** a structure of definite shape. <formatio>

4 ***fossa*** a trench, channel, or hollow place.

5 ***fossula*** a small fossa; a slight depression in the surface of a structure or organ.

6 ***fovea*** pit or depression: a small pit in the surface of a structure or organ.

7 ***foveola*** a small pit or extremely small depression.

8 ***frenulum*** a small fold of integument or mucous membrane that checks, curbs, or limits the movements of an organ or part.

9 ***fundus*** the bottom or base of an organ, or the part of a hollow organ farthest from its mouth.

10 ***funiculus*** cord: a cordlike structure or part.

11 ***galea*** a helmetlike structure or part.

12 ***geniculum*** a sharp, kneelike bend in a small structure or organ, such as a nerve.

13 ***girdle*** an encircling structure or part; anything that encircles a body. <cingulum>

14 ***gland*** an aggregation of cells, specialized to secrete or excrete materials not related to their ordinary metabolic needs. <glandula>

15 ***glans*** a small rounded mass or glandlike body.

16 ***globus*** a spherical structure.

17 ***hamulus*** hook: a long, thin, curved structure.

18 **haustrum** a recess or sacculation. B

19 ***hemisphere*** half of a spherical or spheroid structure. <hemispherium>

20 ***hiatus*** a gap, cleft, or opening.

21 ***hilum*** a depression or pit at the part of an organ where vessels and nerves enter.

22 ***impression*** an indentation or concavity, especially one produced in the surface of one organ by pressure exerted by another. <impressio>

23 ***incisure*** an indentation or depression, usually on the edge of a bone or other structure. Called also *notch*. <incisura>

24 ***insertion*** a place of attachment, as of a muscle to the bone it moves.

25 ***integument*** 1. a covering or investment. <integumentum> 2. the covering of the body, or skin,including its various layers and their appendages; in humans, it comprises the epidermis, dermis, subcutaneous tissue, hair, nails, cutaneous glands, the breast, and mammary glands. <integumentum commune>

26 ***intermediate*** a term denoting the middle of three structures, one of which is situated closer to and the other farther from the median plane of the body or part. <intermedius>

27 ***internal*** something situated nearer to the center of an organ or a cavity. <internus>

28 ***intersection*** a cutting across, or between. <intersectio>

29 ***intumescentia*** an enlargement or swelling.

30 ***investing layer*** a layer of fascia that closely invests a muscle or ligament. <fascia investiens>

31 ***isthmus*** a narrow connection between two larger bodies or parts.

32 ***labyrinth*** a system of intercommunicating cavities or canals. <labyrinthus>

33 ***lacuna*** a small pit or hollow cavity, such as one within or between other body structures.

34 ***lamina*** layer: a thin flat plate or stratum of a composite structure. The term is often used alone to mean the lamina of the vertebral arch.

35 ***ligament*** 1.a band of tissue that connects bones or supports viscera. 2. a double layer of peritoneum extending from one visceral organ to another. 3. cordlike remnants of fetal tubular structures that are nonfunctional after birth. <ligamentum>

36 ***limen*** the beginning point, boundary, or threshold of a structure.

37 ***linea*** line: a streak or narrow ridge on the surface of some structure.

38 ***lingula*** a small tonguelike structure.

39 ***lobule*** a small lobe or one of the primary divisions of a lobe. <lobulus>

40 ***locus*** a site in the body.

41 ***longitudinal*** lengthwise; parallel to the long axis of the body or an organ. <longitudinalis>

42 ***longus*** long; a long structure, as a muscle.

43 ***luminal*** pertaining to the lumen of a tubular structure. <luminalis>

44 ***macula*** a stain, spot, or thickening, especially one distinguishable from its surroundings because of color or some other characteristic.

45 **male genital system** the internal and external reproductive organs in the male. <systema genitale masculinum> B

46 ***malleolus*** a rounded process, such as the protuberance on either side of the ankle joint.

47 ***manubrium*** a handlelike structure or part; often used alone to designate the manubrium of the sternum.

48 ***margin*** a boundary or edge, such as the border of a structure. <margo>

49 **meatus** a passageway in the body, especially one opening on the surface. A

50 ***medius*** in the middle; used in reference to a structure lying between two other structures that are anterior and posterior, superior and inferior, or internal and external in position.

51 ***medulla*** the most interior portion of an organ or structure.

52 ***membrane*** a thin layer of tissue covering a surface, lining a cavity, or dividing a space or organ. <membrana>

53 ***meniscus*** a crescent-shaped structure of the body. Often used alone to designate one of the crescent-shaped disks of fibrocartilage attached to the superior articular surface of the tibia.

A

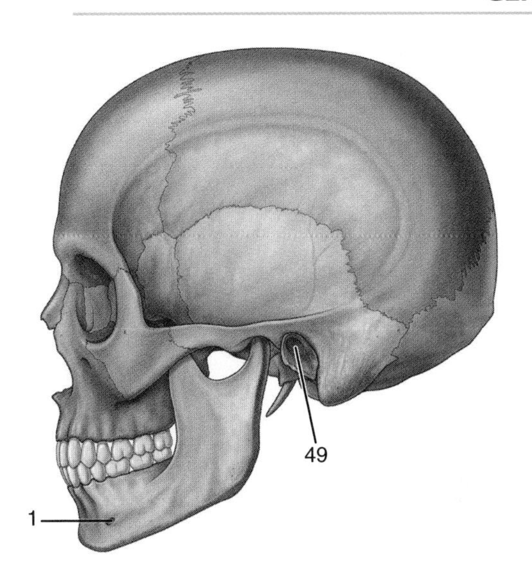

B

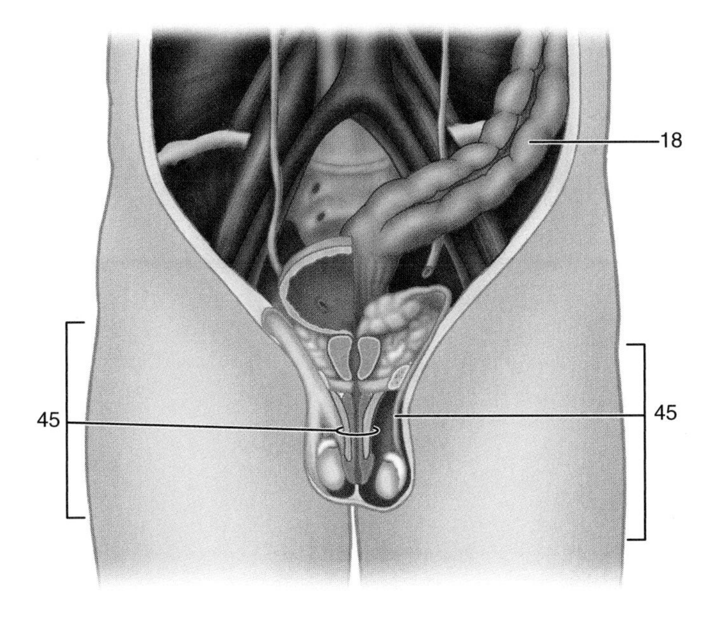

1 **mesentery** 1. the folds of peritoneum that attach the intestines to the abdominal wall. 2. more generally, a membranous fold attaching any of various organs to the body wall. <mesenterium> 2. more generally,a membranous fold attaching any of various organs to the body wall. A

2 *mons* an elevation or eminence.

3 **mucosa** the mucous lining of various tubular structures, facing the lumen, comprising the epithelium, basement membrane, lamina propria mucosae, and lamina muscularis mucosae. <tunica mucosa> B

4 **organ** a somewhat independent part of the body that is arranged according to a characteristic structural plan, and performs a special function or functions; it is composed of various tissues, one of which is primary in function. <organum> D

5 *ostium* an opening, aperture, or orifice.

6 *perichondrium* the layer of dense fibrous connective tissue which invests all cartilage except the articular cartilage of synovial joints.

7 *peripheral* location away from a center or central structure. Called also *peripheralis*. <periphericus>

8 *plexus* a network of lymphatic vessels, nerves, or veins.

9 *pole* either extremity of an axis or of an organ. <polus>

10 *pore* any of various small openings in the body. <porus>

11 *portion* a division of a larger structure. <portio>

12 **process** a prominence or projection. <processus> C

13 *prominence* a small protrusion on another structure or part. <prominentia>

14 *promontory* a projecting eminence or process. <promontorium>

15 **protuberance** a projecting part, or prominence. <protuberantia> C

16 *pulp* any soft, juicy tissue. <pulpa>

17 *pyramid* a part or structure with a pointed or conical shape. <pyramis>

18 *raphe* a seam; the line of union of the halves of any of various symmetrical parts.

19 **recess** any of various cavities, pockets, or pouches. <recessus> D

20 *region* an area on the surface of the body within defined boundaries. <regio>

21 **respiratory system** the tubular and cavernous organs and structures by means of which pulmonary ventilation and gas exchange between ambient air and the blood are brought about; the chief organs involved are the nose, larynx, trachea, bronchi, bronchioles, and lungs. <systema respiratorium> D

22 *retinaculum* a structure that retains an organ or tissue in place.

23 *rima* a cleft, crack, or similar opening.

24 *rostral* toward the oral and nasal region, which may mean superior (for areas of the spinal cord) or anterior (for brain areas). <rostralis>

25 *rostrum* a beaklike appendage or part.

26 *saccus* a baglike structure.

27 *scapular line* an imaginary vertical line on the posterior surface of the body, passing through the inferior angle of the scapula when it is in the anatomical position, i.e., at rest. <linea scapularis>

28 *segment* a part of an organ or other structure set off by natural or arbitrarily established boundaries. <segmentum>

29 *semicanal* a channel which is open on one side. <semicanalis>

30 *septulum* a small separating wall or partition.

31 *septum* a dividing wall or partition.

32 *serosa* the membrane lining the exterior of the walls of various body cavities, reflected over the surfaces of protruding organs; it consists of mesothelium lying upon a connective tissue layer, and it secretes a watery exudate. <tunica serosa>

33 *sinus* a cavity, channel, or space, such as a venous sinus or paranasal sinus.

34 *space* an actual or potential delimited area or open region. <spatium>

35 **spine** a thornlike process or projection. <spina> C

36 *squama* a scale or platelike structure.

37 *stratum* a sheetlike mass of substance.

A

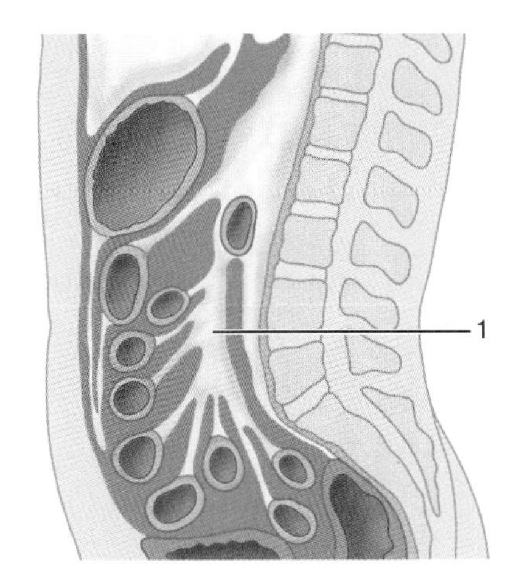

B

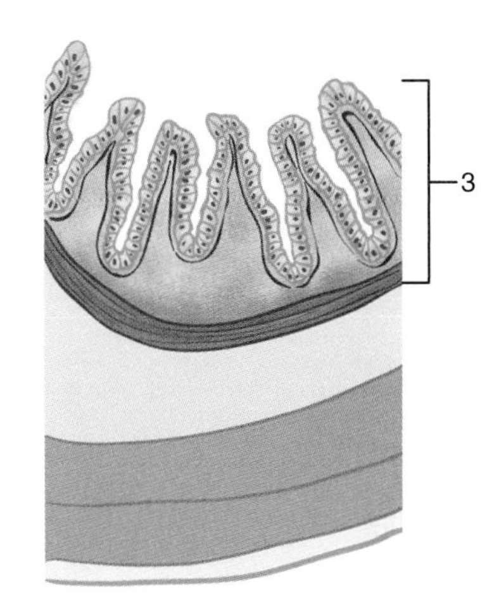

C

D

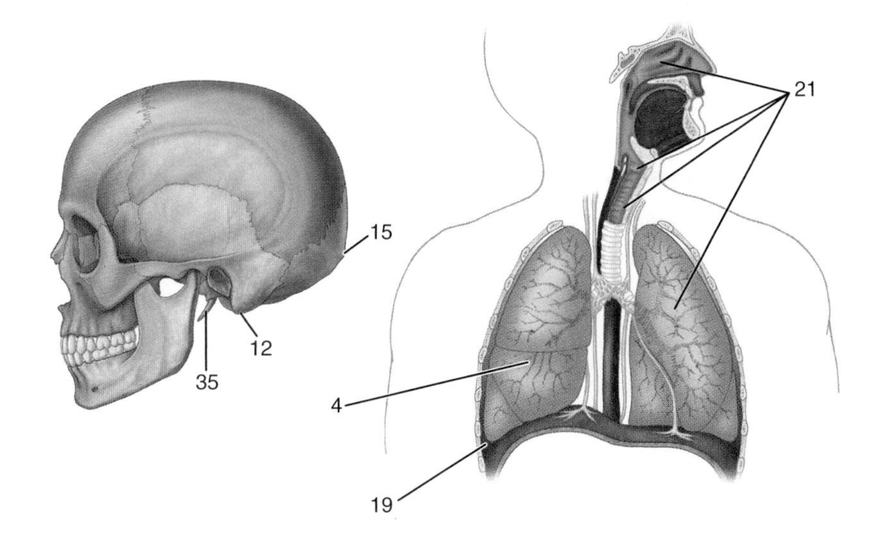

1 **subcutaneous bursa** a synovial sac found beneath the skin. <bursa subcutanea> A

2 ***subfascial bursa*** a synovial sac found beneath a fascial layer. <bursa subfascialis> A

3 ***submuscular bursa*** a synovial sac found beneath a muscle. <bursa submuscularis> B

4 ***subserosa*** a layer of loose areolar tissue underlying the tunica serosa of various organs. <tela subserosa>

5 **subtendinous bursa** a synovial sac found between tendons and bone, tendons and ligaments, and one tendon and another. <bursa subtendinea> C

6 ***sulcus*** a groove, trench, or furrow, especially one of the cerebral sulci.

7 ***superficial*** a structure situated closer than another to the surface of the body. <superficialis>

8 **synovial bursa** a closed synovial sac interposed between surfaces that glide upon each other; it may be simple or multilocular in structure, and subcutaneous, submuscular, subfascial, or subtendinous in location. <bursa synovialis> A

9 ***synovial sheath*** a double-layered, fibrous sheath with synovial fluid present between the layers. <vagina synovialis>

10 ***tegmen*** any covering, or shelter; a covering structure or roof.

11 ***tissue*** a thin weblike layer or membrane. <tela>

12 ***torus*** a bulging projection or swelling.

13 ***trabecula*** a supporting or anchoring strand of connective tissue, such as one extending from a capsule into the substance of the enclosed organ.

14 ***tract*** a collection or bundle, especially of nerve fibers, having the same origin and termination, and serving the same function. <tractus>

15 ***transversalis*** denoting a structure situated at a right angle to the long axis of the body or of an organ.

16 ***triangle*** a three-cornered area or figure; called also *trigone*. <trigonum>

17 ***trochlea*** a pulley-shaped part or structure.

18 ***trunk*** for a major, undivided, usually short portion of a nerve, blood vessel, lymphatic vessel, or duct. <truncus>

19 ***tube*** an elongated hollow cylindrical organ. <tuba>

20 ***tuber*** a swelling or protuberance. Called also *tuberosity*.

21 ***tubercle*** a nodule or small eminence. <tuberculum>

22 **tuberosity** an elevation or protuberance. <tuberositas> A

23 ***tubule*** a small tube. <tubulus>

24 ***tunica*** a membrane or other structure covering or lining a body part or organ. Called also *coat* and *tunic*.

25 ***tunica albuginea*** a dense, white, fibrous sheath enclosing a part or organ.

26 **urinary system** the organs and passageways concerned with the production and excretion of urine, including the kidneys, ureters, urinary bladder, and urethra. <systema urinarium> D

27 ***vallecula*** a depression or furrow.

28 ***valve*** a membranous fold in a canal or passage that prevents reflux of the contents passing through it. <valva>

29 ***valvula*** a small valve; in official terminology the term is now restricted to certain small valves in the body and cusps of heart valves.

30 ***velum*** a veil or veillike structure.

31 ***vertex*** a summit or top.

32 ***vessel*** a canal for carrying fluid, especially one carrying blood, lymph, or spermatozoa. <vas>

33 ***vestibule*** a space or cavity at the entrance to a canal. <vestibulum>

34 ***vestige*** the degenerating remains of any structure that served as a functioning entity in the embryo or fetus. <vestigium>

35 **villus** a small vascular process or protrusion, especially one on the free surface of a membrane. E

36 ***vinculum*** a band or bandlike structure.

37 ***vortex*** a structure arranged in a spiral or whorl.

38 ***wall*** the structure constituting one side of an organ or body cavity. <paries>

39 ***yoke*** a depression or ridge connecting two structures. <jugum>

A

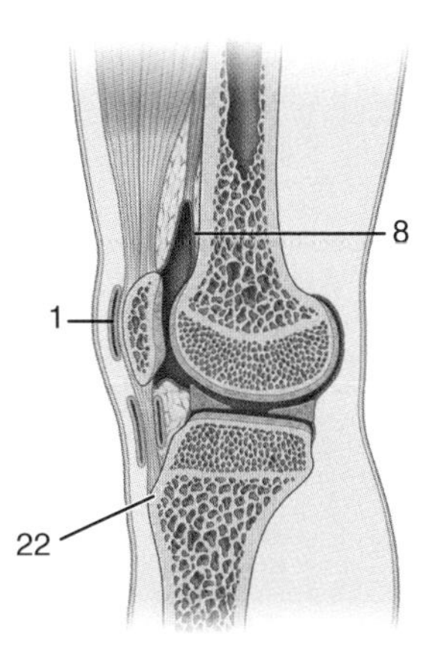

B

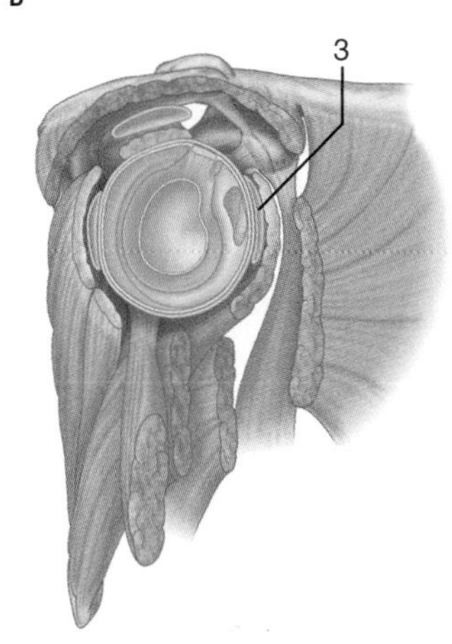

C

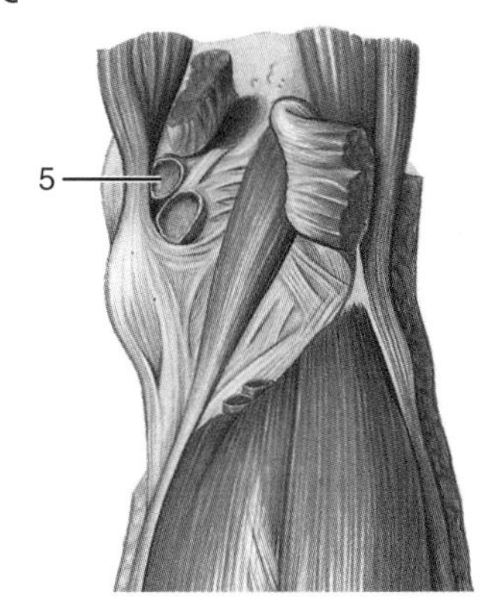

D

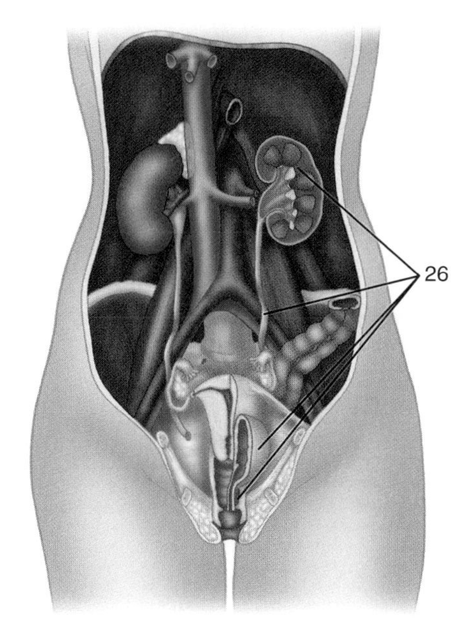

E

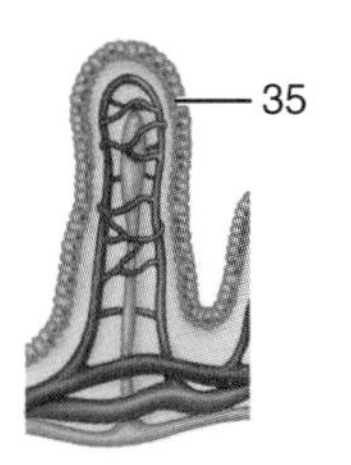

GENERAL LOCATION

1 **anatomical position** the position of the human body, standing erect, with the face directed anteriorly, the upper limbs at the sides and the palms turned anteriorly (supinated), and the feet pointed anteriorly; used as the position of reference in description of site or direction of various structures or parts as established in official anatomical terminology. A

2 ***anterior axillary line*** an imaginary vertical line continuing the line of the anterior axillary fold with the upper limb in the anatomical position; called also *preaxillary line.* <linea axillaris anterior> A

3 **anterior median line** an imaginary vertical line on the anterior surface of the body, dividing the surface equally into right and left sides. <linea mediana anterior> A

4 **anterior** in humans and other bipeds, toward the belly surface of the body; called also *ventral.* A

5 **caudal** 1. pertaining to a tail or cauda. 2. in human anatomy, a synonym of *inferior.* 3. in embryology, denoting a position more toward the tail. <caudalis> A

6 **cranial** 1. pertaining to the cranium. 2. toward the head end of the body; a synonym of *superior* in human anatomy. A

7 **dexter** right: denoting the right-hand one of two similar structures, or the one situated on the right side of the body. A

8 **distal** denoting remoteness from the point of origin or attachment of an organ or part. <distalis> A

9 **dorsal** 1. pertaining to the back. 2. denoting a position more toward the back surface than some other object of reference; a synonym of *posterior* in human anatomy. <dorsalis> A

10 **frontal** 1. pertaining to the forehead. 2. denoting a relationship to the frontal planes. <frontalis> A

11 **frontal planes** planes passing longitudinally through the body from side to side, at right angles to the median plane, and dividing the body into front and back parts. So called because such planes roughly parallel the frontal suture of the skull. Called also *coronal planes* because one of these planes passes through the coronal suture. <plana frontalia> A

12 **horizontal** 1. parallel to the plane of the horizon. 2. denoting relationship to this orientation when the body is in the anatomical (upright) position. <horizontalis> A

13 **horizontal planes** planes at right angles to the long axis of the body. <plana horizontalia> A

14 **inferior** 1. situated below or directed downward. 2. pertaining to the lower surface of an organ or other structure, or to the lower of two (or more) similar structures. A

15 **lateral** denoting a structure situated farther from the median plane of the body or the midline of an organ. <lateralis> A

16 ***mammillary line*** an imaginary vertical line on the anterior surface of the body, passing through the center of the nipple.<linea mammillaris> A

17 **medial** denoting a structure situated nearer to the median plane or the midline of a body or structure. <medialis> A

18 **median** situated in the middle; structures lying in the median plane. <medianus> A

19 ***median axillary line*** an imaginary line halfway between the anterior axillary line and the posterior axillary line, passing through the apex of the axilla; called also *midaxillary line.* <linea axillaris media> A

20 **median plane** the imaginary plane passing longitudinally through the middle of the body from front to back and dividing it into right and left halves. Called also *median sagittal* and *midsagittal plane.* <planum medianum> A

21 ***midclavicular line*** an imaginary vertical line on the anterior surface of the body, passing through the midpoint of the clavicle; called also *midclavicular plane.* <linea medioclavicularis> A

22 ***paramedian planes*** sagittal planes other than the median plane. <plana paramediana> A

23 ***parasternal line*** an imaginary line on the anterior surface of the body midway between the mammillary line and the border of the sternum. <linea parasternalis> A

24 ***posterior axillary line*** an imaginary vertical line continuing the line of the posterior axillary fold with the upper limb in the anatomical position; called also *l. postaxillaris* and *postaxillary line.* <linea axillaris posterior> A

25 **posterior median line** an imaginary vertical line on the posterior surface of the body, dividing the surface equally into right and left sides. <linea mediana posterior> A

26 **posterior** in humans and other bipeds, towards the back surface of the body; called also *dorsal.* A

27 **proximal** denoting proximity to the point of origin or attachment of an organ or part. <proximalis> A

28 **sagittal planes** vertical planes that pass through the body parallel to the median plane (or to the sagittal suture) and divide the body into left and right portions. Included are the *median* and *paramedian planes.* <plana sagittalia> A

29 **sagittal** situated in the plane of or parallel to the sagittal suture, i.e., the median plane of the body. <sagittalis> A

30 **sinister** left: denoting the left-hand one of two similar structures, or the one situated on the left side of the body. A

31 ***sternal line*** an imaginary vertical line on the anterior surface of the body, corresponding to the lateral border of the sternum. <linea sternalis> A

A

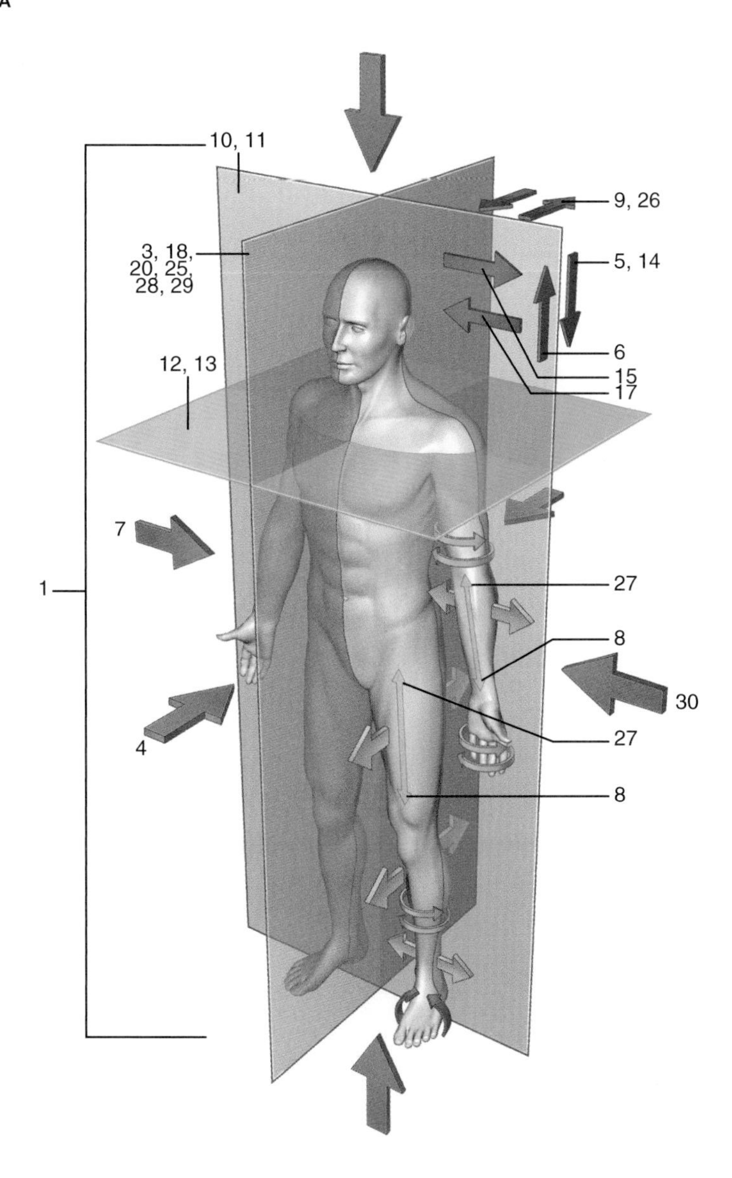

1 **superior** pertaining to a structure occupying a position nearer the vertex. A

2 **transverse** denoting or situated in a position at right angles to a long axis. <transversus> A

3 **transverse planes** horizontal planes of the body dividing the body into superior and inferior portions. <plana transversalia> A

4 **ventral** denoting a position more toward the belly surface than some other object of reference; in human anatomy, a synonym of *anterior*. <ventralis> A

BONES

5 **nutrient canal** one of the freely anastomosing channels of compact bone that contain blood vessels, lymph vessels, and nerves. <canalis nutricius> B

A

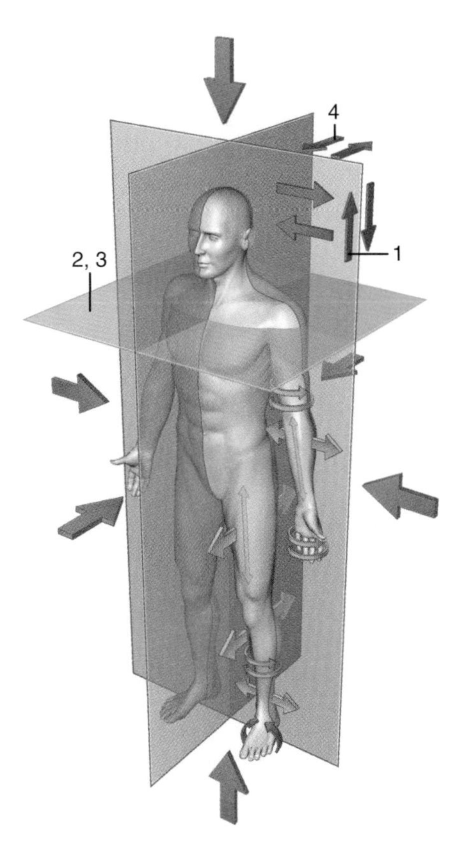

B

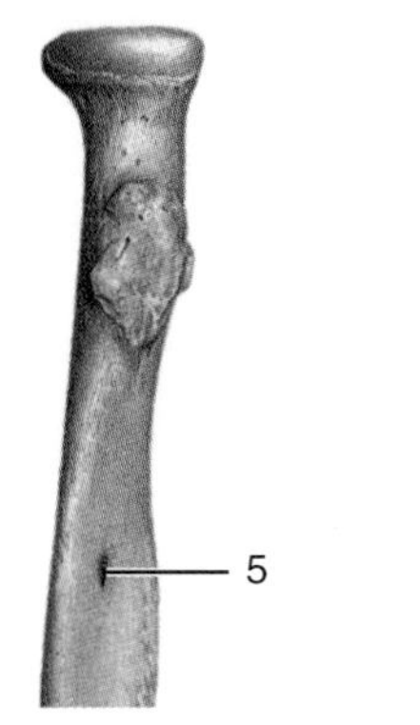

1 ***bone*** 1. the hard form of connective tissue that constitutes the majority of the skeleton of most vertebrates; it consists of an organic component (the cells and matrix) and an inorganic, or mineral, component; the matrix contains a framework of collagenous fibers and is impregnated with the mineral component, chiefly calcium phosphate (85 per cent) and calcium carbonate (10 per cent), which imparts the quality of rigidity to bone. 2. any distinct piece of the osseous framework, or skeleton, of the body. <os> A

2 **bone marrow** the soft material filling the cavities of the bones, made up of a meshwork of connective tissue containing branching fibers, the meshes being filled with marrow cells, which consist variously of fat cells, large nucleated cells or myelocytes, and giant cells called megakaryocytes. <medulla ossium> A

3 ***capitulum*** a little head, or a small eminence on a bone by which it articulates with another bone. A

4 **compact bone** bone substance which is dense and hard. <substantia compacta ossium> A, C

5 **condyle** a rounded projection on a bone, usually for articulation with another. <condylus> A

6 **cortical bone** the substance comprising the hard outer layer of a bone. <substantia corticalis ossium> A, C

7 **diaphysis** the elongated cylindrical portion of a long bone, between the ends or extremities (epiphyses), which are usually articular and wider; it consists of a tube of compact bone that encloses the medullary cavity. Called also *shaft.* A

8 **endosteum** the tissue lining the medullary cavity of a bone. A

9 ***epicondyle*** an eminence on a bone above its condyle. <epicondylus> A

10 ***epiphyseal cartilage*** the cartilage composing the epiphysis prior to ossification. <cartilago epiphysialis> A

11 **epiphyseal disk** the disk or plate of cartilage interposed between the epiphysis and the shaft of the bone during the period of growth; by its growth the bone increases in length. Called also *epiphyseal plate* and *growth disk* or *plate.* <lamina epiphysialis> A, B, C

12 **epiphyseal line** a plane or plate on a long bone, visible as a line, marking the junction of the epiphysis and diaphysis. <linea epiphysialis> A

13 **epiphysis** the expanded articular end of a long bone, developed from a secondary ossification center, which during the period of growth is either entirely cartilaginous or is separated from the shaft by the epiphyseal disk. A

14 ***flat bone*** any bone whose thickness is slight, sometimes consisting of only a thin layer of compact bone, or two layers with intervening spongy bone and marrow; usually bent or curved, rather than flat. <os planum> A

15 ***irregular bone*** a bone that is not readily classified as long, short, or flat; e.g., skull and hip bones and vertebrae. <os irregulare> A

16 **long bone** a bone that has a longitudinal axis of considerable length, consisting of a body or shaft (diaphysis) and an expanded portion (epiphysis) at each end that is usually articular; typically found in the limbs. <os longum> A

17 **medullary cavity** the space in the diaphysis of a long bone containing the marrow; called also *marrow cavity.* <cavitas medullaris> A

18 **metaphysis** the wider part at the extremity of the shaft of a long bone, adjacent to the epiphyseal disk. During development it contains the growth zone and consists of spongy bone; in the adult it is continuous with the epiphysis. A

19 **nutrient artery** any artery that supplies the marrow of a long bone; called also *medullary artery.* <arteria nutricia> B, C

20 **nutrient foramen** any one of the passages that admit the nutrient vessels to the medullary cavity of a bone. <foramen nutricium> C

21 ***ossicle*** a small bone. <ossiculum> A

22 **ossification center** any point at which the process of ossification begins in bones; in a long bone there is a primary center for the diaphysis and a secondary center for the epiphysis. <centrum ossificationis> A

23 **periosteum** a specialized connective tissue covering all bones of the body, and possessing bone-forming potentialities; in adults, it consists of two layers that are not sharply defined, the external layer being a network of dense connective tissue containing blood vessels, and the deep layer composed of more loosely arranged collagenous bundles with spindle-shaped connective tissue cells and a network of thin elastic fibers. A

24 ***phalanx*** any of the bones of the fingers or toes. A

25 ***pneumatized bone*** a bone that contains air-filled cavities or sinuses. <os pneumaticum> A

26 ***red bone marrow*** marrow of developing bone, of the ribs, vertebrae, and many of the smaller bones; it is the site of production of erythrocytes and granular leukocytes. <medulla ossium rubra> A

27 ***short bone*** one whose main dimensions are approximately equal, e.g., one of the bones of the carpus or tarsus. <os breve> A

28 **spongy bone** bone substance made up of thin intersecting lamellae, usually found internal to compact bone; called also *cancellated* or *cancellous bone.* <substantia spongiosa ossium> A

29 ***yellow bone marrow*** ordinary bone marrow of the kind in which the fat cells predominate. <medulla ossium flava> A

A

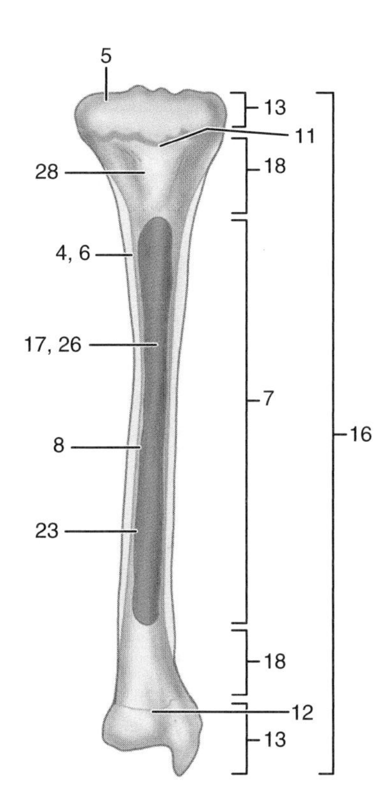

B

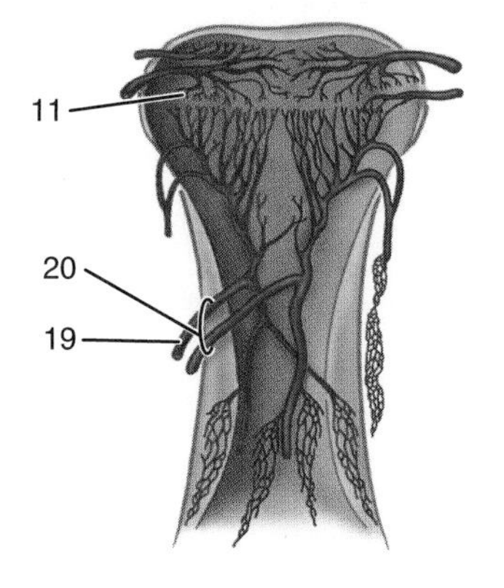

C

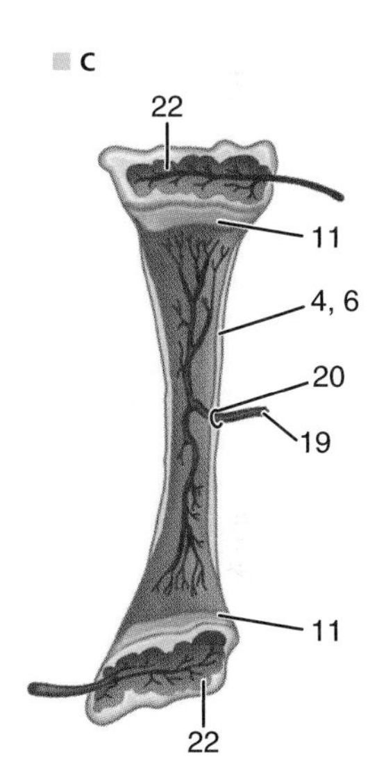

1 **appendicular skeleton** the bones of the upper and lower limbs. <skeleton appendiculare> A

2 **axial skeleton** the bones of the cranium, vertebral column, ribs, and sternum. <skeleton axiale> A

3 ***skeletal system*** the bones *(bony part)* and cartilages *(cartilaginous part)* of the body. <systema skeletale> A

JOINTS

4 ***joint*** the place of union or junction between two or more bones of the skeleton, especially one that allows motion of one or more of the bones. B

5 ***bony joints*** the places of junction between two or more bones of the skeleton, including the fibrous, cartilaginous, and synovial joints. Called also *osseous joints*. <juncturae ossium> B

6 **cartilaginous joint** a type of synarthrosis in which the union of the bony elements is by intervening cartilage; the two types are *synchondrosis* and *symphysis*. <junctura cartilaginea> B

7 ***cotyloid joint*** a type of ball and socket joint. <articulatio cotylica> B

8 ***cranial synchondroses*** the cartilaginous junctions between the bones of the cranium. <synchondroses cranii> B

9 ***ellipsoidal joint*** a modification of the ball-and-socket type of synovial joint in which the articular surfaces are ellipsoid rather than spheroid, e.g., the radiocarpal joint; owing to the arrangement of the muscles and ligaments around the joint, all movements are permitted except rotation about a vertical axis. Called also *condylar* or *condyloid joint*. <articulatio ellipsoidea> B

10 **fibrous joint** a type of synarthrosis in which the union of bony elements is by continuous intervening fibrous tissue, which makes little motion possible; the group includes sutures and syndesmoses. <junctura fibrosa> B

11 ***flat suture*** a type of suture in which there is simple apposition of the contiguous surfaces, with no interlocking of the edges of the participating bones. Called also *false suture*. <sutura plana> B

12 **ginglymus** a type of synovial joint that allows movement in but one plane, forward and backward, as the hinge of a door; called also *hinge joint*. B

13 ***limbous suture*** a type of squamous suture in which there is interlocking of the beveled surfaces of the bones. <sutura limbosa> B

14 **pivot joint** a type of synovial joint that allows a rotary motion in but one plane; a pivot-like process turns within a ring, or a ring turns on a pivot. <articulatio trochoidea> B

15 **plane joint** a type of synovial joint in which the opposed surfaces are flat or only slightly curved; it permits only simple gliding movement, in any direction, within narrow limits imposed by ligaments. Called also *gliding joint* and *arthrodial joint*. <articulatio plana> B

16 **saddle joint** a type of synovial joint in which the articular surface of one bone is concave in one direction and convex in the direction at right angles to the first (concavoconvex), and the articular surface of the second bone is reciprocally convexoconcave; movement is possible along two main axes at right angles to each other. <articulatio sellaris> B

17 ***schindylesis*** a form of articulation in which a thin plate of one bone is received into a cleft in another, as in the articulation of the perpendicular plate of the ethmoid bone with the vomer. Called also *wedge-and-groove joint*. B

18 **serrated suture** a type of suture in which the participating bones are united by interlocking processes resembling the teeth of a saw. <sutura serrata> B

19 ***simple joint*** a type of synovial joint in which only two bones are involved. <articulatio simplex> B

20 **spheroidal joint** a type of synovial joint in which a spheroidal surface on one bone ("ball") moves within a concavity ("socket") on the other bone; it is the most movable type of joint. Called also *ball-and-socket joint, multiaxial joint,* and *polyaxial joint*. <articulatio spheroidea> B

21 **suture** a type of fibrous joint in which the apposed bony surfaces are so closely united by a thin layer of fibrous connective tissue that no movement can occur; found only in the skull. Called also *s. vera* and *true suture*. <sutura> B

22 **symphysis** a type of cartilaginous joint in which the apposed bony surfaces are firmly united by a plate of fibrocartilage; called also *fibrocartilaginous joint*. B

23 ***synarthrosis*** a bony junction that is immovable and is connected by solid connective tissue; the two types are the fibrous joint and the cartilaginous joint. B

24 **synchondrosis** a union between two bones or ossification centers formed by either hyaline cartilage or fibrocartilage; it is usually temporary, the intervening cartilage being converted into bone before adult life. B

25 **syndesmosis** a type of fibrous joint in which the intervening fibrous connective tissue forms an interosseous membrane or ligament. B

26 **synovial joint** a specialized joint permitting more or less free movement, the union of the bony elements being surrounded by an articular capsule enclosing a cavity lined by synovial membrane; called also *diarthrosis* and *diarthrodial joint*. <junctura synovialis> B

27 ***composite joint*** a type of synovial joint in which more than two bones are involved; called also *compound articulation* or *joint*. <articulatio composita> B

28 **squamous suture** <sutura squamosa> B

A

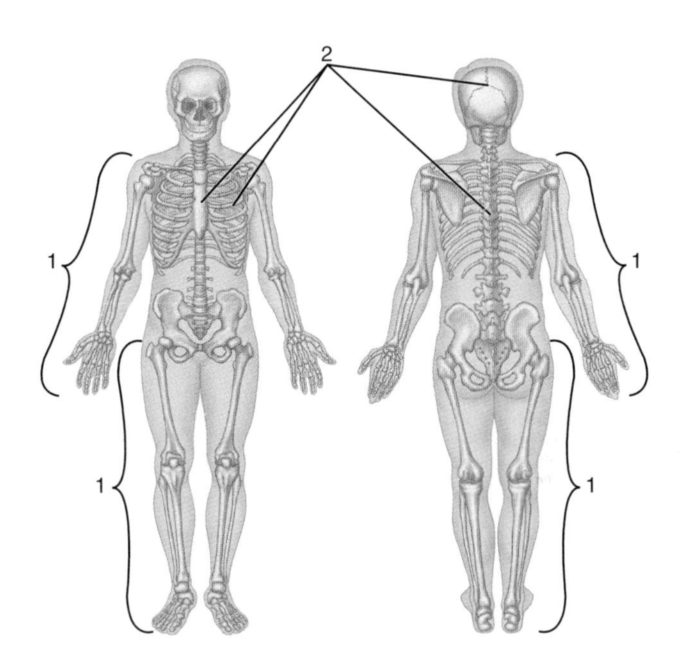

B

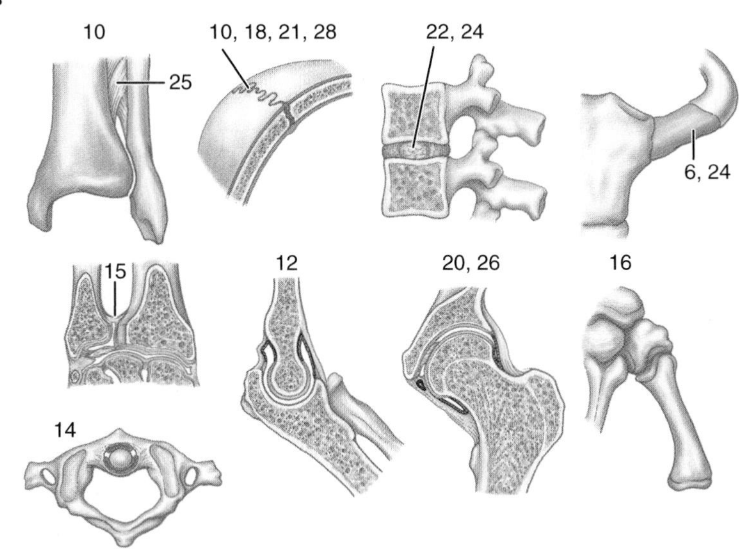

1 **articular cavity** the minute space of a synovial joint, enclosed by the synovial membrane and articular cartilages. <cavitas articularis> A

2 **articular disk** a pad composed of fibrocartilage or dense fibrous tissue found in some synovial joints; it extends into the joint from a marginal attachment at the articular capsule and in some cases completely divides the joint cavity into two separate compartments. Called also *interarticular disk.* <discus articularis> A, E

3 ***capsular ligaments*** thickenings of the fibrous membrane of a joint capsule. <ligamenta capsularia> A

4 ***extracapsular ligaments*** ligaments of a joint capsule that are outside the capsule. <ligamenta extracapsularia> A

5 **fibrous membrane of joint capsule** the outer of the two layers of the joint capsule of a synovial joint, composed of dense white fibrous tissue. <membrana fibrosa capsulae articularis> A

6 **intracapsular ligaments** ligaments within a joint capsule. <ligamenta intracapsularia> D

7 **joint capsule** the saclike envelope that encloses the cavity of a synovial joint by attaching to the circumference of the articular end of each involved bone; it consists of a fibrous membrane and a synovial membrane. Called also *articular capsule* and *synovial capsule.* <capsula articularis> A, D

8 ***synovial fluid*** a transparent alkaline viscid fluid, resembling the white of an egg, secreted by the synovial membrane, and contained in joint cavities, bursae, and tendon sheaths. <synovia> A

9 ***synovial fold*** an extension of the synovial membrane from its free inner surface into the joint cavity. <plica synovialis> A

10 **synovial membrane of joint capsule** the inner of the two layers of the articular capsule of a synovial joint, composed of loose connective tissue and having a free smooth surface that lines the joint cavity. It secretes the synovial fluid. Called also *stratum synoviale capsulae articularis* [TA alternative]. <membrana synovialis capsulae articularis> A

11 ***synovial villi*** slender projections of the synovial membrane from its free inner surface into the joint cavity. <villi synoviales> A

12 **gomphosis** a type of fibrous joint in which a conical process is inserted into a socketlike portion, such as the styloid process in the temporal bone, or the teeth in the dental alveoli. Called also *socket.* B

13 **labrum** a prominent fibrocartilaginous rim around the periphery of certain joints, such as the acetabulum of the hip bone and the glenoid cavity of the scapula. <labrum articulare> C

14 **articular meniscus** a pad, commonly a wedge-shaped crescent of fibrocartilage or dense fibrous tissue, found in some synovial joints; one side forms a marginal attachment at the articular capsule and the other two sides extend into the joint, ending in a free edge. <meniscus articularis> D

15 **bicondylar joint** a condylar joint with a meniscus between the articular surfaces, as in the temporomandibular joint; called also *bicondylar joint.* <articulatio bicondylaris> E

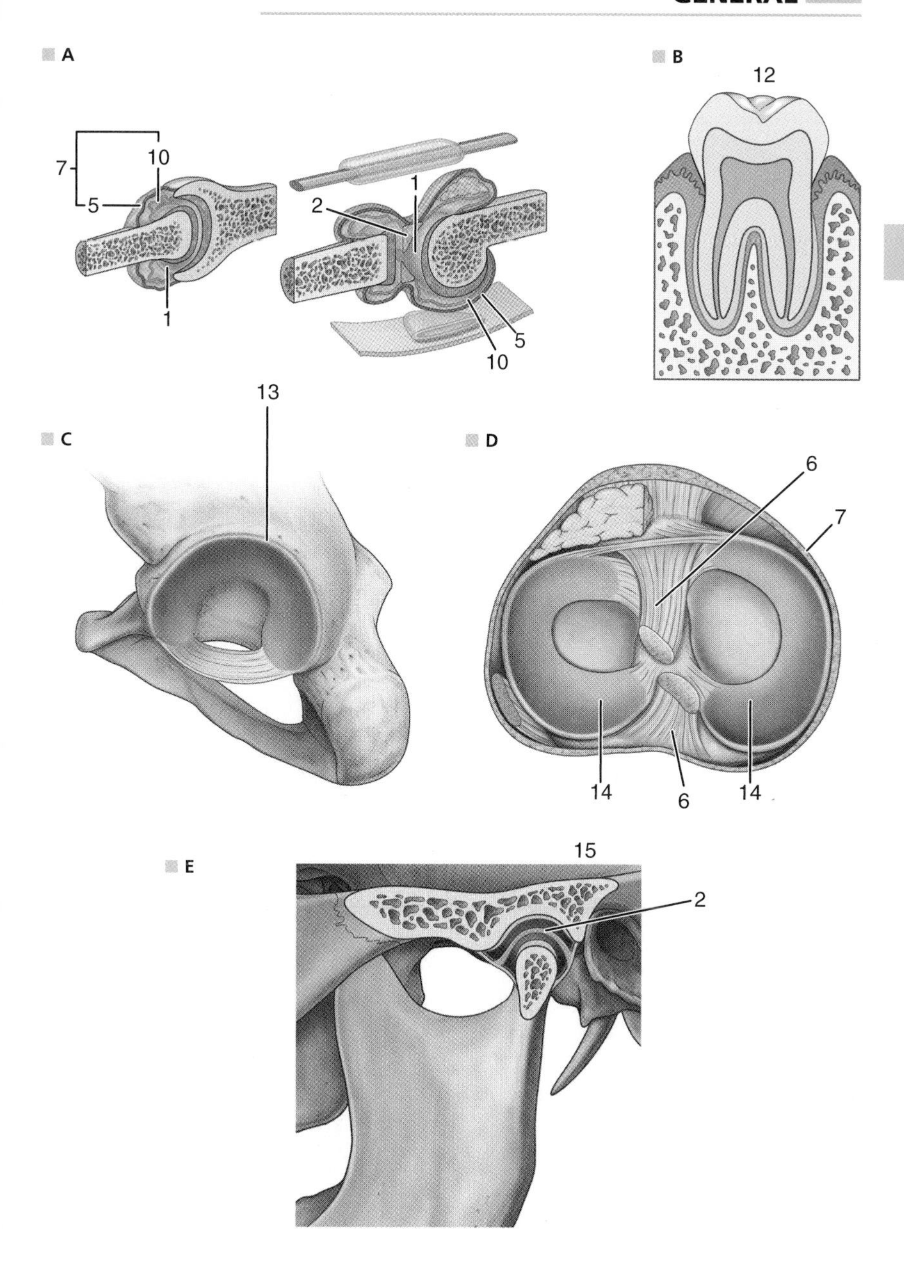
A
7
10
5
1
2
1
5
10
B
12
C
13
D
6
7
14
6
14
E
15
2

MUSCLES

1 **muscle** an organ which by contraction produces the movements of an animal organism; there are two varieties: *striated,* including the skeletal muscles, in which contraction is voluntary, and the cardiac muscle; and *smooth,* or *nonstriated,* including all the involuntary muscles except the heart. Striated muscles are covered with a thin layer of connective tissue *(epimysium)* from which septa *(perimysium)* pass, dividing the muscle into *fasciculi* containing parallel fibers separated by connective tissue septa *(endomysium).* Each fiber consists of sarcoplasm composed of alternate light and dark portions (whence the name *striated muscle*); each contains embedded in it the *myofibrils* and is surrounded by *sarcolemma.* Smooth muscles are composed of elongated, spindle-shaped, nucleated cells arranged parallel to one another and to the long axis of the muscle, and these cells are often grouped into bundles of varying size. The muscles, bundles, and cells are enclosed in an indifferent connective tissue material much as is found in striated muscles. <musculus> A

2 ***aponeurosis*** a white, flattened or ribbonlike tendinous expansion, usually serving to connect a muscle with the parts that it moves. A

3 **belly** a fleshy contractile part of a muscle. <venter> A

4 ***cutaneous muscle*** striated muscle that inserts into the skin, such as the platysma. <musculus cutaneus> A

5 ***dilator*** a muscle that dilates. <musculus dilatator> A

6 **endomysium** the sheath of delicate reticular fibrils which surrounds each muscle fiber. A

7 **epimysium** the fibrous sheath about an entire muscle. A

8 ***extensor*** any muscle that extends a joint. A

9 ***flexor*** 1. causing flexion. 2. any muscle that flexes a joint. A

10 **fusiform muscle** a spindle-shaped muscle in which the fibers are approximately parallel to the long axis of the muscle but converge upon a tendon at either end. <musculus fusiformis> A

11 **head of muscle** the end of a muscle at the site of its attachment to a bone or other fixed structure (origin). <caput musculi> A

12 ***lacertus*** any of various fibrous attachments of muscles. A

13 **levator** a muscle for elevating the organ or structure into which it is inserted. B

14 **multipennate muscle** a muscle in which the fiber bundles converge to several tendons. <musculus multipennatus> A

15 ***muscular system*** all the muscles of the body considered collectively; in official terminology included also are the tendons, bursae, and synovial sheaths. <systema musculare> A

16 ***muscular trochlea*** an anatomical part that serves to change the direction of pull of a tendon; it may be fibrous or bony. <trochlea muscularis> A

17 ***orbicular muscle*** a muscle that encircles a body opening, such as the eye or mouth. <musculus orbicularis> A

18 **pennate muscle** a muscle in which the fibers approach the tendon of insertion from a wide area and are inserted through a large segment of its circumference. Called also *m. bipennatus* [TA alternative] and *bipennate* or *penniform muscle.* <musculus pennatus> A

19 **perimysium** the connective tissue demarcating a fascicle of skeletal muscle fibers. A

20 **semipennate muscle** a muscle in which the fiber bundles approach the tendon of insertion from only one direction and are inserted through only a small segment of its circumference. Called also *unipennate muscle.* <musculus semipennatus> A

21 ***tendinous arch*** a linear thickening of fascia over some part of a muscle, such as that over the soleus or the obturator internus. <arcus tendineus> A

22 **tendinous intersection** a fibrous band that crosses the belly of a muscle and more or less completely divides it into two parts. <intersectio tendinea> A

23 ***quadrate muscle*** a square-shaped muscle. <musculus quadratus>

24 **sphincter muscle** a ringlike muscle that closes a natural orifice; it may be either of smooth muscle or striated muscle. Called also *sphincter.* <musculus sphincter> B

■ A

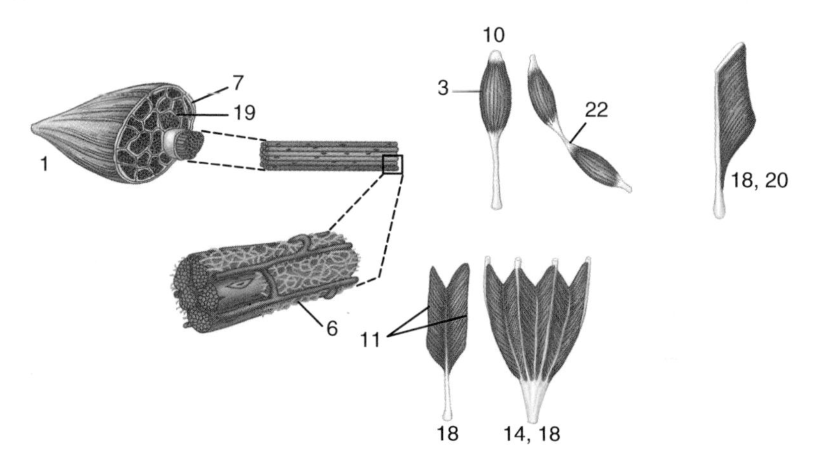

■ B

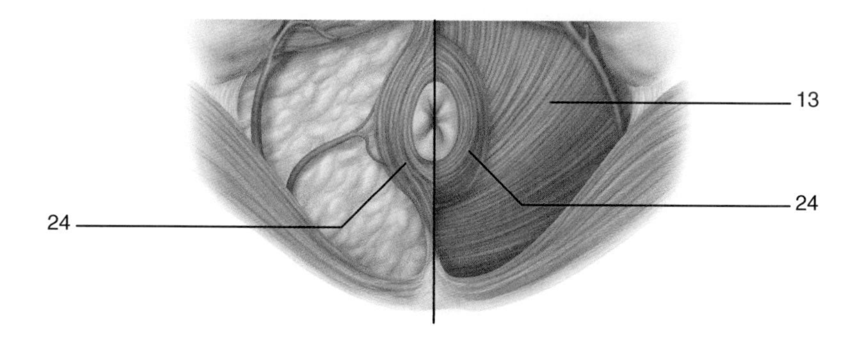

1 **fibrous sheath** the fibrous layer of a tendon sheath. <stratum fibrosum vaginae tendinis> A

2 *mesotendon* the delicate connective tissue sheath attaching a tendon to its fibrous sheath. <mesotendinium> A

3 **synovial sheath** the synovial layer of a tendon sheath. <stratum synoviale vaginae tendinis> A

4 **tendon** a fibrous cord of connective tissue in which the fibers of a muscle end and by which the muscle is attached to a bone or other structure. <tendo> A

5 **tendon sheath** a sheath of tissue that covers a tendon; it has both fibrous and synovial layers. <vagina tendinis> A

NERVES

6 *nervous system* the organ system which, along with the endocrine system, correlates the adjustments and reactions of an organism to internal and environmental conditions. It comprises the central and peripheral nervous systems. <systema nervosum> B

7 **nerve** a cordlike structure made up of a collection of fibers that convey impulses between a part of the central nervous system and some other region of the body. A nerve consists of a connective tissue sheath (epineurium) enclosing bundles of fibers (funiculi or fasciculi); each bundle is in turn surrounded by its own sheath of connective tissue (perineureum), the inner surface of which is formed by a membrane of flattened mesothelial cells. Very small nerves may consist of only one funiculus derived from the parent nerve. Within each such bundle, the individual nerve fibers, which are microscopic in size, are surrounded by interstitial connective tissue (endoneurium). An individual nerve fiber (an axon with its covering sheath) consists of formed elements in a matrix of protoplasm (axoplasm), the entire structure being enclosed in a thin membrane (axolemma). Each nerve fiber is enclosed by a cellular sheath (neurilemma), from which it may or may not be separated by a lipid layer (myelin sheath) derived from neurilemmal cells. <nervus> B

8 **neuron** any of the conducting cells of the nervous system. A typical neuron consists of a cell body, containing the nucleus and the surrounding cytoplasm (perikaryon); several short radiating processes (dendrites); and one long process (the axon), which terminates in twiglike branches (telodendrons) and may have branches (collaterals) projecting along its course. The axon together with its covering or sheath forms the nerve fiber. Called also *nerve cell.* B

9 *articular branch* any of the branches of a mixed (afferent or efferent) peripheral nerve supplying a joint and its associated structures. <ramus articularis>

10 *autonomic branch* any of the branches of the parasympathetic or sympathetic nerves of the autonomic nervous system. <ramus autonomicus>

11 *autonomic nervous system* the portion of the nervous system concerned with regulation of the activity of cardiac muscle, smooth muscle, and glands; usually restricted to the two visceral efferent peripheral components, the sympathetic nervous system and the parasympathetic nervous system. <divisio autonomica systematis nervosi peripherici>

12 *autonomic plexus* any of the extensive networks of nerve fibers and cell bodies associated with the autonomic nervous system; found particularly in the thorax, abdomen, and pelvis, and containing sympathetic, parasympathetic, and visceral afferent fibers. <plexus autonomicus>

13 *central nervous system* the brain and spinal cord. <pars centralis systematis nervosi>

14 *cutaneous branch* a branch of a mixed (afferent or efferent) peripheral nerve innervating a region of the skin. <ramus cutaneus>

15 *interganglionic branches of sympathetic trunk* the branches that interconnect the ganglia of the sympathetic trunk. <rami interganglionares trunci sympathici>

16 *muscular branch* a branch of a peripheral nerve or vessel that supplies muscle. <ramus muscularis>

17 *parasympathetic nervous system* the craniosacral division of the autonomic nervous system, its preganglionic fibers traveling with cranial nerves III, VII, IX, X, and XI, and with the second to fourth sacral anterior roots; it innervates the heart, the smooth muscle and glands of the head and neck, and the thoracic, abdominal, and pelvic viscera. The ganglion cells with which these fibers synapse are in or near the organs innervated. <pars parasympathica divisionis autonomici systematis nervosi>

18 *periarterial plexus* a network of autonomic and sensory nerve fibers in the adventitia of an artery, some of which follow the course of the artery to reach and innervate other structures and some of which innervate the artery itself. <plexus periarterialis>

19 *peripheral nervous system* the peripheral part of the nervous system, consisting of the nerves and ganglia outside the brain and spinal cord. <pars peripherica systematis nervosi>

20 *plexus of spinal nerves* a plexus formed by the intermingling of the fibers of two or more spinal nerves, such as the brachial or lumbosacral plexus. <plexus nervorum spinalium>

21 *sympathetic nervous system* the portion of the autonomic nervous system that receives its fibers of connection with the central nervous system through the thoracolumbar outflow of visceral efferent fibers. These fibers (preganglionic) arise from cells in the thoracic and upper lumbar levels of the spinal cord, leave by way of ventral roots, and, by way of rami communicantes, enter sympathetic trunks, where some synapse with ganglion cells. The fibers (postganglionic) of these ganglion cells return to spinal nerves by way of rami communicantes to supply the blood vessels, smooth muscle, and glands of the trunk and limbs, or go as visceral branches to the blood

vessels, smooth muscles, and glands of the head and neck, and the viscera of the thorax, abdomen, and pelvis. Some preganglionic fibers pass through the sympathetic trunks and synapse in the prevertebral ganglia; postganglionic fibers from those ganglia supply adjacent viscera. <pars sympathica divisionis autonomici systematis nervosi>

22 ***vascular nerves*** the nerve branches that supply the adventitia of the blood vessels. <nervi vasorum>

23 ***vascular plexus*** a plexus of peripheral nerves through which blood vessels receive innervation. <plexus vascularis>

A

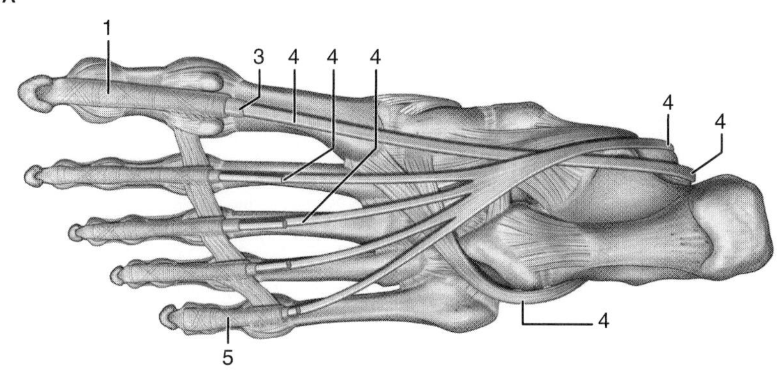

B

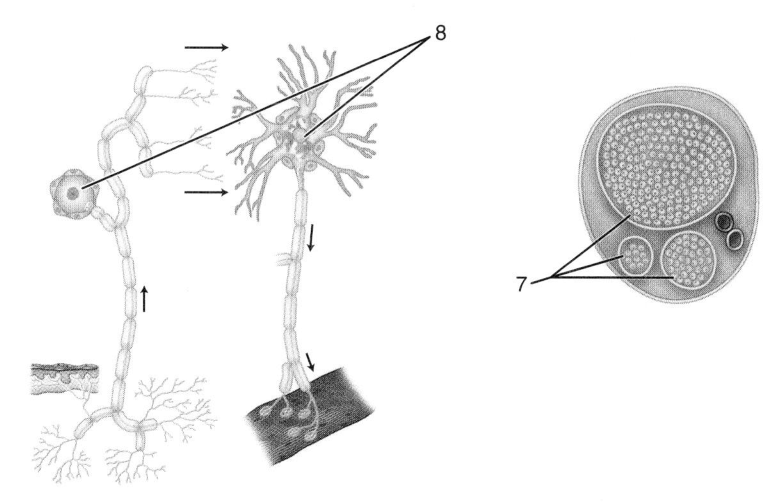

1 **cerebrospinal fluid** the fluid contained within the four ventricles of the brain, the subarachnoid space, and the central canal of the spinal cord; formed by choroid plexuses and brain parenchyma, it circulates through the ventricles into the subarachnoid space and is absorbed into the venous system. <liquor cerebrospinalis> A

2 **dura mater** the outermost, toughest, and most fibrous of the three membranes (meninges) covering the brain and spinal cord; called also *pachymeninx.* A

3 ***ependyma*** the lining membrane of the ventricles of the brain and of the central canal of the spinal cord. A

4 **leptomeninges** the pia mater and arachnoid considered together as one functional unit; the pia-arachnoid. A

5 ***leptomeninx*** singular of *leptomeninges.* NOTE: In official terminology, this term is used as the preferred term when contrasting this structure (arachnoidea mater and pia mater) with the pachymeninx, which is equivalent to the dura mater. A

6 **meninges** the three membranes that envelop the brain and spinal cord: the dura mater, pia mater, and arachnoid. A

7 ***subarachnoid cisterns*** localized enlargements of the subarachnoid space, occurring in areas where the dura mater and arachnoid do not closely follow the contour of the brain with its covering pia mater, and serving as reservoirs of cerebrospinal fluid. <cisternae subarachnoideae> A

8 ***capsule of ganglion*** the laminated connective tissue capsule surrounding a neural ganglion and continuous with epineurium of its associated nerve root. <capsula ganglii> B

9 **endoneurium** the innermost layer of connective tissue in a peripheral nerve, forming an interstitial layer around each individual fiber outside the neurilemma. B

10 **epineurium** the outermost layer of connective tissue of a peripheral nerve, surrounding the entire nerve and containing its supplying blood vessels and lymphatics. B

11 ***mixed nerve*** a nerve composed of both sensory (afferent) and motor (efferent) fibers. <nervus mixtus> B

12 **motor neuron** a peripheral efferent fiber that conducts impulses from the spinal cord or brain to motor end plates or other terminals, resulting in stimulation of muscle contractions. <nervus motorius> B

13 **sensory neuron** a peripheral afferent fiber that conducts impulses from receptors on a sense organ to the termination of its axon in the spinal cord or brain. <nervus sensorius> B

14 ***synapse*** the site of functional apposition between neurons, at which an impulse is transmitted from one neuron to another, usually by a chemical neurotransmitter (e.g., acetylcholine, norepinephrine, etc.) released by the axon terminal of the excited (presynaptic) cell. The neurotransmitter diffuses across the synaptic cleft to bind with receptors on the postsynaptic cell membrane, and thereby effects electrical changes in the postsynaptic cell which result in depolarization (excitation) or hyperpolarization (inhibition). Synapses also occur at sites of apposition between nerve endings and effector organs (e.g., the neuromuscular junction). <synapsis> B

VESSELS

15 ***arterial circle*** a complete or incomplete circle of anastomosing arteries. <circulus arteriosus>

16 ***arterial plexus*** an anastomotic network formed by arteries just before they become arterioles or capillaries. <rete arteriosum>

17 ***articular vascular plexus*** an anastomotic network of blood vessels in or around a joint; called also *articular rete.* <rete vasculosum articulare>

18 ***blood*** the fluid that circulates through the heart, arteries, capillaries, and veins, carrying nutriment and oxygen to the body cells. It consists of the *plasma,* a pale yellow liquid containing the microscopically visible formed elements of the blood: the *erythrocytes,* or red blood corpuscles; the *leukocytes,* or white blood corpuscles; and the *platelets,* or thrombocytes.

19 ***collateral vessel*** a vessel that parallels another vessel, nerve, or other structure. <vas collaterale>

20 ***deep vein*** any deeply situated vein. <vena profunda>

21 ***glomus*** a small, histologically recognizable body, composed of fine arterioles connecting directly with veins, and possessing a rich nerve supply.

22 ***sinusoid*** a form of terminal blood channel consisting of a large, irregular anastomosing vessel, having a lining of reticuloendothelium but little or no adventitia; sinusoids are found in the liver, adrenals, heart, parathyroid, carotid gland, spleen, hemolymph glands, and pancreas. Called also *sinusoidal capillary.* <vas sinusoideum>

23 ***superficial vein*** any superficially situated vein. <vena superficialis>

24 ***vasa vasorum*** the small nutrient arteries and the veins in the walls of the larger blood vessels.

25 ***vascular circle*** a complete or incomplete circle of anastomosing blood vessels. <circulus vasculosus>

26 ***vascular plexus*** a network of intercommunicating blood vessels. <plexus vasculosus>

27 ***vena comitans*** a vein, usually occurring in a pair (venae comitantes), that closely accompanies its homonymous artery and is found especially in the extremities.

28 ***venous plexus*** 1. a network of interconnecting veins. <plexus venosus> 2. an anastomotic network of small veins. <rete venosum>

29 ***venous valve*** any of the small cusps or folds found in the tunica intima of many veins, serving to prevent backflow of blood. <valvula venosa>

30 ***vessels of nerves*** blood vessels supplying the nerves. <vasa nervorum>

A

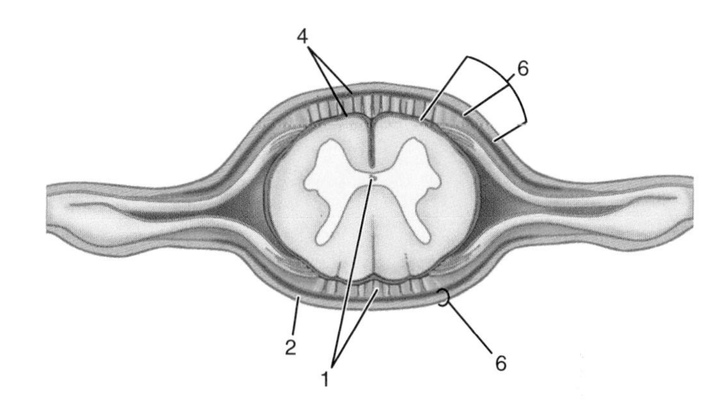

B

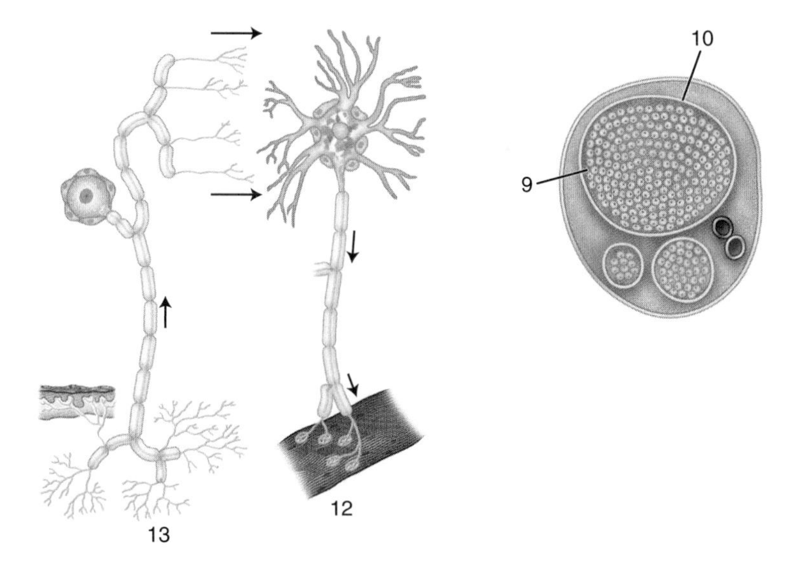

1 **anastomotic vessel** a vessel that serves to interconnect other vessels; such communications are present in the palm of the hand, sole of the foot, base of the brain, and other regions. <vas anastomoticum> A

2 **arteriolovenular anastomosis** arteriolovenular anastomosis: a vessel that directly interconnects an arteriole and a venule and that acts as a shunt to bypass the capillary bed. Called also *a. arteriovenosa* [TA alternative]. <anastomosis arteriolovenularis> A

3 **artery** a vessel through which blood passes away from the heart to various parts of the body. The wall of an artery consists typically of outer, middle, and inner coats. <arteria> A, B

4 **arteriole** a minute arterial branch, especially one just proximal to a capillary. <arteriola> A, C

5 **capillary** any of the minute vessels that connect the arterioles and venules, forming a network in nearly all parts of the body. Their walls act as semipermeable membranes for the interchange of various substances, including fluids, between the blood and tissue fluid. The two principal types are *continuous* and *fenestrated capillaries.* <vas capillare> A, C

6 **venule** any of the small vessels that collect blood from the capillary plexuses and join to form veins. <venula> A, C

7 **vein** a vessel through which blood passes from various organs or parts back to the heart; all veins except the pulmonary veins carry blood low in oxygen. Like arteries, veins have three coats, *inner, middle,* and *outer,* but the coats are not so thick, and they collapse when the vessel is cut. Many veins have valves formed of reduplications of their lining membrane, which prevent the backward flow of blood away from the heart. <vena> A, D

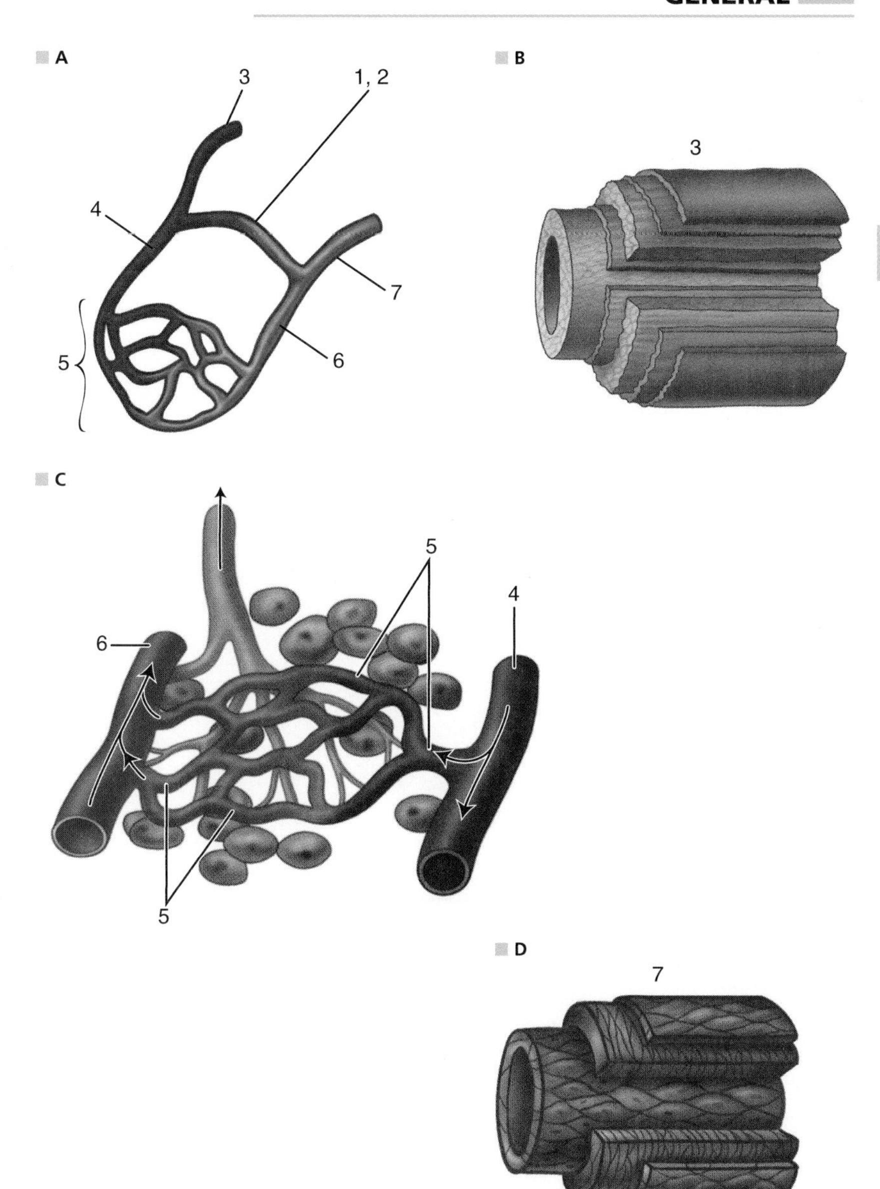
A
3
1, 2
4
7
5
6
B
3
C
5
4
6
5
D
7

LYMPHATICS

1 ***lymphatic system*** the lymphatic vessels and the lymphoid tissue, considered collectively. <systema lymphoideum> A

2 **deep lymphatic vessel** any lymphatic vessel that drains lymph from deep body structures; deep lymphatic vessels accompany the deeply placed blood vessels. <vas lymphaticum profundum> A

3 ***lymph*** a transparent, slightly yellow liquid of alkaline reaction, found in the lymphatic vessels and derived from the tissue fluids. It is occasionally of a light-rose color from the presence of red blood corpuscles, and is often opalescent from particles of fat. Under the microscope, lymph is seen to consist of a liquid portion and of cells, most of which are lymphocytes. Lymph is collected from all parts of the body and returned to the blood via the lymphatic system. <lympha> A

4 ***lymphatic rete*** any of the closed, freely communicating networks formed by the lymphocapillary vessels. <rete lymphocapillare> A

5 **lymphatic vessels** collectively, the lymphocapillary vessels, collecting vessels, and trunks which collect lymph from the tissues and through which the lymph passes to reach the bloodstream. <vasa lymphatica> A

6 ***lymphocapillary vessel*** one of the minute vessels of the lymphatic system, having a caliber greater than a blood capillary; they form closed networks by which they communicate freely with one another. <vas lymphocapillare> A

7 **superficial lymphatic vessel** any lymphatic vessel located under the skin and superficial fascia, in the submucous areolar tissue of the digestive, respiratory, and genitourinary tracts, and in the subserous tissue of the walls of the abdomen and thorax. <vas lymphaticum superficiale> A

8 **capsule of lymph node** the outer layer of a lymph node, composed mainly of collagen fibers with a few fibroblasts and elastin fibers. <capsula nodi lymphoidei> B

9 **cortex of lymph node** the outer portion of the node, consisting mainly of dense lymphatic tissue and follicles. <cortex nodi lymphoidei> B

10 **hilum of lymph node** the indentation on a lymph node where the arteries enter and the veins and efferent lymphatic vessels leave. <hilum nodi lymphoidei> B

11 **lymph node** any of the accumulations of lymphoid tissue organized as definite lymphoid organs, varying from 1 to 25 mm in diameter, situated along the course of lymphatic vessels, and consisting of an outer cortical and an inner medullary part. The lymph nodes are the main source of lymphocytes of the peripheral blood and, as part of the reticuloendothelial system, serve as a defense mechanism by removing noxious agents, such as bacteria and toxins, and probably play a role in antibody production. <nodus lymphoideus> A, B

12 **lymphatic valve** any of the usually doubled cusps in the collecting lymphatic vessels, serving to ensure flow in only one direction. <valvula lymphaticum> B

13 **medulla of lymph node** the central part of a lymph node, comprising cords and sinuses. <medulla nodi lymphoidei> B

14 **trabeculae of lymph node** strands of dense connective tissue radiating out from the capsule through the interior of the node. <trabeculae nodi lymphoidei> B

A

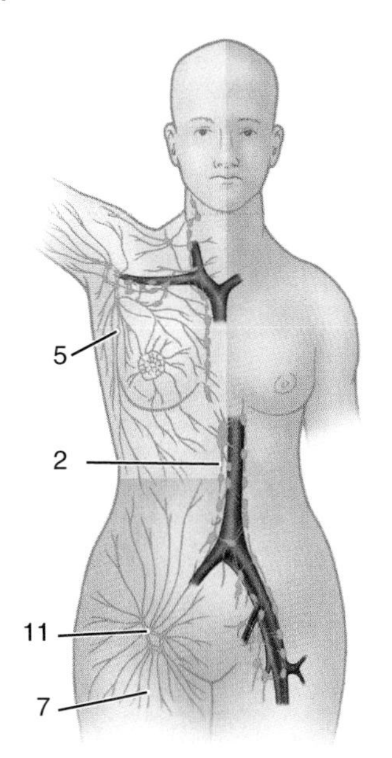

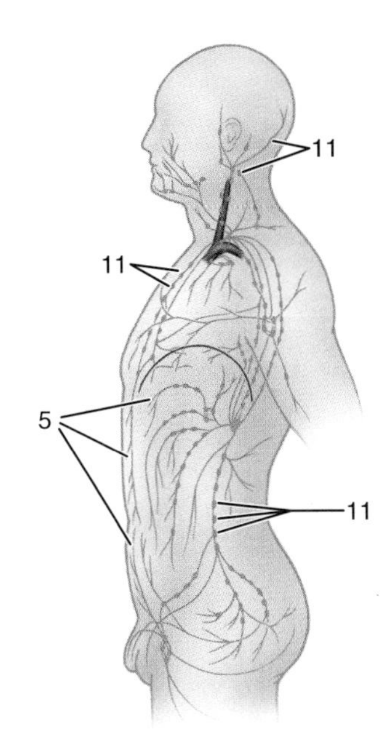

B

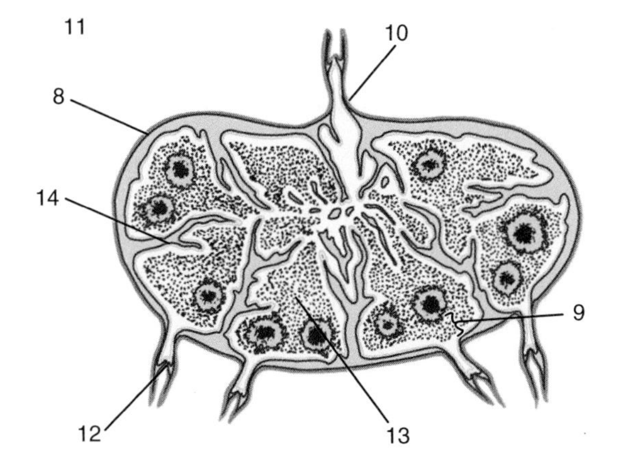

SKIN

1 ***skin*** the outer protective covering of the body, consisting of the epidermis and dermis (or corium), and resting upon the subcutaneous tissues. <cutis> A

2 **arrector muscle of hair** minute smooth muscles of the skin whose contraction causes the hair to stand erect with cutis anserina (goose flesh). *origin,* papillary layer of skin; *insertion,* a hair follicle; *innervation,* sympathetic; *action,* elevate a hair of skin. <musculus arrector pili> A

3 **cutaneous glands** the glands of the skin, including the sweat, sebaceous, and ceruminous glands. <glandulae cutis> A

4 **cutaneous vein** one of the small veins that begin in the papillae of the skin, form subpapillary plexuses, and open into the subcutaneous veins. <vena cutanea> A

5 **dermal papilla** any of the conical extensions of the collagen fibers, the capillary blood vessels, and sometimes the nerves of the dermis into corresponding spaces among the downward- or inward-projecting rete ridges on the under surface of the epidermis. On the forehead and ear these are less prominent; on the face, neck, and pubes the relations are reversed and rete ridges extend inward or downward into spaces among a network of dermal ridges. <papilla dermis> A

6 **dermis** the layer of the skin deep to the epidermis, consisting of a dense bed of vascular connective tissue. A

7 **epidermis** the outermost and nonvascular layer of the skin, derived from the embryonic ectoderm, varying in thickness from 0.07 to 0.12 mm, except on the palms and soles where it may be 0.8 and 1.4 mm, respectively. On the palmar and plantar surfaces, it exhibits maximal cellular differentiation and layering, and comprises, from within outward, five layers: the stratum basale (basal layer); the stratum spinosum (prickle cell or spinous layer); the stratum granulosum (granular layer); the stratum lucidum (clear layer); and the stratum corneum (horny layer). In the thinner epidermis of the general body surface, the stratum lucidum is usually absent. A

8 ***membranous layer of subcutaneous tissue*** the deeper layer of subcutaneous tissue in certain areas of the body, underlying the subcutaneous fat. <stratum membranosum telae subcutaneae> A

9 **papillary layer of dermis** the outer layer of the dermis, characterized by the presence of ridges or papillae protruding into the epidermis. <stratum papillare dermidis> A

10 **reticular layer of dermis** the inner layer of the dermis, consisting chiefly of dense fibrous tissue. <stratum reticulare dermidis> A

11 **sebaceous gland** one of the holocrine glands of the skin that secrete sebum and are situated in the corium. <glandula sebacea> A

12 ***skin ligaments*** bands of connective tissue attaching the dermis to the subcutaneous tissue. <retinacula cutis> A

13 **subcutaneous fat** a layer of adipose tissue underlying the dermis. Called also *pannus.* <panniculus adiposus> A

14 ***subcutaneous tissue*** the layer of loose connective tissue situated just beneath the skin; called also *hypodermis.* <tela subcutanea> A

15 **sweat gland** one of the glands that secrete sweat, situated in the dermis or subcutaneous tissue, and opening by a duct on the surface of the body. There are two types: the ordinary or *eccrine sweat glands* and the *apocrine sweat glands.* Called also *sudoriparous gland.* <glandula sudorifera> A

16 ***cruces pilorum*** crosslike figures formed by the pattern of hair growth, the hairs lying in opposite directions.

17 ***epithelium*** the covering of internal and external surfaces of the body, including the lining of vessels and other small cavities. It consists of cells joined by small amounts of cementing substances. Epithelium is classified into types on the basis of the number of layers deep and the shape of the superficial cells.

18 ***hair streams*** continuous lines formed by the pattern of hair growth on various parts of the body, the hairs lying in the same direction. <flumina pilorum>

19 ***hair whorls*** whorled patterns of hair growth on the body, as that on the crown of the head. <vortices pilorum>

20 ***lanugo*** the fine hair on the body of the fetus; called also *down* and *lanugo hair.*

21 ***tactile elevations*** the small elevations on the skin of the palm and the sole, richly supplied with sensory nerve endings. <toruli tactiles>

22 **body of nail** the large distal, exposed portion of the nail of a digit. <corpus unguis> B

23 **free border of nail** the distal overhanging edge of a nail. <margo liber unguis> B

24 ***hidden border of nail*** the proximal buried edge of a nail. <margo occultus unguis> B

25 **hyponychium** the thickened epidermis underneath the free distal end of the nail. B

26 **lateral border of nail** the edge on either side of the nail. <margo lateralis unguis> B

27 **lunula of nail** the crescentic white area at the base of the nail. <lunula unguis> B

28 ***nail matrix*** the tissue upon which the deep aspect of the nail rests; called also *nail bed.* The term is also used to denote the proximal portion of the nail bed from which growth chiefly proceeds. <matrix unguis> B

29 **nail wall** the fold of skin overlapping the sides and the proximal end of a nail. <vallum unguis> B

30 **eponychium** the narrow band of epidermis that extends from the nail wall onto the nail surface; called also *cuticle* and *perionychium.* B

A

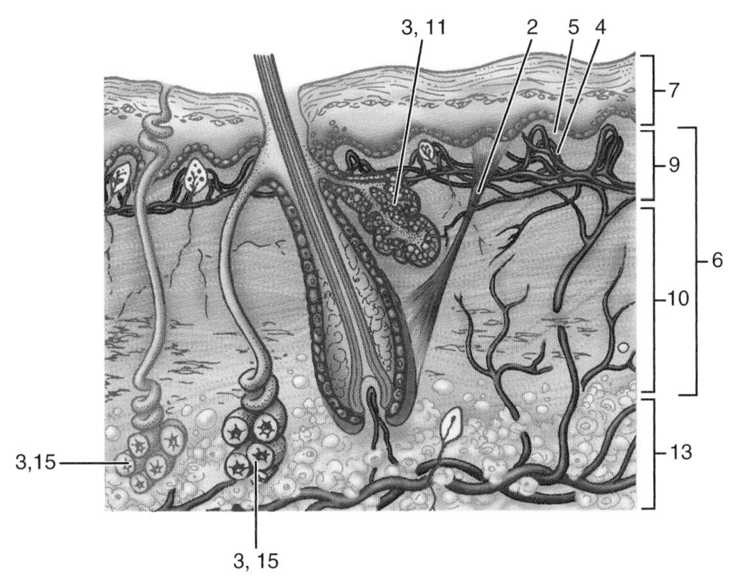

B

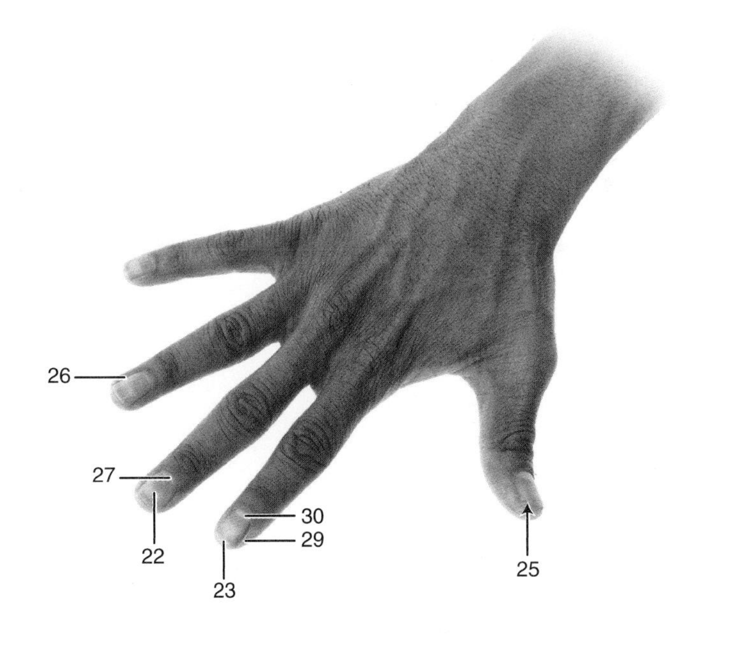

1 **dermal ridges** ridges of the skin produced by the projecting papillae of the dermis on the palm or sole, producing a fingerprint or footprint that is characteristic of the individual. <cristae cutis> A

2 ***skin furrows*** the fine depressions on the surface of the skin between the dermal ridges. Called also *skin sulci.* <sulci cutis> A

3 **hair follicle** one of the tubular invaginations of the epidermis that enclose the hairs, and from which the hairs grow. It is divided into upper and lower segments: the upper comprises the infundibulum, extending from the free surface to the sebaceous gland, and the isthmus, extending from the sebaceous gland to the arrector pili; the lower comprises the stem and the bulb. At the level of the stem, the hair cuticle is surrounded successively by the inner and outer root sheaths, which are enclosed by a dermal sheath. <folliculus pili> B

4 **hair** one of the filamentous appendages growing out of the skin, consisting primarily of keratin. Anatomists distinguish between the hairs on the limbs and trunk (*pili*) and those on the scalp (*capilli*). Each hair consists of a cylindrical shaft that begins with a root, which is contained in a follicle. The base of the root expands to form the hair bulb, which rests upon and encloses the hair papilla. B

5 **cleavage lines** linear clefts in the skin over many parts of the body, indicative of direction of the fibers. They correspond closely to the crease lines in the skin, and assume a characteristic pattern in each body part, although they can vary with body configuration. Called also *Langer* or *tension lines.* <lineae distractiones> C

A

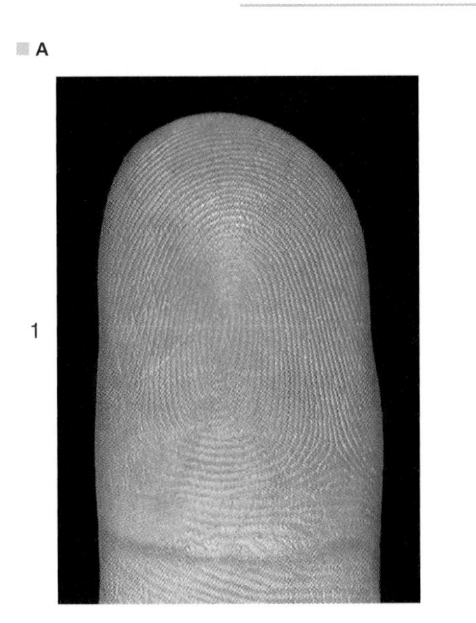

B

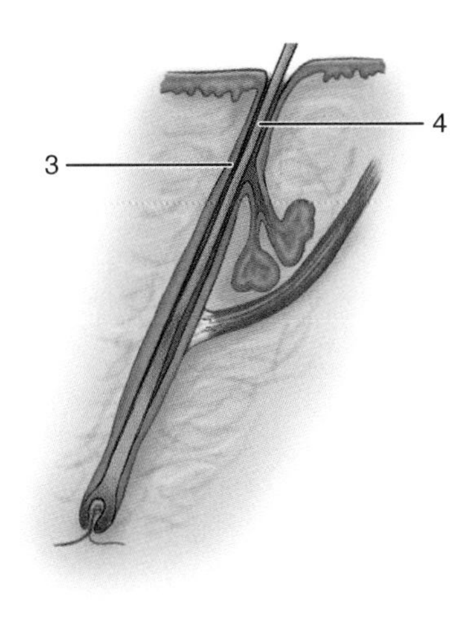

C

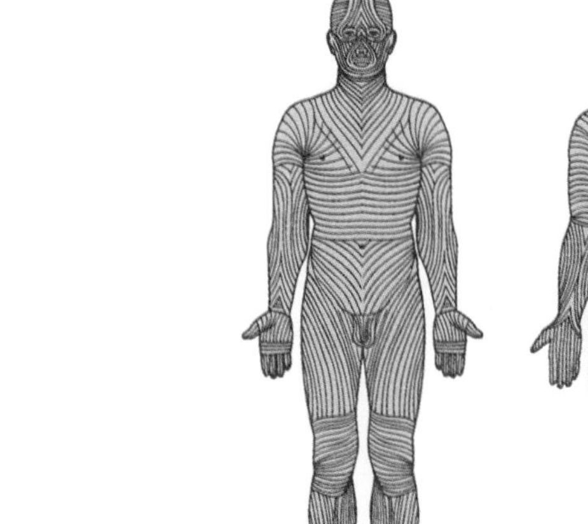
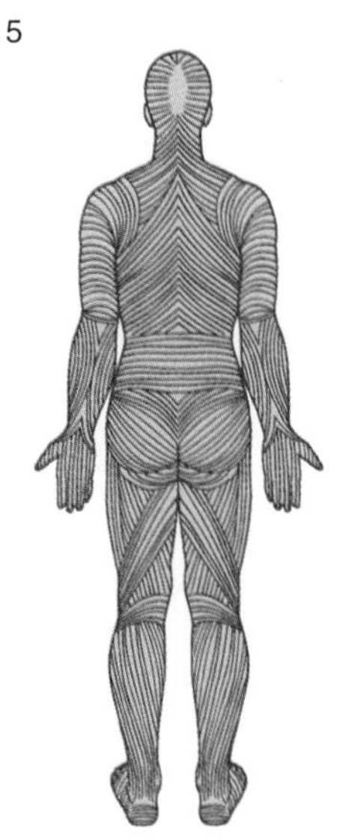

BACK

GENERAL

1 ***back*** the posterior part of the trunk from the neck to the pelvis. <dorsum> A

2 ***regions of back*** the various anatomical regions of the back, including the vertebral, sacral, scapular, infrascapular, and lumbar regions. <regiones dorsales> A

3 **infrascapular region** the region of the back inferior to the scapula and lateral to the inferior thoracic vertebrae. <regio infrascapularis> A

4 **lumbar region** the region of the back lying lateral to the lumbar vertebrae. <regio lumbalis> A

5 **sacral region** the region of the back overlying the sacrum. <regio sacralis> A

6 **scapular region** the region of the back overlying the scapula. <regio scapularis> A

7 **vertebral region** the middle region of the back, overlying the vertebral column. <regio vertebralis> A

8 **lumbar triangle** a small triangular interval between the inferolateral margin of the latissimus dorsi muscle and the external oblique muscle of the abdomen, just superior to the ilium. <trigonum lumbale inferius> B

9 ***superior lumbar triangle*** an inconstant triangle or rhombus bounded by the twelfth rib and the serratus posterior inferior, erector spinae, and internal oblique muscles, and overlapped by the latissimus dorsi and the external oblique, with the thoracolumbar fascia as its floor. When present it is sometimes a site through which abscesses point or hernias occur. <trigonum lumbale superius> B

10 **triangle of auscultation** the area limited by the lower edge of the trapezius muscle, the latissimus dorsi, and the medial margin of the scapula. <trigonum auscultationis> B

BONES

11 **primary curvature of vertebral column** a dorsally convex part of the spinal (vertebral) column. <curvatura primaria columnae vertebralis> C

12 **sacral curvature** the dorsally convex curve formed by the sacrum when seen from the side. Called also *pelvic curvature*. <kyphosis sacralis> C

13 **secondary curvatures of vertebral column** dorsally concave sections of the spinal (vertebral) column. <curvaturae secundariae columnae vertebralis> C

14 **thoracic curvature** the dorsally convex curve formed by the thoracic spinal column when seen from the side. <kyphosis thoracica> C

15 **vertebra prominens** the seventh cervical vertebra, so called because of the length of its spinous process, although it is only the spine that is prominent. (NOTE: the spinous process of the first thoracic vertebra is often more prominent.) C

16 **costal process** a process that projects laterally from the transverse process of a lumbar vertebra and resembles a rib. <processus costiformis> D

17 **inferior articular facet of vertebra** the articulating surface on the inferior articular process of a vertebra. <facies articularis inferior vertebrae> D

18 **inferior articular process of vertebra** a process on either side of a vertebra, arising from the inferior surface of the arch near the junction of the lamina and pedicle; it bears a surface that faces anteriorly and inferiorly, articulating with the superior articular process of the vertebra below. <processus articularis inferior vertebrae> D

19 **lamina of vertebral arch** either of the pair of broad plates of bone flaring out from the pedicles of the vertebral arches and fusing together at the midline to complete the dorsal part of the arch and provide a base for the spinous process. <lamina arcus vertebrae> D

20 **pedicle of vertebral arch** one of the paired parts of the vertebral arch that connect a lamina to the vertebral body. <pediculus arcus vertebrae> D

21 **spinous process of vertebra** a part of the vertebra projecting backward from the arch, giving attachment to muscles of the back. <processus spinosus vertebrae> D

22 **superior articular facet of vertebra** the articulating surface on the superior articular process of a vertebra. <facies articularis superior vertebrae> D

23 **superior articular process of vertebra** a process on either side of a vertebra, arising from the superior surface of the arch near the junction of the lamina and pedicle; it bears a surface that faces posteriorly and superiorly, articulating with the inferior articular process of the vertebra above. <processus articularis superior vertebrae> D

24 **transverse process of vertebra** a process on either side of a vertebra, projecting laterally from the junction between the lamina and the pedicle. <processus transversus vertebrae> D

25 **vertebral arch** the bony arch on the dorsal aspect of a vertebra, composed of the laminae and pedicles. <arcus vertebrae> D

26 **vertebral body** the body of a vertebra, consisting of the centrum, the ossified neurocentral joint and part of the vertebral arches, and the facets for the heads of the ribs. <corpus vertebrae> D

27 **vertebral foramen** the large opening in a vertebra formed by its body and arch. <foramen vertebrale> D

28 ***paravertebral line*** an imaginary line corresponding to the transverse vertebral processes. <linea paravertebralis> D

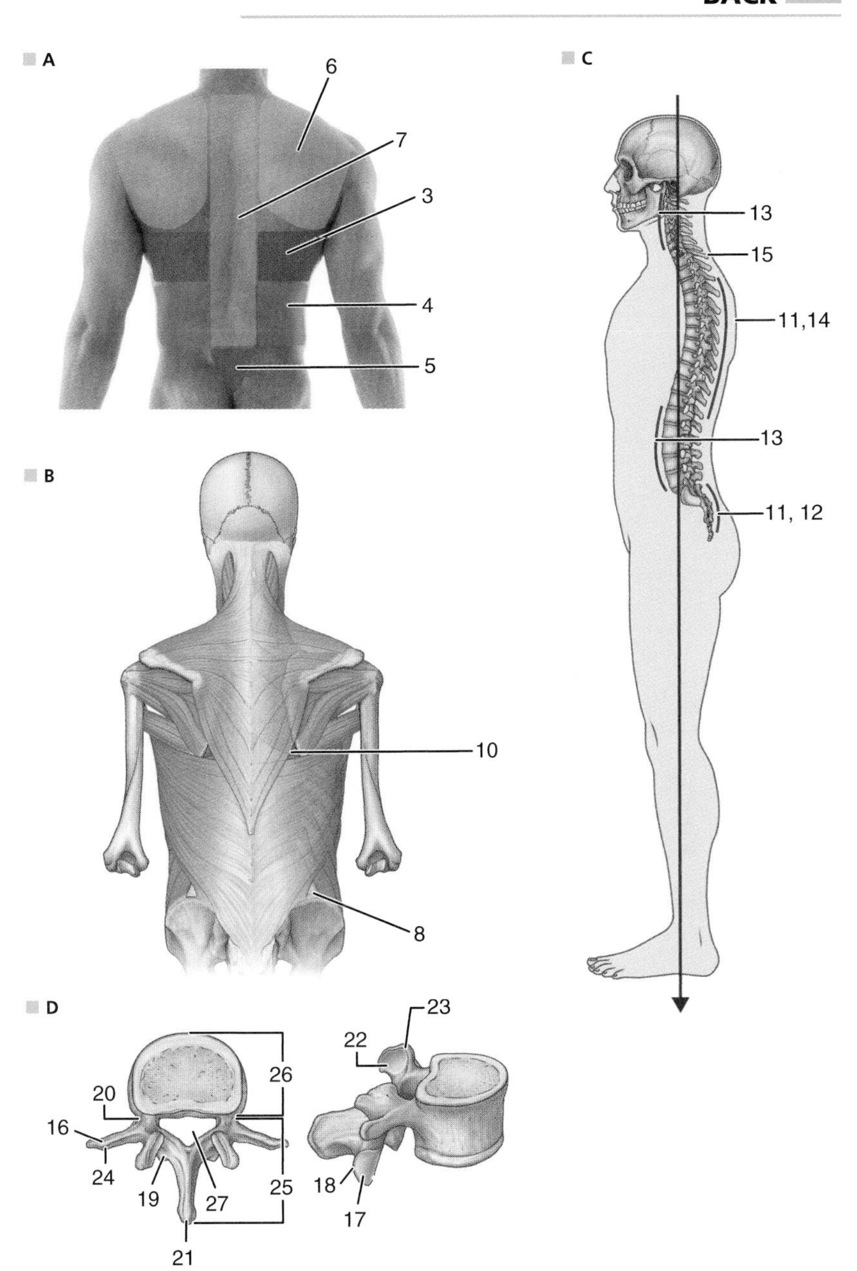
A
6
7
3
4
5
B
10
8
C
13
15
11,14
13
11, 12
D
23
22
26
20
16
24
19
27
25
21
18
17

1 **anterior tubercle of cervical vertebra** a tubercle on the anterior part of the extremity of each transverse process, lying lateral to the posterior tubercle and at a slightly higher level in all except the sixth vertebra, to which are attached the scalenus anterior, longus capitis, and longus colli muscles. <tuberculum anterius vertebrae cervicalis> A

2 **groove for spinal nerve** the groove on the upper surface of each transverse process of a cervical vertebra, extending from the foramen transversarium lateralward and separating the anterior and posterior tubercles. It lodges the anterior branch of a cervical nerve. <sulcus nervi spinalis> A, B

3 **posterior tubercle of cervical vertebra** a tubercle on the posterior part of the extremity of each transverse process of a cervical vertebra, lying lateral to the anterior tubercle and at a slightly lower level in all except the sixth vertebra, to which are attached the splenius, longissimus and iliocostalis cervicis, levator scapulae, and posterior and middle scalene muscles. <tuberculum posterius vertebrae cervicalis> A

4 **uncus of body of cervical vertebra** a hooklike projection found on each side of the superior surface of the third to seventh cervical vertebral bodies. This is a frequent site of formation of spurs (osteophytes), leading to spondylosis uncovertebralis. Called also *uncinate process of cervical vertebra.* <uncus corporis vertebrae cervicalis> B

5 **anterior arch of atlas** the more slender portion joining the lateral masses of the atlas ventrally, constituting about one-fifth of the entire circumference of the atlas. <arcus anterior atlantis> C

6 **anterior tubercle of atlas** the conical eminence on the front of the anterior arch of the atlas. <tuberculum anterius atlantis> C

7 **atlas** the first cervical vertebra, which articulates above with the occipital bone and below with the axis. C

8 **facet for dens** the facet on the inner surface of the anterior arch of the atlas for the articulation of the dens of the axis. <fovea dentis atlantis> C

9 **foramen transversarium** the passage in either process of a cervical vertebra that, in the upper six vertebrae, transmits the vertebral vessels; it is small or may be absent in the seventh. A, C

10 **lateral mass of atlas** the thickened lateral portion of the atlas to which the arches are attached and which bears the articulating surfaces and the transverse process. <massa lateralis atlantis> C

11 **posterior arch of atlas** the slender portion joining the lateral masses of the atlas dorsally, constituting about two-fifths of the entire circumference of the atlas. <arcus posterior atlantis> C

12 **posterior tubercle of atlas** a variable prominence on the posterior surface of the posterior arch of the atlas, which represents a spinous process and gives attachment to the rectus capitis posterior minor muscle. <tuberculum posterius atlantis> C

13 **sulcus for vertebral artery** the groove on the cranial surface of the posterior arch of the atlas; it lodges the vertebral artery and the first spinal nerve. <sulcus arteriae vertebralis atlantis> C

14 **superior articular surface of atlas** the large oval facet on the superior aspect of either lateral mass of the atlas; called also *superior articular facet of atlas.* <facies articularis superior atlantis> C

15 **anterior articular facet of dens** an oval facet on the anterior surface of the dens of the axis, articulating with the fovea dentis of the atlas. <facies articularis anterior dentis> D

16 **axis** the second cervical vertebra; called also *epistropheus, odontoid vertebra,* and *vertebra dentata.* D, E

17 **dens** the toothlike process that projects from the superior surface of the body of the axis, ascending to articulate with the atlas. D, E

18 **apex of dens** the tip of the dens of the axis. <apex dentis> E

19 **posterior articular facet of dens** a smooth groove on the posterior surface of the dens of the axis, which lodges the transverse ligament of the atlas. <facies articularis posterior dentis> E

20 ***median atlantoaxial joint*** a single joint formed by the two articular facets of the dens of the axis, one in relation with the articular facet on the anterior arch of the atlas, the other in relation with the transverse ligament of the atlas. <articulatio atlantoaxialis mediana> F

21 **transverse ligament of atlas** the strong horizontal portion of the cruciform ligament of the atlas. It is attached at each end to the lateral masses of the atlas and curves posteriorly around the dens of the axis. It thus divides the atlantal ring into a smaller anterior division for the dens and a larger posterior division for the spinal cord and related structures. <ligamentum transversum atlantis> F

22 **inferior costal facet** a small facet on the lower edge of the body of a vertebra articulating with the head of a rib. <fovea costalis inferior> G

23 **inferior vertebral notch** the indentation found below each pedicle of a vertebra which, with the indentation located above the pedicle of the vertebra below, forms the intervertebral foramen. <incisura vertebralis inferior> G

24 **superior costal facet** a small facet on the upper edge of the body of a vertebra articulating with the head of a rib. <fovea costalis superior> G

25 **superior vertebral notch** the indentation found above each pedicle of a vertebra which, with the indentation located below the corresponding pedicle of the vertebra above, forms the intervertebral foramen. <incisura vertebralis superior> G

26 **transverse costal facet** a facet on the transverse process of a vertebra for articulation with the tubercle of a rib. <fovea costalis processus transversi> G

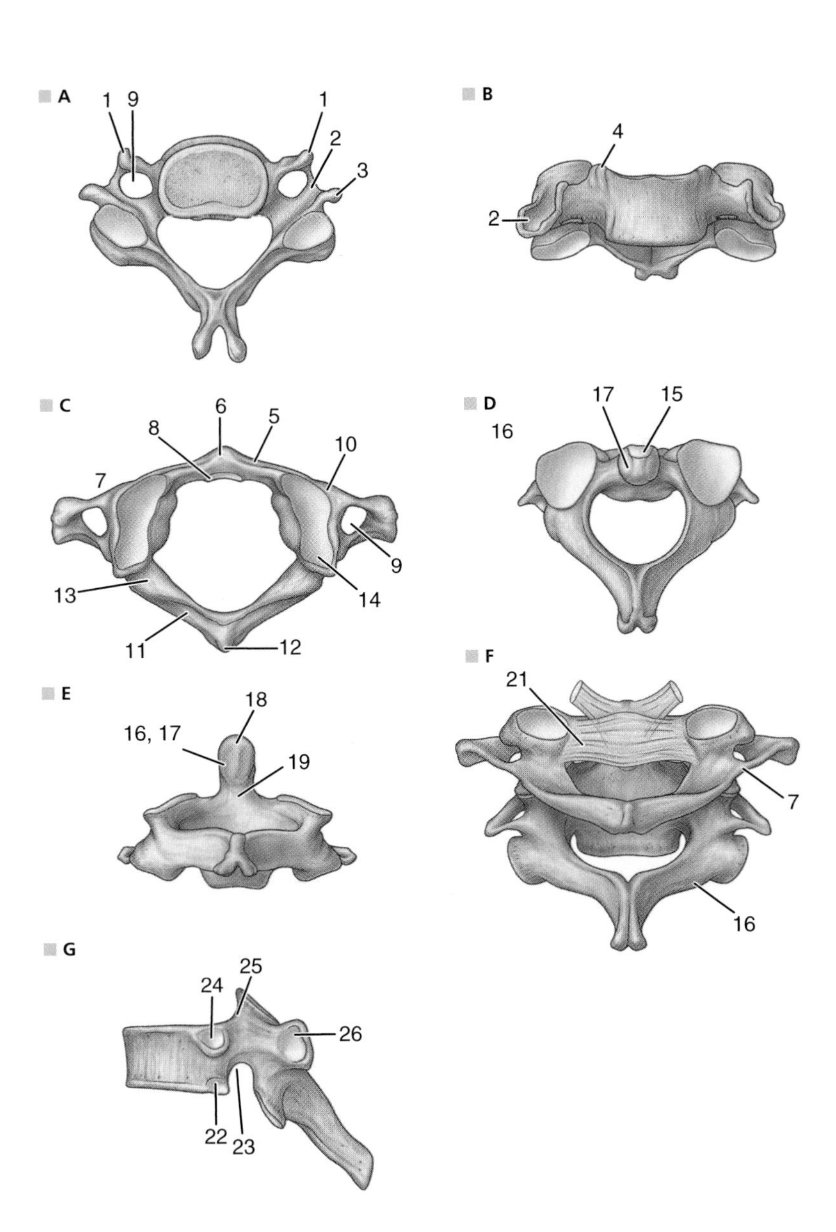
A
1
9
1
2
3
B
4
2
C
6
5
8
10
7
9
13
14
11
12
D
17
15
16
E
18
16, 17
19
F
21
7
16
G
25
24
26
22
23

1 **promontory of sacrum** the prominent anterior border of the pelvic surface of the body of the first sacral vertebra. <promontorium ossis sacri> A

2 **superior articular process of sacrum** either of two processes projecting posteriorly and medially from the first sacral vertebra at the junctions between the body and the alae; they articulate with the inferior articular processes of the fifth lumbar vertebra. <processus articularis superior ossis sacri> A

3 **intervertebral disks** the 23 plates of fibrocartilage found, from the axis to the sacrum, between the bodies of adjacent vertebrae, each consisting of a fibrous ring (anulus fibrosus) enclosing a pulpy center (nucleus pulposus). <disci intervertebrales> B, C

4 **intervertebral foramen** the passage formed by the inferior and superior notches on the pedicles of adjacent vertebrae; it transmits a spinal nerve and vessels. <foramen intervertebrale> B

5 **intervertebral surface** the surface of the vertebral body adjacent to the intervertebral disk, having an elevated rim (the annular apophysis) surrounding a rough center. <facies intervertebralis> B

6 **zygapophyseal joints** the joints between the articular processes of the vertebrae (zygapophyses). <articulationes zygapophysiales> B

7 **anulus fibrosus of intervertebral disk** the circumferential ringlike portion of an intervertebral disk, composed of fibrocartilage and fibrous tissue. <anulus fibrosus disci intervertebralis> C

8 **nucleus pulposus of intervertebral disk** a semifluid mass of fine white and elastic fibers that forms the central portion of an intervertebral disk; it has been regarded as the persistent remains of the embryonic notochord. <nucleus pulposus disci intervertebralis> C

9 **intervertebral symphysis** the union between the vertebral bodies, consisting of the anterior and posterior longitudinal ligaments, and the intervertebral disks. <symphysis intervertebralis> D

10 **vertebral canal** the canal formed by the foramina in the successive vertebrae, which encloses the spinal cord and meninges; called also *spinal canal.* <canalis vertebralis> D

11 ***carotid tubercle*** the large anterior tubercle of the transverse process of the sixth cervical vertebra, which lies lateral to and at a slightly higher level than the posterior tubercle. <tuberculum caroticum> E

12 ***caudal retinaculum*** a fibrous band that extends from the tip of the coccyx to the adjacent skin and thus forms the foveola coccygea. <retinaculum caudale> E

13 **cervical vertebrae** (C1-C7) the upper seven vertebrae, constituting the skeleton of the neck. <vertebrae cervicales> E

14 **coccyx** the small bone caudad to the sacrum in humans, formed by union of four (sometimes five or three) rudimentary vertebrae, and forming the caudal extremity of the vertebral column; called also *tail bone.* <os coccygis> E

15 **lateral part of sacrum** the part or mass of the sacrum on either side lateral to the dorsal and pelvic sacral foramina. <pars lateralis ossis sacri> A

16 **lumbar vertebrae** (L1-L5) the five vertebrae between the thoracic vertebrae and the sacrum. <vertebrae lumbales> E

17 **sacrum** the wedge-shaped bone formed usually by five fused vertebrae that are lodged dorsally between the two hip bones. <os sacrum> E

18 **thoracic vertebrae** (T1-T12) the vertebrae, usually twelve in number, situated between the cervical and the lumbar vertebrae, giving attachment to the ribs and forming part of the posterior wall of the thorax. <vertebrae thoracicae> E

19 **vertebra** any of the thirty-three bones of the spinal (vertebral) column, comprising the seven *cervical,* twelve *thoracic,* five *lumbar,* five *sacral,* and four *coccygeal* vertebrae. <vertebra> E

20 ***inferior articular surface of atlas*** either of the two inferior articular surfaces found on the lateral masses of the atlas. Called also *inferior articular facet of atlas.* <facies articularis inferior atlantis> F

21 **accessory process** a small nodule that projects backward from the posterior surface of the transverse process and lateral to and below the mammillary process of a lumbar vertebra. Such a process also occurs on the tenth, eleventh, and twelfth thoracic vertebrae. <processus accessorius> F

22 **mammillary process** a tubercle on each superior articular process of the lumbar vertebrae and on the tenth, eleventh, and twelfth thoracic vertebrae. <processus mammillaris> F

23 **anterior atlanto-occipital membrane** a single midline ligamentous structure that passes from the anterior arch of the atlas to the anterior margin of the foramen magnum, and corresponds in position with the anterior longitudinal ligament of the vertebral column. Called also *anterior* or *deep atlanto-occipital ligament* and *ligamentum atlantooccipitale anterius.* <membrana atlantooccipitalis anterior> G

24 **atlantooccipital joint** either of two joints, each formed by a superior articular pit of the atlas and a condyle of the occipital bone; called also *craniovertebral, occipitoatlantal,* or *Cruveilhier's joint.* <articulatio atlantooccipitalis> G

25 **lateral atlantoaxial joint** either of a pair of joints, one on each side of the body, formed by the inferior articular surface of the atlas and the superior surface of the axis. <articulatio atlantoaxialis lateralis> G

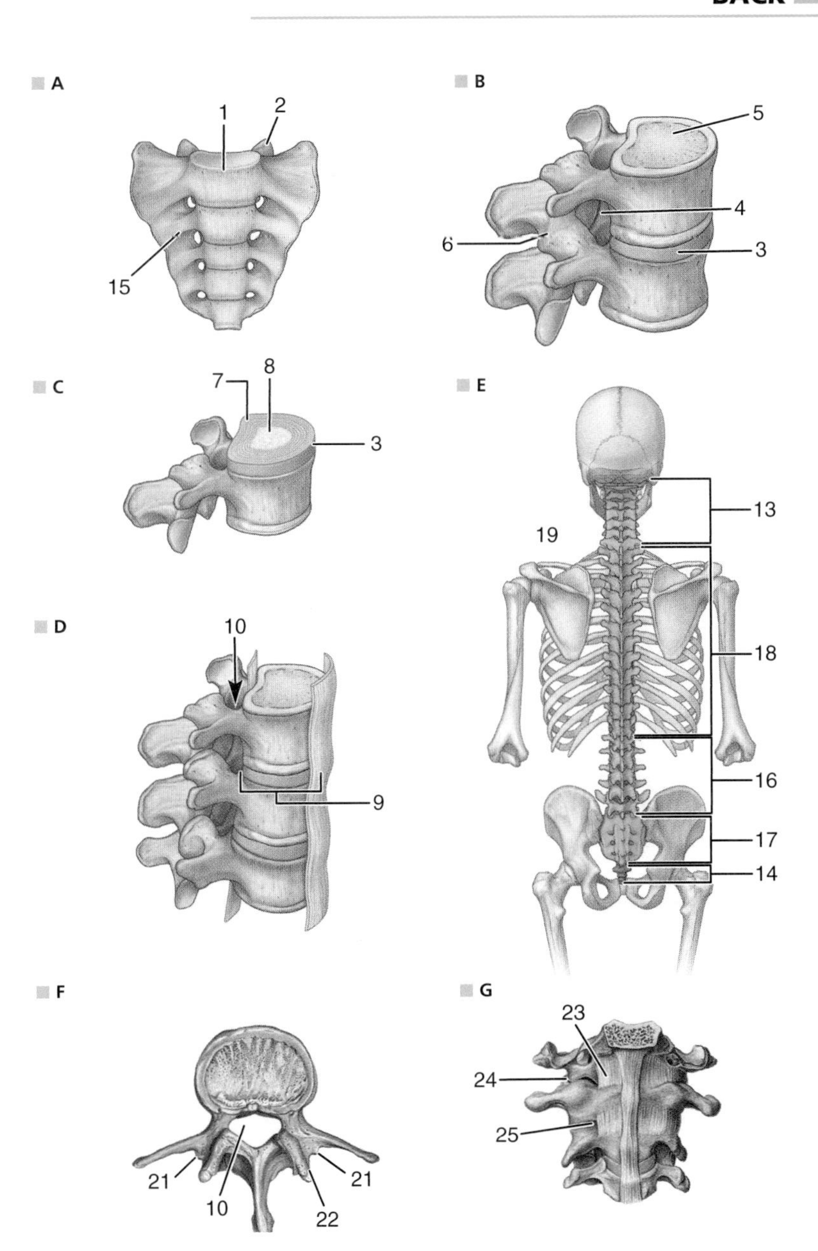
A
1
2
15
B
5
4
6
3
C
7
8
3
E
13
19
18
16
17
14
D
10
9
F
21
10
22
21
G
23
24
25

1 **posterior atlanto-occipital membrane** a single midline ligamentous structure that passes from the posterior arch of the atlas to the posterior margin of the foramen magnum, and corresponds in position with the ligamenta flava. <membrana atlantooccipitalis posterior> A

2 **lateral atlantooccipital ligament** a thickened portion of the articular capsule of the atlantooccipital joint attached to the jugular processes of the occipital bone and to the base of the transverse process of the atlas. <ligamentum atlanto-occipitale laterale> B

3 **tectorial membrane** a strong fibrous band connected cranially with the basilar part of the occipital bone and caudally with the dorsal surface of the bodies of the second and third cervical vertebrae. It is actually the cranial prolongation of the deeper portion of the posterior longitudinal ligament of the vertebral column. <membrana tectoria> C

LIGAMENTS

4 **transverse ligament of atlas** the strong horizontal portion of the cruciform ligament of the atlas. It is attached at each end to the lateral masses of the atlas and curves posteriorly around the dens of the axis. It thus divides the atlantal ring into a smaller anterior division for the dens and a larger posterior division for the spinal cord and related structures. <ligamentum transversum atlantis> D

5 **anterior longitudinal ligament** a single long, fibrous band in the midline, attached to the ventral surfaces of the bodies of the vertebrae; it extends from the occipital bone and the anterior tubercle of the atlas down to the sacrum. <ligamentum longitudinale anterius> E

6 ***anterior sacrococcygeal ligament*** a flat band, homologous with the anterior longitudinal ligament of the vertebral column, that passes from the lower part of the sacrum over onto the anterior part of the coccyx. <ligamentum sacrococcygeum anterius> E

7 ***deep posterior sacrococcygeal ligament*** the terminal portion of the posterior longitudinal ligament of the vertebral column; it helps to unite the dorsal surfaces of the fifth sacral and the coccygeal vertebrae. <ligamentum sacrococcygeum posterius profundum> E

8 **posterior longitudinal ligament** a single midline fibrous band attached to the dorsal surfaces of the bodies of the vertebrae, extending from the occipital bone to the coccyx. <ligamentum longitudinale posterius> E

9 **ligamenta flava** a series of bands of yellow elastic tissue attached to and extending between the ventral portions of the laminae of two adjacent vertebrae, from the junction of the axis and the third cervical vertebra to the junction of the fifth lumbar vertebra and the sacrum. They assist in maintaining or regaining the erect position and serve to close in the spaces between the arches. F

10 **ligamentum nuchae** a broad, fibrous, roughly triangular sagittal septum in the back of the neck, separating the right and left sides. It extends from the tips of the spinous processes of all the cervical vertebrae to attach to the entire length of the external occipital crest. Caudally it is continuous with the supraspinous ligament. G

11 **alar ligaments** two strong bands that pass from the posterolateral part of the tip of the dens of the axis upward and laterally to the condyles of the occipital bone; they limit rotation of the head. <ligamenta alaria> D

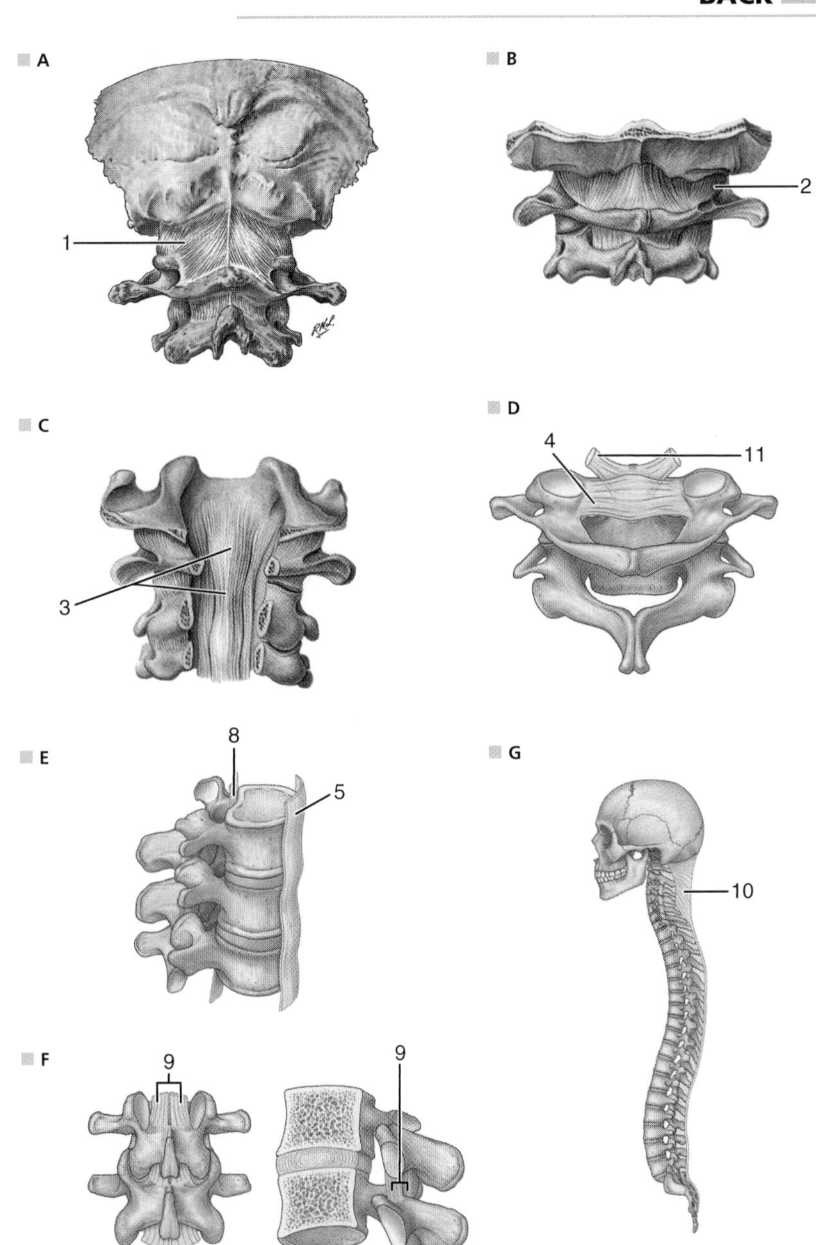
A
B
C
D
E
F
G
1
2
3
4
11
8
5
10
9
9

1 **interspinous ligaments** several fine fibrous membranes that extend from one vertebral spinous process to the next. They extend obliquely from the ligamenta flava ventrally to the supraspinous ligament dorsally, and contain white fibrous and yellow elastic tissue. They are poorly developed or lacking in the cervical region. <ligamenta interspinalia> A

2 ***superficial posterior sacrococcygeal ligament*** a fibrous band continuous with the supraspinous ligament of the vertebral column; attached cranially to the margin of the sacral hiatus, and diverging as it passes caudally to attach to the dorsal surface of the coccyx. <ligamentum sacrococcygeum posterius superficiale> A

3 **supraspinous ligament** a single long, vertical fibrous band passing over and attached to the tips of the spinous processes of the vertebrae from the seventh cervical to the sacrum; it is continuous above with the ligamentum nuchae. <ligamentum supraspinale> A

4 **cruciform ligament of atlas** a ligament in the form of a cross, of which the transverse ligament of the atlas forms the horizontal bar, and the longitudinal fascicles the vertical bar of the cross. <ligamentum cruciforme atlantis> B

5 **apical ligament of dens** a cord of tissue extending from the tip of the dens of the axis to the occipital bone, near the anterior margin of the foramen magnum; it is usually delicate but is sometimes well developed. <ligamentum apicis dentis> C

6 **intertransverse ligaments** several poorly developed fibrous bands that extend from one vertebral transverse process to the next. They consist of fine membranes in the lumbar region and of small cords in the thoracic region, and are lacking in the cervical region. <ligamenta intertransversaria> D

7 ***lateral sacrococcygeal ligament*** a fibrous band, homologous with the intertransverse ligaments, that passes from the transverse process of the first coccygeal vertebra to the lower lateral angle of the sacrum, thus helping to complete the foramen of the fifth sacral nerve. <ligamentum sacrococcygeum laterale> D

MUSCLES

8 ***dorsal muscles*** the muscles of the back. <musculi dorsi>

Superficial Group of Muscles of Back

9 **trapezius muscle** *origin,* occipital bone, ligamentum nuchae, spinous processes of seventh cervical and all thoracic vertebrae; *insertion,* clavicle, acromion, spine of scapula; *innervation,* accessory nerve and cervical plexus; *action,* elevates shoulder, rotates scapula to raise shoulder in abduction of arm, draws scapula backward. <musculus trapezius> E

10 **latissimus dorsi muscle** *origin,* spines of lower thoracic vertebrae, lumbar and sacral vertebrae through attachment to thoracolumbar fascia, iliac crest, lower ribs, inferior angle of scapula; *insertion,* floor of intertubercular sulcus of humerus; *innervation,* thoracodorsal; *action,* adducts, extends, and rotates humerus medially. <musculus latissimus dorsi> E

11 **levator scapulae muscle** *origin,* transverse processes of four upper cervical vertebrae; *insertion,* medial border of scapula; *innervation,* third and fourth cervical; *action,* raises scapula. <musculus levator scapulae> E

12 **rhomboid major muscle** *origin,* spinous processes of second, third, fourth, and fifth thoracic vertebrae; *insertion,* medial margin of scapula; *innervation,* dorsal scapular; *action,* retracts, elevates scapula. <musculus rhomboideus major> E

13 **rhomboid minor muscle** *origin,* spinous processes of seventh cervical to first thoracic vertebrae, lower part of ligamentum nuchae; *insertion,* medial margin of scapula at root of the spine; *innervation,* dorsal scapular; *action,* adducts, elevates scapula. <musculus rhomboideus minor> E

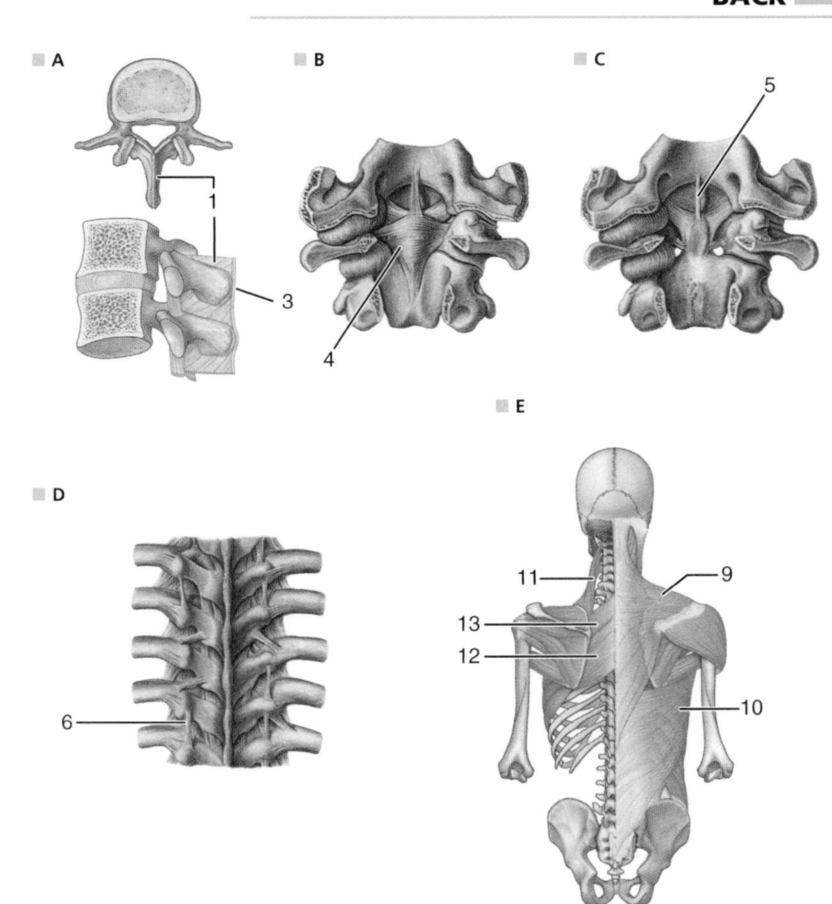
A
B
C
1
3
4
5
D
E
6
11
9
13
12
10

1 **thoracolumbar fascia** the fascia of the back that attaches medially to the spinous processes of the vertebral column for its entire length and blends laterally with the aponeurosis of the transversus abdominis muscle; inferiorly it attaches to the iliac crest and the sacrum. <fascia thoracolumbalis> A

Intermediate Group of Muscles of Back

2 **serratus posterior inferior muscle** *origin,* spines of two lower thoracic and two or three upper lumbar vertebrae; *insertion,* inferior border of four lower ribs; *innervation,* ninth to twelfth thoracic; *action,* lowers ribs in expiration. <musculus serratus posterior inferior> B

3 **serratus posterior superior muscle** *origin,* ligamentum nuchae, spinous processes of upper thoracic vertebrae; *insertion,* second, third, fourth, and fifth ribs; *innervation,* first four thoracic; *action,* raises ribs in inspiration. <musculus serratus posterior superior> B

Deep Group of Muscles of Back

4 **splenius capitis muscle** *origin,* lower half of ligamentum nuchae, spinous processes of seventh cervical and three or four upper thoracic vertebrae; *insertion,* mastoid part of temporal bone, occipital bone; *innervation,* middle and lower cervical; *action,* extends, rotates head. <musculus splenius capitis> C

5 **splenius cervicis muscle** *origin,* spinous processes of third to sixth thoracic vertebrae; *insertion,* transverse processes of two or three upper cervical vertebrae; *innervation,* dorsal branches of lower cervical; *action,* extends, rotates head and neck. <musculus splenius cervicis> C

6 **erector spinae muscle** a name given the fibers of the more superficial of the deep muscles of the back, originating from the sacrum, spines of the lumbar and the eleventh and twelfth thoracic vertebrae, and the iliac crest, which split and insert as the iliocostalis, longissimus, and spinalis muscles. <musculus erector spinae> C

7 **iliocostalis muscle** the lateral division of erector spinae muscle, which includes the *iliocostalis cervicis, iliocostalis thoracis,* and *iliocostalis lumborum muscles.* <musculus iliocostalis> D

8 **iliocostalis cervicis muscle** *origin,* angles of third, fourth, fifth, and sixth ribs; *insertion,* transverse processes of fourth, fifth, and sixth cervical vertebrae; *innervation,* branches of cervical; *action,* extends cervical spine. <musculus iliocostalis cervicis> D

9 **iliocostalis lumborum muscle** *origin,* iliac crest; *insertion,* angles of lower six or seven ribs; *innervation,* branches of thoracic and lumbar; *action,* extends lumbar spine. <pars lumbalis musculi iliocostalis lumborum> D

10 **iliocostalis thoracis muscle** *origin,* upper borders of angles of six lower ribs; *insertion,* angles of six upper ribs and transverse process of seventh cervical vertebra; *innervation,* branches of thoracic; *action,* keeps thoracic spine erect. <pars thoracica musculi iliocostalis lumborum> D

11 **longissimus muscle** the largest element of the erector spinae muscle, which includes the *longissimus capitis, longissimus cervicis,* and *longissimus thoracis muscles.* <musculus longissimus> D

12 **longissimus capitis muscle** *origin,* transverse processes of four or five upper thoracic vertebrae, articular processes of three or four lower cervical vertebrae; *insertion,* mastoid process of temporal bone; *innervation,* branches of cervical; *action,* draws head backward, rotates head. <musculus longissimus capitis> D

13 **longissimus cervicis muscle** *origin,* transverse processes of four or five upper thoracic vertebrae; *insertion,* transverse processes of second to sixth cervical vertebrae; *innervation,* lower cervical and upper thoracic; *action,* extends cervical vertebrae. <musculus longissimus cervicis> D

14 **longissimus thoracis muscle** *origin,* transverse and articular processes of lumbar vertebrae and thoracolumbar fascia; *insertion,* transverse processes of all thoracic vertebrae, nine or ten lower ribs; *innervation,* lumbar and thoracic; *action,* extends thoracic vertebrae. <musculus longissimus thoracis> D

15 **spinalis muscle** the medial division of the erector spinae, including the *spinalis capitis, spinalis cervicis,* and *spinalis thoracis muscles.* <musculus spinalis> D

16 ***spinalis capitis muscle*** *origin,* spines of upper thoracic and lower cervical vertebrae; *insertion,* occipital bone; *innervation,* branches of cervical; *action,* extends head. <musculus spinalis capitis> D

17 ***spinalis cervicis muscle*** *origin,* lower part of ligamentum nuchae, spinous processes of seventh cervical and sometimes two upper thoracic vertebrae; *insertion,* spinous processes of axis and sometimes of second to fourth cervical vertebrae; *innervation,* branches of cervical; *action,* extends vertebral column. <musculus spinalis cervicis> D

18 **spinalis thoracis muscle** *origin,* spinous processes of two upper lumbar and two lower thoracic; *insertion,* spines of upper thoracic vertebrae; *innervation,* branches of thoracic and lumbar; *action,* extends vertebral column. <musculus spinalis thoracis> D

A

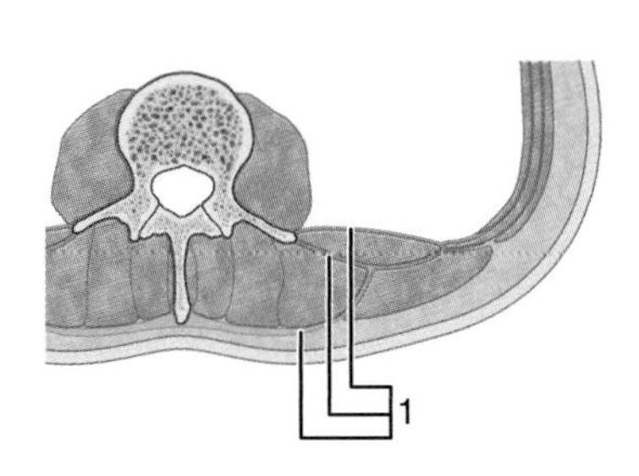

C

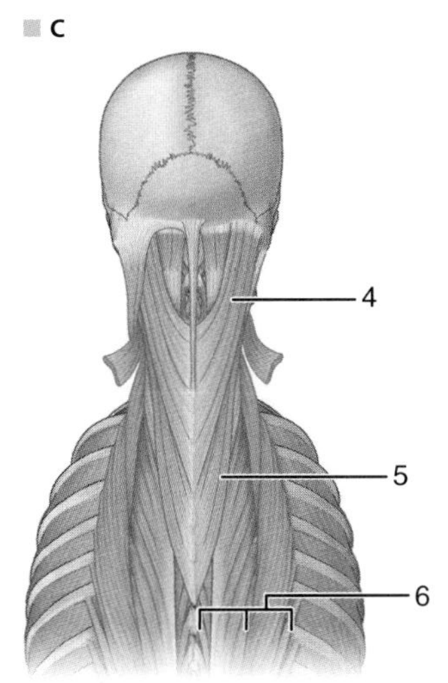

B

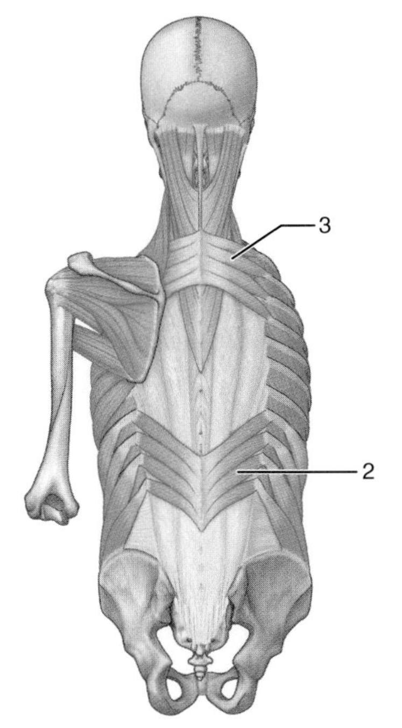

D

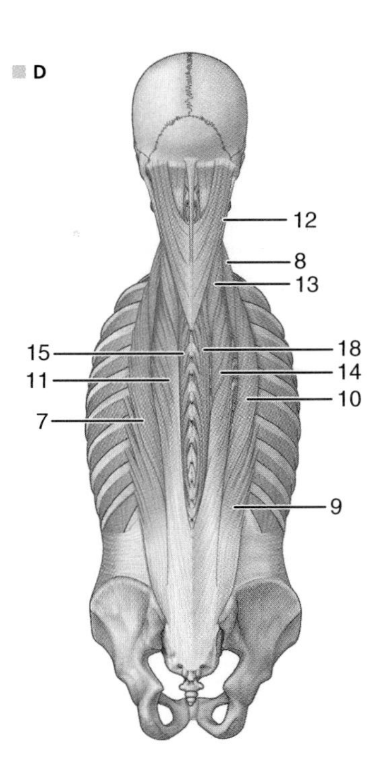

1 **intertransversarii muscles** small muscles passing between the transverse processes of contiguous vertebrae, including the lateral and medial lumbar, thoracic, and anterior, lateral posterior, and medial posterior cervical intertransversarii muscles. <musculi intertransversarii> A

2 ***anterior cervical intertransversarii muscles*** small muscles passing between the anterior tubercles of adjacent cervical vertebrae, innervated by anterior primary rami of spinal nerves, and acting to bend the vertebral column laterally. <musculi intertransversarii anteriores cervicis> A

3 ***lateral lumbar intertransversarii muscles*** small muscles passing between the transverse processes of adjacent lumbar vertebrae, innervated by anterior primary rami of spinal nerves, and acting to bend the vertebral column laterally. <musculi intertransversarii laterales lumborum> A

4 ***lateral posterior cervical intertransversarii muscles*** small muscles passing between the posterior tubercles of adjacent cervical vertebrae, lateral to the posterior medial intertransversarii, innervated by anterior primary rami of spinal nerves, and acting to bend the vertebral column laterally. <musculi intertransversarii posteriores laterales cervicis> A

5 ***medial lumbar intertransversarii muscles*** small muscles passing from the accessory process of one lumbar vertebra to the mammillary process of the contiguous lumbar vertebra, innervated by posterior primary rami of spinal nerves, and acting to bend the vertebral column laterally. <musculi intertransversarii mediales lumborum> A

6 ***medial posterior cervical intertransversarii muscles*** small muscles passing between the posterior tubercles of adjacent cervical vertebrae, close to the vertebral body, innervated by posterior primary rami of spinal nerves, and acting to bend the vertebral column laterally. <musculi intertransversarii posteriores mediales cervicis> A

7 ***thoracic intertransversarii muscles*** poorly developed muscle bundles extending between the anterior tubercles of adjacent thoracic vertebrae, innervated by posterior primary rami of spinal nerves, and acting to bend the vertebral column laterally. <musculi intertransversarii thoracis> A

8 **levatores costarum muscles** (12 on each side): originating from the transverse processes of the seventh cervical and first to eleventh thoracic vertebrae and inserting medial to the angle of a lower rib; innervated by intercostal nerves and aiding in elevation of the ribs in respiration. <musculi levatores costarum> A

9 **levatores costarum breves muscles** the levatores costarum muscles of each side that insert medial to the angle of the rib next below the vertebra of origin. <musculi levatores costarum breves> A

10 **levatores costarum longi muscles** the lower levatores costarum muscles of each side, which have fascicles extending down to the second rib below the vertebra of origin. <musculi levatores costarum longi> A

11 **multifidus muscles** *origin,* sacrum, sacroiliac ligament, and mammillary processes of lumbar, transverse processes of thoracic, and articular processes of cervical vertebrae; *insertion,* spines of contiguous vertebrae above; *innervation,* posterior rami of spinal nerves; *action,* extend, rotate vertebral column. <musculi multifidi> A

12 **rotatores muscles** a series of small muscles deep in the groove between the spinous and transverse processes of the vertebrae. <musculi rotatores> A

13 ***rotatores cervicis muscles*** *origin,* transverse processes of cervical vertebrae; *insertion,* base of spinous process of superjacent vertebrae; *innervation,* spinal nerves; *action,* extend vertebral column and rotate it toward the opposite side. <musculi rotatores cervicis> A

14 ***rotatores lumborum muscles*** *origin,* transverse processes of lumbar vertebrae; *insertion,* base of spinous process of superjacent vertebrae; *innervation,* spinal nerves; *action,* extend vertebral column and rotate it toward the opposite side. <musculi rotatores lumborum> A

15 ***rotatores thoracis muscles*** *origin,* transverse processes of thoracic vertebrae; *insertion,* base of spinous process of superjacent vertebrae; *innervation,* spinal nerves; *action,* extend vertebral column and rotate it toward the opposite side. <musculi rotatores thoracis> A

16 ***semispinalis muscle*** a muscle composed of fibers extending obliquely from the transverse processes of the vertebrae to the spine, except for the semispinalis capitis; it includes the *m. semispinalis capitis, m. semispinalis cervicis,* and *m. semispinalis thoracis.* <musculus semispinalis> A

17 **semispinalis capitis muscle** *origin,* transverse processes of five or six upper thoracic and four lower cervical vertebrae; *insertion,* between superior and inferior nuchal lines of occipital bone; *innervation,* suboccipital, greater occipital, and other branches of cervical; *action,* extends head and rotates to opposite side. <musculus semispinalis capitis> A

18 **semispinalis cervicis muscle** *origin,* transverse processes of five or six upper thoracic vertebrae; *insertion,* spinous processes of second to fifth cervical vertebrae; *innervation,* branches of cervical; *action,* extends, rotates vertebral column. <musculus semispinalis cervicis> A

19 **semispinalis thoracis muscle** *origin,* transverse processes of sixth to tenth thoracic vertebrae; *insertion,* spinous processes of two lower cervical and four upper thoracic vertebrae; *innervation,* spinal nerves; *action,* extends, rotates vertebral column. <musculus semispinalis thoracis> A

20 ***transversospinales muscles*** a general term including the semispinalis, multifidus and rotatores muscles. <musculi transversospinales> A

A

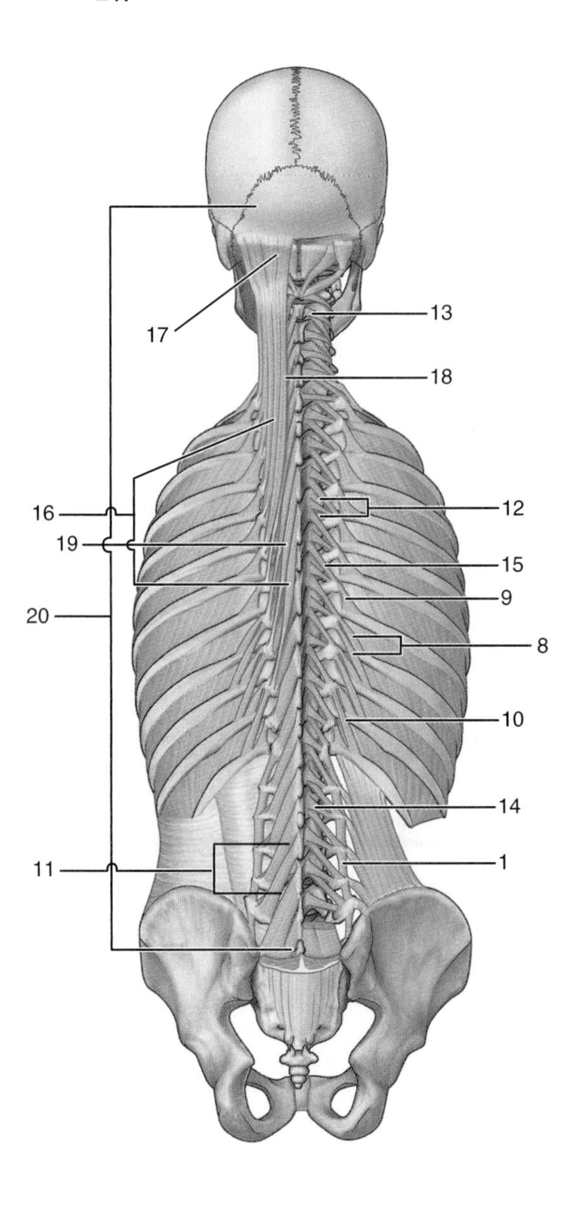

1 **interspinales muscles** short bands of muscle fibers between spinous processes of contiguous vertebrae. <musculi interspinales> A

2 ***interspinales cervicis muscles*** paired bands of muscle fibers extending between spinous processes of contiguous cervical vertebrae, innervated by spinal nerves, and acting to extend the vertebral column. <musculi interspinales cervicis> A

3 ***interspinales lumborum muscles*** paired bands of muscle fibers extending between spinous processes of contiguous lumbar vertebrae, innervated by spinal nerves, and acting to extend the vertebral column. <musculi interspinales lumborum> A

4 ***interspinales thoracis muscles*** paired bands of muscle fibers extending between spinous processes of contiguous thoracic vertebrae, innervated by spinal nerves, and acting to extend the vertebral column. <musculi interspinales thoracis> A

Suboccipital Group of Muscles of Back

5 **obliquus capitis inferior muscle** *origin,* spinous process of axis; *insertion,* transverse process of atlas; *innervation,* dorsal branches of spinal nerves; *action,* rotates atlas and head. <musculus obliquus capitis inferior> B

6 **obliquus capitis superior muscle** *origin,* transverse process of atlas; *insertion,* occipital bone; *innervation,* dorsal branches of spinal nerves; *action,* extends and moves head laterally. <musculus obliquus capitis superior> B

7 **rectus capitis posterior major muscle** *origin,* spinous process of axis; *insertion,* occipital bone; *innervation,* suboccipital and greater occipital; *action,* extends head. <musculus rectus capitis posterior major> B

8 **rectus capitis posterior minor muscle** *origin,* posterior tubercle of atlas; *insertion,* occipital bone; *innervation,* suboccipital and greater occipital; *action,* extends head. <musculus rectus capitis posterior minor> B

9 ***suboccipital muscles*** the muscles situated just below the occipital bone, including the recti capitis posteriores major and minor, the obliquus capitis inferior and superior, the recti capitis anterior and lateral, the splenius capitis, and the longus capitis muscles. <musculi suboccipitales> B

Innervation

10 **posterior cutaneous branch of posterior ramus of thoracic nerve** cutaneous branches of either of the two divisions of a posterior ramus. <ramus cutaneus posterior rami posterioris nervi thoracici> C

11 **suboccipital nerve** *origin,* posterior ramus of first cervical nerve; *distribution,* emerges above posterior arch of atlas and supplies muscles of suboccipital triangle and semispinalis capitis muscle; *modality,* motor. <nervus suboccipitalis> B

12 **lesser occipital nerve** *origin,* superficial cervical plexus (C2-C3); *distribution,* ascends behind the auricle and supplies some of the skin on the side of the head and on the cranial surface of the auricle; *modality,* general sensory. <nervus occipitalis minor> D

13 **third occipital nerve** *origin,* medial branch of posterior ramus of C3; *distribution,* skin of upper part of back of neck and head; *modality,* general sensory. <nervus occipitalis tertius> D

VESSELS

14 ***vertebral vein*** a vein that arises from the suboccipital venous plexus, passes with the vertebral artery through the foramina of the transverse processes of the upper six cervical vertebrae, and opens into the brachiocephalic vein. <vena vertebralis>

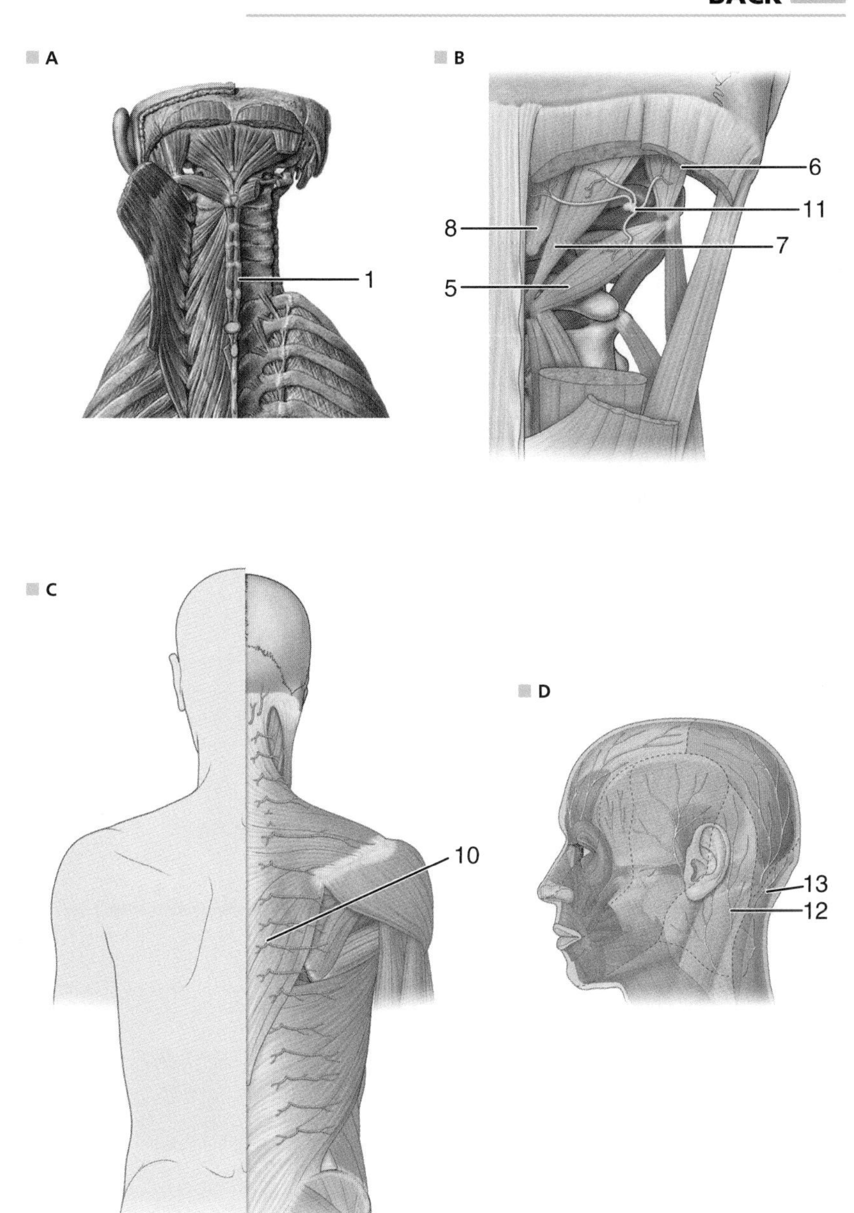
A
B
C
D
1
6
11
8
7
5
10
13
12

1 **superficial branch of transverse cervical artery** a branch that arises from the transverse cervical artery at the anterior border of the levator scapulae muscle, it has ascending and descending branches that supply the levator scapulae, trapezius, and splenius muscles. <ramus superficialis arteriae transversae colli> A

2 **deep branch of transverse cervical artery** *origin,* transverse cervical artery; *branches,* none; *distribution,* rhomboid, latissimus, and trapezius muscles. It is an anatomical variant that is called the dorsal scapular artery when it is a branch of the subclavian artery. <ramus profundus arteriae transversae colli> B

3 **atlantic part of vertebral artery** the third part of the artery, which winds behind the lateral mass of the atlas to lie in a groove on the upper surface of the posterior arch of the atlas. <pars atlantica arteriae vertebralis> C

4 **segmental spinal branches of posterior intercostal artery** branches that arise from each posterior intercostal artery and enter the vertebral canal through the vertebral foramen to supply the vertebrae, spinal cord, and meninges. They may arise as single spinal branches that enter the foramen, and then divide into anterior and posterior branches, or may arise from the dorsal branch as a number of independent spinal branches. <rami spinales rami dorsalis arteriae intercostalis posterioris> D

5 **cervical part of vertebral artery** the second part of the vertebral artery, ascending through the transverse foramina of the upper six cervical vertebrae; it has spinal and muscular branches. <pars transversaria arteriae vertebralis> E

6 ***posterior branch of first posterior intercostal artery*** the posterior branch arising from the first posterior intercostal artery. Its distribution is similar to that of the other posterior intercostals. <ramus dorsalis arteriae intercostalis posterioris primae> F

7 **posterior branch of posterior intercostal artery** a branch arising from a posterior intercostal artery, passing backward with the posterior branch of the corresponding intercostal nerve to supply the posterior thoracic wall; it has several spinal branches and a medial and a lateral cutaneous branch. <ramus dorsalis arteriae intercostalis posterioris> F

8 ***posterior branch of posterior intercostal vein*** a branch corresponding to the posterior branch of the posterior intercostal artery. <ramus dorsalis venae intercostalis posterioris> F

9 ***posterior branch of second posterior intercostal artery*** the posterior branch arising from the second posterior intercostal artery. Its distribution is similar to that of the other posterior intercostals. <ramus dorsalis arteriae intercostalis posterioris secundae> F

10 ***posterior branch of subcostal artery*** a branch supplying back muscles, its distribution being similar to that of the posterior branches of the posterior intercostal arteries. <ramus dorsalis arteriae subcostalis> F

11 **posterior intercostal arteries** eleven pairs (the two highest called *first* and *second*): *origin,* thoracic aorta; *branches,* posterior, collateral, muscular, and lateral cutaneous; *distribution,* thoracic wall. <arteriae intercostales posteriores> F

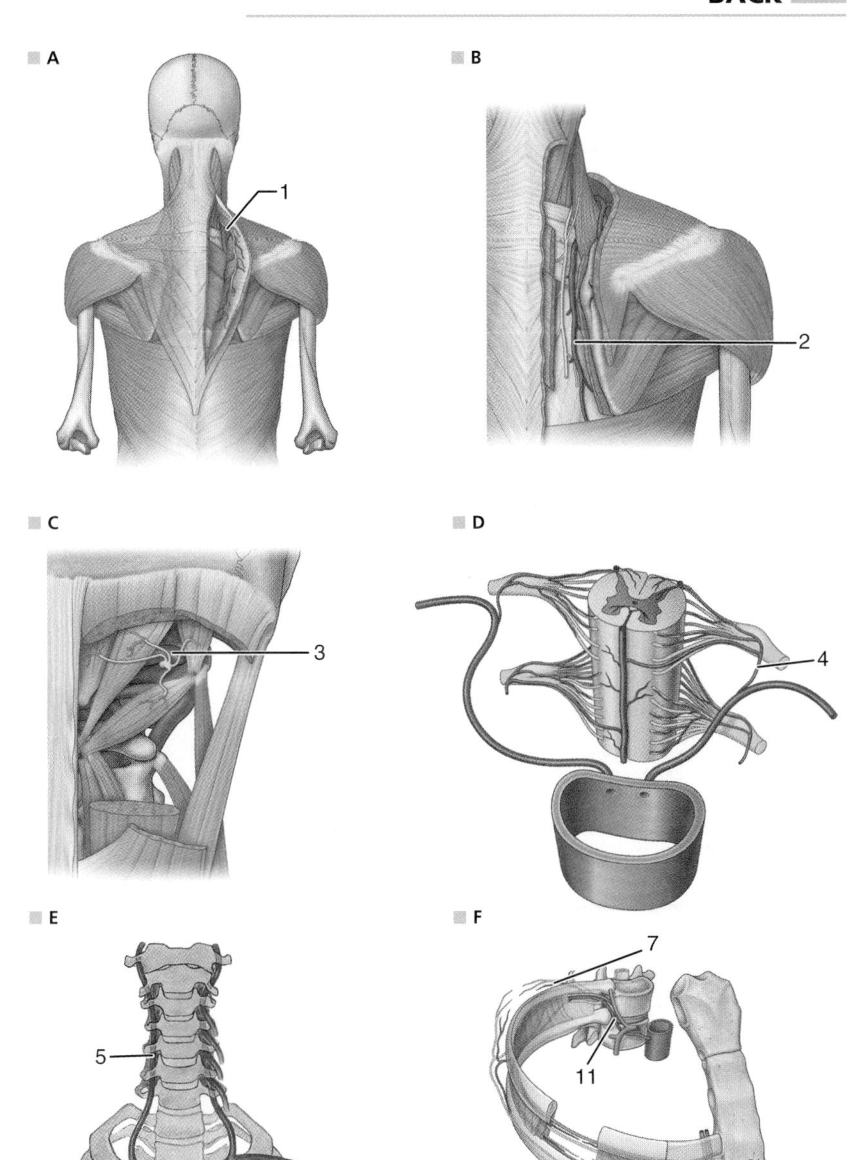
A
1
B
2
C
3
D
4
E
5
F
7
11

1 ***posterior branch of lumbar artery*** the larger of the two branches into which each lumbar artery (four or five) divides; it supplies lumbar back muscles, joints, and skin, and gives off a spinal branch. <ramus dorsalis arteriae lumbalis> A

2 **postcentral branch of posterior branch of posterior intercostal artery** one of the spinal branches of the posterior branch of a posterior intercostal artery; it is the branch that tracks behind the vertebral body. <ramus postcentralis rami dorsalis arteriae intercostalis posterioris> A

3 **prelaminar branch of posterior branch of posterior intercostal artery** one of the spinal branches of the posterior branch of a posterior intercostal artery; it is the branch that tracks in front of the lamina of the vertebral arch. <ramus prelaminaris rami dorsalis arteriae intercostalis posterioris> A

4 **basivertebral veins** venous sinuses in the cancellous tissue of the bodies of the vertebrae, which communicate with the plexus of veins on the anterior surface of the vertebrae and with the anterior internal and anterior external vertebral plexuses. <venae basivertebrales> B

5 ***accessory vertebral vein*** a vein that sometimes arises from a plexus formed around the vertebral artery by the vertebral vein, descends with the vertebral vein, and emerges through the transverse foramen of the seventh cervical vertebra to empty into the brachiocephalic vein. <vena vertebralis accessoria> B

6 **anterior external vertebral venous plexus** the venous plexus formed by the anterior external group of veins of the vertebral column that lies on the anterior aspects of the vertebral bodies. <plexus venosus vertebralis externus anterior> B

7 **anterior internal vertebral venous plexus** the venous plexus formed by the anterior internal group of veins of the vertebral column that lies on the posterior aspects of the vertebral bodies and intervertebral disks, on either side of the posterior longitudinal ligament. <plexus venosus vertebralis internus anterior> B

8 ***anterior vertebral vein*** a small vein accompanying the ascending cervical artery; it arises in a venous plexus adjacent to the more cranial cervical transverse processes, and descends to end in the vertebral vein. <vena vertebralis anterior> B

9 **intervertebral vein** any one of the veins that drain the vertebral plexuses, passing out through the intervertebral foramina and emptying into the regional veins: in the neck, into the vertebral; in the thorax, the intercostal; in the abdomen, the lumbar; and in the pelvis, the lateral sacral veins. <vena intervertebralis> B

10 **posterior external vertebral venous plexus** the venous plexus formed by the posterior external group of veins of the vertebral column that lie on the posterior aspects of the laminae and around the spinous, articular, and transverse processes of the vertebrae. <plexus venosus vertebralis externus posterior> B

11 **posterior internal vertebral venous plexus** the venous plexus formed by the posterior internal group of veins of the vertebral column that lie on either side of the midline in front of the vertebral arches and ligamenta flava. <plexus venosus vertebralis internus posterior> B

12 ***suboccipital venous plexus*** that part of the posterior external vertebral plexus which lies on and in the suboccipital triangle, receives the occipital veins of the scalp, and drains into the vertebral vein. <plexus venosus suboccipitalis> B

13 ***veins of the vertebral column*** a plexiform venous network extending the entire length of the vertebral column, outside or inside the vertebral canal; the anterior and posterior external and anterior and posterior internal groups freely anastomose and end in the intervertebral veins. <venae columnae> B

14 ***descending branch of occipital artery*** a branch that arises from the occipital artery on the obliquus capitis superior muscle and divides into superficial and deep branches, supplying the trapezius and deep neck muscles. <ramus descendens arteriae occipitalis>

A

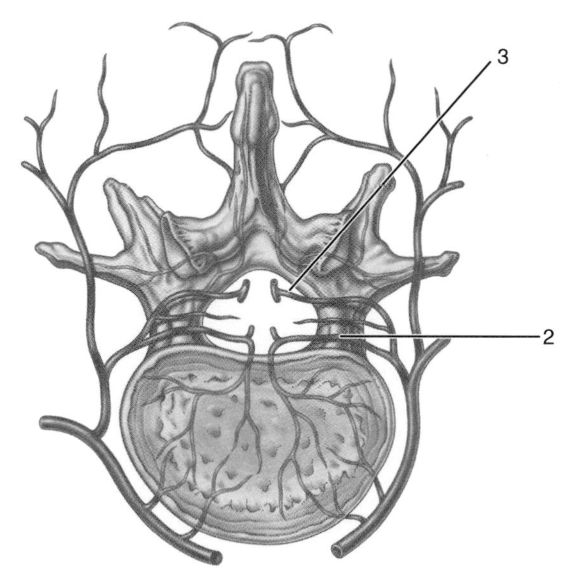

B

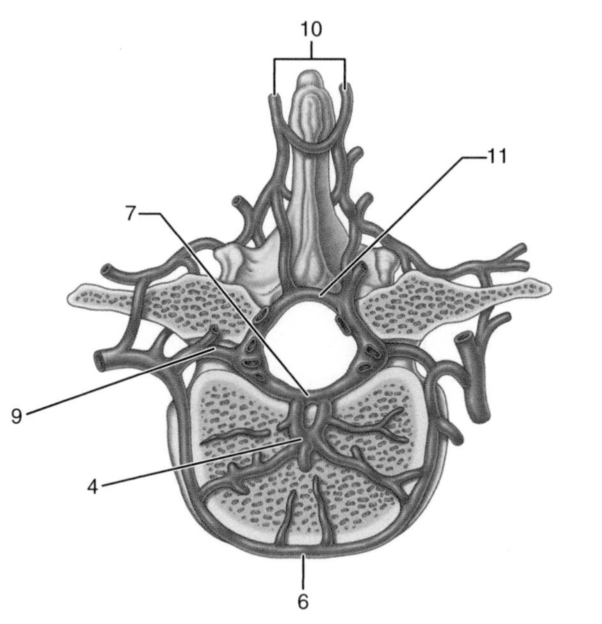

SPINAL CORD

1 **spinal cord** that part of the central nervous system that is lodged in the vertebral canal. It extends from the foramen magnum, where it is continuous with the medulla oblongata, to the upper part of the lumbar region, ending between the twelfth thoracic and third lumbar vertebrae, often at or near the first and second lumbar vertebrae. It is composed of an inner core of *gray substance,* and an outer layer of *white substance,* and is enclosed in three protective membranes: the *dura mater, arachnoid,* and *pia mater.* <medulla spinalis> A

2 **cauda equina** the collection of spinal roots that descend from the lower part of the spinal cord and occupy the vertebral canal below the cord; their appearance resembles the tail of a horse. A

3 **cervical enlargement** the enlargement of the cervical spinal cord at the level of attachment of the nerves to the upper limbs. <intumescentia cervicalis> A

4 **cervical nerves** the eight pairs of nerves (C1-C8) that arise from the cervical segments of the spinal cord and, except for the last pair, leave the vertebral column above the correspondingly numbered vertebra. The anterior branches of the upper four, on either side, unite to form the cervical plexus, and those of the lower four, together with the anterior branch of the first thoracic nerve, form most of the brachial plexus. <nervi cervicales> A

5 ***cervical part of spinal cord*** the part of the cord that is within the cervical part of the vertebral canal and gives rise to the eight pairs of cervical spinal nerves. <pars cervicalis medullae spinalis> A

6 **coccygeal nerve** either of the thirty-first pair of spinal nerves (Co), arising from the coccygeal segment of the spinal cord. <nervus coccygeus> A

7 ***coccygeal part of spinal cord*** the part of the cord that gives rise to the coccygeal spinal nerves. <pars coccygea medullae spinalis> A

8 **conus medullaris** the cone-shaped lower end of the spinal cord, at the level of the upper lumbar vertebrae. A

9 **lumbar nerves** the five pairs of nerves (L1-L5) that arise from the lumbar segments of the spinal cord, each pair leaving the vertebral column below the correspondingly numbered vertebra. The anterior branches of these nerves participate in the formation of the lumbosacral plexus. <nervi lumbales> A

10 ***lumbar part of spinal cord*** the part of the cord that is within the lower thoracic part of the vertebral canal (in adults) and gives rise to the five pairs of lumbar spinal nerves. <pars lumbalis medullae spinalis> A

11 **lumbosacral enlargement** the enlargement of the lumbar spinal cord at the level of attachment of the nerves to the lower limbs. <intumescentia lumbosacralis> A

12 **sacral nerves** the five pairs of nerves (S1-S5) that arise from the sacral segments of the spinal cord; the anterior branches of the first four pairs participate in the formation of the sacral plexus. <nervi sacrales> A

13 ***sacral part of spinal cord*** the part of the cord that is within the lumbar part of the vertebral canal and gives rise to the five pairs of sacral spinal nerves *(segmenta sacralia medullae spinalis [1 to 5]).* <pars sacralis medullae spinalis> A

14 ***segments of spinal cord*** small sections of the spinal cord, to each of which is attached anterior and posterior roots of one of the 31 pairs of spinal nerves; there are 8 cervical, 12 thoracic, 5 lumbar, 5 sacral, and one or more coccygeal segments. <segmenta medullae spinalis> A

15 ***spinal nerves*** the thirty-one pairs of nerves that arise from the spinal cord and pass out between the vertebrae, including the 8 pairs of cervical, 12 of thoracic, 5 of lumbar, 5 of sacral, and 1 of coccygeal nerves. <nervi spinales> A

16 **thoracic nerves** the twelve pairs of spinal nerves (T1-T12) that arise from the thoracic segments of the spinal cord, each pair leaving the vertebral column below the correspondingly numbered vertebra. They innervate the body wall of the thorax and upper abdomen. <nervi thoracici> A

17 ***thoracic part of spinal cord*** the part of the cord that is within the upper three-fourths of the thoracic part of the vertebral canal (in the adult) and gives rise to the twelve pairs of thoracic spinal nerves. <pars thoracica medullae spinalis> A

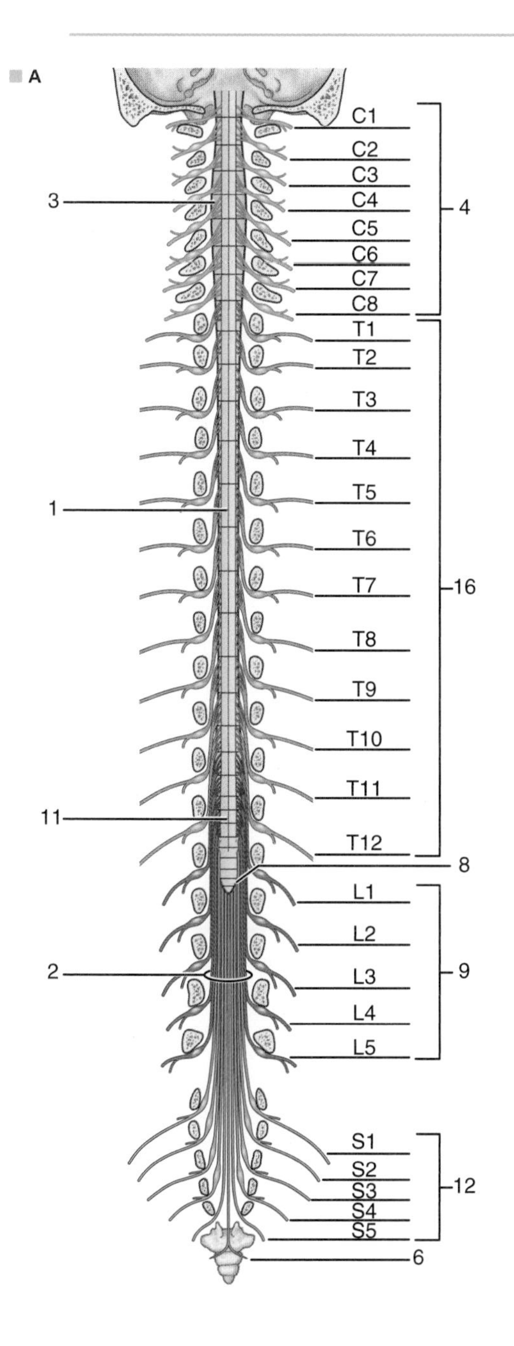
A
C1
C2
C3
C4
C5
C6
C7
C8
T1
T2
T3
T4
T5
T6
T7
T8
T9
T10
T11
T12
L1
L2
L3
L4
L5
S1
S2
S3
S4
S5
1
2
3
4
6
8
9
11
12
16

1 **vertebral column** the columnar assemblage of the vertebrae from the cranium through the coccyx; called also backbone and spine. <columna vertebralis> A

2 ***vertebral nerve*** *origin,* cervicothoracic and vertebral ganglia; *distribution,* ascends with vertebral artery and gives fibers to spinal meninges, cervical nerves, and posterior cranial fossa; *modality,* sympathetic. <nervus vertebralis> See B, p. 51

3 **lateral branch of posterior ramus of cervical nerve** the lateral branch that arises from the posterior ramus of a cervical nerve and supplies adjacent skin and muscles. <ramus lateralis rami posterioris nervi cervicalis> B

4 **lateral branch of posterior ramus of lumbar nerve** a branch that runs inferolaterally from the posterior ramus of each lumbar nerve and innervates adjacent muscle; these branches have upper terminal branches that constitute the superior cluneal nerves and innervate the skin of the buttock. <ramus lateralis rami posterioris nervi lumbalis> B

5 **lateral branch of posterior ramus of thoracic nerve** the lateral of the terminal divisions of the posterior ramus; it supplies first its corresponding levator costae muscle and then the longissimus thoracis and iliocostalis thoracis muscles. The lower branches pierce the latissimus dorsi and supply the skin of the back. <ramus lateralis rami posterioris nervi thoracici> B

6 **medial branch of posterior ramus of cervical nerve** the medial branch that arises from the posterior ramus of a cervical nerve and supplies muscle, periosteum, ligaments, and joints; all except the first and sometimes the sixth through eighth have an eventual cutaneous distribution. <ramus medialis rami posterioris nervi cervicalis> B

7 **medial branch of posterior ramus of lumbar nerve** a branch of the posterior ramus of each lumbar nerve, mainly innervating deep muscle but also helping supply ligaments, periosteum, and joints. <ramus medialis rami posterioris nervi lumbalis> B

8 ***medial branch of posterior ramus of sacral nerve*** a branch arising from the posterior ramus of one of the three upper sacral nerves and innervating the multifidus muscle. <ramus medialis rami posterioris nervi sacralis> B

9 **medial branch of posterior ramus of thoracic nerve** one of the terminal divisions of the posterior ramus of the thoracic nerve, supplying periosteum, ligaments, and joints. Those of the upper thoracic nerves supply the skin of the back and those of the lower nerves supply primarily the erector spinae muscle. <ramus medialis rami posterioris nervi thoracici> B

10 **dural part of filum terminale** the downward prolongation of the spinal dura mater from the lower border of the second sacral vertebra to the first coccygeal vertebral segment. <pars duralis fili terminalis> C

11 **filum terminale** a slender threadlike filament of connective tissue that descends from the conus medullaris to the base of the coccyx; divided into a pial part and a dural part, the dividing line being the lower border of the second sacral vertebra. C

12 **pial part of filum terminale** the prolongation of the spinal pia mater, surrounded by extensions of the dural and arachnoid meninges, from the conus medullaris to the lower border of the second sacral vertebra. <pars pialis fili terminalis> C

A

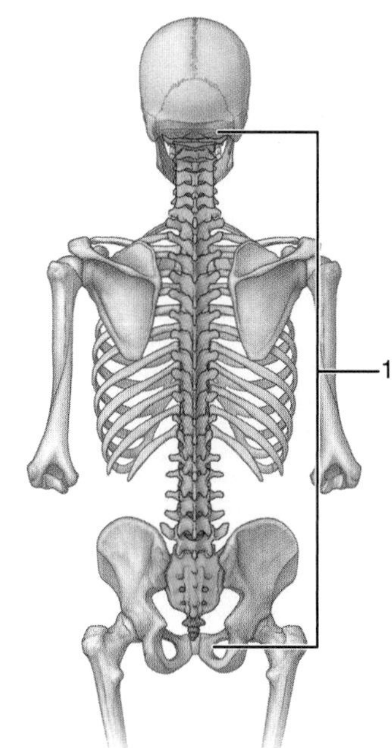

B

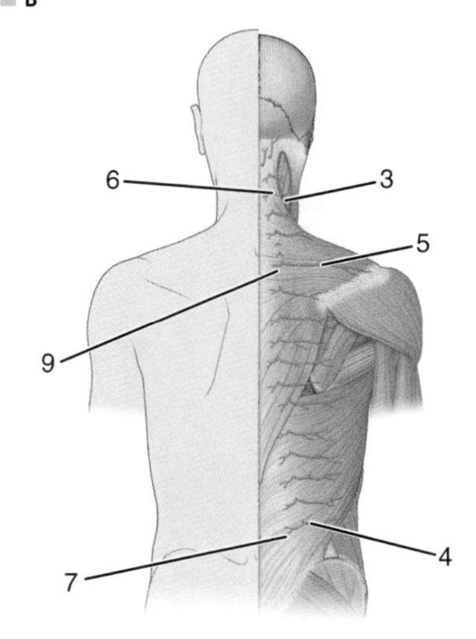

C

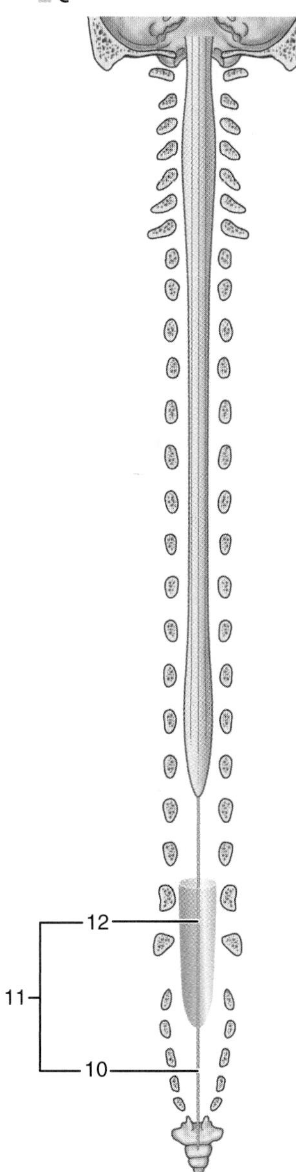

1 **anterior median fissure of spinal cord** the deep longitudinal fissure in the median plane of the anterior aspect of the spinal cord; it contains the anterior spinal artery ensheathed in the linea splendens. <fissura mediana anterior medullae spinalis> A

2 **base of posterior horn of spinal cord** the portion of the posterior horn of gray substance in the spinal cord that is continuous with the lateral horn. <basis cornus posterioris medullae spinalis> A

3 **central canal of spinal cord** a small canal extending throughout the length of the spinal cord, lined by ependymal cells. Above, it continues into the medulla oblongata, where it opens into the fourth ventricle. <canalis centralis medullae spinalis> A

4 **gray matter of spinal cord** the gray nervous tissue composed of nerve cell bodies, unmyelinated nerve fibers, and supportive tissue; it contains fewer myelinated fibers but more nerve cell bodies, unmyelinated nerve fibers, and blood vessels than the white matter. <substantia grisea medullae spinalis> A

5 **head of posterior horn of spinal cord** the oval or fusiform portion of the dorsal horn of gray substance in the spinal cord between the constricted portion (neck) and the apex of the horn. <caput cornus posterioris medullae spinalis> A

6 ***linea splendens*** the sheath for the anterior spinal artery formed by the pia mater in the fissura mediana anterior medullae spinalis. A

7 **neck of posterior horn of spinal cord** the constricted portion of the posterior horn of gray matter in the spinal cord between the base of the horn and the head. <cervix cornus posterioris medullae spinalis> A

8 **posterior median fissure of spinal cord** a shallow vertical groove on the posterior median surface of the spinal cord, separating the two posterior funiculi; the posterior median septum extends out from it. Called also *posterior median sulcus of spinal cord.* <sulcus medianus posterior medullae spinalis> A

9 **posterolateral sulcus of spinal cord** a longitudinal sulcus on the posterolateral surface of the spinal cord; it gives entrance to the posterior nerve roots and separates the lateral and posterior funiculi. <sulcus posterolateralis medullae spinalis> A

10 **white matter of spinal cord** the white substance of the spinal cord, consisting of long myelinated nerve fibers arranged in parallel longitudinal bundles. <substantia alba medullae spinalis> A

11 ***vertebral plexus*** a nerve plexus accompanying the vertebral artery, formed by fibers from the vertebral and cervicothoracic ganglia and carrying sympathetic fibers to the posterior cranial fossa via cranial nerves. <plexus vertebralis> B

12 **anterior radicular artery** one of the spinal branches of the posterior branch of a posterior intercostal artery; it is the branch supplying the anterior root of a particular spinal nerve. <arteria radicularis anterior> B

13 **anterior spinal artery** *origin,* intracranial part of vertebral artery; *branches,* none; *distribution,* the two branches, one from each vertebral artery, unite to form a single vessel, which descends on the anterior midline of the spinal cord, supplying the anterior region of the cord. <arteria spinalis anterior> B

14 **posterior radicular artery** one of the spinal branches of the posterior branch of a posterior intercostal artery; it is the branch supplying the posterior root of a particular spinal nerve. <arteria radicularis posterior> B

15 **posterior spinal artery** *origin,* posterior inferior cerebellar artery (usually) or vertebral artery; *branches,* none; *distribution,* posterior region of spinal cord. <arteria spinalis posterior> B

16 ***radicular branches of spinal branches of vertebral artery*** one type of spinal branch of the cervical part of the vertebral artery; the radicular branches enter the vertebral foramina and follow and supply the roots of individual spinal nerves. <rami radiculares rami spinalum arteriae vertebralis> B

17 ***spinal branch of first posterior intercostal artery*** a vessel arising from the posterior branch of the first posterior intercostal artery, entering the intervertebral foramen to help supply the contents of the vertebral canal. <ramus spinalis arteriae intercostalis posterioris primae> B

18 ***spinal branch of lateral sacral artery*** a vessel arising from a lateral sacral artery and entering the pelvic sacral foramen to help supply the contents of the vertebral canal. <ramus spinalis arteriae sacralis lateralis> B

19 ***spinal branch of lumbar artery*** a branch arising from the posterior branch of a lumbar artery and entering an intervertebral foramen with the spinal nerve to help supply the contents of the vertebral canal. <ramus spinalis arteriae lumbalis> B

20 **spinal branch of posterior branch of posterior intercostal arteries** one of the two branches into which the posterior branch of a posterior intercostal artery divides, passing through the intervertebral foramen with the corresponding spinal nerve to help supply the contents of the vertebral canal. <ramus spinalis rami dorsalis arteriarum intercostalium posteriorum> B

21 ***spinal branch of posterior intercostal vein*** a vessel, the vena comitans of the arterial spinal branch, that emerges from the vertebral canal and contributes to the posterior branch of each posterior intercostal vein. <ramus spinalis venae intercostalis posterioris> B

22 ***spinal branch of second posterior intercostal artery*** a vessel arising from the dorsal branch of the second posterior intercostal artery, entering the intervertebral foramen to help supply the contents of the vertebral canal. <ramus spinalis arteriae intercostalis posterioris secundae> B

A

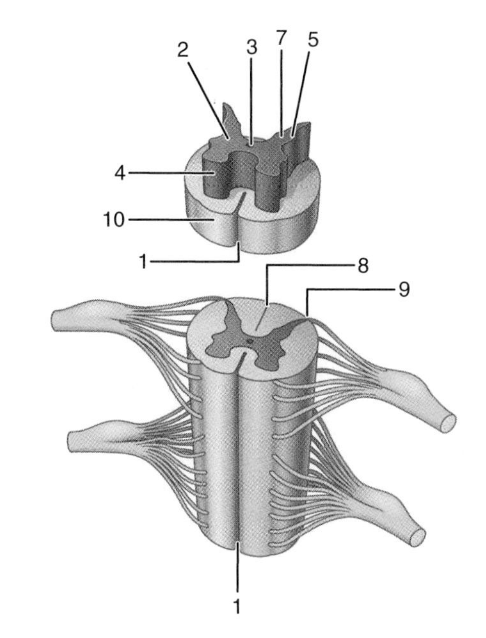

B

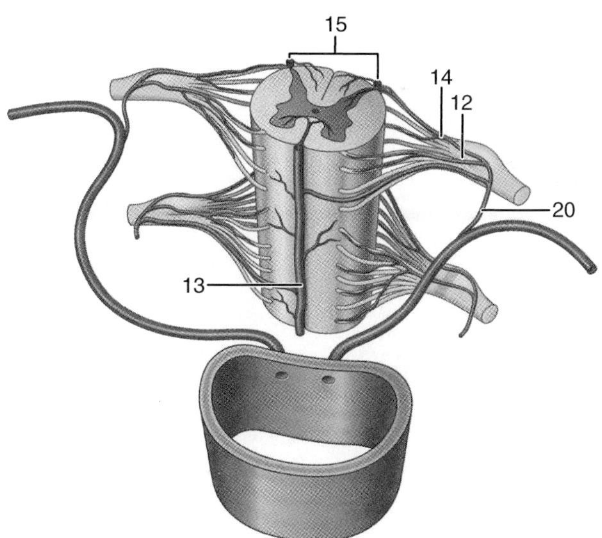

1 ***spinal branch of subcostal artery*** a spinal branch corresponding to those arising from the posterior branches of the posterior intercostal arteries; it enters the vertebral canal to help supply the contents of the canal. <ramus spinalis arteriae subcostalis> A

2 ***spinal branches of vertebral artery*** branches of the transverse part of the vertebral artery; they supply the spinal cord and its meninges, the vertebral bodies, and the intervertebral disks. <rami spinales arteriae vertebralis> A

3 **anterior spinal veins** a group of longitudinal veins forming a plexus on the anterior surface of the spinal cord, comprising a median vein lying anterior to the anterior median fissure and two anterolateral veins lying slightly posterior to the anterior nerve roots; they drain the anterior spinal cord. <venae spinales anteriores> B

4 **posterior spinal veins** a group of longitudinal, usually discontinuous, veins forming a plexus on the posterior surface of the spinal cord, comprising a median vein lying behind the posterior median septum and two posterolateral veins lying posterior to the posterior nerve roots; they drain the posterior spinal cord. <venae spinales posteriores> B

5 **denticulate ligament** either of two symmetrical folds of pia mater of the spinal cord, each beginning in a longitudinal line between the lines of attachment of the anterior and posterior roots. The lateral edge is scalloped and has about 21 pointed processes that extend laterally and fuse with the arachnoid and dura mater. <ligamentum denticulatum> C

6 **spinal arachnoid mater** the arachnoid covering the spinal cord. <arachnoidea mater spinalis> C

7 **spinal dura mater** the dura mater covering the spinal cord; it is separated from the periosteum of the enclosing vertebrae by an epidural space containing blood vessels and fibrous and areolar tissue. <dura mater spinalis> C

8 **spinal pia mater** the pia mater covering the spinal cord and consisting of collagenous fibers, which also form the denticulate ligament, and reticular fibers, which closely invest the cord, form the various septa, and form an investment for the rootlets. <pia mater spinalis> C

9 **anterior ramus of spinal nerve** the anterior and usually larger of the two branches into which each spinal nerve divides almost as soon as it emerges from the intervertebral foramen; the anterior rami supply anterior and lateral parts of the trunk and all parts of the limbs. <ramus anterior nervi spinalis> D

10 ***anterior rami of lumbar nerves*** the anterior branches of the five lumbar spinal nerves; the upper four form the lumbar plexus, and the fifth and a part of the fourth form part of the sacral plexus. <rami anteriores nervorum lumbalium> D

11 ***anterior rami of sacral nerves*** branches of the five sacral spinal nerves; the upper four emerge from the sacrum through the anterior foramina and contribute to the sacral plexus, and the fifth emerges through the sacral hiatus and, with a communication from the fourth, contributes to the coccygeal plexus. <rami anteriores nervorum sacralium> D

12 ***anterior rami of thoracic nerves*** anterior branches of the twelve thoracic spinal nerves. The first eleven (T1—11) are situated between the ribs and are intercostal nerves; the last (T12) is below the ribs and is the subcostal nerve. <rami anteriores nervorum thoracicorum> D

13 ***anterior ramus of coccygeal nerve*** a branch of the last spinal nerve, emerging from the sacral hiatus and contributing to the coccygeal plexus. <ramus anterior nervi coccygei> D

14 **anterior root of spinal nerve** the motor division of each spinal nerve, attached centrally to the spinal cord and joining peripherally with the corresponding posterior (sensory) root to form the nerve before it emerges through the intervertebral foramen. It conveys motor fibers to skeletal muscle and contains preganglionic autonomic fibers at thoracolumbar and sacral levels. There are 31 anterior and 31 posterior roots: 8 cervical, 12 thoracic, 5 lumbar, 5 sacral, and 1 coccygeal. <radix anterior nervi spinalis> D

15 ***lateral branch of posterior ramus of sacral nerve*** a branch that arises from the posterior ramus of one of the three upper sacral nerves and supplies the posterior gluteal skin. <ramus lateralis rami posterioris nervi sacralis> D

16 ***posterior rami of cervical nerves*** branches of the eight cervical spinal nerves; they subdivide into lateral and medial branches. <rami posteriores nervorum cervicalium> D

17 ***posterior rami of lumbar nerves*** branches of the five lumbar spinal nerves; they subdivide into lateral and medial branches. <rami posteriores nervorum lumbalium> D

18 ***posterior rami of sacral nerves*** branches of the five sacral spinal nerves, emerging from the sacrum through the posterior foramina; they subdivide into medial and lateral branches. <rami posteriores nervorum sacralium> D

19 ***posterior rami of thoracic nerves*** branches of the twelve thoracic spinal nerves, they subdivide into lateral and medial and posterior cutaneous branches. <rami posteriores nervorum thoracicorum> D

20 ***posterior ramus of coccygeal nerve*** a branch of the last spinal nerve, helping to innervate the skin over the coccyx. <ramus posterior nervi coccygei> D

21 **posterior ramus of spinal nerve** the posterior and usually smaller of the two rami into which each spinal nerve divides almost as soon as it emerges from the intervertebral foramen; the posterior rami supply the skin, muscles, joints, and bone of the posterior part of the neck and trunk. Most of these rami further subdivide into a medial and a lateral portion. <ramus posterior nervi spinalis> D

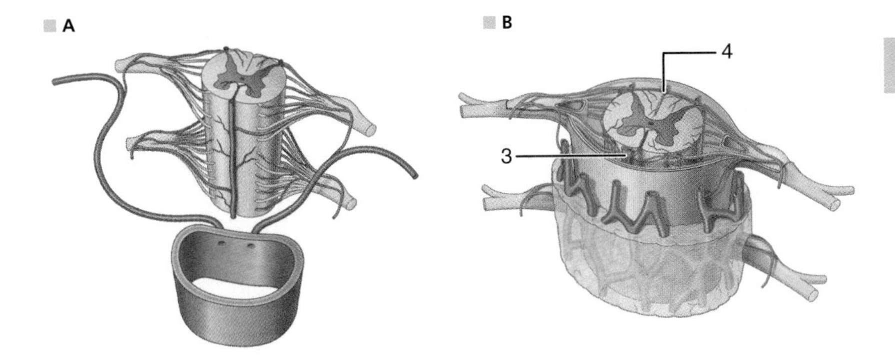
A
B
4
3

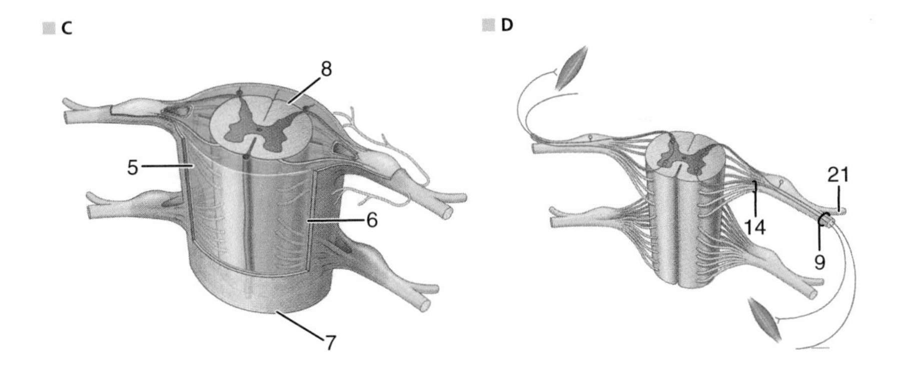
C
8
5
6
7
D
21
14
9

1 **posterior root of spinal nerve** the sensory division of each spinal nerve, attached centrally to the spinal cord and joining peripherally with the anterior (motor) root to form the nerve before it emerges through the intervertebral foramen: each posterior root bears a spinal ganglion and conveys sensory fibers to the spinal cord. There are 31 anterior and 31 posterior nerve roots: 8 cervical, 12 thoracic, 5 lumbar, 5 sacral, and 1 coccygeal. <radix posterior nervi spinalis> A

2 **rootlets of spinal nerve** the threadlike filaments by which the anterior and posterior roots of each spinal nerve are attached to the spinal cord. <fila radicularia nervi spinalis> A

3 **spinal ganglion** the ganglion found on the posterior root of each spinal nerve, composed of the unipolar nerve cell bodies of the sensory neurons of the nerve. <ganglion sensorium nervi spinalis> A

4 **trunk of spinal nerve** the usually very short nerve trunk formed by the anterior and posterior roots of a spinal nerve. <truncus nervi spinalis> A

5 ***rami communicantes of spinal nerves*** branches connecting spinal nerves with sympathetic ganglia, each spinal nerve receiving a gray ramus communicans, and the thoracic and upper lumbar spinal nerves having in addition a white ramus communicans. <rami communicantes nervorum spinalium> B

6 **gray ramus communicans of spinal nerve** one of the two types of communicating nerve branches between a sympathetic ganglion and a spinal nerve; these branches carry postganglionic impulses back to the spinal nerves and then to the periphery, supplying blood vessels, sweat glands, and smooth muscles. <ramus communicans griseus nervi spinalis> B

7 **white ramus communicans of spinal nerve** one of the two types of communicating nerve branches between a sympathetic ganglion and a spinal nerve; the white type are largely myelinated, send impulses from the spinal nerves to and through the ganglia, and are located mainly in the thoracic and upper lumbar region. <ramus communicans albus nervi spinalis> B

A

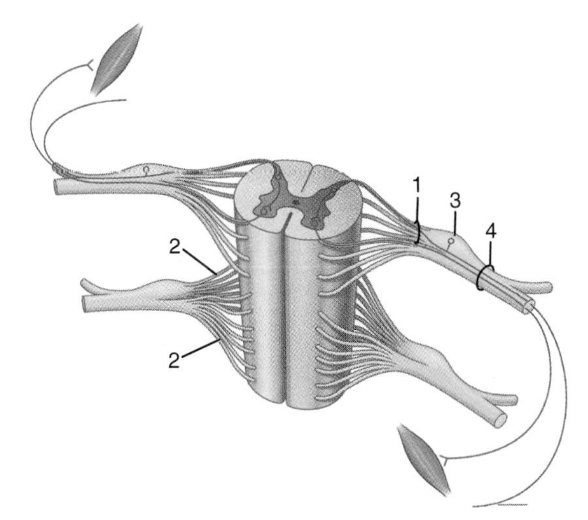

B

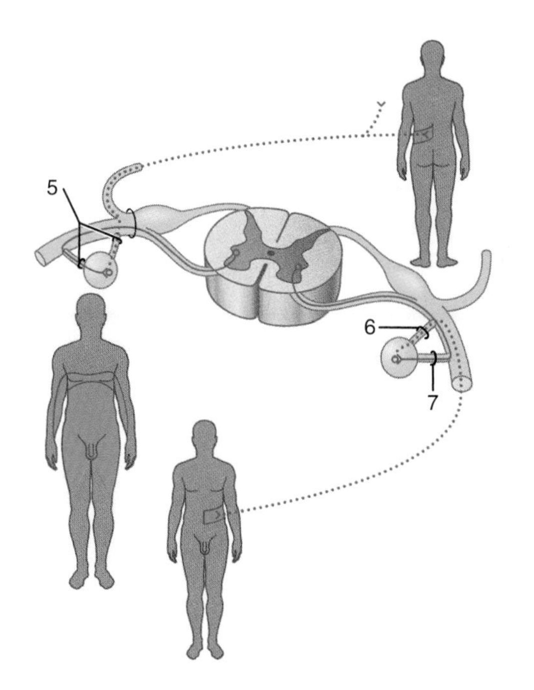

GENERAL

1 **thorax** chest: the part of the body between the neck and the thoracic diaphragm, encased by the ribs. A

2 **inferior thoracic aperture** the irregular opening at the inferior part of the thorax bounded by the twelfth thoracic vertebra, the twelfth ribs, and the curving edge of the costal cartilages as they meet the sternum; called also *thoracic outlet.* <apertura thoracis inferior> A

3 **superior thoracic aperture** the elliptical opening at the superior end of the thorax, bounded by the first thoracic vertebra, the first ribs and cartilage, and the upper margin of the manubrium sterni. Called also *thoracic inlet.* NOTE: In clinical usage, the term "thoracic outlet syndrome" refers to this opening, not to the inferior thoracic aperture. <apertura thoracis superior> A

4 **thoracic cavity** the portion of the body cavity situated between the neck and the diaphragm; called also *pectoral cavity.* <cavitas thoracis> A

BONES

5 **inferior costal facet** a small facet on the lower edge of the body of a vertebra articulating with the head of a rib. <fovea costalis inferior> B

6 **superior costal facet** a small facet on the upper edge of the body of a vertebra articulating with the head of a rib. <fovea costalis superior> B

7 **transverse costal facet** a facet on the transverse process of a vertebra for articulation with the tubercle of a rib. <fovea costalis processus transversi> B

8 **costal arch** the anterior portion of the inferior thoracic aperture, consisting of the costal cartilages of ribs 7 to 10, inclusive; called also *arch of ribs.* <arcus costalis> C

9 **costochondral joints** joints between the lateral extremity of each costal cartilage and the sternal ends of the ribs. <articulationes costochondrales> C

10 **false ribs** the lower five ribs on either side (VIII-XII); the anterior tips of the upper three connect with the costal cartilages of the superiorly adjacent ribs, and the anterior tips of the lower two (XI and XII) ordinarily have no attachment. <costae spuriae> C

11 **first rib** (I), the superior rib on either side. <costa prima> C

12 **floating ribs** the lower two ribs on either side (XI and XII), which ordinarily have no anterior attachment. <costae fluctuantes> C

13 **interchondral joints** the unions, on either side, between the costal cartilages of the upper false ribs, usually ribs seven through ten; called also *intercostal joints.* <articulationes interchondrales> C

14 ***presternal region*** the region of the chest superficial to the sternum. <regio presternalis> C

15 **rib** any one of the paired elastic arches of bone, twelve on either side (I-XII), extending from the thoracic vertebrae toward the median line on the anterior aspect of the trunk; they form the major part of the thoracic skeleton. The upper seven (I-VII) are called true ribs; the lower five (VIII-XII) are called false ribs. <costa> C

16 **second rib** (II), the rib just inferior to the first rib on either side. <costa secunda> C

17 **sternocostal joints** the joints between the costal notches of the sternum and the medial ends of the costal cartilages of the upper seven ribs. <articulationes sternocostales> C

18 **sternocostal synchondrosis of first rib** the joint between the sternum and the first rib, in which the costal cartilage is united directly to the sternum. <synchondrosis sternocostalis costae primae> C

19 ***thoracic joints*** the joints of the thorax considered as a group, including syndesmoses, synchondroses, and the thoracic synovial joints. <juncturae thoracis> C

20 ***thoracic skeleton*** the skeletal framework enclosing the thorax, consisting of the thoracic vertebrae and intervertebral disks, the ribs and costal cartilages, and the sternum; called also *rib cage.* <skeleton thoracis> C

21 ***thoracic synovial joints*** the costovertebral, sternocostal, costochondral, and interchondral joints. <articulationes thoracis> C

22 **true ribs** the upper seven ribs on either side (I-VII), which are connected to the sides of the sternum by their costal cartilages. <costae verae> C

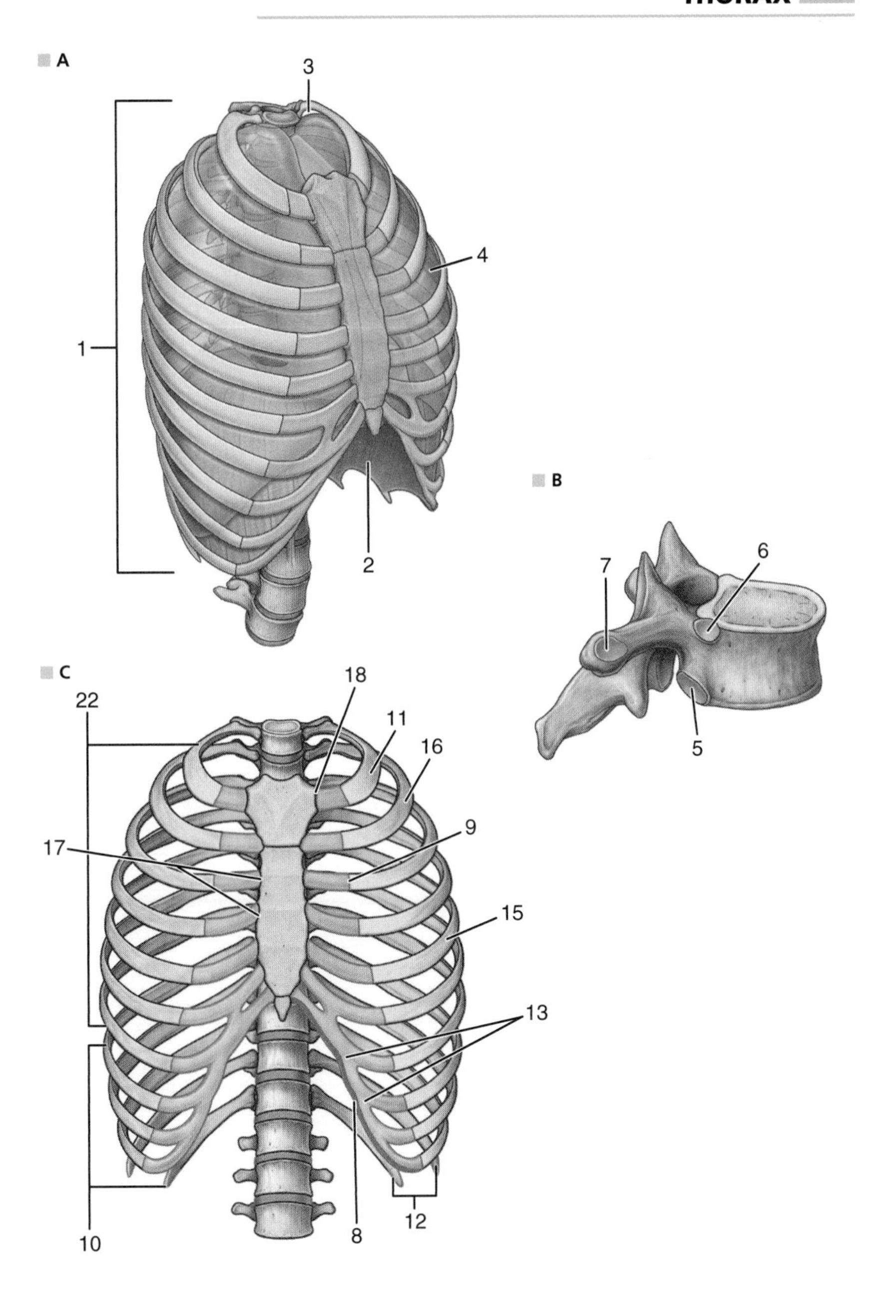
A
1
2
3
4
B
5
6
7
C
8
9
10
11
12
13
15
16
17
18
22

THORAX

1 **articular facet of head of rib** the surface on the head of a rib where it articulates with the body of a vertebra. Typically it is divided into two facets by a transverse crest, the lower facet articulating with the corresponding vertebra, and the upper facet with the superjacent vertebra. The articular surfaces of the heads of the first, tenth, eleventh, and twelfth ribs generally consist of only one facet. Called also *posterior* or *superior costal facet.* <facies articularis capitis costae> A

2 **body of rib** the part of a rib extending between its posteriorly placed tubercle and its anterior extremity; called also *shaft of rib.* <corpus costae> A

3 **costal angle** a prominent line on the external surface of a rib, slightly anterior to the tubercle, where the rib is bent in two directions and at the same time twisted on its long axis; called also *angle of rib.* <angulus costae> A

4 **costal cartilage** a bar of hyaline cartilage by which the anterior extremity of a rib is attached to the sternum in the case of the true ribs, or to the superiorly adjacent ribs in the case of the upper false ribs. <cartilago costalis> A

5 **costal groove** a sulcus that follows the inferior and internal surface of a rib anteriorly from the tubercle, gradually becoming less distinct; it lodges the intercostal vessels and nerves. <sulcus costae> A

6 **crest of head of rib** a horizontal crest dividing the articular surface of the head of the rib into two facets, for articulation with the depression on the bodies of two adjacent vertebrae. <crista capitis costae> A

7 **crest of neck of rib** a crest on the superior border of the neck of a rib, giving attachment to the anterior costotransverse ligament. <crista colli costae> A

8 **head of rib** the posterior end of a rib, which articulates with the body of a vertebra. <caput costae> A

9 **neck of rib** the part of a rib extending from the head to the tubercle. <collum costae> A

10 **tubercle of rib** a small eminence on the posterior surface of a rib where the neck and body join; it protrudes inferiorly and posteriorly, and bears on its medial part a surface that articulates with the transverse process of the corresponding vertebra. <tuberculum costae> A

11 **articular facet of tubercle of rib** the convex facet on the costal tubercle that articulates with the transverse process of a vertebra; called also *anterior* or *inferior costal facet.* <facies articularis tuberculi costae> A

12 **groove for subclavian artery** a transverse groove on the cranial surface of the first rib, just posterior to the anterior scalene tubercle; it lodges the subclavian artery; called also *subclavian s.* and *s. subclavius.* <sulcus arteriae subclaviae> B

13 **scalene tubercle** the tubercle on the cranial surface of the first rib for the insertion of the anterior scalene muscle. <tuberculum musculi scaleni anterioris> B

14 ***tuberosity for serratus anterior muscle*** a roughened, raised area on the second rib that gives attachment to a slip of the serratus anterior muscle. <tuberositas musculi serrati anterioris>

15 **body of sternum** the second or principal portion of the sternum, located between the manubrium above and the xiphoid process below. <corpus sterni> C

16 **clavicular notch of sternum** either of two oval surfaces, one on each side of the superior border of the manubrium of the sternum, where it articulates with the clavicle. <incisura clavicularis sterni> C

17 **costal notches of sternum** the facets on the sternum, seven on each lateral edge, for articulation with the costal cartilages. <incisurae costales sterni> C

18 **jugular notch of sternum** the notch on the upper border of the sternum between the clavicular notches; called also *suprasternal notch.* <incisura jugularis sterni> C

19 **manubriosternal joint** the joint uniting the manubrium with the body of the sternum, which begins as a synchondrosis (manubriosternal synchondrosis) and later becomes a symphysis. <symphysis manubriosternalis> C

20 **manubrium of sternum** the cranial portion of the sternum, which articulates with the clavicles and the first two pairs of ribs. <manubrium sterni> C

21 **sternal angle** the angle formed on the anterior surface of the sternum at the junction of its body and manubrium. <angulus sterni> C

22 ***sternal membrane*** the thick fibrous membrane that envelopes the sternum; it is formed by the intermingling of fibers of the radiate sternocostal ligaments, the periosteum, and the tendinous origin of the pectoralis major. <membrana sterni> C

23 **sternal synchondrosis** the cartilaginous union between the manubrium and the body of the sternum. <synchondrosis sternalis> C

24 **sternum** a longitudinal unpaired plate of bone forming the middle of the anterior wall of the thorax; it articulates above with the clavicles and along its sides with the cartilages of the first seven ribs. Its three parts are the manubrium, body, and xiphoid process. C

25 ***suprasternal bones*** ossicles occasionally occurring in the ligaments of the sternoclavicular articulation. <ossa suprasternalia> C

26 **xiphisternal joint** the joint between the xiphoid process and the body of the sternum. <synchondrosis xiphisternalis> C

27 **xiphoid process** the pointed process of cartilage, supported by a core of bone, connected with the lower end of the body of the sternum. <processus xiphoideus> C

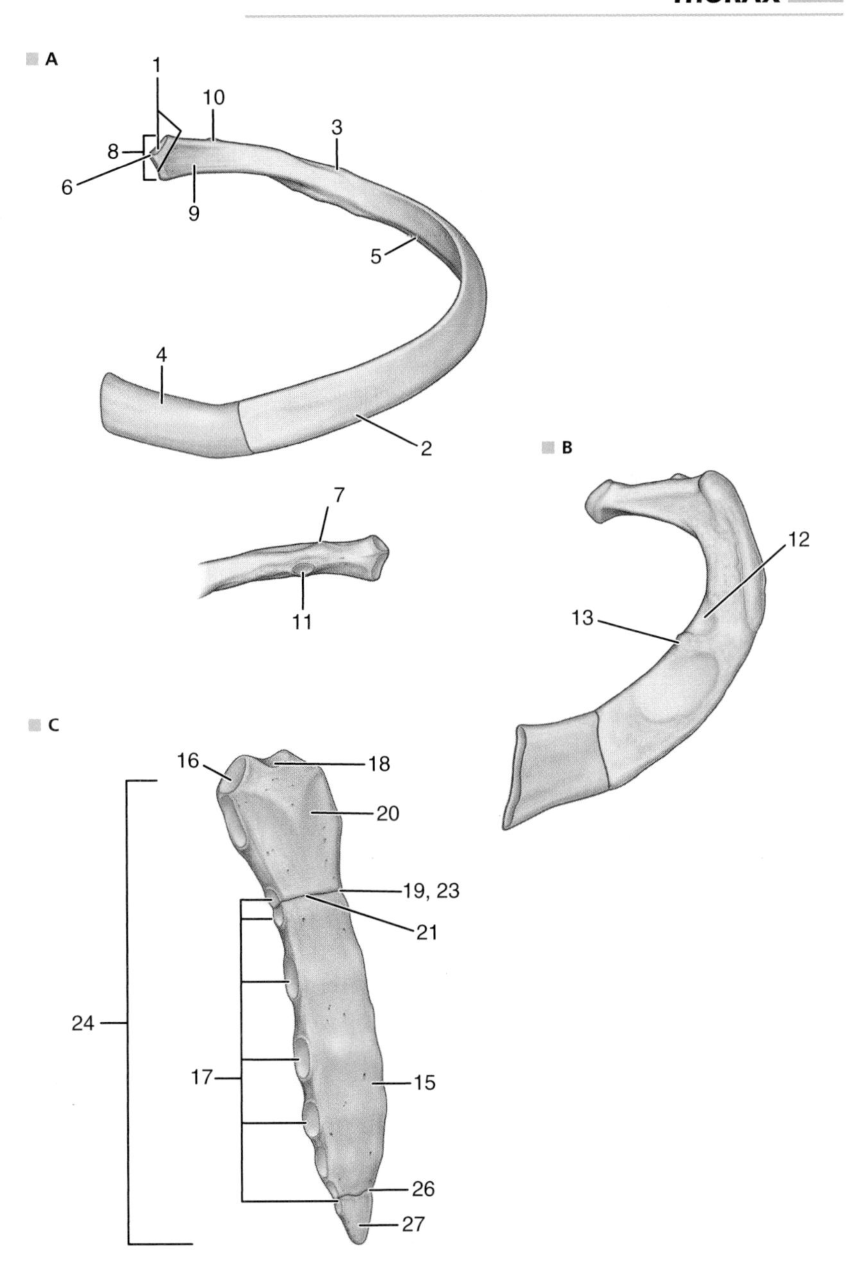
A
1
10
3
8
6
9
5
4
2
7
11
B
12
13
C
16
18
20
19, 23
21
24
17
15
26
27

1 **intra-articular ligament of head of rib** a horizontal band of fibers attached to the crest separating the two articular facets on the head of the rib, and to the intervertebral disk, thus dividing the joint of the head of the rib into two cavities. It is lacking in the joints of the first, tenth, eleventh, and twelfth ribs. <ligamentum capitis costae intraarticulare> A

2 **joint of head of rib** the joint of the head of the rib with the bodies of two vertebrae. <articulatio capitis costae> A

3 **lateral costotransverse ligament** a fibrous band that passes transversely from the posterior surface of the tip of a transverse process of a vertebra to the nonarticular part of the tubercle of the corresponding rib. <ligamentum costotransversarium laterale> A

4 **superior costotransverse ligament** a strong band of fibers ascending from the crest of the neck of a rib to the transverse process of the vertebra above; it may be divided into a stronger anterior portion and a weaker posterior portion. It is lacking for the first rib. <ligamentum costotransversarium superius> A

5 **costotransverse joint** one of the two types of joints between ribs and vertebrae, being that of the tubercle of the rib with the transverse process of a vertebra. It is lacking for the eleventh and twelfth ribs. <articulatio costotransversaria> A

6 **costotransverse ligament** short fibers that connect the dorsal surface of the neck of a rib with the anterior surface of the transverse process of the corresponding vertebra. <ligamentum costotransversarium> A

7 **costovertebral joints** the joints between the ribs and vertebrae; there are costotransverse joints and joints of heads of ribs. <articulationes costovertebrales> A

8 **radiate ligament of head of rib** fibers that from their attachment on the anterior surface of the head of a rib radiate medially, in a fanlike manner, to attach to the two adjacent vertebrae and to the intervertebral disk between them. <ligamentum capitis costae radiatum> B

9 **costotransverse foramen** the narrow space between the posterior surface of the neck of a rib and the anterior surface of the transverse process of the corresponding vertebra. <foramen costotransversarium> A, B

10 **costoxiphoid ligaments** inconstant strandlike bands that pass obliquely from the anterior surface of the seventh and sometimes from the sixth costal cartilage to the anterior surface of the xiphoid process of the sternum. Some bands may also be present on the posterior surface. <ligamenta costoxiphoidea> C

11 **intra-articular sternocostal ligament** a horizontal fibrocartilaginous plate in the center of the second sternocostal joint, which joins the tip of the costal cartilage to the fibrous junction between the manubrium and the body of the sternum, and thus divides the joint into two parts. <ligamentum sternocostale intraarticulare> C

12 **radiate sternocostal ligaments** fibrous bands attached to the sternal end of a costal cartilage, radiating from there out onto the anterior part of the sternum. <ligamenta sternocostalia radiata> C

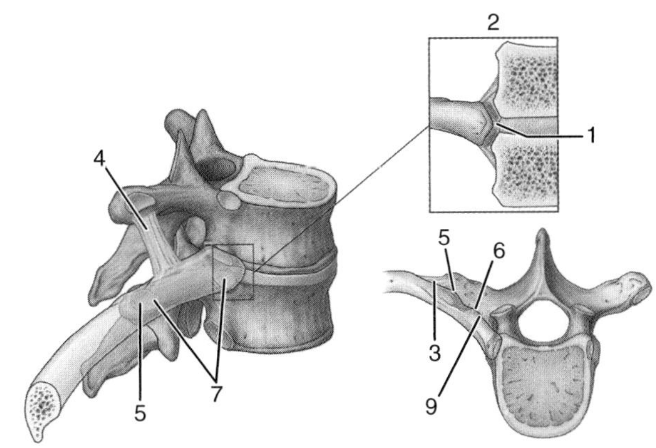
A
2
1
4
5
6
3
7
5
9

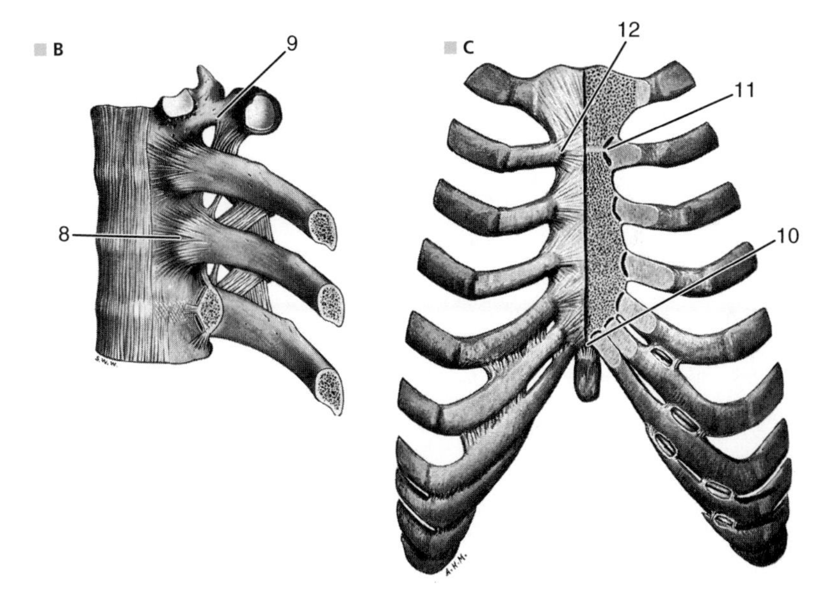
B
9
8
C
12
11
10

PECTORAL REGION

1 *pectoral region* the aspect of the thorax, or chest, bounded by the pectoralis major muscle and including the lateral pectoral, mammary, and inframammary regions. <regio pectoralis>

2 **infrasternal angle** the angle on the anteroinferior surface of the thorax, the apex of which is the xiphisternal joint, and the sides of which are the seventh, eighth, and ninth costal cartilages; it partially delimits two sides of the triangular epigastric region on the anterior body surface; called also *subcostal angle.* <angulus infrasternalis> A

3 **mammary region** the region of the anterior aspect of the thorax, around the mammary gland. <regio mammaria> B

4 **areola** the darkened ring surrounding the nipple of a breast. <areola mammae> B

5 ***areolar venous plexus*** a venous plexus in the areola around the nipple, formed by branches of the internal thoracic veins and draining into the lateral thoracic vein. <plexus venosus areolaris> B

6 **axillary process of mammary gland** the superolateral part of the mammary gland that extends toward the axilla. <processus axillaris glandulae mammariae> B

7 **breast** the modified cutaneous, glandular structure on the anterior aspect of the thorax that contains, in the female, the elements that secrete milk for nourishment of the young. <mamma> B

8 **inframammary region** the region of the anterior aspect of the thorax situated inferior to either breast and superior to the inferior border of the twelfth rib. <regio inframammaria> B

9 **nipple** the pigmented projection on the anterior surface of the mammary gland, surrounded by the areola; the lactiferous ducts open onto it. <papilla mammaria> B

10 **corpus mammae** the essential mass of the mammary gland, exclusive of the glandular elements, which is thickest beneath the nipple and thinner toward the periphery. C

11 **areolar glands** sebaceous glands of the mammary areola; called also *Montgomery glands.* <glandulae areolares> D

12 **lactiferous ducts** channels conveying the milk secreted by the lobes of the breast to and through the nipples; called also *mammary ducts.* <ductus lactiferi> D

13 **lactiferous sinuses** enlargements in the lactiferous ducts just before they open onto the mammary papilla. <sinus lactiferi> D

14 **lobes of mammary gland** the major subdivisions of the secreting portion of the mammary gland, each drained by a single lactiferous duct and further subdivided into lobules. <lobi glandulae mammariae> C

15 **mammary gland** the specialized accessory gland of the skin of female mammals that secretes milk. It is a compound tubuloalveolar gland composed of 15 to 25 lobes arranged radially about the nipple and separated by connective and adipose tissue, each lobe having its own excretory (lactiferous) duct opening on the nipple. The lobes are subdivided into lobules, with the alveolar ducts and alveoli being the secretory portion of the gland. Called also *lactiferous gland.* <glandula mammaria> C

16 **suspensory ligaments of breast** fibrous processes, extending from the corpus mammae to the dermis, homologous with the retinacula cutis of other regions of the body. <ligamenta suspensoria mammaria> C

A

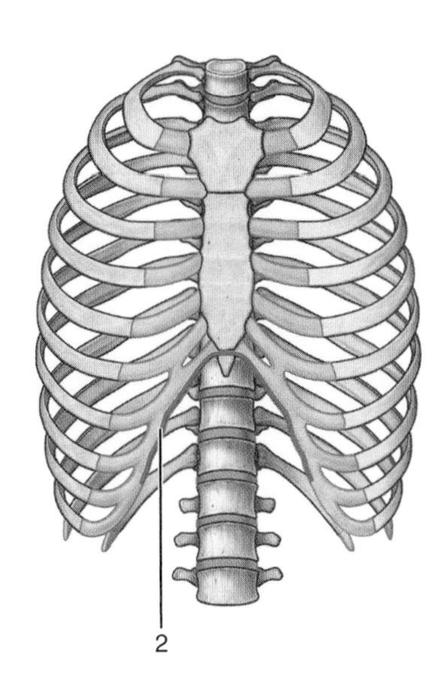

B

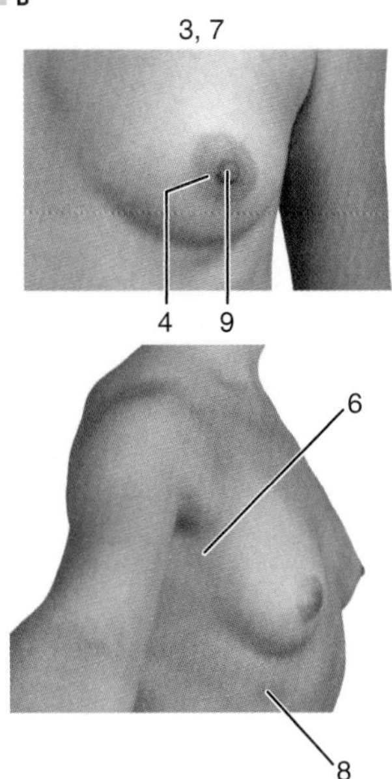

C

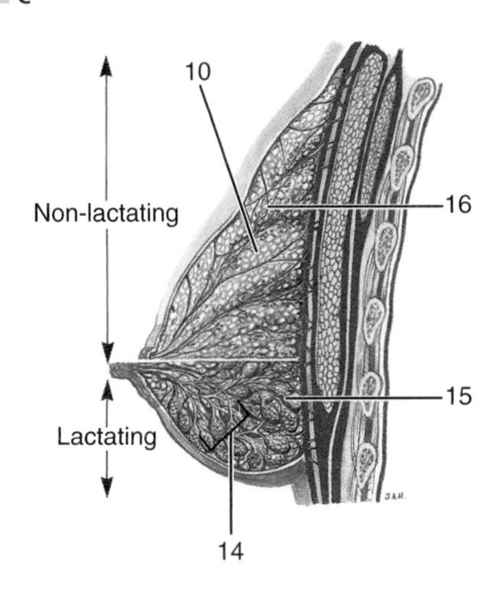

D

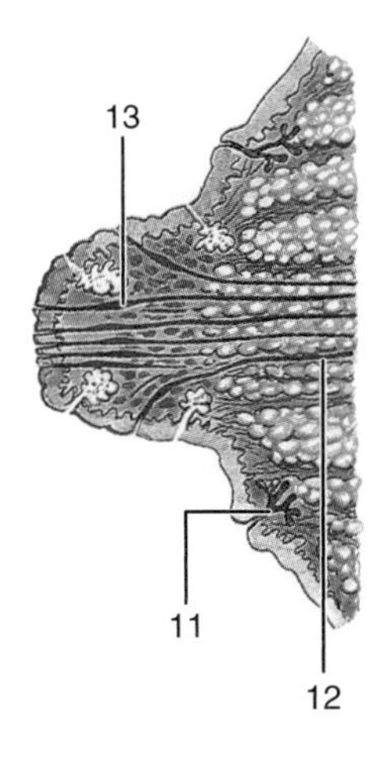

1 ***interpectoral lymph nodes*** small inconstant lymph nodes that may occur between the mammary gland and apical lymph nodes. <nodi lymphoidei interpectorales> A

2 ***paramammary lymph nodes*** lymph nodes on the lateral mammary gland that drain into the axillary lymph nodes. <nodi lymphoidei paramammarii> A

3 **parasternal lymph nodes** nodes located along the course of the internal thoracic artery, which drain the mammary gland, abdominal wall, and diaphragm. <nodi lymphoidei parasternales> A

4 **pectoral axillary lymph nodes** four or five axillary lymph nodes along the inferior border of the pectoralis minor muscle near the lateral thoracic artery; they receive lymph from the skin and muscles of the anterior and lateral thoracic walls and mammary gland and drain into the central and apical nodes. Called also *anterior axillary lymph nodes.* <nodi lymphoidei axillares pectorales> A

5 **accessory breast** a mammary gland present in excess of the normal number, generally found along the line of the embryonic mammary ridge; called also *accessory mammary gland.* <mamma accessoria> B

6 **lateral mammary branches of lateral thoracic artery** branches that supply the mammary gland. <rami mammarii laterales arteriae thoracicae lateralis> C

7 **lateral thoracic artery** *origin,* axillary artery; *branches,* mammary branches; *distribution,* pectoral muscles, mammary gland. Called also *external mammary artery.* <arteria thoracica lateralis> C

8 ***lateral thoracic vein*** a large vein accompanying the lateral thoracic artery and draining into the axillary vein. <vena thoracica lateralis> C

9 ***pectoral veins*** collective term for branches of the subclavian vein that drain the pectoral region. <venae pectorales> C

10 **pectoralis minor muscle** *origin,* third, fourth, and fifth ribs; *insertion,* coracoid process of scapula; *innervation,* lateral and medial pectoral; *action,* draws shoulder forward and downward, raises third, fourth, and fifth ribs in forced inspiration. <musculus pectoralis minor> D

11 **clavipectoral triangle** the triangular region separating the upper border of the pectoralis minor muscle from the clavicle, which contains the clavipectoral fascia. <trigonum clavipectorale> E

12 **infraclavicular fossa** the triangular region of the thorax or chest just inferior to the clavicle, between the deltoid and pectoralis major muscles. <fossa infraclavicularis> E

13 ***pectoral fascia*** the sheet of fascia investing the pectoralis major muscle. <fascia pectoralis> E

14 **pectoralis major muscle** *origin,* clavicle, sternum, six upper costal cartilages, aponeurosis of external oblique. These origins are reflected in the subdivision of the muscle into clavicular, sternocostal, and abdominal parts; *insertion,* crest of intertubercular groove of humerus; *innervation,* medial and lateral pectoral; *action,* adducts, flexes, rotates arm medially. <musculus pectoralis major> E

15 **abdominal part of pectoralis major muscle** the portion of the muscle that originates from the aponeurosis of the external oblique. <pars abdominalis musculi pectoralis majoris> E

16 **clavicular part of pectoralis major muscle** the portion of the muscle that originates from the clavicle. <pars clavicularis musculi pectoralis majoris> E

17 **sternocostal part of pectoralis major muscle** the portion of the muscle that originates from the sternum and the ribs. <pars sternocostalis musculi pectoralis majoris> E

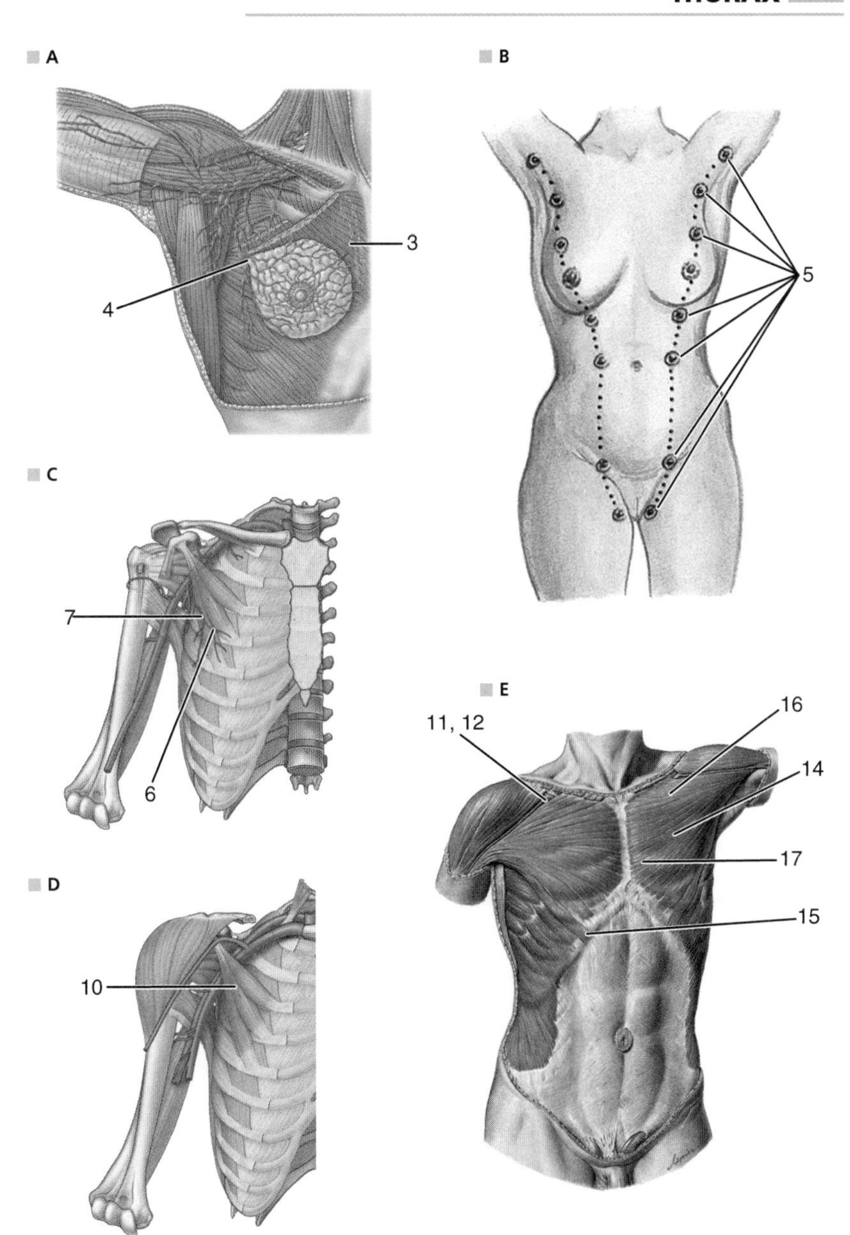
A
B
C
D
E
3
4
5
7
6
10
11, 12
16
14
17
15

THORACIC WALL

1 **endothoracic fascia** the extrapleural fascial sheet beneath the serous lining of the thoracic cavity. <fascia endothoracica> A

2 ***suprapleural membrane*** the strengthened portion of the endothoracic fascia attached to the inner part of the first rib and the transverse process of the seventh cervical vertebra. <membrana suprapleuralis> A

3 ***thoracic fascia*** the deep fascia that covers the outside of the thoracic cavity. <fascia thoracica> A

4 **pulmonary groove** a large vertical groove in the posterior part of the thoracic cavity, one on either side of the bodies of the vertebrae posterior to the level of their anterior surface, lodging the posterior, bulky portion of the lung. <sulcus pulmonalis> B

5 **external intercostal membrane** any of the aponeurotic bands parallel with, and perhaps replacing, the fibers of the external intercostal muscles in the spaces between the costal cartilages, from the anterior tips of the ribs medially to the sternum. <membrana intercostalis externa> C

6 **intercostal space** the space intervening between two adjacent ribs. <spatium intercostale> C

7 ***internal intercostal membrane*** any of the aponeurotic bands parallel with, and perhaps replacing, the fibers of the internal intercostal muscles in the spaces between the ribs, from the angles of the ribs medially to the vertebral column. <membrana intercostalis interna> C

MUSCLES OF THORAX

8 **external intercostal muscles** (11 on each side): *origin,* inferior border of rib; *insertion,* superior border of rib below; *innervation,* intercostal; *action,* primarily elevate ribs in inspiration, also active in expiration. <musculi intercostales externi> C

9 **innermost intercostal muscles** the layer of muscle fibers separated from the internal intercostal muscles by the intercostal nerves. <musculi intercostales intimi> C

10 **internal intercostal muscles** (11 on each side): *origin,* inferior border of rib and costal cartilage; *insertion,* superior border of rib and costal cartilage below; *innervation,* intercostal; *action,* draw ribs together in respiration and expulsive movements, also act in inspiration. <musculi intercostales interni> C

11 ***sternalis muscle*** a band occasionally found parallel to the sternum on the sternocostal origin of the pectoralis major. <musculus sternalis> C

12 **subcostales muscles** *origin,* inner surface of ribs: *insertion,* inner surface of first, second, third rib below; *innervation,* intercostal; *action,* draw adjacent ribs together, depress ribs. <musculi subcostales> D

13 **transversus thoracis muscle** *origin,* mediastinal surface of sternum and of xiphoid process; *insertion,* cartilages of second to sixth ribs; *innervation,* intercostal; *action,* draws ribs downward. <musculus transversus thoracis> D

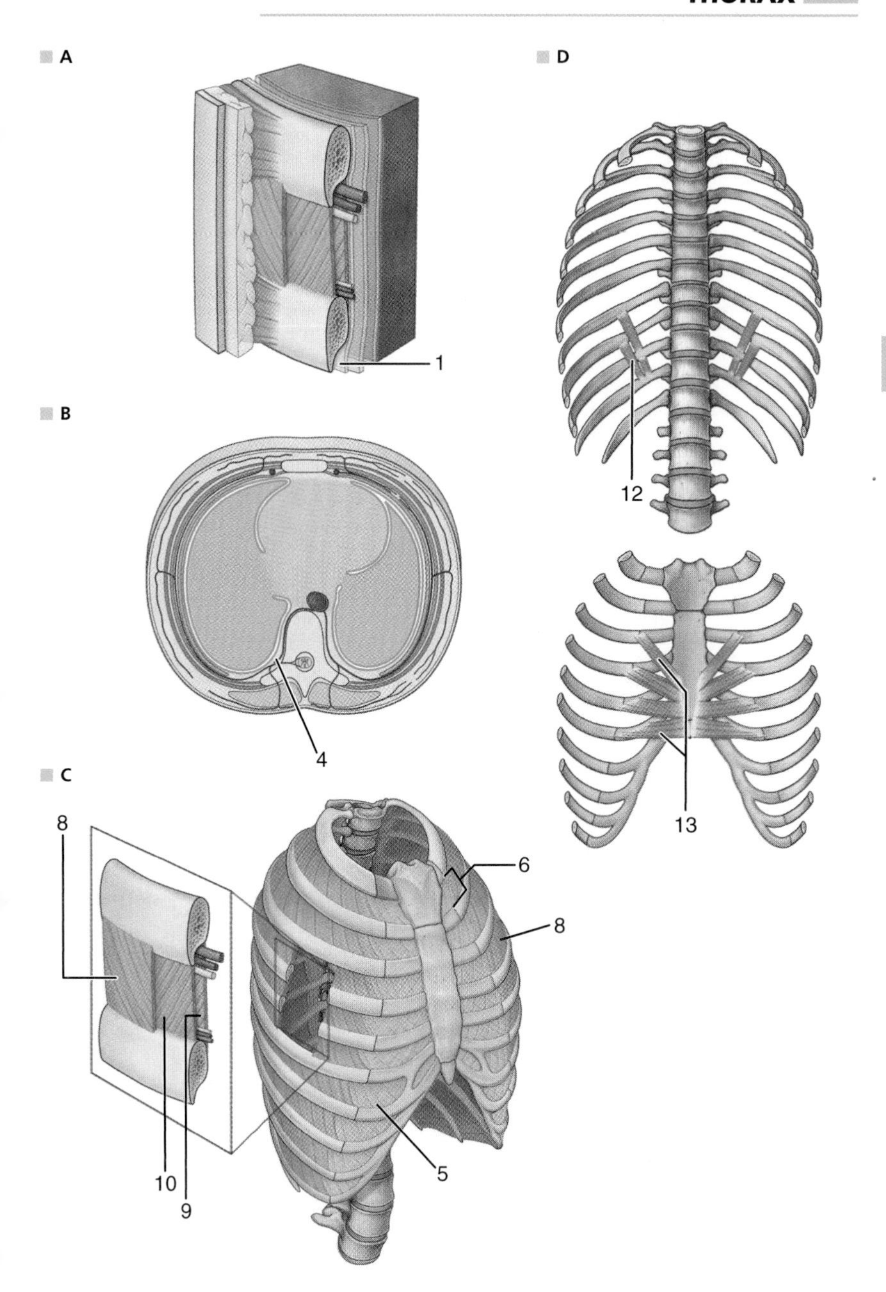
A
1
B
4
C
8
6
8
10
9
5
D
12
13

NERVES OF THORAX

1 **anterior pectoral cutaneous branch of intercostal nerve** a branch from the intercostal nerve that contributes to innervation of the skin in the anteromedial thoracic region; in the breast region it sends further branches (medial mammary branches). Its modality is general sensory. <ramus cutaneus anterior pectoralis nervi intercostalis> A

2 **intercostal nerves** anterior rami of the first eleven thoracic spinal nerves, situated between the ribs. The first three send branches to the brachial plexus as well as to the thoracic wall; the fourth, fifth, and sixth supply only the thoracic wall; and the seventh through eleventh are thoracoabdominal in distribution. The primary anterior division of the twelfth thoracic nerve is subcostal rather than intercostal in position and is known as the subcostal nerve. <nervi intercostales> A

3 **lateral mammary branches of lateral pectoral cutaneous branch of intercostal nerve** branches that the supply the mammary gland; *modality,* general sensory. <rami mammarii laterales rami cutanei lateralis pectoralis nervi intercostalis> A

4 **lateral pectoral cutaneous branch of intercostal nerve** a branch from the intercostal that, through its anterior and posterior subdivisions, innervates the skin of the lateral and posterior body wall; it has a general sensory modality. Those of the fourth through sixth anterior thoracic branches send further branches (lateral mammary branches). <ramus cutaneus lateralis pectoralis nervi intercostalis> A

5 **medial mammary branches of anterior pectoral cutaneous branch of intercostal nerve** further branches of the anterior pectoral cutaneous branch that innervate the skin of the mammary region; *modality,* general sensory. <rami mammarii mediales rami cutanei anterioris pectoralis nervi intercostalis> A

6 ***muscular branches of intercostal nerves*** branches that supply numerous muscles of the lateral and ventral regions of the thorax and abdomen. <rami musculares nervorum intercostalium> A

VESSELS OF THORAX

7 **anterior intercostal branches of internal thoracic artery** twelve branches, two in each of the upper six intercostal spaces, that supply the intercostal spaces and the pectoralis major muscle. Within each space both branches run laterally, the upper anastomosing with the posterior intercostal artery, the lower with the collateral branch of that artery. <rami intercostales anteriores arteriae thoracicae internae> A

8 **anterior intercostal veins** the twelve paired venae comitantes of the anterior thoracic arteries, which drain into the internal thoracic veins. <venae intercostales anteriores> A

9 **collateral branch of posterior intercostal artery** a branch helping supply the thoracic wall, arising from the posterior intercostal arteries near the angle of the rib and running forward in the lower part of the corresponding intercostal space. <ramus collateralis arteriarum intercostalis posterioris> A

10 ***intercostal lymph nodes*** lymph nodes in the back of the thorax, along the intercostal vessels. <nodi lymphoidei intercostales> A

11 ***lateral costal branch of internal thoracic artery*** an occasional branch passing inferolaterally behind the ribs, supplying ribs and costal cartilages, and anastomosing with the posterior intercostal arteries. <ramus costalis lateralis arteriae thoracicae internae> A

12 **lateral cutaneous branch of posterior branch of posterior intercostal artery** a branch from the posterior branch of each posterior intercostal artery that supplies the posterolateral aspect of the thorax. <ramus cutaneus lateralis rami dorsalis arteriae intercostalis posterioris> A

13 **lateral cutaneous branch of posterior intercostal artery** a branch arising from a posterior intercostal artery, supplying the anterolateral thoracic wall. The branches of the third through fifth posterior intercostal arteries give off small mammary branches. <ramus cutaneus lateralis arteriae intercostalis posterioris> A

14 **lateral mammary branches of lateral cutaneous branch of posterior intercostal artery** branches that arise from each of the third, fourth, and fifth posterior intercostal arteries to supply the pectoral muscles, skin, and mammary tissue. <rami mammarii laterales rami cutanei lateralis arteriae intercostalis posterioris> A

15 **medial cutaneous branch of posterior branch of posterior intercostal artery** a branch of the posterior branch of each posterior intercostal artery that supplies the skin adjacent to the vertebral column. <ramus cutaneus medialis rami dorsalis arteriae intercostalis posterioris> A

16 **medial mammary branches of perforating branch of internal thoracic artery** branches that arise from the second, third, and fourth perforating branches of the internal thoracic artery and help supply the mammary gland. <rami mammarii mediales rami perforantium arteriae thoracicae internae> A

17 **perforating branches of internal thoracic artery** six branches, one in each of the upper six intercostal spaces, supplying the pectoralis major muscle and adjacent skin; the second, third, and fourth branches give off mammary branches. <rami perforantes arteriae thoracicae internae> A

18 **posterior intercostal arteries** for the first two, see *first posterior intercostal artery* and *second posterior intercostal artery;* there are nine other pairs (III-–XI): *origin,* thoracic aorta; *branches,* dorsal, collateral, muscular, and lateral cutaneous; *distribution,* thoracic wall. <arteriae intercostales posteriores> A

A

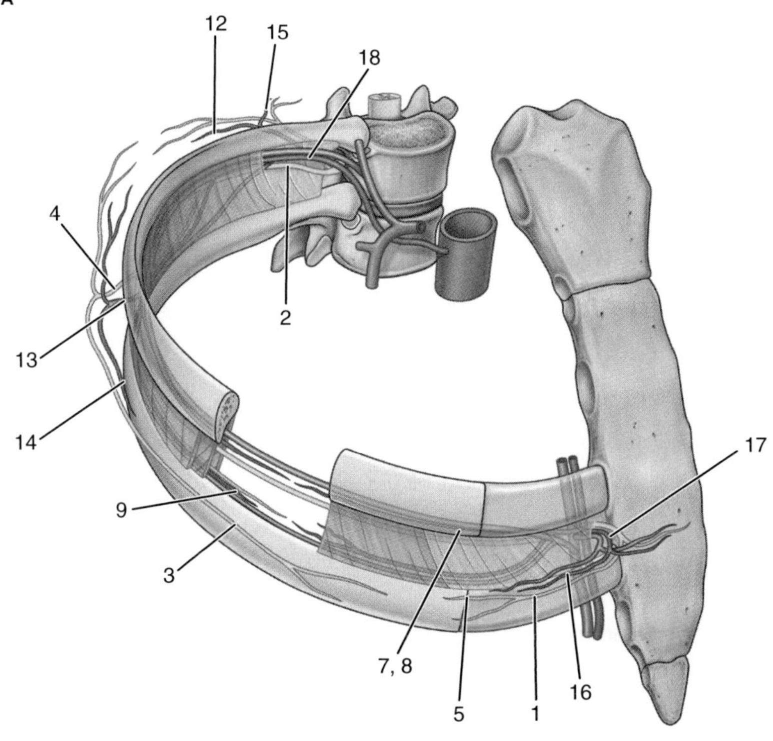

1 ***bronchial branches of internal thoracic artery*** small, variable branches of the internal thoracic artery, with distribution to the bronchi and trachea. Called also *anterior bronchial arteries.* <rami bronchiales arteriae thoracicae internae> A

2 ***first posterior intercostal artery*** *origin,* highest intercostal artery; *branches,* dorsal and spinal branches; *distribution,* upper thoracic wall. <arteria intercostalis posterior prima> A

3 **supreme intercostal artery** *origin,* costocervical trunk; *branches,* first and second posterior intercostal arteries; *distribution,* upper thoracic wall. Called also *superior intercostal artery.* <arteria intercostalis suprema> A

4 **internal thoracic artery** *origin,* subclavian artery; *branches,* mediastinal, thymic, bronchial, tracheal, sternal, perforating, medial mammary, lateral costal, and anterior intercostal branches, pericardiacophrenic, musculophrenic, and superior epigastric arteries; *distribution,* anterior thoracic wall, mediastinal structures, diaphragm. Called also *internal mammary artery.* <arteria thoracica interna> A

5 ***mediastinal branches of internal thoracic artery*** branches that supply areolar tissue, pericardium, lymph nodes, and the thymus in the anterior and superior mediastina. Called also *anterior mediastinal arteries.* <rami mediastinales arteriae thoracicae internae> A

6 **musculophrenic artery** *origin,* internal thoracic artery; *branches,* none; *distribution,* diaphragm, abdominal and thoracic walls. <arteria musculophrenica> A

7 ***second posterior intercostal artery*** *origin,* supreme intercostal artery; *branches,* dorsal and spinal branches; *distribution,* upper thoracic wall. <arteria intercostalis posterior secunda> A

8 ***sternal branches of internal thoracic artery*** branches that supply the sternum and the transversus thoracis muscle. <rami sternales arteriae thoracicae internae> A

9 **subcostal artery** *origin,* thoracic aorta; *branches,* posterior and spinal branches; *distribution,* upper posterior abdominal wall. <arteria subcostalis> A

10 **superior epigastric artery** *origin,* internal thoracic artery; *branches,* none; *distribution,* abdominal wall, diaphragm. <arteria epigastrica superior> A

11 ***tracheal branches of internal thoracic artery*** branches supplying the trachea. <rami tracheales arteriae thoracicae internae> A

12 **internal thoracic veins** two veins formed by junction of the venae comitantes of the internal thoracic artery of either side; each continues along the artery to open into the brachiocephalic vein. Called also *internal mammary veins.* <venae thoracicae internae> B

13 **musculophrenic veins** the venae comitantes of the musculophrenic artery, draining blood from parts of the diaphragm and from the wall of the thorax and abdomen and emptying into the internal thoracic veins. <venae musculophrenicae> B

14 **right superior intercostal vein** a common trunk formed by union of the second, third, and sometimes fourth posterior intercostal veins, which drains into the azygos vein. <vena intercostalis superior dextra> B

15 **superior epigastric veins** the venae comitantes of the superior epigastric artery, which open into the internal thoracic vein. <venae epigastricae superiores> B

16 **subclavian vein** the vein that continues the axillary as the main venous stem of the upper member, follows the subclavian artery, and joins with the internal jugular vein to form the brachiocephalic vein. <vena subclavia> C

PLEURAL CAVITY
Pleura

17 **pleural cavity** the potential space between the parietal and visceral pleurae. Called also *pleural space.* <cavitas pleuralis> D

18 **pleural recesses** the spaces where the different portions of the pleura join at an angle and which are never completely filled by lung tissue. <recessus pleurales> D

19 **visceral pleura** the portion of the pleura investing the lungs and lining their fissures, completely separating the different lobes. <pleura visceralis> D

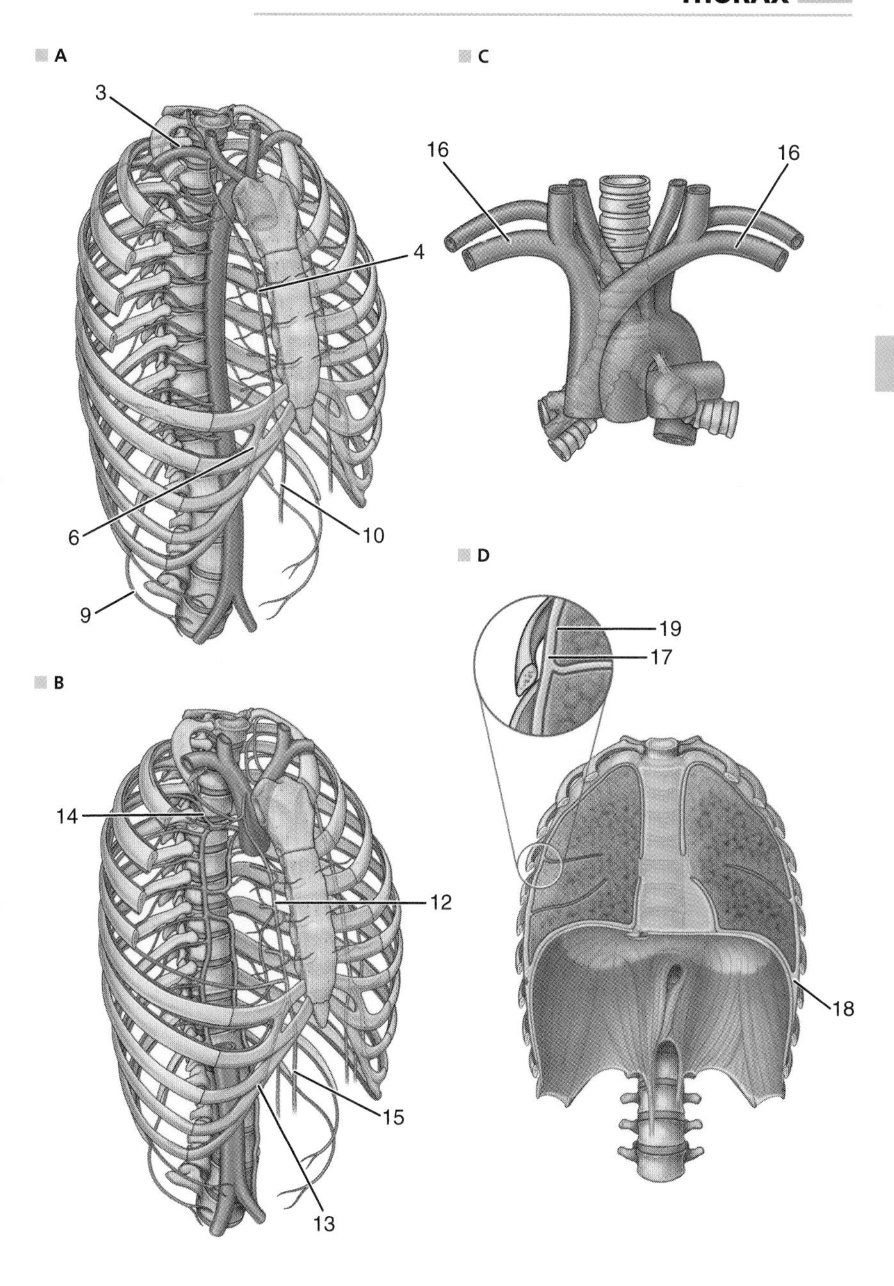
A
3
4
6
10
9
B
14
12
15
13
C
16
16
D
19
17
18

1 **cupula of pleura** the domelike roof of the pleural cavity on either side, extending up through the superior aperture of the thorax. <cupula pleurae> A

2 **parietal pleura** the portion of the pleura lining the walls of the thoracic cavity. <pleura parietalis> A

3 ***phrenicopleural fascia*** the fascial layer on the upper surface of the diaphragm, beneath the pleura. <fascia phrenicopleuralis> A

4 ***pleura*** the serous membrane investing the lungs (visceral pleura) and lining the thoracic cavity (parietal pleura), completely enclosing a potential space known as the pleural cavity. There are two pleurae, right and left, entirely distinct from each other. The pleura is moistened with a serous secretion which facilitates the movements of the lungs in the chest. A

5 **pulmonary ligament** a vertical pleural fold associated with the lung, extending from the hilum down to the base on the medial surface of the lung; on the left it forms the posterior boundary of the cardiac impression. <ligamentum pulmonale> A, C

6 **costodiaphragmatic recess** the pleural recess situated at the junction of the costal and diaphragmatic pleurae. <recessus costodiaphragmaticus pleuralis> B

7 **costomediastinal recess** a wedge-shaped space, not completely filled with lung tissue, along the line at which the costal pleura meets the mediastinal pleura in front. <recessus costomediastinalis pleuralis> B

8 ***phrenicomediastinal recess*** the pleural recess situated at the line of junction of the diaphragmatic and mediastinal pleurae. <recessus phrenicomediastinalis pleuralis> B

Lungs

9 **root of lung** the attachment of either lung, comprising the structures entering and emerging at the hilum. <radix pulmonis> C

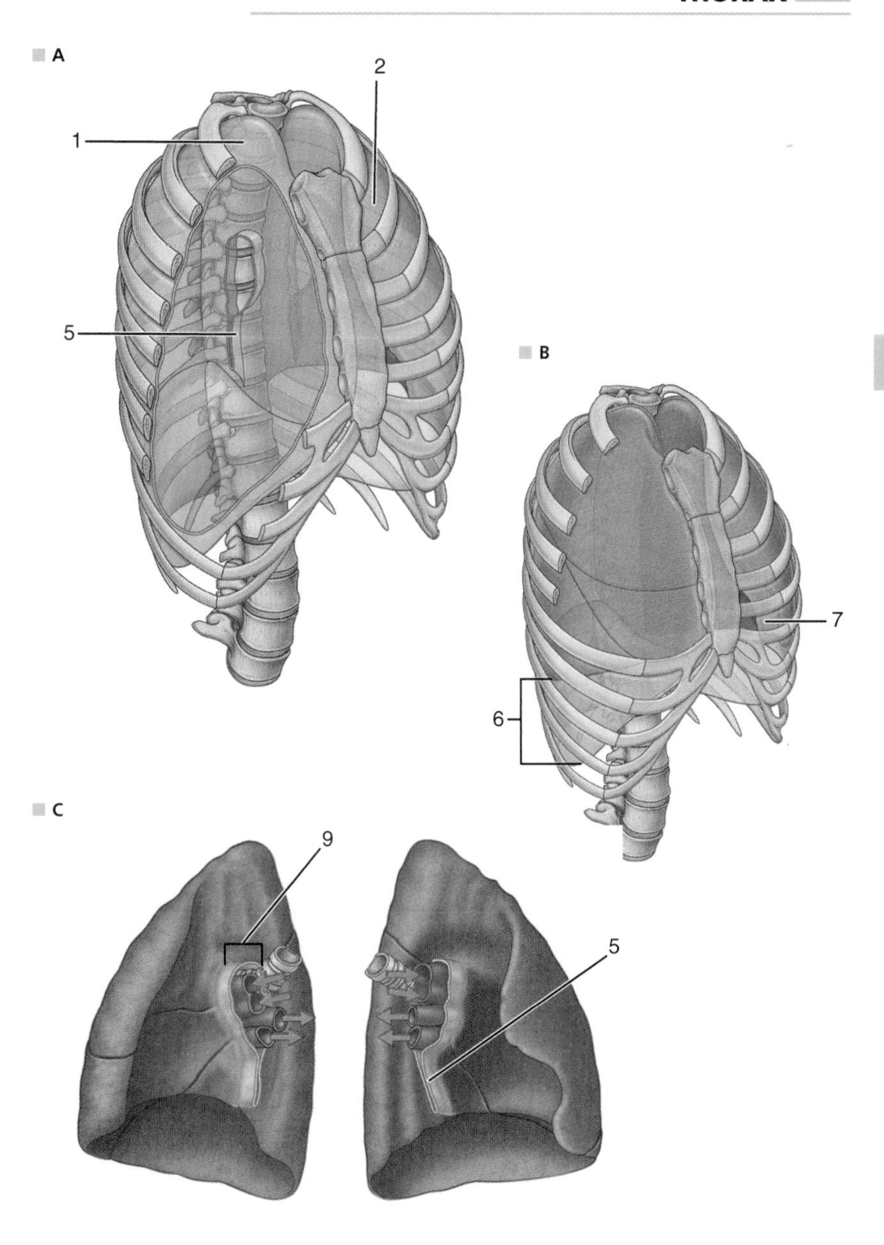
A
1
2
5
B
7
6
C
9
5

1 **horizontal fissure of right lung** the cleft that extends forward from the oblique fissure in the right lung, separating the upper and middle lobes. <fissura horizontalis pulmonis dextri> A

2 **inferior lobe of right lung** the inferior of the three lobes of the right lung; it has five bronchopulmonary segments. <lobus inferior pulmonis dextri> A

3 **middle lobe of right lung** the middle of the three lobes of the right lung; it has two bronchopulmonary segments. <lobus medius pulmonis dextri> A

4 **superior lobe of right lung** the superior of the three lobes of the right lung; it has three bronchopulmonary segments. <lobus superior pulmonis dextri> A

5 **cardiac impression of lung** the indentation on the medial surface of either lung produced by the heart and pericardium. <impressio cardiaca pulmonis> B

6 **interlobar surface of lung** the surface of a lung lying within a fissure (oblique or horizontal). <facies interlobaris pulmonis> A

7 **apex of lung** the rounded upper extremity of either lung, extending upward as high as the first thoracic vertebra. <apex pulmonis> A, B

8 **diaphragmatic surface of lung** the surface of each lung that is adjacent to the diaphragm. <facies diaphragmatica pulmonis> A, B

9 **hilum of lung** the depression on the mediastinal surface of the lung where the bronchus and the blood vessels and nerves enter. <hilum pulmonis> A, B

10 **inferior border of lung** the border of the lung that extends in a curve behind the sixth costal cartilage, the upper margin of the eighth rib in the axillary line, the ninth or tenth rib in the scapular line, and passes medially to the eleventh costovertebral joint. <margo inferior pulmonis> A, B

11 **lung** either of the pair of organs of respiration, one on the right and one on the left of the thorax, which are separated from each other by the heart and mediastinal structures. The right lung is composed of upper, middle, and lower lobes, and the left, of upper and lower lobes. Each lung consists of an external serous coat (the visceral layer of the pleura), subserous areolar tissue, and lung parenchyma. The latter is made up of lobules, which are bound together by connective tissue. <pulmo> A, B

12 **mediastinal surface of lung** the surface of each lung lying medially to the vertebral column and mediastinum; it contains the cardiac impression. <facies mediastinalis pulmonis> A, B

13 **base of lung** the portion of each lung that is directed toward the diaphragm. <basis pulmonis> A, B, C

14 **cardiac notch of left lung** a notch in the anterior border of the left lung. <incisura cardiaca pulmonis sinistri> B

15 **lingula of left lung** a projection from the lower portion of the upper lobe of the left lung, just beneath the cardiac notch, between the cardiac impression and the inferior margin. <lingula pulmonis sinistri> B

16 **inferior lobe of left lung** the inferior of the two lobes of the left lung; it has four or five bronchopulmonary segments. <lobus inferior pulmonis sinistri> B

17 **superior lobe of left lung** the superior of the two lobes of the left lung; it has four bronchopulmonary segments. <lobus superior pulmonis sinistri> B

18 **anterior border of lung** the anterior border of either lung, which descends from behind the sternum, slightly lateral to the midline, and curves laterally to meet the inferior margin. <margo anterior pulmonis> A, B

19 **costal surface of lung** the convex surface of each lung in close adaptation to the curvatures of the ribs and the costal cartilages, which joins the mediastinal surface at the anterior and posterior borders and the diaphragmatic surface at the inferior border. It is related behind to the sides of the vertebral bodies. <facies costalis pulmonis> A, B

20 **oblique fissure of lung** the cleft that separates the lower from the middle and upper lobes in the right lung or the upper from the lower lobe in the left lung. <fissura obliqua pulmonis> A, B

21 **vertebral part of costal surface of lung** the part of the costal surface of each lung related behind to the sides of the vertebral bodies. <pars vertebralis faciei costalis pulmonis> A, B

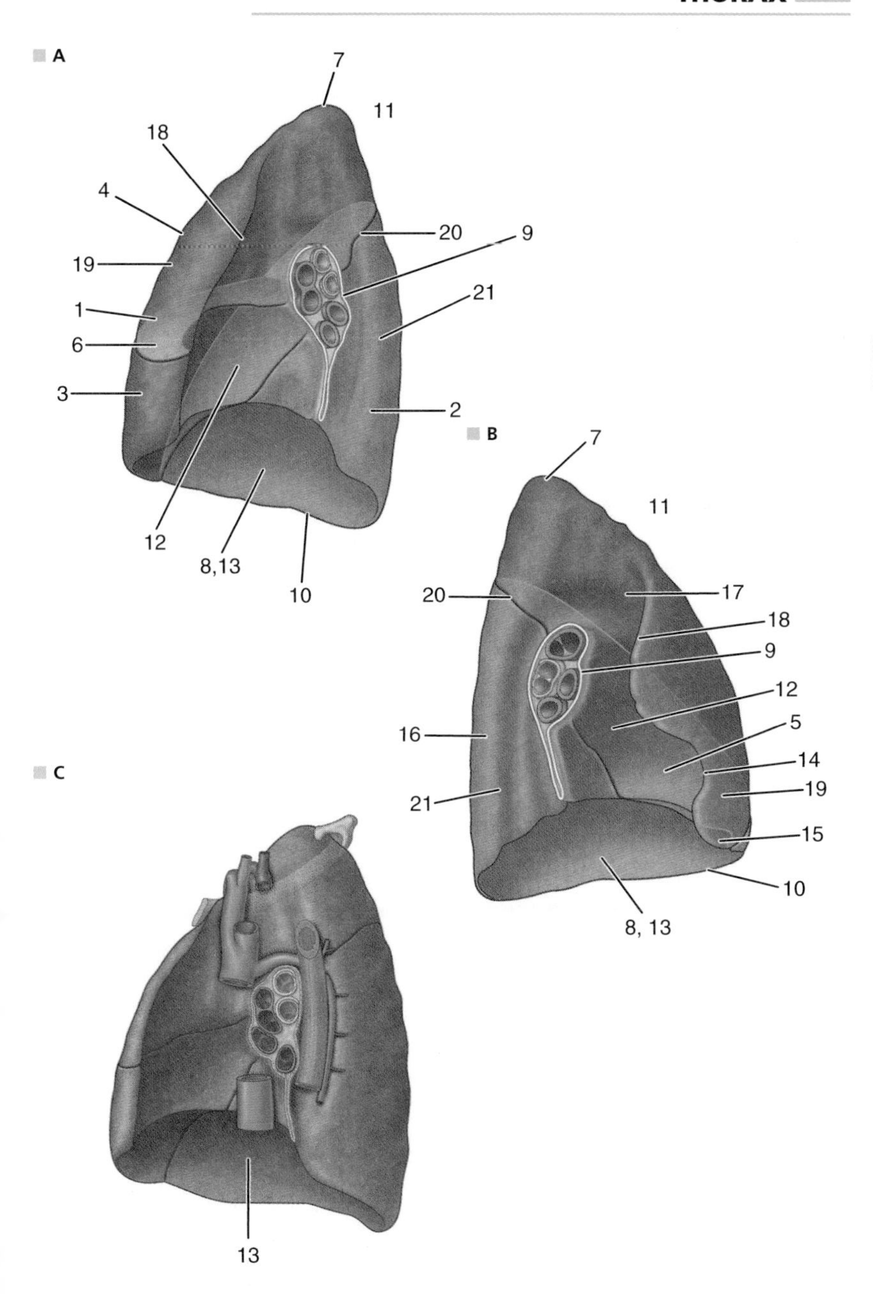
A
7
11
18
4
20
9
19
21
1
6
3
2
12
8,13
10
B
7
11
20
17
18
9
12
5
16
14
19
21
15
10
8, 13
C
13

1 ***bronchial glands*** seromucous glands in the mucosa and submucosa of the bronchial walls. <glandulae bronchiales> A

2 ***bronchial tree*** the bronchi and their branching structures. <arbor bronchialis> A

3 ***bronchiole*** one of the finer subdivisions of the branched bronchial tree, 1 mm or less in diameter, differing from the bronchi in having no cartilage plates and having cuboidal epithelial cells. <bronchiolus> A

4 ***bronchus*** any of the larger air passages of the lungs, having an outer fibrous coat with irregularly placed plates of hyaline cartilage, an interlacing network of smooth muscle, and a mucous membrane of columnar ciliated epithelial cells. A

5 **left primary bronchus** one of the two main branches into which the trachea divides, passing to the left lung. Called also *left main bronchus.* <bronchus principalis sinister> A

6 **lobar bronchi** passages arising from the primary bronchi and passing to the lobes of the right and left lungs. There are three right and two left lobar bronchi, which divide into the segmental bronchi. <bronchi lobares> A

7 ***mucosa of bronchi*** the mucous membrane (tunica mucosa) lining the bronchi. <tunica mucosa bronchiorum> A

8 **right primary bronchus** one of the two main branches into which the trachea divides, passing to the right lung. Called also *right main bronchus.* <bronchus principalis dexter> A

9 ***submucosa of bronchi*** the layer of tissue underlying the tunica mucosa of the bronchi. <tela submucosa bronchiorum> A

10 **segmental bronchi** air passages arising from the lobar bronchi and passing to the different segments of the two lungs, where they further subdivide into smaller and smaller passages (bronchioles). The segmental bronchi are designated by roman numerals, with the three right lobar bronchi divided into ten segmental bronchi and the two left lobar bronchi into eight or nine, depending on the system of classification. <bronchi segmentales> A, B

11 **bronchopulmonary segments** the smaller subdivisions of the lobes of the lungs, separated by connective tissue septa and supplied by branches of the respective lobar bronchi; they are also designated by roman numerals. <segmenta bronchopulmonalia> C

A

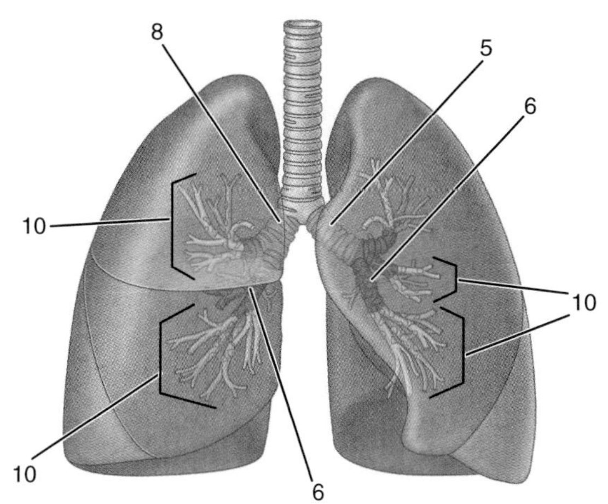
8
5
6
10
10
10
10
6

B

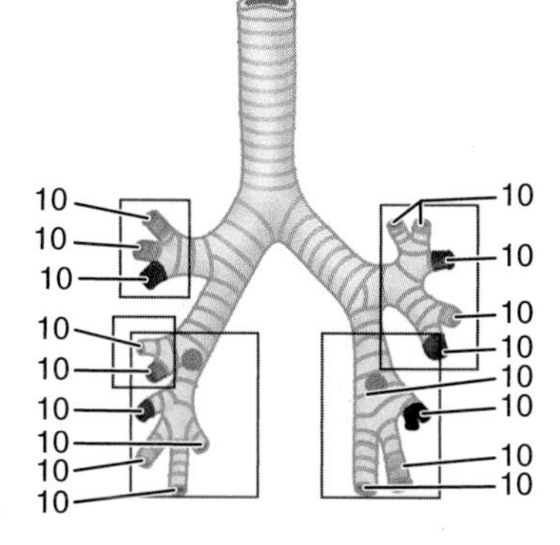
10
10
10
10
10
10
10
10
10
10
10
10
10
10
10
10
10
10
10

C

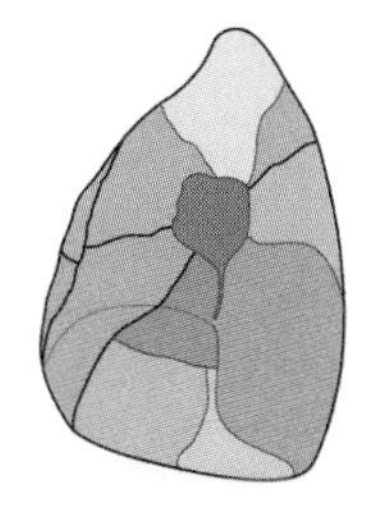

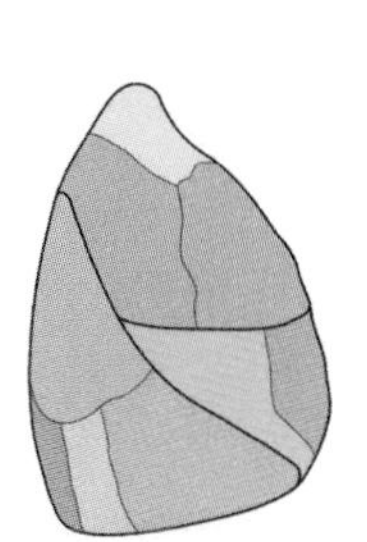

11

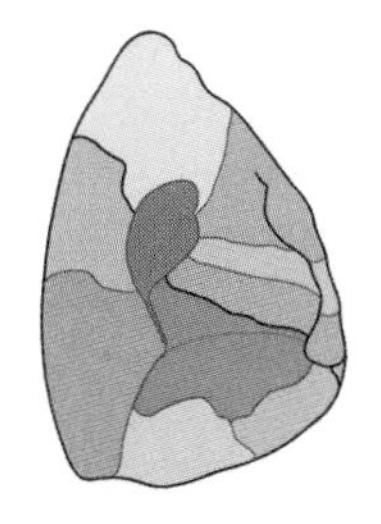

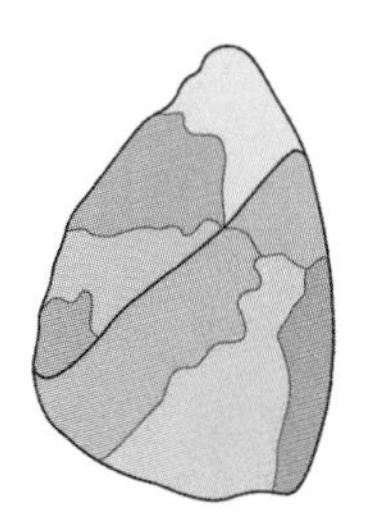

1 **bifurcation of pulmonary trunk** the site of the division of the pulmonary trunk into right and left pulmonary arteries. <bifurcatio trunci pulmonalis> A

2 **left pulmonary artery** *origin,* pulmonary trunk; *branches, of superior lobe:* apical, ascending and descending, posterior, lingular (inferior and superior), *of inferior lobe:* superior, anterior basal, lateral basal, medial basal, posterior basal; *distribution,* left lung. <arteria pulmonalis sinistra> A

3 **right pulmonary artery** *origin,* pulmonary trunk; *branches, of superior lobe:* apical, anterior ascending and descending, posterior ascending and descending, *of medial lobe:* medial and lateral, *of inferior lobe:* anterior basal, lateral basal, medial basal; posterior basal; *distribution,* right lung. <arteria pulmonalis dextra> A

4 **bronchial branches of vagus nerve** branches of the vagus that supply the bronchi and the pulmonary vessels, both directly and by way of the anterior and posterior parts of the pulmonary plexus; there are two or three short anterior branches and numerous longer posterior branches; modality, parasympathetic and visceral afferent. <rami bronchiales nervi vagi> B

5 **pulmonary plexus** a nerve plexus formed by several strong trunks of the vagus nerve that are joined at the root of the lung by branches from the sympathetic trunk and cardiac plexus. The plexus is often described as having anterior and posterior parts; filaments from each accompany the blood vessels and bronchi into the lungs. <plexus pulmonalis> B

6 **pulmonary thoracic branches of thoracic ganglia** branches from the second to the fourth, fifth, or sixth thoracic ganglia to the posterior pulmonary plexus, sometimes going on to follow intercostal arteries to the hilum of the lung. <rami pulmonales thoracici gangliorum thoracicorum> B

7 **bronchomediastinal trunk** either of the two lymphatic trunks (right and left) that drain the pulmonary, bronchopulmonary, tracheobronchial, tracheal, and parasternal lymph nodes: that on the right side into the right lymphatic duct or subclavian vein, and that on the left into the thoracic duct or the subclavian vein. <truncus bronchomediastinalis> C

8 **bronchopulmonary lymph nodes** lymph nodes embedded in the root of the lung, mainly at the hilum, that drain into the tracheobronchial lymph nodes; called also *hilar lymph nodes.* <nodi lymphoidei bronchopulmonales> C

9 **inferior tracheobronchial lymph nodes** nodes in the angle of the bifurcation of the trachea, receiving lymph from adjacent structures. <nodi lymphoidei tracheobronchiales inferiores> C

10 ***intrapulmonary lymph nodes*** nodes located along the larger bronchi within the lung substance, through which lymph from the lung drains. Called also *pulmonary lymph nodes.* <nodi lymphoidei intrapulmonales> C

11 **paratracheal lymph nodes** lymph nodes on either side of the esophagus and trachea, extending upward into the neck, which receive lymph from the esophagus, trachea, and tracheobronchial lymph nodes. <nodi lymphoidei paratracheales> C

12 **superior tracheobronchial lymph nodes** nodes between the trachea and the bronchus on either side, receiving lymph from adjacent structures. <nodi lymphoidei tracheobronchiales superiores> C

MEDIASTINUM

13 **anterior mediastinum** the division of the inferior mediastinum bounded posteriorly by the pericardium, anteriorly by the sternum, and on each side by the pleura. It contains loose areolar tissue and lymphatic vessels. <mediastinum anterius> D

14 **inferior mediastinum** the three inferior portions of the mediastinum, comprising the anterior, middle, and superior mediastinum. <mediastinum inferius> D

15 **mediastinum** the mass of tissues and organs separating the two pleural sacs, between the sternum anteriorly and the vertebral column posteriorly and from the thoracic inlet superiorly to the diaphragm inferiorly. It is divided into a superior region and an inferior region that comprises anterior, middle, and posterior parts. D

16 **middle mediastinum** the division of the inferior mediastinum containing the heart enclosed in its pericardium, the ascending aorta, the superior vena cava, the bifurcation of the trachea into bronchi, the pulmonary arteries and veins, the phrenic nerves, a large portion of the roots of the lungs, and the arch of the azygos vein. <mediastinum medium> D

17 **posterior mediastinum** the division of the inferior mediastinum bounded posteriorly by the vertebral column, anteriorly by the pericardium, and on each side by the pleurae. It contains the descending aorta, parts of the greater and lesser azygos and superior intercostal veins, the thoracic duct, the esophagus, the vagus nerves, and the greater splanchnic nerves. <mediastinum posterius> D

18 **superior mediastinum** the division of the mediastinum extending from the pericardium to the root of the neck, and containing the esophagus and the trachea posteriorly, the thymus or its remains anteriorly, and the great vessels related to the heart and pericardium, the thoracic duct, and the vagus nerves in between. <mediastinum superius> D

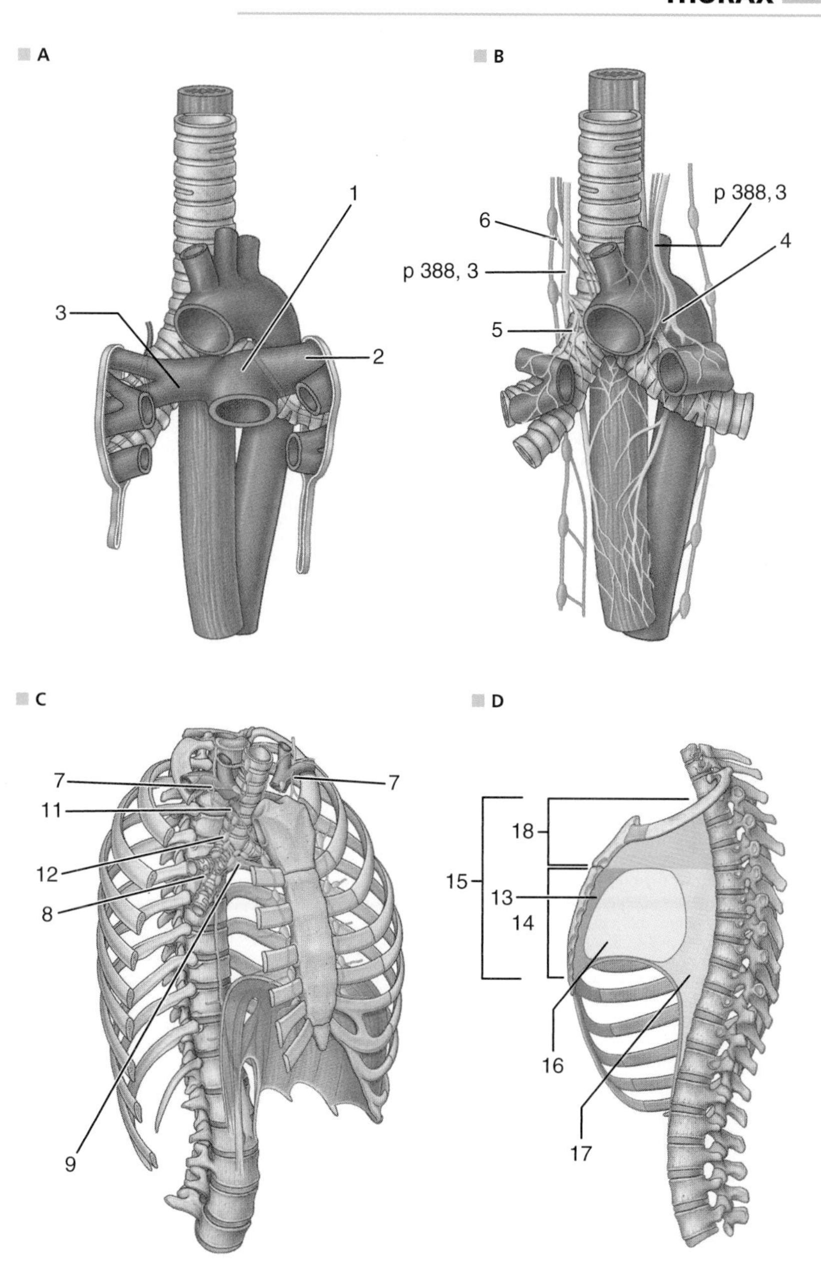
A
1
2
3
B
6
p 388, 3
p 388, 3
4
5
C
7
7
11
12
8
9
D
15
18
13
14
16
17

Superior Mediastinum

1 **anterior mediastinal lymph nodes** nodes along the great vessels of the superior mediastinum and on the anterior part of the diaphragm, receiving lymph from adjacent structures. <nodi lymphatici mediastinales anteriores> A

2 **annular ligaments of trachea** circular horizontal ligaments that join the tracheal cartilages together. <ligamenta anularia tracheae> D

3 **bifurcation of trachea** the site of division of the trachea into the right and left primary bronchi. <bifurcatio tracheae> B

4 **carina of trachea** a projection of the lowest tracheal cartilage, forming a prominent semilunar ridge running anteroposteriorly between the orifices of the two bronchi. <carina tracheae> B

5 ***mucosa of trachea*** the mucous membrane (tunica mucosa) lining the trachea. <tunica mucosa tracheae> B

6 **trachea** the cartilaginous and membranous tube descending from the larynx and branching into the right and left main bronchi. It is kept patent by a series of about twenty transverse horseshoe-shaped cartilages. Called also *windpipe.* B, C, D, E

7 **tracheal cartilages** the 16 to 20 incomplete rings which, held together and enclosed by a strong, elastic, fibrous membrane, constitute the wall of the trachea. Called also *tracheal rings.* <cartilagines tracheales> D

8 ***tracheal glands*** mucous glands in the elastic submucous coat between the cartilaginous rings and on the posterior wall of the trachea. <glandulae tracheales> B

9 ***submucosa of esophagus*** the submucous layer of the esophagus. <tela submucosa oesophagi> C

10 ***adventitia of esophagus*** the tunica adventitia of the esophagus. <tunica adventitia oesophagi> C, D

11 **lobe of thymus** either of the two chief parts (right or left) of the thymus, which meet in the midline. <lobus thymi> E

12 ***lobules of thymus*** the smaller subdivisions of the lobes of the thymus, separated by fibrous trabeculae. <lobuli thymi> E

13 ***medulla of thymus*** the central portion of each lobule of the thymus; it contains many more reticular cells and far fewer lymphocytes than does the surrounding cortex. <medulla thymi> E

14 **thymic branches of internal thoracic artery** branches distributed to the thymus gland in the anterior mediastinum; called also *thymic arteries.* <rami thymici arteriae thoracicae internae> E

15 ***thymic veins*** small branches from the thymus gland that open into the left brachiocephalic vein. <venae thymicae> E

16 **thymus** a bilaterally symmetric lymphoid organ consisting of two pyramidal lobes situated in the anterior superior mediastinum. It develops as an outgrowth of the epithelium of the third branchial pouch, which is invaded by lymphoid stem cells that migrate via the blood from the yolk sac and later from the bone marrow. Each lobe is surrounded by a fibrous capsule from which septa penetrate to divide the parenchyma into lobules; each lobule consists of an outer zone, the cortex, relatively rich in lymphocytes (thymocytes), and an inner zone, the medulla, relatively rich in epithelial cells. The thymus is the site of production of T lymphocytes. Precursor cells migrate into the outer cortex, where they actively proliferate. As they mature and acquire T cell surface markers they move through the inner cortex, where approximately 90 per cent die; the remainder move on to the medulla, become mature T cells, and enter the circulation. The thymus reaches its maximal development at about puberty and then undergoes a gradual process of involution (replacement of parenchyma by fat and fibrous tissue), resulting in a slow decline of immune function throughout adulthood. E

17 **brachiocephalic vein** either of the two veins that drain blood from the head, neck, and upper limbs, and unite to form the superior vena cava. Each is formed at the root of the neck by union of the ipsilateral internal jugular and subclavian veins. The right vein *(v. brachiocephalica dextra)* passes almost vertically downward in front of the brachiocephalic artery, and the left vein *(v. brachiocephalica sinistra)* passes from left to right behind the upper part of the sternum. Each vein receives the vertebral, deep cervical, deep thyroid, and internal thoracic veins. The left vein also receives intercostal, thymic, tracheal, esophageal, phrenic, mediastinal, and pericardiac branches, as well as the thoracic duct; and the right vein receives the right lymphatic duct. Called also *innominate vein.* <vena brachiocephalica> E

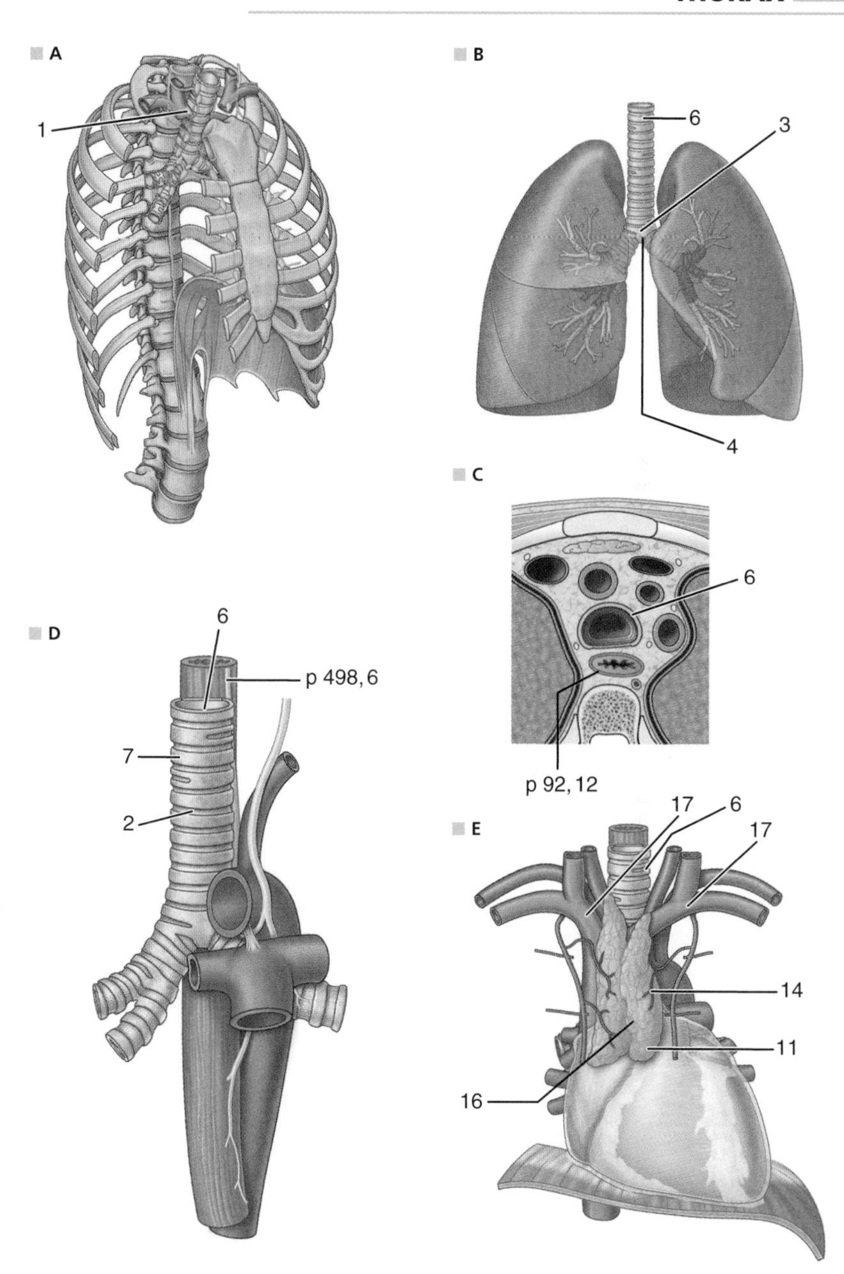
A
1
B
6
3
4
C
6
p 92, 12
D
6
p 498, 6
7
2
E
17
6
17
14
11
16

1 ***ductus arteriosus*** a fetal blood vessel connecting the left pulmonary artery directly to the descending aorta; called also *arterial canal, duct of Botallo,* and *pulmoaortic canal.* A

2 **ligamentum arteriosum** a short, thick, strong fibromuscular cord extending from the pulmonary artery to the arch of the aorta; it is the remains of the ductus arteriosus. A

3 ***lymph node of ligamentum arteriosum*** the lowest anterior mediastinal lymph node situated anterior to the ligamentum arteriosum. <nodus lymphoideus ligamenti arteriosi> A

4 ***mediastinal veins*** numerous small branches that drain blood from the anterior mediastinum into the brachiocephalic vein, azygos vein, or the superior vena cava. <venae mediastinales> A

5 **subclavian artery** *origin,* brachiocephalic trunk (right), arch of aorta (left); *branches,* vertebral, internal thoracic arteries, thyrocervical and costocervical trunks; *distribution,* neck, thoracic wall, spinal cord, brain, meninges, upper limb. <arteria subclavia> A, C, D

6 ***tracheal veins*** small branches that drain blood from the trachea into the brachiocephalic vein. <venae tracheales> A

7 ***highest intercostal vein*** the first posterior intercostal vein of either side, which passes over the apex of the lung and ends in the brachiocephalic, vertebral, or superior intercostal vein. <vena intercostalis suprema> B

8 **left superior intercostal vein** the common trunk formed by union of the second, third, and sometimes fourth posterior intercostal veins, which crosses the arch of the aorta and joins the left brachiocephalic vein. <vena intercostalis superior sinistra> B

9 ***esophageal branches of recurrent laryngeal nerve*** branches that help innervate the esophagus; modality, visceral afferent and general sensory. <rami oesophagei nervi laryngei recurrentis> C

10 **thoracic part of esophagus** the part of the esophagus located in the thoracic region, posterior to the trachea and pericardium and anterior to the vertebral column. <pars thoracica oesophagi> C

11 **thoracic part of trachea** the part of the trachea that lies posteriorly in the superior mediastinum, separated from the upper four thoracic vertebrae by the esophagus. <pars thoracica tracheae> C

12 **esophagus** the musculomembranous passage extending down from the pharynx and emptying into the stomach at the cardiac opening, passing through the diaphragm at the esophageal hiatus. It is divided anatomically into three parts, *cervical, thoracic,* and *abdominal.* A, C, D

13 ***mucosa of esophagus*** the mucous membrane (tunica mucosa) lining the esophagus. <tunica mucosa oesophagi> C

14 ***muscular coat of esophagus*** the muscular layer of the esophagus. <tunica muscularis oesophagi> C

15 ***muscularis mucosae of esophagus*** the muscular layer of the mucous membrane (tunica mucosa) of the esophagus. <lamina muscularis mucosae oesophagi> C

16 ***serosa of esophagus*** the serous coat or layer (tunica serosa) of the esophagus. <tunica serosa oesophagi> C

17 **thoracic constrictions of esophagus** narrowing of the thoracic esophagus where it is compressed by the aortic arch and the left main bronchus. <constrictio partis thoracicae oesophagi> D

Anterior Mediastinum

18 ***prepericardial lymph nodes*** lymph nodes situated between the pericardium and sternum. <nodi lymphoidei prepericardiaci>

19 ***sternopericardial ligaments*** two (superior and inferior) or more fibrous bands that attach the pericardium to the dorsal surface of the sternum. Called also *ligaments of Luschka.* <ligamenta sternopericardiaca>

Middle Mediastinum

20 **pulmonary trunk** the vessel arising from the conus arteriosus of the right ventricle, extending upward obliquely to divide into the right and left pulmonary arteries beneath the arch of the aorta, and conveying unaerated blood toward the lungs. <truncus pulmonalis> A

21 **superior vena cava** the venous trunk draining blood from the head, neck, upper extremities, and chest; it begins by union of the two brachiocephalic veins, passes directly downward, and empties into the right atrium of the heart. <vena cava superior> A

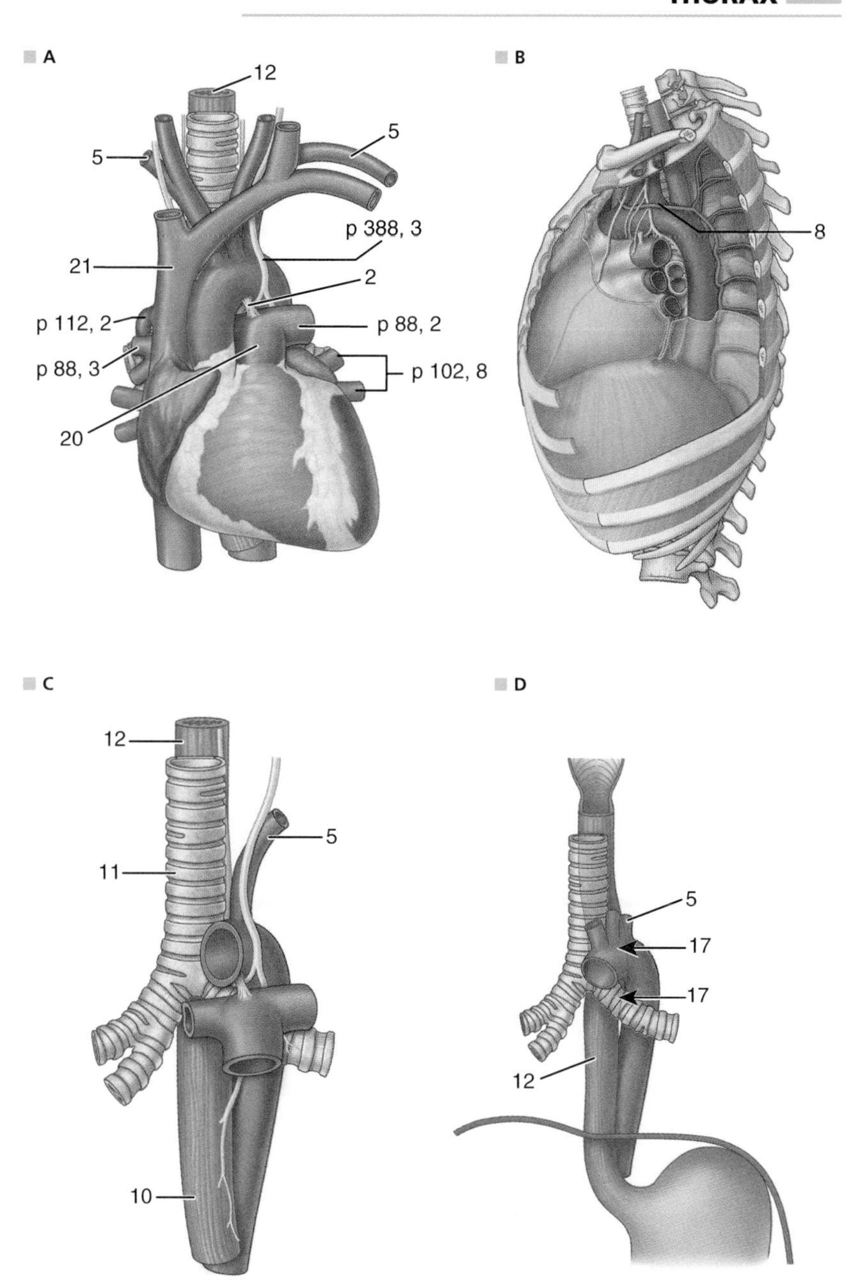
A
12
5
5
p 388, 3
21
2
p 112, 2
p 88, 2
p 88, 3
p 102, 8
20
B
8
C
12
5
11
10
D
5
17
17
12

Pericardium

1 **pericardium** the fibroserous sac that surrounds the heart and the roots of the great vessels, comprising an external layer of fibrous tissue and an inner serous layer. The base of the pericardium is attached to the central tendon of the diaphragm. A, B

2 **fibrous pericardium** the external layer of the pericardium, consisting of fibrous tissue. <pericardium fibrosum> A

3 ***fold of left vena cava*** a triangular fold of visceral pericardium occurring between the left superior pulmonary vein and the left pulmonary artery and enclosing the ligament of the left vena cava. <plica venae cavae sinistrae> A

4 ***ligament of left vena cava*** a remnant of one of the venous trunks opening into the embryonic primordial atrium, extending from the left intercostal vein to the oblique vein of left atrium; it is enclosed by the fold of left vena cava. <ligamentum venae cavae sinistrae> A

5 **parietal layer of serous pericardium** the outer of the two layers of the serous pericardium, lining the fibrous pericardium. Called also *parietal pericardium.* <lamina parietalis pericardii serosi> A

6 **pericardial cavity** the potential space between the parietal layer and the visceral layer (epicardium) of the serous pericardium. <cavitas pericardiaca> A

7 ***serous pericardium*** the inner serous portion of the pericardium, consisting of two layers, the parietal layer, and the visceral layer, or epicardium, which is reflected onto the roots of the great vessels and the heart. <pericardium serosum> A

8 **visceral layer of serous pericardium** the inner layer of the serous pericardium; it is in contact with the heart and the roots of the great vessels; called also *epicardium* and *visceral pericardium.* <lamina visceralis pericardii serosi> A

9 ***accessory phrenic nerves*** an inconstant contribution of the fifth cervical nerve to the phrenic nerve; when present, they run a separate course to the root of the neck or into the thorax before joining the phrenic nerve. <nervi phrenici accessorii> B

10 ***lateral pericardial lymph nodes*** lymph nodes accompanying the pericardiacophrenic artery. <nodi lymphoidei pericardiaci laterales> B

11 **pericardiac branch of phrenic nerve** a branch arising from the phrenic or accessory phrenic nerve and supplying the pericardium; *modality*, general sensory. <ramus pericardiacus nervi phrenici> B

12 **pericardiacophrenic artery** *origin,* internal thoracic artery; *branches,* none; *distribution,* pericardium, diaphragm, pleura. <arteria pericardiacophrenica> B

13 **pericardiacophrenic veins** small veins that drain blood from the pericardium and diaphragm into the left brachiocephalic vein. <venae pericardiacophrenicae> B

14 ***pericardial veins*** numerous small branches that drain blood from the pericardium into the brachiocephalic, inferior thyroid, and azygos veins, and the superior vena cava. <venae pericardiacae> B

15 **phrenic nerve** *origin,* cervical plexus (C4-C5); *branches,* pericardial and phrenicoabdominal branches; *distribution,* pleura, pericardium, diaphragm, peritoneum, and sympathetic plexuses; *modality,* general sensory and motor. <nervus phrenicus> B

16 ***phrenicoabdominal branches of phrenic nerve*** branches of the phrenic or accessory phrenic nerve that supply the diaphragm; *modality,* general sensory and motor. <rami phrenicoabdominales nervi phrenici> B

17 **oblique pericardial sinus** a recess of serous pericardium that passes upward behind the left atrium and between the left and right pulmonary veins. <sinus obliquus pericardii> C

18 **transverse pericardial sinus** a passage behind the aorta and pulmonary trunk and in front of the atria; it is lined by serous pericardium. <sinus transversus pericardii> C

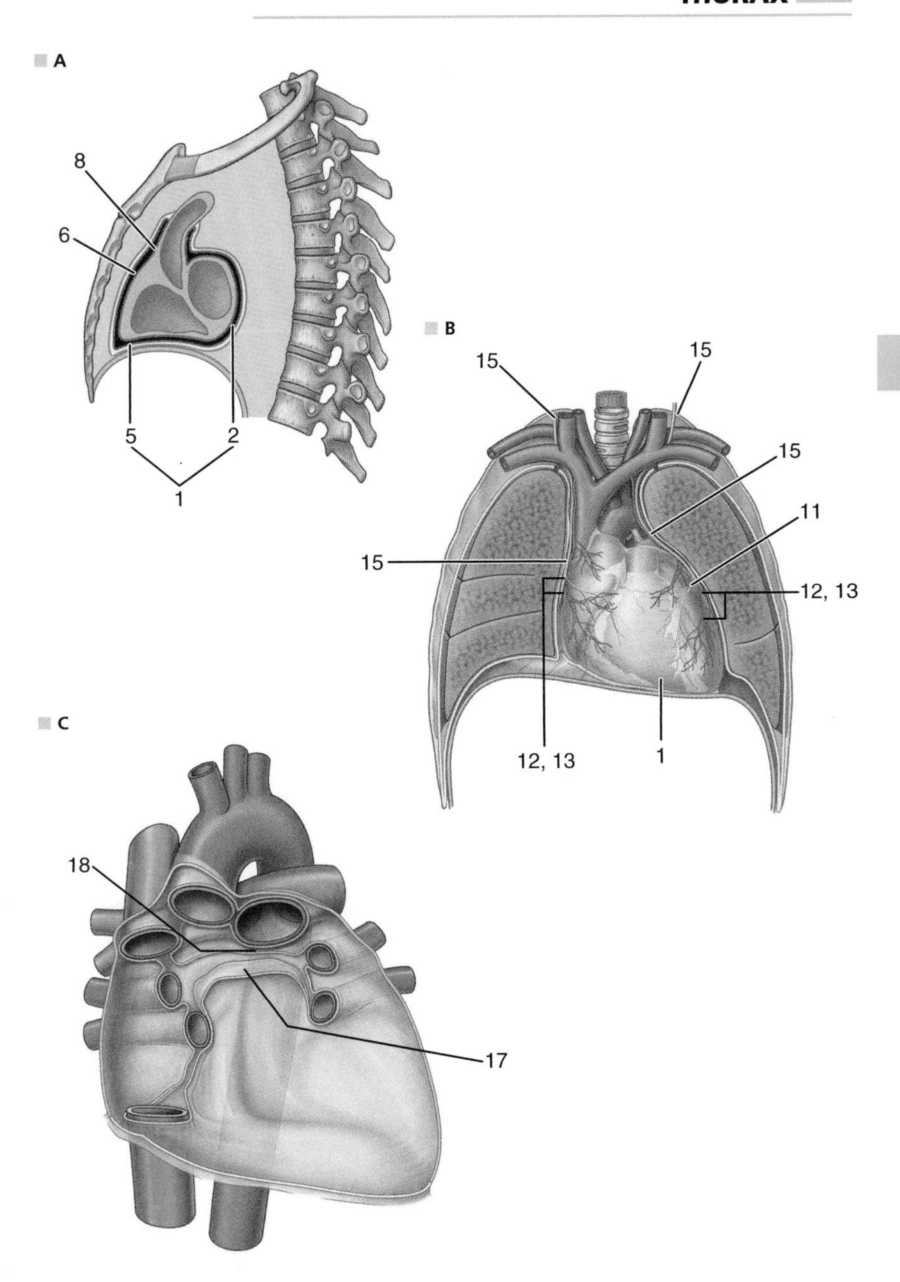
A
8
6
5
2
1
B
15
15
15
11
15
12, 13
12, 13
1
C
18
17

Heart

1 ***heart*** the viscus of cardiac muscle that maintains the circulation of the blood.. It is divided into four cavities-two atria and two ventricles. The left atrium receives oxygenated blood from the lungs. From there the blood passes to the left ventricle, which forces it via the aorta through the arteries to supply the tissues of the body. The right atrium receives the blood after it has passed through the tissues and given up much of its oxygen. The blood then passes to the right ventricle, and then to the lungs, to be oxygenated. There are four major valves: left atrioventricular (mitral); right atrioventricular (tricuspid); aortic; and pulmonary. The heart tissue itself is nourished by the blood in the coronary arteries. <cor> A

2 ***endocardium*** the endothelial lining membrane of the cavities of the heart and the connective tissue bed on which it lies. This subendothelial connective tissue contains varying amounts of elastic and collagen fibers and smooth muscle cells.

3 ***myocardium*** the middle and thickest layer of the heart wall, composed of cardiac muscle.

4 **anterior interventricular sulcus** a groove on the anterior surface of the heart marking the position of the interventricular septum and the line of separation between the ventricles; called also *anterior interventricular groove.* <sulcus interventricularis anterior> A

5 **anterior surface of heart** the convex surface of the heart, which in general is directed anteriorly and somewhat superiorly, being formed mainly by the right ventricle, and to a lesser degree by the left ventricle and the atria; called also *sternocostal surface of heart.* <facies sternocostalis cordis> A

6 **apex of heart** the blunt rounded extremity of the heart formed by the left ventricle; it is directed anteriorly, inferiorly, and to the left. <apex cordis> A

7 **coronary sulcus** a groove on the external surface of the heart, separating the atria from the ventricles; portions of it are occupied by the major arteries and veins of the heart; called also *atrioventricular groove.* <sulcus coronarius cordis> A

8 **notch of cardiac apex** a slight notch found at the site where the anterior and posterior interventricular sulci become continuous and cross the right margin of the heart. <incisura apicis cordis> A

9 **right border of heart** the margin of the heart formed by the wall of the profile of the right atrium, running from the apex to the right, and marking the junction of the anterior and diaphragmatic cardiac surfaces; seen as a surface except in two-dimensional representations. Called also *acute margin, inferior border,* or *right surface of heart.* <margo dexter cordis> A

10 ***right/left pulmonary surface of heart*** the surface of the heart that faces either of the lungs. <facies pulmonalis dextra/sinistra cordis> A

11 ***base of heart*** a poorly delimited region of the heart, formed, in general, by the atria and the area occupied by the roots of the great vessels. It lies opposite the middle thoracic vertebrae, its exact position varying with heart action, and is directed superiorly, posteriorly, and to the right. <basis cordis> B

12 **diaphragmatic surface of heart** the surface of the heart (within the pericardium) that rests on the diaphragm and is directed inferiorly and somewhat posteriorly; it is formed by the two ventricles, the left ventricle contributing a little more than the right. <facies diaphragmatica cordis> B

13 **posterior interventricular sulcus** a groove on the diaphragmatic surface of the heart marking the position of the interventricular septum and the line of separation between the ventricles; called also *posterior interventricular groove.* <sulcus interventricularis posterior> B

14 **cusp** a tapering projection or structure, applied especially to one of the triangular segments of a cardiac valve. <cuspis> C

15 **left fibrous trigone of heart** a thickened and irregularly triangular portion of the fibrous skeleton of the base of the heart, located between the left atrioventricular fibrous ring and the left posterior margin of the aortic orifice. <trigonum fibrosum sinistrum cordis> C

16 **right fibrous trigone of heart** a thickened, irregularly triangular portion of the fibrous skeleton of the base of the heart, located between the right and left atrioventricular fibrous rings, posterior to the aortic orifice. <trigonum fibrosum dextrum cordis> C

17 **right/left fibrous ring of heart** one of the dense fibrous rings that surround the right and left atrioventricular orifices. To these rings, either directly or indirectly, are attached the atrial and ventricular muscle fibers. The rings form part of the cardiac skeleton. Called also *Lower's rings.* <anulus fibrosus dexter/sinister cordis> C

18 ***tendon of infundibulum*** a collagenous band connecting the posterior surface of the pulmonary valve and the muscular infundibulum to the root of the aorta; called also *conus ligament.* <tendo infundibuli> C

19 **vortex of heart** the whorled arrangement of muscle fibers at the apex in the left ventricle of the heart, through which the more superficial fibers pass to the interior of the left ventricle toward the base. <vortex cordis> D

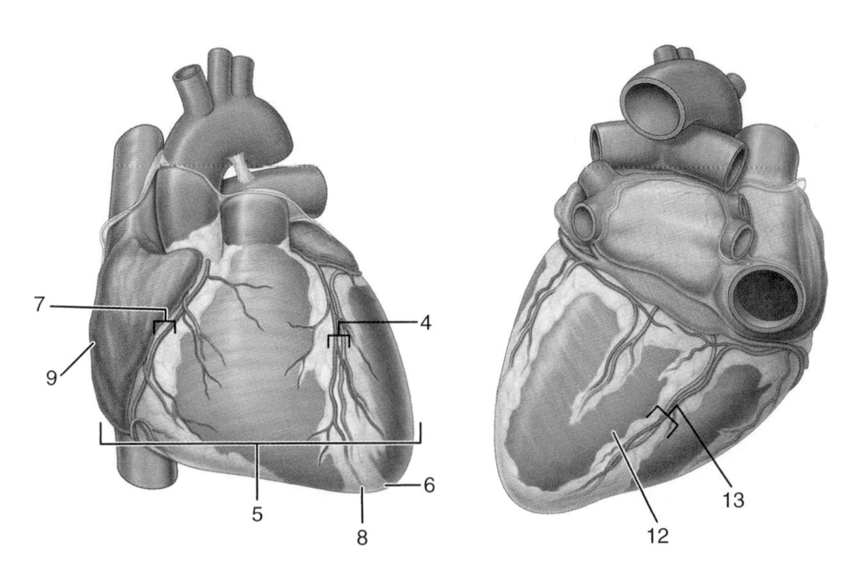
A
B
7
4
9
5
6
8
12
13

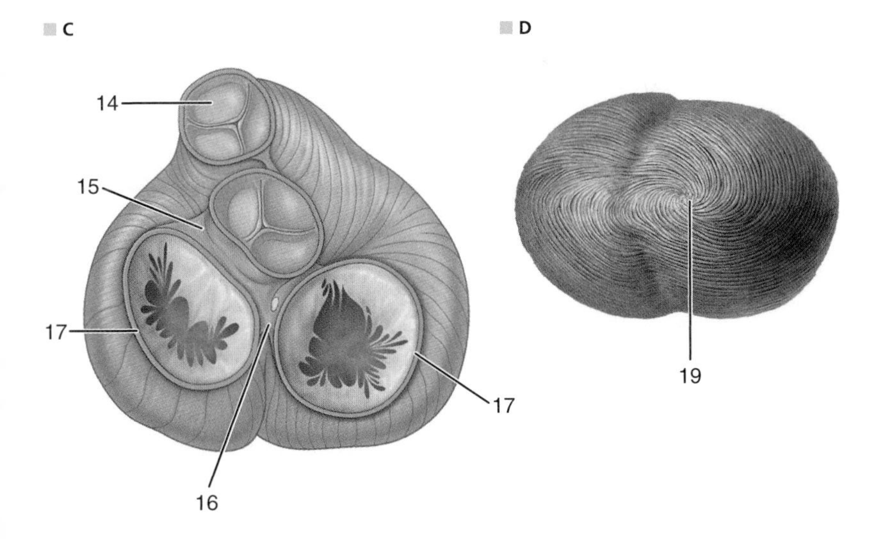
C
D
14
15
17
17
16
19

Right Atrium

1 **right atrium** the atrium of the right side of the heart; it receives blood from the superior and the inferior venae cavae, and delivers it to the right ventricle. <atrium dextrum> A

2 **sulcus terminalis cordis** a shallow groove on the external surface of the right atrium of the heart between the superior and inferior venae cavae; it represents the junction of the sinus venosus with the primitive atrium in the embryo, and corresponds to a ridge on the internal surface, the crista terminalis. B

3 **atrioventricular septum** the portion of the membranous part of the interventricular septum between the right atrium and left ventricle. <septum atrioventriculare cordis> C

4 **crista terrminalis of right atrium** a ridge on the internal surface of the right atrium of the heart, located to the right of the orifices of the superior and inferior venae cavae, and separating the sinus venarum cavarum from the atrium proper and auricle. The pectinate muscles of the right atrium are attached at this crest. It corresponds to a groove on the external surface, the sulcus terminalis. Called also *taenia terminalis.* <crista terminalis atrii dextri> C

5 ***foramen ovale of heart*** the aperture in the fetal heart that provides a communication between the atria. <foramen ovale cordis> C

6 **fossa ovalis** an oval depression in the right atrium on the lower part of the interatrial septum, within a triangular area bounded by the openings of the two venae cavae and the coronary sinus. It represents the remains of the fetal foramen ovale. <fossa ovalis cordis> C

7 **interatrial septum** the wall that separates the atria of the heart. <septum interatriale cordis> C

8 ***intervenous tubercle*** a more or less distinct ridge across the inner surface of the right atrium between the openings of the venae cavae. <tuberculum intervenosum> C

9 **limbus of fossa ovalis** the prominent rounded margin of the fossa ovalis; it represents the edge of the second septum of the fetal heart. Called also *border of oval fossa.* <limbus fossae ovalis> C

10 **opening of coronary sinus** an opening in the wall of the right atrium, situated between the opening of the inferior vena cava and the atrioventricular opening, the lower part of which is covered by the valve of the coronary sinus. <ostium sinus coronarii> C

11 **opening of inferior vena cava** the opening of the inferior vena cava into the right atrium of the heart; it is accompanied by a valve that, in the adult, is usually rudimentary. <ostium venae cavae inferioris> C

12 **opening of superior vena cava** the opening of the superior vena cava into the right atrium of the heart; it is unaccompanied by a valve. <ostium venae cavae superioris> C

13 **pectinate muscles of right atrium** small ridges of muscle fibers projecting from the inner walls of the right auricle of the heart and extending in the right atrium to the crista terminalis. <musculi pectinati atrii dextri> C

14 **right atrioventricular orifice** the opening between the right atrium and the right ventricle of the heart, guarded by the tricuspid (right atrioventricular) valve; called also *tricuspid orifice.* <ostium atrioventriculare dextrum> C

15 **right auricle** the ear-shaped appendage of the right atrium of the heart. <auricula dextra> B

16 ***sinus of venae cavae*** the portion of the right atrium bounded medially by the interatrial septum and laterally by the crista terminalis, and into which the superior and inferior venae cavae empty; it is often called the *sinus venosus* because it develops from that embryonic structure. <sinus venarum cavarum> C

17 ***sinus venosus*** the common venous receptacle in the embryonic heart, attached to the posterior wall of the primordial atrium; it receives the umbilical and vitelline veins and the common cardinal veins. C

18 **valve of coronary sinus** a fold of endocardium along the right and inferior margins of the opening of the coronary sinus into the right atrium of the heart; it covers the lower part of the sinus and prevents regurgitation into the sinus during atrial contractions. <valvula sinus coronarii> C

19 ***valve of foramen ovale*** a crescentic ridge on the left side of the interatrial septum, representing the edge of what was the first septum of the fetal heart. <valvula foraminis ovalis> C

20 **valve of inferior vena cava** the variably sized crescentic fold of endocardial tissue, enclosing a few muscle fibers, that is attached to the anterior margin of the opening of the inferior vena cava into the right atrium of the heart. Rudimentary in the adult, in the fetus it directs blood flow from the inferior vena cava into the left atrium via the foramen ovale. <valvula venae cavae inferioris> C

Right Ventricle

21 **right ventricle** the cavity of the heart that propels the blood through the pulmonary trunk and arteries into the lungs. <ventriculus dexter cordis> A

22 **tricuspid valve** the valve between the right atrium and right ventricle of the heart; it usually has three cusps (anterior, posterior, and septal), but additional small cusps may be present. Called also *right atrioventricular valve.* <valva atrioventricularis dextra> B

23 **anterior cusp of pulmonary valve** the anterior cusp of the valve of the pulmonary trunk. <valvula semilunaris anterior valvae trunci pulmonalis> B

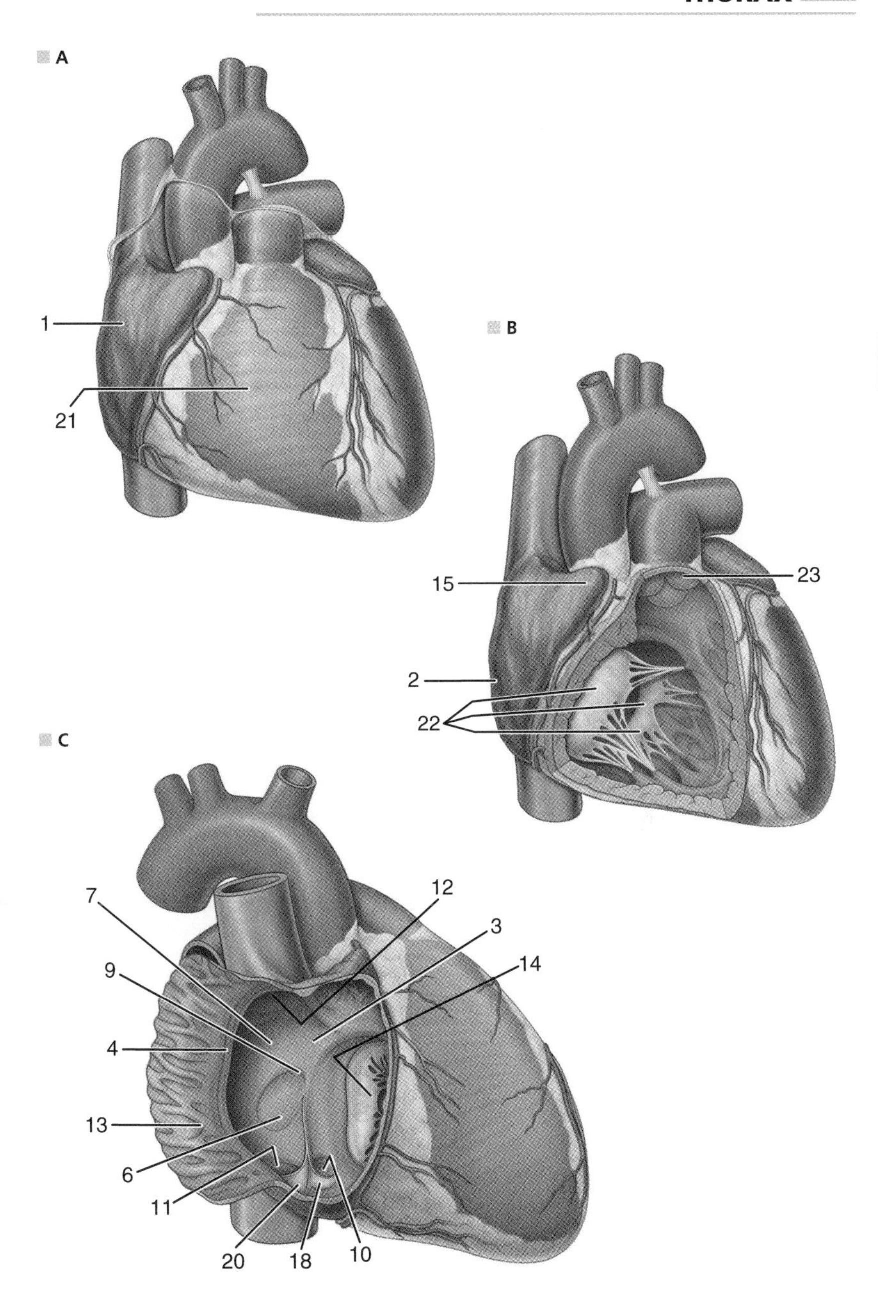
A
1
21
B
15
23
2
22
C
7
12
3
9
14
4
13
6
11
20
18
10

1 **anterior cusp of tricuspid valve** the anterior of the cusps of the tricuspid (right atrioventricular) valve. <cuspis anterior valvae atrioventricularis dextrae> A

2 **anterior papillary muscle of right ventricle** the papillary muscle arising from the anterior wall of the right ventricle. <musculus papillaris anterior ventriculi dextri> A

3 **chordae tendineae** the tendinous cords that connect each cusp of the two atrioventricular valves to appropriate papillary muscles in the heart ventricles. The cords are of varying lengths and thicknesses and are frequently branched. <chordae tendineae cordis> A

4 **conus arteriosus** the anterosuperior portion of the right ventricle of the heart, which is delimited from the rest of the ventricle by the supraventricular crest and which joins the pulmonary trunk, thus forming the outflow tract for blood in the right ventricle. Called also *infundibulum.* A

5 **interventricular septum** the partition that separates the left ventricle from the right ventricle, consisting of a muscular and a membranous part; called also *ventricular s.* <septum interventriculare cordis> A

6 **left semilunar cusp of pulmonary valve** the left cusp of the valve of the pulmonary trunk. <valvula semilunaris sinistra valvae trunci pulmonalis> A

7 ***membranous part of interventricular septum*** the very small, completely membranous area of the interventricular septum of the heart; situated near the root of the aorta, it can be viewed between the opposed margins of the right and posterior semilunar valves of the aorta. <pars membranacea septi interventricularis> A

8 **muscular part of interventricular septum** the thick muscular partition forming the greater part of the septum between the ventricles of the heart. <pars muscularis septi interventricularis> A

9 **opening of pulmonary trunk** the opening of the pulmonary trunk, guarded by the pulmonary valve. <ostium trunci pulmonalis> A

10 **papillary muscles** conical muscular projections from the walls of the cardiac ventricles, attached to the cusps of the atrioventricular valves by the chordae tendineae. There is an anterior and a posterior papillary muscle in each ventricle, as well as a group of small papillary muscles on the septum in the right ventricle. <musculi papillares> A

11 **posterior cusp of tricuspid valve** the posterior of the cusps of the tricuspid (right atrioventricular) valve. <cuspis posterior valvae atrioventricularis dextrae> A

12 **posterior papillary muscle of right ventricle** the papillary muscle arising from the diaphragmatic wall of the right ventricle. <musculus papillaris posterior ventriculi dextri> A

13 **pulmonary valve** a valve composed of three semilunar cusps or segments, guarding the pulmonary orifice in the right ventricle of the heart; it prevents backflow of blood into the right ventricle. <valva trunci pulmonalis> A, B

14 **right semilunar cusp of pulmonary valve** the right cusp of the valve of the pulmonary trunk. <valvula semilunaris dextra valvae trunci pulmonalis> A

15 **septal cusp of tricuspid valve** the cusp that is attached to the membranous interventricular septum. <cuspis septalis valvae atrioventricularis dextrae> A

16 **septomarginal trabecula** a bundle of muscle at the apical end of the right ventricle of the heart, connecting the base of the anterior papillary muscle to the interventricular septum; it usually contains a branch of the atrioventricular bundle. It has been thought to prevent ventricular overdistention and thus is called also *moderator band.* <trabecula septomarginalis> A

17 **supraventricular crest** a ridge on the inner surface of the right ventricle of the heart, marking off the conus arteriosus, or outflow tract, from the remainder of the right ventricle, the inflow tract. <crista supraventricularis> A

18 **trabeculae carneae** irregular bundles and bands of muscle projecting from a great part of the interior of the walls of the ventricles of the heart. They occur as three types: as simple muscular ridges, as bundles attached at both ends but free in the middle, or as papillary muscles, projecting from the heart wall and attaching to the chordae tendineae. <trabeculae carneae cordis> A

19 **lunulae of semilunar cusps of pulmonary valve** small thinned areas in the cusps of the pulmonary valve, one on each side of the nodule of the cusp. <lunulae valvularum semilunarium valvae trunci pulmonalis> B

20 **nodules of semilunar cusps of pulmonary valve** small fibrous tubercles, one at the center of the free margin of each semilunar cusp. <noduli valvularum semilunarium valvae trunci pulmonalis> B

21 **sinus of pulmonary trunk** a slight dilatation between the wall of the pulmonary trunk and each of the semilunar cusps of the pulmonary trunk valve. <sinus trunci pulmonalis> B

A

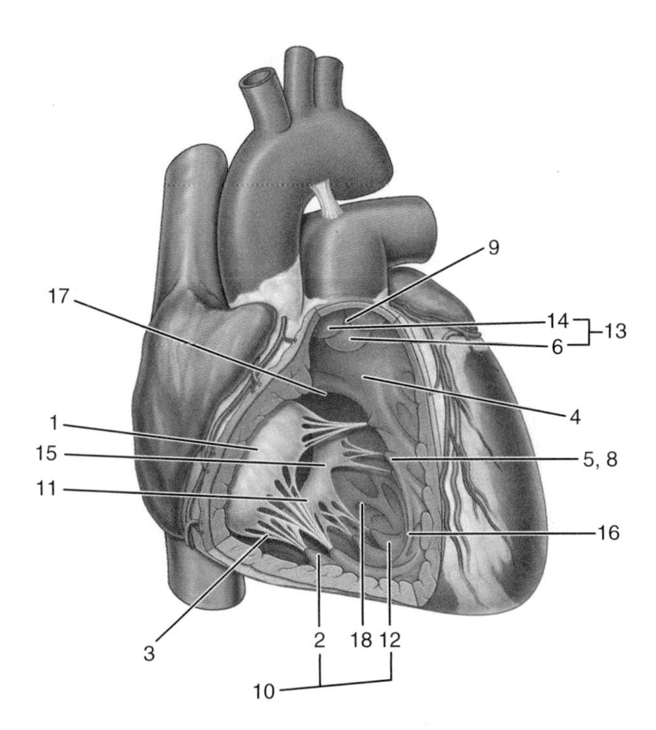

B

13

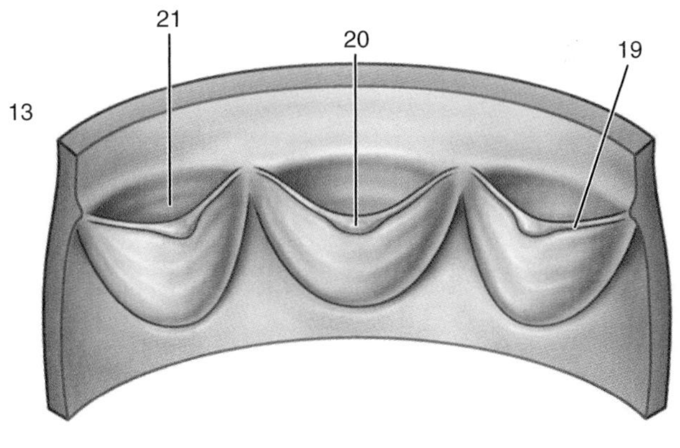

Left Atrium

1 **left atrium** the atrium of the left side of the heart; it receives blood from the pulmonary veins, and delivers it to the left ventricle. <atrium sinistrum> A

2 **left atrioventricular orifice** the opening between the left atrium and the left ventricle of the heart, guarded by the mitral (left atrioventricular) valve; called also *mitral orifice*. <ostium atrioventriculare sinistrum> A

3 **openings of pulmonary veins** the openings of the pulmonary veins (usually four) into the left atrium of the heart; they are unaccompanied by valves. <ostia venarum pulmonalium> A

4 **pectinate muscles of left atrium** small ridges of muscle fibers projecting from the inner walls of the left auricle of the heart. <musculi pectinati atrii sinistri> A

5 **left auricle** the ear-shaped appendage of the left atrium of the heart. <auricula sinistra> B

6 **left inferior pulmonary vein** the vein that returns blood from the lower lobe of the left lung (from the superior segmental and common basal veins) to the left atrium of the heart. <vena pulmonalis sinistra inferior> B

7 **left superior pulmonary vein** the vein that returns blood from the upper lobe of the left lung (from the apicoposterior, anterior segmental, and lingular veins) to the left atrium of the heart. <vena pulmonalis sinistra superior> B

8 **pulmonary veins** the four veins, right and left superior and right and left inferior, that return aerated blood from the lungs to the left atrium of the heart. <venae pulmonales> B

9 **right inferior pulmonary vein** the vein that returns blood from the lower lobe of the right lung (from the superior segmental and common basal veins) to the left atrium of the heart. <vena pulmonalis dextra inferior> B

10 **right superior pulmonary vein** the vein that returns blood from the upper and middle lobes of the right lung (from the middle lobe, apical, anterior segmental, and posterior segmental veins) to the left atrium of the heart. <vena pulmonalis dextra superior> B

Left Ventricle

11 **left ventricle** the cavity of the heart that propels the blood out through the aorta into the systemic arteries. <ventriculus sinister cordis> B

12 **anterior cusp of mitral valve** the anterior of the cusps of the mitral (left atrioventricular) valve. <cuspis anterior valvae atrioventricularis sinistrae> B

13 **anterior papillary muscle of left ventricle** the conical muscular projection arising from the anterior wall of the left ventricle and attached to the anterior cusp of the mitral valve. <musculus papillaris anterior ventriculi sinistri> B

14 **aortic orifice** the opening between the left ventricle and the ascending aorta; guarded by the aortic valve. <ostium aortae> B

15 **aortic vestibule** a space within the left ventricle of the heart at the root of the aorta. <vestibulum aortae> B

16 ***commissural cusps*** two small cusps that form the two outer of the three scallops constituting the posterior cusp of the mitral (left atrioventricular) valve. <cuspides commisurales> B

17 **mitral valve** the valve between the left atrium and left ventricle of the heart; it usually has two cusps (anterior and posterior), but additional small cusps may be present. Called also *left atrioventricular valve*. <valva atrioventricularis sinistra> A

18 **posterior cusp of mitral valve** the posterior of the cusps of the mitral (left atrioventricular) valve; the term is sometimes used to denote the entire three-scalloped region posterior to the anterior cusp but at other times is restricted to the central scallop, with the two outer scallops called the commissural cusps. <cuspis posterior valvae atrioventricularis sinistrae> B

19 **posterior papillary muscle of left ventricle** the conical muscular projection arising from the posterior wall of the left ventricle and attached to the posterior cusp of the mitral valve. <musculus papillaris posterior ventriculi sinistri> B

20 **aortic sinus** a dilatation between the aortic wall and each of the semilunar cusps of the aortic valve; from two of these sinuses the coronary arteries take origin. <sinus aortae> C

21 **aortic valve** a valve composed of three semilunar cusps or segments, guarding the aortic orifice in the left ventricle of the heart; it prevents backflow into the left ventricle. <valva aortae> B, C

22 ***bulb of aorta*** the enlargement of the aorta at its point of origin from the heart, where the bulges of the aortic sinuses occur. <bulbus aortae> C

23 **left semilunar cusp of aortic valve** the left cusp of the aortic valve. <valvula semilunaris sinistra valvae aortae> C

24 **lunulae of semilunar cusps of aortic valve** small thinned areas in the cusps of the aortic valve, one on each side of the nodule of the cusp. <lunulae valvularum semilunarium valvae aortae> C

25 **nodules of semilunar cusps of aortic valve** small fibrous tubercles, one at the center of the free margin of each semilunar cusp of the valve. <noduli valvularum semilunarium valvae aortae> C

26 **posterior semilunar cusp of aortic valve** the posterior cusp of the aortic valve. <valvula semilunaris posterior valvae aortae> C

27 **right semilunar cusp of aortic valve** the right cusp of the aortic valve. <valvula semilunaris dextra valvae aortae> C

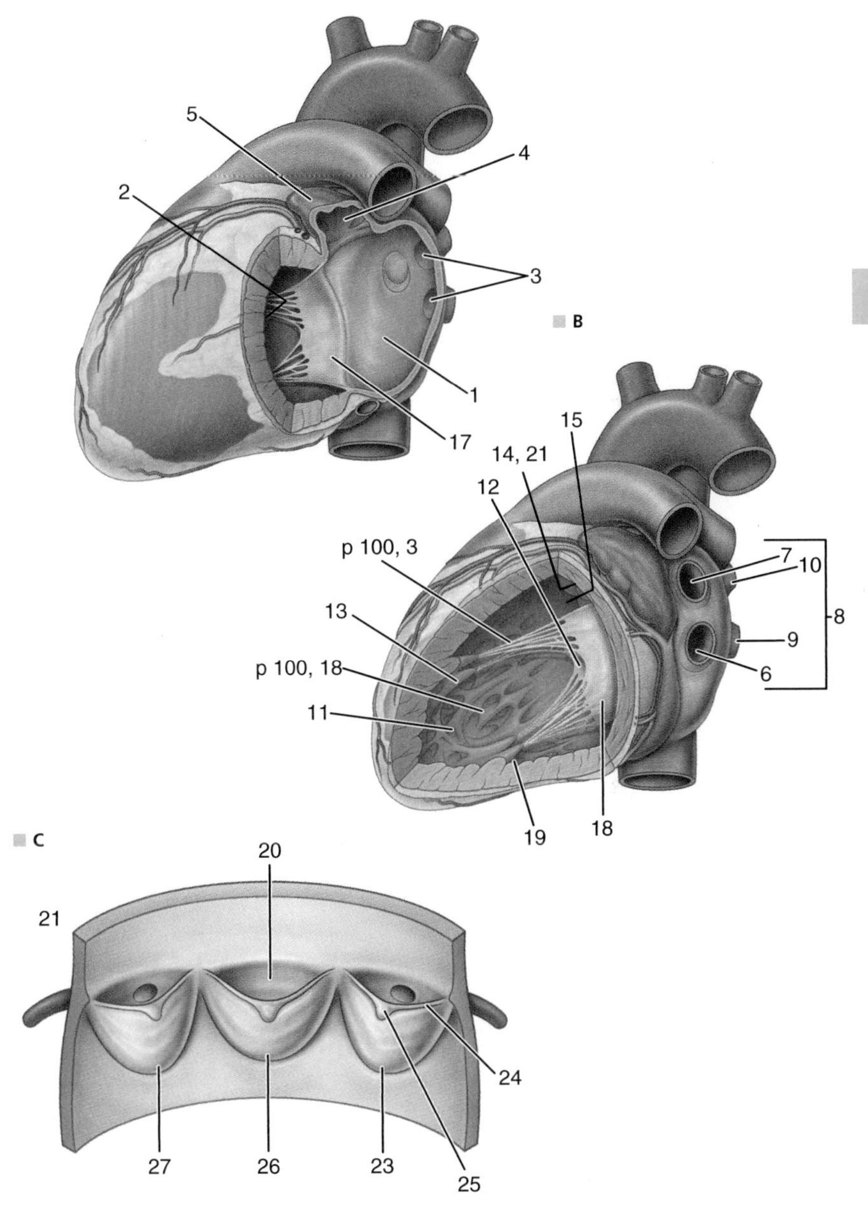
A
5
4
2
3
1
17
B
15
14, 21
12
p 100, 3
13
p 100, 18
11
7
10
8
9
6
19
18
C
20
21
24
27
26
23
25

Coronary Vasculature

1 **anterior interventricular branch of left coronary artery** the branch of the left coronary artery that runs to the apex of the heart in the anterior interventricular sulcus, supplying the ventricles and most of the interventricular septum; called also *anterior interventricular artery.* <ramus interventricularis anterior arteriae coronariae sinistrae> A

2 **aorta** the main trunk from which the systemic arterial system proceeds. It arises from the left ventricle of the heart; passes upward *(ascending aorta),* bends over *(arch of aorta),* and then proceeds downward *(descending aorta).* At about the level of the fourth lumbar vertebra it divides into the two common iliac arteries. A

3 ***aortic isthmus*** a narrowed portion of the aorta, especially noticeable in the fetus, at the point where the ductus arteriosus is attached. <isthmus aortae> A

4 **arch of aorta** the continuation of the ascending aorta, giving rise to the brachiocephalic trunk, and the left common carotid and left subclavian arteries; it continues as the thoracic aorta. <arcus aortae> A

5 **ascending aorta** the proximal portion of the aorta, arising from the left ventricle, and giving origin to the right and left coronary arteries before continuing as the arch of the aorta. <pars ascendens aortae> A

6 ***atrial anastomotic branch of circumflex branch of left coronary artery*** a branch of the circumflex branch of the left coronary artery that passes the interatrial septum to anastomose with the right coronary artery; called also *atrial anastomotic artery.* <ramus atrialis anastomoticus rami circumflexi arteriae coronariae sinistrae> A

7 ***atrial branches of circumflex branch of left coronary artery*** branches distributed to the left atrium, consisting of anterior, lateral, and posterior groups. <rami atriales rami circumflexi arteriae coronariae sinistrae> A

8 ***atrial branches of right coronary artery*** branches of the right coronary artery, consisting of anterior and lateral branches chiefly distributed to the right atrium, and usually a single posterior branch distributed to the right and left atria. <rami atriales arteriae coronariae dextrae> A

9 ***atrioventricular branches of circumflex branch of left coronary artery*** small recurrent branches distributed to the atria and ventricles. <rami atrioventriculares rami circumflexi arteriae coronariae sinistrae> A

10 ***atrioventricular nodal branch of circumflex branch of left coronary artery*** a branch of the left coronary artery occasionally found supplying the atrioventricular node. <ramus nodi atrioventricularis rami circumflexi arteriae coronariae sinistrae> A

11 ***atrioventricular nodal branch of right coronary artery*** a branch of the right coronary artery usually arising opposite the origin of the posterior interventricular artery and inserting into the atrioventricular node. Called also *atrioventricular nodal artery.* <ramus nodi atrioventricularis arteriae coronariae dextrae> A

12 **circumflex branch of left coronary artery** a branch that curves around to the back of the left ventricle in the coronary sulcus, supplying the left ventricle and left atrium. Called also *circumflex artery.* <ramus circumflexus arteriae coronariae sinistrae> A

13 ***conus arteriosus branch of left coronary artery*** a small branch of the anterior interventricular branch of the left coronary artery, which supplies the conus arteriosus, and anastomoses with the right conus artery (branch) of the right coronary artery; called also *conus artery* and *left conus artery.* <ramus coni arteriosi arteriae coronariae sinistrae> A

14 ***conus arteriosus branch of right coronary artery*** the first ventricular branch of the right coronary artery, which supplies the conus arteriosus, and anastomoses with the left conus artery (branch) of the anterior interventricular branch of the left coronary artery; called also *conus artery, right conus artery,* and *third conus artery.* <ramus coni arteriosi arteriae coronariae dextrae> A

15 ***intermediate atrial branch of circumflex branch of left coronary artery*** a branch that distributes along the left atrium above the coronary sulcus; called also *left intermediate atrial artery.* <ramus atrialis intermedius rami circumflexi arteriae coronariae sinistri> A

16 ***intermediate atrial branch of right coronary artery*** a branch of the right coronary artery arising opposite to the marginal branch and ascending to over the right atrium; called also *right intermediate atrial artery.* <ramus atrialis intermedius arteriae coronariae dextrae> A

17 ***interventricular septal branches of left coronary artery*** branches of the anterior interventricular branch of the left coronary artery that supply about the anterior two-thirds of the interventricular septum; called also *anterior septal arteries.* <rami interventriculares septales arteriae coronariae sinistrae> A

18 ***interventricular septal branches of right coronary artery*** numerous relatively small branches of the posterior interventricular branch of the right coronary artery that supply about the posterior one-third of the interventricular septum. Called also *posterior septal arteries.* <rami interventriculares septales arteriae coronariae dextrae> A

19 **left coronary artery** *origin,* left aortic sinus; *branches,* anterior interventricular and circumflex rami; *distribution,* left ventricle, left atrium. <arteria coronaria sinistra> A

A

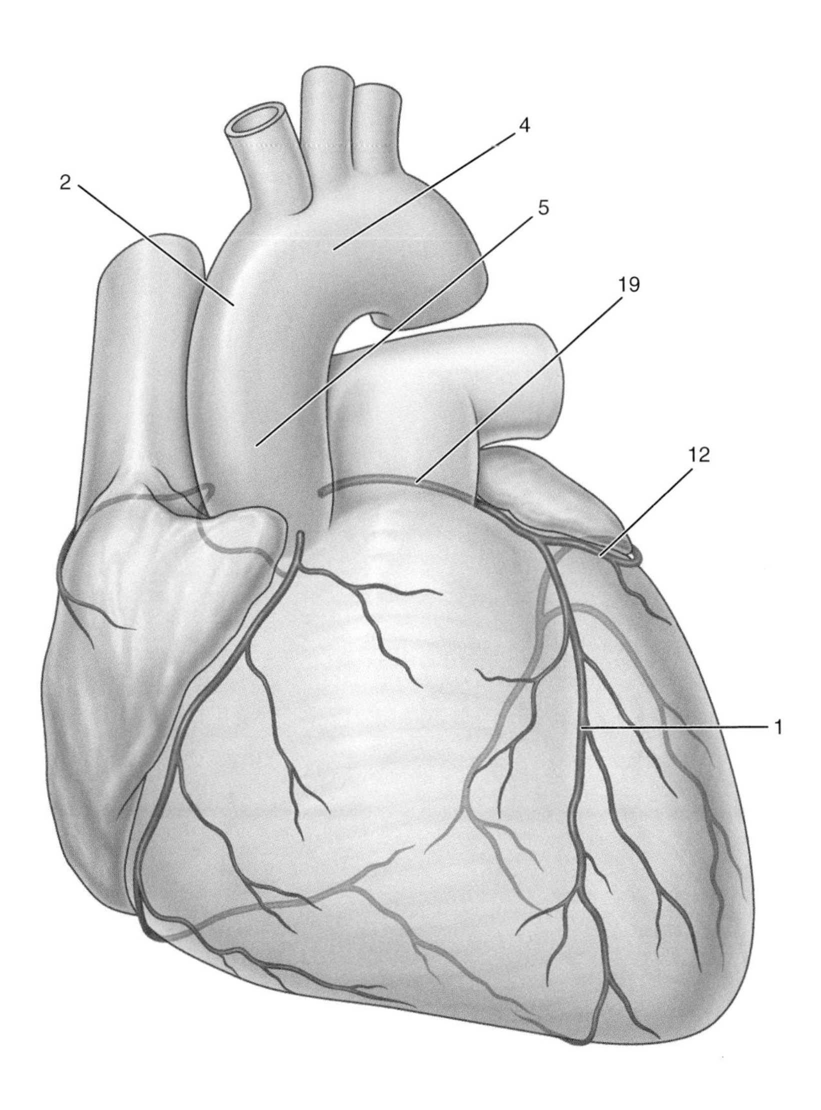

1 **left marginal branch of circumflex branch of left coronary artery** a branch that follows the left margin of the heart and supplies the left ventricle. Called also *left marginal artery.* <ramus marginalis sinister rami circumflexi arteriae coronariae sinistrae> A

2 **posterior interventricular branch of right coronary artery** a branch running toward the apex of the heart in the posterior interventricular sulcus, supplying the diaphragmatic surface of the ventricles and part of the interventricular septum. <ramus interventricularis posterior arteriae coronariae dextrae> A

3 ***posterior left ventricular branch of circumflex branch of left coronary artery*** an interventricular continuation of the circumflex branch; there may be two or three vessels. <ramus posterior ventriculi sinistri rami circumflexi arteriae coronariae sinistrae> A

4 **right coronary artery** *origin,* right aortic sinus; *branches,* conus artery and atrial, atrioventricular node, intermediate atrial, posterior interventricular, right marginal, and sinoatrial node branches; *distribution,* right ventricle, right atrium. <arteria coronaria dextra> A

5 **right marginal branch of right coronary artery** a branch that passes toward the apex of the heart along the acute margin of the heart and ramifies over the right ventricle; called also *right marginal artery.* <ramus marginalis dexter arteriae coronariae dextrae> A

6 ***right posterolateral branch of right coronary artery*** an inconstant branch of the right coronary artery. <ramus posterolateralis dexter arteriae coronariae dextrae> A

7 **sinuatrial nodal branch of right coronary artery** a branch of the right coronary artery that supplies the right atrium, encircles the base of the superior vena cava, and inserts into the sinuatrial node. Called also *nodal artery* or *sinuatrial nodal artery.* <ramus nodi sinuatrialis arteriae coronariae dextrae> A

8 **sinuatrial nodal branch of circumflex branch of left coronary artery** a branch of the left coronary artery occasionally found supplying the sinuatrial node. <ramus nodi sinuatrialis rami circumflexi arteriae coronariae sinistrae> B

9 **anterior interventricular vein** the portion of the great cardiac vein ascending in the anterior interventricular sulcus and emptying into the left coronary vein. <vena interventricularis anterior> C

10 **anterior veins of right ventricle** small veins that drain blood from the anterior aspect of the right ventricle, ascend in subepicardial tissue to cross the right part of the atrioventricular sulcus, and empty into the right atrium. Called also *anterior cardiac veins.* <venae ventriculi dextri anteriores> C

11 ***cardiac veins*** the veins of the heart, which drain blood from the various tissues making up the organ. <venae cordis> C

12 **coronary sinus** the terminal portion of the great cardiac vein, which lies in the coronary sulcus between the left atrium and ventricle, and empties into the right atrium between the opening of the inferior vena cava and the atrioventricular orifice. <sinus coronarius> C

13 ***foramina of smallest cardiac veins*** minute openings in the walls of the heart, through which small veins, the venae cardiacae minimae, empty their blood directly into the heart; they are most numerous in the right atrium and ventricle, occasional in the left atrium, and rare in the left ventricle. <foramina venarum minimarum atrii dextri> C

14 **great cardiac vein** a vein that collects blood from the anterior surface of the ventricles, follows the anterior longitudinal sulcus, and empties into the coronary sinus. <vena cardiaca magna> C

15 ***left atrial veins*** inconstant smallest cardiac veins (venae cardiacae minimae) emptying into the left atrium of the heart. <venae atriales sinistrae> C

16 **middle cardiac vein** a vein that collects blood from the diaphragmatic surface of the ventricles, follows the posterior longitudinal sulcus, and empties into the coronary sinus. <vena cardiaca media> C

17 ***oblique vein of left atrium*** a small vein from the left atrium that opens into the coronary sinus. Called also *vein of Marshall.* <vena obliqua atrii sinistri> C

18 **posterior vein of left ventricle** the vein that drains blood from the posterior surface of the left ventricle into the coronary sinus. <vena ventriculi sinistri posterior> C

19 ***right atrial veins*** the smallest cardiac veins (venae cardiacae minimae) emptying into the right atrium of the heart. <venae atriales dextrae> C

20 **right marginal vein** a vein ascending along the right margin of the heart, draining adjacent parts of the right ventricle and opening into the right atrium or anterior cardiac veins. <vena marginalis dextra> C

21 ***right ventricular veins*** the smallest cardiac veins (venae cardiacae minimae) emptying into the right ventricle of the heart. <venae ventriculares dextrae> C

22 **small cardiac vein** a vein that collects blood from both parts of the right heart, follows the coronary sulcus to the left, and opens into the coronary sinus. <vena cardiaca parva> C

23 ***smallest cardiac veins*** numerous small veins arising in the muscular walls and draining independently into the cavities of the heart, and most readily seen in the atria; called also *thebesian veins* and *venae cordis minimae.* <venae cardiacae minimae> C

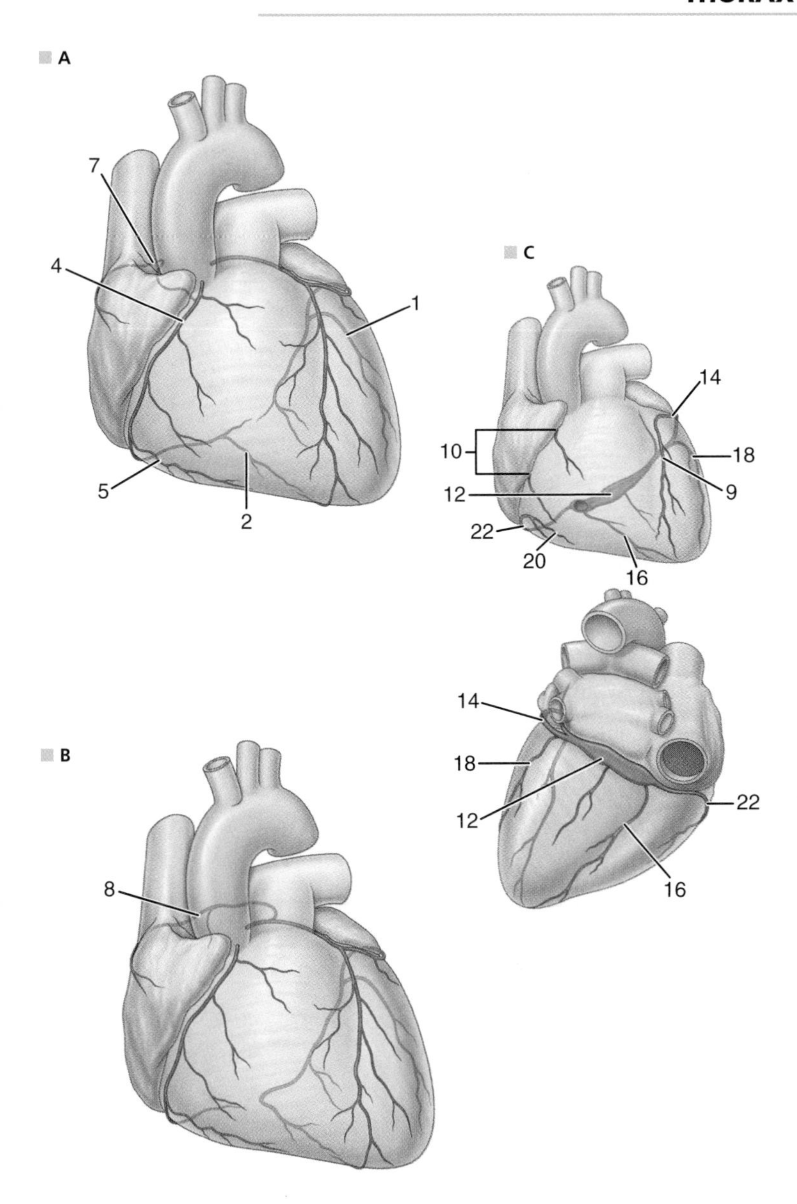
A
7
4
1
5
2
C
14
10
18
12
9
22
20
16
14
18
12
22
16
B
8

Cardiac Conduction System

1 ***cardiac conduction system*** a system of specialized muscle fibers that generate and rapidly transmit cardiac impulses and serve to coordinate contractions, comprising the sinoatrial node, atrioventricular node, bundle of His and its right and left bundle branches, and the subendocardial branches of Purkinje fibers. <complexus stimulans cordis> A, B

2 **atrioventricular bundle** a small band of atypical cardiac muscle fibers originating in the atrioventricular node and comprising the trunk of the bundle of His and the left and right bundle branches. <fasciculus atrioventricularis> A

3 **atrioventricular node** a small area of specialized cardiac muscle cells and fibers that receives the cardiac impulses from the sinoatrial node and passes them on toward the ventricles, introducing a delay in impulse conduction. It is located in the right atrium between the tricuspid valve and the orifice of the coronary sinus, is composed of a meshwork of (Purkinje) fibers continuous with the atrial muscle fibers and the bundle of His, and is supplied by a branch of th e right coronary artery. <nodus atrioventricularis> A

4 **right crus of atrioventricular bundle** a discrete group of fascicles arising from the trunk of the bundle at the superior end of the muscular part of the interventricular septum; it descends to be distributed to the right ventricle of the heart as a terminal network of Purkinje fibers that become continuous with the ventricular cardiac muscle fibers. Called also *right bundle branch.* <crus dextrum fasciculi atrioventricularis> A

5 **sinoatrial node** a microscopic collection of atypical cardiac muscle fibers (Purkinje fibers) at the superior end of the sulcus terminalis, at the junction of the superior vena cava and the right atrium. The cardiac rhythm normally takes its origin in this node, which thus is known also as the *cardiac pacemaker.* Called also *sinuatrial* or *sinus node.* <nodus sinuatrialis> A

6 **subendocardial branches** small ramifications of the conducting system of the heart (Purkinje fibers), which form a plexus in the papillary muscles and ventricles. Called also *Purkinje network.* <rami subendocardiales> A

7 **left crus of atrioventricular bundle** a dispersed array of fascicles arising from the trunk of the bundle at the superior end of the muscular part of the interventricular septum; they continue as a flattened sheet, generally diverging into anterior and posterior limbs, and descend to be distributed to the papillary muscles of the left ventricle of the heart as a terminal network of Purkinje fibers that become continuous with the ventricular cardiac muscle fibers. Called also *left bundle branch.* <crus sinistrum fasciculi atrioventricularis> A, B

8 **cardiac plexus** the plexus around the base of the heart, chiefly in the epicardium. It is formed by cardiac branches from the vagus nerves and the sympathetic trunks and ganglia, contains visceral afferent fibers, and shows subdivisions related to the arch of the aorta, right and left atria, and right and left coronary arteries. The cardiac plexus is continuous with the right and left pulmonary plexuses. <plexus cardiacus> C, D

9 ***inferior cardiac branches of the vagus nerve*** branches (sometimes called cervicothoracic) arising from the vagi and from the recurrent laryngeal nerves at the thoracic inlet, and joining cervicothoracic sympathetic cardiac nerves, the combined nerves passing to the cardiac plexus; *modality,* parasympathetic and visceral afferent. <rami cardiaci cervicales inferiores nervi vagi>

10 **superior cervical cardiac branches of vagus nerve** variable branches arising from the vagus in the cervical region and usually joining the cervical sympathetic cardiac nerves. The conjoined nerves then descend in front of, or behind, the arch of the aorta to the cardiac plexus; modality, parasympathetic and visceral afferent. <rami cardiaci cervicales superiores nervi vagi> C, D

11 **thoracic cardiac branches of the vagus nerve** branches which arise in the thorax from the right and left vagus and left recurrent laryngeal nerves. They go directly to the posterior walls of the atria, to the coronary plexuses, and to the anterior pulmonary plexuses. <rami cardiaci thoracici nervi vagi> C, D

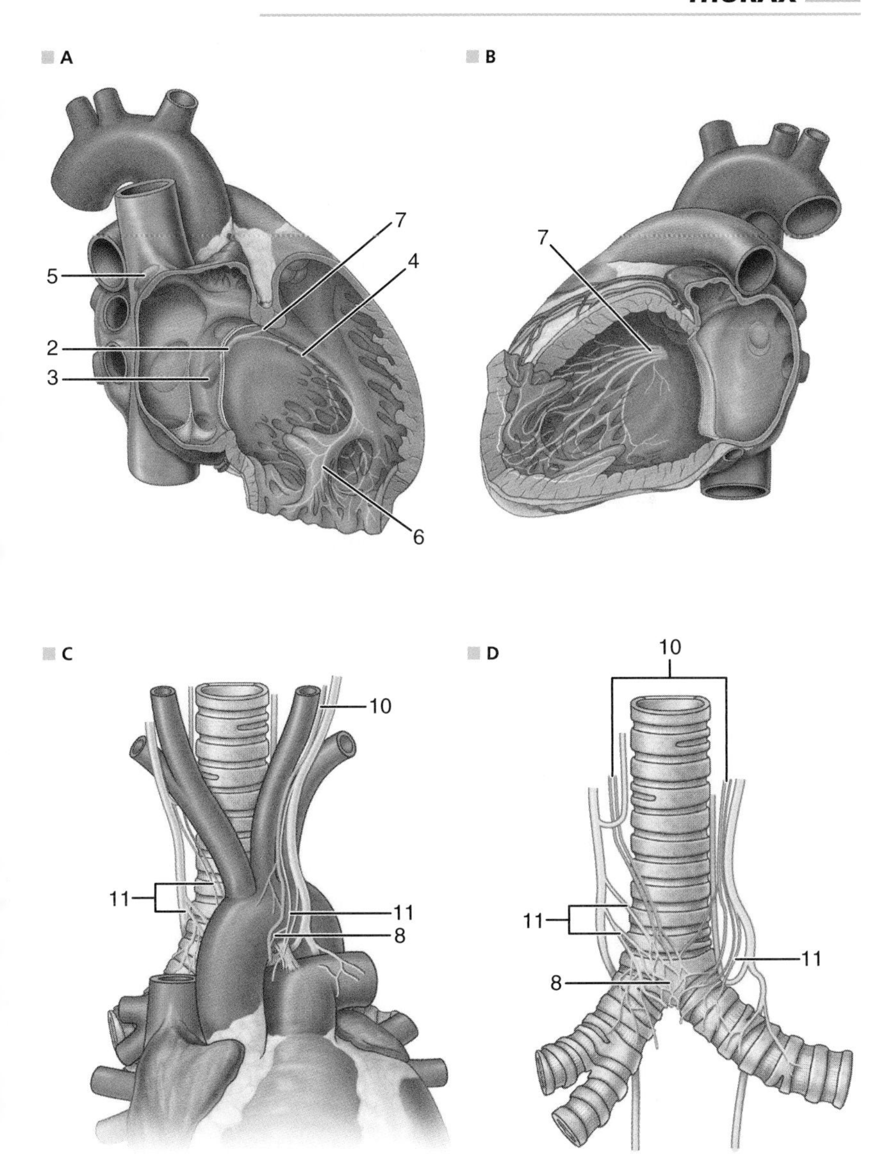
A
7
4
5
2
3
6
B
7
C
10
11
11
8
D
10
11
11
8

Posterior Mediastinum

1 **arch of thoracic duct** see *thoracic duct.* <arcus ductus thoracici> A

2 ***posterior mediastinal lymph nodes*** a group of lymph nodes situated behind the pericardium in relation to the esophagus and thoracic aorta, which receive lymph from the esophagus, pericardium, diaphragm, and lungs and pass efferent vessels mainly to the thoracic duct and sometimes to the tracheobronchial lymph nodes. <nodi lymphatici mediastinales posteriores> A

3 ***prevertebral lymph nodes*** lymph nodes situated in back of the thoracic aorta. <nodi lymphoidei prevertebrales> A

4 **thoracic duct** the largest lymph channel in the body, which collects lymph from the portions of the body below the diaphragm and from the left side of the body above the diaphragm; it begins in the abdomen *(abdominal part)* at the junction of the intestinal, lumbar, and descending intercostal trunks (which consists of a plexus, or the *cisterna chyli*) at about the level of the second lumbar vertebra, enters the thorax through the aortic hiatus of the diaphragm *(thoracic part)*, ascends to cross the posterior mediastinum, and enters the neck *(cervical part)*, where it forms a downward arch across the subclavian artery, and ends at the junction of the subclavian and internal jugular veins. <ductus thoracicus> A

5 **thoracic part of thoracic duct** <pars thoracica ductus thoracici> A

6 **arch of azygos vein** an arch formed by the azygos vein above the root of the right lung. <arcus venae azygou> B

7 ***broncho-esophageus muscle*** a name given muscular fasciculi that arise from the wall of the left bronchus and reinforce muscles of the esophagus. <musculus bronchooesophageus> C

8 ***esophageal branches of thoracic ganglia*** sympathetic nerve branches from the thoracic ganglia, helping to innervate the thoracic and abdominal esophagus. <rami oesophageales gangliorum thoracicorum> C

9 **esophageal plexus** a plexus surrounding the esophagus formed by branches of the left and right vagi and sympathetic trunks and containing also visceral afferent fibers from the esophagus; it is subdivided into anterior and posterior parts. <plexus oesophageus> C

10 ***esophageal veins*** small veins that drain blood from the esophagus into the hemiazygos and azygos veins, or into the left brachiocephalic vein. <venae oesophageales> C

11 ***juxtaesophageal lymph nodes*** posterior mediastinal lymph nodes situated on both sides of the esophagus. <nodi lymphoidei juxtaoesophageales> C

12 ***pleuroesophageus muscle*** a bundle of smooth muscle fibers usually connecting the esophagus with the left mediastinal pleura. <musculus pleurooesophageus> C

13 **bronchial branches of thoracic aorta** branches arising from the thoracic aorta to supply the bronchi and lower trachea, and passing along the posterior sides of the bronchi to ramify about the respiratory bronchioles; distributed also to adjacent lymph nodes, pulmonary vessels, and pericardium, and to part of the esophagus. Called also *bronchial arteries.* <rami bronchiales partis thoracicae aortae> D

14 **descending aorta** the continuation of the aorta from the arch of the aorta, in the thorax, to the point of its division into the common iliac arteries, in the abdomen; divided anatomically into thoracic and abdominal parts. <pars descendens aortae> D

15 **esophageal branches of thoracic aorta** branches, usually two, that arise from the front of the aorta to supply the esophagus. <rami oesophageales partis thoracicae aortae> D

16 **mediastinal branches of thoracic aorta** small vessels supplying connective tissue and lymph nodes in the posterior mediastinum. <rami mediastinales partis thoracicae aortae> D

17 ***pericardial branches of thoracic aorta*** small branches from the aorta distributed to the surface of the pericardium. <rami pericardiaci partis thoracicae aortae> D

18 **thoracic aorta** the proximal portion of the descending aorta, which proceeds from the arch of the aorta and gives rise to the bronchial, esophageal, pericardiac, and mediastinal branches, and the superior phrenic, posterior intercostal III to XI, and subcostal arteries; it is continuous through the diaphragm with the abdominal aorta. Called also *aorta thoracica* [TA alternative]. <pars thoracica aortae> D

19 ***thoracic aortic plexus*** a plexus around the thoracic aorta formed by filaments from the sympathetic trunks and vagus nerves, and from which fine twigs accompany branches of the aorta. It is continuous below with the celiac plexus and the abdominal aortic plexus. <plexus aorticus thoracicus> D

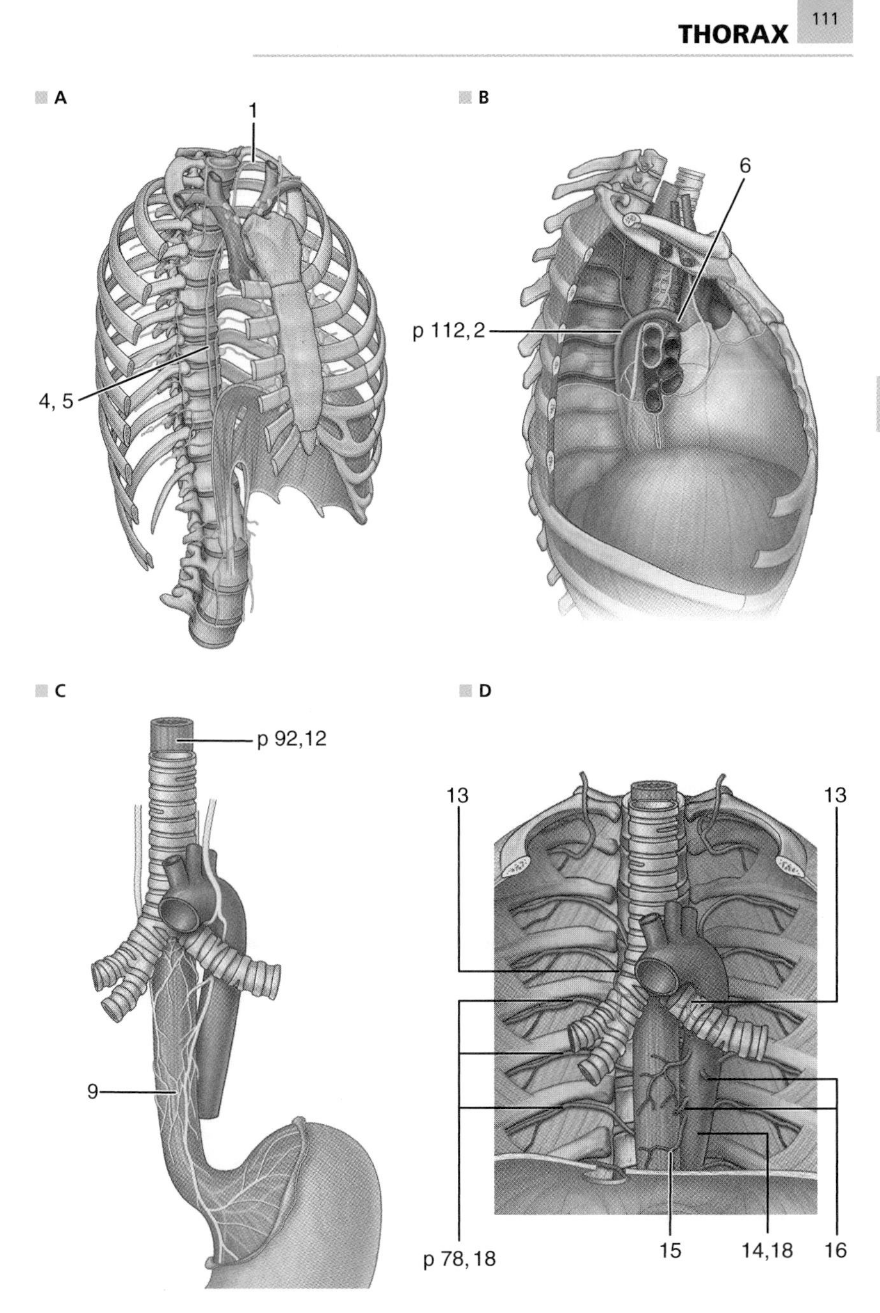
A
1
4, 5
B
6
p 112, 2
C
p 92,12
9
D
13
13
p 78, 18
15
14,18
16

1 **accessory hemiazygos vein** the descending intercepting trunk for the upper, often the fourth through the eighth, left posterior intercostal veins. It lies on the left side and at the eighth thoracic vertebra joins the hemiazygos vein or crosses to the right side to join the azygos vein directly; above, it may communicate with the left superior intercostal vein. <vena hemiazygos accessoria> A

2 **azygos vein** an intercepting trunk for the right intercostal veins as well as a connecting branch between the superior and inferior venae cavae: it arises from the ascending lumbar vein, passes up in the posterior mediastinum to the level of the fourth thoracic vertebra, where it arches over the root of the right lung and empties into the superior vena cava. <vena azygos> A

3 **hemiazygos vein** an intercepting trunk for the lower left posterior intercostal veins; it arises from the ascending lumbar vein, passes up on the left side of the vertebrae to the eighth thoracic vertebra, where it may receive the accessory branch, and crosses over the vertebral column to open into the azygos vein. <vena hemiazygos> A

4 ***lymph node of arch of azygos vein*** a lymph node sometimes present on the azygos vein at the point where it arches over the root of the lung. <nodus lymphoideus arcus venae azygos> A

5 **posterior intercostal veins** the veins that accompany the corresponding intercostal arteries and drain the intercostal spaces posteriorly; the first ends in the brachiocephalic or the vertebral vein, the second and third join the superior intercostal vein, and the fourth to eleventh join the azygos vein on the right and the hemiazygos or accessory hemiazygos vein on the left. <venae intercostales posteriores> A

6 **greater splanchnic nerve** *origin,* thoracic sympathetic trunk and fifth through tenth thoracic ganglia; *distribution,* descending through the diaphragm or its aortic openings, ends in celiac ganglia and plexuses, with a splanchnic ganglion commonly occurring near the diaphragm; *modality,* preganglionic sympathetic and visceral afferent. <nervus splanchnicus major> B

7 **least splanchnic nerve** *origin,* last ganglion of sympathetic trunk or lesser splanchnic nerve; *distribution,* aorticorenal ganglion and adjacent plexus; *modality,* sympathetic and visceral afferent. <nervus splanchnicus imus> B

8 **lesser splanchnic nerve** *origin,* ninth and tenth thoracic ganglia of sympathetic trunk; *branches,* renal ramus; *distribution,* pierces the diaphragm, joins the aorticorenal ganglion and celiac plexus, and communicates with the renal and superior mesenteric plexuses; *modality,* preganglionic sympathetic and visceral afferent. <nervus splanchnicus minor> B

9 ***splanchnic thoracic ganglion*** a small ganglion formed on the greater thoracic splanchnic nerve near the twelfth thoracic vertebra. <ganglion thoracicum splanchnicum> B

10 **sympathetic trunk** two long nerve strands, one on each side of the vertebral column, extending from the base of the skull to the coccyx. Interconnected by nerve strands, each has cervical, thoracic, lumbar, and sacral sympathetic ganglia. These receive preganglionic fibers from thoracic and upper lumbar anterior roots by way of rami communicantes, send postganglionic fibers to anterior roots by rami communicantes, and give branches to prevertebral plexuses and adjacent viscera and blood vessels. <truncus sympathicus> B

11 ***thoracic cardiac branches*** branches of the second through fourth or fifth thoracic ganglia of the sympathetic trunk, supplying the heart and having a sympathetic (accelerator) modality as well as a visceral afferent one (chiefly for pain). Called also *thoracic cardiac nerves.* <rami cardiaci thoracici> B

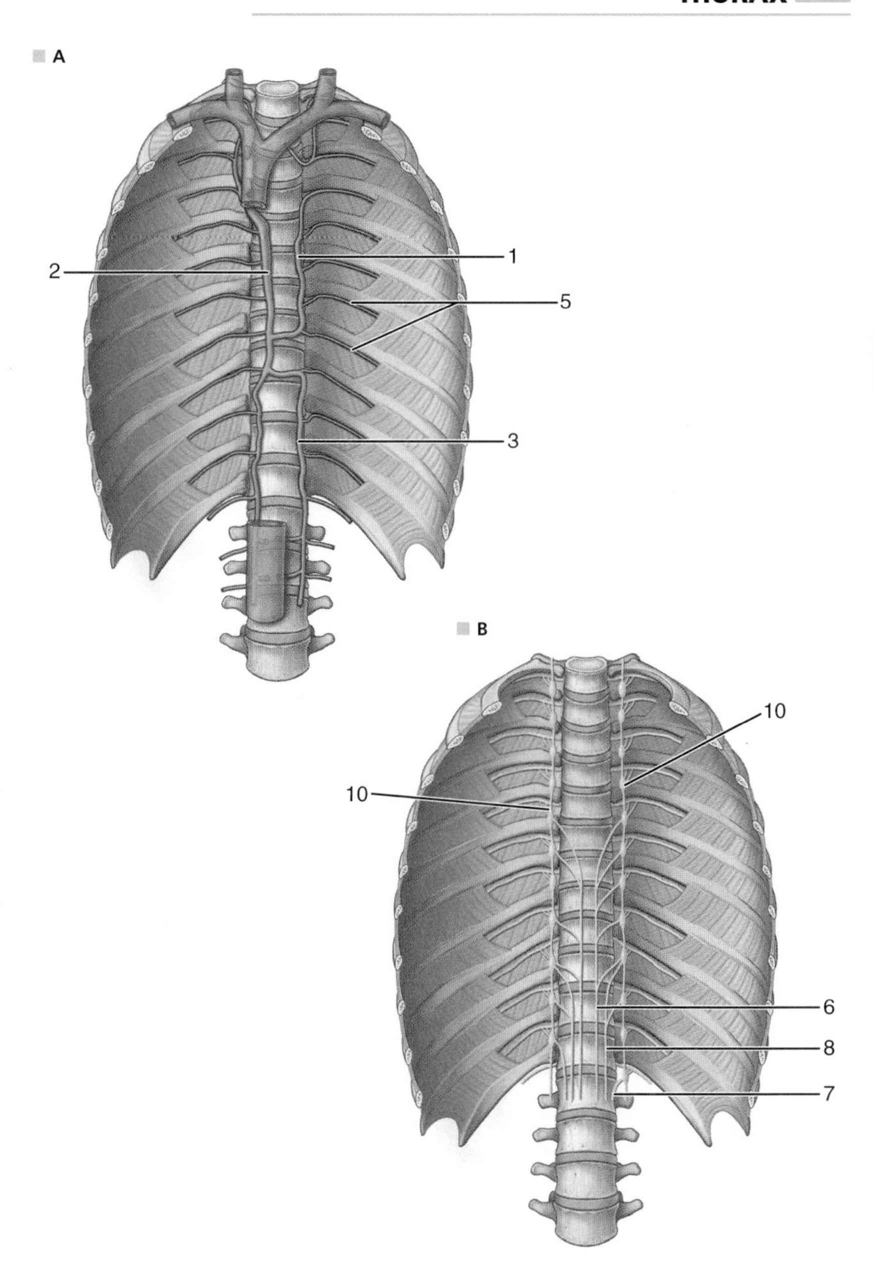
A
2
1
5
3
B
10
10
6
8
7

ABDOMEN

GENERAL

1 **abdomen** that portion of the body that lies between the thorax and the pelvis; it contains a cavity *(abdominal cavity)* separated by the diaphragm from the thoracic cavity above, and by the plane of the pelvic inlet from the pelvic cavity below, and lined with a serous membrane, the peritoneum. This cavity contains the abdominal viscera and is enclosed by a wall *(abdominal wall)* formed by the abdominal muscles, vertebral column, and the ilia. Called also *belly* and *venter.* A

2 **abdominal cavity** the body cavity located inferior to the diaphragm and superior to the pelvis, forming the superior and major part of the abdominopelvic cavity. <cavitas abdominis> A

3 **abdominopelvic cavity** the space within the trunk between the diaphragm and the inferior boundary of the lesser pelvis; it is divided into the abdominal and pelvic cavities. <cavitas abdominis et pelvis> A

ABDOMINAL WALL

4 **interspinous plane** a horizontal plane transecting the trunk at the level of the anterior superior iliac spines. <planum interspinale> D

5 **semilunar line** a curved line along the lateral border of each rectus abdominis muscle, corresponding to the meeting of the aponeuroses of the internal oblique and transverse abdominal muscles. <linea semilunaris> B

6 **intertubercular plane** a horizontal plane transecting the trunk at the level of the iliac tubercles. <planum intertuberculare> C, D

7 **subcostal plane** a horizontal plane transecting the trunk at the level of the inferior margins of the tenth costal cartilages. <planum subcostale> C, D

8 **supracristal plane** a horizontal plane transecting the trunk at the summits of the iliac crests at the level of the fourth lumbar spinous process; called also *interiliac plane.* <planum supracristale> C

9 **transpyloric plane** a horizontal plane half way between the superior margins of the manubrium sterni and the pubic symphysis, which usually does not correspond to the level of the pylorus. <planum transpyloricum> C

10 **abdominal regions** the various anatomical regions of the abdomen, which is divided into nine regions by four imaginary lines projected onto the anterior wall; two of the lines pass horizontally around the body (the upper at the level of the cartilages of the ninth ribs, the lower at the tops of the crests of the ilia), and two extend vertically on each side of the body from the cartilage of the eighth rib to the center of the inguinal ligament. <regiones abdominales> D

11 **epigastric region** the upper middle region of the abdomen, located within the infrasternal angle. <epigastrium> D

12 **flank** either side of the body between the ribs and the pelvis, constituting the middle row left or right abdominal region. <latus> D

13 **groin** the junctional region between the abdomen and thigh on either side of the body, together constituting the inferolateral abdominal regions. <inguen> D

14 **hypochondrium** either of the superolateral regions of the abdomen, lateral to the epigastric region, overlying the costal cartilages. D

15 **pubic region** the middle portion of the most inferior region of the abdomen. <hypogastrium> D

16 **umbilical region** the central region of the abdomen, surrounding the umbilicus. <umbilicus> D

17 **umbilicus** navel; the cicatrix marking the site of attachment of the umbilical cord in the fetus. D

A

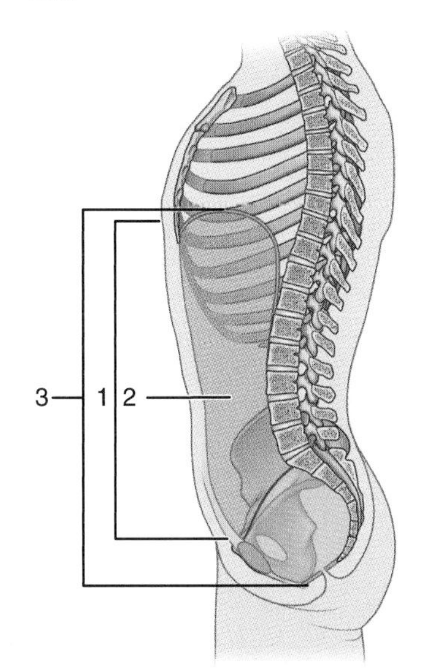

B

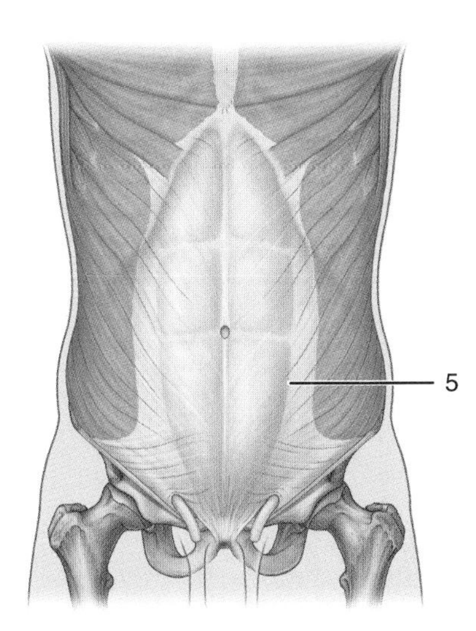

C

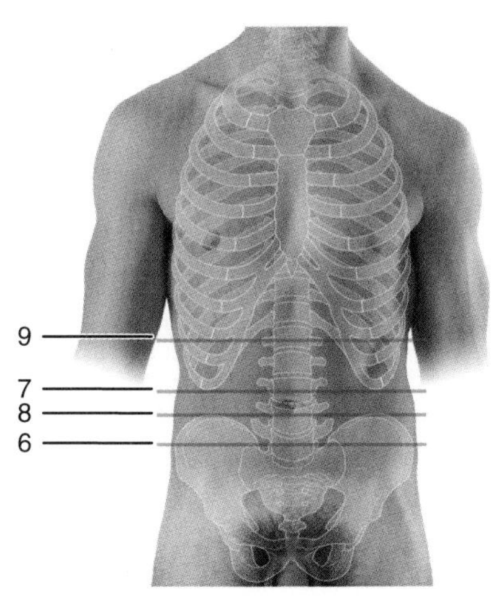

D

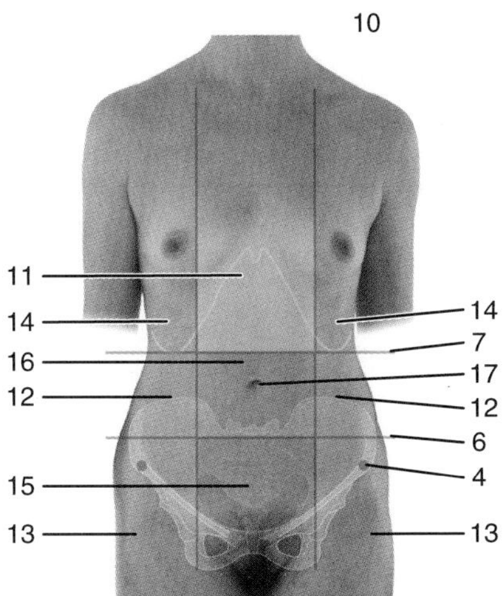

Layers of Abdominal Wall

1 ***abdominal fascia*** an inclusive term for the fascia that forms part of the general layer lining the walls of the abdominal cavity and investing the abdominal organs; it is subdivided into visceral abdominal, parietal abdominal, and extraperitoneal fascia. <fascia abdominis> A

2 **external oblique muscle of abdomen** *origin,* lower eight ribs at costal cartilages; *insertion,* crest of ilium, linea alba through rectus sheath; *innervation,* seventh to twelfth intercostal; *action,* flexes and rotates vertebral column, compresses abdominal viscera. <musculus obliquus externus abdominis> A, C, D

3 **fatty layer of subcutaneous tissue of abdomen** the more superficial layer of the subcutaneous tissue of the abdominal wall. Called also *Camper's fascia.* <panniculus adiposus telae subcutaneae abdominis> A

4 **internal oblique muscle of abdomen** *origin,* inguinal ligament, iliac crest, thoracolumbar fascia; *insertion,* inferior three or four costal cartilages, linea alba, conjoined tendon to pubis; *innervation,* seventh to twelfth intercostal, first lumbar; *action,* flexes and rotates vertebral column, compresses abdominal viscera. <musculus obliquus internus abdominis> A, D, E

5 **membranous layer of subcutaneous tissue of abdomen** the deeper layer of subcutaneous tissue of the skin of the abdomen. Called also *Scarpa's fascia.* <stratum membranosum telae subcutaneae abdominis> A

6 ***retroinguinal space*** the subdivision of the extraperitoneal space, in the inguinal region, bounded by the peritoneum above and the transversalis fascia below. <spatium retroinguinale>

7 **subcutaneous tissue of abdomen** the loose connective tissue just below the skin of the abdomen, consisting of an adipose layer overlying a membranous layer. <tela subcutanea abdominis> A

8 **subserosa of peritoneum** the subserous layer of the peritoneum, a layer of loose areolar tissue. <tela subserosa peritonei> A

9 **transversalis fascia** part of the inner investing layer of the abdominal wall, continuous with the fascia of the other side behind the rectus abdominis muscle and its sheath, as well as with the diaphragmatic fascia, iliopsoas fascia, and parietal pelvic fascia. <fascia transversalis> A, D, E

10 **transversus abdominis muscle** *origin,* cartilages of six lower ribs, thoracolumbar fascia, iliac crest, inguinal ligament; *insertion,* linea alba through rectus sheath, conjoined tendon to pubis; *innervation,* lower intercostals, iliohypogastric, ilioinguinal; *action,* compresses abdominal viscera. <musculus transversus abdominis> A, D

11 **extraperitoneal fascia** the thin layer of areolar connective tissue separating the parietal peritoneum from the abdominal walls. <fascia extraperitonealis> A, B

12 **extraperitoneal space** the space between the parietal peritoneum and the transversalis fascia, containing the extraperitoneal fascia around various organs; it is subdivided into the retroperitoneal, retropubic, and retroinguinal spaces. <spatium extraperitoneale> A, B

13 **inguinal ligament** a fibrous band running from the anterior superior iliac spine to the pubic tubercle. Called also *inguinal arch.* <ligamentum inguinale> C

14 **linea alba** the tendinous median line on the anterior abdominal wall between the two rectus muscles, formed by the decussating fibers of the aponeuroses of the three flat abdominal muscles. C, D, E

15 **anterior lamina of rectus sheath** the portion of the muscle sheath lying anterior to the rectus abdominis muscle, formed by aponeuroses of the internal and external oblique muscles superior to the arcuate line and by the aponeuroses of the internal oblique and transversus muscles inferior to the arcuate line. <lamina anterior vaginae musculi recti abdominis> C, D

16 **arcuate line of sheath of rectus abdominis** a crescentic line marking the termination of the posterior layer of the sheath of the rectus abdominis muscle, just inferior to the level of the iliac crest. <linea arcuata vaginae musculi recti abdominis> E

17 **pyramidalis muscle** *origin,* anterior aspect of pubis, anterior pubic ligament; *insertion,* linea alba; *innervation,* last thoracic; *action,* tenses abdominal wall. <musculus pyramidalis> E

18 **tendinous intersections of rectus abdominis muscle** three or more fibrous bands that cross the front of the rectus abdominis muscle, fusing with the anterior layer of its sheath. <intersectiones tendineae musculi recti abdominis> E

19 **posterior lamina of sheath of rectus abdominis** the portion of the muscle sheath lying posterior to the rectus abdominis muscle, formed by the transversus abdominis muscle and its aponeurosis at the level of the xiphoid process; inferior to the xiphoid process, as far as the arcuate line, it is formed by the aponeuroses of the internal oblique and the transversus muscles. <lamina posterior vaginae musculi recti abdominis> D, E

20 **rectus abdominis muscle** origin, pubic crest and symphysis; *insertion*, xiphoid process, cartilages of fifth, sixth, and seventh ribs; *innervation*, branches of lower thoracic; *action*, flexes lumbar vertebrae, supports abdomen. <musculus rectus abdominis> D, E

21 **rectus sheath** a sheath formed by the aponeuroses of other abdominal muscles, within which the rectus abdominis can move. <vagina musculi recti abdominis> D

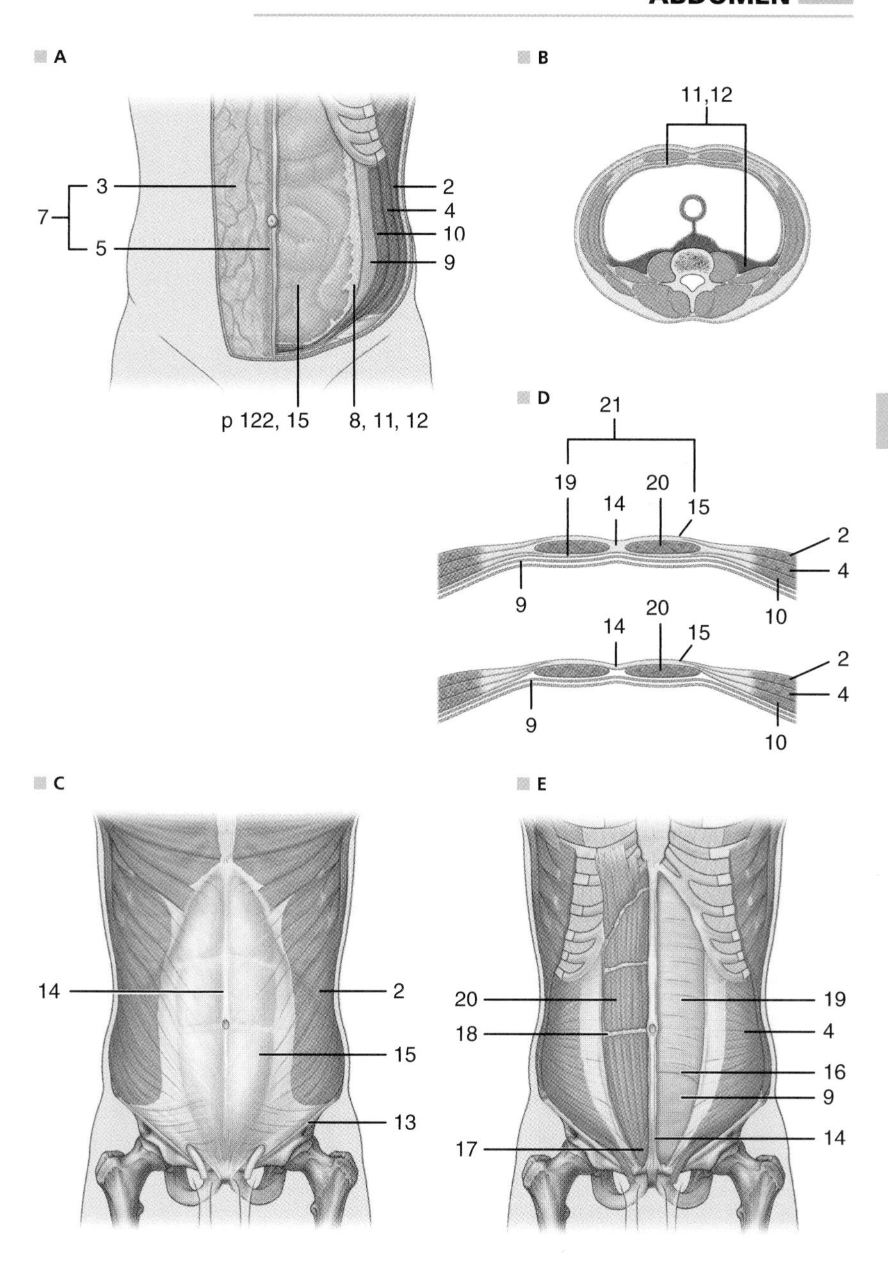
A
7
3
5
2
4
10
9
p 122, 15
8, 11, 12
B
11,12
D
21
19
20
14
15
2
4
9
10
20
14
15
2
4
9
10
C
14
2
15
13
E
20
18
17
19
4
16
9
14

1 **retroperitoneal space** the subdivision of the extraperitoneal space between the posterior parietal peritoneum and the posterior abdominal wall, containing the kidneys, suprarenal glands, ureters, duodenum, ascending and descending colon, pancreas, and the large vessels and nerves. Called also *retroperitoneum.* <spatium retroperitoneale> A

2 ***interfoveolar ligament*** a thickening in the transversalis fascia on the medial side of the deep inguinal ring; it is connected above to the transversus muscle and below to the inguinal ligament. <ligamentum interfoveolare> B

3 ***iliopubic tract*** a thickened band of tissue that strengthens the lower part of the deep inguinal ring and forms the base of the internal spermatic fascia. <tractus iliopubicus> B

4 **inguinal triangle** the triangular area on the anteroinferior abdominal wall bounded by the rectus abdominis muscle, the inguinal ligament, and the inferior epigastric vessels: the site in which a direct inguinal hernia begins. <trigonum inguinale> B

5 **lateral inguinal fossa** the depression on the inside of the anterior abdominal wall lateral to the lateral umbilical fold; called also *external inguinal fossa.* <fossa inguinalis lateralis> C

6 **lateral umbilical fold** a laterally placed fold of peritoneum on either side of the inferior part of the anterior abdominal wall, overlying the inferior epigastric vessels; called also *epigastric fold.* <plica umbilicalis lateralis> C

7 **medial inguinal fossa** the depression on the inside of the anterior abdominal wall between the medial and lateral umbilical folds; called also *internal inguinal fossa.* <fossa inguinalis medialis> C

8 **medial umbilical fold** the fold of peritoneum that covers the obliterated umbilical artery. <plica umbilicalis medialis> C

9 **median umbilical fold** the fold of peritoneum that covers the median umbilical ligament; called also *urachal fold.* <plica umbilicalis mediana> C

10 **median umbilical ligament** a fibrous cord, the remains of the partially obliterated urachus, extending from the urinary bladder to the umbilicus; it is situated in and produces the median umbilical fold. <ligamentum umbilicale medianum> C

11 **supravesical fossa** the depression on the inside of the anterior abdominal wall between the median and the medial umbilical fold. <fossa supravesicalis> C

12 ***umbilical fascia*** a thickening of the transversalis fascia extending along the median umbilical ligament downward from the umbilicus. <fascia umbilicalis> C

13 ***umbilical ring*** the aperture in the abdominal wall through which the umbilical cord communicates with the fetus. After birth it is felt for some time as a distinct fibrous ring surrounding the umbilicus; these fibers later shrink progressively. Called also *umbilical canal.* <anulus umbilicalis> C

Nerves of Abdominal Wall

14 **anterior abdominal cutaneous branch of intercostal nerve** a branch from the intercostal nerve that contributes to innervation of the skin in the anteromedial abdominal region. Its modality is general sensory. <ramus cutaneus anterior abdominalis nervi intercostalis> D

15 **lateral abdominal cutaneous branch of intercostal nerve** a branch from the intercostal nerve that, through its anterior and posterior subdivisions, innervates the skin of the lateral and posterior body wall; it has a general sensory modality. <ramus cutaneus lateralis abdominalis nervi intercostalis> D

16 **anterior cutaneous branch of iliohypogastric nerve** a branch that runs forward between the internal and external oblique muscles and innervates the skin over the pubis; *modality,* general sensory. <ramus cutaneus anterior nervi iliohypogastrici> E

17 **femoral branch of genitofemoral nerve** a branch arising by division of the genitofemoral nerve above the inguinal ligament; entering the femoral sheath, it turns forward and supplies the skin of the femoral triangle; *modality,* general sensory. <ramus femoralis nervi genitofemoralis> E

18 ***genital branch of genitofemoral nerve*** a branch arising from the genitofemoral nerve above the inguinal ligament; entering the inguinal canal through the deep inguinal ring, it supplies the cremaster and continues to the skin of the scrotum or of the labium majus, and that of the adjacent area of the thigh; modality, general sensory and motor. <ramus genitalis nervi genitofemoralis> E

19 **iliohypogastric nerve** *origin,* lumbar plexus-L1 (sometimes T12); *branches,* lateral and anterior cutaneous branches; *distribution,* the skin above the pubis and over the lateral side of the buttock, and occasionally the pyramidalis; *modality,* motor and general sensory. Called also *iliopubic nerve.* <nervus iliohypogastricus> E

20 **ilioinguinal nerve** *origin,* lumbar plexus-L1 (sometimes T12); accompanies the spermatic cord through the inguinal canal; *branches,* anterior scrotal or labial branches; *distribution,* skin of scrotum or labia majora, and adjacent part of thigh; *modality,* general sensory. <nervus ilioinguinalis> E

21 **lateral cutaneous branch of iliohypogastric nerve** a nerve branch distributed to the skin over the side of the buttock; *modality,* general sensory. <ramus cutaneus lateralis nervi iliohypogastrici> E

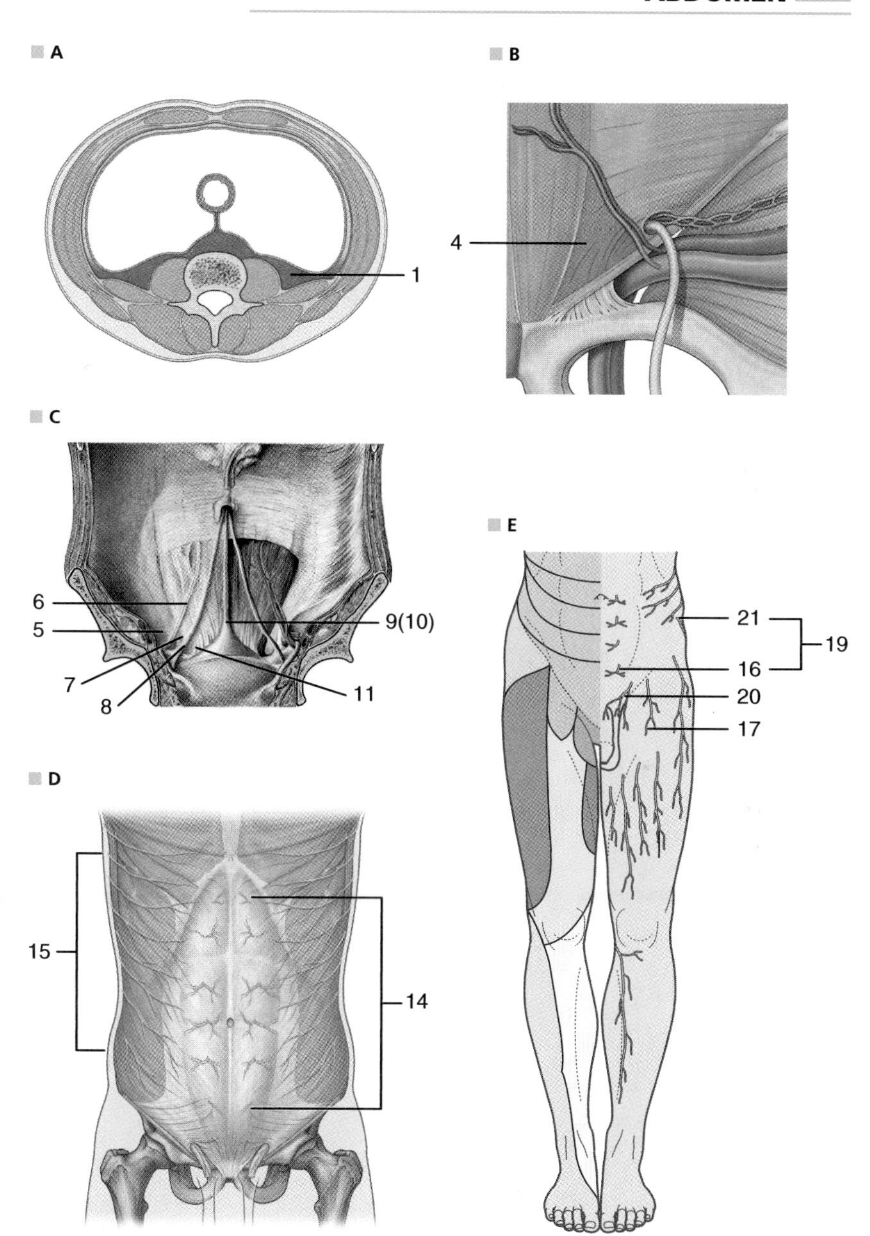
A
1
B
4
C
6
5
9(10)
7
8
11
D
15
14
E
21
19
16
20
17

Vessels of Abdominal Wall

1 **ascending branch of deep circumflex iliac artery** a branch leaving the deep circumflex iliac artery near the anterior superior iliac spine, rising between and distributing to the transversus abdominis and internal oblique muscles. <ramus ascendens arteriae circumflexae ilium profundae> A

2 **deep circumflex iliac artery** *origin,* external iliac artery; *branches,* ascending branches; *distribution,* iliac region, abdominal wall, groin. <arteria circumflexa ilium profunda> A

3 ***deep circumflex iliac vein*** a common trunk formed from the venae comitantes of the homonymous artery and emptying into the external iliac vein. <vena circumflexa ilium profunda> A

4 **inferior epigastric artery** *origin,* external iliac; *branches,* pubic branch, cremasteric artery, artery of round ligament of uterus; *distribution,* abdominal wall. <arteria epigastrica inferior> A

5 ***inferior epigastric vein*** a vein that accompanies the inferior epigastric artery and opens into the external iliac vein. <vena epigastrica inferior> A

6 ***pubic branch of inferior epigastric artery*** a branch that arises from the inferior epigastric artery near the deep inguinal ring and descends on the back of the pubis, anastomosing through an obturator branch with the pubic branch of the obturator artery. <ramus pubicus arteriae epigastricae inferioris> A

7 ***subcutaneous veins of abdomen*** the superficial veins of the abdominal wall. <venae subcutaneae abdominis> A

8 **superficial circumflex iliac artery** *origin,* femoral artery; *branches,* none; *distribution,* groin, abdominal wall. <arteria circumflexa ilium superficialis> A

9 **superficial epigastric artery** *origin,* femoral; *branches,* none; *distribution,* abdominal wall, groin. <arteria epigastrica superficialis> A

10 ***superficial epigastric vein*** a vein that follows its homonymous artery and opens into the great saphenous or the femoral vein. <vena epigastrica superficialis> A

11 **superior epigastric artery** *origin,* internal thoracic artery; *branches,* none; *distribution,* abdominal wall, diaphragm. <arteria epigastrica superior> A

12 ***superior epigastric veins*** the venae comitantes of the superior epigastric artery, which open into the internal thoracic vein. <venae epigastricae superiores> A

13 ***thoracoepigastric veins*** long, longitudinal, superficial veins in the anterolateral subcutaneous tissue of the trunk, which empty superiorly into the lateral thoracic and inferiorly into the femoral vein. <venae thoracoepigastricae> A

INGUINAL REGION (GROIN)

Inguinal Canal

14 **lacunar ligament** a small triangular membrane with its base just medial to the femoral ring; one side is attached to the inguinal ligament and the other to the pectineal line of the pubis. <ligamentum lacunare> B

15 **pectineal ligament** a strong aponeurotic lateral continuation of the lacunar ligament along the pectineal line of the pubis; called also Cooper's ligament. <ligamentum pectineum> B

16 ***reflex inguinal ligament*** a triangular band of fibers arising from the lacunar ligament and the pubic bone and passing diagonally upward and medially behind the superficial abdominal ring and in front of the conjoint tendon to the linea alba. <ligamentum inguinale reflexum> B

17 **deep inguinal ring** an aperture in the transversalis fascia for the spermatic cord or for the round ligament; called also *internal inguinal ring.* <anulus inguinalis profundus> B

18 **inguinal canal** the passage superficial to the deep inguinal ring, transmitting the spermatic cord in the male and the round ligament in the female. <canalis inguinalis> C, D

19 **superficial inguinal ring** an opening in the aponeurosis of the external oblique muscle for the spermatic cord or for the round ligament; called also *external inguinal ring.* <anulus inguinalis superficialis> C, D

20 **lateral crus of superficial inguinal ring** the part of the superficial inguinal ring that blends with the inguinal ligament as it goes to the pubic tubercle. <crus laterale anuli inguinalis superficialis> D

21 **medial crus of superficial inguinal ring** the part of the superficial inguinal ring that is attached to the pubic symphysis and that blends with the fundiform ligament of the penis. <crus mediale anuli inguinalis superficialis> D

22 **conjoint tendon** the united tendons of the transversus abdominis and internal oblique muscles going to the linea alba and pectineal line of the pubic bone; called also *inguinal falx.* <falx inguinalis> E

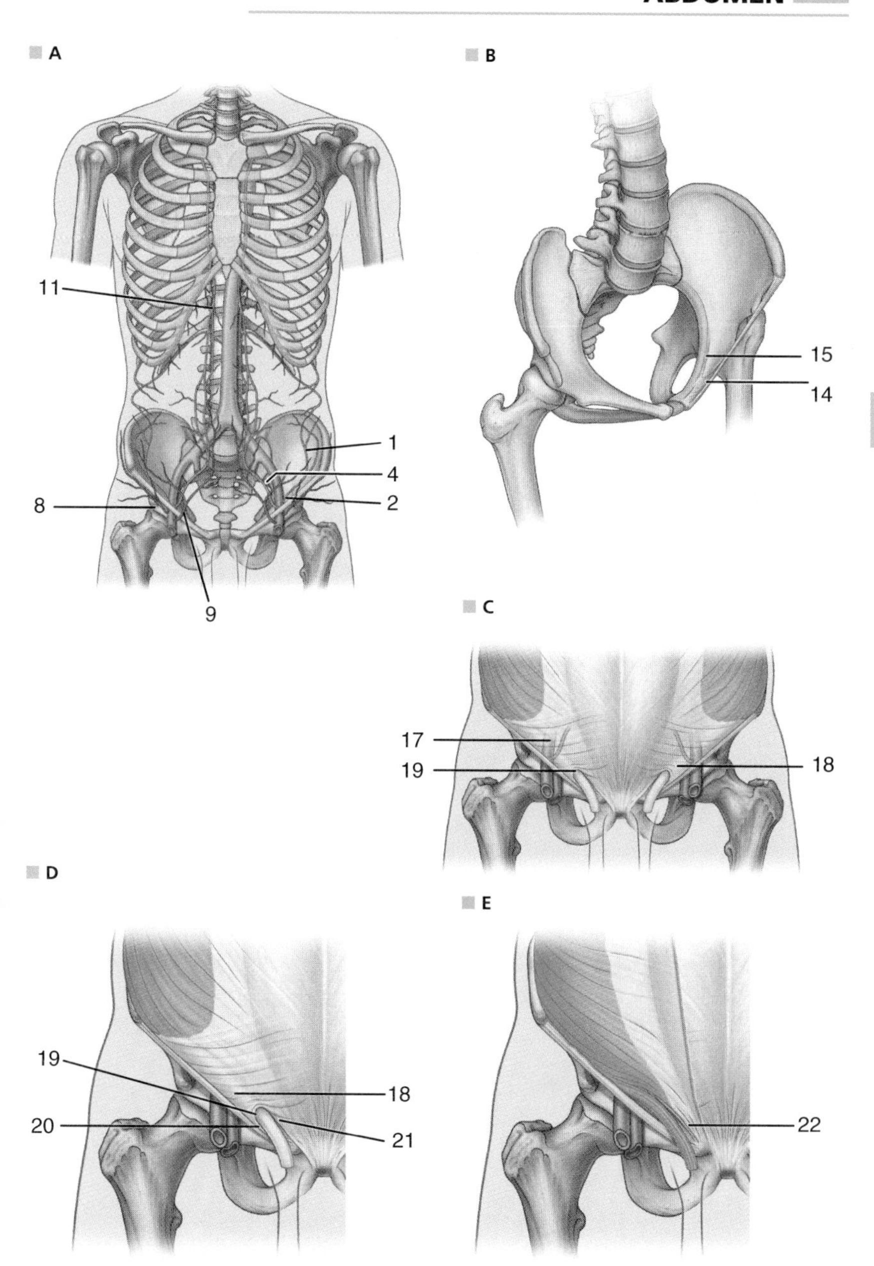
A
11
1
4
8
2
9
B
15
14
C
17
19
18
D
19
18
20
21
E
22

Contents of Inguinal Region

1 **cremaster muscle** *origin,* inferior margin of internal oblique muscle of abdomen; *insertion,* pubic tubercle; *innervation,* genital branch of genitofemoral; *action,* elevates testis. <musculus cremaster> A

2 **cremasteric artery** *origin,* inferior epigastric; *branches,* none; *distribution,* cremaster muscle, coverings of spermatic cord. Called also *external spermatic artery.* <arteria cremasterica> A

3 **cremasteric fascia** the thin covering of the spermatic cord formed by the investing fascia of the cremaster muscle; it is adjacent to the external surface of the internal spermatic fascia. Called also *Cooper's fascia.* <fascia cremasterica> A

4 **external spermatic fascia** the thin outer covering of the spermatic cord, which is continuous with the investing fascia of the external oblique muscle. <fascia spermatica externa> A

5 **internal spermatic fascia** the thin innermost covering of the spermatic cord, derived from the transversalis fascia of the abdominal wall. <fascia spermatica interna> A

6 **spermatic cord** a structure that extends from the abdominal inguinal ring to the testis, comprising the ductus deferens, testicular artery, pampiniform plexus, and nerves, as well as various other vessels; it is enclosed by various coverings. <funiculus spermaticus> A

Hernias of Inguinal Region

7 **direct inguinal hernia** an inguinal hernia that emerges between the inferior epigastric artery and the edge of the rectus muscle. B

8 **indirect inguinal hernia** an inguinal hernia that leaves the abdomen through the deep inguinal ring and passes down obliquely through the inguinal canal, lateral to the inferior epigastric artery. C

PERITONEUM AND PERITONEAL CAVITY

9 **peritoneal cavity** the potential space of capillary thinness between the parietal and the visceral peritoneum; it is normally empty except for a thin serous fluid that keeps the surfaces moist. Called also *greater peritoneal cavity.* <cavitas peritonealis> D

10 **peritoneum** the serous membrane lining the abdominopelvic walls *(parietal peritoneum)* and investing the viscera *(visceral peritoneum).* A strong, colorless membrane with a smooth surface, it forms a double-layered sac that is closed in the male and is continuous with the mucous membrane of the uterine tubes in the female. The potential space between the parietal and visceral peritoneum is called the *peritoneal cavity.* D

11 ***ascending mesocolon*** the peritoneum attaching the ascending colon to the posterior abdominal wall, usually obliterated when the ascending colon becomes retroperitoneal. <mesocolon ascendens> D

12 ***descending mesocolon*** the peritoneum attaching the descending colon to the posterior abdominal wall; it is usually absent because the descending colon is ordinarily retroperitoneal. <mesocolon descendens> D

13 **inferior omental recess** the lower portion of the omental bursa, including its extension down into the greater omentum. It is bounded in front by the posterior wall of the stomach, and behind by the pancreas, the transverse colon and its mesocolon, the left suprarenal gland, and part of the left kidney. <recessus inferior bursae omentalis> D

14 ***mesocolon*** the process of the peritoneum by which the colon is attached to the posterior abdominal wall. It is divided into ascending, transverse, descending, and sigmoid or pelvic portions, according to the segment of the colon to which it gives attachment. D

15 **parietal peritoneum** the peritoneum that lines the abdominal and pelvic walls and the inferior surface of the diaphragm. <peritoneum parietale> E

16 **superior omental recess** a long, narrow peritoneal pocket leading from the vestibule upward toward the liver, between the inferior vena cava on the right, the esophagus on the left, the gastrohepatic ligament in front, and the diaphragm behind. <recessus superior bursae omentalis> E

17 **transverse mesocolon** the peritoneum attaching the transverse colon to the posterior abdominal wall. <mesocolon transversum> D

18 ***visceral abdominal fascia*** the fascia that invests the abdominal viscera. <fascia abdominis visceralis> E

19 **visceral peritoneum** a continuation of the parietal peritoneum reflected at various places over the viscera, forming a complete covering for the stomach, spleen, liver, intestines from the distal duodenum to the upper end of the rectum, uterus, and ovaries; it also partially covers some other abdominal organs. It holds the viscera in position by its folds, including the *mesenteries,* the *omenta,* and the *ligaments* of the liver, spleen, stomach, kidneys, bladder, and uterus. The potential space between the visceral and the parietal peritoneum is the *peritoneal cavity.* <peritoneum viscerale> E

Omenta, Mesenteries, and Ligaments

20 **omental bursa** a serous peritoneal cavity situated behind the stomach, the lesser omentum, and part of the liver and in front of the pancreas and duodenum. It communicates with the general peritoneal cavity (greater sac) through the epiploic foramen and sometimes is continuous with the cavity of the greater omentum. Called also *lesser peritoneal cavity* or *sac.* <bursa omentalis> D

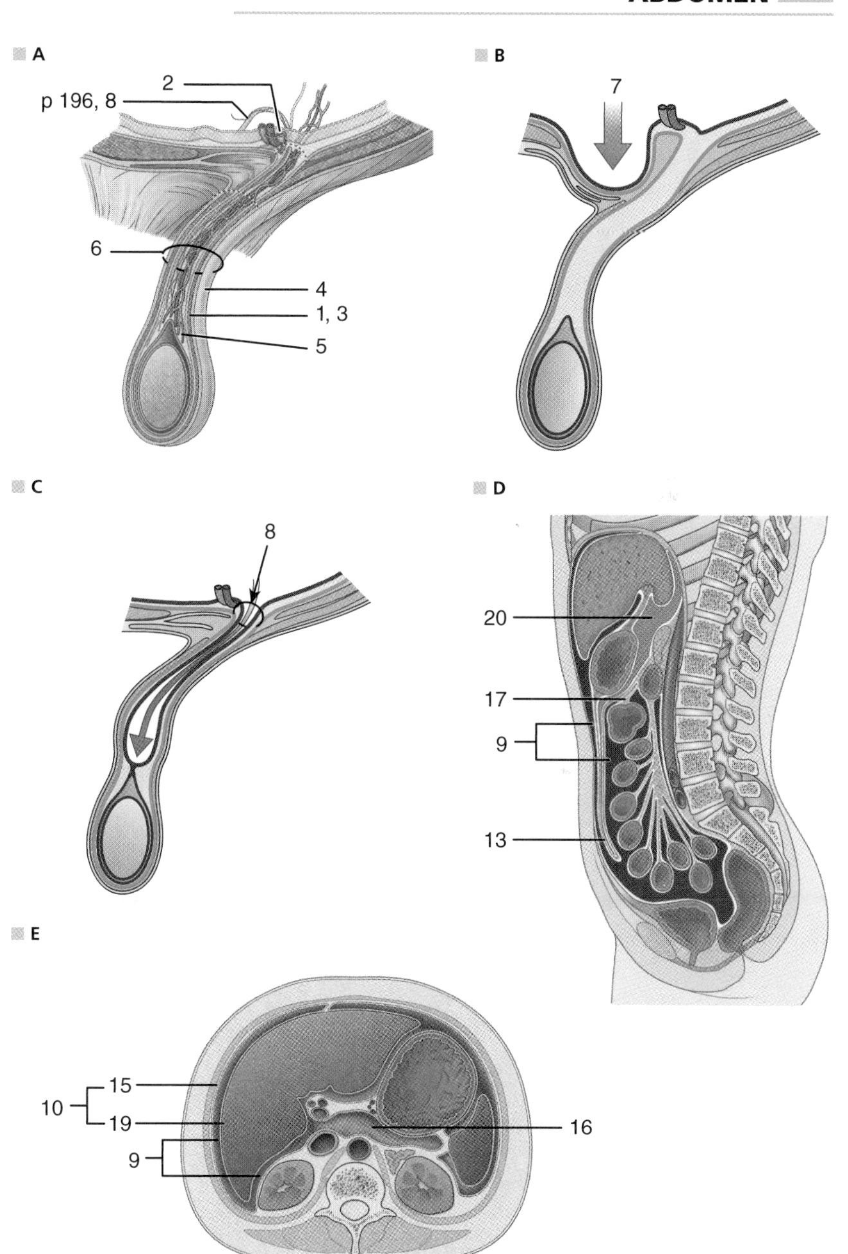
A
2
p 196, 8
6
4
1, 3
5
B
7
C
8
D
20
17
9
13
E
15
10
19
9
16

1 **splenic recess** an extension of the omental bursa to the left behind the gastrosplenic ligament almost to the spleen. <recessus splenicus> A

2 **splenorenal ligament** a fold in the greater omentum that connects the kidney and the concave surface of the spleen; this and the phrenicosplenic ligament are the same structure in some persons. <ligamentum splenorenale> A

3 **vestibule of omental bursa** that part of the omental bursa posterior to the lesser omentum and adjacent to the omental foramen. <vestibulum bursae omentalis> A

4 **greater omentum** a prominent peritoneal fold suspended from the greater curvature of the stomach and passing inferiorly a variable distance in front of the intestines; it is attached to the anterior surface of the transverse colon. <omentum majus> B

5 **mesoappendix** the peritoneal fold attaching the appendix to the mesentery of the ileum. C

6 **omental foramen** the opening connecting the greater and lesser peritoneal sacs, situated below and behind the porta hepatis. Called also *epiploic foramen.* <foramen omentale> A, C, D

7 **gastrocolic ligament** a peritoneal fold, part of the greater omentum, that extends from the greater curvature of the stomach to the transverse colon. <ligamentum gastrocolicum> D

8 **gastrosplenic ligament** a peritoneal fold extending from the greater curvature of the stomach to the hilum of the spleen. <ligamentum gastrosplenicum> A

9 ***hepatocolic ligament*** an occasional fold of peritoneum, an extension of the lesser omentum to the right, passing from the visceral surface of the liver near the gallbladder to the right colic flexure. <ligamentum hepatocolicum> D

10 **hepatoduodenal ligament** a peritoneal fold that passes from the porta hepatis to the superior part of the duodenum. It is continuous on the left with the hepatogastric ligament, and on the right it forms one of the borders of the omental foramen. It contains the hepatic artery, portal vein, bile duct, nerves, and lymphatics. <ligamentum hepatoduodenale> D

11 **hepatogastric ligament** a peritoneal fold, part of the lesser omentum, that passes from the visceral surface of the liver to the lesser curvature of the stomach. <ligamentum hepatogastricum> D

12 ***hepatopancreatic fold*** a crescentic fold of peritoneum formed by the hepatic artery as it runs forward from the posterior abdominal wall to the lesser omentum. <plica hepatopancreatica> D

13 **lesser omentum** a peritoneal fold joining the lesser curvature of the stomach and the first part of the duodenum to the porta hepatis. <omentum minus> D

14 **gastropancreatic fold** a crescentic fold of peritoneum formed by the left gastric artery as it runs from the posterior abdominal wall to the lesser curvature of the stomach. <plica gastropancreatica> E

15 **gastrophrenic ligament** a fold of peritoneum continuous with the gastrosplenic ligament, extending from the right inferior surface of the diaphragm to the cardiac part of the stomach. <ligamentum gastrophrenicum> E

16 ***hepatorenal ligament*** a fold of peritoneum that passes from the posterior part of the visceral surface of the liver to the anterior surface of the right kidney; it forms the right margin of the omental foramen. <ligamentum hepatorenale> E

17 **inferior duodenal fold** a thin fold of peritoneum that bounds the inferior duodenal recess. <plica duodenalis inferior> E

18 **inferior duodenal recess** a pocket in the peritoneum on the left side of the ascending part of the duodenum, bounded by the inferior duodenal fold. <recessus duodenalis inferior> E

19 **inferior ileocecal recess** a peritoneal pocket situated behind the ileocecal fold, above the vermiform appendix below the ileum, and medial to the cecum. <recessus ileocaecalis inferior> E

20 **intersigmoidal recess** a shallow peritioneal pocket running downward and to the left at the base of the sigmoid mesocolon. <recessus intersigmoideus> E

21 **paracolic gutters** small, shallow, and variable peritoneal pockets found lateral to the descending colon; called also paracolic recesses. <sulci paracolici> E

22 ***paraduodenal fold*** an occasionally found peritoneal fold containing a branch of the left colic artery. <plica paraduodenalis> E

23 ***paraduodenal recess*** a pocket occasionally found in the peritoneum behind a fold containing a branch of the left colic artery. <recessus paraduodenalis> E

24 **phrenicocolic ligament** a peritoneal fold that passes from the left colic flexure to the adjacent costal portion of the diaphragm. <ligamentum phrenicocolicum> D

25 ***phrenicosplenic ligament*** a fold in the greater omentum that connects the diaphragm with the concave surface of the spleen; this and the splenorenal ligament are the same structure in some persons. <ligamentum phrenicosplenicum> E

26 **retrocecal recess** a peritoneal pocket extending upward behind the cecum and sometimes behind the colon. <recessus retrocaecalis> C

27 **retroduodenal recess** an occasional peritoneal pocket extending behind the horizontal and ascending parts of the duodenum. <recessus retroduodenalis> E

28 **superior duodenal fold** a fold of peritoneum covering the inferior mesenteric vein and the ascending branch of the left colic artery. <plica duodenalis superior> E

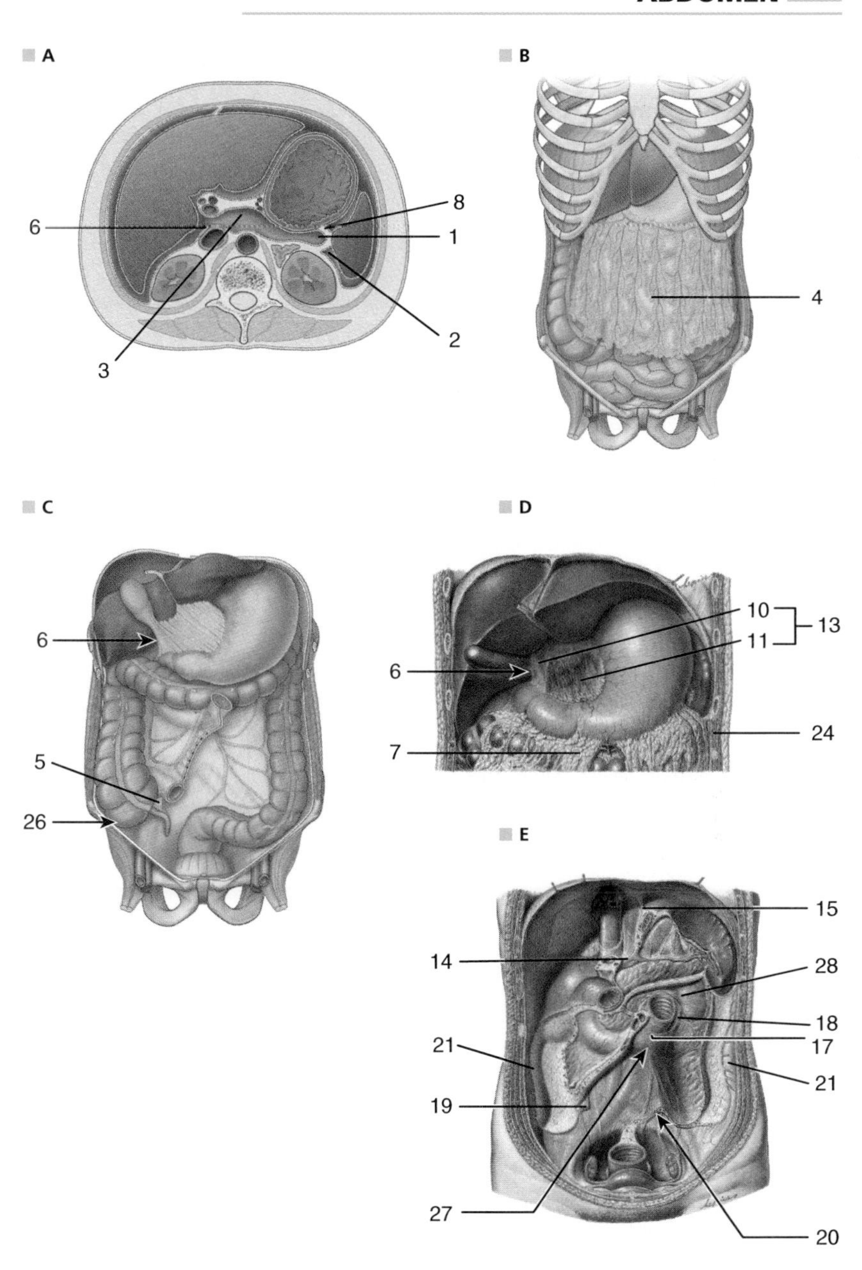
A
B
C
D
E
6
8
1
2
3
4
6
5
26
10
11
13
6
24
7
15
14
28
18
17
21
21
19
27
20

1 **superior duodenal recess** a peritoneal pocket behind the superior duodenal fold. <recessus duodenalis superior> A

2 **superior ileocecal recess** a peritoneal pocket situated behind and below the vascular cecal fold, above the ileum and medial to the lower end of the ascending colon. <recessus ileocaecalis superior> A

3 **vascular cecal fold** the fold of peritoneum that covers the anterior cecal vessels, forming the superior ileocecal recess. <plica caecalis vascularis> A

4 ***cecal folds*** the folds of peritoneum on either side of the retrocecal recess, which may connect the cecum to the abdominal wall. <plicae caecales> A

5 ***celiacoduodenal part of suspensory muscle of duodenum*** a band of smooth muscle that passes from the terminal duodenum to join the phrenicoceliac part and end in connective tissue that attaches to the celiac trunk. <pars coeliacoduodenalis musculi suspensorii duodeni> A

6 ***phrenicoceliac part of suspensory muscle of duodenum*** a band of skeletal muscle that passes from the right crus of the diaphragm to join the celiacoduodenal part and attach to the celiac trunk. <pars phrenicocoeliaca musculi suspensorii duodeni> A

7 ***suspensory muscle of duodenum*** a flat band of smooth muscle originating from the left crus of the diaphragm, and continuous with the muscular coat of the duodenum at its junction with the jejunum. Called also *ligament of Treitz.* <musculus suspensorius duodeni> A

8 **root of mesentery** the line of attachment of the mesentery to the posterior abdominal wall, extending from the duodenojejunal flexure near the second lumbar vertebra diagonally downward to the upper border of the right sacroiliac joint. <radix mesenterii> A, B

9 **sigmoid mesocolon** the peritoneum attaching the sigmoid colon to the posterior abdominal wall. <mesocolon sigmoideum> B

10 **hepatorenal recess** a peritoneal pouch between the liver and the kidney. <recessus hepatorenalis> C

11 **subhepatic recesses** peritoneal pockets located beneath the liver. <recessus subhepatici> C

12 **subphrenic recesses** peritoneal pockets located beneath the diaphragm. <recessus subphrenici> C

ABDOMINAL VISCERA
Organs of Abdominal Viscera

13 **hidden part of duodenum** a portion of the duodenum that is concealed by the overlying transverse colon, ascending colon, and root of the mesentery. <pars tecta duodeni> D

14 **phrenic constriction** the narrowing in the esophagus where it crosses the diaphragm at the esophageal hiatus. <constrictio phrenica> E

15 **angular incisure** the lowest point on the lesser curvature of the stomach, marking the junction of the cranial two-thirds and caudal one-third of the stomach. <incisura angularis> F

16 ***anterior wall of stomach*** the part of the stomach wall that is directed toward the anterior surface of the body. <paries anterior gastris> F

17 **body of stomach** that part of the stomach between the fundus and the pyloric part. <corpus gastricum> F

18 **cardia** the part of the stomach immediately adjacent to and surrounding the cardiac opening where the esophagus connects to the stomach; it contains the cardiac glands, and does not have any parietal cells or chief cells. Called also cardiac part of stomach. F

19 **cardiac orifice** the orifice between the esophagus and the cardia (cardiac part of stomach). <ostium cardiacum> F

20 **cardial notch** a notch at the junction of the esophagus and the greater curvature of the stomach. <incisura cardialis> F

21 ***esophageal glands*** the mucous glands in the submucous layer of the esophagus. <glandulae oesophageae> F

22 ***fornix gastricus*** a term used in radiographic anatomy to refer to the arch of the fundus of the stomach. F

23 **fundus of stomach** that part of the stomach to the left and above the level of the entrance of the esophagus. <fundus gastricus> F

24 **greater curvature of stomach** the left or lateral and inferior border of the stomach, marking the inferior junction of the anterior and posterior surfaces. <curvatura major gastris> F

25 **lesser curvature of stomach** the right or medial border of the stomach, marking the superior junction of the anterior and posterior surfaces. <curvatura minor gastris> F

26 ***posterior wall of stomach*** the part of the stomach wall that is directed toward the posterior surface of the body. Called also *posterior gastric wall* and *posterior stomach wall.* <paries posterior gastris> F

27 **pyloric antrum** the dilated portion of the pyloric part of the stomach, distal to the body of the stomach and proximal to the pyloric canal. <antrum pyloricum> F

28 **pyloric canal** the short, narrow part of the stomach extending from the gastroduodenal junction to the pyloric antrum. <canalis pyloricus> F

29 **pyloric part of stomach** the caudal third of the stomach, consisting of the pyloric antrum and canal, and distinguished by the presence of the pyloric glands and by the absence of parietal cells. <pars pylorica gastris> F

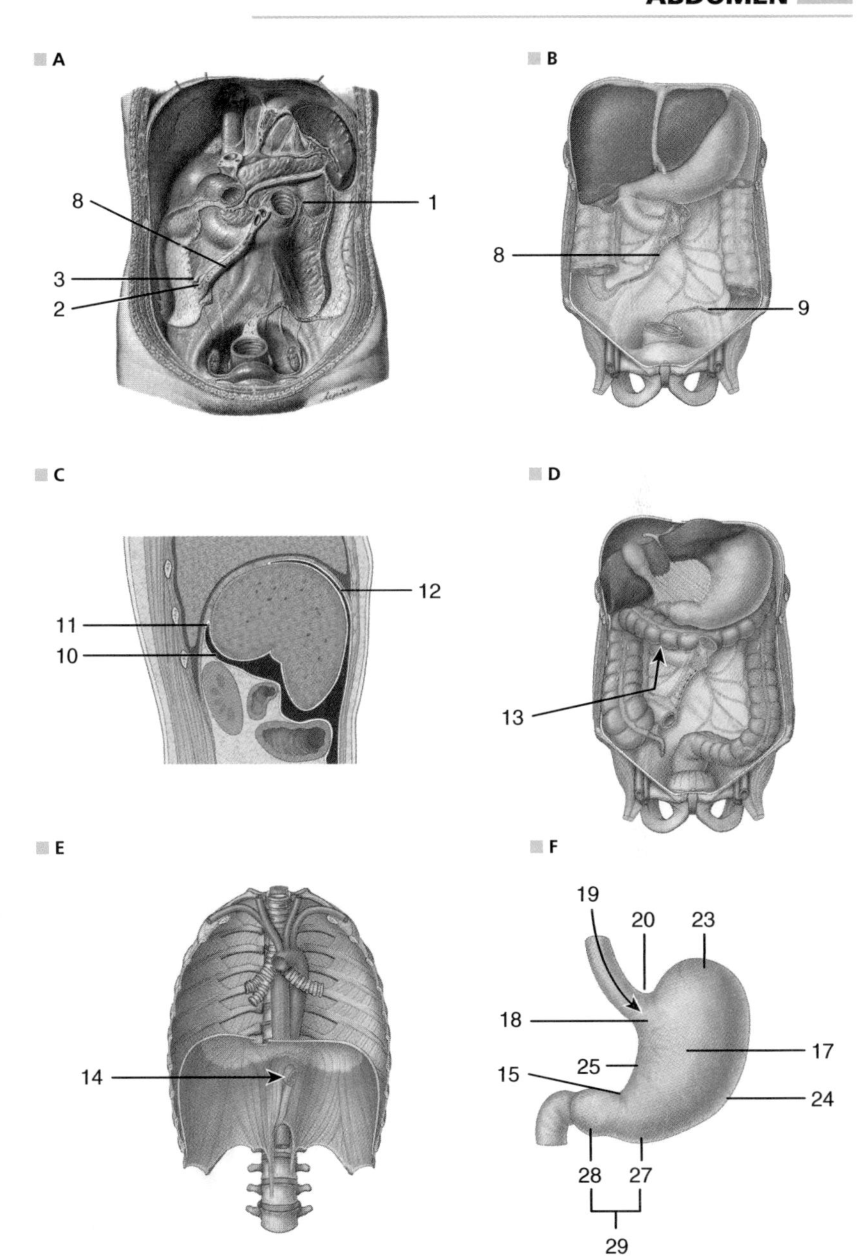
A
8
1
3
2
B
8
9
C
12
11
10
D
13
E
14
F
19
20
23
18
17
25
15
24
28
27
29

1 **stomach** the musculomembranous expansion of the alimentary tract between the esophagus and the duodenum, divided into the *cardia, fundus, pyloric part,* and *body.* The upper concave surface or edge is the *lesser curvature,* and the lower convex edge is the *greater curvature.* The stomach wall has four coats or layers. Gastric glands, found in the mucous membrane, secrete gastric juice into the cavity of the stomach; it contains hydrochloric acid, pepsin, and various other digestive enzymes. Food mixed with this juice forms a semifluid substance called chyme that is suitable for further digestion by the intestine. <gaster> A

2 **abdominal part of esophagus** the part of the esophagus below the diaphragm, joining the stomach. <pars abdominalis oesophagi> A

3 **pylorus** the most distal part of the stomach, surrounded by a strong band of circular muscle, and through which the stomach contents are emptied into the duodenum. The term is variously used to mean either the entire pyloric part of the stomach or just a section of it, such as the pyloric antrum, canal, or opening. A

4 **circular layer of muscular coat of colon** the inner layer of circularly coursing fibers in the muscular coat (tunica muscularis) of the colon. <stratum circulare tunicae muscularis coli> C

5 **longitudinal layer of muscular coat of colon** the outer layer of the muscular coat (tunica muscularis) of the colon, consisting of longitudinally coursing fibers; it is thick in the regions of the three teniae coli and very thin between them. <stratum longitudinale tunicae muscularis coli> C

6 **muscular coat of colon** the muscular layer of the colon, consisting of sublayers or strata of longitudinally and circularly coursing fibers. <tunica muscularis coli> C

7 **circular layer of muscular coat of stomach** the layer of circularly coursing fibers in the muscular coat (tunica muscularis) of the stomach. <stratum circulare tunicae muscularis gastris> D

8 **longitudinal layer of muscular coat of stomach** the layer of longitudinally coursing fibers in the muscular coat (tunica muscularis) of the stomach. <stratum longitudinale tunicae muscularis gastris> D

9 **muscular coat of stomach** the muscular layer of the stomach, composed of sublayers or strata of longitudinal, circular, and oblique fibers. <tunica muscularis gastris> D

10 **muscularis mucosae of stomach** the muscular layer of the mucous membrane (tunica mucosa) of the stomach. <lamina muscularis mucosae gastris> D

11 ***oblique fibers of muscular coat*** the inner obliquely coursing fibers of the muscular layer or coat of the stomach; called also *oblique fibers of stomach.* <fibrae obliquae tunicae muscularis> D

12 **serosa of stomach** the serous coat or layer (tunica serosa) of the stomach. <tunica serosa gastris> D

13 **submucosa of stomach** the submucous layer of the stomach. <tela submucosa gastris> D

14 ***subserosa of stomach*** the subserous layer of the stomach. <tela subserosa gastris> D

15 ***areae gastricae*** small patches of gastric mucosa, 1 to 5 mm in diameter, separated by the villous folds and containing the gastric pits. B

16 ***gastric canal*** the longitudinal grooved channel formed by the more or less regular ridges along the lesser curvature of the stomach. <canalis gastricus> B

17 **gastric folds** the series of folds in the mucous membrane of the stomach; they are oriented chiefly longitudinally and partially disappear when the stomach is distended. Called also *gastric rugae* and *rugae of stomach.* <plicae gastricae> B

18 **gastric glands** the secreting glands of the stomach, including the fundic, cardiac, and pyloric glands; sometimes used specifically to denote the fundic glands. <glandulae gastricae> B

19 **gastric pits** the numerous pits in the gastric mucosa marking the openings of the gastric glands. <foveolae gastricae> B

20 **mucosa of stomach** the mucous membrane (tunica mucosa) lining the stomach. <tunica mucosa gastris> D

21 **pyloric orifice** the orifice between the pylorus of the stomach and the duodenum. <ostium pyloricum> B

22 **pyloric sphincter** a thickening of the circular muscle of the stomach around its opening into the duodenum. <musculus sphincter pyloricus> B

23 **villous folds of stomach** a fine network of furrows demarcating the gastric areas. <plicae villosae gastris> B

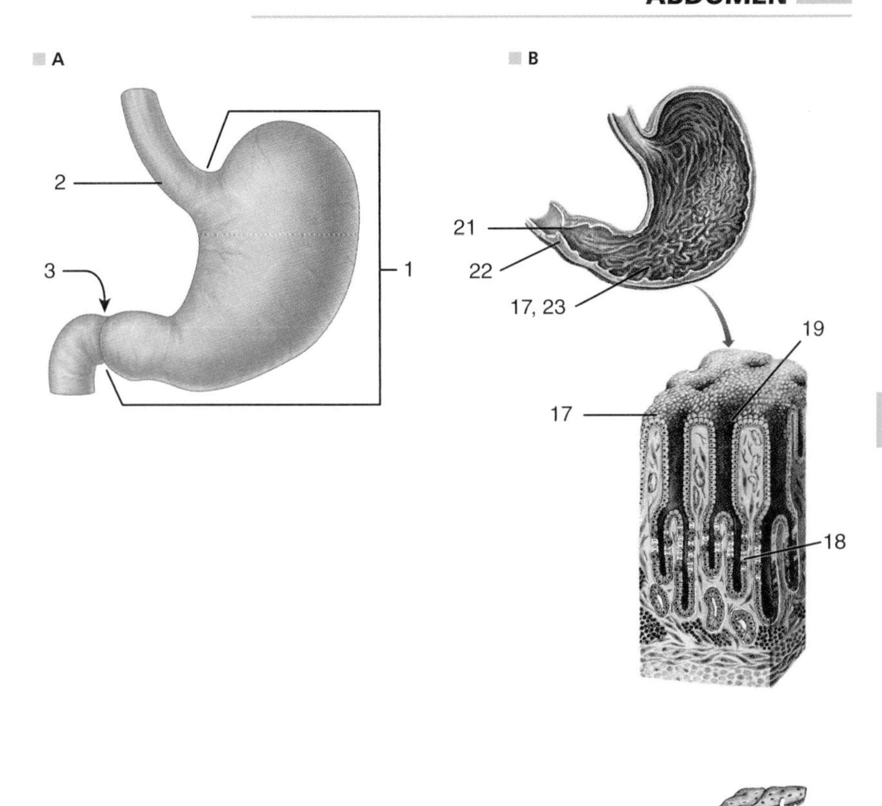
A
B
2
3
1
21
22
17, 23
19
17
18

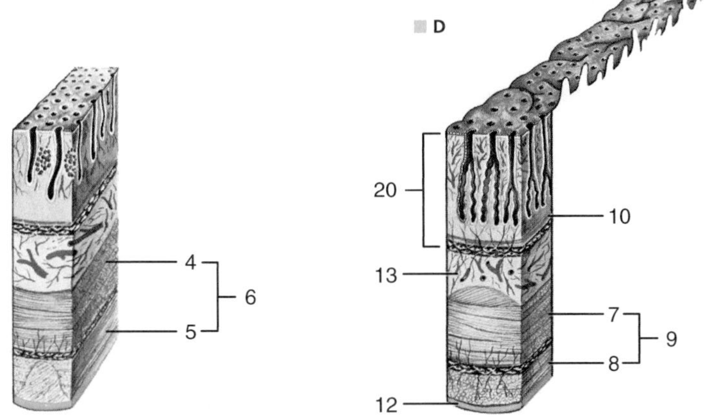
C
D
4
5
6
20
10
13
7
8
9
12

1 **ascending part of duodenum** the terminal part of the duodenum, which goes upwards to end at the duodenojejunal flexure. <pars ascendens duodeni> A

2 **descending part of duodenum** the part between the superior and inferior parts, into which the bile and pancreatic ducts open. <pars descendens duodeni> A

3 ***duodenal glands*** tubuloacinar glands in the submucous layer of the duodenum that open into the intestinal glands; they secrete urogastrone. Called also *Brunner's glands.* <glandulae duodenales> A

4 **duodenojejunal flexure** the bend in the small intestine at the junction between the duodenum and jejunum; the suspensory muscle of the duodenum attaches to this point. <flexura duodenojejunalis> A

5 **duodenum** the first or proximal portion of the small intestine, much shorter than the following portions, extending from the pylorus to the jejunum; so called because its length is about 12 finger breadths. A

6 **horizontal part of duodenum** the third, more horizontal or transverse, part of the duodenum, located between the descending and ascending parts, crossing from right to left anterior to the third lumbar vertebra. <pars horizontalis duodeni> A

7 **inferior duodenal flexure** the bend where the descending part of the duodenum ends and the horizontal part begins. <flexura duodeni inferior> A

8 **superior duodenal flexure** the bend where the superior part of the duodenum ends and the descending part begins. <flexura duodeni superior> A

9 **superior part of duodenum** the most proximal part of the duodenum, adjacent to the pylorus; it forms the superior duodenal flexure. <pars superior duodeni> A

10 **ileum** the distal and longest portion of the small intestine, extending from the jejunum to the cecum. B

11 **jejunum** the second section of the small intestine, extending from the duodenum to the ileum. B

12 **frenulum of ileal orifice** a fold formed by the joined extremities of the ileal orifice, extending partly around the lumen of the colon. <frenulum ostii ilealis> C

13 **ileal orifice** the opening at the junction of the ileum and cecum, which is rounded at the left (anterior) end and pointed at the right (posterior) end. It has two lips, *ileocecal* and *ileocolic.* In a cadaver this is the site of the so-called *ileocecal valve;* in the living individual the ileum here forms a conical projection called the *ileal papilla.* <ostium ileale> C

14 **ileal papilla** the conical projection formed by the terminal ileum at its junction with the cecum, with the ileum extending into the large intestine; this structure in the living individual corresponds to the so-called *ileocecal valve* in the cadaver. <papilla ilealis> C

15 ***ileocecal fold*** a fold of peritoneum at the left border of the cecum, extending from the ileum above to the appendix below. <plica ileocaecalis> C

16 **ileocecal lip of ileal orifice** the inferior of the two lips forming the ileal orifice. <labrum ileocaecalis ostii ilealis> C

17 **ileocolic lip of ileal orifice** the superior of the two lips forming the ileal orifice. <labrum ileocolicum ostii ilealis> C

18 ***opening of vermiform appendix*** the orifice between the vermiform appendix and the cecum. <ostium appendicis vermiformis> C

19 **terminal part of ileum** the part just before the ileum meets the cecum at the ileal orifice and ileal papilla. <pars terminalis ilei> C

20 ***aggregated lymphoid nodules of vermiform appendix*** oval elevated areas of lymphoid tissue occupying the greater part of the submucosa of the vermiform appendix. <noduli lymphoidei aggregati appendicis vermiformis> D

21 **ascending colon** the portion of the colon between the cecum and the right colic flexure. <colon ascendens> D

22 **cecum** the first part of the large intestine, forming a dilated pouch into which open the ileum, colon, and vermiform appendix. C, D

23 **colon** the part of the large intestine extending from the cecum to the rectum. D

24 **descending colon** the portion of the colon between the left colic flexure and the sigmoid colon at the pelvic brim; the part of it that lies in the left part of the iliac fossa is sometimes called the *iliac colon.* <colon descendens> D

25 **haustra of colon** sacculations in the wall of the colon produced by adaptation of its length to that of the taeniae coli, or by the arrangement of the circular muscle fibers. <haustra coli> D

26 **large intestine** the distal portion of the intestine, about 1.5 meters (5 feet) long, extending from its junction with the small intestine to the anus; it comprises the cecum, colon, rectum, and anal canal. <intestinum crassum> D

27 **left colic flexure** the bend in the large intestine at which the transverse colon becomes the descending colon. <flexura coli sinistra> D

28 **omental appendices** peritoneum-covered tabs of fat, 2 to 10 cm long, attached in rows along the taeniae coli; called also *epiploic appendices.* <appendices omentales> D

29 **rectum** the distal portion of the large intestine, beginning anterior to the third sacral vertebra as a continuation of the sigmoid colon and ending at the anal canal. D

30 **right colic flexure** the bend in the large intestine at which the ascending colon becomes the transverse colon. <flexura coli dextra> D

A

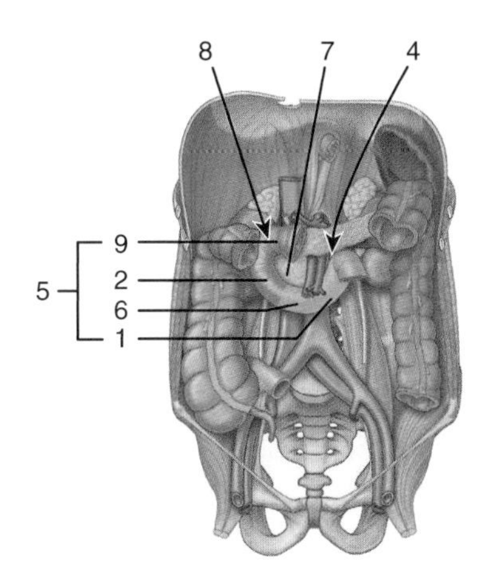
8
7
4
9
5
2
6
1

B

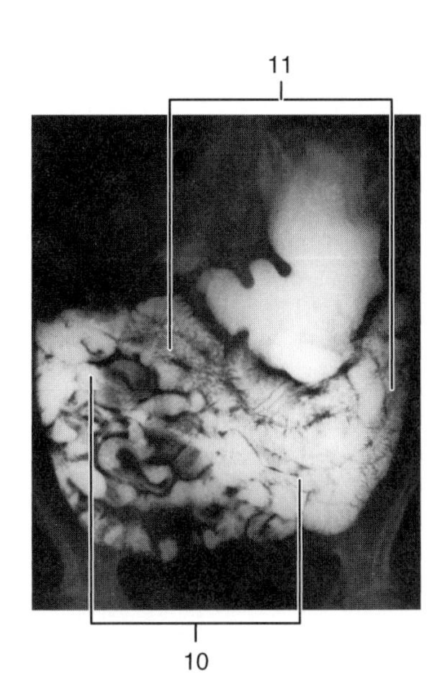
11
10

C

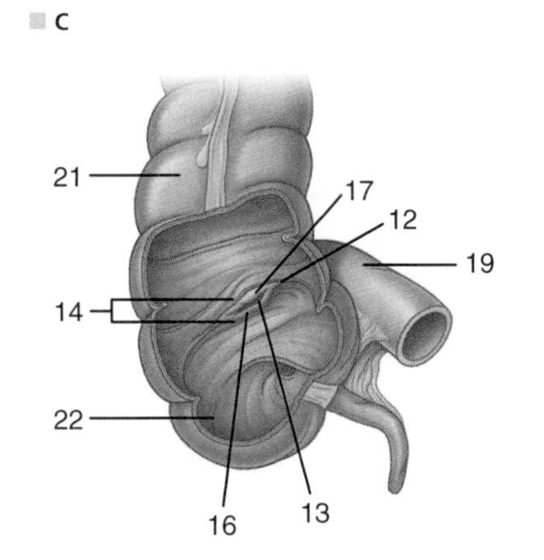
21
17
12
19
14
22
16
13

D

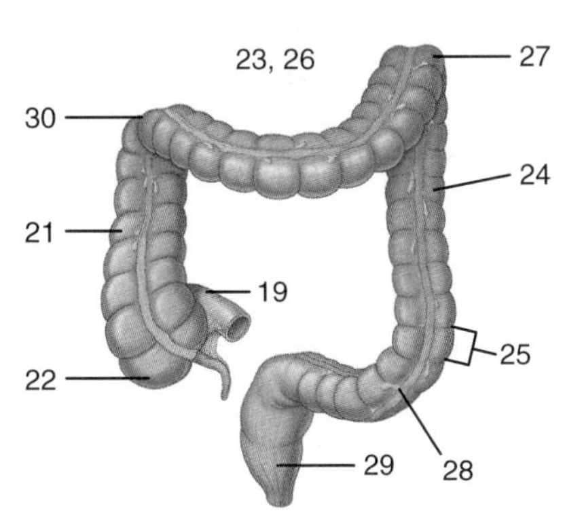
23, 26
27
30
24
21
19
25
22
29
28

1 ***semilunar folds of colon*** crescentic folds in the wall of the colon, projecting into the lumen between the haustra. <plicae semilunares coli> A

2 **sigmoid colon** the S-shaped part of the colon that lies in the pelvis, extending from the pelvic brim to the third segment of the sacrum; it is continuous above with the descending colon and below with the rectum. Called also *sigmoid flexure.* <colon sigmoideum> A

3 **transverse colon** the portion of the colon that runs transversely across the upper part of the abdomen, between the right and left colic flexures. <colon transversum> A

4 ***transverse folds of rectum*** permanent transverse folds in the rectum, usually three in number (two on the left and one on the right), involving the tunica mucosa and tela submucosa, and the circular layer of the tunica muscularis. <plicae transversae recti> A

5 **vermiform appendix** a wormlike diverticulum of the cecum, varying in length from 7 to 15 cm, and measuring about 1 cm in diameter. Called also *appendix.* <appendix vermiformis> A

6 **taeniae coli** three thickened bands, about 0.6 cm wide and one-sixth shorter than the colon, formed by the longitudinal fibers in the tunica muscularis of the large intestine; they extend from the root of the vermiform appendix to the rectum, where the fibers spread out and form a continuous layer encircling the tube. A

7 **circular layer of muscular coat of rectum** the inner layer of circularly coursing fibers in the muscular coat (tunica muscularis) of the rectum. <stratum circulare tunicae muscularis recti> B

8 **longitudinal layer of muscular coat of rectum** the outer layer of longitudinally coursing fibers of the muscular coat (tunica muscularis) of the rectum. <stratum longitudinale tunicae muscularis recti> B

9 **muscular coat of rectum** the muscular layer of the rectum, having two sublayers or strata, an outer longitudinal and an inner circular one. <tunica muscularis recti> B

10 ***aggregated lymphoid nodules of small intestine*** oval elevated areas of lymphoid tissue on the mucosa of the small intestine, composed of many lymphoid follicles closely packed together; called also *Peyer's patches.* <noduli lymphoidei aggregati intestini tenuis> C

11 ***circular folds*** the permanent transverse folds of the luminal surface of the small intestine, involving both the mucosa and submucosa. Called also *Kerckring's folds* or *valves.* <plicae circulares> C

12 **small intestine** the proximal portion of the intestine, smaller in caliber than the large intestine, and about 6 meters (20 feet) long, extending from the pylorus to the cecum; it comprises the duodenum, jejunum, and ileum. <intestinum tenue> C

13 **circular layer of muscular coat of small intestine** the inner layer of circularly coursing fibers in the muscular coat (tunica muscularis) of the small intestine. <stratum circulare tunicae muscularis intestini tenuis> D

14 **intestinal glands** simple tubular glands in the mucous membrane of either the small intestine, opening between the bases of the villi and containing argentaffin cells; or the large intestine. Called also *crypts* or *glands of Lieberkühn.* <glandulae intestinales> D

15 **intestinal villi** the multitudinous threadlike projections that cover the surface of the mucosa of the small intestine and serve as the sites of absorption of fluids and nutrients. <villi intestinales> D

16 **longitudinal layer of muscular coat of small intestine** the outer layer of the muscular coat (tunica muscularis) of the small intestine, consisting of longitudinally coursing fibers. <stratum longitudinale tunicae muscularis intestini tenuis> D

17 **mucosa of small intestine** the mucous membrane (tunica mucosa) lining the small intestine. <tunica mucosa intestini tenuis> D

18 **muscular coat of small intestine** the muscular layer of the small intestine, having two sublayers or strata, an inner circular and an outer longitudinal one. <tunica muscularis intestini tenuis> D

19 **muscularis mucosae of small intestine** the muscular layer of the mucous membrane (tunica mucosa) of the small intestine. <lamina muscularis mucosae intestini tenuis> D

20 **serosa of small intestine** the serous coat or layer (tunica serosa) of the small intestine. <tunica serosa intestini tenuis> D

21 **submucosa of small intestine** the submucous layer of the small intestine. <tela submucosa intestini tenuis> D

22 ***subserosa of small intestine*** the subserous layer of the small intestine. <tela subserosa intestini tenuis> D

23 **mucosa of large intestine** the mucous membrane (tunica mucosa) lining the large intestine. <tunica mucosa intestini crassi> E

24 **muscular coat of large intestine** the muscular layer of the large intestine. <tunica muscularis intestini crassi> E

25 **muscularis mucosae of large intestine** the muscular layer of the mucous membrane (tunica mucosa) of the large intestine. <lamina muscularis mucosae intestini crassi> E

26 ***serosa of large intestine*** the serous coat or layer (tunica serosa) of the large intestine. <tunica serosa intestini crassi> E

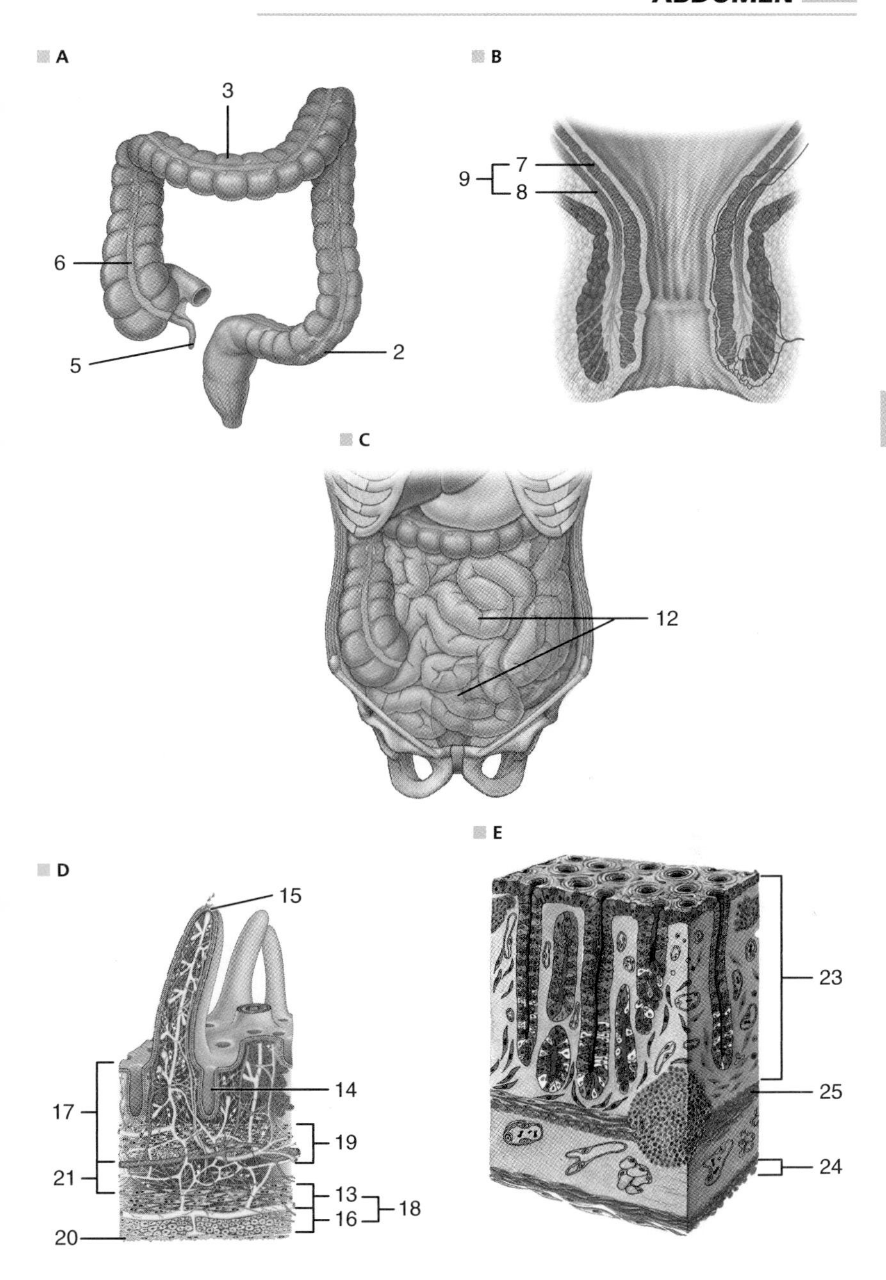
A
3
6
5
2
B
9
7
8
C
12
D
15
14
17
19
21
13
16
18
20
E
23
25
24

1 **taenia libera** the thickened band formed by anterior longitudinal muscle fibers of the large intestine, almost equidistant from the taenia mesocolica and the taenia omentalis. A

2 **taenia mesocolica** the thickened band of longitudinal muscle fibers of the large intestine along the site of attachment of the mesocolon. A

3 **taenia omentalis** the band of longitudinal muscle fibers of the large intestine along the site of attachment of the greater omentum. A

4 **diaphragmatic surface of liver** the surface of the liver that is in contact with the diaphragm, divided anatomically into superior, anterior, right, and posterior parts. <facies diaphragmatica hepatis> B

5 **posterior part of diaphragmatic surface of liver** the part directed toward the posterior surface of the body. <pars posterior faciei diaphragmaticae hepatis> B

6 **superior part of diaphragmatic surface of liver** the most superior part of the diaphragmatic surface, projecting convexly upwards. <pars superior faciei diaphragmaticae hepatis> B

7 **anterior part of diaphragmatic surface of liver** the part directed toward the anterior surface of the body. <pars anterior faciei diaphragmaticae hepatis> B, C

8 ***cardiac impression of liver*** a depression on the superior part of the diaphragmatic surface of the liver, corresponding to the position of the heart. <impressio cardiaca hepatis> C

9 **falciform ligament of liver** a sickle-shaped sagittal fold of peritoneum that helps to attach the liver to the diaphragm and extends from the coronary ligament of the liver to the umbilicus, forming the boundary between the right and left lobes. <ligamentum falciforme hepatis> C

10 ***fibrous appendix of liver*** a fibrous band at the left extremity of the liver, being the atrophied remnant of formerly more extensive liver tissue. <appendix fibrosa hepatis> C

11 ***fibrous capsule of liver*** the fibroelastic layer that surrounds the liver beneath the peritoneum; at the porta hepatis it is continuous with the perivascular fibrous capsule. <tunica fibrosa hepatis> C

12 ***interlobular bile ducts*** small channels between the hepatic lobules, draining into the biliary ductules. <ductus biliferi interlobulares> C

13 **liver** a large gland of a dark red color found in the upper part of the abdomen on the right side. Its domed upper surface fits closely against and is adherent to the inferior surface of the right diaphragmatic dome, and it has a double blood supply from the hepatic artery and the portal vein. Its traditional anatomic divisions have been four lobes: left, right, caudate, and quadrate lobes. Based on the internal blood supply and biliary drainage, a newer anatomic system divides the liver into three regions, the left liver, right liver, and posterior liver, with the latter being identical to the caudate lobe. The liver is made up of thousands of minute lobules, which are its functional units. C

14 **right part of diaphragmatic surface of liver** the portion that is directed toward the right side of the body, near the ribs on that side. <pars dextra faciei diaphragmaticae hepatis> C

15 ***serosa of liver*** the serous coat or layer (tunica serosa) of the liver. <tunica serosa hepatis> C

16 ***subserosa of liver*** the subserous layer of the liver. <tela subserosa hepatis> C

17 **fissure for ligamentum teres** the fissure or fossa on the visceral surface of the liver lodging the ligamentum teres in the adult, forming part of the boundary between the right and left lobes of the liver. Called also *fissure of round ligament.* <fissura ligamenti teretis> D

18 **fissure for ligamentum venosum** a fossa on the posterior part of the diaphragmatic surface of the liver, lodging the ligamentum venosum in the adult. <fissura ligamenti venosi> D

19 **groove for vena cava** a groove on the posterior part of the diaphragmatic surface of the liver, separating the right lobe from the caudate lobe and lodging the inferior vena cava. <sulcus venae cavae> D

20 **notch for ligamentum teres** a notch in the inferior border of the liver, occupied by the ligamentum teres in the adult; called also *interlobar* or *umbilical notch.* <incisura ligamenti teretis> D

21 **porta hepatis** the transverse fissure on the visceral surface of the liver where the portal vein and hepatic artery enter the liver and the hepatic ducts leave. D

22 **bare area of liver** a large part of the diaphragmatic surface of the liver that lacks a peritoneal covering; its boundaries are formed by the coronary ligament of the liver and the left and right triangular ligaments. <area nuda hepatis> E

23 **caudate lobe of liver** a small lobe of the liver, bounded on the right by the inferior vena cava, which separates it from the right lobe, and on the left by the attachment of the hepatogastric ligament, which separates it from the left lobe. <lobus caudatus hepatis> D, E

24 ***caudate process*** the right of the two processes seen on the caudate lobe of the liver. <processus caudatus> E

25 **colic impression of liver** a variable concavity in the right liver, where it is in contact with the right flexure of the colon. <impressio colica hepatis> E

26 **coronary ligament of liver** the line of reflection of the peritoneum from the diaphragmatic surface of the liver to the under surface of the diaphragm. <ligamentum coronarium hepatis> E

27 **duodenal impression of liver** a concavity on the right lobe of the liver where it is in contact with the descending part of the duodenum. <impressio duodenalis hepatis> E

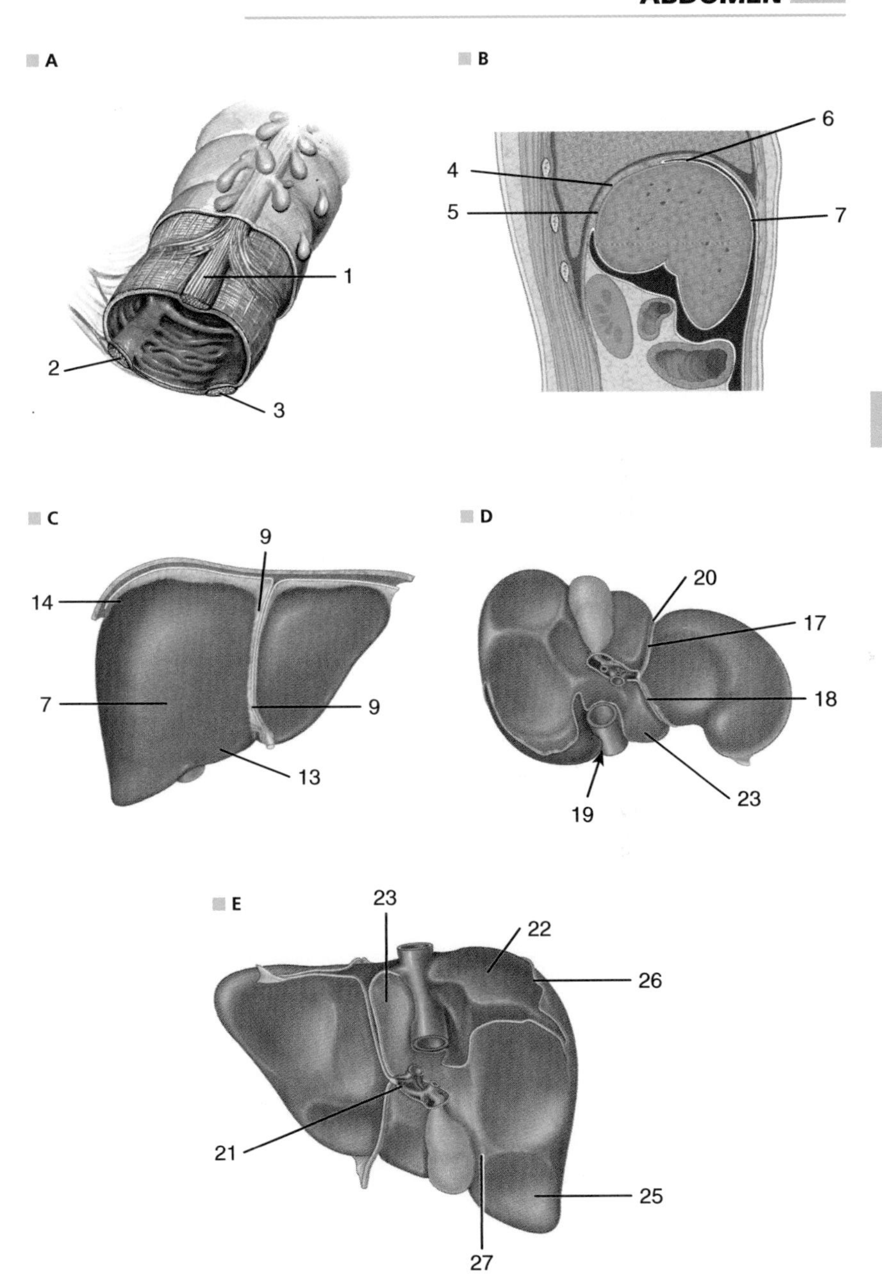
A
1
2
3
B
4
5
6
7
C
9
14
7
9
13
D
20
17
18
23
19
E
23
22
26
21
25
27

1 **esophageal impression of liver** a concavity on the left lobe of the liver corresponding to the position of the abdominal part of the esophagus. <impressio oesophagea hepatis> A

2 **fossa for gallbladder** the fossa on the visceral surface of the liver that lodges the gallbladder and helps to separate the right and left lobes. <fossa vesicae biliaris> A

3 **gastric impression of liver** a large concavity in the left liver where it is in contact with the anterior surface of the stomach. <impressio gastrica hepatis> A

4 ***hepatic ligaments*** the ligaments of the liver, including the coronary, falciform, and right and left triangular ligaments. <ligamenta hepatis> A

5 **inferior border of liver** the anteroinferior edge of the liver, separating the diaphragmatic and visceral surfaces. <margo inferior hepatis> A

6 ***left duct of caudate lobe*** the left duct draining the caudate lobe (posterior part) of the liver. <ductus lobi caudati sinister> A

7 **left lobe of liver** in traditional anatomic terminology, the smaller of the two main lobes of the liver. Anteriorly, it is separated from the right lobe by the falciform ligament. Posteroinferiorly, it is separated from the caudate and quadrate lobes by the attachment of the gastrohepatic ligament and the ligamentum teres. <lobus hepatis sinister> A

8 ***left part of liver*** the part that receives blood from the left branches of the hepatic portal vein and hepatic artery proper and whose bile flows out through the left hepatic duct. Called also *left liver.* <pars hepatis sinistra> A

9 **left triangular ligament of liver** a triangular extension of the left extremity of the coronary ligament, which helps to attach the left liver to the diaphragm. <ligamentum triangulare sinistrum hepatis> A

10 **ligamentum venosum of liver** a fibrous cord, the remains of the fetal ductus venosus, lying in the fissure of the venous ligament. <ligamentum venosum> A

11 **omental tuberosity of liver** the rounded prominence on the posteroinferior surface of the left liver, just cranial to the lesser curvature of the stomach. <tuber omentale hepatis> A

12 **papillary process** the left of the two processes seen on the caudate lobe of the liver. <processus papillaris> A

13 ***posterior liver*** a term used to refer to the posterior region that is not part of either the left part or the right part; this part is coextensive with the caudate lobe. <pars posterior hepatis> A

14 **quadrate lobe of liver** a small lobe of the liver bounded on the right by the gallbladder, which separates it from the right lobe, and on the left by the ligamentum teres, which separates it from the left lobe. <lobus quadratus hepatis> A

15 **renal impression of liver** the concavity on the right liver where it is in contact with the right kidney. <impressio renalis hepatis> A

16 ***right duct of caudate lobe*** the right duct draining the caudate lobe (posterior part) of the liver. <ductus lobi caudati dexter> A

17 **right lobe of liver** in traditional anatomic terminology, the largest of the four lobes of the liver. Anteriorly, it is separated from the left lobe by the falciform ligament. Posteroinferiorly, it is separated from the caudate lobe by the inferior vena cava and from the quadrate lobe by the gallbladder. Sometimes the term has been extended to include the caudate and quadrate lobes. <lobus hepatis dexter> A

18 ***right part of liver*** in newer anatomic terminology, the part that receives blood from the right branches of the hepatic portal vein and hepatic artery proper and whose bile flows out through the right hepatic duct. Called also *right liver.* <pars hepatis dextra> A

19 **right triangular ligament of liver** the pointed right extremity of the coronary ligament of the liver, extending to attach to the diaphragm. <ligamentum triangulare dextrum hepatis> A

20 **round ligament of liver** a fibrous cord, the remains of the left umbilical vein, extending from the porta hepatis (where it is attached to the left branch of the portal vein) out through the fissure for the ligamentum teres and the falciform ligament to the umbilicus. <ligamentum teres hepatis> A

21 **suprarenal impression of liver** a small concavity on the right liver, superior to the renal impression, caused by contact with the right adrenal (suprarenal) gland. <impressio suprarenalis hepatis> A

22 ***visceral surface of liver*** the posteroinferior surface of the liver, which is in contact with various abdominal viscera. Called also *inferior surface of liver.* <facies visceralis hepatis> A

23 **hepatic lobules** the small vascular units composing the substance of the liver, each of which is polygonal, with a central vein at its center and portal canals peripherally at the corners. <lobuli hepatis> B

24 **anterior border of body of pancreas** the pancreatic border that bounds the anterosuperior and anteroinferior surfaces. <margo anterior corporis pancreatis> C

25 **anteroinferior surface of body of pancreas** the anterior and inferior surface of the body of the pancreas. <facies anteroinferior corporis pancreatis> C

26 **anterosuperior surface of body of pancreas** the anterior and superior surface of the body of the pancreas. <facies anterosuperior corporis pancreatis> C

27 **body of pancreas** the triangularly prismatic central portion of the pancreas, extending from the neck on the right to the tail on the left. <corpus pancreatis> C

A

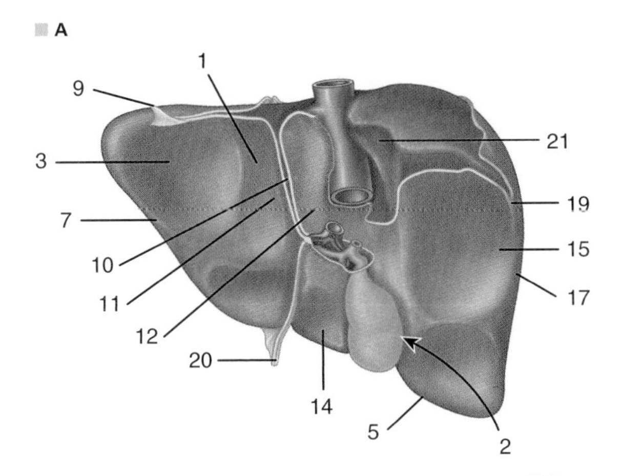

B

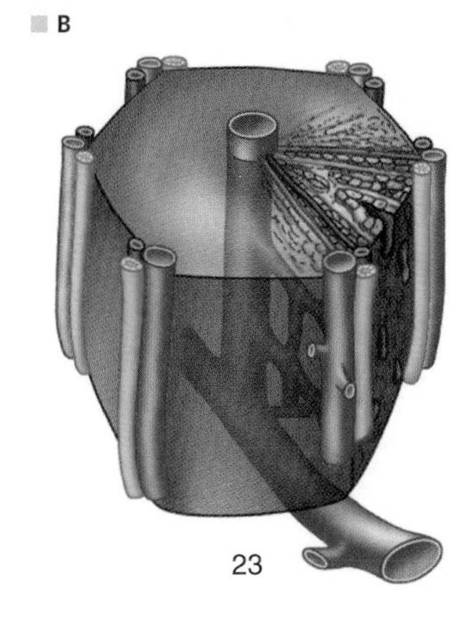

C

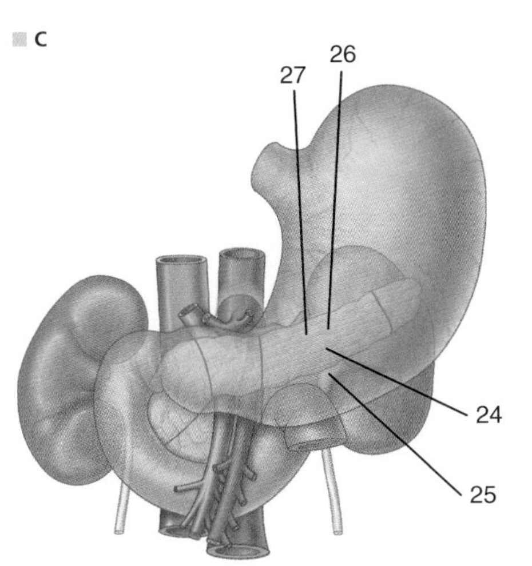

1 **head of pancreas** the discoidal mass forming the enlarged right extremity of the pancreas, lying in a flexure of the duodenum. <caput pancreatis> A

2 ***islets of Langerhans*** irregular microscopic structures scattered throughout the pancreas and constituting its endocrine part (the *endocrine pancreas*). They are composed of at least four types of cells: the *alpha cells,* which secrete glucagon; the *beta cells,* which are the most abundant and secrete insulin; the *delta cells,* which secrete somatostatin; and the *PP cells,* which secrete pancreatic polypeptide. A

3 **neck of pancreas** a constricted portion marking the junction of the head and body of the pancreas. <collum pancreatis> A

4 ***omental eminence of body of pancreas*** a rounded prominence chiefly on the anterior surface of the neck of the pancreas. <tuber omentale corporis pancreatis> A

5 **pancreas** a large, elongated, racemose gland situated transversely behind the stomach, between the spleen and the duodenum. Its right extremity or *head* is larger and directed downward; its left extremity or *tail* is transverse and terminates close to the spleen. It is subdivided into lobules by septa that extend into the gland from the thin, areolar tissue that forms an indefinite capsule. The endocrine part *(endocrine pancreas)* consists of *islets of Langerhans.* The exocrine part *(exocrine pancreas)* consists of pancreatic acini, secretory units that produce and secrete into the duodenum pancreatic juice, which contains enzymes essential to protein digestion. A

6 **pancreatic notch** a notch at the junction of the left half of the head of the pancreas and the neck of the pancreas. <incisura pancreatis> A

7 ***posterior surface of body of pancreas*** the pancreatic surface directed toward the posterior part of the body. <facies posterior corporis pancreatis> A

8 **superior border of body of pancreas** the pancreatic edge that bounds the anterosuperior and posterior surfaces. <margo superior corporis pancreatis> A

9 **tail of pancreas** the left extremity of the pancreas, usually in contact with the medial aspect of the spleen and the junction of the transverse colon and descending colon. <cauda pancreatis> A

10 **uncinate process of pancreas** the left and caudal part of the head of the pancreas, which hooks around behind the pancreatic vessels; called also *lesser pancreas.* <processus uncinatus pancreatis> A

11 **accessory pancreatic duct** a small inconstant duct draining part of the head of the pancreas into the minor duodenal papilla. <ductus pancreaticus accessorius> A

12 **duodenal ampulla** the superior part of the duodenum, often seen only radiographically after a barium meal. <ampulla duodeni> A

13 ***longitudinal fold of duodenum*** a mucosal ridge running longitudinally on the inner surface of the medial wall of the descending part of the duodenum. <plica longitudinalis duodeni> A

14 **major duodenal papilla** a small elevation at the site of the opening of the conjoined common bile duct and pancreatic duct into the lumen of the duodenum. <papilla duodeni major> A

15 **minor duodenal papilla** a small elevation at the site of the opening of the accessory pancreatic duct into the lumen of the duodenum. <papilla duodeni minor> A

16 **pancreatic duct** the main excretory duct of the pancreas, which usually unites with the common bile duct before entering the duodenum at the major duodenal papilla. <ductus pancreaticus> A

17 ***sphincter of bile duct*** an annular sheath of muscle that invests the bile duct within the wall of the duodenum. <musculus sphincter ductus choledochi> A

18 ***sphincter of hepatopancreatic ampulla*** muscle fibers investing the hepatopancreatic ampulla in the wall of the duodenum. <musculus sphincter ampullae hepatopancreaticae> A

19 ***sphincter of pancreatic duct*** a sphincter that surrounds the pancreatic duct just above the hepatopancreatic ampulla. <musculus sphincter ductus pancreatici> A

20 ***anterior branch of right hepatic duct*** the branch of the right hepatic duct draining anterior segments of the right liver. <ramus anterior ductus hepatici dextri> B

21 **bile duct** the duct formed by union of the common hepatic duct and the cystic duct; it empties into the duodenum at the major duodenal papilla, along with the pancreatic duct. <ductus choledochus> A, B

22 **body of gallbladder** the central portion of the gallbladder, extending from the fundus to the neck. <corpus vesicae biliaris> B

23 **common hepatic duct** the duct formed by union of the right and left hepatic ducts, which in turn joins the cystic duct to form the common bile duct. <ductus hepaticus communis> B

24 **cystic duct** the passage connecting the neck of the gallbladder and the common bile duct. <ductus cysticus> B

25 ***folds of tunica mucosa of gallbladder*** the folds in the mucosa that bound the polygonal spaces, giving the interior a honeycombed appearance. <plicae mucosae vesicae biliaris> B

26 **fundus of gallbladder** the inferior dilated portion of the gallbladder. <fundus vesicae biliaris> B

27 **gallbladder** the pear-shaped reservoir for the bile, found within a fossa on the visceral surface of the liver. From its neck, the cystic duct projects to join the common bile duct. <vesica biliaris> B

A

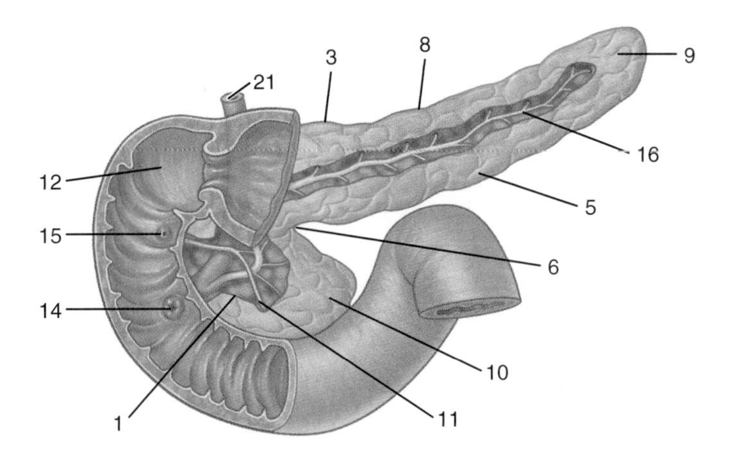

B

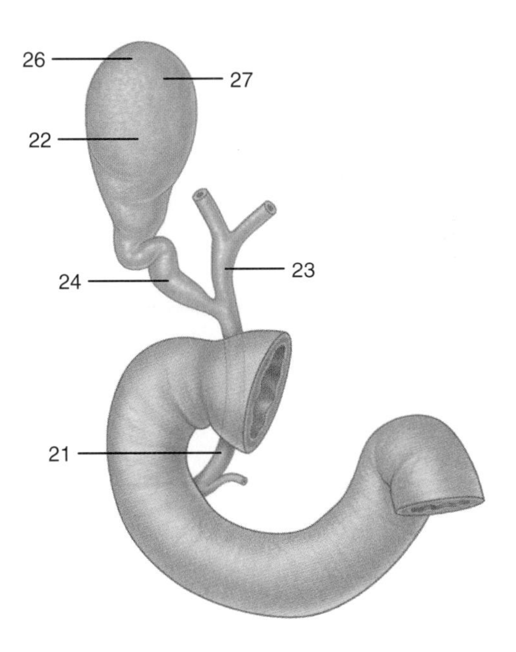

1 ***glands of bile duct*** tubuloacinar glands in the mucosa of the bile ducts and the neck of the gallbladder; called also *biliary glands.* <glandulae ductus choledochi> A

2 **infundibulum of gallbladder** the tapering part of the gallbladder, ending at the neck. <infundibulum vesicae biliaris> A

3 ***lateral branch of left hepatic duct*** the branch of the left hepatic duct draining lateral segments of the left lobe of liver. <ramus lateralis ductus hepatici sinistri> A

4 **left hepatic duct** the duct that drains the left liver and part of the posterior liver (caudate lobe). It has lateral and medial branches draining from specific segments of the liver. <ductus hepaticus sinister> A, B

5 ***medial branch of left hepatic duct*** the branch of the left hepatic duct draining medial segments of the left liver. <ramus medialis ductus hepatici sinistri> A

6 ***mucosa of gallbladder*** the mucous membrane (tunica mucosa) lining the gallbladder. <tunica mucosa vesicae biliaris> A

7 ***muscular coat of gallbladder*** the muscular layer of the gallbladder. <tunica muscularis vesicae biliaris> A

8 **neck of gallbladder** the upper constricted portion of the gallbladder, between the body and the cystic duct. <collum vesicae biliaris> A

9 ***posterior branch of right hepatic duct*** the branch of the right hepatic duct draining posterior segments of the right liver. <ramus posterior ductus hepatici dextri> A

10 **right hepatic duct** the duct that drains the right liver and part of the posterior liver (caudate lobe). It has anterior and posterior branches draining from specific segments of the liver. <ductus hepaticus dexter> A, B

11 ***serosa of gallbladder*** the serous coat or layer (tunica serosa) of the gallbladder. <tunica serosa vesicae biliaris> A

12 ***spiral fold*** a spirally arranged elevation in the mucosa of the first part of the cystic duct. <plica spiralis> A

13 ***subserosa of gallbladder*** the subserous layer of the gallbladder. <tela subserosa vesicae biliaris> A

14 ***accessory spleen*** a connected or detached outlying portion, or exclave, of the spleen; called also *lien accessorius* [TA alternative], *splenculus, spleneolus, splenulus,* and *splenunculus.* <splen accessorius> C

15 **anterior extremity of spleen** the lower pole of the spleen, which is situated anterior to the upper pole. <extremitas anterior splenis> C

16 ***capsule of spleen*** the fibroelastic coat of the spleen. <capsula splenis> C

17 **colic impression of spleen** the concave surface of the spleen in contact with the colon. <facies colica splenis> C

18 **diaphragmatic surface of spleen** the convex posterolateral surface of the spleen, directed toward the diaphragm. <facies diaphragmatica splenis> C

19 **gastric impression of spleen** the concave surface of the spleen in contact with the stomach. <facies gastrica splenis> C

20 **hilum of spleen** the fissure on the gastric surface of the spleen where the vessels and nerves enter. <hilum splenicum> C

21 **inferior border of spleen** a straight margin of the spleen somewhat less prominent than the superior margin, separating the renal surface from the diaphragmatic surface; called also *obtuse* or *posterior margin of spleen.* <margo inferior splenis> C

22 **posterior extremity of spleen** the uppermost pole of the spleen, situated somewhat posterior to the lower pole. <extremitas posterior splenis> C

23 **renal impression of spleen** the concave surface of the spleen in contact with the left kidney. <facies renalis splenis> C

24 ***serosa of spleen*** the serous coat or layer (tunica serosa) of the spleen. <tunica serosa splenis> C

25 **spleen** a large glandlike but ductless organ in the upper part of the abdominal cavity on the left side lateral to the cardiac end of the stomach; it has a flattened oblong shape and is about 125 mm long, the largest structure in the lymphoid system. It has a purple color and a pliable consistency, and is distinguished by two types of tissue: red pulp and white pulp. The spleen disintegrates red blood cells and sets free hemoglobin, which the liver converts into bilirubin; it serves as a reservoir of blood; and it produces lymphocytes and plasma cells. During fetal life and in the newborn it gives rise to new red blood cells. <splen> C

26 ***splenic pulp*** the dark, reddish-brown substance that fills up the interspaces of the sinuses of the spleen; called also *red pulp.* <pulpa splenica> C

27 ***splenic sinus*** a dilated venous sinus not lined by ordinary endothelial cells, found in the splenic pulp. <sinus splenicus> C

28 ***splenic trabeculae*** fibrous bands that pass into the spleen from the tunica fibrosa and form the supporting framework of the organ. <trabeculae splenicae> C

29 **superior border of spleen** a somewhat sharp, convex line, sometimes serrated, between the gastric and diaphragmatic surfaces of the spleen. <margo superior splenis> C

30 **visceral surface of spleen** the surface of the spleen which comes in contact with various other viscera, including the colon, kidney, and stomach. <facies visceralis splenis> C

31 **cystohepatic triangle** the triangle formed by the liver superiorly, the cystic duct inferiorly, and the common hepatic duct medially; this space usually contains the cystic artery. <trigonum cystohepaticum> B

A

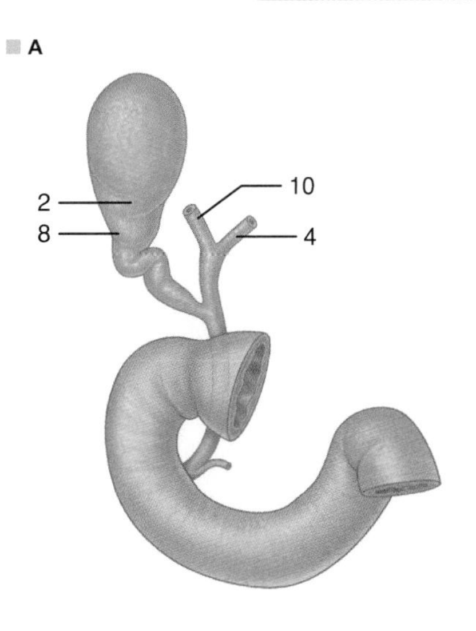

C

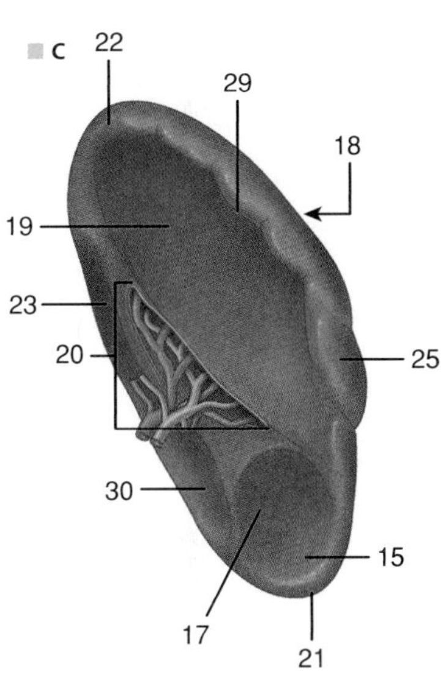

B

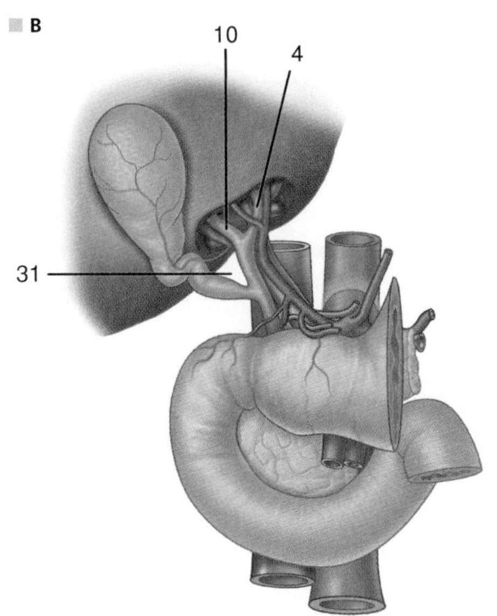

Nerves of Abdominal Viscera

1 ***gastric plexuses*** subdivisions of the celiac portion of the prevertebral plexuses, accompanying the gastric arteries and branches and supplying nerve fibers to the stomach. <plexus gastrici> A

2 ***hepatic plexus*** a subdivision of the celiac plexus accompanying the hepatic artery to the liver. <plexus hepaticus> A

3 ***inferior mesenteric plexus*** a subdivision of the aortic plexus accompanying the inferior mesenteric artery. <plexus mesentericus inferior> A

4 ***pancreatic plexus*** a subdivision of the celiac plexus, accompanying pancreatic arteries. <plexus pancreaticus> A

5 ***superior mesenteric plexus*** a subdivision of the celiac plexus accompanying the superior mesenteric artery. <plexus mesentericus superior> A

6 ***superior rectal plexus*** a plexus accompanying the superior rectal artery to the rectum, derived from the inferior mesenteric and hypogastric plexuses. Called also *superior hemorrhoidal p.* <plexus rectalis superior> A

7 ***suprarenal plexus*** a subdivision of the celiac plexus, in proximity with and supplying nerve fibers to an adrenal (suprarenal) gland. <plexus suprarenalis> A

8 **enteric plexus** a plexus of autonomic nerve fibers within the wall of the digestive tube, and made up of the submucosal, myenteric, and subserosal plexuses; it contains visceral afferent fibers, sympathetic postganglionic fibers, parasympathetic preganglionic and postganglionic fibers, and parasympathetic postganglionic cell bodies. <plexus entericus> B

9 **myenteric plexus** that part of the enteric plexus within the tunica muscularis. <plexus myentericus> B

10 **submucosal plexus** the part of the enteric plexus that is situated in the submucosa. <plexus submucosus> B

11 ***subserosal plexus*** the part of the enteric plexus situated deep to the serosal surface of the tunica serosa. <plexus subserosus> B

12 **anterior gastric branches of anterior vagal trunk** branches arising from the anterior trunk of the vagus near the cardiac end of the stomach, innervating the anterior aspect of the lesser curvature and the anterior surface of the stomach almost to the pylorus; *modality,* parasympathetic and visceral afferent. <rami gastrici anteriores trunci vagalis anterioris> C

13 **anterior vagal trunk** a nerve trunk (or trunks) formed by fibers from both left and right vagus nerves, collected from the anterior part of the esophageal plexus; it descends through the esophageal hiatus of the diaphragm to supply branches to the anterior surface of the stomach. <truncus vagalis anterior> C

14 **hepatic branches of anterior vagal trunk** branches (sometimes only one) arising from the anterior vagal trunk, contributing to the hepatic plexus, and helping innervate the liver, gallbladder, pancreas, pylorus, and duodenum; modality, parasympathetic and visceral afferent. <rami hepatici trunci vagalis anterioris> C

15 **posterior gastric branches of posterior vagal trunk** branches arising from the posterior vagal trunk near the cardiac end of the stomach and innervating the cardiac orifice and fundus, the posterior aspect of the lesser curvature, and the posterior surface of the stomach to the pyloric antrum; *modality,* parasympathetic and visceral afferent. <rami gastrici posteriores trunci vagalis posterioris> C

16 **posterior vagal trunk** a nerve trunk or trunks formed by fibers from both left and right vagus nerves, collected from the posterior part of the esophageal plexus; it descends through the esophageal hiatus of the diaphragm to supply branches to the posterior surface of the stomach. <truncus vagalis posterior> C

Vessels of Abdominal Viscera

17 **omental branches of left gastro-omental artery** branches that supply the stomach and greater omentum. <rami omentales arteriae gastroomentalis sinistrae> D

18 **omental branches of right gastro-omental artery** branches that supply the greater omentum. <rami omentales arteriae gastroomentalis dextrae> D

19 ***cardiac lymphatic ring*** a chain of lymph nodes (paracardial lymph nodes) around the cardiac opening of the stomach. <anulus lymphaticus cardiae>

20 ***perivascular fibrous capsule*** the connective tissue sheath that accompanies the vessels and ducts through the porta hepatis ; it is continuous with the fibrous coat. Called also *Glisson's capsule* and *hepatobiliary capsule.* <capsula fibrosa perivascularis>

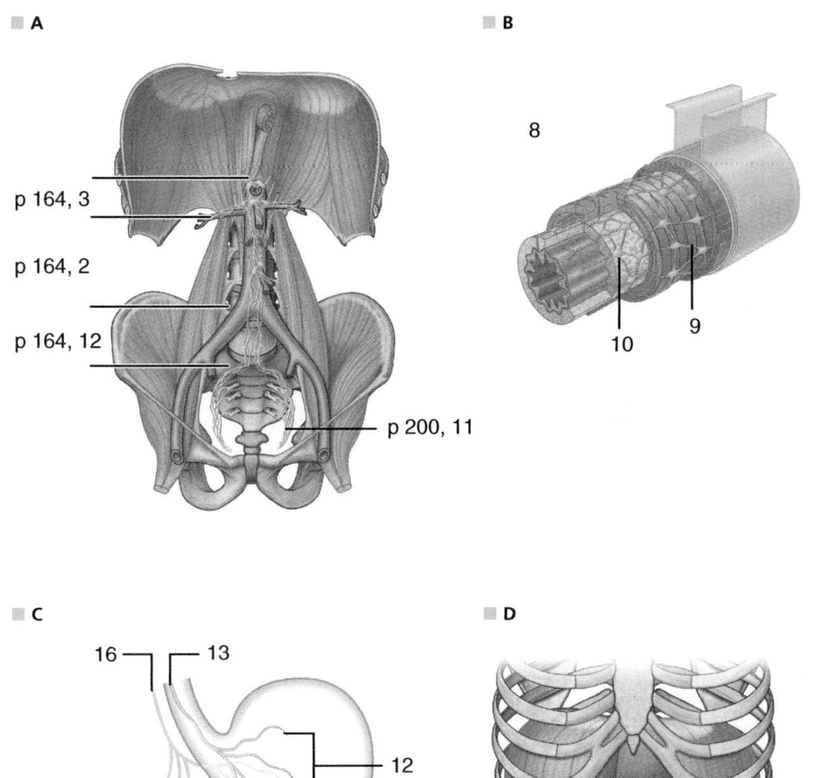
A
p 164, 3
p 164, 2
p 164, 12
p 200, 11
B
8
10
9

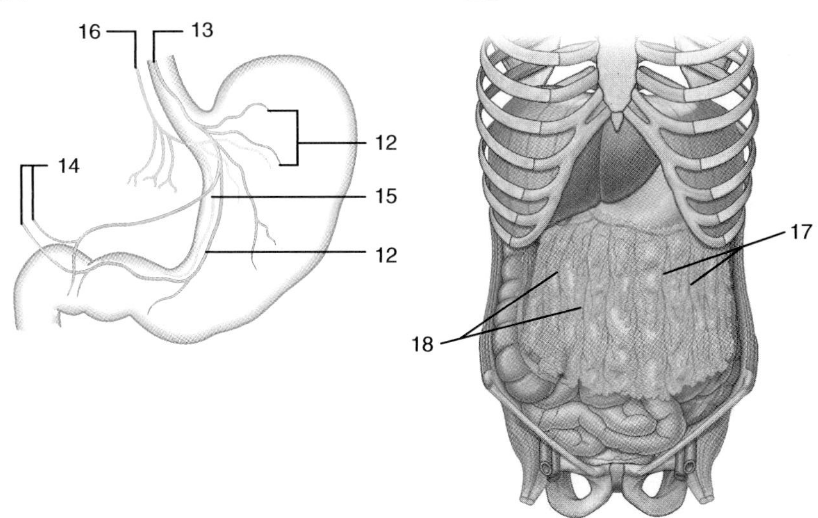
C
16
13
12
14
15
12
D
17
18

1 **left gastric artery** *origin,* celiac; *branches,* esophageal; *distribution,* esophagus, lesser curvature of stomach. Called also *left coronary artery of stomach.* <arteria gastrica sinistra> A, B

2 **left gastro-omental artery** *origin,* splenic artery; *branches,* gastric, omental; *distribution,* stomach, greater omentum. Called also *left inferior gastric artery.* <arteria gastroomentalis sinistra> A, B

3 **celiac trunk** the arterial trunk that arises from the abdominal aorta, gives off the left gastric, common hepatic, and splenic arteries, and supplies the esophagus, stomach, duodenum, spleen, pancreas, liver, and gallbladder. <truncus coeliacus> B

4 **common hepatic artery** *origin,* celiac trunk; *branches,* right gastric, gastroduodenal, proper hepatic; *distribution,* stomach, pancreas, duodenum, liver, gallbladder, greater omentum. <arteria hepatica communis> A, B

5 **esophageal branches of left gastric artery** branches that supply the esophagus; called also *inferior esophageal arteries.* <rami oesophageales arteriae gastricae sinistrae> A

6 ***gastric branches of left gastro-omental artery*** vessels that supply both surfaces of the stomach. <rami gastrici arteriae gastroomentalis sinistrae> A

7 ***gastric branches of right gastro-omental artery*** vessels that supply both surfaces of the stomach. <rami gastrici arteriae gastroomentalis dextrae> A

8 **gastroduodenal artery** *origin,* common hepatic artery; *branches,* supraduodenal and posterior superior pancreaticoduodenal arteries; *distribution,* stomach, duodenum, pancreas, greater omentum. <arteria gastroduodenalis> A, B

9 **hepatic artery proper** *origin,* common hepatic artery; *branches,* right and left branches; *distribution,* liver, gallbladder. <arteria hepatica propria> A, B

10 ***retroduodenal arteries*** *origin,* first branch of gastroduodenal; *branches,* none; *distribution,* bile duct, duodenum, head of pancreas. <arteriae retroduodenales> A

11 **right gastric artery** *origin,* common hepatic artery; *branches,* none; *distribution,* lesser curvature of stomach. Called also *pyloric artery.* <arteria gastrica dextra> A

12 **right gastro-omental artery** *origin,* gastroduodenal artery; *branches,* gastric, omental; *distribution,* stomach, greater omentum. Called also *right inferior gastric artery.* <arteria gastroomentalis dextra> A

13 ***supraduodenal artery*** *origin,* gastroduodenal; *branches,* duodenal; *distribution,* superior part of duodenum. <arteria supraduodenalis> A

14 **anterior branch of inferior pancreaticoduodenal artery** a branch that passes in front of the head of the pancreas and then ascends to anastomose with the anterior superior pancreaticoduodenal artery; it supplies the head of the pancreas and adjoining parts of the duodenum. <ramus anterior arteriae pancreaticoduodenalis inferioris> B

15 **anterior superior pancreaticoduodenal artery** *origin,* gastroduodenal artery; *branches,* pancreatic and duodenal; *distribution,* pancreas and duodenum. <arteria pancreaticoduodenalis superior anterior> B

16 **artery to tail of pancreas** *origin,* splenic; *branches and distribution,* supplies branches to tail of pancreas, and accessory spleen (if present). <arteria caudae pancreatis> B

17 ***dorsal pancreatic artery*** *origin,* splenic; *branches,* inferior pancreatic; *distribution,* neck and body of pancreas. <arteria pancreatica dorsalis> B

18 ***duodenal branches of anterior superior pancreaticoduodenal artery*** vessels that supply the duodenum. <rami duodenales arteriae pancreaticoduodenalis superioris anterioris> B

19 ***duodenal branches of posterior superior pancreaticoduodenal artery*** vessels supplying the duodenum. <rami duodenales arteriae pancreaticoduodenalis superioris posterioris> B

20 ***great pancreatic artery*** *origin,* splenic artery; *branches and distribution,* right and left branches anastomose with other pancreatic arteries. <arteria pancreatica magna> B

21 ***inferior pancreatic artery*** *origin,* dorsal pancreatic; *branches,* none; *distribution,* body and tail of pancreas. <arteria pancreatica inferior> B

22 **inferior pancreaticoduodenal arteries** *origin,* superior mesenteric artery; *branches,* anterior, posterior; *distribution,* pancreas, duodenum. <arteriae pancreaticoduodenales inferiores> B

23 ***pancreatic branches of posterior superior pancreaticoduodenal artery*** vessels that supply the pancreas. <rami pancreatici arteriae pancreaticoduodenalis superioris posterioris> B

24 **pancreatic branches of splenic artery** branches that supply the pancreas, arising from the splenic artery during its tortuous course along the superior border of the body of the pancreas. <rami pancreatici arteriae splenicae> B

25 ***pancreatic branches of the anterior superior pancreaticoduodenal artery*** vessels that supply the pancreas. <rami pancreatici arteriae pancreaticoduodenalis superioris anterioris> B

26 **posterior branch of inferior pancreaticoduodenal artery** a branch that ascends behind the head of the pancreas, which it sometimes pierces, and anastomoses with the posterior superior pancreaticoduodenal artery; it supplies the head of the pancreas and adjoining parts of the duodenum. <ramus posterior arteriae pancreaticoduodenalis inferioris> B

27 ***posterior gastric artery*** *origin,* splenic artery; *branches,* none; *distribution,* posterior gastric wall. <arteria gastrica posterior> B

28 **posterior superior pancreaticoduodenal artery** *origin,* gastroduodenal artery; *branches,* pancreatic and duodenal; *distribution,* pancreas, duodenum. <arteria pancreaticoduodenalis superior posterior> B

A

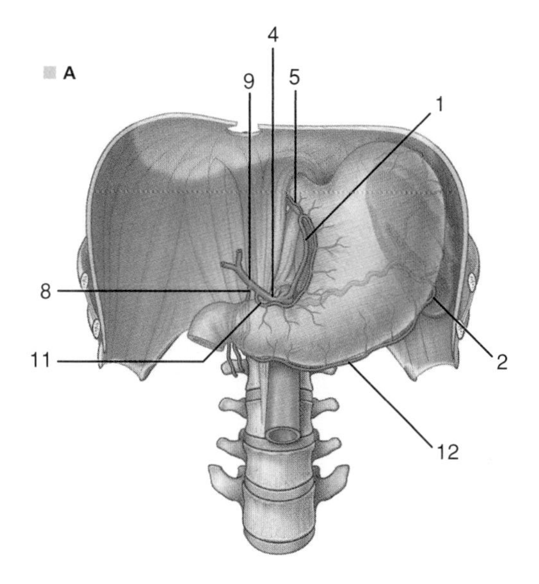

B

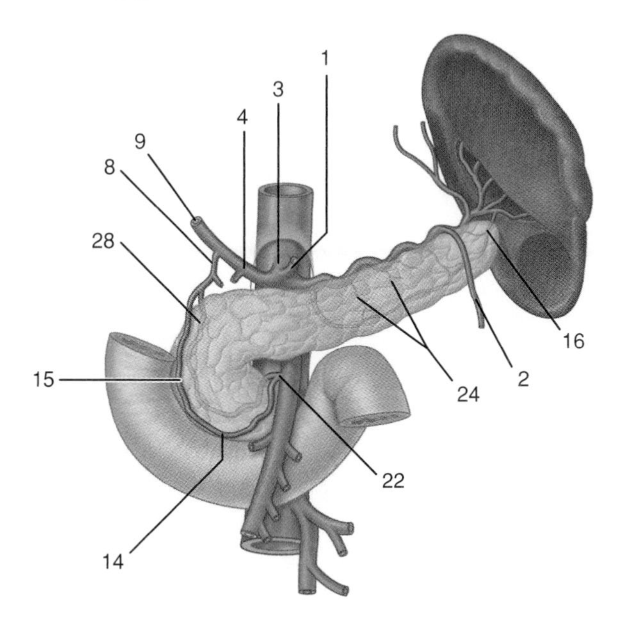

1 ***prepancreatic artery*** an arterial arch between the neck and uncinate process of the pancreas, formed by the right branch of the dorsal branch of the splenic artery and a branch from the anterior superior pancreaticoduodenal artery. <arteria prepancreatica> A

2 **short gastric arteries** *origin,* splenic; *branches,* none; *distribution,* upper part of stomach. <arteriae gastricae breves> A, B

3 **splenic branches of splenic artery** the terminal branches of the splenic artery, which follow the trabeculae of the spleen. <rami splenici arteriae splenicae> A

4 **splenic artery** *origin,* celiac trunk; *branches,* pancreatic and splenic branches, prepancreatic, left gastroomental, and short gastric arteries; *distribution,* spleen, pancreas, stomach, greater omentum. <arteria splenica> A, B, C

5 ***anterior segmental artery of liver*** *origin,* right branch of hepatic artery proper; *branches,* none; *distribution,* anterior segment of right liver. <arteria segmenti anterioris hepatici> C

6 ***artery of caudate lobe*** either of two branches, one from the right and one from the left hepatic artery, supplying twigs to the caudate lobe of the liver. <arteria lobi caudati> C

7 **cystic artery** *origin,* right branch of proper hepatic artery; *branches,* none; *distribution,* gallbladder. <arteria cystica> C

8 ***interlobular arteries of liver*** arteries originating from the right or left branch of the proper hepatic artery, forming a plexus outside each hepatic lobule and supplying the walls of the interlobular veins and the accompanying bile ducts. <arteriae interlobulares hepatis> C

9 ***lateral segmental artery of liver*** *origin,* hepatic artery proper; *branches,* none; *distribution,* lateral segment of left liver. <arteria segmenti lateralis hepatici> C

10 **left branch of hepatic artery proper** a branch that supplies the left liver. <ramus sinister arteriae hepaticae propriae> B, C

11 ***medial segmental artery of liver*** *origin,* left branch of hepatic artery proper; *branches,* none; *distribution,* medial segment of left liver. <arteria segmenti medialis hepatici> C

12 **right branch of hepatic artery proper** the right of the two branches into which the hepatic artery proper normally divides; it supplies the right liver, and a branch, the cystic artery, supplies the gallbladder. <ramus dexter arteriae hepaticae propriae> B, C

13 ***posterior segmental artery of liver*** *origin,* right branch of hepatic artery proper; *branches,* none; *distribution,* posterior segment of right liver. <arteria segmenti posterioris hepatici> C

14 ***caudate branches of transverse part of hepatic portal vein*** the branches of the left branch of the portal vein that drain the caudate lobe (posterior part of liver). <rami lobi caudati partis transversae venae portae hepatis> C

15 **anterior cecal artery** *origin,* ileocolic; *branches,* none; *distribution,* cecum. <arteria caecalis anterior> D

16 **appendicular artery** *origin,* ileocolic artery; *branches,* none; *distribution,* vermiform appendix. Called also *vermiform artery.* <arteria appendicularis> D

17 ***appendicular lymph nodes*** lymph nodes situated along the appendicular artery and in the mesoappendix that drain into the ileocolic lymph nodes. <nodi lymphoidei appendiculares> D

18 ***appendicular vein*** the vena comitans of the appendicular artery; it opens into the ileocolic vein. <vena appendicularis> D

19 ***colic branch of ileocolic artery*** a branch that passes upward on the ascending colon and anastomoses with the right colic artery; called also *ascending ileocolic artery.* <ramus colicus arteriae ileocolicae> D

20 **ileal arteries** *origin,* superior mesenteric; *branches,* none; *distribution,* ileum. <arteriae ileales> D

21 **ileal branch of ileocolic artery** a branch that passes upward and to the left of the lower ileum and anastomoses with the end of the superior mesenteric artery. <ramus ilealis arteriae ileocolicae> D

22 **ileocolic artery** *origin,* superior mesenteric; *branches,* anterior and posterior cecal and appendicular arteries and colic (ascending) and ileal branches; *distribution,* ileum, cecum, vermiform appendix, ascending colon. Called also *inferior right colic artery.* <arteria ileocolica> D

23 **jejunal arteries** *origin,* superior mesenteric; *branches,* none; *distribution,* jejunum. <arteriae jejunales> D

24 **marginal artery of colon** a continuous vessel running along the inner perimeter of the large intestine from the ileocolic junction to the rectum, formed by branches from the superior and inferior mesenteric arteries and giving rise to straight arteries that supply the intestinal wall. Called also *marginal artery of Drummond.* <arteria marginalis coli> D

25 **middle colic artery** *origin,* superior mesenteric artery; *branches,* none; *distribution,* transverse colon. <arteria colica media> D

26 **posterior cecal artery** *origin,* ileocolic; *branches,* none; *distribution,* cecum. <arteria caecalis posterior> D

27 **right colic artery** *origin,* superior mesenteric artery; *branches,* none; *distribution,* ascending colon. <arteria colica dextra> D

28 **superior mesenteric artery** *origin,* abdominal aorta; *branches,* inferior pancreaticoduodenal, jejunal, ileal, ileocolic, right colic, and middle colic arteries; *distribution,* small intestine, proximal half of colon. <arteria mesenterica superior> D

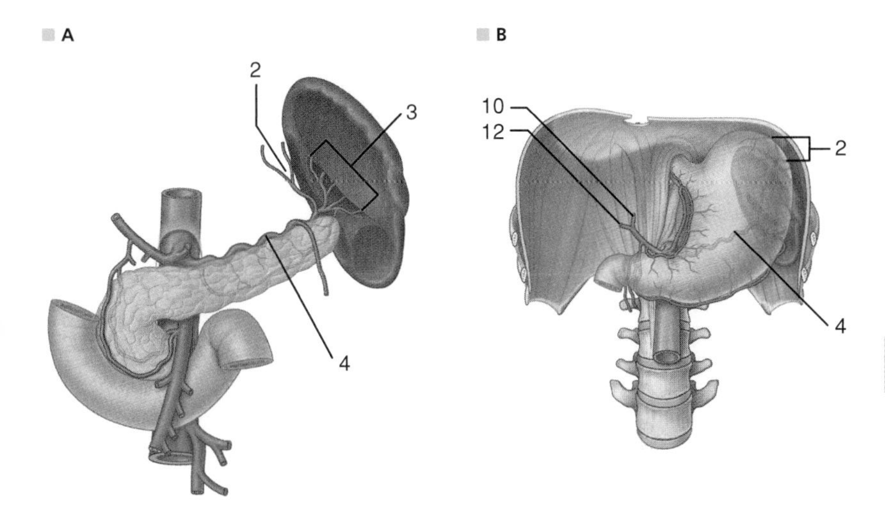
A
2
3
4
B
10
12
2
4

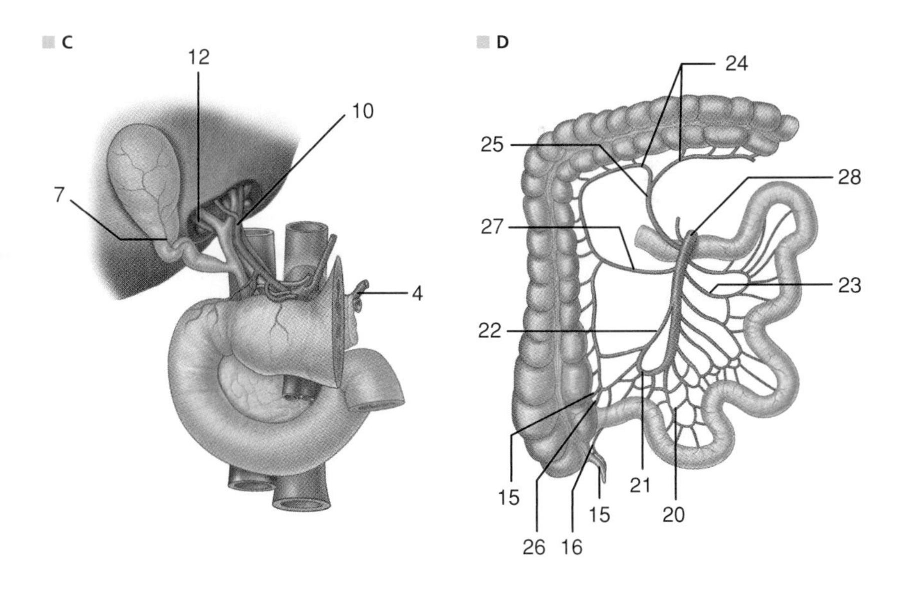
C
12
10
7
4
D
24
25
28
27
23
22
15
26
16
15
21
20

1 **left colic artery** *origin,* inferior mesenteric; *branches,* none; *distribution,* descending colon. <arteria colica sinistra> A

2 **sigmoid arteries** *origin,* inferior mesenteric artery; *branches,* none; *distribution,* sigmoid colon. <arteriae sigmoideae> A

3 **superior rectal artery** *origin,* inferior mesenteric artery; *branches,* none; *distribution,* rectum. <arteria rectalis superior> A

4 **inferior mesenteric artery** *origin,* abdominal aorta; *branches,* left colic, sigmoid, and superior rectal arteries; *distribution,* descending colon, rectum. <arteria mesenterica inferior> A

5 ***cystic vein*** a small vein that returns the blood from the gallbladder to the right branch of the portal vein, within the substance of the liver. <vena cystica> B

6 **left branch of hepatic portal vein** a branch distributed to the left liver. <ramus sinister venae portae hepatis> C

7 ***pancreatic veins*** numerous branches from the pancreas which open into the splenic and the superior mesenteric veins. <venae pancreaticae> B

8 ***pancreaticoduodenal veins*** four veins that drain blood from the pancreas and duodenum, closely following the homonymous arteries. A superior and an inferior vein originate from both an anterior and a posterior venous arcade. The anterior superior vein joins the right gastro-omental vein; the posterior superior vein joins the portal vein. The anterior and posterior inferior veins join, sometimes as one trunk and other times singly, the uppermost jejunal vein or the superior mesenteric vein. <venae pancreaticoduodenales> B

9 ***prepyloric vein*** a vein that passes upward over the anterior surface of the junction between the pylorus and the duodenum and empties into the right gastric vein. Called also *Mayo's vein.* <vena prepylorica> B

10 **right branch of hepatic portal vein** a branch distributed to the right liver. <ramus dexter venae portae hepatis> C

11 ***right gastric vein*** the vena comitans of the right gastric artery, emptying into the portal vein. <vena gastrica dextra> C

12 **inferior mesenteric vein** a vein that follows the distribution of its homonymous artery and empties into the splenic vein. <vena mesenterica inferior> B, C

13 **superior mesenteric vein** a vein that follows the distribution of its homonymous artery and joins with the splenic vein to form the hepatic portal vein. <vena mesenterica superior> B, C

14 **ileal veins** veins draining blood from the ileum into the superior mesenteric vein. <venae ileales> C

15 **ileocolic vein** a vein that follows the distribution of its homonymous artery and empties into the superior mesenteric vein. <vena ileocolica> C

16 **jejunal veins** veins draining blood from the jejunum into the superior mesenteric vein. <venae jejunales> C

17 **left colic vein** a vein that follows the left colic artery and opens into the inferior mesenteric vein. <vena colica sinistra> C

18 **left gastric vein** the vena comitans of the left gastric artery, emptying into the portal vein. <vena gastrica sinistra> C

19 **left gastro-omental vein** a vein that follows the distribution of its homonymous artery and empties into the splenic vein; called also *left epiploic vein* and *left gastroepiploic vein.* <vena gastroomentalis sinistra> C

20 **middle colic vein** a vein that follows the distribution of the middle colic artery and empties into the superior mesenteric vein. <vena colica media> C

21 **right colic vein** a vein that follows the distribution of the right colic artery and empties into the superior mesenteric vein. <vena colica dextra> C

22 **right gastro-omental vein** a vein that follows the distribution of its homonymous artery and empties into the superior mesenteric vein; called also *right epiploic vein* and *right gastroepiploic vein.* <vena gastroomentalis dextra> C

23 **short gastric veins** small vessels draining the left portion of the greater curvature of the stomach and emptying into the splenic vein. <venae gastricae breves> C

24 **sigmoid veins** veins from the sigmoid colon that empty into the inferior mesenteric vein. <venae sigmoideae> C

25 **superior rectal vein** the vein that drains the upper part of the rectal plexus into the inferior mesenteric vein and thus establishes connection between the portal system and the systemic circulation. <vena rectalis superior> C

26 ***transverse part of left branch of hepatic portal vein*** <pars transversa rami sinistri venae portae hepatis> C

27 **portal vein** a short thick trunk formed by union of the superior mesenteric and the splenic veins behind the neck of the pancreas; it passes upward to the right end of the porta hepatis, where it divides into successively smaller branches, following the branches of the hepatic artery, until it forms a capillarylike system of sinusoids that permeates the entire substance of the liver. Called also *hepatic portal vein.* <vena portae hepatis> B, C

28 **splenic vein** the vein formed by union of several branches at the hilum of the spleen, passing from left to right to the neck of the pancreas, where it joins the superior mesenteric vein to form the portal vein. <vena splenica> B, C

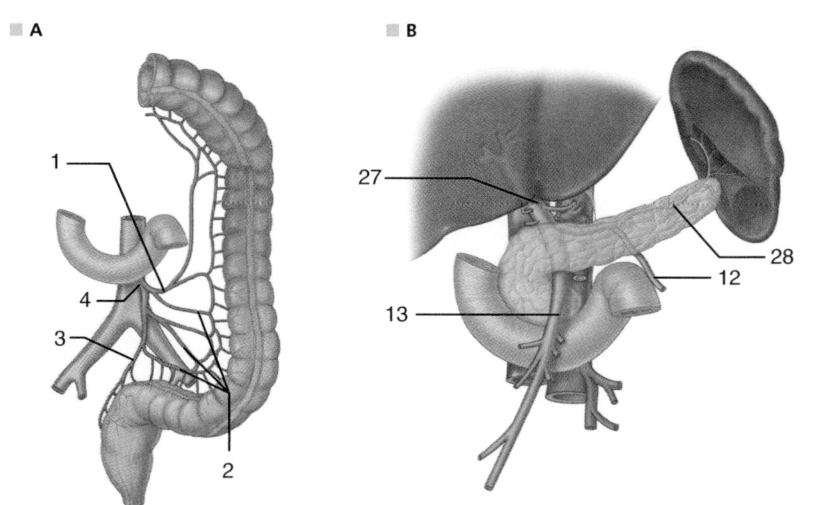
A
1
4
3
2
B
27
28
12
13

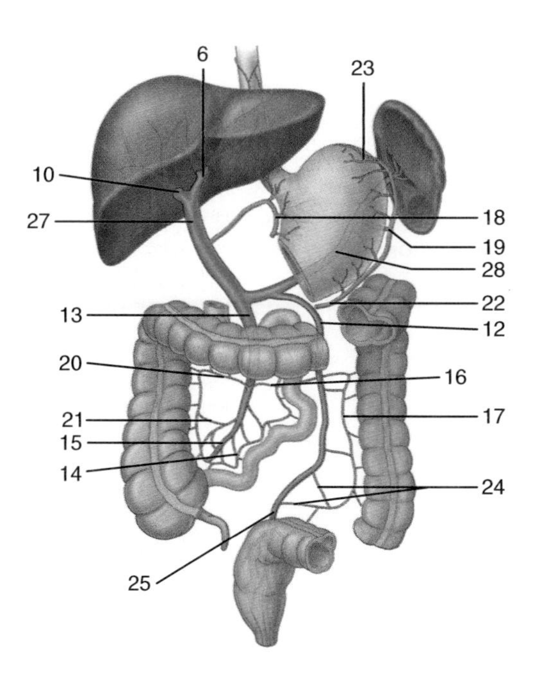
C
6
23
10
27
18
19
28
22
13
12
20
16
21
17
15
14
24
25

1 **hepatic veins** veins that receive blood from the central veins of the liver. The upper group usually consists of three large veins (left, middle, and right hepatic veins), and the lower group consists of six to twenty small veins, which come from the right and caudate lobes; all are contiguous with the hepatic tissue and valveless, and open into the inferior vena cava on the posterior aspect of the liver. <venae hepaticae> A

2 ***interlobular veins of liver*** the veins that arise as tributaries of the hepatic veins. <venae interlobulares hepatis> A

3 ***intermediate hepatic vein*** the large hepatic vein that drains the central veins in the middle part of the liver and empties into the inferior vena cava. Called also *middle hepatic vein.* <vena hepatica intermedia> A

4 **left hepatic vein** the large hepatic vein that drains the central veins in the left side of the liver and empties into the inferior vena cava. <vena hepatica sinistra> A

5 **paraumbilical veins** veins that communicate with the portal vein and anastomose with the superior and inferior epigastric and the superior vesical veins in the region of the umbilicus. They form a part of the collateral circulation of the portal vein in the event of hepatic obstruction. Called also *veins of Sappey.* <venae paraumbilicales> A

6 **right hepatic vein** the large hepatic vein that drains the central veins in the right side of the liver and empties into the inferior vena cava. <vena hepatica dextra> A

7 ***umbilical part of left branch*** the part of the left branch of the hepatic portal vein that passes from the hilum of the liver to the umbilicus. <pars umbilicalis rami sinistri venae portae hepatis> A

8 ***umbilical vein*** the vein formed from the left umbilical vein after degeneration of the right umbilical vein; it carries the blood from the placenta to the ductus venosus. <vena umbilicalis> A

POSTERIOR ABDOMINAL REGION

Muscles of Posterior Abdominal Region

9 ***iliac part of iliopsoas fascia*** the part of the fascia that invests the iliacus muscle. <pars iliaca fasciae iliopsoas> B

10 **iliacus muscle** *origin,* iliac fossa and base of sacrum; *insertion,* greater psoas tendon and lesser trochanter of femur; *innervation,* femoral; *action,* flexes thigh, trunk on limb. <musculus iliacus> B

11 ***iliopsoas fascia*** a strong fascia covering the inner surface of the iliacus and psoas muscles; it has both an iliac part and a psoas part. <fascia iliopsoas> B

12 **lumbocostal ligament** a strong fascial band that passes from the twelfth rib to the tips of the transverse processes of the first and second lumbar vertebrae. <ligamentum lumbocostale> B

13 **psoas major muscle** *origin,* lumbar vertebrae; *insertion,* lesser trochanter of femur; *innervation,* second and third lumbar; *action,* flexes thigh or trunk. <musculus psoas major> B,C

14 **psoas minor muscle** *origin,* last thoracic and first lumbar vertebrae; *insertion,* pectineal line, iliopectineal eminence, iliac fascia; *innervation,* first lumbar; *action,* flexes trunk. Absent in 40 to 50 per cent of persons. <musculus psoas minor> B

15 ***psoas part of iliopsoas fascia*** the part of the fascia that invests the psoas major muscle. <pars psoatica fasciae iliopsoas> B

16 **quadratus lumborum muscle** *origin,* iliac crest, thoracolumbar fascia; *insertion,* twelfth rib, transverse processes of four upper lumbar vertebrae; *innervation,* first and second lumbar and twelfth thoracic; *action,* flexes lumbar vertebrae laterally, fixes last rib. <musculus quadratus lumborum> B,C

17 **aortic hiatus** the opening in the diaphragm through which the aorta and thoracic duct pass. <hiatus aorticus> C

18 **caval opening** the opening in the diaphragm that transmits the inferior vena cava and some branches of the right vagus nerve. <foramen venae cavae> C

19 **central tendon of diaphragm** the cloverleaf-shaped aponeurosis, immediately below the pericardium, onto which the diaphragmatic fibers converge to insert. <centrum tendineum diaphragmatis> C

20 **costal part of diaphragm** the part of the diaphragm arising from the inner surfaces of the ribs and their cartilages. <pars costalis diaphragmatis> C

21 ***diaphragm*** the musculomembranous partition separating the abdominal and thoracic cavities, and serving as a major thoracic muscle. <diaphragma> C

22 **esophageal hiatus** the opening in the diaphragm for the passage of the esophagus and the vagus nerves. <hiatus oesophageus> C

23 **lateral arcuate ligament** the ligamentous arch, formed by the fascia of the quadratus lumborum muscle, constituting part of the lumbar portion of the diaphragm. <ligamentum arcuatum laterale> C

24 **left crus of diaphragm** a fibromuscular band arising from the superior two or three lumbar vertebrae, and ascending along with the right crus, to insert into the central tendon of the diaphragm. <crus sinistrum diaphragmatis> C

25 **lumbar part of diaphragm** the portion of the diaphragm that arises from the lumbar vertebrae, comprising the right and left diaphragmatic crura, the right crus arising from the superior three or four vertebrae, and the left from the superior two or three. <pars lumbalis diaphragmatis> C

26 ***lumbocostal triangle*** a triangular opening of variable size between the lateral arcuate ligament and the costal part of the diaphragm. <trigonum lumbocostale> C

A

B

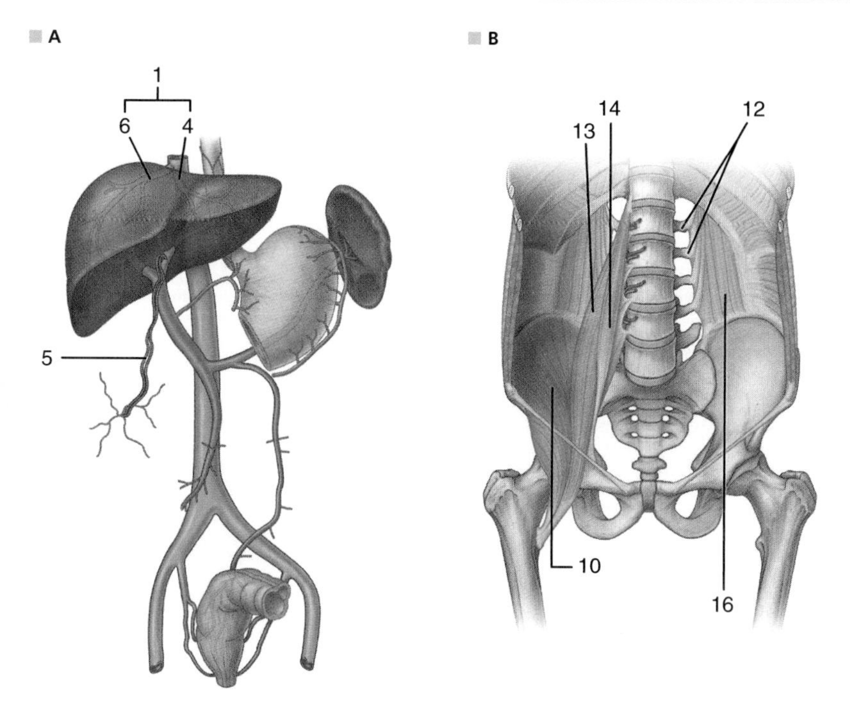

C

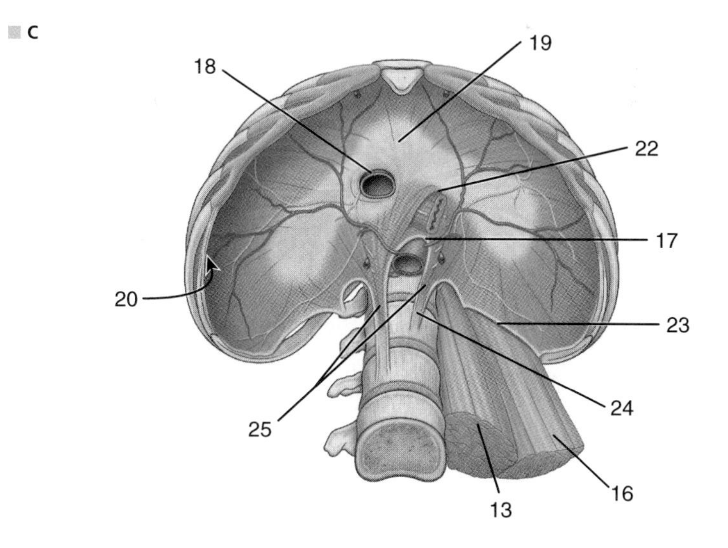

1 **medial arcuate ligament** the ligamentous arch, formed by the fascia of the psoas muscle, constituting part of the lumbar portion of the diaphragm. <ligamentum arcuatum mediale> A

2 **median arcuate ligament** the ligamentous arch across the anterior surface of the aorta, interconnecting the crura of the diaphragm. <ligamentum arcuatum medianum> A

3 **right crus of diaphragm** a fibromuscular band arising from the superior three or four lumbar vertebrae, and ascending along with the left crus, to insert into the central tendon of the diaphragm. <crus dextrum diaphragmatis> A

4 **sternal part of diaphragm** the portion of the diaphragm that arises from the inner aspect of the xiphoid process of the sternum. <pars sternalis diaphragmatis> A

5 ***sternocostal triangle*** a triangular opening between the costal and sternal parts of the diaphragm; beyond this point the internal thoracic vessels become the superior epigastric vessels. <trigonum sternocostale> A

ORGANS OF POSTERIOR ABDOMINAL REGION

Kidneys of Posterior Abdominal Region

6 **adipose capsule of kidney** the investment of fat surrounding the fibrous capsule of the kidney and continuous at the hilum with the fat in the renal sinus; called also *perinephric* or *perirenal fat.* <capsula adiposa renis> B

7 **anterior surface of kidney** the kidney surface that is directed toward the viscera and covered by peritoneum. <facies anterior renis> B

8 **fibrous capsule of kidney** the connective tissue investment of the kidney, which continues through the hilum to line the renal sinus. <capsula fibrosa renis> B

9 **lateral border of kidney** the convex narrow border of the kidney. <margo lateralis renis> B

10 **medial border of kidney** the concave border of the kidney, which contains the hilum. <margo medialis renis> B

11 **pararenal fat body** a large mass of fat lying dorsal to the renal fascia; called also *paranephric* or *pararenal fat.* <corpus adiposum pararenale> B

12 **posterior surface of kidney** the kidney surface directed toward the posterior body wall, and not covered by peritoneum. <facies posterior renis> B

13 **renal fascia** a thin membranous sheath that encloses the kidney, perirenal fat, and adrenal gland, formed by condensation of fibroareolar tissue. Some authorities distinguish between *anterior renal fascia* and *posterior renal fascia,* which join laterally to form the *lateroconal fascia.* Called also *Gerota's capsule* or *fascia, perinephric fascia,* and *perirenal fascia.* <fascia renalis> B

14 ***adventitia of renal pelvis*** the tunica adventitia of the renal pelvis. <tunica adventitia pelvis renalis> C

15 ***anterior inferior segment of kidney*** one of the renal segments (see *renal segments*). <segmentum anterius inferius renis> C

16 ***anterior superior segment of kidney*** one of the renal segments (see *renal segments*). <segmentum anterius superius renis> C

17 **cortical cortex** part of the renal cortex consisting of a narrow peripheral zone where the renal columns do not have visible renal corpuscles. <cortex corticis> C

18 ***cortical labyrinth*** a network of tubules and blood vessels in the renal cortex. <labyrinthus corticis> C

19 **cribriform area** the tip of a renal pyramid, which is perforated by 10-25 openings for the papillary ducts. <area cribrosa papillae renalis> C

20 **hilum of kidney** the indented area on the medial margin of the kidney where the vessels, nerves, and ureter enter. <hilum renale> C

21 **inferior extremity of kidney** the lower, smaller end of the kidney; called also *inferior* or *lower pole of kidney.* <extremitas inferior renis> C

22 ***inferior segment of kidney*** the renal segment located most inferiorly (see *renal segments*). <segmentum inferius renis> C

23 **kidney** either of the two organs in the lumbar region that filter the blood, excreting the end-products of body metabolism in the form of urine, and regulating the concentrations of hydrogen, sodium, potassium, phosphate, and other ions in the extracellular fluid. <ren> C

24 **major renal calices** the two or more larger subdivisions of the renal pelvis, into which the minor calices open. <calices renales majores> C

25 ***medullary rays*** cortical extensions of bundles of tubules from the renal pyramids. <radii medullares> C

26 **minor renal calices** a variable number of smaller subdivisions of the renal pelvis that enclose the renal pyramids and open into the major calices. <calices renales minores> C

27 ***mucosa of renal pelvis*** the mucous membrane (tunica mucosa) lining the renal pelvis. <tunica mucosa pelvis renalis> C

28 ***muscular layer of renal pelvis*** the muscular layer or coat of the renal pelvis. <tunica muscularis pelvis renalis> C

29 ***openings of papillary ducts*** minute openings in the summit of each renal papilla, the orifices of the collecting tubules. <foramina papillaria renis> C

30 ***posterior segment of kidney*** the renal segment located most posteriorly (see *renal segments*). <segmentum posterius renis> C

31 **renal columns** inward extensions of the cortical structure of the kidney between the renal pyramids. <columnae renales> C

32 **renal cortex** the outer part of the substance of the kidney, composed mainly of glomeruli and convoluted tubules. <cortex renalis> C

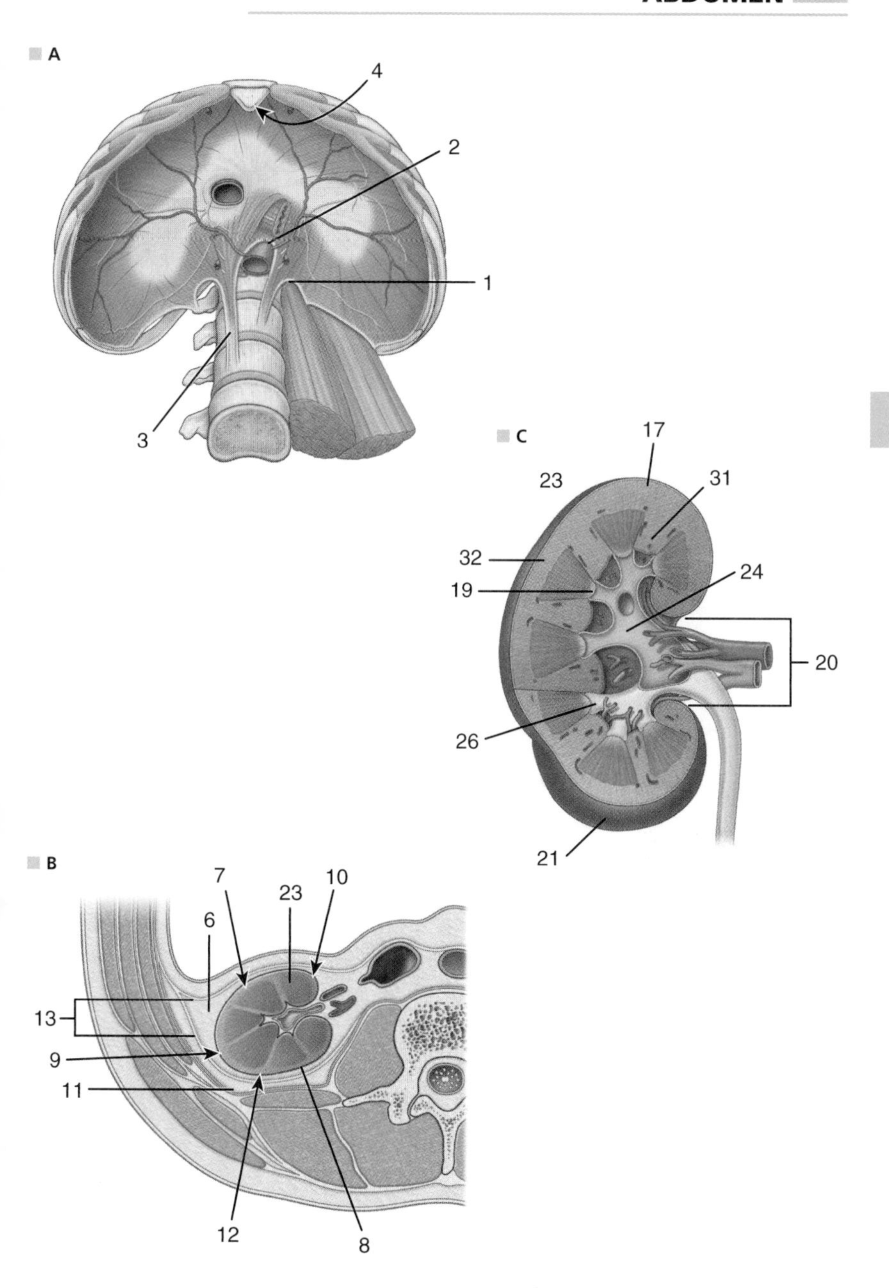
A
4
2
1
3
C
17
23
31
32
24
19
20
26
21
B
7
10
23
6
13
9
11
12
8

1 **renal lobes** the units of the kidney, each consisting of a pyramid and its surrounding cortical substance; the division of the kidney into lobes is more distinctly marked in infants than in adults. <lobi renales> A

2 **renal papilla** the blunted apex of a renal pyramid, found in the inner zone of the renal medulla and projecting into a renal sinus. <papilla renalis> A

3 **renal pelvis** the expansion from the upper end of the ureter into which the renal calices open; ordinarily lodged within the renal sinus, under certain conditions a large part of it may be outside the kidney (extrarenal pelvis). <pelvis renalis> A

4 **renal pyramids** the conical masses that make up the substance of the renal medulla; they contain collecting ducts and the straight arterioles of kidney. <pyramides renales> A

5 ***renal segments*** subdivisions of the kidney that have independent blood supply from branches of the renal artery: *superior segment, anterior superior segment, anterior inferior segment, inferior segment,* and *posterior segment.* <segmenta renalia> A

6 **renal sinus** a cavity within the substance of the kidney, occupied by the renal pelvis, calices, vessels, nerves, and fat. <sinus renalis> A

7 **superior extremity of kidney** the upper, larger end of the kidney; called also *superior* or *upper pole of kidney.* <extremitas superior renis> B

8 ***superior segment of kidney*** the renal segment located most superiorly (see *renal segments*). <segmentum superius renis> A

9 **anterior branch of renal artery** a branch supplying the anterior, superior, and inferior segments of the kidney. <ramus anterior arteriae renalis> A

10 ***anterior inferior segmental artery of kidney*** *origin,* anterior branch of renal artery; *branches,* none; *distribution,* anterior inferior segment of kidney. <arteria segmenti anterioris inferioris renalis> B

11 ***anterior superior segmental artery of kidney*** *origin,* anterior branch of renal artery; *branches,* none; *distribution,* anterior superior segment of kidney. <arteria segmenti anterioris superioris renalis> B

12 ***capsular branches of intrarenal arteries*** branches that supply the renal capsule. <rami capsulares arteriarum intrarenalium> B

13 ***capsular branches of renal artery*** branches that supply the renal capsule. Called also *capsular arteries,* and *perirenal arteries.* <rami capsulares arteriae renalis> B

14 ***inferior segmental artery of kidney*** *origin,* anterior branch of renal artery; *branches,* none; *distribution,* inferior segment of kidney. <arteria segmenti inferioris renalis> B

15 **posterior branch of renal artery** a branch supplying the posterior segment of the kidney. <ramus posterior arteriae renalis> A

16 ***posterior segmental artery of kidney*** *origin,* posterior branch of renal artery; *branches,* none; *distribution,* posterior segment of kidney. <arteria segmenti posterioris renalis> B

17 **renal artery** *origin,* abdominal aorta; *branches,* ureteral branches, inferior suprarenal artery; *distribution,* kidney, suprarenal gland, ureter. <arteria renalis> B, C

18 **renal veins** two veins, one from each kidney, that receive blood from the interlobar veins, with the left also receiving blood from the left testicular (or ovarian), left suprarenal, and (sometimes) inferior phrenic veins; they empty into the inferior vena cava at the level of the second lumbar vertebra. <venae renales> B

19 **superior segmental artery of kidney** *origin,* anterior branch of renal artery; *branches,* none; *distribution,* superior segment of kidney. <arteria segmenti superioris renalis> C

20 **abdominal part of ureter** that portion of the ureter extending from the kidney to the linea terminalis of the pelvis. <pars abdominalis ureteris> C

21 ***adventitia of ureter*** the tunica adventitia of the ureter. <tunica adventitia ureteris> C

22 ***mucosa of ureter*** the mucous membrane (tunica mucosa) lining the ureter. <tunica mucosa ureteris> C

23 ***muscular layer of ureter*** the muscular layer or coat of the ureter. <tunica muscularis ureteris> C

24 **ureter** the fibromuscular tube that conveys the urine from the kidney to the bladder. It begins with the renal pelvis and empties into the base of the bladder. It is 40 to 46 cm long and is divided into abdominal and pelvic parts. C

25 **ureteral branches of renal artery** branches that supply the upper portion of the ureter. <rami ureterici arteriae renalis> C

A

B

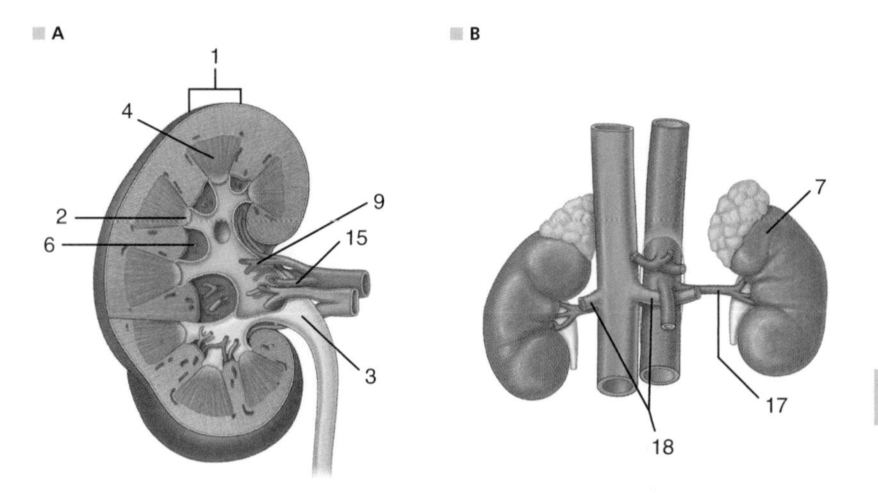

C

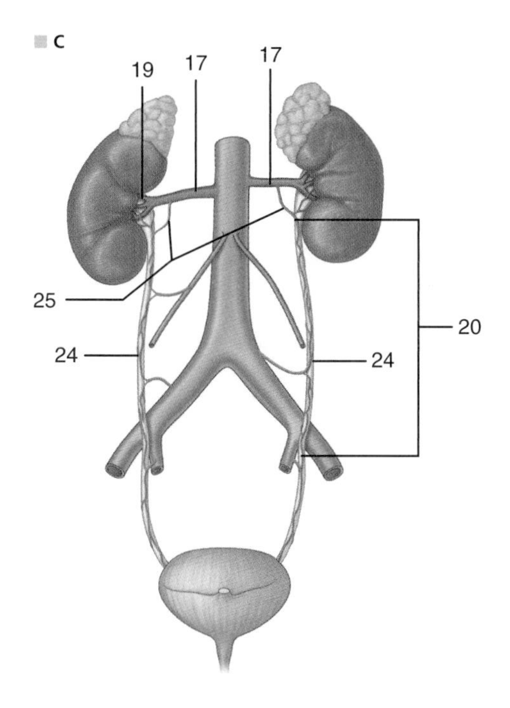

1 **afferent glomerular arteriole** a branch of an interlobular artery that goes to a renal glomerulus. <arteriola glomerularis afferens> A

2 **efferent glomerular arteriole** an arteriole that arises from a renal glomerulus and breaks up into capillaries to supply renal tubules. <arteriola glomerularis efferens> A

3 ***rete mirabile*** a vascular network formed by division of an artery or a vein into a large number of smaller vessels that subsequently reunite into a single vessel, occurring only in the arterioles that supply the glomeruli of the kidney. A

4 **arcuate arteries of kidney** *origin,* interlobar artery; *branches,* interlobular artery and straight arterioles; *distribution,* parenchyma of kidney. <arteriae arcuatae renis> B

5 **arcuate veins of kidney** a series of complete arches across the bases of the renal pyramids; they are formed by union of the interlobular veins and the straight venules and drain into the interlobar veins. <venae arcuatae renis> B

6 ***cortical radiate arteries*** arteries originating from the arcuate arteries of the kidney and distributed to the renal glomeruli. Called also *interlobular arteries of kidney.* <arteriae corticales radiatae> B

7 **inner stripe of outer zone of renal medulla** in the outer zone of the renal medulla, the part next to the inner zone and away from the cortex. <stria interna zonae externae medullae renalis> B

8 **inner zone of renal medulla** the part of the renal medulla farthest in from the cortex. <zona interna medullae renalis> B

9 **interlobar arteries of kidney** *origin,* lobar branches of segmental arteries; *branches,* arcuate arteries; *distribution,* parenchyma of kidney. <arteriae interlobares renis> B

10 **interlobar veins of kidney** veins that drain the arcuate veins, pass down between the renal pyramids, and unite to form the renal vein. <venae interlobares renis> B

11 ***interlobular veins of kidney*** veins that collect blood from the capillary network of the renal cortex and empty into the arcuate veins. <venae interlobulares renis> B

12 ***intrarenal arteries*** the arteries within the kidney, including the interlobar, arcuate, and cortical radiate (interlobular) arteries, and the straight arterioles. <arteriae intrarenales> B

13 ***intrarenal veins*** the veins within the kidney, including the interlobar, arcuate, interlobular, and stellate veins, and the straight venules. Called also *renal veins.* <venae intrarenales> B

14 **outer stripe of outer zone of renal medulla** in the outer zone of the renal medulla, the part next to the cortex. <stria externa zonae externae medullae renalis> B

15 **outer zone of renal medulla** the part of the renal medulla nearest to the cortex. It is subdivided into the *inner stripe* and the *outer stripe.* <zona externa medullae renalis> B

16 ***perforating radiate arteries*** small arteries that are continuations of the cortical radiate arteries and perforate the renal capsule. <arteriae perforantes radiatae> B

17 **renal medulla** the inner part of the substance of the kidney, organized grossly into renal pyramids. <medulla renalis> B

18 ***stellate veins of kidney*** veins on the surface of the kidney that collect blood from the superficial parts of the renal cortex and empty into the interlobular veins. <venae stellatae renis> B

19 ***straight arterioles of kidney*** branches of the arcuate arteries of the kidney arising from the efferent glomerular arterioles and passing down to the renal pyramids; called also *straight arteries of kidney.* Also sometimes *false straight arterioles* to distinguish them from straight direct branches from the arcuate and interlobular arteries that are called *true straight arterioles.* <arteriolae rectae renis> B

20 ***straight venules of kidney*** venules that drain the papillary part of the kidney and empty into the arcuate veins. <venulae rectae renis> B

A

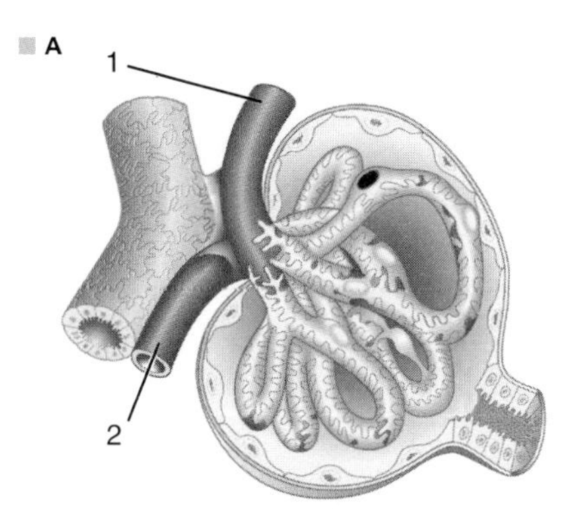

B

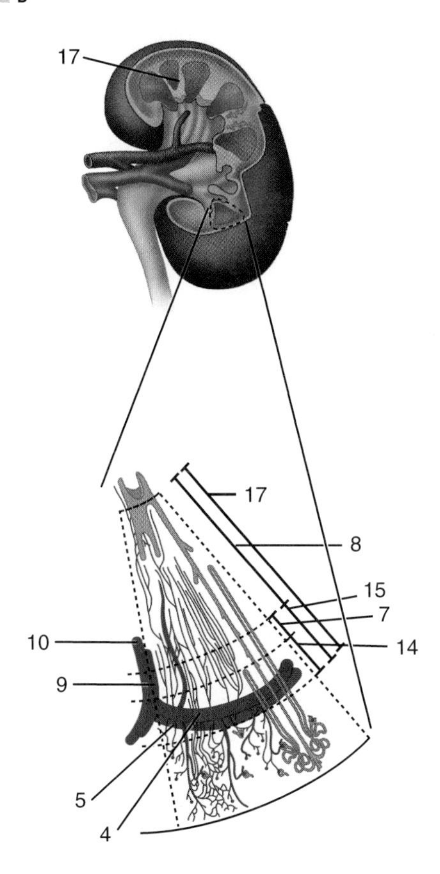

Suprarenal Glands of Posterior Abdominal Region

1 **adrenal cortex** the outer, firm yellowish layer that comprises the larger part of the adrenal gland; it secretes, in response to release of corticotropin by the pulmonary gland, many steroid hormones. Called also *suprarenal cortex.* <cortex glandulae suprarenalis> A

2 **adrenal medulla** the inner, reddish brown, soft part of the adrenal gland; it synthesizes, stores, and releases catecholamines. Called also *suprarenal medulla.* <medulla glandulae suprarenalis> A

3 **central vein of suprarenal gland** the large single vein into which the various veins within the substance of the gland empty, and which continues at the hilum as the suprarenal vein. <vena centralis glandulae suprarenalis> A

4 ***accessory suprarenal glands*** adrenal glandular tissue, usually either cortical or medullary, found in the abdomen or pelvis. Called also *adrenal* or *suprarenal rests.* <glandulae suprarenales accessoriae> B

5 **anterior surface of suprarenal gland** the anterior surface of the suprarenal gland, where the hilum is located. <facies anterior glandulae suprarenalis> B

6 **hilum of suprarenal gland** the depression on the anterior surface of the gland where the suprarenal vein enters it. <hilum glandulae suprarenalis> B

7 **inferior suprarenal artery** *origin,* renal artery; *branches,* none; *distribution,* suprarenal gland. Called also *inferior capsular artery.* <arteria suprarenalis inferior> B

8 **left suprarenal vein** the vein that returns blood from the left suprarenal gland to the left renal vein. <vena suprarenalis sinistra> B

9 **medial border of suprarenal gland** the border that with the superior border divides the anterior from the posterior surface. <margo medialis glandulae suprarenalis> B

10 **middle suprarenal artery** *origin,* abdominal aorta; *branches,* none; *distribution,* suprarenal gland. Called also *middle capsular artery.* <arteria suprarenalis media> B

11 ***posterior surface of the suprarenal gland*** the posterior surface of the suprarenal gland, which borders the peritoneum. <facies posterior glandulae suprarenalis> B

12 **renal surface of suprarenal gland** the surface directed toward the kidney, being separated from it by a layer of fat; called also *inferior margin of suprarenal gland.* <facies renalis glandulae suprarenalis> B

13 **right suprarenal vein** a vein that drains the right suprarenal gland into the inferior vena cava. <vena suprarenalis dextra> B

14 **superior border of suprarenal gland** the superior border, which with the medial border divides the anterior from the posterior surface. <margo superior glandulae suprarenalis> B

15 **superior suprarenal arteries** *origin,* inferior phrenic artery; *branches,* none; *distribution,* suprarenal gland. <arteriae suprarenales superiores> B

16 **suprarenal gland** a flattened body found in the retroperitoneal tissues at the cranial pole of the kidney. It consists of two components of different embryologic origin, the cortex and medulla. Called also *adrenal gland.* <glandula suprarenalis> B

Lymphatics of Abdomen

17 ***abdominal part of thoracic duct*** <pars abdominalis ductus thoracici> C

18 **celiac lymph nodes** a few nodes along the celiac trunk, which receive lymph from the stomach, spleen, duodenum, liver, and pancreas. <nodi lymphoidei coeliaci> C

19 ***central superior mesenteric lymph nodes*** the middle group of superior mesenteric nodes, situated along the ileal and jejunal branches of the superior mesenteric artery. <nodi lymphoidei mesenterici superiores centrales> C

20 **cisterna chyli** a dilated portion of the thoracic duct at its origin in the lumbar region; it receives several lymph-collecting vessels, including the intestinal, lumbar, and descending intercostal trunks. C

21 ***colic lymph nodes*** a subgroup of the mesocolic lymph nodes, situated along the right, middle, and left colic arteries. <nodi lymphoidei colici dextri/medii/sinistri> C

22 **common iliac lymph nodes** the four to six lymph nodes grouped at the sides and dorsal to the common iliac vessels, comprising five groups: medial, intermediate, lateral, subaortic, and promontory; they receive afferent vessels from the lateral and internal iliac lymph nodes and send efferent vessels to the lateral aortic lymph nodes. <nodi lymphoidei iliaci communes> C

23 ***cystic lymph node*** a hepatic lymph node situated in the curve of the neck of the gallbladder at the junction of the cystic and common hepatic ducts. <nodus lymphoideus cysticus> C

24 **external iliac lymph nodes** the eight to ten nodes along the external iliac vessels, comprising five groups: medial, intermediate, lateral, interiliac, and obturator lymph nodes; they receive afferent vessels from the inguinal lymph nodes, deep part of the abdominal wall below the umbilicus, and some pelvic viscera and send efferent vessels to the common iliac lymph nodes. <nodi lymphoidei iliaci externi> C

25 ***foraminal lymph node*** a hepatic lymph node situated along the upper part of the common bile duct. <nodus lymphoideus foraminalis> C

26 ***hepatic lymph nodes*** a variable number of lymph nodes situated along the common hepatic artery and hepatic artery proper and the bile ducts that receive lymph from the stomach, spleen, duodenum, liver, and pancreas; two are fairly common: the cystic node and the foraminal node. <nodi lymphoidei hepatici> C

A

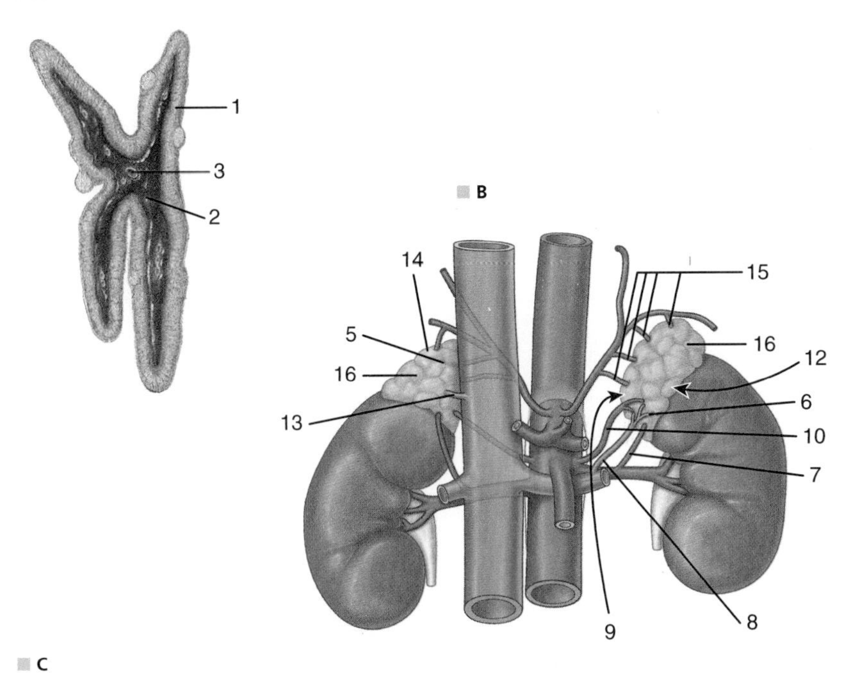

C

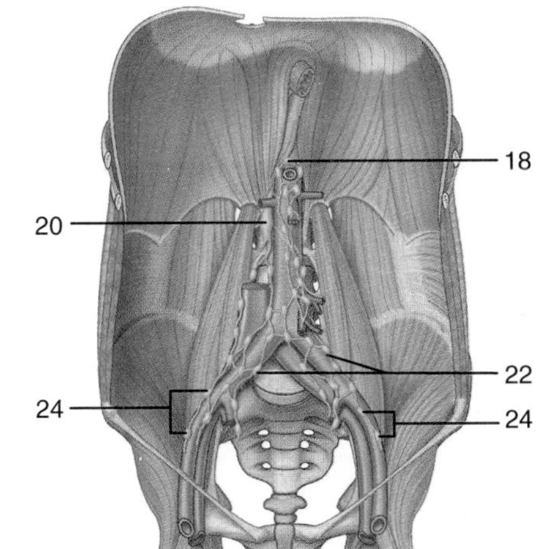

1 ***ileocolic lymph nodes*** nodes in the region of the ileocolic junction, draining adjacent structures and draining into the superior mesenteric lymph node. <nodi lymphoidei ileocolici> A

2 ***inferior epigastric lymph nodes*** lymph nodes along the deep epigastric vessels, receiving lymph from the lower abdominal wall. <nodi lymphoidei epigastrici inferiores> A

3 **inferior mesenteric lymph nodes** nodes situated along the inferior mesenteric vessels and receiving lymph from the adjacent region; they comprise two groups: the sigmoid and superior rectal lymph nodes. <nodi lymphoidei mesenterici inferiores> A

4 ***inferior pancreatic lymph nodes*** lymph nodes associated with the inferior pancreatic artery. <nodi lymphoidei pancreatici inferiores> A

5 ***inferior pancreaticoduodenal lymph nodes*** lymph nodes situated along the inferior pancreaticoduodenal artery. <nodi lymphoidei pancreaticoduodenales inferiores> A

6 ***inferior phrenic lymph nodes*** parietal lymph nodes accompanying the inferior vessels of the diaphragm. <nodi lymphoidei phrenici inferiores> A

7 ***interiliac lymph nodes*** the external iliac lymph nodes situated between the external and internal iliac vessels and the obturator artery. <nodi lymphoidei interiliaci> A

8 ***intermediate common iliac lymph nodes*** the common iliac lymph nodes situated between the common iliac vessels. <nodi lymphoidei iliaci communes intermedii> A

9 ***intermediate external iliac lymph nodes*** the external iliac lymph nodes situated between the external iliac vessels. <nodi lymphoidei iliaci externi intermedii> A

10 ***intermediate lacunar lymph node*** a lymph node situated between the external iliac vessels at the lacuna vasorum. <nodus lymphoideus lacunaris intermedius> A

11 ***intermediate lumbar lymph nodes*** the chain of lumbar lymph nodes that lie in the median plane, between the left and right lumbar lymph nodes. <nodi lymphoidei lumbales intermedii> A

12 **intestinal trunks** short lymphatic trunks that leave the gastrointestinal tract and participate in formation of the thoracic duct. <trunci intestinales> A

13 ***juxtaintestinal mesenteric lymph nodes*** the mesenteric lymph nodes situated close to the wall of the intestine between the branches of the jejunal and ileal arteries; they drain into the superior mesenteric lymph node. <nodi lymphoidei mesenterici juxtaintestinales> A

14 **lateral aortic lymph nodes** two chains (right and left) of the left lumbar group that are on the left side of the aorta and drain the suprarenal glands, kidneys, ureters, testes, ovaries, pelvic viscera (except the intestines), and posterior abdominal wall. <nodi lymphoidei aortici laterales> A

15 ***lateral caval lymph nodes*** a group of lymph nodes of the right lumbar group that are on the right side of the inferior vena cava. <nodi lymphoidei cavales laterales> A

16 **lateral common iliac lymph nodes** the common iliac lymph nodes situated on the lateral aspect of the common iliac vessels. <nodi lymphoidei iliaci communes laterales> A

17 **lateral external iliac lymph nodes** the external iliac lymph nodes situated on the lateral aspect of the external iliac vessels. <nodi lymphoidei iliaci externi laterales> A

18 ***lateral lacunar lymph node*** a lymph node situated on the lateral aspect of the external iliac vessels at the lacuna vasorum. <nodus lymphoideus lacunaris lateralis> A

19 ***lateral vesicular lymph nodes*** the paravesicular lymph nodes situated in relation to the lateral umbilical ligament. <nodi lymphoidei vesicales laterales> A

20 **left lumbar lymph nodes** the chain of lumbar lymph nodes situated at the side of the abdominal aorta on the psoas major muscle, comprising three groups: right and left lateral aortic, preaortic, and retroaortic (postaortic) lymph nodes. <nodi lymphoidei lumbales sinistri> A

21 **lumbar trunk** either of the two lymphatic trunks (right and left) that drain lymph upward from the lumbar lymph nodes and help form the thoracic duct. <truncus lumbalis> A

22 **medial common iliac lymph nodes** the common iliac lymph nodes situated on the medial aspect of the common iliac vessels. <nodi lymphoidei iliaci communes mediales> A

23 **medial external iliac lymph nodes** the external iliac lymph nodes situated on the medial aspect of the external iliac vessels. <nodi lymphoidei iliaci externi mediales> A

24 ***medial lacunar lymph node*** a lymph node situated on the medial aspect of the external iliac vessels at the lacuna vasorum. <nodus lymphoideus lacunaris medialis> A

25 ***mesocolic lymph nodes*** lymph nodes situated in the mesocolon, comprising two groups: paracolic and colic (right, middle, and left colic); they drain through the superior mesenteric lymph node. <nodi lymphoidei mesocolici> A

26 ***pancreatic lymph nodes*** nodes found along the pancreatic arteries that drain lymph from the pancreas to the pancreaticosplenic lymph nodes. <nodi lymphoidei pancreatici> A

27 ***paracolic lymph nodes*** a subgroup of the mesocolic lymph nodes, situated along the medial borders of the ascending and descending colon and along the mesenteric borders of the transverse and sigmoid colon. <nodi lymphoidei paracolici> A

A

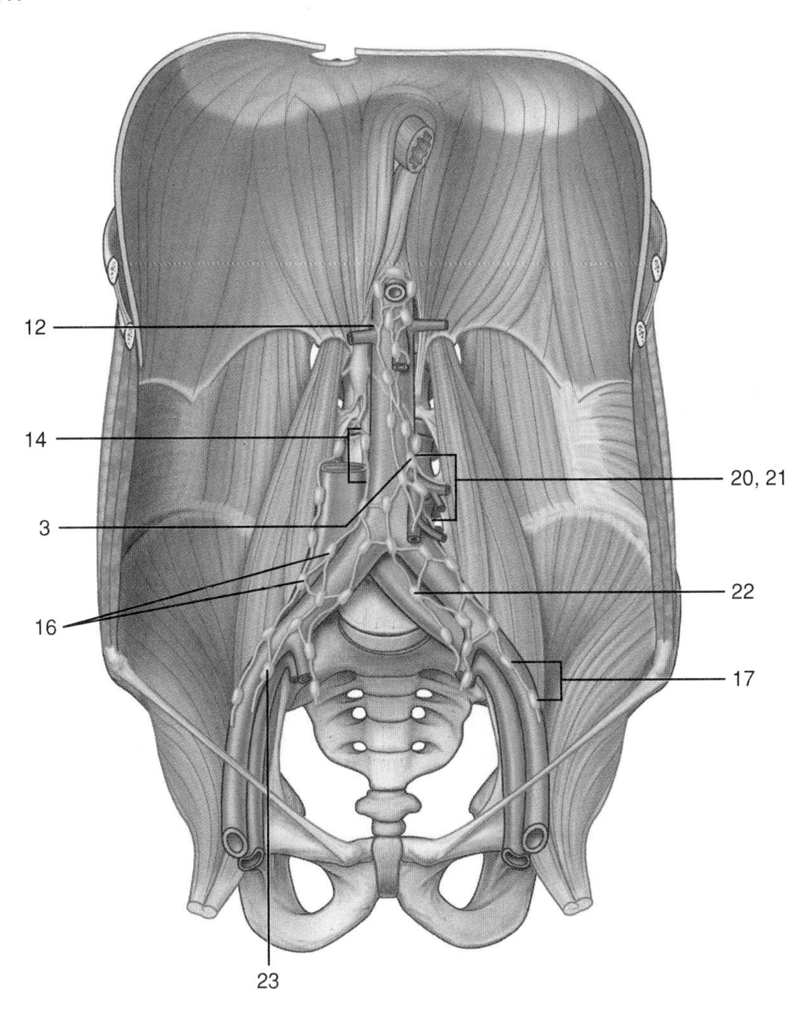

1 ***parietal abdominal lymph nodes*** the lymph nodes that drain the abdominal walls, comprising the left, intermediate, and lumbar lymph nodes, inferior phrenic lymph nodes, and inferior epigastric lymph nodes. <nodi lymphoidei abdominis parietales> A

2 ***parietal lymph nodes*** lymph nodes that receive lymph from the walls of a body cavity. <nodi lymphoidei parietales> A

3 **preaortic lymph nodes** a group of lymph nodes of the left lumbar group that is in front of the aorta and drains the abdominal part of the alimentary canal and its derivatives. <nodi lymphoidei preaortici> A

4 ***precaval lymph nodes*** a group of lymph nodes of the right lumbar group that is in front of the inferior vena cava. <nodi lymphoidei precavales> A

5 ***prececal lymph nodes*** lymph nodes situated in front of the cecum that drain into the anterior ileocolic lymph nodes. <nodi lymphoidei precaecales> A

6 ***promontorial common iliac lymph nodes*** the common iliac lymph nodes situated in front of the sacral promontory. <nodi lymphoidei iliaci communes promontorii> A

7 ***pyloric lymph nodes*** lymph nodes found anterior to the head of the pancreas, receiving lymph from the pyloric part of the stomach. They are subdivided into three groups: suprapyloric, subpyloric, and retropyloric nodes. <nodi lymphoidei pylorici> A

8 ***retroaortic lymph nodes*** a group of lymph nodes of the left lumbar group, situated behind the aorta and formed by peripheral nodes of the right and left lateral aortic lymph nodes. Called also postaortic lymph nodes. <nodi lymphoidei retroaortici> A

9 ***retrocaval lymph nodes*** a group of lymph nodes of the right lumbar group situated behind the inferior vena cava. Called also *postcaval lymph nodes.* <nodi lymphoidei retrocavales> A

10 ***retrocecal lymph nodes*** lymph nodes situated in back of the cecum that drain into the posterior ileocecal lymph nodes. <nodi lymphoidei retrocaecales> A

11 ***retropyloric lymph nodes*** pyloric lymph nodes situated posterior to the pylorus. <nodi lymphoidei retropylorici> A

12 **right lumbar lymph nodes** the chain of lumbar lymph nodes situated partly in front of the vena cava and partly behind it on the psoas major muscle, comprising three groups: lateral caval, precaval, and retrocaval (postcaval) lymph nodes. <nodi lymphoidei lumbales dextri> A

13 ***right/left gastric lymph nodes*** a few nodes along the right and left gastric arteries that receive lymph from the stomach, spleen, duodenum, liver, and pancreas. <nodi lymphoidei gastrici dextri/sinistri> A

14 ***right/left gastro-omental lymph nodes*** lymph nodes situated in the greater omentum along the pyloric half of the greater curvature of the stomach in association with the right and left gastro-omental arteries. <nodi lymphoidei gastroomentales dextri/sinistri> A

15 ***sigmoid lymph nodes*** a group of lymph nodes of the inferior mesenteric group, situated along the sigmoid arteries. <nodi lymphoidei sigmoidei> A

16 ***solitary lymphatic or lymphoid nodules*** small concentrations of lymphoid tissue scattered throughout the mucosa and submucosa of the small and large intestines. <noduli lymphoidei solitarii> A

17 ***splenic lymph nodes*** lymph nodes in the capsule and larger trabeculae of the spleen that drain into adjacent lymph nodes. <nodi lymphoidei splenici> A

18 ***splenic lymphoid nodules*** aggregations of lymphatic tissue that ensheath the arteries in the spleen. Called also malpighian bodies or corpuscles of spleen and *white pulp.* <noduli lymphoidei splenici> A

19 ***subaortic common iliac lymph nodes*** the common iliac lymph nodes situated below the bifurcation of the aorta. <nodi lymphoidei iliaci communes subaortici> A

20 ***subpyloric lymph nodes*** pyloric lymph nodes located inferior to the pylorus. <nodi lymphoidei subpylorici> A

21 **superior mesenteric lymph nodes** mesenteric lymph nodes situated along the superior mesenteric artery and draining various other groups of nodes in the region. <nodi lymphoidei mesenterici superiores> A

22 ***superior pancreatic lymph nodes*** lymph nodes associated with the superior pancreatic artery. <nodi lymphoidei pancreatici superiores> A

23 ***superior pancreaticoduodenal lymph nodes*** lymph nodes situated along the superior pancreaticoduodenal artery. <nodi lymphoidei pancreaticoduodenales superiores> A

24 ***superior phrenic lymph nodes*** several nodes on the thoracic surface of the diaphragm, receiving lymph from the intercostal spaces, pericardium, diaphragm, and liver; called also *diaphragmatic lymph nodes.* <nodi lymphoidei phrenici superiores> A

25 ***superior rectal lymph nodes*** a group of lymph nodes of the inferior mesenteric group, situated along the superior rectal artery. <nodi lymphoidei rectales superiores> A

26 ***suprapyloric lymph node*** a pyloric lymph node located superior to the duodenum on the right gastric artery. <nodus lymphoideus suprapyloricus> A

A

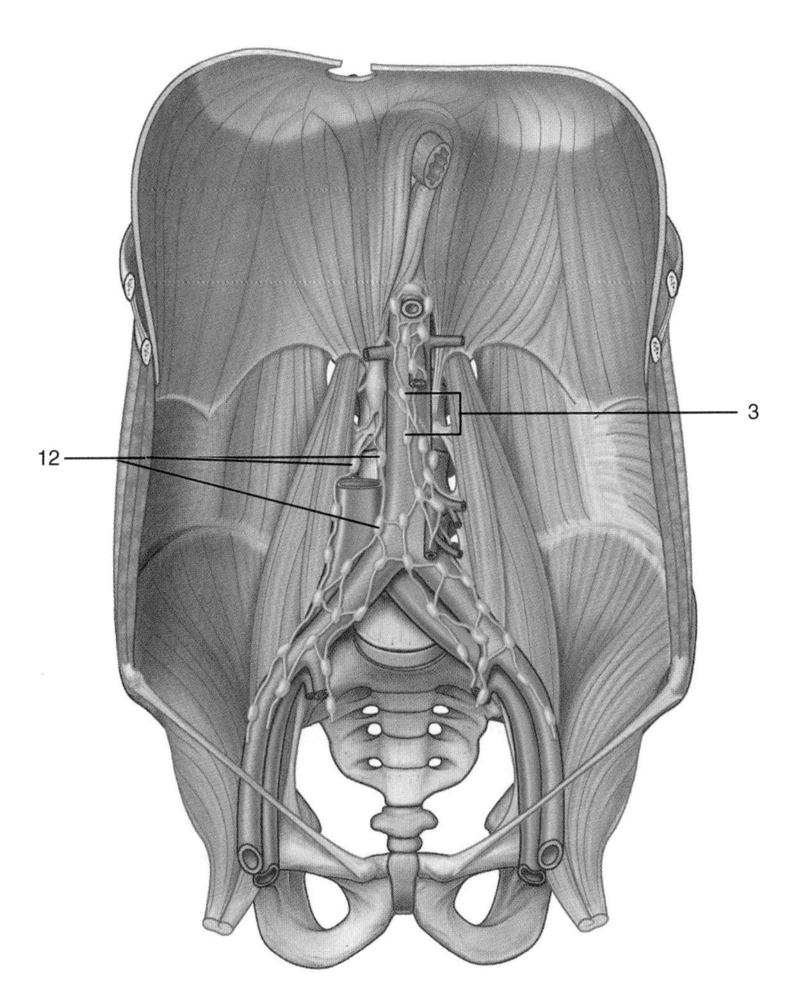

Nerves of Posterior Abdominal Region

1 **celiac branches of vagus nerve** branches that arise from both the anterior and posterior vagal trunks and join the celiac plexus; modality, parasympathetic and visceral afferent. Called also *celiac nerves.* <rami coeliaci nervi vagi> A

2 **abdominal aortic plexus** an unpaired plexus composed of interconnecting bundles of fibers that arise from the celiac and superior mesenteric plexuses and descend along the aorta. Receiving branches from the lumbar splanchnic nerves, it becomes the superior hypogastric plexus below the bifurcation of the aorta. Branches of the plexus are distributed along the adjacent branches of the aorta. <plexus aorticus abdominalis> B

3 **celiac plexus** a prevertebral plexus that lies on the front and sides of the aorta at the origins of the celiac trunk and superior mesenteric and renal arteries. It contains the paired celiac ganglia, the superior mesenteric ganglion (or ganglia), and small unnamed ganglionic masses. Branches of the plexus extend along all of the adjacent arteries. Called also *solar plexus.* <plexus coeliacus> B

4 ***iliac plexus*** any of the plexuses derived chiefly from the aortic plexus and accompanying the common iliac arteries. <plexus iliacus> B

5 **intermesenteric plexus** the part of the aortic plexus that is located between the origins of the superior and inferior mesenteric arteries. <plexus intermesentericus> B

6 **lumbar splanchnic nerves** *origin,* lumbar ganglia or sympathetic trunk; *distribution,* upper nerves join celiac and adjacent plexuses, middle ones go to intermesenteric and adjacent plexuses, and lower ones descend to superior hypogastric plexus; *modality,* preganglionic sympathetic and visceral afferent. <nervi splanchnici lumbales> B

7 ***para-aortic bodies*** exclaves of glandular cells of sympathetic origin (chromaffin cells) found near the sympathetic ganglia along the aorta in the abdominal cavity; they serve as chemoreceptors responsive to oxygen, carbon dioxide, and hydrogen ion concentration that help to control respiration. <corpora paraaortica> B

8 ***renal branch of lesser splanchnic nerve*** a branch from the lesser splanchnic nerve to the aorticorenal ganglion; *modality,* sympathetic preganglionic fibers and visceral afferent. <ramus renalis nervi splanchnici minoris> B

9 ***renal branches of vagus nerve*** branches passing from the vagal trunks via the celiac plexus to the kidney; *modality,* parasympathetic and visceral afferent. <rami renales nervi vagi> B

10 **renal plexus** a subdivision of the celiac plexus, accompanying the renal artery. <plexus renalis> B

11 ***splenic plexus*** a subdivision of the celiac plexus, which accompanies the splenic artery. <plexus splenicus> B

12 **superior hypogastric plexus** the downward continuation of the abdominal aortic plexus; it lies in front of the upper part of the sacrum, just below the bifurcation of the aorta, receives fibers from the lower lumbar splanchnic nerves, and divides into the right and left hypogastric nerves. Called also *presacral nerve.* <plexus hypogastricus superior> B

13 **sympathetic trunk** two long nerve strands, one on each side of the vertebral column, extending from the base of the skull to the coccyx. Interconnected by nerve strands, each has cervical, thoracic, lumbar, and sacral sympathetic ganglia. These receive preganglionic fibers from thoracic and upper lumbar anterior roots by way of rami communicantes, send postganglionic fibers to anterior roots by rami communicantes, and give branches to prevertebral plexuses and adjacent viscera and blood vessels. <truncus sympathicus> B

14 ***testicular plexus*** a subdivision of the aortic plexus accompanying the testicular arteries. <plexus testicularis> B

15 ***ureteric plexus*** a plexus supplying the ureter and derived from the renal and hypogastric plexuses. <plexus uretericus> B

16 **genitofemoral nerve** *origin,* lumbar plexus (L1-L2); *branches,* genital and femoral branches; *modality,* general sensory and motor. <nervus genitofemoralis> C

17 ***lumbar plexus*** a plexus formed by the anterior branches of the second to fifth lumbar nerves in the psoas major muscle (the branches of the first lumbar nerve often are included). The lower division of the fourth lumbar nerve joins the fifth, and the lumbosacral trunk thus formed becomes part of the sacral plexus. The branches of the first lumbar nerve are the ilioinguinal and iliohypogastric nerves; branches of the plexus proper are the genitofemoral, lateral femoral cutaneous, obturator, and femoral nerves. <plexus lumbalis> C

18 **lumbosacral trunk** a trunk formed by union of the lower division of the anterior branch of the fourth lumbar nerve with the anterior branch of the fifth lumbar nerve; it descends to the sacral plexus. <truncus lumbosacralis> C

19 **subcostal nerve** *origin,* anterior ramus of twelfth thoracic nerve; *distribution,* skin of lower abdomen and lateral side of gluteal region, parts of transversus, oblique, and rectus muscles, and usually the pyramidalis muscle, and adjacent peritoneum; *modality,* general sensory and motor. <nervus subcostalis> C

Vessels of Posterior Abdominal Region

20 **ureteral branches of testicular artery** branches distributed to the ureter. <rami ureterici arteriae testicularis> D

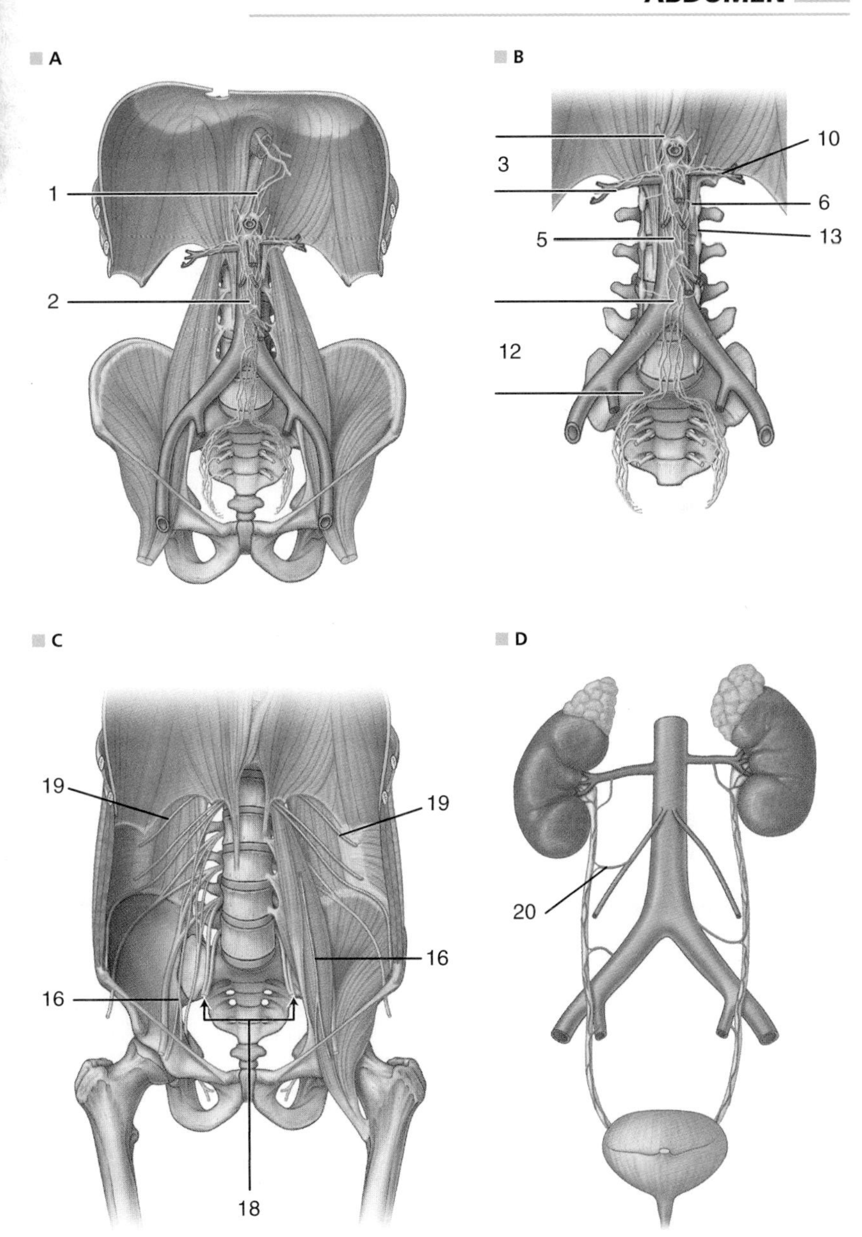
A
1
2
B
3
10
6
5
13
12
C
19
19
16
16
18
D
20

1 **abdominal aorta** the distal part of the descending aorta, which is the continuation of the thoracic part and gives rise to the inferior phrenic, lumbar, median sacral, superior and inferior mesenteric, middle suprarenal, renal, and testicular or ovarian arteries, and celiac trunk. <pars abdominalis aortae> A

2 **arteriae lumbales imae** *origin,* median sacral; *branches,* none; *distribution,* sacrum, gluteus maximus muscle. Called also *fifth lumbar arteries.* A

3 **bifurcation of aorta** the site on the left side of the body of the fourth lumbar vertebra, where the abdominal aorta divides into the right and left common iliac arteries. <bifurcatio aortae> A

4 **common iliac artery** *origin,* abdominal aorta; *branches,* internal and external iliac; *distribution,* pelvis, abdominal wall, lower limb. <arteria iliaca communis> A

5 **external iliac artery** *origin,* common iliac; *branches,* inferior epigastric, deep circumflex iliac; *distribution,* abdominal wall, external genitalia, lower limb. <arteria iliaca externa> A

6 **inferior phrenic artery** *origin,* abdominal aorta; *branches,* superior suprarenal; *distribution,* diaphragm, suprarenal gland. Called also *great phrenic artery.* <arteria phrenica inferior> A

7 **internal iliac artery** *origin,* continuation of common iliac; *branches,* iliolumbar, obturator, superior gluteal, inferior gluteal, umbilical, inferior vesical, uterine, middle rectal, and internal pudendal arteries; *distribution,* wall and viscera of pelvis, buttock, reproductive organs, medial aspect of thigh. Called also *hypogastric artery.* <arteria iliaca interna> A

8 **lumbar arteries** *origin,* abdominal aorta; *branches,* dorsal and spinal branches; *distribution,* posterior abdominal wall, renal capsule. <arteriae lumbales> A

9 **ovarian artery** *origin,* abdominal aorta; *branches,* ureteral, tubal; *distribution,* ureter, ovary, uterine tube. Called also *tubo-ovarian artery.* <arteria ovarica> A

10 ***superior phrenic arteries*** *origin,* thoracic aorta; *branches,* none; *distribution,* upper surface of vertebral portion of diaphragm. <arteriae phrenicae superiores> A

11 **testicular artery** *origin,* abdominal aorta; *branches,* ureteral, epididymal; *distribution,* ureter, epididymis, testis. <arteria testicularis> A

12 **external iliac vein** the continuation of the femoral vein from the inguinal ligament to the sacroiliac joint, where it joins with the internal iliac vein to form the common iliac vein. <vena iliaca externa> B

13 **inferior phrenic veins** veins that follow the homonymous arteries, the one on the right entering the inferior vena cava, and the one on the left entering the left suprarenal or renal vein or the inferior vena cava. <venae phrenicae inferiores> B

14 **inferior vena cava** the venous trunk for the lower extremities and for the pelvic and abdominal viscera; it begins at the level of the fifth lumbar vertebra by union of the common iliac veins, passes upward on the right of the aorta, and empties into the right atrium of the heart. <vena cava inferior> B, C

15 **internal iliac vein** a short trunk formed by union of parietal branches; it extends from the greater sciatic notch to the brim of the pelvis, where it joins the external iliac vein to form the common iliac vein. <vena iliaca interna> B

16 **left ovarian vein** a vein that drains the left pampiniform plexus of the broad ligament and empties into the left renal vein. <vena ovarica sinistra> B

17 **left testicular vein** a vein that drains the left pampiniform plexus and empties into the left renal vein. <vena testicularis sinistra> B

18 **right ovarian vein** a vein that drains the right pampiniform plexus of the broad ligament and empties into the inferior vena cava. <vena ovarica dextra> B

19 **right testicular vein** a vein that drains the right pampiniform plexus and empties into the inferior vena cava. <vena testicularis dextra> B

20 ***superior phrenic veins*** small veins on the superior surface of the diaphragm that drain into the azygos and hemiazygos veins. <venae phrenicae superiores> B

21 **ascending lumbar vein** an ascending intercepting vein for the lumbar veins of either side; it begins in the lateral sacral veins and passes up the spine to the first lumbar vertebra, where by union with the subcostal vein it becomes on the right side the azygos vein, and on the left side, the hemiazygos vein. <vena lumbalis ascendens> C

22 **common iliac vein** a vein that arises at the sacroiliac joint by union of the external iliac and internal iliac veins, and passes upward to the right side of the fifth lumbar vertebra where it unites with its fellow of the opposite side to form the inferior vena cava. <vena iliaca communis> B, C

23 **lumbar veins** the veins, four or five on each side, that accompany the corresponding lumbar arteries and drain the posterior wall of the abdomen, vertebral canal, spinal cord, and meninges; the first four usually end in the inferior vena cava, although the first may end in the ascending lumbar vein; the fifth is a tributary of the iliolumbar or of the common iliac vein; and all are generally united by the ascending iliac vein. <venae lumbales> C

24 **subcostal vein** the vena comitans of the subcostal artery on the left or right side; it joins the ascending lumbar vein to form the azygos vein on the right or the hemiazygos vein on the left. <vena subcostalis> C

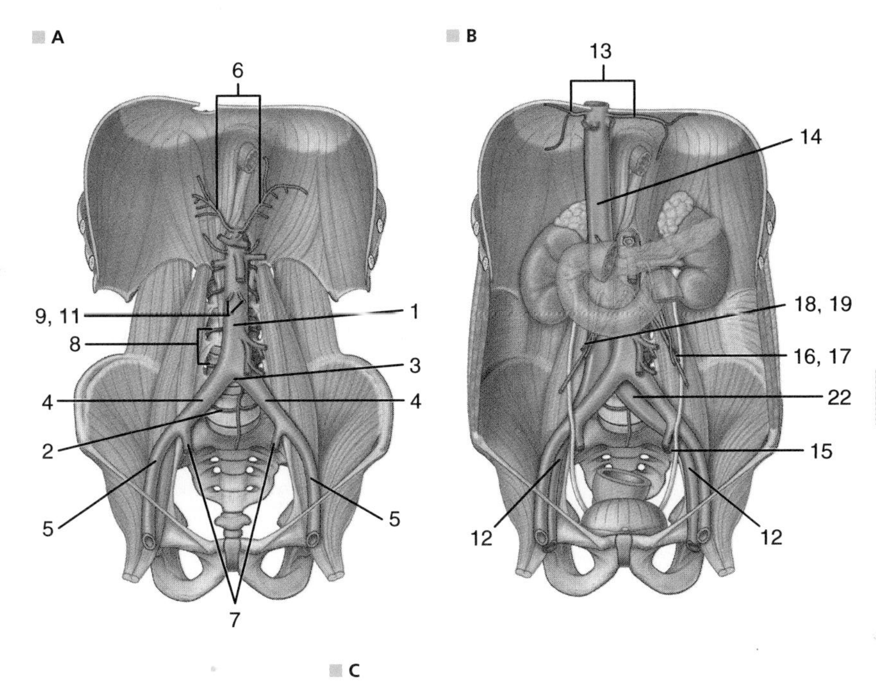
A
6
9, 11
1
8
3
4
4
2
5
5
7
B
13
14
18, 19
16, 17
22
15
12
12

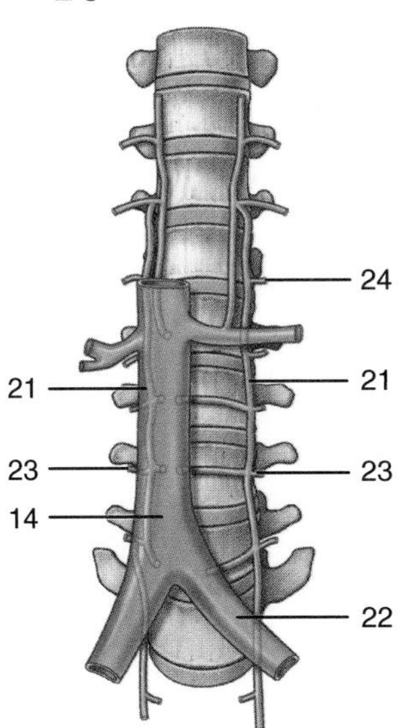
C
24
21
21
23
23
14
22

PELVIS AND PERINEUM

GENERAL

1 **pelvis** the inferior portion of the trunk of the body, bounded anteriorly and laterally by the two hip bones and posteriorly by the sacrum and coccyx. The pelvis is divided by a plane passing through the terminal lines into the *pelvis major* superiorly and the *pelvis minor* inferiorly. The superior boundary of the pelvic cavity is the *inlet*, and the inferior boundary of the pelvis minor is the *outlet*. The outlet is closed by the coccygeus and levator ani muscles and the perineal fascia, which form the *floor of the pelvis*. The inlet and outlet each have three important diameters: an anteroposterior (conjugate), an oblique, and a transverse, the relations of which determine types variously classified by different authors. A

2 **anatomical conjugate of pelvis** the anteroposterior diameter of the pelvic inlet, measured from the superior margin of the pubic symphysis to the sacrovertebral angle; called also anteroposterior or conjugate diameter of pelvis; anatomical, internal, or true conjugate diameter; and anatomical, internal, or true conjugate. <conjugata anatomica pelvis> B

3 **diagonal conjugate of pelvis** a diameter of the pelvic inlet; the distance from the posterior surface of the pubis to the tip of the sacral promontory. Called also *diagonal conjugate diameter.* <conjugata diagonalis pelvis> C

4 ***external conjugate of pelvis*** the distance from the depression under the last lumbar spine to the upper margin of the pubis; called also *external conjugate diameter.* <conjugata externa pelvis> B

5 **interspinous distance** the greatest width between the anterior superior iliac spines; called also *interspinous diameter.* <distantia interspinosa> B

6 **oblique diameter of pelvis** the oblique diameter across the pelvic inlet, measured from one sacroiliac joint to the iliopubic eminence of the other side. Designated right or left depending on which sacroiliac joint is used for reference; the left is uniformly 0.5 cm shorter than the right. <diameter obliqua pelvis> B

7 **transverse diameter of pelvis** the greatest distance from side to side across the pelvic inlet. <diameter transversa pelvis> B

8 **pelvic inclination** the angle between the plane of the superior aperture of the minor pelvis and the horizontal plane, when the body is in the erect position; called also *pelvic incline.* <inclinatio pelvis> C

9 **axis of pelvis** an imaginary curved line through the pelvis minor at right angles to the plane of the superior aperture, the plane of the cavity, and the plane of the inferior aperture at their central points. <axis pelvis> C

10 **coccygeal foveola** a dermal pit near the tip of the coccyx, indicative of the site of attachment of the embryonic neural tube to the skin; called also *postanal dimple* or *pit.* <foveola coccygea> D

11 **pelvic cavity** the space within the walls of the pelvis, forming the inferior and lesser part of the abdominopelvic cavity. <cavitas pelvis> A

12 **greater pelvis** the part of the pelvis superior to a plane passing through the iliopectineal lines. Called also *false pelvis.* A

13 **lesser pelvis** the part of the pelvis inferior to a plane passing through the iliopectineal lines. Called also *true pelvis.* A

14 **perineum** the *area* inferior to the pelvic diaphragm (pelvic floor); it is bounded anteriorly by the pubic symphysis, laterally by the ischial tuberosities, and posteriorly by the coccyx. A

15 **greater sciatic foramen** a hole converted from the major sciatic notch by the sacrotuberal and sacrospinal ligaments. <foramen ischiadicum majus> E

16 **lesser sciatic foramen** a hole converted from the minor sciatic notch by the sacrotuberal and sacrospinal ligaments. <foramen ischiadicum minus> E

17 **obturator canal** an opening within the obturator membrane for the passage of the obturator vessels and nerve; its boundaries are the edge of the obturator membrane, together with the obturator groove of the pubic bone. <canalis obturatorius> E

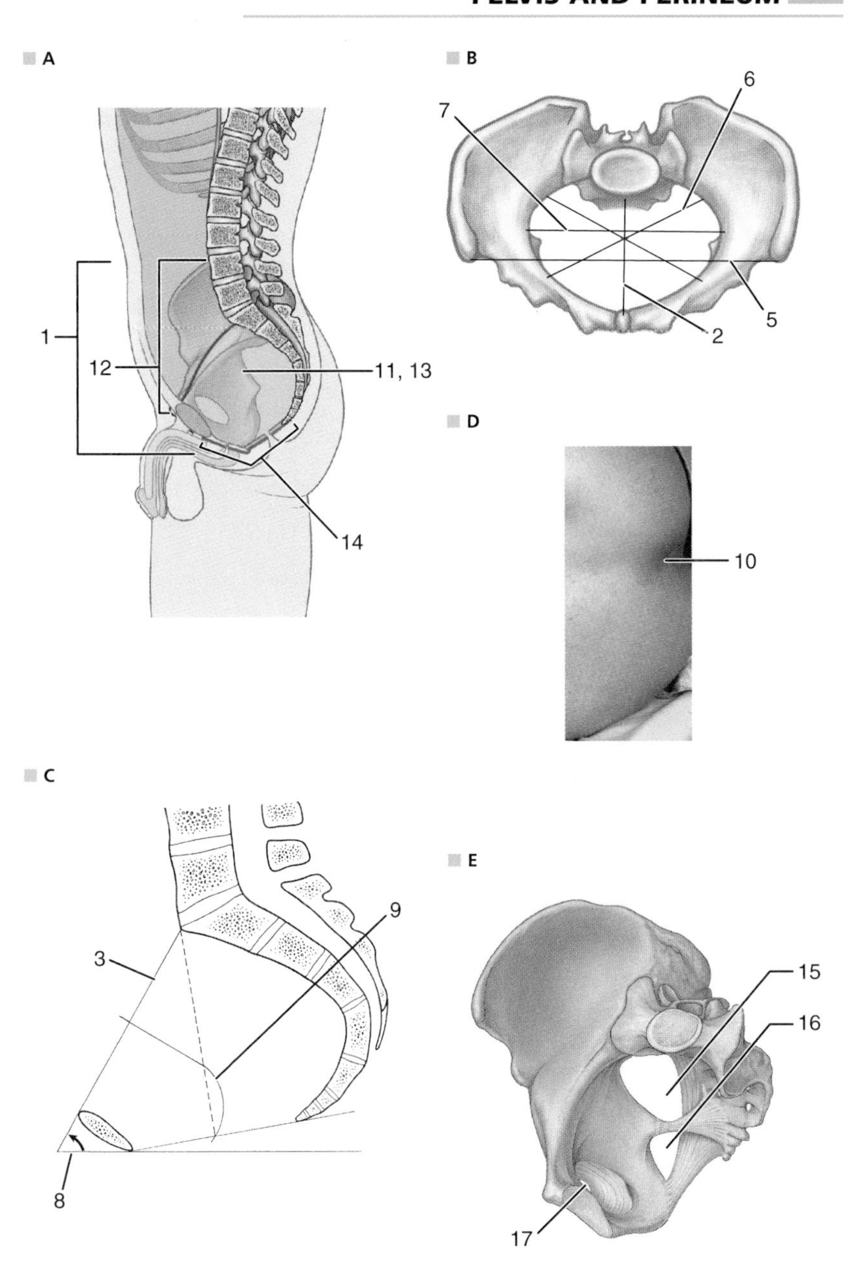
A
B
D
C
E
1
12
11, 13
14
7
6
2
5
10
3
9
8
15
16
17

1 **pelvic diaphragm** the portion of the floor of the pelvis formed by the coccygei and levatores ani muscles and their fasciae. <diaphragma pelvis> A

2 **pelvic inlet** the superior aperture of the pelvis minor, bounded by the crest and pecten of the pubic bones, the arcuate lines of the ilia, and the anterior margin of the base of the sacrum. <apertura pelvis superior> B

3 **rectouterine fold** a crescentic fold of peritoneum extending from the rectum to the base of the broad ligament on either side, forming the rectouterine pouch. <plica rectouterina> C

4 **rectouterine pouch** a sac or recess formed by a fold of the peritoneum dipping down between the rectum and the uterus. <excavatio rectouterina> C

5 **vesicouterine pouch** the space between the bladder and the uterus in the female peritoneal cavity. <excavatio vesicouterina> C

6 **rectovesical pouch** the space between the rectum and the bladder in the peritoneal cavity in the male; in females the corresponding cavity is divided by the uterus and broad ligament into the *rectouterine* and *vesicouterine* pouches. <excavatio rectovesicalis> E

7 **pelvic outlet** the inferior, very irregular aperture of the pelvis minor, bounded by the coccyx, the sacrotuberous ligaments, part of the ischium, the sides of the pubic arch, and the pubic symphysis. <apertura pelvis inferior> D

8 **pararectal fossa** either of two cavities formed by folds of peritoneum, one on either side of the rectum, varying in size according to distention of the rectum; in males this is continuous with the rectovesical pouch and in females it is continuous with the rectouterine pouch. Called also *pararectal pouch.* <fossa pararectalis> E

9 **paravesical fossa** the fossa formed by the peritoneum on each side of the urinary bladder; in females it is the lateral part of the vesicouterine pouch. <fossa paravesicalis> E

10 **anal cleft** the cleft between the buttocks on which the anus opens. Called also *intergluteal, gluteal,* or *natal cleft.* <crena analis> F

11 **hip** the area of the body lateral to and including the hip joint. <coxa> F

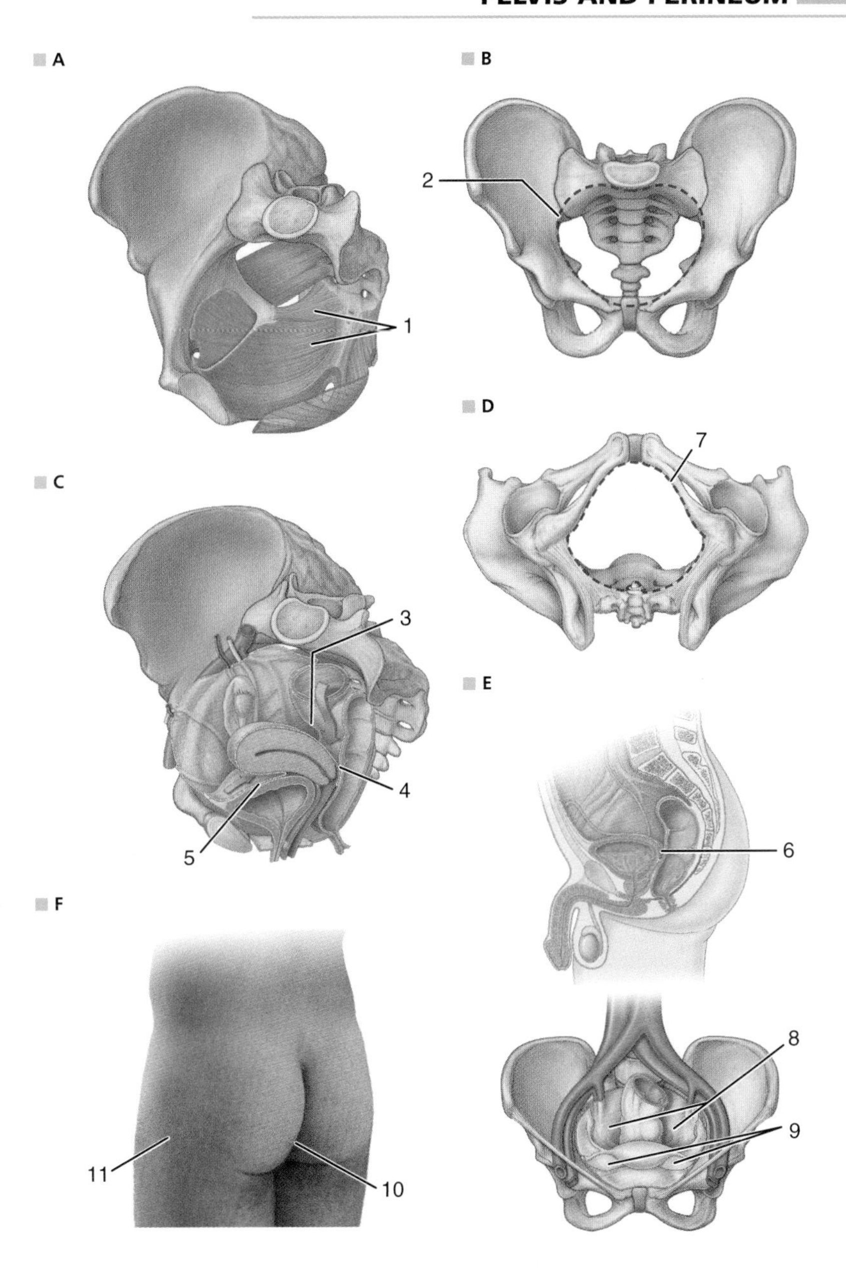
A
B
C
D
E
F
1
2
3
4
5
6
7
8
9
10
11

1 **transverse vesical fold** a transverse fold of the urogenital peritoneum that extends from the bladder onto the pelvic wall when the bladder is empty. <plica vesicalis transversa> A

2 ***parietal pelvic fascia*** the fascia on the wall of the pelvis that covers the muscles passing from the interior of the pelvis to the thigh. <fascia pelvis parietalis> B

3 ***pelvic fascia*** an inclusive term for the fascia that forms part of the general layer lining the walls of the pelvis and invests the pelvic organs. <fascia pelvis> B

4 **presacral fascia** a layer of parietal pelvic fascia between the sacrum and the rectum; the superior and inferior hypogastric plexuses are imbedded in it. <fascia presacralis> B

5 **rectoprostatic fascia** a membranous partition separating the rectum from the prostate and urinary bladder; this structure in the male corresponds to the rectovaginal fascia in the female. <fascia rectoprostatica> B

6 ***rectosacral fascia*** the fusion of the inferior part of the presacral fascia with the visceral fascia on the posterior aspect of the rectum. <fascia rectosacralis> B

7 **retropubic space** the part of the extraperitoneal space between the inferior aspect of the apex of the bladder, the transversalis fascia, and the posterosuperior aspect of the pubic symphysis, extending along the sides of the bladder to the lateral ligaments and limited inferiorly by the puboprostatic ligaments. <spatium retropubicum> B

8 **visceral pelvic fascia** the fascia that covers the organs and vessels of the pelvis. <fascia pelvis visceralis> B

9 **coccygeal body** an oval structure consisting of irregular masses of spherical or polyhedral epithelioid cells grouped around a dilated, sinusoidal capillary vessel, occurring anterior to, or immediately inferior to, the apex of the coccyx, at the termination of the median sacral vessels. <glomus coccygeum> C

10 **rectovaginal fascia** the membranous partition between the rectum and the vagina; this structure in the female corresponds to the rectoprostatic fascia in the male. <fascia rectovaginalis> C

11 **superior fascia of pelvic diaphragm** the fascia on the upper surface of the levator ani and coccygeus muscles. <fascia superior diaphragmatis pelvis> D

12 **urogenital peritoneum** the peritoneum lining the urogenital structures in the lower pelvis. <peritoneum urogenitale> D

13 **obturator fascia** the part of the parietal fascia of the pelvis that covers the obturator internus muscle. <fascia obturatoria> E

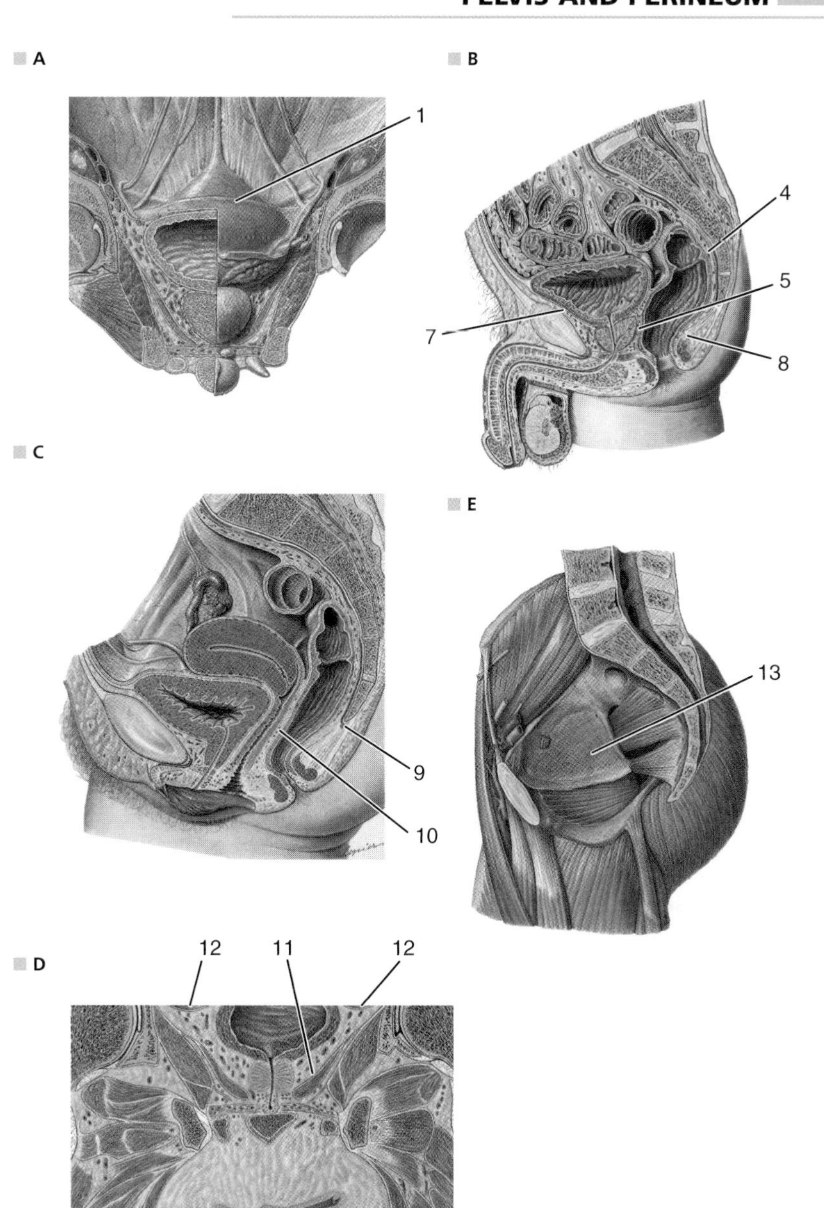
A
B
C
D
E
1
4
5
7
8
9
10
11
12
12
13

BONES

1 **anterior inferior iliac spine** a blunt bony process projecting forward from the lower part of the anterior margin of the ilium, just above the acetabulum. <spina iliaca anterior inferior> A

2 ***anterior obturator tubercle*** a small spur sometimes present on the margin of the obturator foramen, projecting from the superior ramus of the pubis. <tuberculum obturatorium anterius> A

3 **anterior superior iliac spine** a blunt bony projection on the anterior border of the ilium, forming the anterior end of the iliac crest. <spina iliaca anterior superior> A

4 **arcuate line of ilium** the iliac portion of the terminal line, limiting the ala of the ilium inferiorly on its medial surface. <linea arcuata ossis ilii> A

5 **auricular surface of ilium** a somewhat ear-shaped area on the sacropelvic surface of the ilium, which articulates with the auricular surface of the sacrum to form the sacroiliac joint. <facies auricularis ossis ilii> A

6 **greater sciatic notch** the large notch on the posterior border of the pelvic bone, where the posterior borders of the ilium and the ischium become continuous. <incisura ischiadica major> A

7 **iliac crest** the thickened, expanded upper border of the ilium. <crista iliaca> A

8 **iliac fossa** a large, smooth concave area occupying much of the inner surface of the ala of the ilium, especially anteriorly; from it arises the iliacus muscle. <fossa iliaca> A

9 **iliac tuberosity** a roughened area on the sacropelvic surface of the ilium, between the iliac crest and the auricular surface, for the attachment of muscles and ligaments. <tuberositas iliaca> A

10 **iliopubic eminence** a diffuse enlargement just anterior to the acetabulum, marking the junction of the ilium with the superior ramus of the pubis; called also *iliopectineal eminence* or *tubercle* and *iliopubic tubercle*. <eminentia iliopubica> A

11 **inferior ramus of pubic bone** the short flattened bar of bone that projects from the body of the pubic bone in a posteroinferolateral direction to meet the ramus of the ischium. <ramus inferior ossis pubis> A

12 **internal lip of iliac crest** the inner margin of the iliac crest. <labium internum cristae iliacae> A

13 **ischial spine** a strong process of bone projecting backward and medialward from the posterior border of the ischium, on a level with the lower border of the acetabulum and serving to separate the greater and lesser sciatic notches. <spina ischiadica> A

14 **ischial tuberosity** a large elongated mass on the inferior part of the posterior margin of the body of the ischium, to which several muscles are attached. <tuber ischiadicum> A

15 **ischiopubic ramus** the inferior ramus of the pubis and the adjacent part of the ramus of the ischium, considered as a unit. <ramus ischiopubicus> A

16 **lesser sciatic notch** the notch on the posterior border of the ischium just inferior to the ischiadic spine. <incisura ischiadica minor> A

17 **linea terminalis** a line on the inner surface of either pelvic bone, extending from the sacroiliac joint to the iliopubic eminence anteriorly, and marking the plane separating the pelvis major from the pelvis minor. <linea terminalis pelvis> A

18 **obturator foramen** the large opening between the pubic bone and the ischium. <foramen obturatum> A

19 **pectineal line** the anterior border of the superior ramus of the pubis, beginning at the pubic tubercle and continuing to the iliopubic eminence. <pecten ossis pubis> A

20 **posterior inferior iliac spine** a blunt bony projection from the posterior border of the ilium, corresponding to the posterior lower extremity of the auricular surface and the posterior upper extremity of the greater sciatic notch. <spina iliaca posterior inferior> A

21 ***posterior obturator tubercle*** a small protuberance often present on the margin of the obturator foramen, projecting from the free edge of the acetabular fossa near the junction of the pubis and ischium. <tuberculum obturatorium posterius> A

22 **posterior superior iliac spine** a blunt bony projection on the posterior border of the ilium, forming the posterior end of the iliac crest. <spina iliaca posterior superior> A

23 **pubic crest** the thick, rough, anterior border of the body of the pubic bone. <crista pubica> A

24 **pubic tubercle** a prominent tubercle situated at the lateral end of the pubic crest and at the medial end of the superior border of the superior ramus of the pubic bone; it is the anterior medial terminal of the obturator crest and of the pecten of the pubic bone. <tuberculum pubicum ossis pubis> A

25 **ramus of ischium** the flattened bar of bone that projects from the inferior end of the body of the ischium in an anterosuperomedial direction to meet the inferior ramus of the pubis. It forms part of the border of the obturator foramen. <ramus ossis ischii> A

26 **sacropelvic surface of ilium** an irregular area on the inner surface of the ala of the ilium, posterior to the iliac fossa; it contains the iliac tuberosity and the auricular surface. <facies sacropelvica ossis ilii> A

27 **superior ramus of pubic bone** the bar of bone projecting from the body of the pubic bone in a posterosuperolateral direction to the iliopubic eminence and forming part of the acetabulum. <ramus superior ossis pubis> A

28 **symphysial surface of pubic bone** the rough, ovoid, medial surface of the body of the pubic bone, by which it articulates at the pubic symphysis with its fellow of the opposite side. <facies symphysialis ossis pubis> A

A

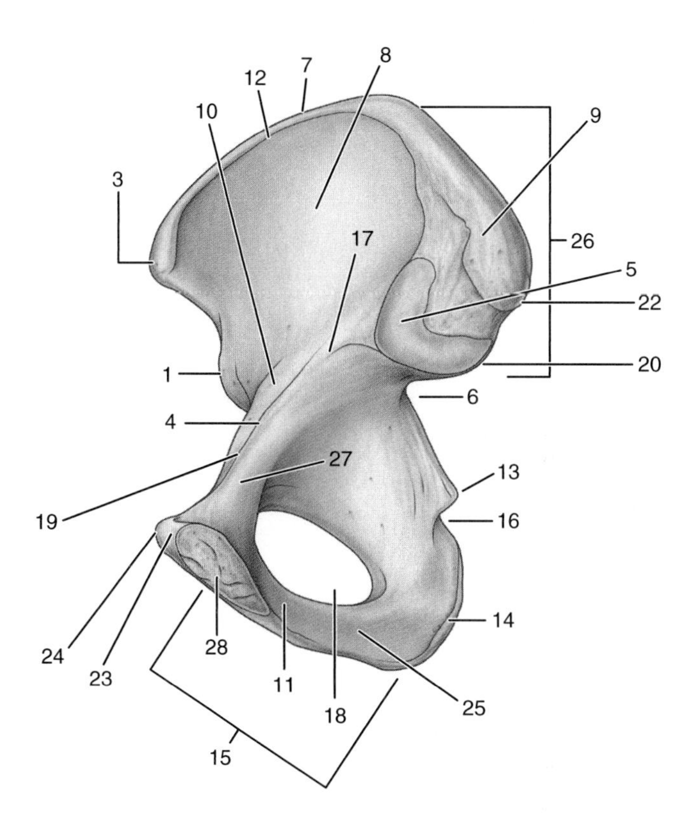

PELVIS AND PERINEUM

1 **ala of ilium** the expanded superior portion of the ilium which forms the lateral boundary of the pelvis major. <ala ossis ilii> A

2 **body of ilium** the inferior portion of the ilium, which forms roughly the superior two-fifths of the acetabulum. <corpus ossis ilii> A

3 **body of ischium** the thick, irregular, prismatic part of the ischium. Its superior end participates in the acetabulum, and from its inferior end the ramus of the ischium projects. <corpus ossis ischii> A

4 **body of pubis** the irregular mass of the pubic bone that lies alongside the median plane, articulating with the similar portion of the opposite pubic bone. From it extend the superior and inferior rami of the pubic bone. <corpus ossis pubis> A

5 **external lip of iliac crest** the outer margin of the iliac crest. <labium externum cristae iliacae> A

6 **gluteal surface of ilium** the large external, or posterior, surface of the ala of the ilium, on which are located the three gluteal lines. <facies glutea ossis ilii> A

7 **tuberculum of iliac crest** a prominence on the iliac crest about 5 cm behind the anterior superior iliac spine. <tuberculum iliacum> A

8 **intermediate line of iliac crest** the area between the internal and external lips of the iliac crest. <linea intermedia cristae iliacae> A

9 **ischium** the inferior posterior portion of the pelvic bone; it is a separate bone in early life. <os ischii> A

10 ***obturator crest*** the inferior border of the superior ramus of the pubic bone, a strong ridge of bone beginning near the pubic tubercle and extending to the anterior part of the gap in the rim of the acetabulum, forming part of the circumference of the obturator foramen, and giving attachment to the obturator membrane. <crista obturatoria> A

11 **pubic bone** the anterior inferior part of the pelvic bone on either side, articulating with its fellow in the anterior midline at the pubic symphysis; it is a separate bone in early life. <os pubis> A

12 **ala of sacrum** the upper surface of the lateral part of the sacrum. <ala ossis sacri> B

13 **anterior sacral foramina** the eight openings (four on each side) on the pelvic surface of the sacrum for the anterior rami of the sacral nerves. <foramina sacralia anteriora> B

14 **apex of sacrum** the inferior end of the body of the fifth sacral vertebra, which articulates with the coccyx. <apex ossis sacri> B

15 **base of sacral bone** the cranial surface of the sacrum; its lateral portions consist of the alae of the sacrum, and its middle portion is the upper surface of the body of the first sacral vertebra, which articulates with the fifth lumbar vertebra. <basis ossis sacri> B

16 **coccygeal cornu** either of the cranial pair of rudimentary articular processes of the coccyx that articulate with the cornua of the sacrum. <cornu coccygeum> C

17 **pelvic surface of sacrum** the smooth, concave, anteroinferiorly directed surface of the sacrum that helps form the posterior wall of the pelvis. <facies pelvica ossis sacri> B

18 **transverse lines of sacrum** four transverse ridges on the pelvic surface of the sacrum, running between the pairs of pelvic sacral foramina, marking the positions of the former intervertebral disks. <lineae transversae ossis sacri> B

19 ***intervertebral foramina of sacrum*** the four short, forked tunnels in each lateral wall of the sacral canal, connecting it with the pelvic and dorsal sacral foramina. <foramina intervertebralia ossis sacri> D

20 **lateral sacral crest** either of two series of tubercles lateral to the posterior sacral foramina, representing the transverse processes of the sacral vertebrae. <crista sacralis lateralis> D

21 **medial sacral crest** either of two indefinite crests just medial to the posterior sacral foramina, formed by fusion of the articular processes of the sacral vertebrae. <crista sacralis medialis> D

22 **median sacral crest** a median ridge on the dorsal surface of the sacrum, formed by the remnants of the spinous processes of the upper four sacral vertebrae. <crista sacralis mediana> D

23 **posterior sacral foramina** the eight openings (four on each side) on the posterior surface of the sacrum for the posterior rami of the sacral nerves. Called also *dorsal sacral foramina*. <foramina sacralia posteriora> D

24 **posterior surface of sacrum** the markedly convex and rough posterior surface of the sacrum, which gives origin to the sacrospinalis and multifidus muscles. <facies dorsalis ossis sacri> D

25 **sacral canal** the continuation of the vertebral canal through the sacrum. <canalis sacralis> D

26 **sacral cornu** either of the two hook-shaped processes extending downward from the arch of the last sacral vertebra. <cornu sacrale> D

27 **sacral hiatus** the opening at the inferior end of the sacral canal formed by failure of the laminae of the fifth and sometimes the fourth sacral vertebrae to meet in the midline. <hiatus sacralis> D

28 **auricular surface of sacrum** the broad irregular surface on the superior half of the lateral aspect of the sacrum, which articulates with the ilium. <facies auricularis ossis sacri> E

29 **sacral tuberosity** a roughened area on the lateral part of the sacrum, on the posterior surface between the lateral sacral crest and the auricular surface, which gives attachment to the sacroiliac ligaments. <tuberositas ossis sacri> E

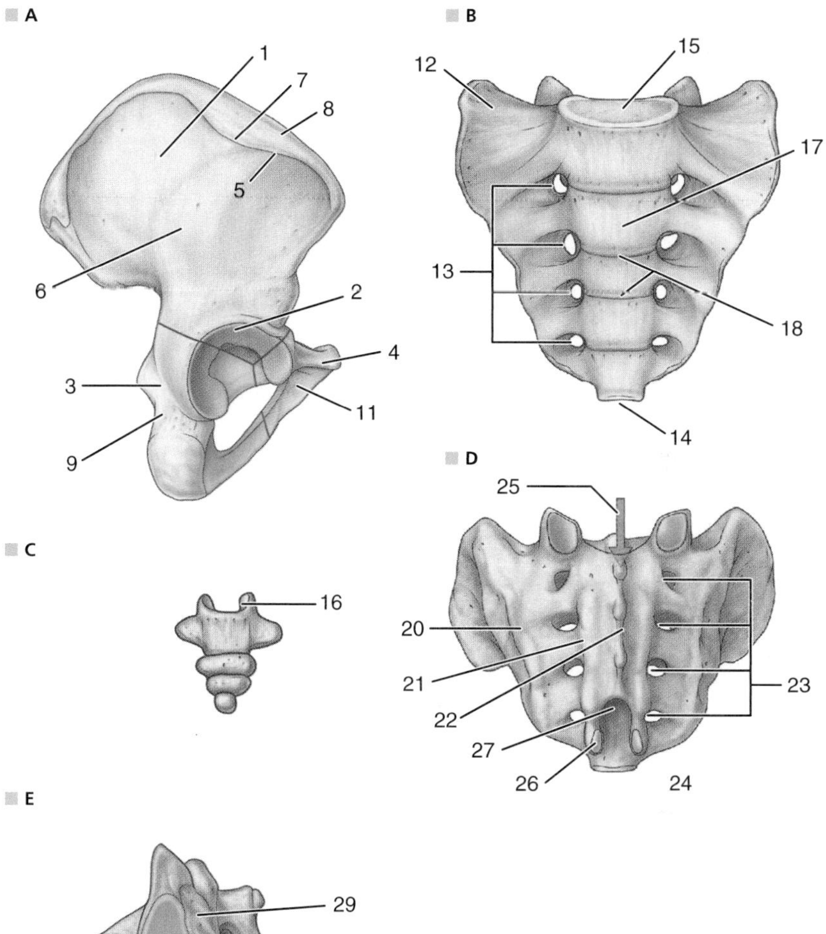

E

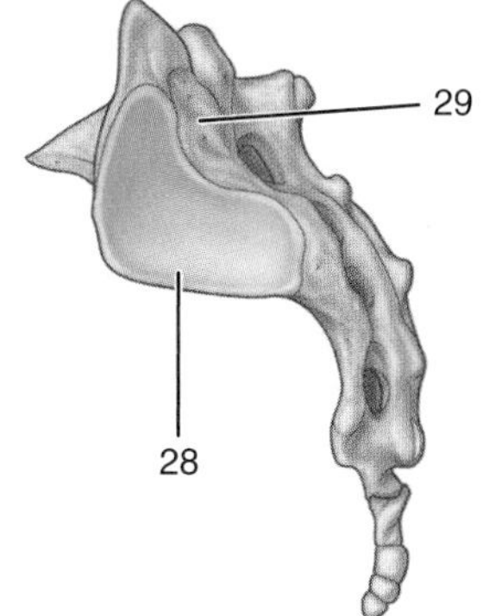

1 ***intercristal diameter*** the distance between the middle points of the iliac crests. <distantia intercristalis> A

2 ***pelvic girdle*** the encircling bony structure supporting the lower limbs, comprising the two pelvic bones, articulating with each other and with the sacrum, to complete the essentially rigid bony ring. <cingulum pelvicum> A

3 **pubic arch** the arch formed by the conjoined rami of the ischial and pubic bones of the two sides of the body. <arcus pubicus> A

4 **subpubic angle** the apex of the pubic arch; the angle formed at the point of meeting of the conjoined rami of the ischial and pubic bones of the two sides of the body. Called also *subpubic arch* and *arch of pelvis.* <angulus subpubicus> A

JOINTS

5 **interpubic disk** a midline plate of fibrocartilage interposed between the symphysial surfaces of the pubic bones, these surfaces being covered by a thin layer of hyaline cartilage. <discus interpubicus> A

6 **lumbosacral joint** the joint between the sacrum and the lumbar vertebrae. <articulatio lumbosacralis> A

7 **sacrococcygeal joint** the joint between the coccyx and sacrum. <articulatio sacrococcygea> A

8 **joints of pelvic girdle** the sacroiliac joint and the pubic symphysis. <juncturae cinguli pelvici> B

9 **pubic symphysis** the joint formed by union of the bodies of the pubic bones in the median plane by a thick mass of fibrocartilage. <symphysis pubica> B

10 **sacroiliac joint** the joint formed between the auricular surfaces of the sacrum and ilium. <articulatio sacroiliaca> A, B

11 ***anterior sacrococcygeal ligament*** a flat band, homologous with the anterior longitudinal ligament of the vertebral column, that passes from the lower part of the sacrum over onto the anterior part of the coccyx. <ligamentum sacrococcygeum anterius>

12 ***deep posterior sacrococcygeal ligament*** the terminal portion of the posterior longitudinal ligament of the vertebral column; it helps to unite the posterior surfaces of the fifth sacral and the coccygeal vertebrae. <ligamentum sacrococcygeum posterius profundum>

13 ***lateral sacrococcygeal ligament*** a fibrous band, homologous with the intertransverse ligaments, that passes from the transverse process of the first coccygeal vertebra to the lower lateral angle of the sacrum, thus helping to complete the foramen of the fifth sacral nerve. <ligamentum sacrococcygeum laterale>

14 ***superficial posterior sacrococcygeal ligament*** a fibrous band continuous with the supraspinous ligament of the vertebral column; attached superiorly to the margin of the sacral hiatus, and diverging as it passes inferiorly to attach to the posterior surface of the coccyx. <ligamentum sacrococcygeum posterius superficiale>

LIGAMENTS

15 **iliolumbar ligament** a strong band that passes from the transverse processes of the fourth and fifth lumbar vertebrae to the internal lip of the adjacent portion of the iliac crest. <ligamentum iliolumbale> A

16 **anterior sacroiliac ligament** any of numerous thin fibrous bands passing from the anterior margin of the auricular surface of the sacrum to the adjacent portions of the ilium. <ligamentum sacroiliacum anterius> A, B

17 **interosseous sacroiliac ligament** any of numerous short, strong bundles connecting the tuberosities and adjacent surfaces of the sacrum and the ilium. <ligamentum sacroiliacum interosseum> C

18 **posterior sacroiliac ligament** any of numerous strong bands that pass from the ilium to the sacrum. The *long posterior sacroiliac ligament* is more superficial and connects the posterior superior iliac spine with the second, third, and fourth articular tubercles of the sacrum. The *short posterior sacroiliac ligament* is deeper and more nearly horizontal and connects the tuberosity of the ilium with the first and second tubercles on the posterior of the sacrum. <ligamentum sacroiliacum posterius> C

19 **inferior pubic ligament** a thick archlike band of fibers situated along the inferior margin of the pubic symphysis. Its fibers are attached to the medial borders of the inferior rami of the pubic bones and thus it rounds out and forms the summit of the pubic arch. <ligamentum pubicum inferius> D

20 **superior pubic ligament** fibers that pass transversely across the superior margin of the pubic symphysis; attached to the bones and to the interpubic disk, they extend laterally as far as the pubic tubercle. <ligamentum pubicum superius> D

21 **obturator membrane** a strong membrane that fills the obturator foramen except superiorly at the obturator groove, where a deficiency is left, the *obturator canal.* <membrana obturatoria> E

22 **sacrospinous ligament** one of the long vertical fibrous bands attached by the apex to the spine of the ischium and by the base to the lateral margins of the sacrum. Called also *sacrospinal ligament.* <ligamentum sacrospinale> E

23 **sacrotuberous ligament** a large, flat band that is attached below to the ischial tuberosity, spreads out as it ascends, and is attached to the lateral margins of the sacrum and the coccyx and to the posterior inferior iliac spine. <ligamentum sacrotuberale> E

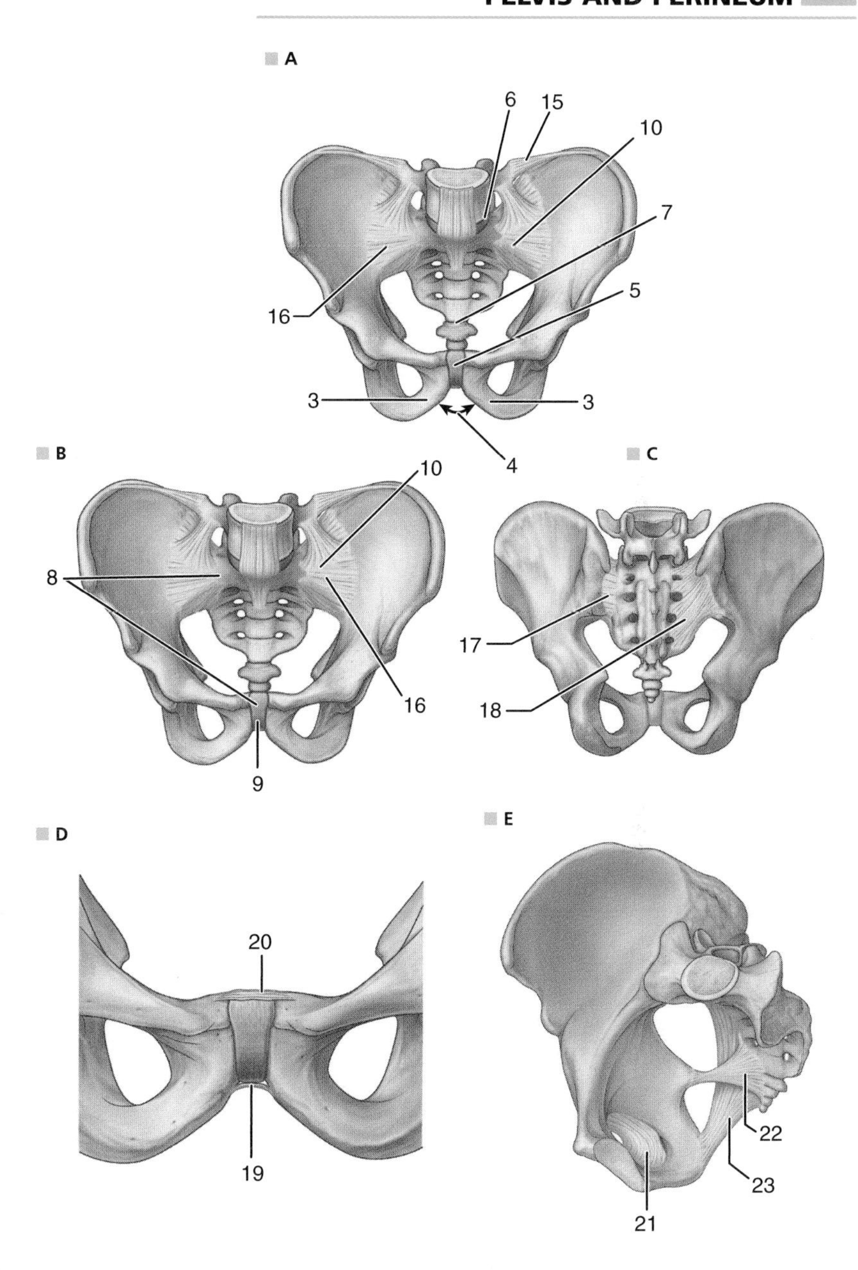
A
6
15
10
7
5
16
3
3
4
B
10
8
16
9
C
17
18
D
20
19
E
22
23
21

1 **tendinous arch of levator ani muscle** a linear thickening of the fascia covering the obturator internus muscle, from which levator ani takes origin. <arcus tendineus musculi levatoris ani> A

2 ***tendinous arch of pelvic fascia*** a thickening of the superior fascia, extending from the ischial spine to the posterior part of the body of the pubis. <arcus tendineus fasciae pelvis> A

3 **transverse perineal ligament** a fibrous band in males that spans the subpubic angle just behind the deep dorsal vein of the penis, formed by thickening of the anterior boundary of the perineal membrane. <ligamentum transversum perinei> B

PELVIS

Muscles of Pelvis

4 **obturator internus muscle** *origin,* pelvic surface of pelvic bone and obturator membrane, margin of obturator foramen, ramus of ischium, inferior ramus of pubis; *insertion,* greater trochanter of femur; *innervation,* fifth lumbar, first and second sacral; *action,* rotates thigh laterally. <musculus obturatorius internus> C

5 **piriformis muscle** *origin,* ilium, second to fourth sacral vertebrae; *insertion,* upper border of greater trochanter of femur; *innervation,* first and second sacral; *action,* rotates thigh laterally. <musculus piriformis> C

6 **coccygeus muscle** *origin,* ischial spine; *insertion,* lateral border of lower part of sacrum, upper coccyx; *innervation,* third and fourth sacral; *action,* supports and raises coccyx. Called also *ischiococcygeus muscle.* <musculus ischiococcygeus> A

7 **iliococcygeus muscle** the posterior portion of the levator ani, which originates as far anteriorly as the obturator canal and inserts on the side of the coccyx and the anococcygeal ligament; *innervation,* third and fourth sacral; *action,* helps to support pelvic viscera and resist increases in intra-abdominal pressure. <musculus iliococcygeus> A

8 **levator ani muscle** a name applied collectively to important muscular components of the pelvic diaphragm, including the pubococcygeus, puborectalis, and iliococcygeus muscles. <musculus levator ani> A

9 **pubococcygeus muscle** the anterior portion of the levator ani, originating anterior to the obturator canal; *insertion,* anococcygeal body and side of coccyx; *innervation,* pudendal; *action,* helps support pelvic viscera and resist increases in intraabdominal pressure. <musculus pubococcygeus> A

10 ***puboprostaticus muscle*** a part of the anterior portion of the pubococcygeus muscle, inserted in the prostate and the tendinous center of the perineum; innervated by the pudendal nerve, it supports and compresses the prostate and is involved in control of micturition. Called also *levator muscle of prostate* or *levator prostatae muscle.* <musculus puboprostaticus> A

11 **puborectalis muscle** a portion of the levator ani having a more lateral origin from the pubic bone, and continuous posteriorly with the corresponding muscle of the opposite side; *innervation,* third and fourth sacral; *action,* helps support pelvic viscera and resist increases in intra-abdominal pressure. <musculus puborectalis> A

12 ***pubovaginalis muscle*** a part of the anterior portion of the pubococcygeus muscle, which is inserted into the urethra and vagina; innervated by the pudendal nerve, it is involved in control of micturition. <musculus pubovaginalis> A

13 **compressor urethrae muscle** compressor muscle of urethra (occurs only in females): *origin,* ischiopubic ramus on each side; *insertion,* blends with its partner on the other side anterior to the urethra below the external urethral sphincter; *innervation,* perineal branches of pudendal nerve; *action,* accessory urethral sphincter. <musculus compressor urethrae> D

14 **iliopectineal bursa** a bursa between the iliopsoas tendon and the iliopectineal eminence; called also *subiliac bursa.* <bursa iliopectinea> E

15 ***anorectoperineal muscles*** bands of smooth muscle fibers extending from the anorectal flexure to the membranous part of urethra in the male. <musculi anorectoperineales>

16 ***pubovesicalis muscle*** smooth muscle fibers extending from the neck of the urinary bladder to the pubis. <musculus pubovesicalis>

17 ***rectococcygeus muscle*** smooth muscle fibers originating on the anterior surface of the second and third coccygeal vertebrae and inserting on the posterior surface of the rectum, innervated by autonomic nerves, and acting to retract and elevate the rectum. <musculus rectococcygeus>

18 ***rectouterinus muscle*** a band of fibers in the female, running between the cervix of the uterus and the rectum, in the rectouterine fold. <musculus rectouterinus>

19 ***rectovesicalis muscle*** a band of fibers in the male, connecting the longitudinal musculature of the rectum with the external muscular coat of the bladder. <musculus rectovesicalis>

A

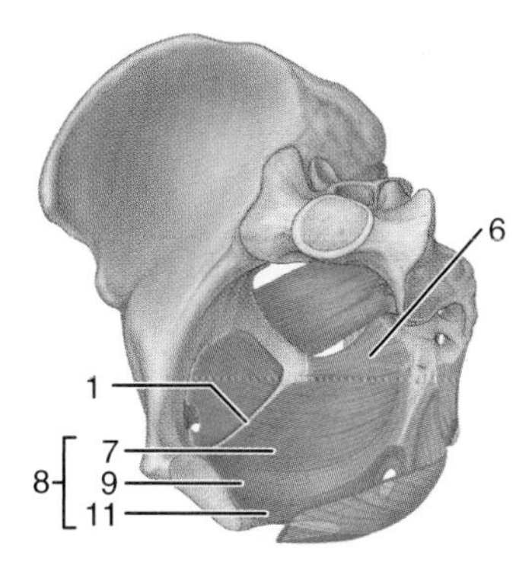

B

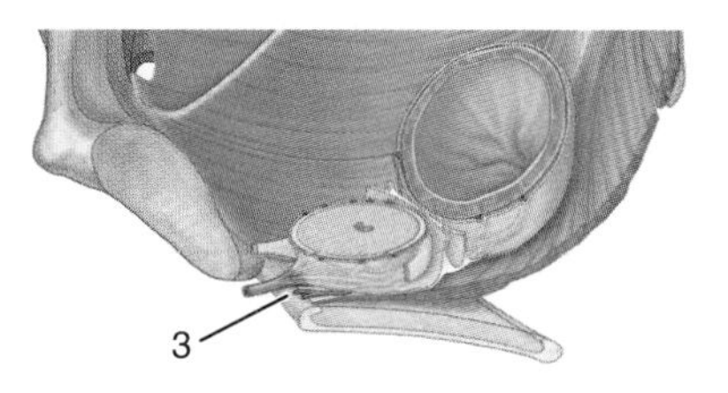

C

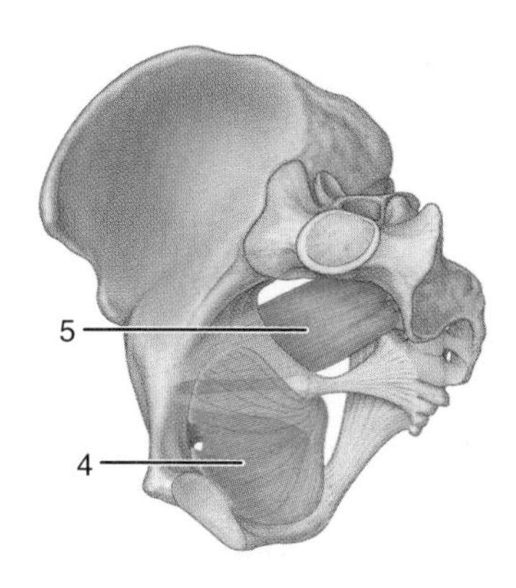

D

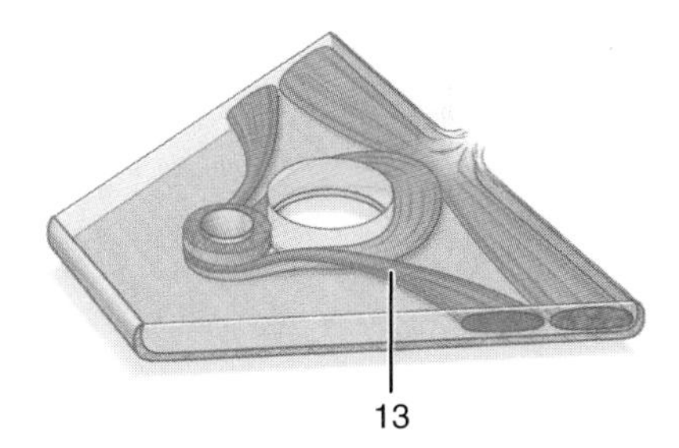

E

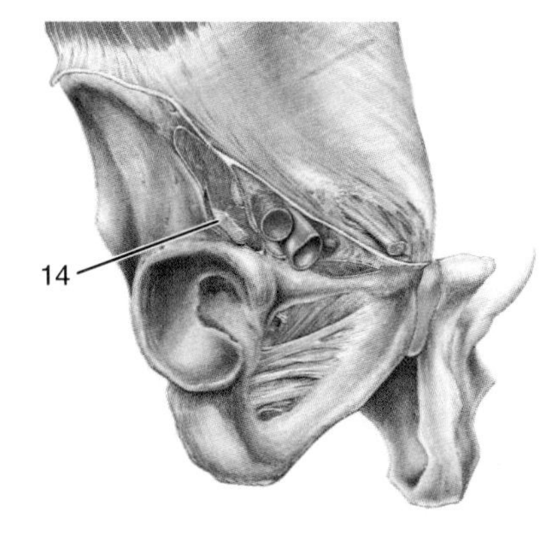

Viscera

Gastrointestinal Viscera

1 **anorectal flexure** the bend at the inferior end of the rectum, where it deviates inferiorly and posteriorly. Called also *perineal flexure*. <flexura anorectalis> A

2 **inferior right lateral flexure of rectum** the fourth bend of the rectum, where it deviates laterally to the right. <flexura inferodextra lateralis recti> A

3 **intermediate left lateral flexure of rectum** the third bend of the rectum, where it deviates laterally to the left. <flexura intermediosinistra lateralis recti> A

4 ***lateral flexures of rectum*** the three lateral bends in the rectum (*inferior right, intermediate left,* and *superior right*). <flexurae laterales recti> A

5 **rectal ampulla** the dilated portion of the rectum just proximal to the anal canal. <ampulla recti> A

6 **sacral flexure of rectum** the first bend in the rectum, where it deviates posteriorly. <flexura sacralis recti> A

7 **superior right lateral flexure of rectum** the second bend in the rectum, where it deviates laterally to the right. <flexura superodextra lateralis recti> A

8 **anal canal** the terminal portion of the alimentary canal, about 4 cm long, extending from the anorectal junction to the anocutaneous line, containing the anus. <canalis analis> A

9 **anal columns** vertical ridges or folds of mucous membrane at the upper half of the anal canal; called also *rectal columns*. <columnae anales> B

10 **anal pecten** the corrugated epithelium within the anal transitional zone between the pectinate line and the anocutaneous line; called also *pecten*. <pecten analis> B

11 **anal sinuses** furrows, with pouchlike recesses at the lower end, separating the anal columns. <sinus anales> B

12 **anal valves** archlike folds of mucous membrane connecting the inferior ends of the anal columns. <valvulae anales> B

13 **anocutaneous line** the sinuous line marking the outer end of the anal pecten, the junction where the anal canal lined with stratified squamous epithelium ends and typical external skin begins. Called also *white line*. <linea anocutanea> B

14 **anorectal junction** the site at which the rectum becomes continuous with the anal canal; called also *anorectal line*. <junctio anorectalis> B

15 **anus** the distal or terminal orifice of the digestive tract, located in the anal cleft. B

16 **internal anal sphincter** a thickening of the circular lamina of the tunica muscularis at the inferior end of the rectum. <musculus sphincter ani internus> B

17 **intersphincteric groove** an indistinct groove in the anal canal, forming the lower border of the anal pecten, marking the change between the subcutaneous part of the external anal sphincter and the border of the internal anal sphincter. <sulcus intersphinctericus> B

18 **pectinate line** the wavy line forming the interior end of the anal pecten just below the anorectal junction. <linea pectinata> B

Urinary Viscera

19 **apex of bladder** the superior area of the urinary bladder, opposite to the fundus; it is at the junction of the superior and inferolateral surfaces of the bladder, and from it the median umbilical ligament (urachus) extends to the umbilicus. <apex vesicae urinariae> C

20 **body of urinary bladder** the central part of the bladder, between the apex and the fundus. <corpus vesicae urinariae> C

21 **urinary bladder** a musculomembranous sac in the anterior part of the pelvic cavity, serving as a reservoir for urine; it receives the excretory products of the kidneys through the ureters and expels them through the urethra. <vesica urinaria> C

22 ***deep trigonal muscle*** the deep layer of trigonal muscle, which is continuous with the detrusor muscle. <musculus trigoni vesicae urinariae profundus> D

23 **detrusor muscle** the bundles of smooth muscle fibers forming the muscular coat of the urinary bladder, which are arranged in a longitudinal and a circular layer and, on contraction, serve to expel urine. Called also *detrusor urinae* and *detrusor urinae muscle*. <musculus detrusor vesicae urinariae> D

24 **internal urethral orifice** the opening between the bladder and the urethra at one corner of the bladder trigone. <ostium urethrae internum> D

25 **interureteric fold** a fold of mucous membrane extending across the fundus of the bladder between the two ureteral orifices, produced by a transverse bundle of muscle fibers; it marks one side of the trigone of the bladder. Called also *interureteric crest* or *ridge*. <plica interureterica> D

26 ***intramural part of ureter*** the short distal portion of the ureter after it bends to run obliquely through the wall of the bladder. <pars intramuralis ureteris> D

27 **mucosa of urinary bladder** the mucous membrane (tunica mucosa) lining the urinary bladder, representing the innermost stratum of the bladder wall. <tunica mucosa vesicae urinariae> D

28 **muscular layer of urinary bladder** muscular layer or coat of urinary bladder: the smooth muscle coat of the bladder wall. <tunica muscularis vesicae urinariae> D

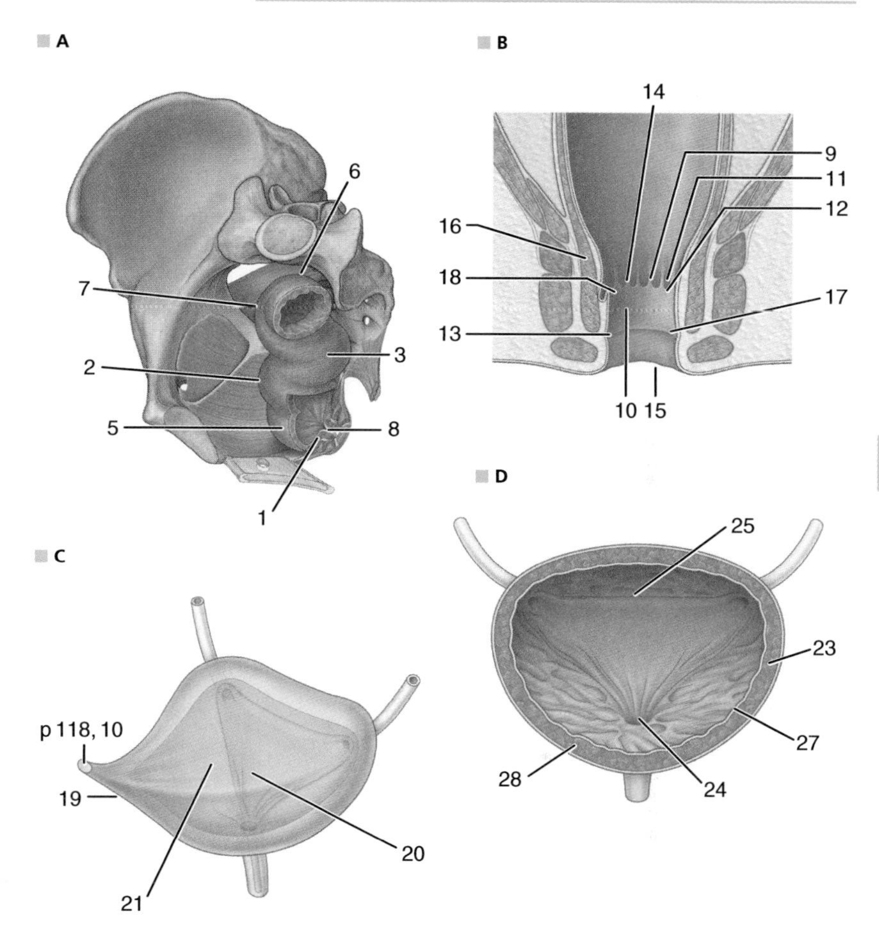
A
6
7
2
3
5
8
1
B
14
9
11
12
16
18
17
13
10 15
C
p 118, 10
19
20
21
D
25
23
27
28
24

1 **neck of urinary bladder** a constricted portion of the bladder, formed by the meeting of its inferolateral surfaces proximal to the internal urethral orifice. <cervix vesicae urinariae> A

2 **serosa of urinary bladder** the serous coat or layer (tunica serosa) of the bladder wall, facing exteriorly to the pelvic cavity. <tunica serosa vesicae urinariae> A

3 **submucosa of urinary bladder** the submucous layer of the wall of the urinary bladder. <tela submucosa vesicae urinariae> A

4 **subserosa of urinary bladder** the subserous layer of the urinary bladder. <tela subserosa vesicae urinariae> A

5 ***superficial trigonal muscle*** the superficial layer of the trigonal muscles, continuous proximally with the muscles of the ureteral wall. <musculus trigoni vesicae urinariae superficialis> A

6 ***trigonal muscles*** a submucous sheet of smooth muscle at the bladder trigone, continuous with ureteral muscles above and with those of the proximal urethra below. Its superficial and deep layers are morphologically distinct. <musculi trigoni vesicae urinariae> A

7 **trigone of urinary bladder** a smooth triangular portion of the mucous membrane in the fundus of the bladder; two of its corners are the ureteral orifices, with the interureteric fold between them, and its third corner is the internal urethral orifice. <trigonum vesicae urinariae> A

8 **ureteric orifice** the opening of a ureter into the bladder; the two orifices are at the lateral angles of the trigone of the bladder. <ostium ureteris> A

9 ***uvula of urinary bladder*** a rounded elevation at the bladder neck, seen mainly in adult males, formed by convergence of many fibers of the trigonal muscle as they pass through the encircling internal sphincter muscle to terminate in the urethra. <uvula vesicae urinariae> A

10 **female urethra** a canal about 3.7 cm long, extending from the neck of the bladder above the anterior vaginal wall to the urinary meatus. <urethra feminina> B

11 **fundus of urinary bladder** an area formed by the posterior surface of the bladder and containing the lowest section (trigone of bladder) where urine empties out through the internal urethral orifice. Called also *base of bladder.* <fundus vesicae urinariae> B, C

12 ***urethral crest of female urethra*** a prominent longitudinal fold in the female urethra, consisting of mucosa bulging inward along the posterior wall of the urethra. <crista urethralis urethrae femininae> B

13 **urethral glands of female urethra** numerous small mucous glands in the mucosa of the female urethra, some of which on either side are drained by the inconstant paraurethral ducts opening into the vestibule. Called also *Skene's glands.* <glandulae urethrales urethrae femininae> B

14 ***urethral lacunae of female urethra*** small depressions or pits in the mucous membrane of the urethra, with their openings usually directed distally; some are openings of ducts of the urethral glands. <lacunae urethrales urethrae femininae> B

15 **distal part of prostatic urethra** the portion of the prostatic urethra between the seminal colliculus and where it becomes the membranous part. <pars distalis partis prostaticae urethrae masculinae> C

16 **internal urethral sphincter** a circular layer of smooth muscle fibers surrounding the internal urethral orifice in males, innervated by the vesical nerve, and acting to close the internal orifice of the urethra. No such structure exists in females. Called also *preprostatic sphincter.* <musculus sphincter urethrae internus> C

17 **male urethra** a canal extending from the neck of the bladder to the urinary meatus, measuring about 20 cm in length, and presenting a double curve when the penis is flaccid; it is divided into four parts: the *spongy, membranous, prostatic,* and *preprostatic.* <urethra masculina> C

18 **membranous part of urethra** a short portion of the urethra between the prostatic and spongiosa parts, traversing the deep perineal space. Called also *intermediate part of urethra* and membranous urethra. <pars intermedia urethrae masculinae> C

19 ***mucosa of female urethra*** the mucous membrane (tunica mucosa) lining the female urethra. <tunica mucosa urethrae femininae> C

20 ***mucosa of male urethra*** the mucous membrane (tunica mucosa) lining the male urethra. <tunica mucosa urethrae masculinae> C

21 ***muscular layer of female urethra*** the muscular layer or coat of the female urethra, consisting of smooth muscle. An inner sublayer runs circularly and an outer sublayer runs longitudinally down the length of the urethra. <tunica muscularis urethrae femininae> C

22 ***muscular layer of male urethra*** the muscular layer or coat of the male urethra, consisting of smooth muscle. An inner sublayer runs circularly, especially in the preprostatic part, and an outer sublayer runs longitudinally down the length of the urethra. <tunica muscularis urethrae masculinae> C

23 **navicular fossa** the lateral expansion of the urethra in the glans penis. <fossa navicularis urethrae> C

24 **preprostatic part of urethra** the short, most proximal, part of the urethra, running almost vertically down from the bladder to where it enters the prostate. Called also *intramural part of urethra* and *preprostatic urethra.* <pars intramuralis urethrae masculinae> C

A

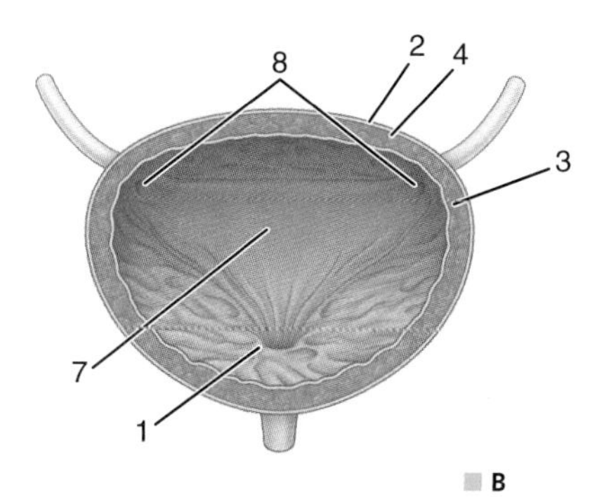

B

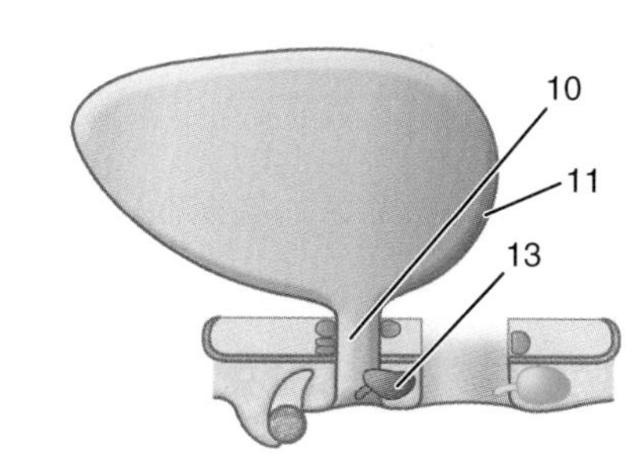

C

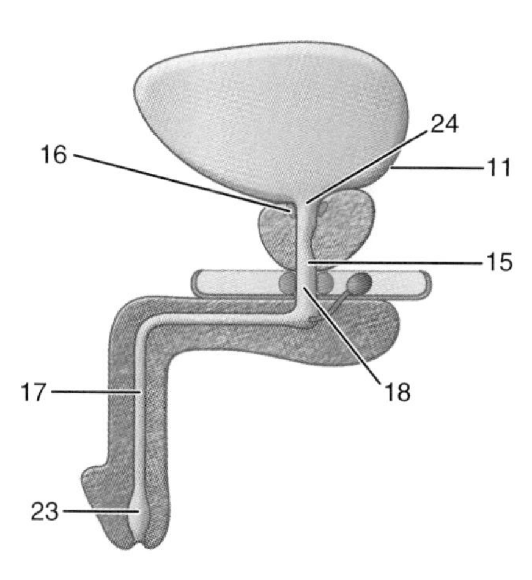

1 **prostatic part of urethra** the part of the urethra that passes through the prostate; called also *prostatic urethra.* <pars prostatica urethrae masculinae> A

2 **proximal part of prostatic urethra** the first portion of the prostatic urethra, up to and including the seminal colliculus. <pars proximalis partis prostaticae urethrae masculinae> A

3 ***spongy layer of female urethra*** a thin layer of spongiose erectile tissue, located just beneath the mucous membrane and containing a plexus of large veins. <tunica spongiosa urethrae femininae>

4 **spongy part of male urethra** the portion of the urethra within the corpus spongiosum of the penis. Called also *spongy urethra.* <pars spongiosa urethrae masculinae> A

5 ***urethral glands of male urethra*** mucous glands in the wall of the male urethra; called also *Littre's glands.* <glandulae urethrales urethrae masculinae> A

6 ***urethral lacunae of male urethra*** small depressions or pits in the mucous membrane of the urethra, with their openings usually directed distally; some are openings of ducts of the urethral glands. <lacunae urethrales urethrae masculinae> A

7 ***valve of navicular fossa*** a fold of mucous membrane occasionally found in the roof of the navicular fossa of the male urethra. <valvula fossae navicularis> A

8 **urethral crest of male urethra** a median elevation along the posterior wall of the male urethra, lying between the prostatic sinuses. <crista urethralis urethrae masculinae> B

9 **pelvic part of ureter** the portion of the ureter that extends from the linea terminalis of the pelvis to the urinary bladder. <pars pelvica ureteris> C

10 **external urethral orifice (female)** the urinary meatus in the female, where the urethra opens into the vestibule; it is surrounded by a sphincter of striated muscle derived from the bulbocavernosus muscle. <ostium ureteris ostium urethrae externum femininae> D

11 **external urethral orifice (male)** the slitlike urinary meatus in the male, where the urethra opens on the tip of the glans penis. <urethra ostium urethrae externum masculinae> E

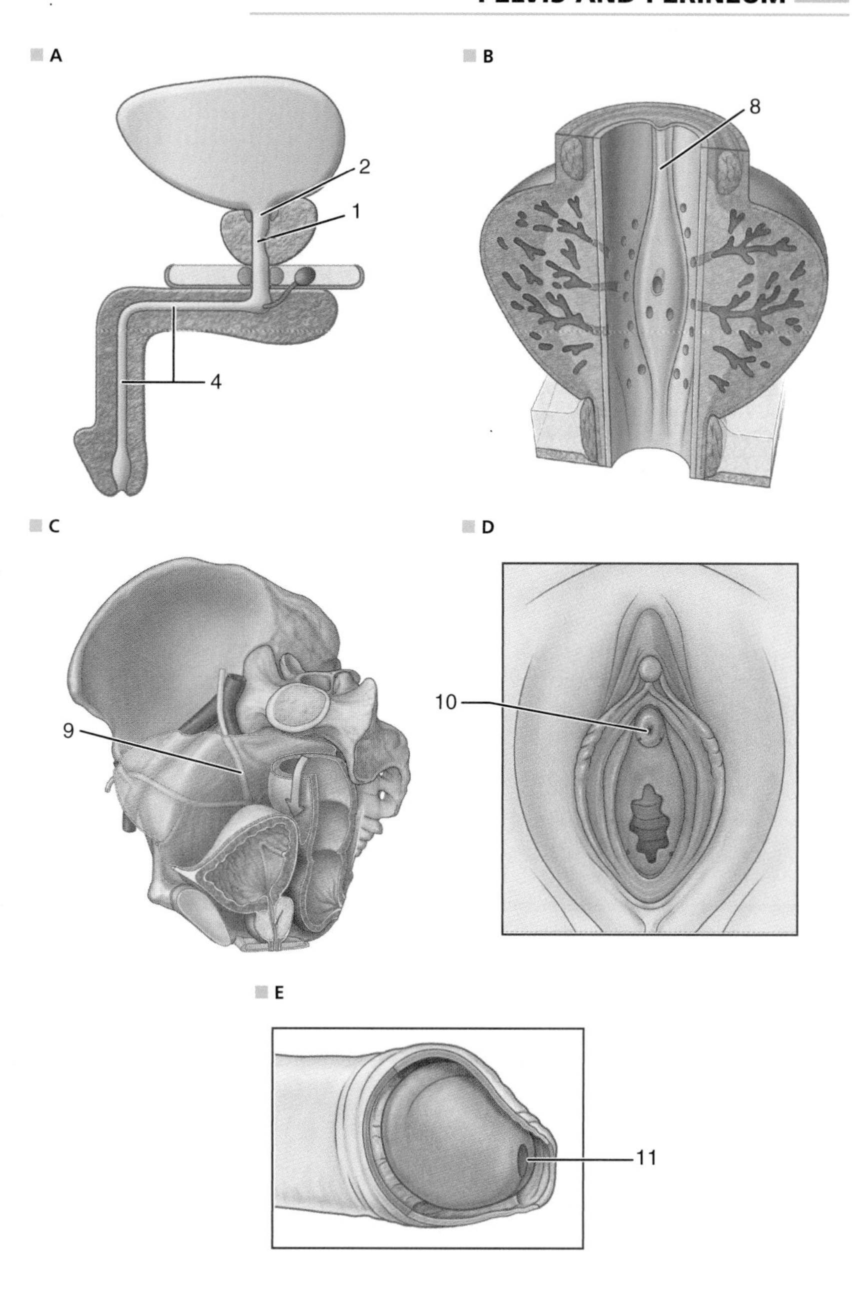
A
2
1
4
B
8
C
9
D
10
E
11

Female Reproductive System

1 ***lateral pubovesical ligament*** the lateral branch of the pubovesical ligament in the female, extending from the bladder neck to the tendinous arch of the pelvic fascia. <ligamentum laterale pubovesicale> A

2 ***medial pubovesical ligament*** the medial branch of the pubovesical ligament in the female, a forward continuation of the tendinous arch of the pelvic fascia to the pubis. <ligamentum mediale pubovesicale> A

3 **pubovesical ligament** a ligament that extends from the bladder neck to the inferior aspect of the pubic bones. In the female it is divided into two distinct parts (lateral and medial). In the male it is parallel and medial to the puboprostatic ligament and is sometimes called also *medial puboprostatic ligament.* <ligamentum pubovesicale> A

4 **body of uterus** that part of the uterus above the isthmus and below the orifices of the uterine tubes. <corpus uteri> B

5 **cervical canal of uterus** the part of the uterine cavity that lies within the cervix. <canalis cervicis uteri> B

6 ***cervical glands of uterus*** mucus-secreting glands located in clefts in the wall of the uterine cervix. <glandulae cervicales uteri> B

7 **cervix of uterus** the lower and narrow end of the uterus, between the isthmus and the uterine ostium. <cervix uteri> B

8 **external os of uterus** the external opening of the cervix of the uterus into the vagina. <ostium uteri> B

9 **fundus of uterus** the part of the uterus above the orifices of the uterine tubes. <fundus uteri> B

10 **horn of uterus** either of the bluntly rounded superior lateral extremities of the body of the uterus that marks the entrance of the uterine tube. <cornu uteri> B

11 ***internal female genital organs*** the various organs in the female that are concerned with reproduction, including the ovaries, uterine tubes, uterus, and vagina. <organa genitalia feminina interna> B

12 **isthmus of uterus** the constricted part of the uterus between the cervix and the body. <isthmus uteri> B

13 **margin of uterus** either border of the uterus (right or left), at the upper portion of which the uterine tube is attached; called also *lateral margin of uterus.* <margo uteri> B

14 ***mucosa of uterus*** the mucous membrane (tunica mucosa) lining the uterus, whose thickness and structure vary with the phase of the menstrual cycle. Called also *endometrium.* <tunica mucosa uteri> B

15 ***mucosa of vagina*** the mucous membrane (tunica mucosa) lining the vagina. <tunica mucosa vaginae> B

16 **muscular layer of uterus** muscular layer or coat of uterus: the smooth muscle layer, which forms the mass of the organ; called also *myometrium* and *mesometrium.* <tunica muscularis uteri> B

17 **muscular layer of vagina** the muscular layer or coat of the vagina. <tunica muscularis vaginae> B

18 ***palmate folds*** a system of folds on the anterior and posterior walls of the cervical canal of the uterus, consisting of a median longitudinal ridge and shorter elevations extending laterally and upward. <plicae palmatae> B

19 ***serosa of uterus*** the serous coat or layer (tunica serosa) of the uterus; called also *perimetrium.* <tunica serosa uteri> B

20 ***spongy layer of vagina*** a thin layer of spongiose erectile tissue, located between the muscular layer and the mucous membrane of the vagina and containing a large plexus of blood vessels. <tunica spongiosa vaginae> B

21 ***subserosa of uterus*** the subserous layer of the uterus. <tela subserosa uteri> B

22 **supravaginal portion of cervix** the part of the cervix uteri that does not protrude into the vagina. <portio supravaginalis cervicis> B

23 **uterine cavity** the flattened space within the uterus, communicating on either side at the cornu with the uterine tubes and below with the vagina. <cavitas uteri> B

24 ***uterine glands*** simple tubular glands throughout the entire thickness and extent of the endometrium, which become enlarged during the premenstrual period. <glandulae uterinae> B

25 **uterine ostium** the point at which the cavity of the uterine tube becomes continuous with that of the uterus. <ostium uterinum tubae uterinae> B

26 **uterine part of uterine tube** the proximal part of the uterine tube, located within the wall of the uterus. <pars uterina tubae uterinae> B

27 **uterus** the hollow muscular organ in which the blastocyst normally becomes embedded and in which the developing embryo and fetus is nourished. In the nongravid adult, it is a pear-shaped structure, about 8 cm in length, consisting of a fundus, body, isthmus, and cervix. Its cavity opens into the vagina below, and into the uterine tube on either side at the cornu. It is supported by direct attachment to the vagina and by indirect attachment to various other nearby pelvic structures. Called also *metra.* B

28 **vaginal portion of cervix** the part of the uterine cervix that protrudes into the vagina and is lined with stratified squamous epithelium. <portio vaginalis cervicis> B

A

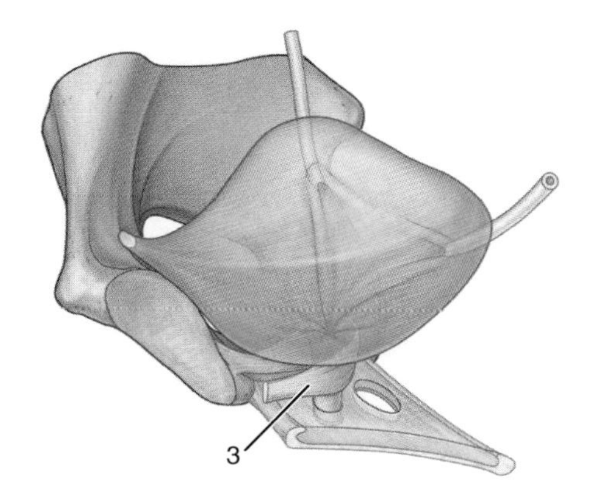

B

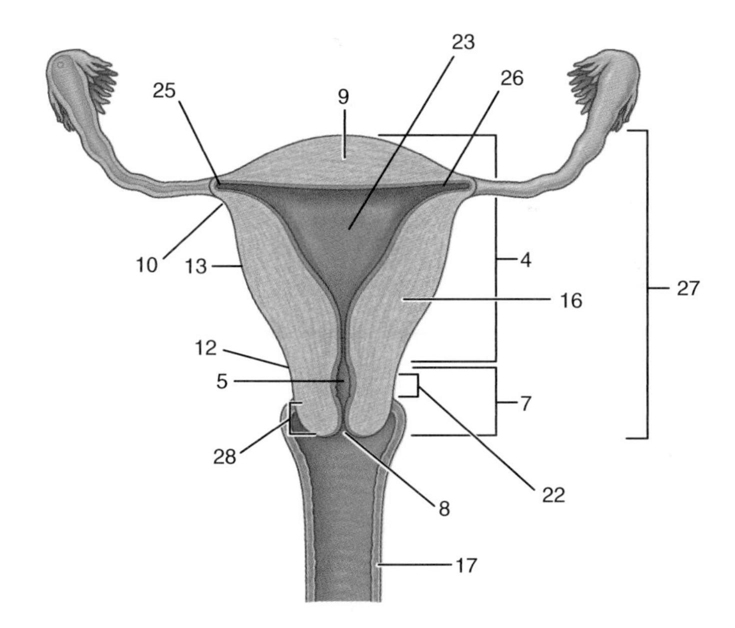

1 **abdominal opening of uterine tube** the funnel-shaped opening by which the uterine tube communicates with the pelvic cavity. <ostium abdominale tubae uterinae> A

2 **ampulla of uterine tube** the thin-walled, almost muscle-free, midregion of the uterine tube; its mucosa is greatly plicated. <ampulla tubae uterinae> B

3 **fimbriae of uterine tube** the numerous divergent fringelike processes on the distal part of the infundibulum of the uterine tube. <fimbriae tubae uterinae> A

4 **infundibulum of uterine tube** the funnel-like dilation at the distal end of the uterine tube. <infundibulum tubae uterinae> A

5 **isthmus of uterine tube** the narrow part of the uterine tube at its junction with the uterus. <isthmus tubae uterinae> B

6 **ligament of ovary** a musculofibrous cord in the broad ligament, joining the ovary to the upper part of the lateral margin of the uterus just below the attachment of the uterine tube. <ligamentum ovarii proprium> B

7 ***mucosa of uterine tube*** the mucous membrane (tunica mucosa) lining the uterine tube, arranged in longitudinal rugae or folds, and continuous with the mucous lining of the uterus. Called also *endosalpinx.* <tunica mucosa tubae uterinae> B

8 ***muscular layer of uterine tube*** the muscular layer or coat of the uterine tube. <tunica muscularis tubae uterinae> B

9 ***ovarian fimbria*** the longest of the processes that make up the fimbriae of the uterine tube, extending along the free border of the mesosalpinx; it is fused to the ovary, so that the ostium of the tube relates to the ovary. <fimbria ovarica> B

10 **round ligament of uterus** a fibromuscular band in the female that is attached to the uterus near the attachment of the uterine tube, passing then along the broad ligament, out through the inguinal ring, and into the labium majus. <ligamentum teres uteri> B

11 ***serosa of uterine tube*** the serous coat or layer (tunica serosa) of the uterine tube. <tunica serosa tubae uterinae> B

12 ***subserosa of uterine tube*** the subserous layer of the uterine tube. <tela subserosa tubae uterinae> B

13 **tubal extremity of ovary** the upper end of the ovary, related to the free end of the uterine tube. <extremitas tubaria ovarii> B

14 ***tubal folds of uterine tube*** the folds of the mucous lining of the uterine tube, which are high and complex in the ampulla. <plicae tubariae tubae uterinae> B

15 **uterine extremity of ovary** the lower end of the ovary, directed toward the uterus; called also *pelvic extremity of ovary.* <extremitas uterina ovarii> B

16 **uterine tube** a long slender tube that extends from the upper lateral cornu of the uterus to the region of the ovary of the same side; it is attached to the broad ligament by the mesosalpinx, and consists of an ampulla, infundibulum, isthmus, uterine part, and two ostia. Called also *fallopian tube, oviduct,* and *salpinx.* <tuba uterina> A

17 **anterior lip of external os of uterus** the anterior projection of the cervix into the vagina; it is shorter and thicker than the posterior lip. <labium anterius ostii uteri> C, D

18 **anterior part of vaginal fornix** the anterior part of the fornix of the vagina; called also *anterior fornix.* <pars anterior fornicis vaginae> C, D

19 ***anterior vaginal column*** a well-marked longitudinal ridge on the anterior wall of the vagina. <columna rugarum anterior vaginae> C

20 **anterior wall of vagina** the wall of the vagina that is intimately associated with the posterior wall of the bladder and urethra. <paries anterior vaginae> C

21 **posterior lip of external os of uterus** the posterior projection of the cervix into the vagina. <labium posterius ostii uteri> C, D

22 **posterior part of vaginal fornix** the posterior part of the fornix of the vagina; called also *posterior fornix.* <pars posterior fornicis vaginae> C, D

23 ***posterior vaginal column*** a well-marked longitudinal ridge on the posterior wall of the vagina. <columna rugarum posterior vaginae> C

24 **posterior wall of vagina** the vaginal wall intimately associated with the anterior wall of the rectum. <paries posterior vaginae> C

25 ***vaginal columns*** well-marked longitudinal ridges on either the anterior or posterior vaginal wall. <columnae rugarum vaginae> C

26 **vaginal fornix** the recess formed between the vaginal wall and the vaginal part of the cervix; sometimes subdivided into anterior, posterior, and lateral fornices, depending on the relation of the recess to the wall of the vagina; called also *fundus of vagina.* <fornix vaginae> D

27 ***vaginal rugae*** small transverse folds of the mucous membrane of the vagina extending outward from the columns. <rugae vaginales> C

28 **lateral part of vaginal fornix** the lateral part of the fornix of the vagina; called also *lateral fornix.* <pars lateralis fornicis vaginae> D

A

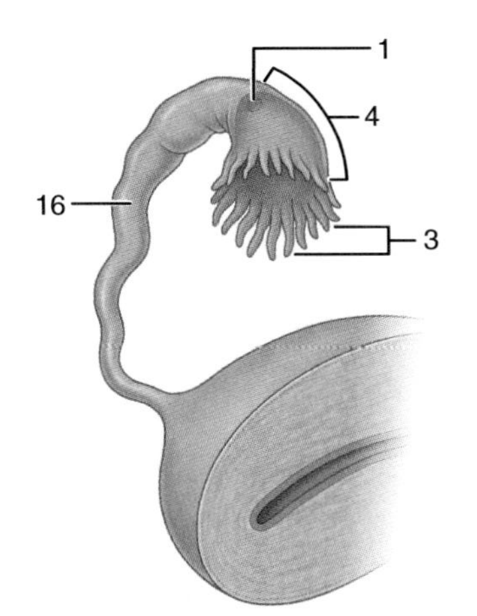

B

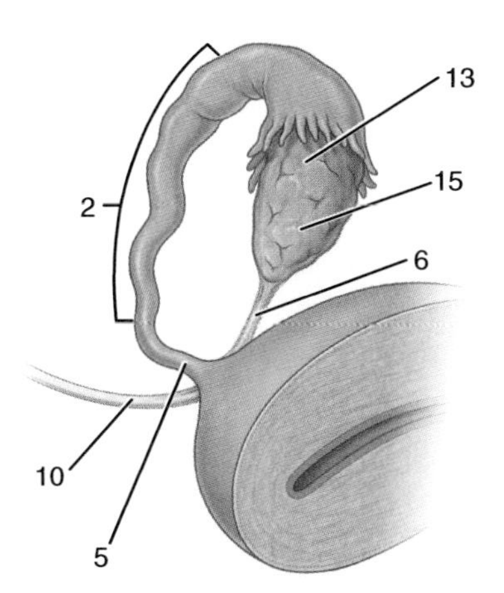

C

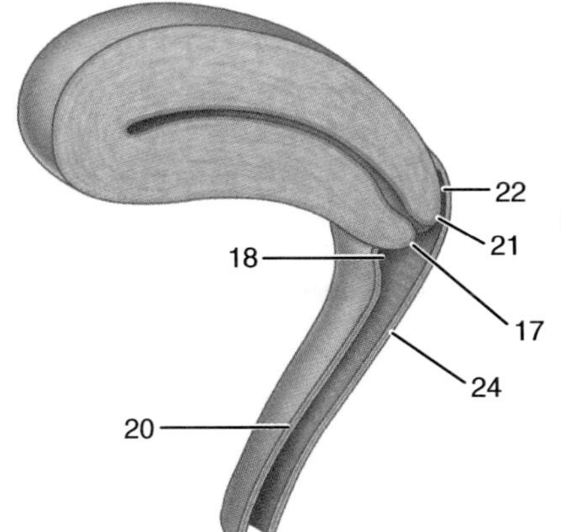

D

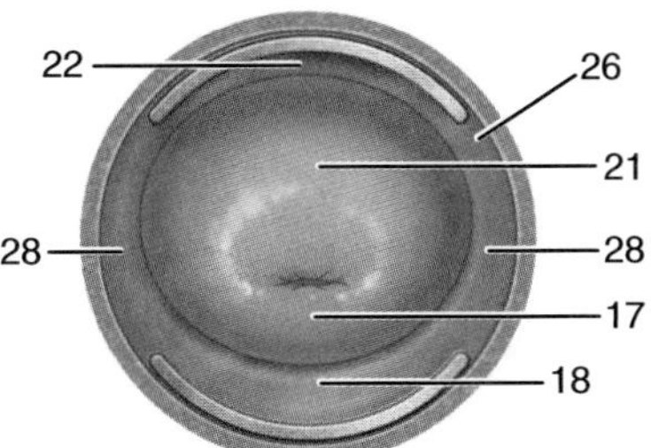

1 **cardinal ligament** part of a thickening of the visceral pelvic fascia beside the cervix and vagina, passing laterally to merge with the upper fascia of the pelvic diaphragm. Called also *transverse cervical ligament.* <ligamentum cardinale> A

2 **intestinal surface of uterus** the convex posterior surface of the uterus, adjacent to the intestine. <facies intestinalis uteri> B

3 ***lateral surface of ovary*** the surface of the ovary in contact with the lateral pelvic wall. <facies lateralis ovarii> B

4 **medial surface of ovary** the side of the ovary in contact with the fimbriated end of the uterine tube and the intestine. <facies medialis ovarii> B

5 **suspensory ligament of ovary** the portion of the broad ligament lateral to and above the ovary; it contains the ovarian vessels and nerves and passes upward over the iliac vessels. <ligamentum suspensorium ovarii> B

6 ***urethral carina of vagina*** the column of rugae in the lower part of the anterior wall of the vagina, immediately beneath the urethra. <carina urethralis vaginae> B

7 **vagina** the genital canal in the female, extending from the vulva to the cervix of the uterus, which receives the penis in copulation. B

8 **vesical surface of uterus** the flat anterior surface of the uterus, adjacent to the urinary bladder. <facies vesicalis uteri> B

9 **corpus albicans** white fibrous tissue that replaces the regressing corpus luteum in the human ovary in the latter half of pregnancy, or soon after ovulation when pregnancy does not supervene. C

10 **corpus luteum** a yellow glandular mass in the ovary formed by an ovarian follicle that has matured and discharged its oocyte. If the oocyte has been fertilized, the corpus luteum increases in size and persists for several months *(true corpus luteum)*; if fertilization has not taken place, it degenerates and shrinks *(false corpus luteum)*. The corpus luteum secretes progesterone. C

11 **cortex of ovary** the dense layer of compact stroma forming the peripheral zone around the ovarian medulla, in which the ovarian follicles are embedded. <cortex ovarii> C

12 **epoöphoron** a vestigial structure associated with the ovary, consisting of a more cranial group of mesonephric tubules and a corresponding portion of the mesonephric duct. Spelled also *epoophoron.* C

13 ***longitudinal duct of epoöphoron*** a closed rudimentary duct lying parallel to the uterine tube into which the transverse ducts of the epoöphoron open; it is a remnant of the part of the mesonephros that participates in formation of the reproductive organs. <ductus longitudinalis epoöphori> C

14 **ovarian medulla** the loose fibroelastic tissue and mass of contorted blood vessels that forms the core of the ovary. <medulla ovarii> C

15 **ovarian stroma** the fibrous tissue and smooth muscle composing the framework of the ovary. <stroma ovarii> C

16 ***transverse ductules of epoöphoron*** the vestigial remains of the mesonephric ducts, which open into the longitudinal duct of the epoöphoron. <ductuli transversi epoöphori> C

17 **tunica albuginea of ovary** the layer of dense connective tissue beneath the germinal epithelium of the ovary. <tunica albuginea ovarii> C

18 ***vesicular appendices of epoöphoron*** small pedunculated structures attached to the uterine tubes near their fimbriated end, being remnants of the mesonephric ducts; called also *morgagnian cyst.* <appendices vesiculosae epoöphori> C

19 **vesicular ovarian follicles** growing ovarian follicles that are nearing maturity and among whose cells fluid has begun to accumulate, leading to formation of a single cavity called the follicular antrum. The oocyte is eccentrically located in the cumulus oophorus, a hillock of follicle cells. Called also *antral follicles* and *graafian follicles* or *vesicles.* <folliculi ovarici vesiculosi> C

A

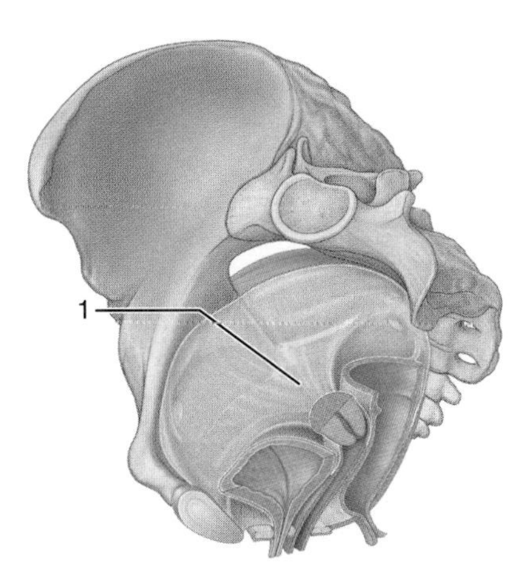

B

C

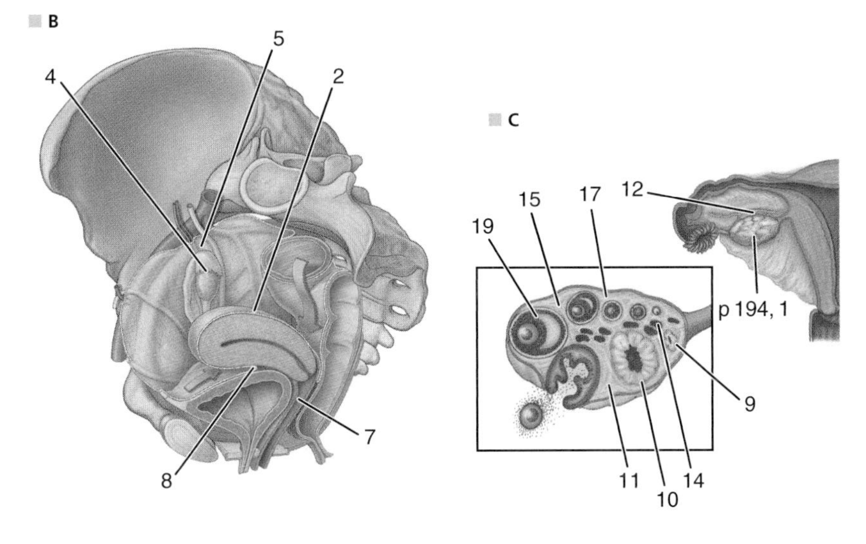

1 **ovary** the female gonad, one of the two sexual glands in which the oocytes are formed. It is a flat oval body along the lateral wall of the pelvic cavity, attached to the posterior surface of the broad ligament. It consists of a cortex and medulla; the former is composed of stroma and ovarian follicles in various stages of maturation, and the latter, occupying a small area in the hilar region, receives the vessels and nerves. The ovarium is covered by specialized epithelium continuous with the peritoneum at the mesovarium. <ovarium> A

2 **free border of ovary** the broad, convex border of the ovary, opposite the mesovarial margin. <margo liber ovarii> B

3 **hilum of ovary** the point on the mesovarial border of the ovary where the vessels and nerves enter. <hilum ovarii> B

4 **mesovarial border of ovary** the border of the ovary that is attached to the broad ligament by means of the mesovarium. <margo mesovaricus ovarii> B

5 ***paracervix*** the inferior part of the parametrium. B

6 ***parametrium*** the extension of the subserous coat of the supracervical portion of the uterus laterally between the layers of the broad ligament. B

7 **broad ligament of uterus** a broad fold of peritoneum extending from the side of the uterus to the wall of the pelvis; it is divided into the *mesometrium, mesosalpinx,* and *mesovarium.* <ligamentum latum uteri> A

8 **ovarian fossa** a shallow pouch on the posterior surface of the broad ligament, in which the ovary is located. <fossa ovarica> A

9 **mesometrium** the portion of the broad ligament below the mesovarium, composed of the layers of peritoneum that separate to enclose the uterus. B, C

10 **mesosalpinx** the part of the broad ligament of the uterus above the mesovarium, composed of layers that enclose the uterine tube. B, C

11 **mesovarium** the portion of the broad ligament of the uterus between the mesometrium and mesosalpinx; it is drawn out to enclose and hold the ovary in place. C

12 ***paroöphoron*** an inconstantly present small group of coiled tubules between the layers of the mesosalpinx, being a remnant of the excretory part of the mesonephros. C

Male Reproductive System

13 **puboprostatic ligament** a thickening of the superior fascia of the pelvic diaphragm in the male that, laterally, extends from the prostate to the tendinous arch of the pelvic fascia and, medially, is a forward continuation of the tendinous arch to the pubis. <ligamentum puboprostaticum> D

14 **capsule of prostate** the fibroelastic capsule that surrounds the prostate and contains an extensive plexus of veins. <capsula prostatica> E

15 **muscular tissue of prostate** the muscular stroma of the prostate, which is intimately blended with the fibrous capsule and permeates the glandular substance. <substantia muscularis prostatae> E

16 **parenchyma of prostate** glandular substance consisting of 30 to 50 small compound tubulosaccular or tubuloalveolar glands, making up the bulk of the prostate; their excretory ducts empty into the prostatic urethra. The parenchyma is surrounded by muscular substance and permeated by muscular strands. Called also *glandular substance of prostate.* <parenchyma prostatae> E

17 **prostatic ductules** minute ducts from the prostate gland that open on either side into or near the prostatic sinuses on the posterior wall of the urethra. Called also *prostatic ducts.* <ductuli prostatici> E

18 **prostatic sinus** the posterolateral recess between the seminal colliculus and the wall of the urethra, where the prostatic ductules empty into the urethra. <sinus prostaticus> E

19 **prostatic utricle** the remains in the male of the lower part of the müllerian (paramesonephric) duct; it is a small blind pouch arising in the parenchyma of the prostate and opening onto the seminal colliculus. <utriculus prostaticus> E

20 **seminal colliculus** a prominent portion of the urethral crest on which are the opening of the prostatic utricle and, on either side of it, the orifices of the ejaculatory ducts. <colliculus seminalis> E

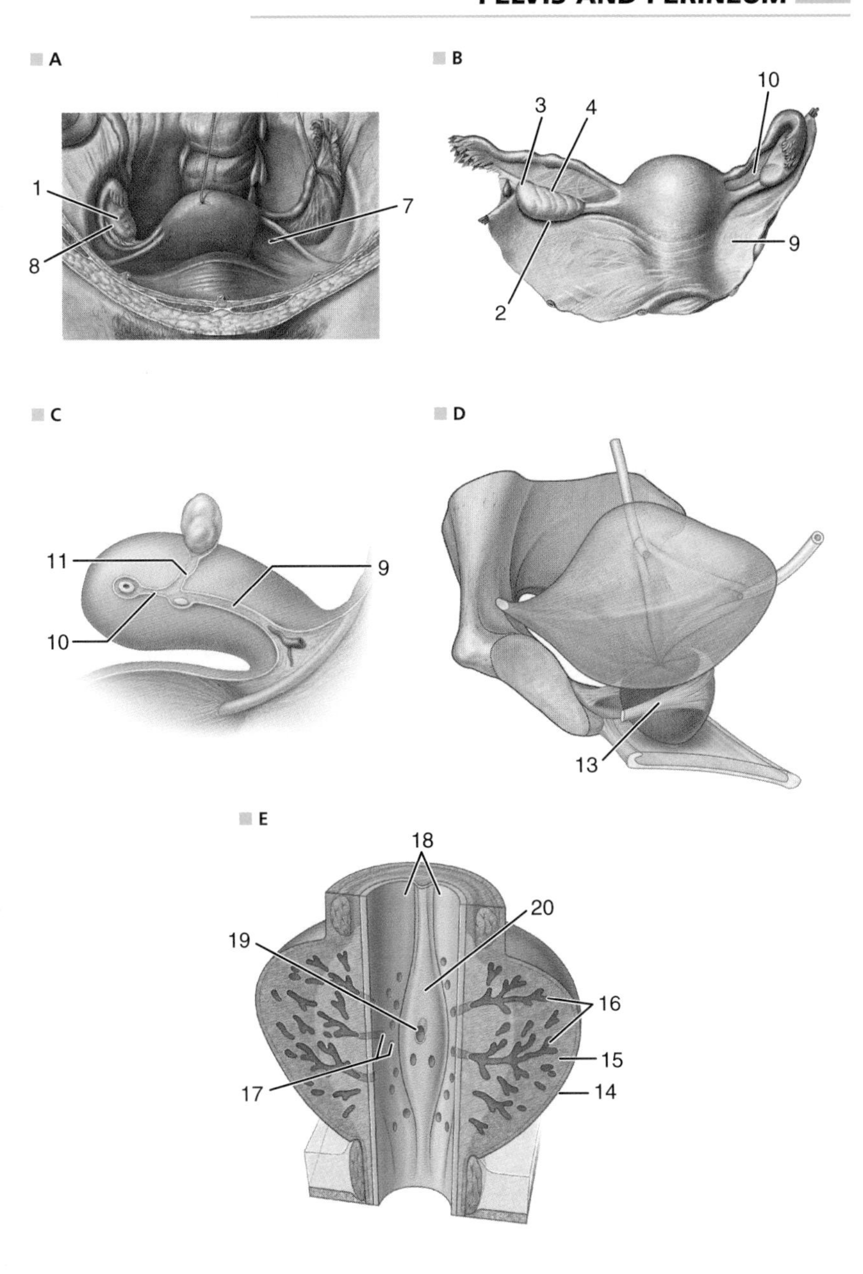
A
B
C
D
E
1
2
3
4
7
8
9
10
9
10
11
9
10
13
14
15
16
17
18
19
20

PELVIS AND PERINEUM

1. ***adventitia of ductus deferens*** the tunica adventitia of the ductus deferens. <tunica adventitia ductus deferentis> A
2. ***adventitia of seminal vesicle*** the adventitious coat (tunica adventitia) of a seminal vesicle. <tunica adventitia glandulae vesiculosae> A
3. **ampulla of ductus deferens** the enlarged and tortuous distal end of the ductus deferens. <ampulla ductus deferentis> A
4. **anterior surface of prostate** the surface towards the pubic symphysis, separated from the symphysis by adipose tissue and the prostatic venous plexus. <facies anterior prostatae> A
5. **apex of prostate** the lower portion of the prostate, located just superior to the pubococcygeus muscle. <apex prostatae> A
6. **base of prostate** the broad upper part of the prostate, in contact with the lower surface of the urinary bladder. <basis prostatae> A
7. ***diverticula of ampulla*** sacculations in the wall of the ampulla of the ductus deferens. <diverticula ampullae ductus deferentis> A
8. **ductus deferens** the excretory duct of the testis, which unites with the excretory duct of the seminal vesicle to form the ejaculatory duct. It has been divided into four parts: scrotal, funicular, inguinal, and pelvic. Called also *vas deferens, spermatic duct,* and *testicular duct.* A
9. ***excretory duct of seminal vesicle*** the duct that drains the seminal vesicle and unites with the ductus deferens to form the ejaculatory duct. <ductus excretorius glandulae vesiculosae> A
10. ***funicular part of ductus deferens*** a middle part of the ductus deferens, where it is within the spermatic cord. <pars funicularis ductus deferentis> A
11. **inferolateral surface of prostate** either of the surfaces on the convex end of the gland away from the base, separated from the superior fascia of the pelvic diaphragm by a venous plexus. <facies inferolateralis prostatae> A
12. ***inguinal part of ductus deferens*** a middle part of the ductus deferens, where it is within the inguinal canal. <pars inguinalis ductus deferentis> A
13. ***internal male genital organs*** the internal organs in the male that are concerned with reproduction, including the testis, epididymis, ductus deferens, seminal vesicle, ejaculatory duct, prostate, and bulbourethral gland. <organa genitalia masculina interna> A
14. ***lateral surface of testis*** the surface of the testis that is directed away from its fellow of the opposite side. <facies lateralis testis> A
15. **medial surface of testis** the surface of the testis that is directed toward its fellow of the opposite side. <facies medialis testis> A
16. ***mucosa of ductus deferens*** the mucous membrane (tunica mucosa) lining the ductus deferens. <tunica mucosa ductus deferentis> A
17. ***mucosa of seminal vesicle*** the mucous membrane (tunica mucosa) lining the seminal vesicle. <tunica mucosa glandulae vesiculosae> A
18. ***muscular coat of ductus deferens*** the muscular layer of the ductus deferens. <tunica muscularis ductus deferentis> A
19. ***muscular coat of seminal vesicle*** the muscular layer of the seminal vesicle.. <tunica muscularis glandulae vesiculosae> A
20. ***paradidymis*** a body made up of a few convoluted tubules in the anterior part of the spermatic cord, considered to be a remnant of the mesonephros. A
21. **pelvic part of ductus deferens** the distal part of the ductus deferens, where it is within the pelvic cavity and terminates at the ampulla. <pars pelvica ductus deferentis> A
22. **posterior surface of prostate** the surface towards the anterior wall of the rectum, separated from it by fascia. <facies posterior prostatae> A
23. ***prostatic fascia*** the reflection of the superior fascia of the pelvic diaphragm onto the prostate. <fascia prostatae> A
24. **scrotal part of ductus deferens** the initial part of the ductus deferens, which is within the scrotum. <pars scrotalis ductus deferentis> A
25. **seminal vesicle** either of the paired, sacculated pouches attached to the posterior part of the urinary bladder; the duct of each joins the ipsilateral ductus deferens to form the ejaculatory duct. Called also *seminal gland.* <glandula vesiculosa> A
26. **prostate** a gland in the male that surrounds the bladder neck and urethra. It consists of a median lobe and two lateral lobes, and is made up partly of glandular matter whose ducts empty into the prostatic part of the urethra, and partly of muscular fibers that encircle the urethra. The prostate contributes to the seminal fluid a secretion that liquefies the coagulated semen. <prostata> A
27. **testis** the male gonad; either of the paired egg-shaped glands normally situated in the scrotum. Each testis is surrounded by an outer tunica vaginalis and an inner tunica albuginea, and is composed of lobules. Specialized interstitial cells *(Leydig cells)* secrete testosterone. Called also *orchis* and *testicle.* A, B
28. ***aberrant ductules*** blind vestiges of mesonephric tubules near the epididymis; see *inferior aberrant ductule* and *superior aberrant ductule.* <ductuli aberrantes> B
29. **body of epididymis** the middle part of the epididymis, formed by the convolutions of the duct of epididymidis. <corpus epididymidis> B
30. **duct of epididymis** the single tube into which the coiled ends of the efferent ductules of the testis open, the convolutions of which make up the greater part of the epididymis. <ductus epididymidis> B

A

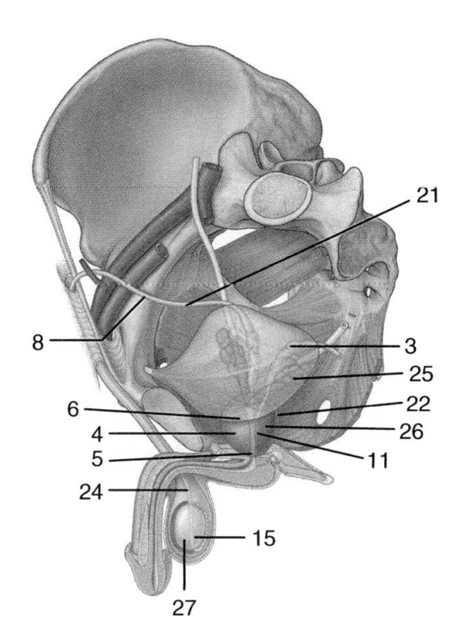

B

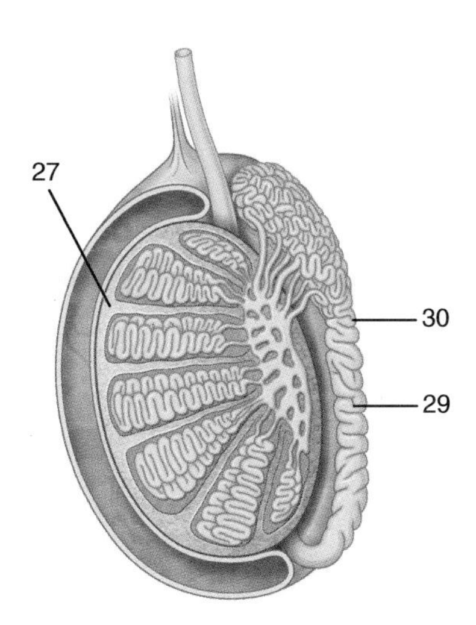

1 **efferent ductules of testis** ductules entering the head of the epididymis from the rete testis. <ductuli efferentes testis> A

2 **epididymis** the elongated cordlike structure along the posterior border of the testis, whose elongated coiled duct provides for storage, transit, and maturation of spermatozoa and is continuous with the ductus deferens. It consists of a head, body, and tail. A, B

3 **head of epididymis** the upper part of the epididymis, in which are found the straight and coiled portions of the efferent ductules of the testis. <caput epididymidis> A

4 ***inferior aberrant ductule*** a narrow, coiled tube often connected with the first part of the ductus deferens, or with the lower part of the duct of the epididymis. <ductulus aberrans inferior> A

5 ***lobules of epididymis*** the wedge-shaped parts of the head of the epididymis, each comprising a single efferent ductule of the testis. <lobuli epididymidis> A

6 **lobules of testis** the pyramidal subdivisions of the testicular substance, each with its base against the tunica albuginea and its apex at the mediastinum, and composed largely of seminiferous tubules. <lobuli testis> A

7 **mediastinum of testis** the partial septum of the testis, formed near its posterior border by fibrous tissue that is continuous with the tunica albuginea. <mediastinum testis> A

8 **parenchyma of testis** the functional tissue of the testis, consisting of the seminiferous tubules, Leydig cells, and Sertoli cells. <parenchyma testis> A

9 **parietal layer of tunica vaginalis of testis** the outer layer of the tunica vaginalis, separated from the visceral layer by a cavity. <lamina parietalis tunicae vaginalis testis> A

10 **rete testis** a network of channels formed by the straight seminiferous tubules, traversing the mediastinum testis and draining into the efferent ductules. A

11 **seminiferous tubules** the numerous delicate, contorted canals within each lobule of the testis; their epithelial linings contain Sertoli cells and germ cells. Called also *convoluted seminiferous tubules.* <tubuli seminiferi contorti> A

12 **septa of testis** connective tissue lamellae from the inner surface of the tunica albuginea, uniting to form the mediastinum of testis. <septula testis> A

13 **straight tubules** the straight terminal portion of the seminiferous tubules; they join to form the rete testis. <tubuli seminiferi recti> A

14 ***superior aberrant ductule*** a narrow tube of variable length that lies in the epididymis and is connected with the rete testis; called also *ductus aberrans.* <ductulus aberrans superior> A

15 **tail of epididymis** the lower part of the epididymis, where the duct of epididymidis is continuous with the ductus deferens. <cauda epididymidis> A

16 **tunica albuginea of testis** the dense, white, inelastic tissue immediately covering the testis, beneath the visceral layer of the tunica vaginalis. <tunica albuginea testis> A

17 **tunica vaginalis** the serous membrane covering the front and sides of the testis and epididymis, composed of visceral and parietal layers. <tunica vaginalis testis> A

18 ***tunica vasculosa*** the vascular coat immediately surrounding the testis, underlying the tunica albuginea. <tunica vasculosa testis> A

19 **visceral layer of tunica vaginalis of testis** the inner part of the tunica vaginalis, firmly attached to the testis and epididymis. <lamina visceralis tunicae vaginalis testis> A

20 **anterior border of testis** the rounded free border of the testis. <margo anterior testis> B

21 **appendix of epididymis** a remnant of the mesonephros sometimes found on the head of the epididymis; called also *appendage of epididymis.* <appendix epididymidis> B

22 **appendix of testis** the remnant of part of the müllerian duct on the upper end of the testis; called also *morgagnian cyst.* <appendix testis> B

23 **inferior extremity of testis** the lower end of the testis, which is attached to the tail of the epididymis. Called also *inferior* or *lower pole of testis.* <extremitas inferior testis> B

24 **inferior ligament of epididymis** a strand of fibrous tissue, covered with a reflection of the tunica vaginalis, which connects the lower end of the body of the epididymis with the testis. <ligamentum epididymidis inferius> B

25 **posterior border of testis** the border of the testis that is attached to the epididymis and the lower end of the ductus deferens. <margo posterior testis> B

26 **sinus of epididymis** a long, slitlike serous pocket between the upper part of the testis and the overlying epididymis. <sinus epididymidis> B

27 **superior extremity of testis** the upper end of the testis, which is attached to the head of the epididymis. Called also *superior* or *upper pole of testis.* <extremitas superior testis> B

28 **superior ligament of epididymis** a strand of fibrous tissue, covered with a reflection of the tunica vaginalis, which connects the upper end of the body of the epididymis with the testis. <ligamentum epididymidis superius> B

A

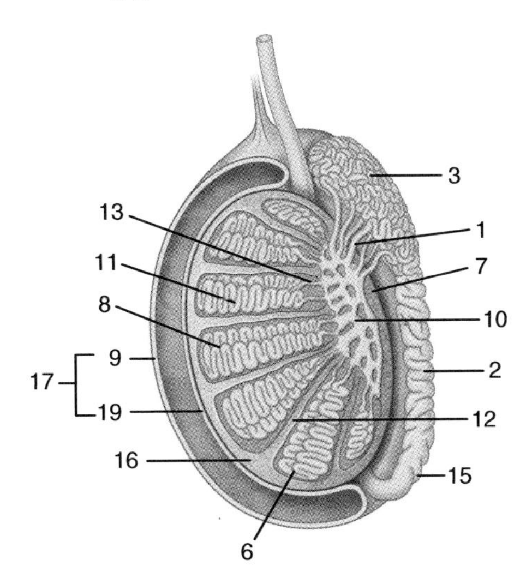
3
13
1
11
7
8
10
9
2
17
19
12
16
15
6

B

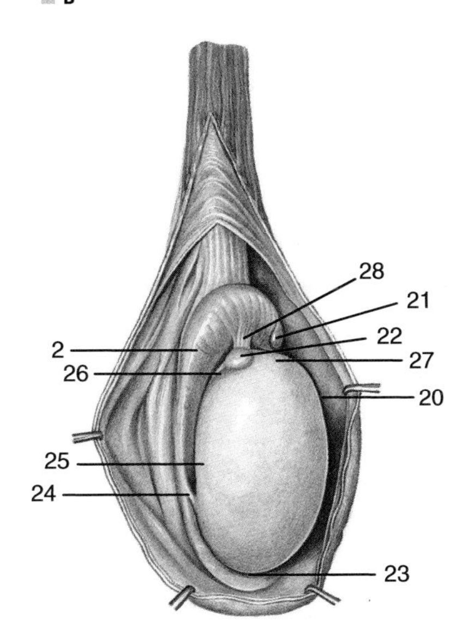
28
21
22
2
27
26
20
25
24
23

1 **isthmus of prostate** the commissure on the base of the prostate, between the right and the left lateral lobe. <isthmus prostatae> A

2 **lobes of prostate** the paired halves (right and left) of the prostate, separated by a more or less distinct median sulcus; called also *lateral lobes of prostate gland.* <lobi prostatae dexter et sinister> A

3 **median lobe of prostate** an inconstant but normal enlargement of the isthmus of the prostate. <lobus medius prostatae> A

Nerves of Pelvis

4 **anococcygeal nerve** *origin,* coccygeal plexus; *distribution,* sacrococcygeal joint, coccyx, skin over the coccyx; *modality,* general sensory. <nervus anococcygeus> B

5 **coccygeal plexus** a small plexus formed by the anterior branches of the coccygeal and the fifth sacral nerves, and a communication from the fourth sacral nerve, and giving off the anococcygeal nerves. <plexus coccygeus> B

6 **nerve to obturator internus** *origin,* anterior branches of anterior rami of L5, S1-S2; *distribution,* superior gemellus and obturator internus muscles; *modality,* general sensory and motor. <nervus musculi obturatorii interni> B

7 **hypogastric nerve** a nerve trunk situated on either side (right and left), interconnecting the superior and inferior hypogastric plexuses. <nervus hypogastricus> C

8 ***parasympathetic root of pelvic ganglia*** *origin,* sacral plexus S2-S4; *distribution,* leaving the sacral plexus, they enter the inferior hypogastric plexus and supply the pelvic organs; *modality,* preganglionic parasympathetic and visceral afferent. Called also *pelvic splanchnic nerves.* <radix parasympathica gangliorum pelvicorum> C

9 **pelvic part of parasympathetic nervous system** the part of the parasympathetic nervous system that includes branches from the second to fourth anterior sacral roots. <pars pelvica partis parasympatheticae systematis nervosi autonomici> C

10 ***deferential plexus*** the subdivision of the inferior hypogastric plexus that supplies nerve fibers to the ductus deferens. Called also *plexus of ductus deferens.* <plexus deferentialis> D

11 **inferior hypogastric plexus** the plexus formed on each side anterior to the lower part of the sacrum, formed by the junction of the hypogastric and pelvic splanchnic nerves; branches are given off to the pelvic organs. <plexus hypogastricus inferior> D

12 ***inferior rectal plexus*** a plexus accompanying the inferior rectal artery, derived chiefly from the inferior rectal nerve. <plexus rectalis inferior> D

13 ***middle rectal plexus*** a subdivision of the inferior hypogastric plexus, in proximity with and supplying nerve fibers to the rectum. <plexus rectalis medius> D

14 **prostatic plexus** a subdivision of the inferior hypogastric plexus that supplies nerve fibers to the prostate and adjacent organs. <plexus prostaticus> D

15 **sacral splanchnic nerves** *origin,* sacral part of sympathetic trunk; *distribution,* pelvic organs and blood vessels via inferior hypogastric plexus; *modality,* preganglionic sympathetic and visceral afferent. <nervi splanchnici sacrales> C, D

16 ***uterovaginal plexus*** the subdivision of the inferior hypogastric plexus that supplies nerve fibers to the uterus, ovary, vagina, urethra, and erectile tissue of the vestibule. <plexus uterovaginalis> D

17 ***vaginal nerves*** *origin,* uterovaginal plexus; *distribution,* vagina; *modality,* sympathetic and parasympathetic. <nervi vaginales> D

18 ***vesical plexus*** the subdivision of the inferior hypogastric plexus that supplies sympathetic nerve fibers to the urinary bladder and parts of the ureter, ductus deferens, and seminal vesicle. <plexus vesicalis> D

19 ***ovarian plexus*** a subdivision of the aortic plexus, accompanying the ovarian arteries. <plexus ovaricus>

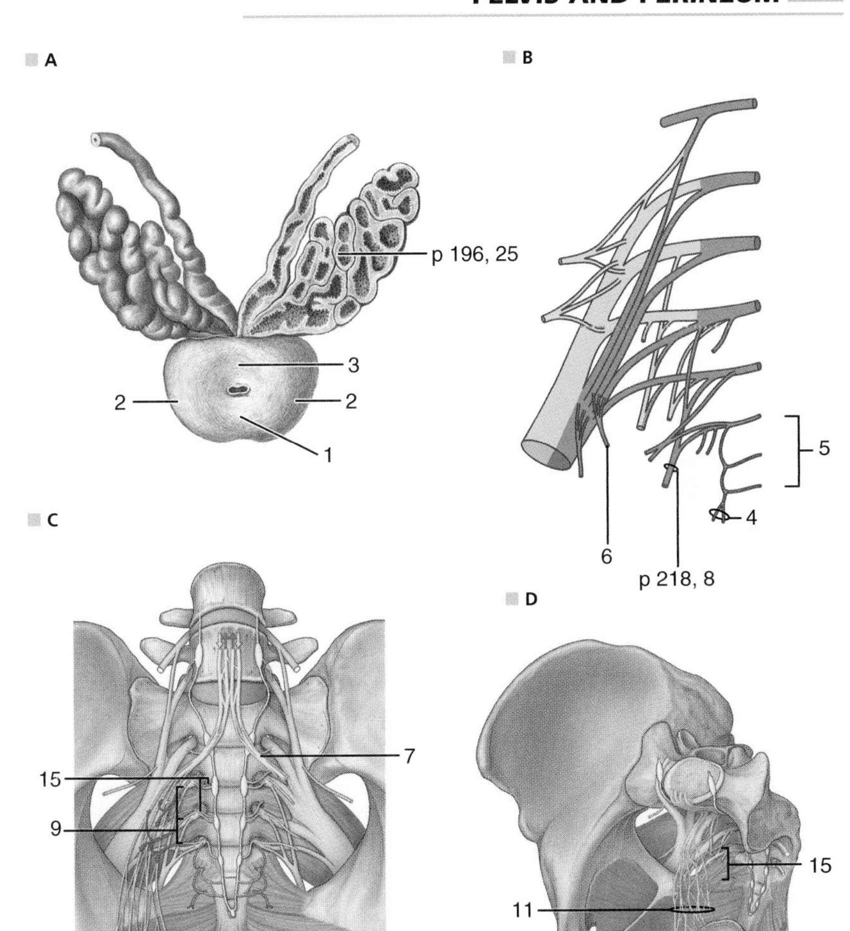
A
p 196, 25
3
2
2
1
B
5
4
6
p 218, 8
C
7
15
9
D
15
11
14

Vessels of Pelvis

1 **artery of ductus deferens** *origin,* umbilical artery; *branches,* ureteral artery; *distribution,* ureter, ductus deferens, seminal vesicles, testes. Called also *deferential artery.* <arteria ductus deferentis> A

2 **pampiniform plexus** 1. in the male, a plexus of veins from the testicle and the epididymis, constituting part of the spermatic cord. 2. in the female, a plexus of ovarian veins in the broad ligament. <plexus pampiniformis> A

3 ***ureteral branches of artery of ductus deferens*** branches that supply the lower portion of the ureter. <rami ureterici arteriae ductus deferentis> A

4 **left ovarian vein** a vein that drains the left pampiniform plexus of the broad ligament and empties into the left renal vein. <vena ovarica sinistra> B

5 **left testicular vein** a vein that drains the left pampiniform plexus and empties into the left renal vein. <vena testicularis sinistra> B

6 **iliac branch of iliolumbar artery** one of the two branches into which the iliolumbar artery divides in the iliac fossa; it supplies the iliacus muscle and sends a large nutrient branch to the ilium. <ramus iliacus arteriae iliolumbalis> C

7 **iliolumbar artery** *origin,* internal iliac; *branches,* iliac and lumbar branches, lateral sacral arteries; *distribution,* pelvic muscles and bones, fifth lumbar segment, sacrum. <arteria iliolumbalis> C

8 ***iliolumbar vein*** a vein that follows the distribution of the iliolumbar artery and opens into the internal iliac or the common iliac vein, or it may divide to end in both. <vena iliolumbalis> C

9 **lateral sacral arteries** usually two, a superior and an inferior, on each side; *origin,* posterior trunk of internal iliac artery; *branches,* spinal branches; *distribution,* structures about coccyx and sacrum. <arteriae sacrales laterales> C

10 ***lateral sacral veins*** veins that follow the homonymous arteries, help to form the lateral sacral plexus, and empty into the internal iliac vein or the superior gluteal veins. <venae sacrales laterales> C

11 **lumbar branch of iliolumbar artery** a branch that arises from the iliolumbar artery in the iliac fossa and ascends to supply the psoas and quadratus lumborum muscles, sending a spinal branch through the intervertebral foramen just above the sacrum. <ramus lumbalis arteriae iliolumbalis> C

12 **spinal branch of iliolumbar artery** a branch that passes through the intervertebral foramen between the fifth lumbar vertebra and the sacrum to help supply the contents of the vertebral canal. <ramus spinalis arteriae iliolumbalis> C

13 **inferior gluteal artery** *origin,* internal iliac; *branches,* sciatic; *distribution,* buttock, back of thigh. <arteria glutea inferior> D

14 **inferior vesical artery** *origin,* internal iliac; *branches,* prostatic; *distribution,* bladder, prostate, seminal vesicles, lower ureter. <arteria vesicalis inferior> D

15 ***lateral sacral branches of median sacral artery*** branches that anastomose with the lateral sacral arteries laterally. <rami sacrales laterales arteriae sacralis medianae> D

16 **median sacral artery** *origin,* continuation of abdominal aorta; *branches,* lowest lumbar artery; *distribution,* sacrum, coccyx, rectum. <arteria sacralis mediana> D

17 **middle rectal artery** *origin,* internal iliac artery; *branches,* vaginal; *distribution,* rectum, prostate, seminal vesicles, vagina. <arteria rectalis media> D

18 **obturator artery** *origin,* internal iliac; *branches,* pubic, acetabular, anterior, and posterior branches; *distribution,* pelvic muscles, hip joint. <arteria obturatoria> D

19 **occluded part of umbilical artery** the portion of an umbilical artery that atrophies at birth when the placental circulation ceases, becoming the medial umbilical ligament. <pars occlusa arteriae umbilicalis> D

20 **patent part of umbilical artery** the proximal section of the fetal umbilical cord, which remains patent in the adult, although reduced in size. <pars patens arteriae umbilicalis> D

21 ***prostatic branches of inferior vesical artery*** branches that supply the prostate and communicate with corresponding vessels on the opposite side. <rami prostatici arteriae vesicalis inferioris> D

22 **superior vesical arteries** *origin,* umbilical artery; *branches,* none; *distribution,* bladder, urachus, ureter. <arteriae vesicales superiores> D

23 **umbilical artery** *origin,* internal iliac artery; *branches,* deferential, superior vesical arteries; *distribution,* ductus deferens, seminal vesicles, testes, urinary bladder, ureter. <arteria umbilicalis> D

24 ***vaginal branches of middle rectal artery*** branches of the middle rectal artery that supply the vagina. <rami vaginales arteriae rectalis mediae> D

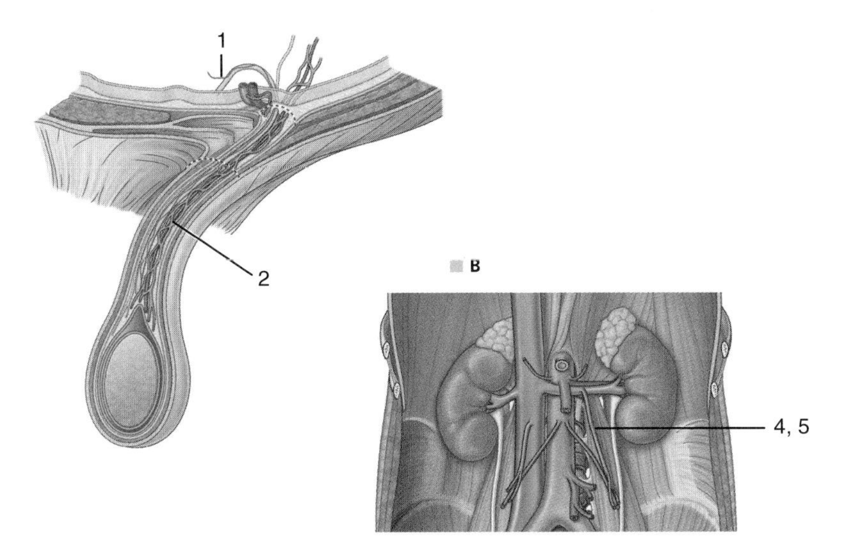
A
1
2
B
4, 5

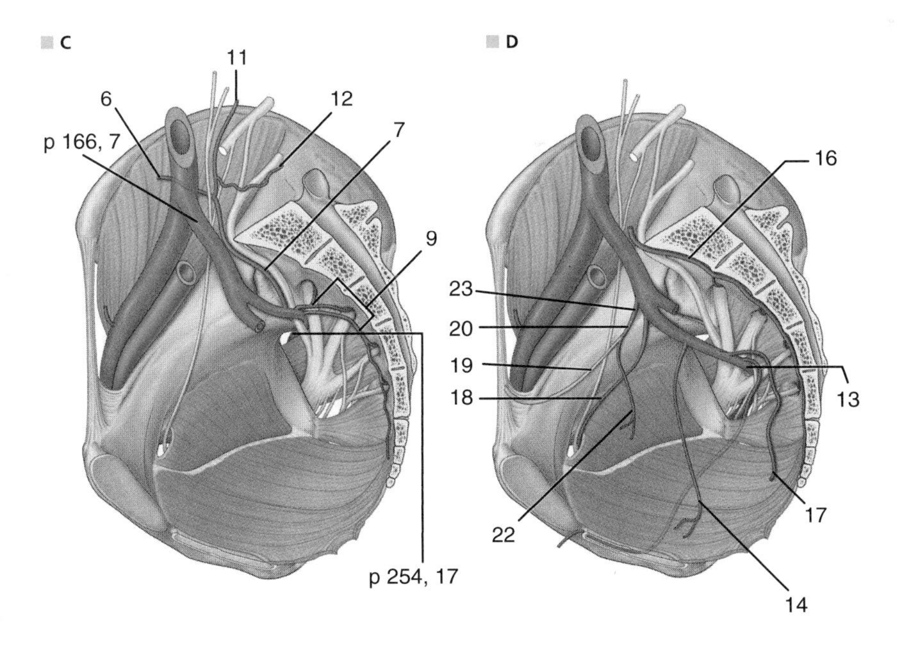
C
11
6
12
p 166, 7
7
9
p 254, 17
D
16
23
20
19
18
13
22
17
14

1 **median sacral vein** a vein that follows the median sacral artery and opens into the common iliac vein. <vena sacralis mediana> A

2 **middle rectal veins** veins that drain the rectal plexus and empty into the internal iliac and superior rectal veins. <venae rectales mediae> A

3 **obturator veins** veins that drain the hip joint and the regional muscles, enter the pelvis through the obturator canal, and empty into the internal iliac or the inferior epigastric vein, or both. <venae obturatoriae> A

4 **prostatic venous plexus** a venous plexus around the prostate gland, receiving the deep dorsal vein of the penis and draining through the vesical plexus and the prostatic veins. <plexus venosus prostaticus> A

5 ***sacral venous plexus*** the plexus on the pelvic surface of the sacrum that receives the sacral intervertebral veins, anastomoses with neighboring lumbar and pelvic veins, and drains into the middle and lateral sacral veins. <plexus venosus sacralis> A

6 ***vesical veins*** veins passing from the vesical plexus to the internal iliac vein. <venae vesicales> A

7 ***vesical venous plexus*** a venous plexus surrounding the upper part of the urethra and the neck of the bladder, communicating with the vaginal plexus in the female and with the prostatic plexus in the male. <plexus venosus vesicalis> A

8 **rectal venous plexus** a venous plexus that surrounds the lower part of the rectum and drains into the rectal veins. <plexus venosus rectalis> B

9 **urethral artery** *origin,* internal pudendal artery; *branches,* none; *distribution,* urethra. <arteria urethralis> C

10 **internal iliac lymph nodes** nodes grouped around the origins of the branches of the internal iliac vessels, comprising two groups: superior and inferior gluteal and sacral lymph nodes; they receive afferent vessels from the pelvic viscera, perineum, and buttocks and send efferent vessels to the common iliac lymph nodes. <nodi lymphoidei iliaci interni> D

11 ***artery of round ligament of uterus*** *origin,* inferior epigastric artery; *branches,* none; *distribution,* round ligament of uterus. <arteria ligamenti teretis uteri> E

12 **helicine branches of uterine artery** the exceedingly tortuous terminal branches of the uterine artery in the uterine muscle. Called also *helicine arteries.* <rami helicini arteriae uterinae> E

13 **ovarian branch of uterine artery** the terminal branch of the uterine artery, which supplies the ovary and anastomoses with the ovarian artery. <ramus ovaricus arteriae uterinae> E

14 **tubal branch of uterine artery** a branch that supplies the uterine tube and the round ligament. <ramus tubarius arteriae uterinae> E

15 ***tubal branches of ovarian artery*** branches distributed to the uterine tubes. <rami tubarii arteriae ovaricae> E

16 **uterine artery** *origin,* internal iliac artery; *branches,* ovarian and tubal branches, vaginal artery; *distribution,* uterus, vagina, round ligament of uterus, uterine tube, ovary. Called also *fallopian artery.* <arteria uterina> E

17 ***uterine venous plexus*** the venous plexus around the uterus, draining into the internal iliac veins by way of the uterine veins. <plexus venosus uterinus> E

18 **vaginal artery** *origin,* uterine artery; *branches,* none; *distribution,* vagina, fundus of bladder. <arteria vaginalis> E

19 **vaginal branches of uterine artery** two median longitudinal vessels formed by anastomosis of branches of the uterine and vaginal arteries, one of which descends in front of and the other behind the vagina. Called also *azygos arteries of vagina.* <rami vaginales arteriae uterinae> E

20 ***vaginal venous plexus*** a venous plexus in the walls of the vagina, which drains into the internal iliac veins by way of the internal pudendal veins. <plexus venosus vaginalis> E

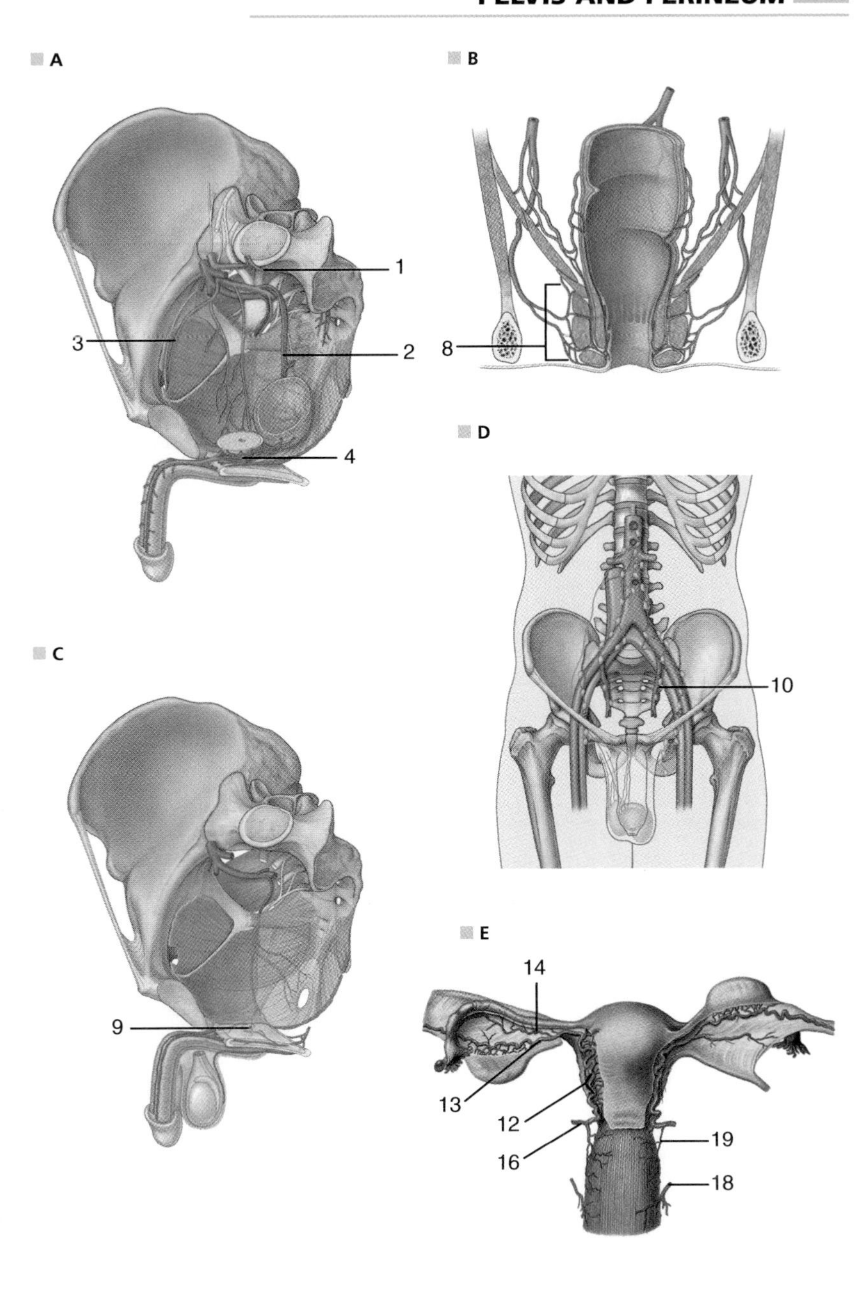
A
1
3
2
4
B
8
D
10
C
9
E
14
13
12
19
16
18

1 ***uterine veins*** veins that drain the uterine plexus into the internal iliac veins. <venae uterinae> A

2 **obturator lymph nodes** the external iliac lymph nodes situated in the obturator canal. <nodi lymphoidei obturatorii> A

3 **pararectal lymph nodes** lymph nodes situated around the rectum, embedded in its muscular coat; they drain into the inferior mesenteric, sacral, internal iliac, common iliac, and superficial inguinal nodes. Called also *anorectal lymph nodes*. <nodi lymphoidei pararectales> A

4 **parauterine lymph nodes** lymph nodes situated around the uterus, consisting of superficial (beneath the peritoneum) and deep (in the substance of the uterine wall) nodes: they drain into the lumbar, external and internal iliac, sacral, and superficial inguinal lymph nodes. <nodi lymphoidei parauterini> A

5 ***paravaginal lymph nodes*** lymph nodes situated around the vagina; they drain into the external and internal iliac, common iliac, and superficial inguinal lymph nodes. <nodi lymphoidei paravaginales> A

6 ***paravesicular lymph nodes*** lymph nodes situated around the urinary bladder, comprising three groups: perivesicular, postvesicular, and lateral vesicular lymph nodes; they drain into the external and internal iliac lymph nodes and, in association with some lymph nodes from the prostate, into the sacral and common iliac lymph nodes. <nodi lymphoidei paravesicales> A

7 **parietal pelvic lymph nodes** the lymph nodes that drain the wall of the pelvis, including the common iliac, external iliac, and internal iliac lymph nodes. <nodi lymphoidei pelvis parietales> A

8 ***prevesicular lymph nodes*** the paravesicular lymph nodes situated in front of the urinary bladder. <nodi lymphoidei prevesicales> A

9 ***pubic branch of obturator artery*** a branch that ascends on the pelvic surface of the ilium, anastomosing with its fellow of the other side and with the pubic branch of the inferior epigastric artery. <ramus pubicus arteriae obturatoriae> A

10 ***retrovesicular lymph nodes*** the paravesicular lymph nodes situated in back of the urinary bladder. Called also *postvesicular lymph nodes*. <nodi lymphoidei retrovesicales> A

11 ***sacral lymph nodes*** the internal iliac lymph nodes situated along the lateral and median sacral vessels; they receive lymph from the rectum and posterior pelvic wall. <nodi lymphoidei sacrales> A

12 ***ureteral branches of ovarian artery*** branches distributed to the ureter. <rami ureterici arteriae ovaricae> A

13 ***visceral pelvic lymph nodes*** the lymph nodes that drain the pelvic viscera, including the paravesicular, parauterine, paravaginal, and pararectal lymph nodes. <nodi lymphoidei pelvis viscerales> A

14 **accessory obturator artery** a name given to the obturator artery when it arises from the inferior epigastric instead of the internal iliac artery. <arteria obturatoria accessoria> B

15 ***obturator branch of inferior epigastric artery*** a vessel connecting the pubic branches of the inferior epigastric and the obturator arteries. The obturator artery is sometimes replaced by an accessory obturator that arises from the inferior epigastric artery by way of this communication. <ramus obturatorius rami pubici arteriae epigastricae inferioris> B

16 **anterior scrotal veins** veins that collect blood from the anterior aspect of the scrotum and drain into the external pudendal vein. <venae scrotales anteriores> C

PERINEUM
General

17 **anococcygeal body** a fibrous band connecting the posterior fibers of the anal sphincter to the coccyx. <corpus anococcygeum> D

18 **perineal body** the fibromuscular mass in the median plane of the perineum where converge and attach the bulbospongiosus and external anal sphincter muscles, the two levatores ani, and the two deep and the two superficial transverse perineal muscles. <corpus perineale> D

19 ***perineal fascia*** fascia that invests the ischiocavernous, bulbospongiosus, and superficial transverse perineal muscles; it is attached to the ischiopubic rami, perineal membrane, and perineal body, and anteriorly it is continuous with the suspensory ligament of the penis or clitoris. <fascia perinei> D

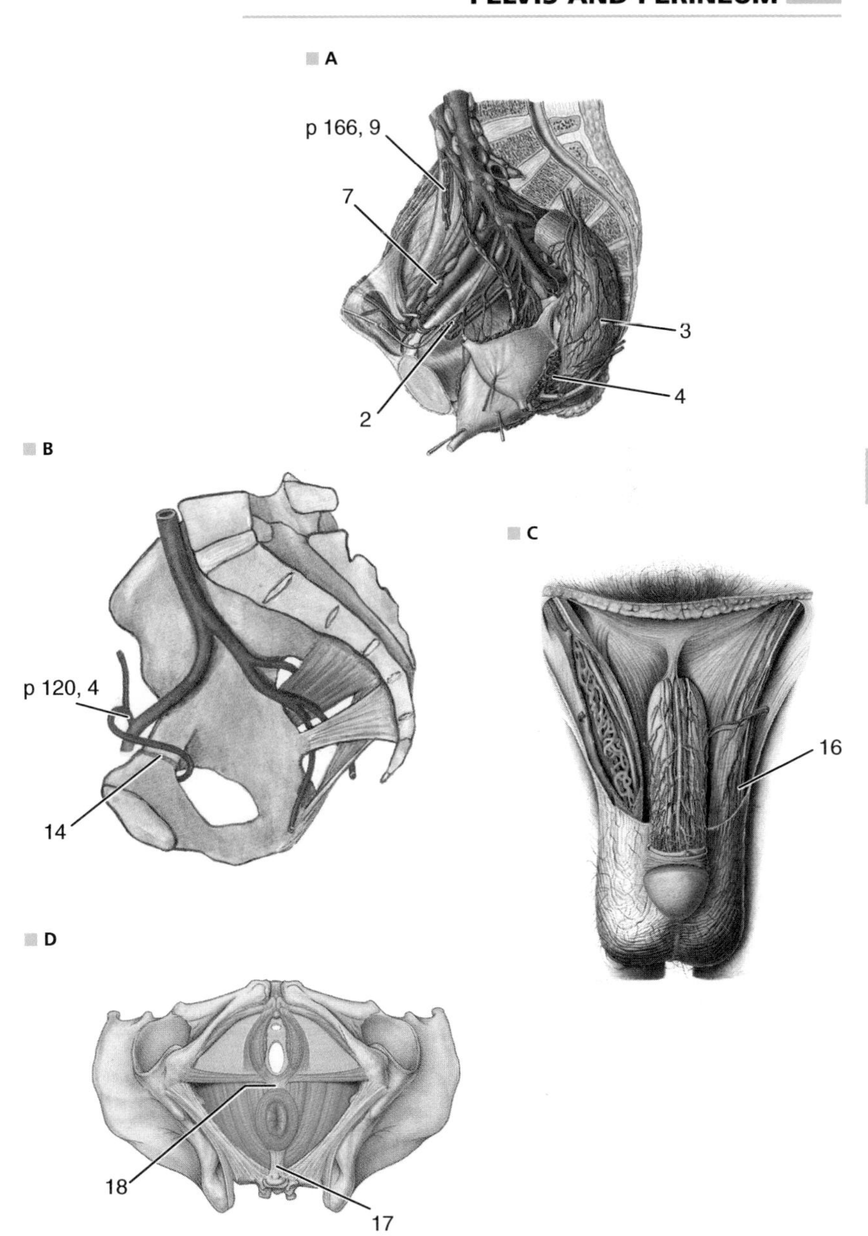
A
p 166, 9
7
3
4
2
B
p 120, 4
14
C
16
D
18
17

PELVIS AND PERINEUM

1 **deep perineal pouch** the region superior (deep) to the perineal membrane and extending into the pelvis; called also deep perineal space. <saccus profundus perinei> A, D

2 **perineal membrane** the triangular fibrous membrane stretched horizontally between the ischiopubic rami, attached at its base to the perineal body; its apex thickens to form the transverse perineal ligament. Called also *inferior fascia of urogenital diaphragm.* <membrana perinei> A

3 **anal triangle** the portion of the perineal region surrounding the anus; called also *anal region.* <regio analis> B

4 **urogenital triangle** the portion of the perineal region surrounding the urogenital organs. Called also *genitourinary region* and *urogenital region.* <regio urogenitalis> B

5 ***perineal region*** the region inferior to the pelvic floor and overlying the pelvic outlet, including the anal and urogenital triangles. <regio perinealis> B, C

6 **perineal raphe** a ridge along the median line of the perineum that runs forward from the anus; in the male, it is continuous with the raphe of the scrotum and the raphe of the penis. Called also *median r. of perineum.* <raphe perinei> C

7 **membranous layer of subcutaneous tissue of perineum** the deeper layer of subcutaneous tissue in the perineal area, where there is no subcutaneous fat. <stratum membranosum telae subcutaneae perinei> D

8 **subcutaneous perineal pouch** a potential space between the membranous layer of the perineal subcutaneous tissue and the superficial layer of the fascia of the perineal muscles. <saccus subcutaneus perinei> D

9 ***subcutaneous tissue of penis*** the loose external layer of fascial tissue of the penis, continuous with the tunica dartos and the subcutaneous tissue of perineum. Called also *superficial penile fascia.* <tela subcutanea penis> D

10 ***subcutaneous tissue of perineum*** the subcutaneous tissue of the urogenital region, consisting almost entirely of the membranous layer with little overlying fat. Called also *superficial perineal fascia.* <tela subcutanea perinei> D

11 **superficial perineal pouch** the region between the perineal membrane and the superficial perineal fascia; called also *superficial perineal compartment* or *space.* <compartimentum superficiale perinei> D

12 **fat body of ischioanal fossa** a pad of fat found in the ischioanal fossa. <corpus adiposum fossae ischioanalis> E

13 **inferior fascia of pelvic diaphragm** the fascia that covers the lower surface of the coccygeus and levator ani muscles, forming the medial wall of the ischiorectal fossa. <fascia inferior diaphragmatis pelvis> E

14 **ischioanal fossa** the potential space between the pelvic diaphragm and the skin inferior to it; an anterior recess extends a variable distance, sometimes reaching the retropubic space. Called also *perineal fossa.* <fossa ischioanalis> E

15 **pudendal canal** the tunnel in the special fascial sheath through which the pudendal vessels and nerve pass; it is intimately related to the obturator fascia. <canalis pudendalis> E

A

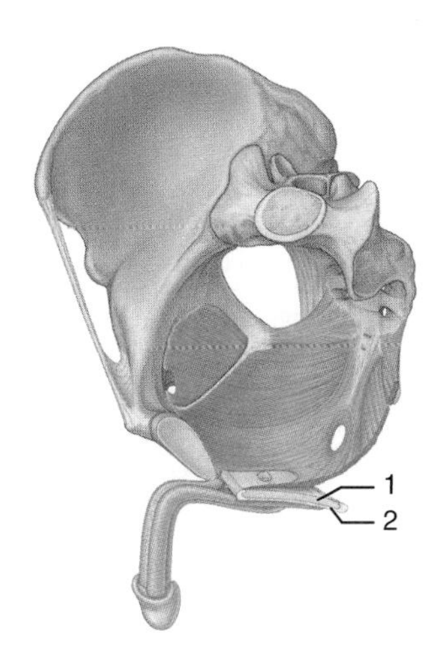

B

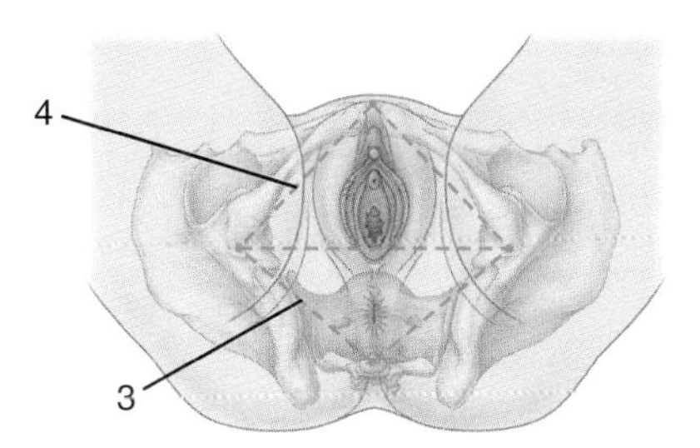

C

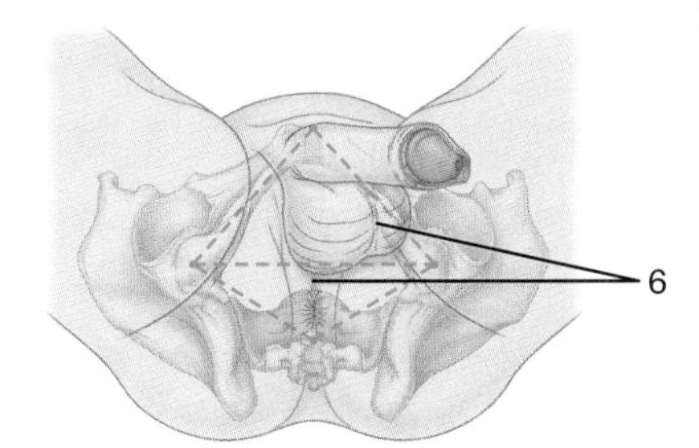

E

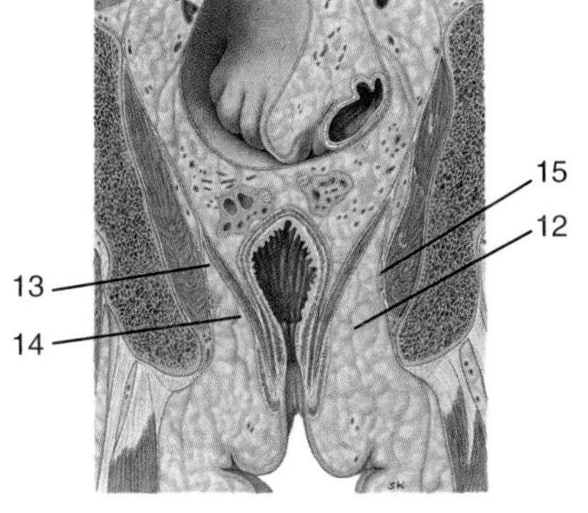

D

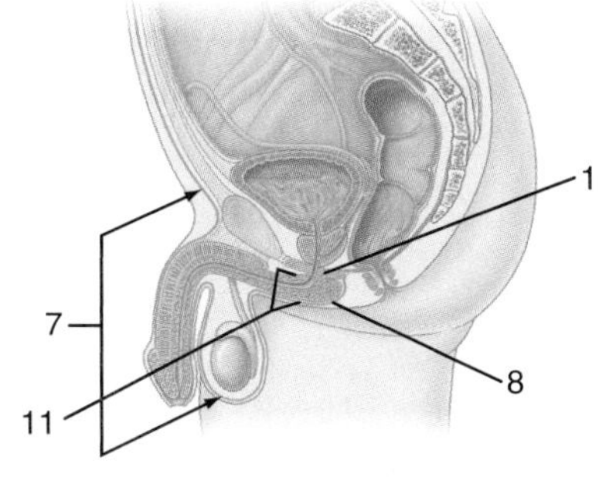

Female Surface Perineum

1 ***external female genital organs*** the external genitalia of the female, comprising the pudendum, clitoris, and urethra. <organa genitalia feminina externa> A, B

2 ***pudendum*** that portion of the female genitalia comprising the mons pubis, labia majora, labia minora, vestibule of the vagina, bulb of the vestibule, greater and lesser vestibular glands, and vaginal orifice. Commonly used to denote the entire external female genitalia (i.e., to include the clitoris and urethra as well). <pudendum femininum> A, B

3 ***fascia of clitoris*** the dense fibrous tissue that encloses the two corpora cavernosa of the clitoris. <fascia clitoridis> C

4 ***fundiform ligament of clitoris*** a broad elastic band of fascial fibers that arises from the linea alba above the pubic symphysis and then passes down to the clitoris, where it divides and passes around the clitoris and fuses with the fascia of the clitoris. <ligamentum fundiforme clitoridis> C

5 **suspensory ligament of clitoris** a strong fibrous band that comes from the external deep investing fascia and attaches the root of the clitoris to the linea alba, pubic symphysis, and arcuate pubic ligament. <ligamentum suspensorium clitoridis> C

6 **mons pubis** the rounded fleshy prominence over the pubic symphysis. A

7 **anterior commissure of labia** the junction of the two labia majora anteriorly, at the lower border of the pubic symphysis. <commissura labiorum anterior> B

8 **clitoris** a small, elongated, erectile body at the anterior angle of the pudendal fissure in the female; it is homologous with the penis in the male. A

9 **frenulum of clitoris** the tissue fold on the under surface of the clitoris formed by union of the two medial parts of the labia minora. <frenulum clitoridis> B

10 **frenulum of labia minora** the posterior union of the labia minora, anterior to the posterior commissure; called also fourchette. <frenulum labiorum pudendi> B

11 **glans of clitoris** erectile tissue at the end of the clitoris, which is continuous with the intermediate part of the vestibulovaginal bulbs. <glans clitoridis> B

12 **hymen** the membranous fold which partially or wholly occludes the external orifice of the vagina. B

13 **hymenal caruncles** small elevations of the mucous membrane encircling the vaginal orifice, being relics of the torn hymen. <carunculae hymenales> B

14 **labium majus** an elongated fold running downward and backward from the mons pubis in the female, one on either side of the median pudendal cleft. <labium majus pudendi> B

15 **labium minus** a small fold of skin, one on each side, running backward from the clitoris between the labium majus and the opening of the vagina. <labium minus pudendi> B

16 ***lesser vestibular glands*** small mucous glands opening upon the vestibular mucous membrane between the urethral and the vaginal orifice. <glandulae vestibulares minores> B

17 **paraurethral ducts of female urethra** inconstantly present ducts in the female, which drain a group of the urethral glands into the vestibule; called also *Skene's ducts* or *tubules*. <ductus paraurethrales urethrae femininae> B

18 **posterior commissure of labia** the apparent junction of the labia majora posteriorly, formed by the forward projection of the perineal body into the pudendal cleft. <commissura labiorum posterior> B

19 **prepuce of clitoris** a fold formed by the union of the labia minora anterior over the clitoris and united with the glans of clitoris. <preputium clitoridis> B

20 ***pudendal fissure*** the cleft between the labia majora in which the urethra and vagina open. <rima pudendi> B

21 **vaginal opening** the external opening of the vagina, situated just posterior to the external urethral orifice. <ostium vaginae> B

22 **vestibular fossa** the part of the vestibule between the opening of the vagina and the frenulum of the pudendal labia. <fossa vestibuli vaginae> B

23 **vestibule of vagina** the space between the labia minora into which the urethra and vagina open. <vestibulum vaginae> B

A

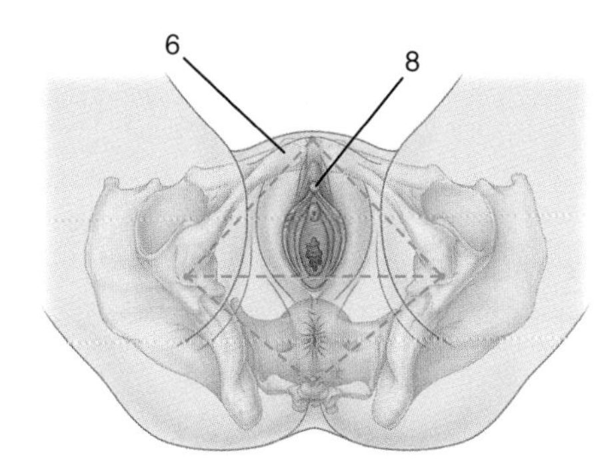

B

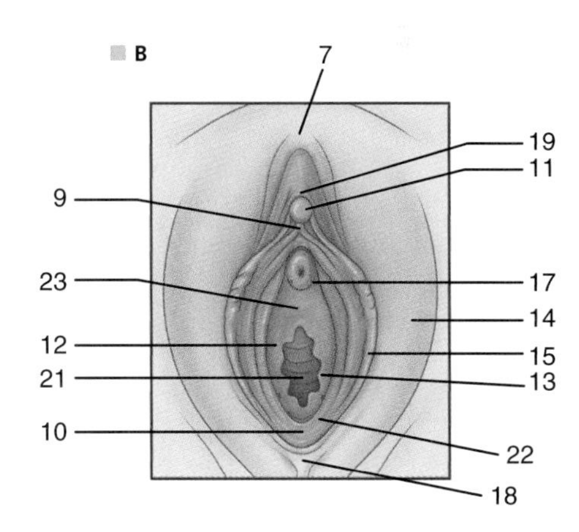

C

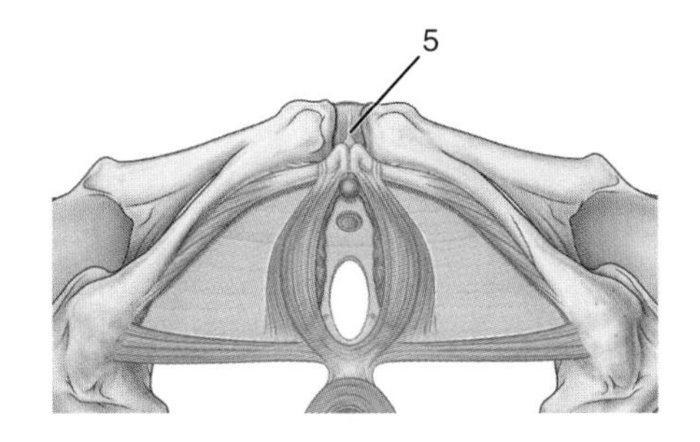

Male Surface Perineum

1 ***external male genital organs*** the external genitalia in the male, comprising the penis, scrotum, and urethra. <organa genitalia masculina externa> A

2 **body of penis** the free part of the penis from the root to the glans, consisting chiefly of the paired corpora cavernosa of penis and the single corpus spongiosum of penis; called also *shaft of penis.* <corpus penis> A

3 **dorsum of penis** the anterior, more extensive surface of the dependent penis, opposite the urethral surface. <dorsum penis> A

4 ***penis*** the male organ of copulation and of urinary excretion, comprising a *root, body,* and *glans.* The penis is homologous with the clitoris in the female. A

5 ***raphe of penis*** a narrow dark streak or ridge continuous posteriorly with the raphe of scrotum and extending forward for a variable distance along the midline of the urethral surface of the penis; in the newborn it may extend to the tip of the glans. <raphe penis> A

6 **raphe of scrotum** a ridge along the surface of the scrotum in the median line, dividing it into nearly equal lateral parts; it is continuous with the perineal raphe and the raphe of the penis. <raphe scroti> A

7 **root of penis** the proximal, attached portion of the penis, consisting of the diverging crura of the corpora cavernosa and the bulb. <radix penis> B

8 ***septum of scrotum*** a fibromuscular partition in the median plane, dividing the scrotum into two nearly equal parts. <septum scroti> A

9 **urethral surface of penis** the surface of the penis overlying the urethra, and opposite the dorsum penis. <facies urethralis penis> A

10 **scrotum** the pouch that contains the testes and their accessory organs. It is composed of skin, the tunica dartos, the spermatic, cremasteric, and infundibuliform fasciae, and the tunica vaginalis of the testis. A

11 **bulb of penis** the enlarged proximal part of the corpus spongiosum found between the two crura of the penis. <bulbus penis> B

12 **fascia of penis** the firm inner fascial layer that surrounds the corpora cavernosa and the corpus spongiosum collectively. <fascia penis> B

13 ***paraurethral ducts of male urethra*** the ducts of the urethral glands situated in the spongy portion of the male urethra. <ductus paraurethrales urethrae masculinae> B

14 **fundiform ligament of penis** a broad elastic band of fascial fibers that arises from the linea alba and from the intercrural fibers just above the pubic symphysis and then passes down to the penis, where it divides and passes around the penis and on into the scrotum. <ligamentum fundiforme penis> C

15 **suspensory ligament of penis** a strong fibrous band that comes from the external deep investing fascia and attaches the root of the penis to the linea alba, pubic symphysis, and arcuate pubic ligament. <ligamentum suspensorium penis> C

16 **corona of glans penis** the rounded proximal border of the glans penis, separated from the corpora cavernosa by the neck of the glans. <corona glandis penis> D

17 **frenulum of prepuce** the fold on the lower surface of the glans penis that connects it with the prepuce. <frenulum preputii penis> D

18 **glans penis** the cap-shaped expansion of the corpus spongiosum at the end of the penis, covered with mucous membrane and ensheathed by the prepuce; called also *balanus.* D

19 **neck of glans penis** the constricted portion between the corona of the glans penis and the corpora cavernosa. <collum glandis penis> D

20 **prepuce of penis** the fold of skin covering the glans penis; called also *foreskin.* <preputium penis> D

21 ***preputial glands*** small sebaceous glands of the corona of the penis and the inner surface of the prepuce, which secrete smegma; called also *Littre's* or *Tyson's glands.* <glandulae preputiales> D

22 ***dartos muscle*** the nonstriated muscle fibers of the tunica dartos, the deeper layers of which help to form the septum of the scrotum. Called also *dartos.* <musculus dartos> A

23 ***tunica dartos*** the thin layer of subcutaneous tissue underlying the skin of the scrotum, consisting mainly of nonstriated muscle fibers (*dartos muscle,* which is sometimes used to refer to the entire tunica dartos). Called also *dartos* and *dartos fascia of scrotum.* A

A

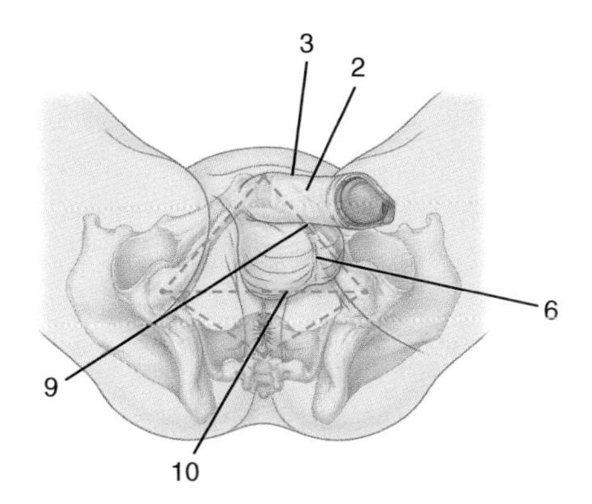

B

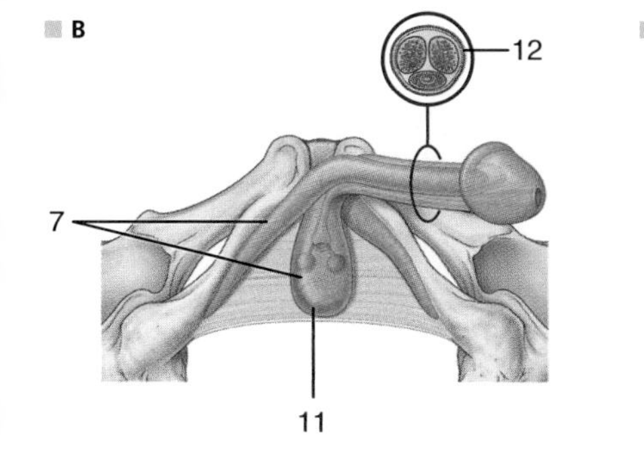

C

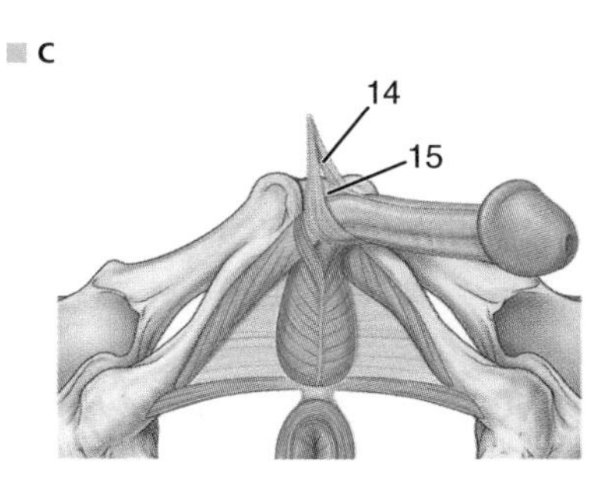

D

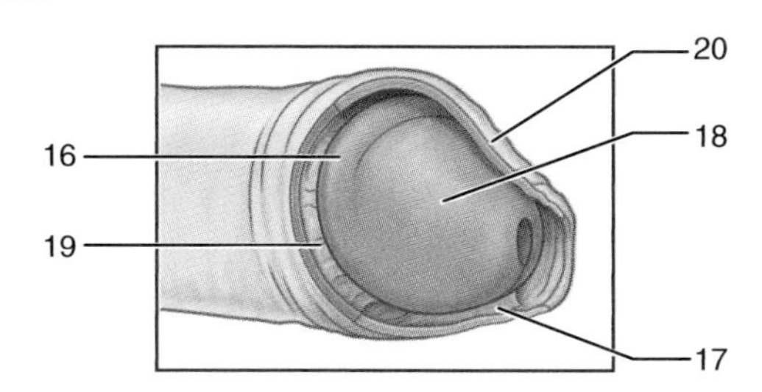

Erectile Tissues

1 **body of clitoris** the main part of the clitoris, formed by the two fused corpora cavernosa, which are embedded anteriorly in the floor of the vestibule of the vagina. <corpus clitoridis> A

2 **bulb of vestibule of vagina** a body consisting of paired elongated masses of erectile tissue, one on either side of the vaginal opening, united anteriorly by a narrow median band, which then expands slightly to form the glans of clitoris. <bulbus vestibuli vaginae> A

3 ***commissure of bulbs of vestibule*** a narrow median band spanning the vaginal orifice to unite the bulbs of the vestibule. <commissura bulborum vestibuli> A

4 **corpus cavernosum of clitoris** a column of erectile tissue on either side (right and left), the two fusing to form the body of the clitoris. <corpus cavernosum clitoridis> A

5 **crus of clitoris** the continuation of each corpus cavernosum of the clitoris, diverging posteriorly to be attached to the ischiopubic rami. <crus clitoridis> A

6 **septum of corpora cavernosa of clitoris** an incomplete fibrous septum between the two lateral halves of the clitoris. <septum corporum cavernosorum clitoridis> A

7 **caverns of corpora cavernosa of penis** the dilatable spaces within the corpora cavernosa of the penis, which fill with blood and become distended with erection. <cavernae corporum cavernosorum penis> B

8 **caverns of corpus spongiosum** the dilatable spaces within the corpus spongiosum of the penis, which fill with blood and become distended with erection. <cavernae corporis spongiosi penis> B

9 **corpus cavernosum of penis** one of the columns of erectile tissue forming the dorsum and sides of the penis. <corpus cavernosum penis> B

10 **corpus spongiosum** the column of erectile tissue that forms the urethral surface of the penis, and in which the urethra is found; its distal expansion forms the glans penis. <corpus spongiosum penis> B

11 **crus of penis** the continuation of each corpus cavernosum of the penis, diverging posteriorly to be attached to the pubic arch. <crus penis> B

12 ***septum of glans*** an incomplete fibrous septum in the median plane of the glans penis, especially below the urethra. <septum glandis penis> B

13 **septum of penis** the fibrous sheet between the two corpora cavernosa of the penis, formed by union of the tunicae albugineae of the two sides. <septum penis> B

14 ***trabeculae of corpora cavernosa*** numerous bands and cords of fibromuscular tissue traversing the interior of the corpora cavernosa of the penis, attached to the tunica albuginea and septum and creating the cavernous spaces that become filled with blood during erection. <trabeculae corporum cavernosorum penis> B

15 ***trabeculae of corpus spongiosum*** numerous bands and cords of fibromuscular tissue traversing the interior of the corpus spongiosum of the penis, creating the cavernous spaces that give the structure its spongy character. <trabeculae corporis spongiosi penis> B

16 **tunica albuginea of corpora cavernosa** the dense, white, fibroelastic sheath that encloses the corpora cavernosa of the penis. Its superficial, longitudinal fibers form a tunic surrounding both corpora, and the deep circularly coursing fibers surround them separately, uniting medially to form the septum of the penis. <tunica albuginea corporum cavernosorum> B

17 **tunica albuginea of corpus spongiosum** the dense, white, fibroelastic sheath that encloses the corpus spongiosum of the penis. <tunica albuginea corporis spongiosi> B

Glands of Perineum

18 **greater vestibular gland** either of two small reddish yellow bodies in the vestibular bulbs, one on each side of the vaginal orifice; they are homologues of the bulbourethral glands in the male. Called also *Bartholin's gland.* <glandula vestibularis major> A

19 **bulbourethral gland** either of two glands embedded in the substance of the external sphincter of the male urethra, just posterior to the membranous part of the urethra; they are homologues of the greater vestibular glands in the female. Called also *Cowper's gland.* <glandula bulbourethralis> B

20 **duct of bulbourethral gland** a duct passing from the bulbourethral gland through the perineal membrane into the bulb of the penis, where it enters the spongy part of the urethra. <ductus glandulae bulbourethralis> B

A

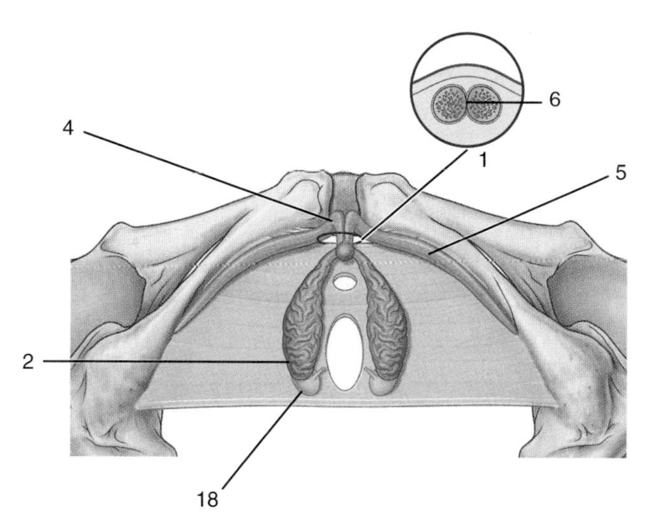

B

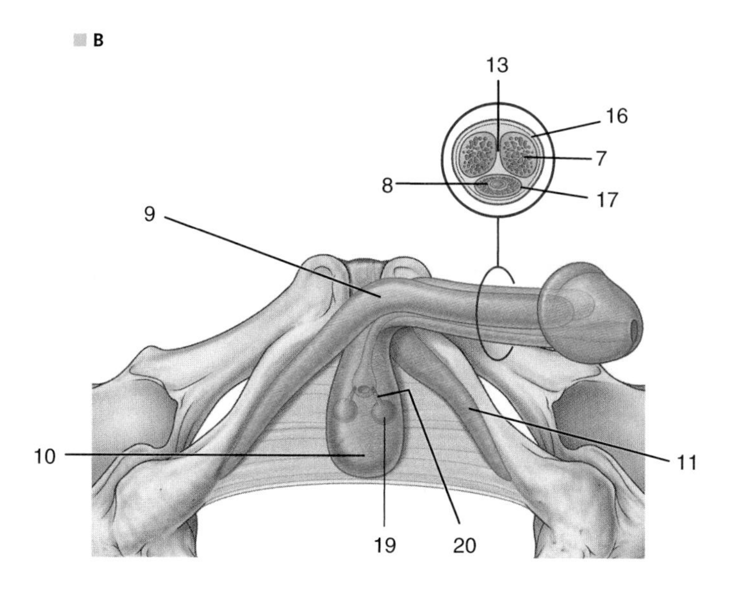

Muscles of Perineum

1 ***perineal muscles*** the muscles participating in formation of the perineum. <musculi perinei> A

2 **bulbospongiosus muscle** *origin,* central point of perineum, median raphe of bulb; *insertion,* fascia of penis (male) or clitoris (female); *innervation,* pudendal; *action,* constricts spongy urethra in males and vaginal orifice in females, contributes to erection of penis or clitoris. Called also *bulbocavernosus* or *bulbocavernous muscle.* <musculus bulbospongiosus> A

3 **ischiocavernosus muscle** *origin,* ramus of ischium; *insertion,* crus of penis or clitoris; *innervation,* perineal; *action,* maintains erection of penis or clitoris. <musculus ischiocavernosus> A

4 **superficial transverse perineal muscle** *origin,* ramus of ischium; *insertion,* perineal body; *innervation,* perineal; *action,* fixes perineal body. <musculus transversus perinei superficialis> A

5 **external urethral sphincter (female)** *origin,* ramus of pubis; *insertion,* median raphe behind and in front of urethra; *innervation,* perineal; *action,* compresses the central part of the urethra. <musculus sphincter urethrae externus urethrae femininae> B

6 **sphincter urethrovaginalis muscle** *origin,* perineal body; *insertion,* passes lateral to the vagina on either side and blends with its partner anterior to the urethra below the compressor urethrae muscle; *innervation,* perineal branches of the pudendal nerve; *action,* accessory urethral sphincter and may facilitate closing the vagina. <musculus sphincter urethrovaginalis> B

7 **deep transverse perineal muscle** *origin,* ramus of ischium; *insertion,* perineal body; *innervation,* perineal; *action,* fixes perineal body. <musculus transversus perinei profundus> C

A

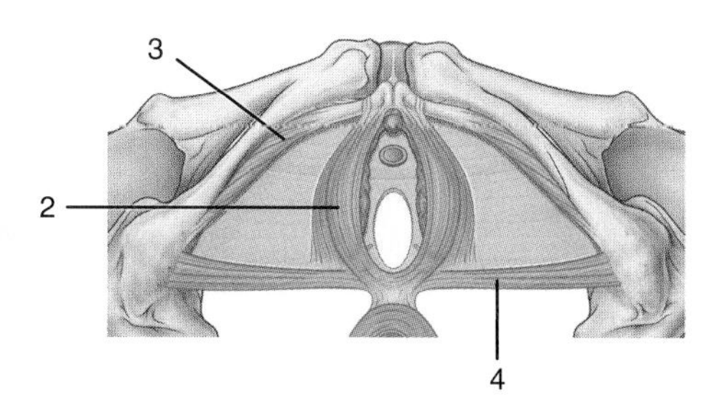

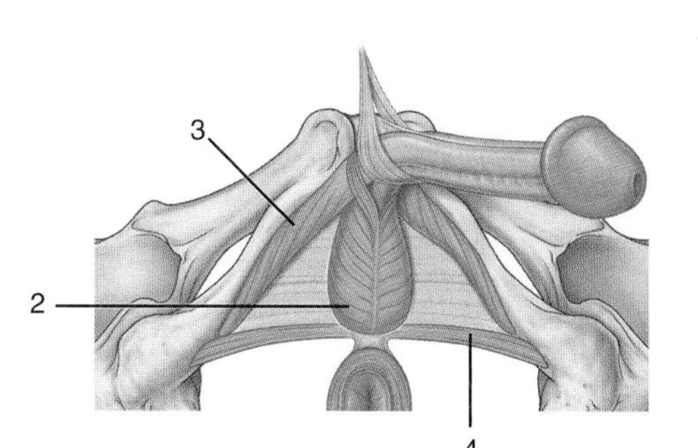

B

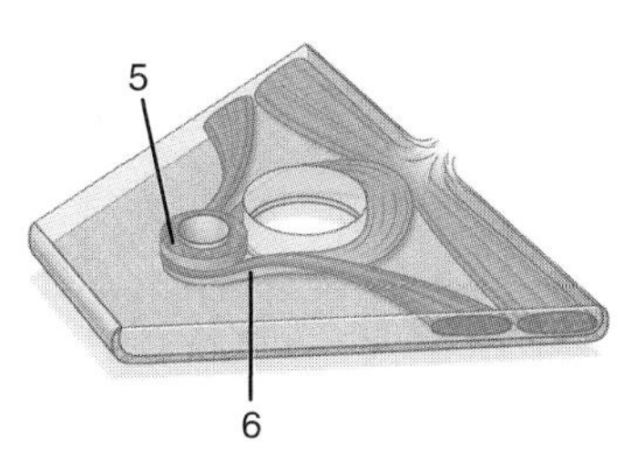

C

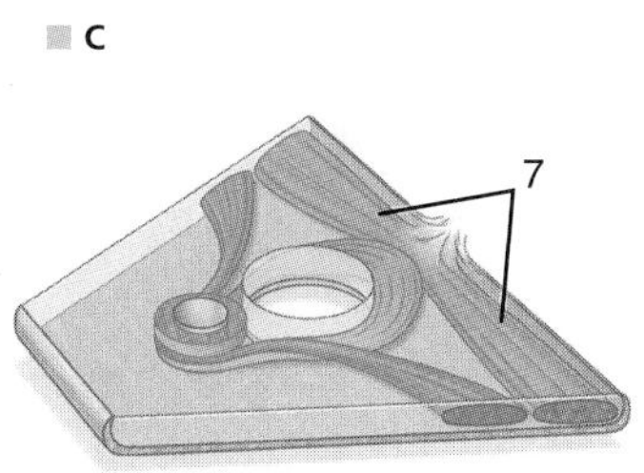

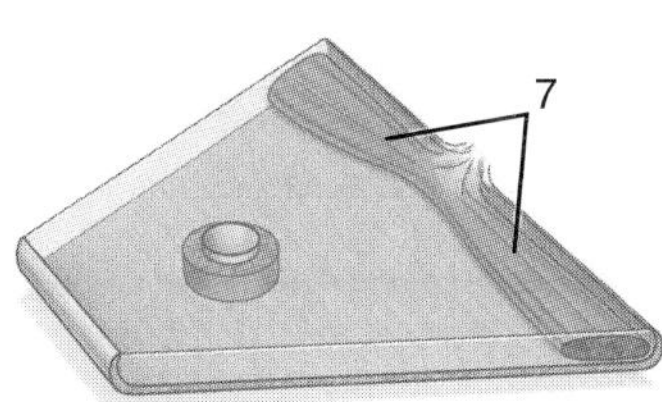

1 **external urethral sphincter (male)** *origin,* ramus of pubis; *insertion,* median raphe behind and in front of urethra; *innervation,* perineal; *action,* compresses the membranous part of the urethra. <musculus sphincter urethrae externus urethrae masculinae> A

2 **deep part of external anal sphincter** the part of the sphincter that surrounds the upper part of the anal canal. <pars profunda musculi sphincteris ani externus> B

3 **external anal sphincter** *origin,* tip of coccyx, anococcygeal ligament; *insertion,* perineal body; *innervation,* inferior rectal and perineal branch of fourth sacral; *action,* closes anus. <musculus sphincter ani externus> B

4 **subcutaneous part of external anal sphincter** the part of the sphincter that surrounds the outermost portion of the anal canal. <pars subcutanea musculi sphincteris ani externus> B

5 **superficial part of external anal sphincter** the part of the sphincter that lies just deep to the subcutaneous part, extending farther toward the rectum. <pars superficialis musculi sphincteris ani externus> B

Nerves of Perineum

6 **cavernous nerves of penis** *origin,* prostatic plexus; *distribution,* erectile tissue of penis; *modality,* sympathetic, parasympathetic, and visceral afferent. <nervi cavernosi penis> C

7 **inferior rectal nerves** *origin,* pudendal nerve, or independently from sacral plexus; *distribution,* external anal sphincter, skin around anus, and lining of anal canal up to pectinate line; *modality,* general sensory and motor. <nervi anales inferiores> E

8 **pudendal nerve** *origin,* sacral plexus (S2-S4); *branches,* enters the pudendal canal, gives off the inferior anal nerve, and then divides into the perineal nerve and dorsal nerve of the penis (clitoris); *distribution,* muscles, skin, and erectile tissue of perineum; *modality,* general sensory, motor, and parasympathetic. <nervus pudendus> E

9 **dorsal nerve of penis** *origin,* pudendal nerve; *distribution,* deep transverse perineal and urethral sphincter muscles; corpus cavernosum, skin, prepuce, and glans of penis; *modality,* general sensory and motor. <nervus dorsalis penis> D

10 ***cavernous nerves of clitoris*** *origin,* uterovaginal plexus; *distribution,* erectile tissue of clitoris; *modality,* parasympathetic, sympathetic, and visceral afferent. <nervi cavernosi clitoridis> E

11 ***dorsal nerve of clitoris*** *origin,* pudendal nerve; *distribution,* deep transverse perineal and urethral sphincter muscles; corpus cavernosum, skin, prepuce, and glans of clitoris; *modality,* general sensory and motor. <nervus dorsalis clitoridis> E

12 **perineal nerves** *origin,* pudendal nerve in the pudendal canal; *branches,* muscular branches and posterior scrotal or labial nerves; *distribution,* muscular branches supply the bulbospongiosus, ischiocavernosus, superficial transverse perineal muscles and bulb of the penis and, in part, the external anal sphincter and levator ani; the scrotal (labial) nerves supply the scrotum or labium majus; *modality,* general sensory and motor. <nervi perineales> E

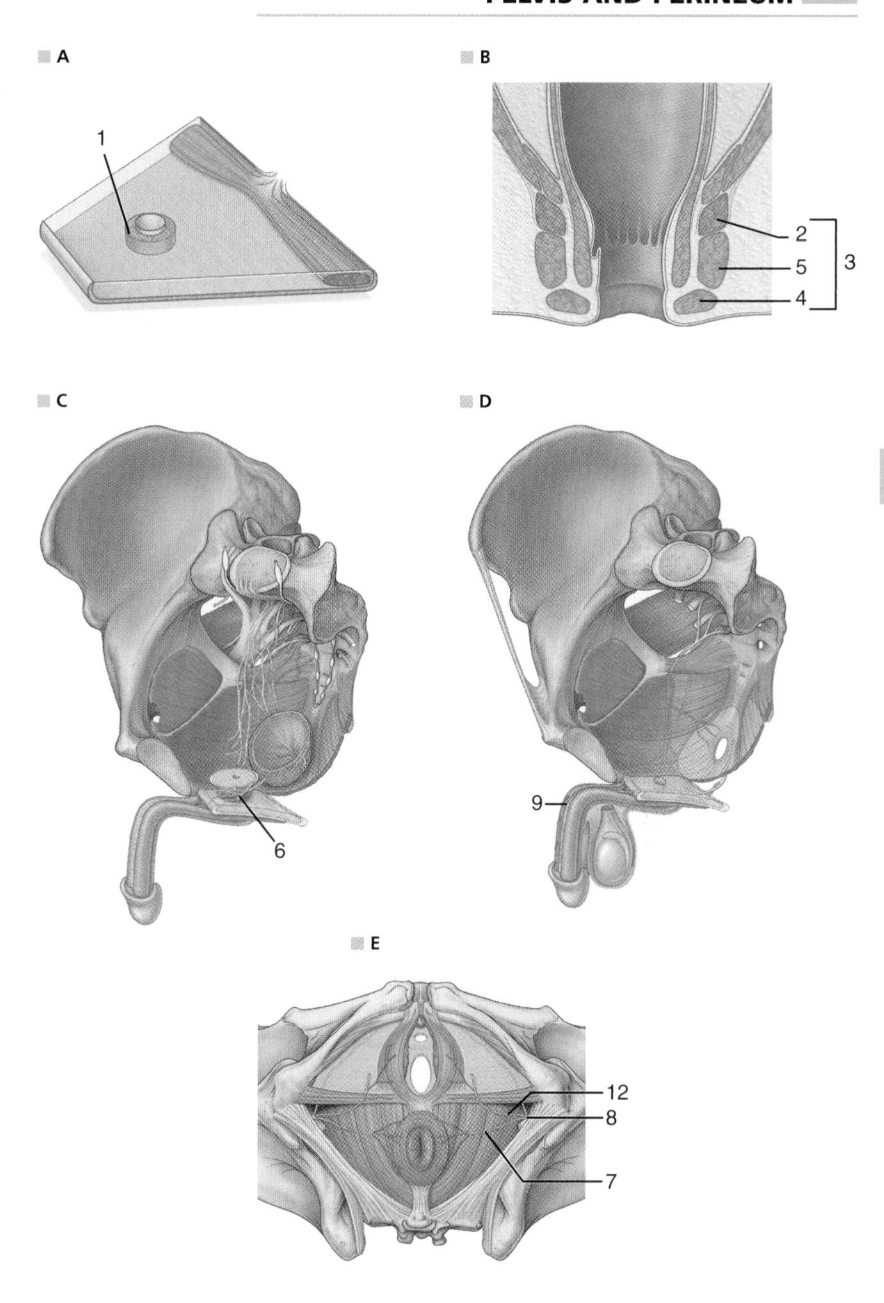
A
1
B
2
5
4
3
C
6
D
9
E
12
8
7

1 **anterior scrotal nerves** *origin,* ilioinguinal nerve; *distribution,* skin of anterior scrotal region; *modality,* general sensory. <nervi scrotales anteriores> A

2 **posterior scrotal nerves** *origin,* perineal nerves; *distribution,* skin of scrotum; *modality,* general sensory. <nervi scrotales posteriores> B

3 ***anterior labial nerves*** *origin,* ilioinguinal nerve; *distribution,* skin of anterior labial region of labia majora, and adjacent part of thigh; *modality,* general sensory. <nervi labiales anteriores> C

4 **posterior labial nerves** *origin,* perineal nerves; *distribution,* labium majus; *modality,* general sensory. <nervi labiales posteriores> C

Vessels of Perineum

5 **artery of bulb of penis** *origin,* internal pudendal artery; *branches,* none; *distribution,* bulbourethral gland, bulb of penis. Called also *bulbourethral artery.* <arteria bulbi penis> D

6 ***helicine arteries of penis*** helicine arteries arising from the vessels of the penis, whose engorgement causes erection of the organ. <arteriae helicinae penis> D

7 ***inferior rectal artery*** *origin,* internal pudendal artery; *branches,* none; *distribution,* rectum, anal canal. Called also *inferior hemorrhoidal artery.* <arteria rectalis inferior> D

8 ***perineal artery*** *origin,* internal pudendal artery; *branches,* none; *distribution,* perineum, skin of external genitalia. <arteria perinealis> D

A

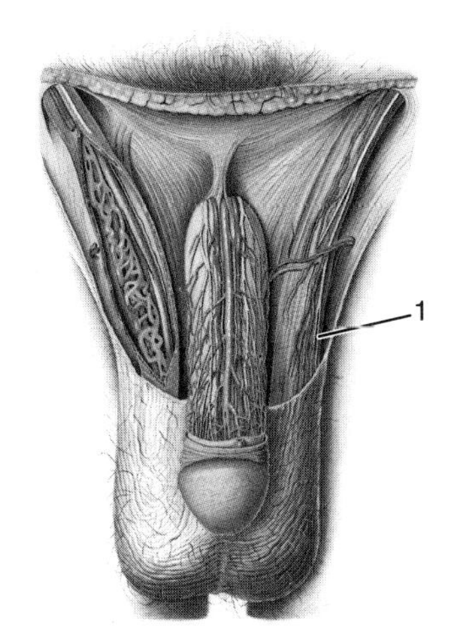

B

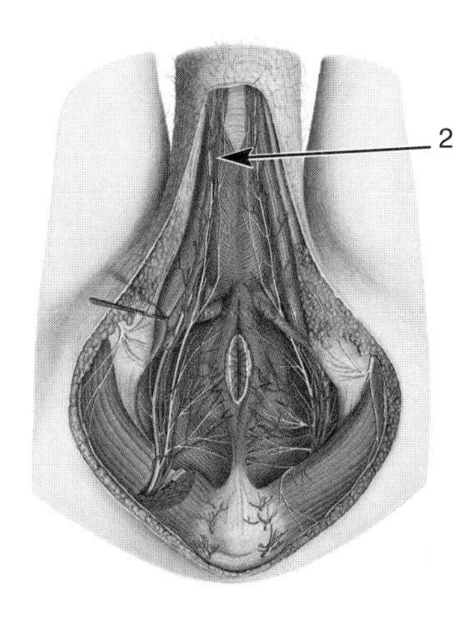

C

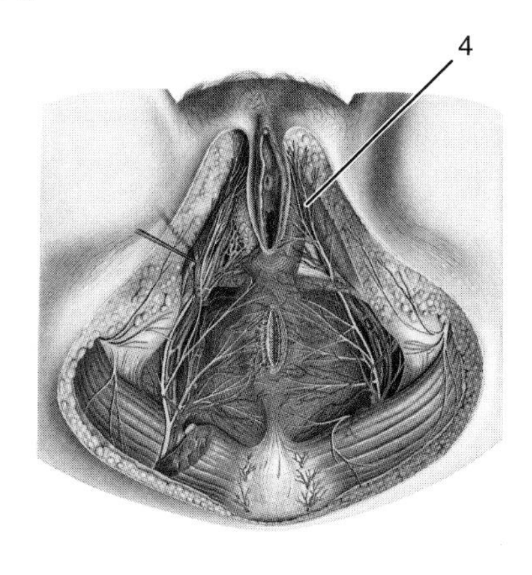

D

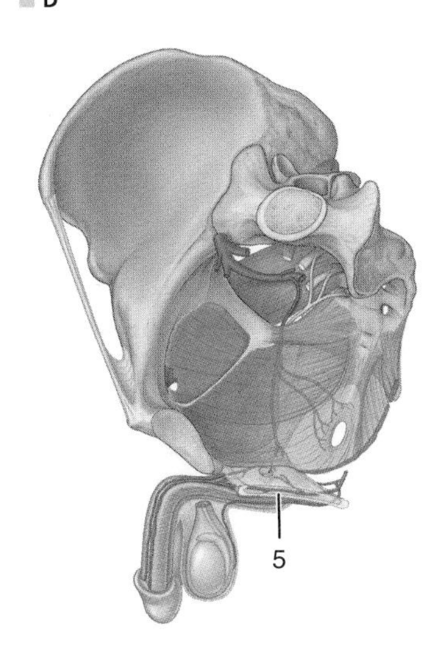

1 **dorsal artery of penis** *origin,* internal pudendal artery; *branches,* none; *distribution,* glans, corona, and prepuce of penis. <arteria dorsalis penis> A

2 **internal pudendal artery** *origin,* internal iliac artery; *branches,* posterior scrotal or posterior labial branches and inferior rectal, perineal, urethral arteries, artery of bulb of penis or vestibule, deep artery of penis or clitoris, dorsal artery of penis or clitoris; *distribution,* external genitalia, anal canal, perineum. <arteria pudenda interna> B, C

3 **deep veins of penis** small veins that empty into the deep dorsal vein of the penis. <venae profundae penis> A

4 **inferior rectal veins** veins that drain the rectal plexus into the internal pudendal vein. <venae rectales inferiores> C

5 ***deep dorsal vein of clitoris*** a vein that follows the course of the dorsal artery of clitoris and opens into the vesical plexus. <vena dorsalis profunda clitoridis>

6 **deep dorsal vein of penis** a vein lying subfascially in the midline of the penis between the dorsal arteries; it begins in small veins around the corona of glans, is joined by the deep veins of the penis as it passes proximally, and passes between the arcuate pubic and transverse perineal ligaments where it divides into a left and right vein to join the prostatic plexus. <vena dorsalis profunda penis> A, D

7 **internal pudendal vein** a vein that follows the course of the internal pudendal artery, and drains into the internal iliac vein. <vena pudenda interna> C, D

8 **cavernous veins of penis** veins that return the blood from the corpora cavernosa to the deep veins and the dorsal vein of the penis. <venae cavernosae penis> A

9 **deep artery of penis** *origin,* internal pudendal artery; *branches,* none; *distribution,* corpus cavernosum penis. <arteria profunda penis> A

10 **superficial dorsal veins of penis** veins that collect blood subcutaneously from the penis and drain into the external pudendal vein. <venae dorsales superficiales penis> A

A

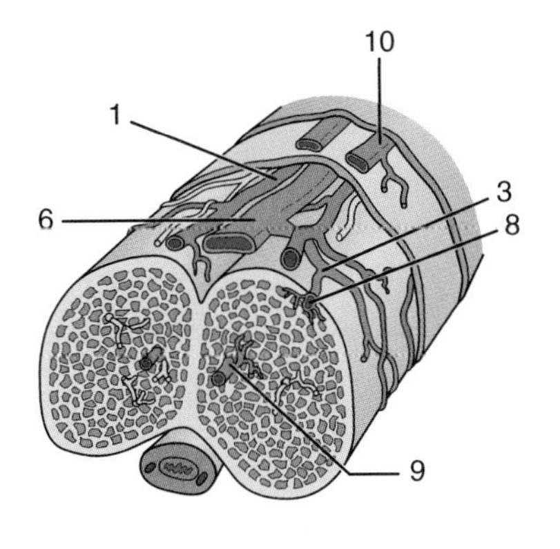

B

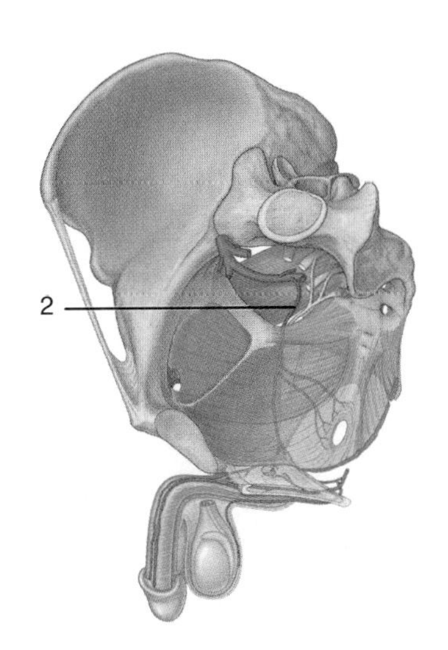

C

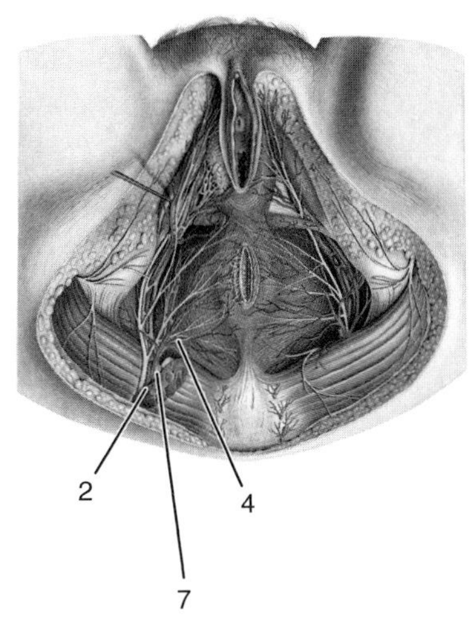

D

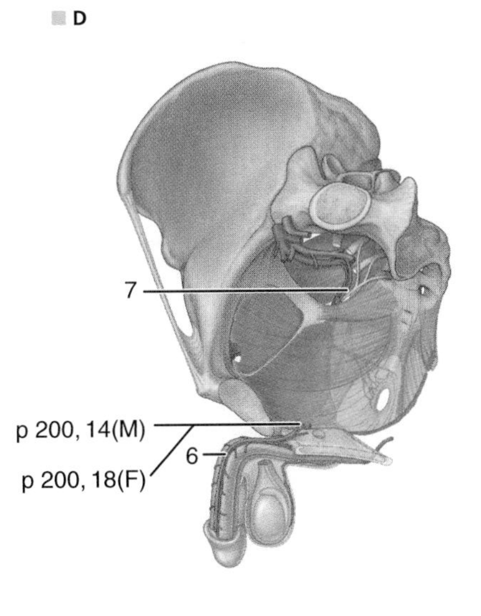

1 **anterior scrotal branches of deep external pudendal artery** branches arising from the deep external pudendal artery and supplying the anterior scrotal region; called also *anterior scrotal arteries.* <rami scrotales anteriores arteriae pudendae externae profundae> A

2 **posterior scrotal branches of internal pudendal artery** two branches arising from the internal pudendal artery in the anterior part of the ischiorectal fossa, helping to supply the ischiocavernosus and bulbospongiosus muscles, and distributed to the scrotum. Called also *posterior scrotal arteries.* <rami scrotales posteriores arteriae pudendae internae> B

3 ***posterior scrotal veins*** small branches from the posterior aspect of the scrotum that open into the vesical venous plexus. <venae scrotales posteriores> B

4 ***vein of bulb of penis*** a vein draining blood from the bulb of the penis into the internal pudendal vein. <vena bulbi penis> B

5 ***anterior labial branches of deep external pudendal artery*** branches that arise from the deep external pudendal artery and supply the labium majus; called also *anterior labial arteries of vulva.* <rami labiales anteriores arteriae pudendae externae profundae> C

6 **artery of bulb of vestibule** *origin,* internal pudendal artery; *branches,* none; *distribution,* bulb of vestibule of vagina, greater vestibular glands. <arteria bulbi vestibuli> C

7 ***deep artery of clitoris*** *origin,* internal pudendal artery; *branches,* none; *distribution,* clitoris. <arteria profunda clitoridis> C

8 ***deep veins of clitoris*** small veins of the clitoris that drain into the vesical venous plexus. <venae profundae clitoridis> C

9 **dorsal artery of clitoris** *origin,* internal pudendal artery; *branches,* none; *distribution,* clitoris. <arteria dorsalis clitoridis> C

10 **posterior labial branches of internal pudendal artery** two branches arising from the internal pudendal artery in the anterior part of the ischiorectal fossa, helping to supply the ischiocavernosus and bulbospongiosus muscles, and supplying the labium majus and labium minus. Called also *posterior labial arteries of vulva.* <rami labiales posteriores arteriae pudendae internae> C

11 ***posterior labial veins*** small branches from the labia which open into the vesical venous plexus; they are homologues of the posterior scrotal veins in the male. <venae labiales posteriores> C

12 ***superficial dorsal veins of clitoris*** veins that collect blood subcutaneously from the clitoris and drain into the external pudendal vein. <venae dorsales superficiales clitoridis> C

13 **vein of bulb of vestibule** a vein draining blood from the bulb of the vestibule of the vagina into the internal pudendal vein. <vena bulbi vestibuli> C

A

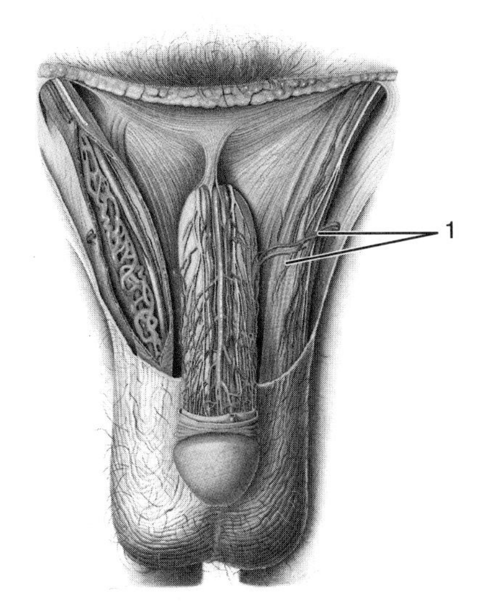

B

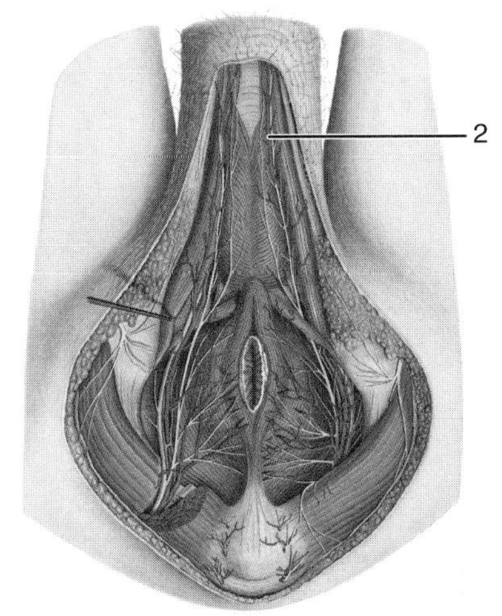

C

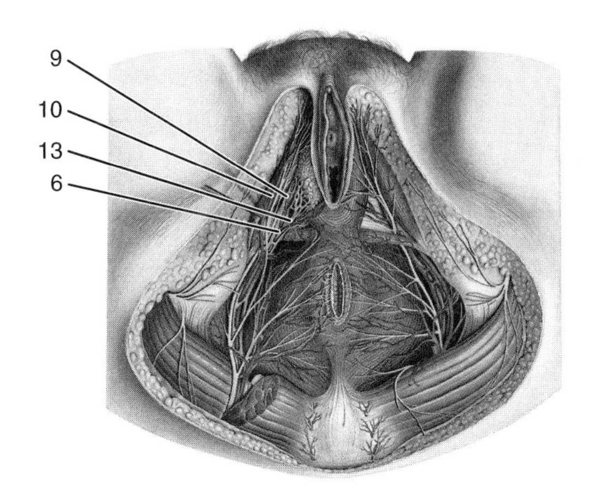

LOWER LIMB

GENERAL

1 **ankle region** the region overlying the tarsal bones. <regio tarsalis> A

2 **anterior region of knee** called also *r. genualis anterior.* <regio genus anterior> A

3 **anterior region of leg** called also *r. cruralis anterior* and *anterior crural region.* <regio cruris anterior> A

4 **anterior region of thigh** called also *r. femoralis anterior.* <regio femoris anterior> A

5 **anterior talocrural region** the anterior aspect of the lower limb between the leg and foot. <regio talocruralis anterior> A

6 **buttocks** the prominences formed by the gluteal muscles on the lower part of the back. <nates> A

7 **calf** the fleshy mass formed chiefly by the gastrocnemius muscle on the posterior aspect of the leg below the knee. <sura> A

8 **foot region** the region of the lower limb distal to the leg region. <regio pedis> A

9 **gluteal region** the region overlying the gluteal muscles. <regio glutealis> A

10 **heel region** the region of the foot overlying the calcaneus. <regio calcanea> A

11 **hip region** the region of the lower limb overlying the hip joint. <regio coxae> A

12 **knee region** the region of the lower limb overlying the knee joint, comprising the anterior and posterior regions. <regio genus> A

13 **knee** the site of articulation between the thigh (femur) and leg. <genu> A

14 **leg region** the lower leg, including the anterior region of the leg, the posterior region of the leg, the anterior talocrural region, and the posterior talocrural region. <regio cruris> A

15 **lower limb** the limb extending from the gluteal region to the foot. It is specialized for weight-bearing and locomotion. <membrum inferius> A

16 **posterior part of knee** <poples> A

17 **posterior region of knee** including the popliteal fossa. <regio genus posterior> A

18 **posterior region of leg** called also *posterior crural region.* <regio cruris posterior> A

19 **posterior region of thigh** <regio femoris posterior> A

20 **posterior talocrural region** the posterior aspect of the lower limb between the leg and foot; called also *posterior ankle region.* <regio talocruralis posterior> A

21 **sural region** the region on the posterior aspect of the leg around the calf. <regio surae> A

22 **deep veins of lower limb** veins that drain the lower limb, found accompanying homonymous arteries, and anastomosing freely with the superficial veins; the principal deep veins are the femoral and popliteal veins. <venae profundae membri inferioris> B

23 **great saphenous vein** the longest vein in the body, extending from the dorsum of the foot to just below the inguinal ligament, where it opens into the femoral vein. It drains the foot and leg through many tributaries, including the medial marginal, accessory saphenous, superficial external pudendal, superficial circumflex iliac, superficial epigastric, and numerous cutaneous veins. <vena saphena magna> B

24 **superficial veins of lower limb** veins that drain the lower limb, found immediately beneath the skin, and anastomosing freely with the deep veins; the principal superficial veins are the great and small saphenous veins. <venae superficiales membri inferioris> B

25 ***veins of lower limb*** veins that drain the thigh, leg, and foot, divided into *superficial* and *deep veins.* <venae membri inferioris> B

26 **deep fascia of leg** the investing fascia of the leg. <fascia cruris> C

27 **fascia lata** the external investing fascia of the thigh. C

28 **femoral region** the region overlying the femur. <regio femoris> D

29 **pelvic girdle** the encircling bony structure supporting the lower limbs, comprising the two pelvic bones, articulating with each other and with the sacrum, to complete the essentially rigid bony ring. <cingulum pelvicum> D

30 **tarsus** the bony region of the foot that articulates with the leg. Called also *ankle.* D

31 **lateral retromalleolar region** the region posterior to the lateral malleolus of the fibula. <regio retromalleolaris lateralis> E

32 **medial retromalleolar region** the region posterior to the medial malleolus of the tibia. <regio retromalleolaris medialis> E

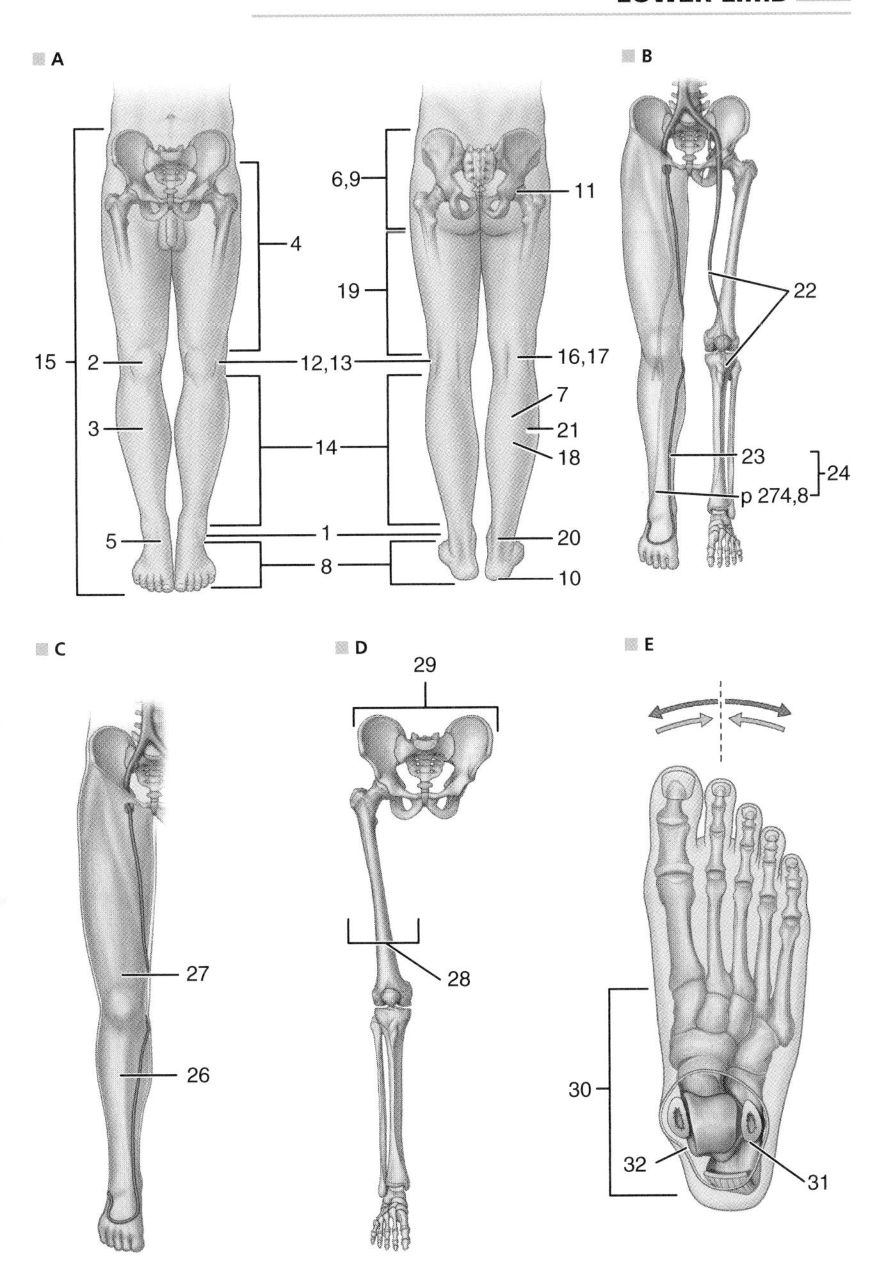
A
6,9
11
4
19
22
15
2
12,13
16,17
7
3
14
21
18
23
24
p 274,8
5
1
20
8
10
B
C
D
E
29
27
28
26
30
32
31

LOWER LIMB

BONES

1 **tuberosity of cuboid bone** a transverse ridge on the lower surface of the cuboid bone over which the tendon of the fibularis longus muscle plays. <tuberositas ossis cuboidei> A

2 **groove for popliteus** a smooth, well-marked groove separated from the lateral condyle of the femur by a prominent lip and extending superiorly and posteriorly to the posterior extremity of the lateral condyle; it lodges the tendon of the popliteus muscle. <sulcus popliteus femoris> B

3 **calcaneal process of cuboid bone** a process projecting posteriorly from the inferomedial angle of the cuboid bone that supports the anterior calcaneus. <processus calcaneus ossis cuboidei> A

4 **distal tuberosity of toes** a roughened, raised bony mass on the plantar surface of the tip of a distal phalanx of the foot. <tuberositas phalangis distalis pedis> C

5 **ilium** the expansive superior portion of the hip bone; it is a separate bone in early life. <os ilium> D

6 **ischium** the inferior dorsal portion of the hip bone; it is a separate bone in early life. <os ischii> D

7 **pubic bone** the anterior inferior part of the pelvic bone on either side, articulating with its fellow in the anterior midline at the pubic symphysis; it is a separate bone in early life. Called also *pubis.* <os pubis> D

8 **gluteal surface of ilium** the large external, or posterior, surface of the ala of the ilium, on which are located the three gluteal lines. <facies glutea ossis ilii> E

9 **anterior gluteal line** the middle of three rough curved lines on the gluteal surface of the ala of the ilium; it begins from the iliac crest about 2.5 cm posterior to the anterior superior iliac spine and arches more or less posteriorly to the greater sciatic notch. <linea glutea anterior> F

10 **inferior gluteal line** a rough curved line, often indistinct, on the gluteal surface of the ala of the ilium; it runs from the notch between the anterior superior and anterior inferior iliac spines posteriorly to the anterior part of the greater sciatic notch. <linea glutea inferior> F

11 **posterior gluteal line** a rough curved line on the gluteal surface of the ala of the ilium; it begins from the iliac crest about 5 cm anterior to the posterior superior iliac spine and runs downward to the greater sciatic notch. <linea glutea posterior> F

12 **acetabular fossa** a rough nonarticular area in the floor of the acetabulum above the acetabular notch. <fossa acetabuli> G

13 **acetabular margin** the peripheral margin of the acetabulum to which the labrum acetabulare is attached. <limbus acetabuli> G

14 **acetabular notch** a notch in the inferior portion of the lunate surface of the acetabulum. <incisura acetabuli> G

15 **acetabulum** the large cup-shaped cavity on the lateral surface of the hip bone in which the head of the femur articulates. G

16 **lunate surface of acetabulum** the articular portion of the acetabulum. <facies lunata acetabuli> G

17 **supra-acetabular groove** a sulcus located posterosuperior to the margin of the acetabulum, which is the site of attachment of the reflected head of the rectus femoris muscle. <sulcus supra-acetabularis> G

18 **greater trochanter** a broad, flat process at the upper end of the lateral surface of the femur, to which several muscles are attached. <trochanter major> H

19 **intertrochanteric line** a line running obliquely downward and medially from the tubercle of the femur, winding around the medial side of the body of the bone. <linea intertrochanterica> H

20 **lesser trochanter** a short conical process projecting medially from the lower part of the posterior border of the base of the neck of the femur. <trochanter minor> H

21 **neck of femur** the heavy column of bone connecting the head of the femur and the shaft. <collum femoris> H

22 ***subcutaneous trochanteric bursa*** a bursa between the greater trochanter of the femur and the skin. <bursa trochanterica subcutanea> H

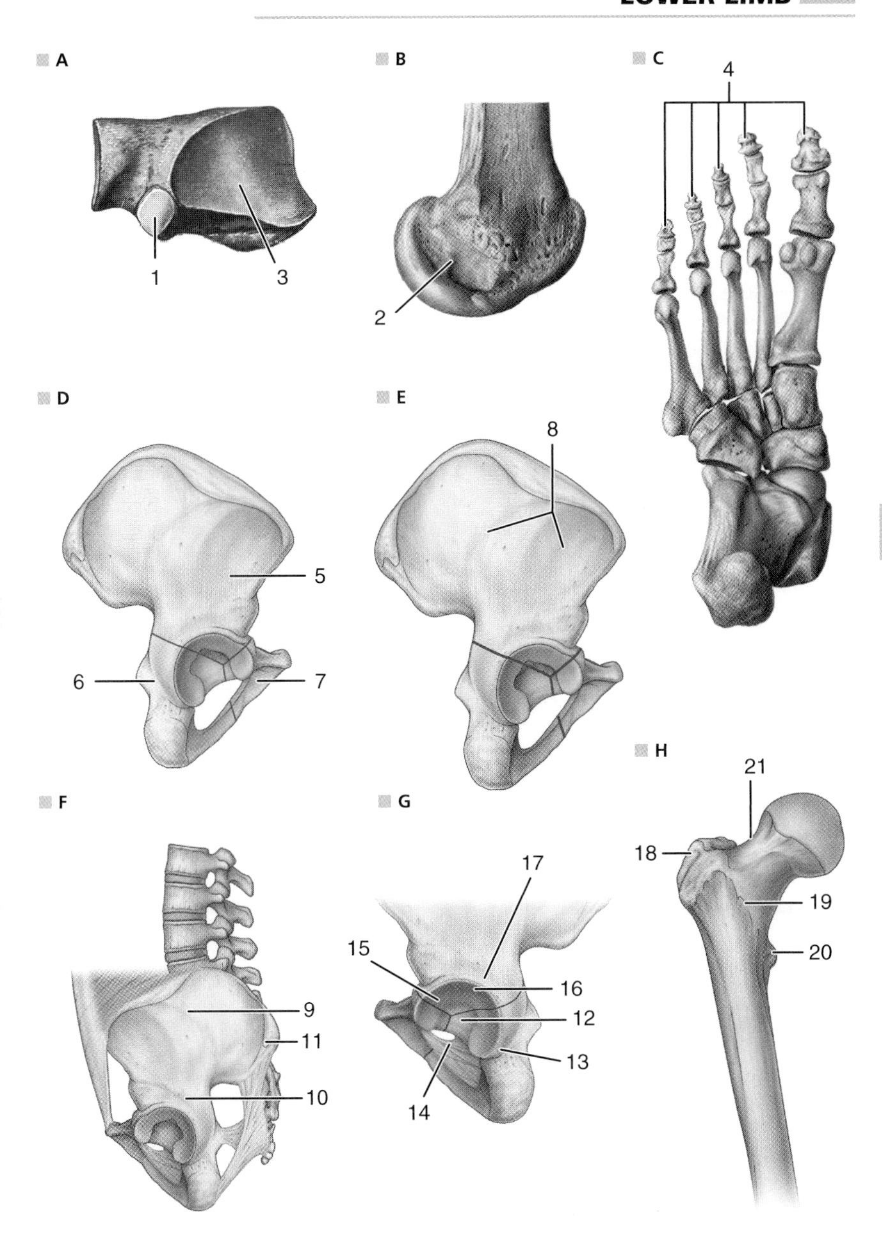
A
1
3
B
2
C
4
D
5
6
7
E
8
F
9
11
10
G
17
15
16
12
13
14
H
21
18
19
20

1 **fovea of femoral head** a depression in the head of the femur where the ligamentum of head of femur is attached. <fovea capitis femoris> A

2 **trochanteric fossa** a deep depression on the medial surface of the greater trochanter that receives the insertion of the tendon of the obturator externus muscle. <fossa trochanterica> A

3 **gluteal tuberosity** an elevation on the upper part of the shaft of the femur for attachment of the gluteus maximus muscle. When the tuberosity is unusually prominent, it may be called the *third trochanter.* <tuberositas glutea femoris> B

4 **intertrochanteric crest** a prominent ridge running obliquely downward and medialward from the summit of the greater trochanter on the posterior surface of the neck of the femur to the lesser trochanter. <crista intertrochanterica> B

5 **lateral lip of linea aspera** the distinct outer part of the linea aspera that becomes continuous with the gluteal tuberosity and ends at the greater trochanter above and with the lateral supracondylar line below. Called also *lateral margin of linea aspera.* <labium laterale lineae asperae femoris> B

6 **linea aspera** the broad, thickened ridge that forms the posterior border of the femur and has distinct lateral and medial lips. B

7 **medial lip of linea aspera** the distinct inner part of the linea aspera that becomes continuous with the intertrochanteric line above and the medial supracondylar line below. <labium medialis lineae asperae femoris> B

8 **quadrate tubercle** an elevation just above the center of the intertrochanteric crest that gives attachment to the quadratus femoris muscle. <tuberculum quadratum femoris> B

9 **pectineal line** a line running down the posterior surface of the shaft of the femur, giving attachment to the pectineus muscle. <linea pectinea> C

10 **adductor tubercle of femur** a small projection from the upper part of the medial epicondyle of the femur, to which the tendon of the adductor magnus muscle is attached. <tuberculum adductorium femoris> D

11 **body of femur** the main part or shaft of the femur. <corpus femoris> D

12 **femur** the bone that extends from the pelvis to the knee, being the longest and largest bone in the body; its head articulates with the acetabulum of the hip bone, and distally, the femur, along with the patella and tibia, forms the knee joint. Called also *thigh bone.* D

13 **head of femur** the proximal end of the femur, articulating with the acetabulum on the hip bone. <caput femoris> D

14 **lateral epicondyle of femur** a projection from the distal end of the femur, above the lateral condyle, for the attachment of collateral ligaments of the knee. <epicondylus lateralis femoris> D

15 **medial epicondyle of femur** a projection from the distal end of the femur, above the medial condyle, for the attachment of collateral ligaments of the knee. <epicondylus medialis femoris> D

16 **patellar surface of femur** the smooth anterior continuation of the condyles that forms the surface of the femur articulating with the patella. <facies patellaris femoris> D

17 **intercondylar fossa of femur** the posterior depression between the condyles of the femur. <fossa intercondylaris femoris> E

18 **intercondylar line** a transverse ridge separating the floor of the intercondylar fossa from the popliteal surface of the femur and giving attachment to the posterior portion of the capsular ligament of the knee. <linea intercondylaris> E

19 **lateral condyle of femur** the lateral of the two surfaces at the distal end of the femur that articulate with the superior surfaces of the head of the tibia. <condylus lateralis femoris> E

20 **lateral supracondylar line** a slight ridge on the lower third of the posterior surface of the femur that is continuous above with the lateral lip of the linea aspera and descends to the lateral epicondyle. <linea supracondylaris lateralis> E

21 **medial condyle of femur** the medial of the two surfaces at the distal end of the femur that articulate with the superior surfaces of the head of the tibia. <condylus medialis femoris> E

22 **medial supracondylar line** an indistinct ridge on the lower third of the posterior surface of the femur that is continuous above with the medial lip of the linea aspera, being interrupted at its upper end to allow passage of the femoral artery, and descends to the adductor tubercle. <linea supracondylaris medialis> E

23 **popliteal surface of femur** the triangular lower third of the posterior surface of the femur, between the medial and lateral supracondylar lines, which forms the superior part of the floor of the popliteal fossa. <facies poplitea femoris> E

24 **anterior surface of patella** the slightly convex, longitudinally striated front surface of the patella, which is perforated by small openings for the nutrient vessels. <facies anterior patellae> F

25 **apex of patella** the inferiorly directed blunt point of the patella, to which the patellar ligament is attached. <apex patellae> F

26 **base of patella** the superior border of the patella, to which the tendon of the quadriceps femoris muscle is attached. <basis patellae> F

27 **articular surface of patella** the posterior surface of the patella, which is largely covered by a thick cartilaginous layer. <facies articularis patellae> G

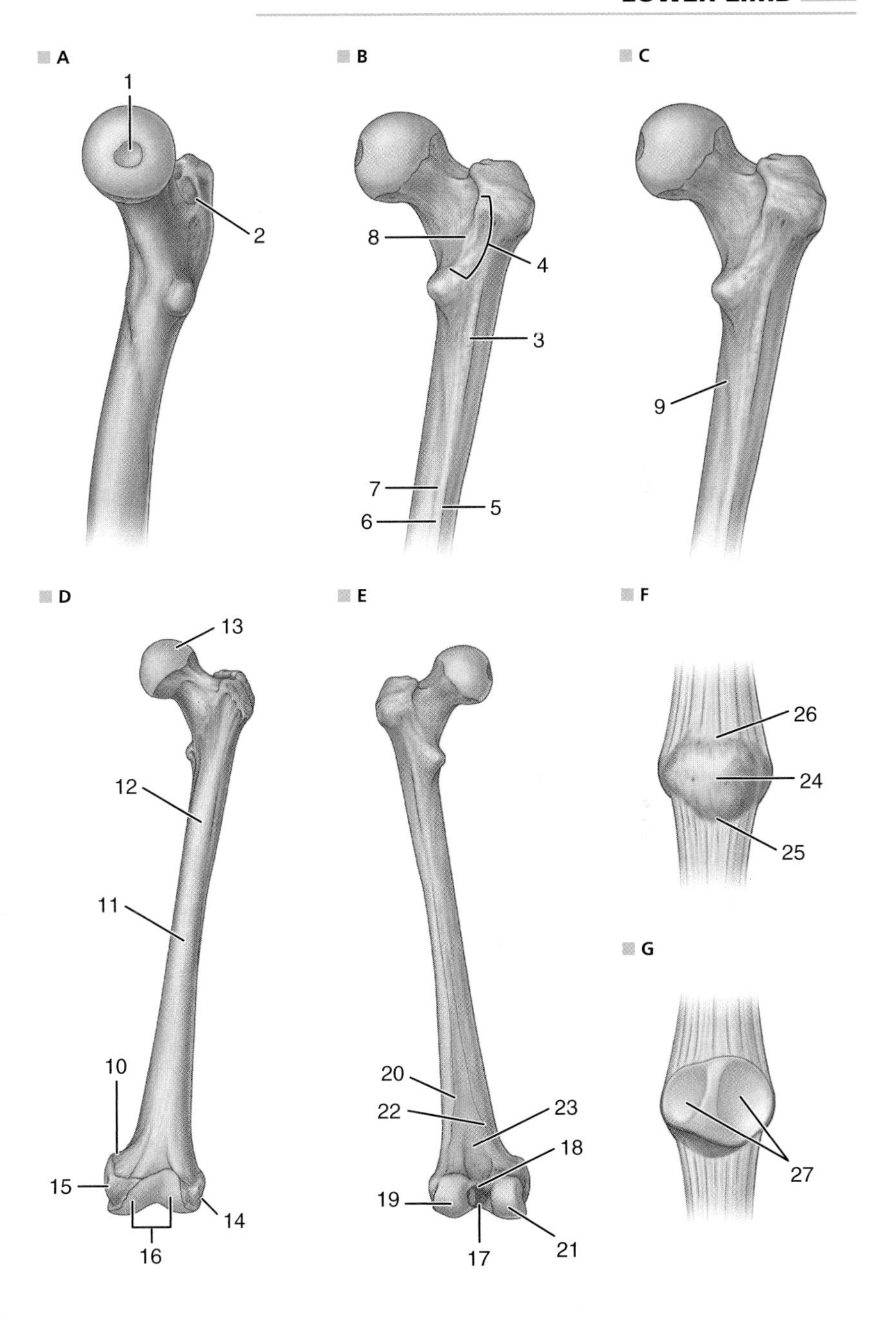
A
1
2
B
8
4
3
7
5
6
C
9
D
13
12
11
10
15
14
16
E
20
22
23
18
19
17
21
F
26
24
25
G
27

1 **anterior intercondylar area of tibia** the broad area between the superior articular surfaces of the tibia; called also *anterior intercondylar fossa of tibia*. <area intercondylaris anterior tibiae> A

2 **posterior intercondylar area of tibia** a deep notch separating the condyles on the posterior surface of the tibia. <area intercondylaris posterior tibiae> A

3 ***superior articular surface of tibia*** the surface on the proximal end of the tibia that articulates with the condyles of the femur. <facies articularis superior tibiae> A

4 **intercondylar eminence** an eminence on the proximal extremity of the tibia, surmounted on either side by a prominent tubercle, on to the sides of which the articular facets are prolonged. <eminentia intercondylaris> A

5 **lateral condyle of tibia** the lateral articular eminence on the proximal end of the tibia. <condylus lateralis tibiae> B

6 **lateral intercondylar tubercle** a lateral spur projecting upward from the intercondylar eminence at the proximal end of the tibia. <tuberculum intercondylare laterale> B

7 **medial condyle of tibia** the medial articular eminence on the proximal end of the tibia. <condylus medialis tibiae> B

8 **medial intercondylar tubercle** a medial spur projecting upward from the intercondylar eminence at the proximal end of the tibia. <tuberculum intercondylare mediale> B

9 **tibial tuberosity** a longitudinally elongated, raised and roughened area on the anterior crest of the tibia, located just distal to the intercondylar eminence, and giving attachment to the patellar ligament. <tuberositas tibiae> B

10 **anterior border of tibia** the prominent anteromedial margin of the body of the tibia, separating the medial and lateral surfaces. <margo anterior tibiae> B

11 **lateral surface of tibia** the surface of the body of the tibia between the interosseous and anterior borders. <facies lateralis tibiae> B

12 **medial surface of tibia** the slightly convex surface of the body of the tibia between the anterior and medial borders. <facies medialis tibiae> B

13 **posterior surface of tibia** the surface of the body of the tibia between the medial and interosseous borders; in the proximal third it presents the soleal line. <facies posterior tibiae> B

14 **fibular articular surface of tibia** the surface on the posteroinferior aspect of the lateral condyle of the tibia that articulates with the head of the fibula. <facies articularis fibularis tibiae> C

15 **soleal line** a line extending from the fibular facet downward and inward across the posterior surface of the tibia, giving attachment to fibers of the soleus muscle. <linea musculi solei> C

16 **apex of head of fibula** a process pointing upward on the posterior surface of the head of the fibula, giving attachment to the arcuate popliteal ligament of the knee joint and part of the biceps tendon. <apex capitis fibulae> D

17 **articular facet of head of fibula** the medial surface of the head of the fibula, which articulates with the lateral condyle of the tibia. <facies articularis capitis fibulae> D

18 **head of fibula** the proximal extremity of the fibula. <caput fibulae> D

19 **neck of fibula** the portion of the fibula between the head and shaft. <collum fibulae> D

20 **anterior border of fibula** the anterolateral border of the body of the fibula. <margo anterior fibulae> D

21 **lateral surface of fibula** the area between the anterior and posterior borders of the body of the fibula. <facies lateralis fibulae> D

22 **medial surface of fibula** the narrow area on the body of the fibula between the interosseous and anterior borders. <facies medialis fibulae> D

23 **posterior border of fibula** the posterolateral margin of the body of the fibula. <margo posterior fibulae> D

24 **posterior surface of fibula** the large area between the posterior and interosseous borders of the body of the fibula, presenting the medial crest. <facies posterior fibulae> D

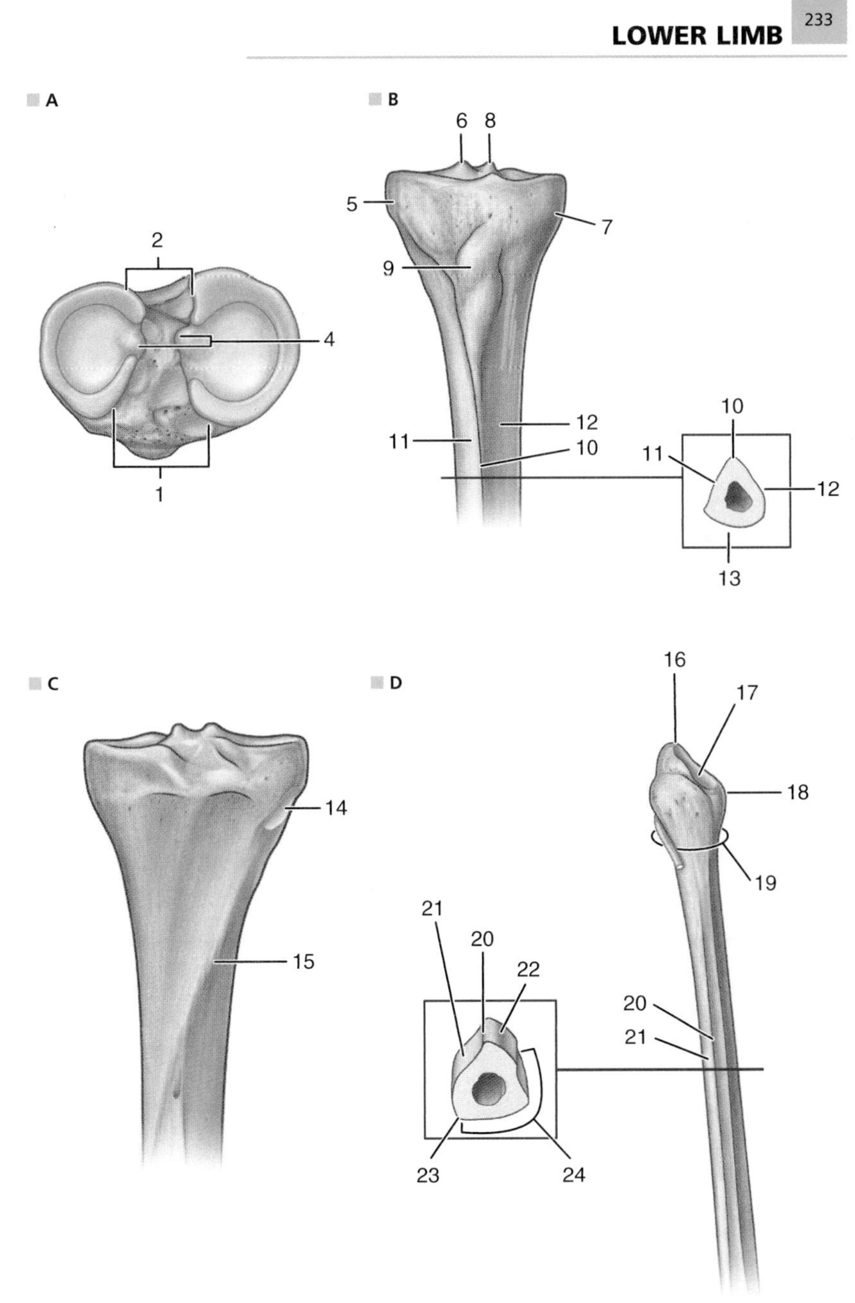
A
2
4
1
B
6
8
5
7
9
11
12
10
10
11
12
13
C
14
15
D
16
17
18
19
21
20
22
20
21
23
24

1 **patella** a triangular sesamoid bone, about 5 cm in diameter, situated at the front of the knee in the tendon of insertion of the quadriceps extensor femoris muscle. Called also *knee cap.* A

2 **pelvic bone** the bone comprising the ilium, ischium, and pubis. Called also *hip bone.* <os coxae> A

3 **tibia** the shin bone: the inner and larger bone of the leg below the knee; it articulates with the femur and head of the fibula above and with the talus below. Called also *shin bone.* A

4 ***tibial*** pertaining to the tibia. <tibialis> A

5 **body of fibula** the principal part or shaft of the fibula. <corpus fibulae> B

6 **body of tibia** the long central part of the tibia. <corpus tibiae> B

7 **fibula** the outer and smaller of the two bones of the leg, which articulates proximally with the tibia and distally is joined to the tibia in a syndesmosis. A

8 ***fibular*** pertaining to the fibula. Called also *peroneal.* <fibularis> B

9 **interosseous border of fibula** a prominent ridge medial to the anterior border of the fibula, connected with a similar ridge on the tibia by a strong, wide fibrous sheet, the interosseous membrane. <margo interosseus fibulae> B

10 **interosseous border of tibia** the prominent lateral border of the body of the tibia, which separates the posterior and lateral surfaces and gives attachment to the interosseous membrane. <margo interosseus tibiae> B

11 **medial border of tibia** the border that extends between the medial condyle and medial malleolus of the tibia, separating the medial and posterior surfaces. <margo medialis tibiae> B

12 **medial crest of fibula** the long crest on the posterior surface of the body of the fibula, which separates the origin of the tibialis posterior muscle from that of the flexor hallucis longus muscle. <crista medialis fibulae> B

13 **lateral malleolus** the process on the lateral side of the distal end of the fibula, forming, with the medial malleolus, the mortise in which the talus articulates. <malleolus lateralis> C

14 **malleolar sulcus of fibula** a groove on the posterior surface of the lateral malleolus of the fibula, which lodges the tendons of the fibularis muscles. <sulcus malleolaris fibulae> C

15 **malleolar sulcus of tibia** a short longitudinal groove on the posterior surface of the medial malleolus of the tibia, which lodges the tendons of the tibialis posterior muscle and the digitorum flexor longus muscle. <sulcus malleolaris tibiae> C

16 **medial malleolus** the process on the medial side of the distal end of the tibia, forming, with the lateral malleolus, the mortise in which the talus articulates. <malleolus medialis> C

17 ***subcutaneous bursa of lateral malleolus*** a bursa between the lateral malleolus and the skin. <bursa subcutanea malleoli lateralis> C

18 ***subcutaneous bursa of medial malleolus*** a bursa between the medial malleolus and the skin. <bursa subcutanea malleoli medialis> C

19 **fibular notch of tibia** a depression on the lateral surface of the lower end of the tibia, which articulates with the lower end of the fibula. <incisura fibularis tibiae> C

20 **inferior articular surface of tibia** the surface on the distal end of the tibia where it articulates with the talus. <facies articularis inferior tibiae> C

21 **interosseous membrane of leg** a thin aponeurotic lamina attached to the interosseous borders of the tibia and fibula, deficient for a short distance at the proximal end of the bones; it separates the muscles on the anterior and posterior parts of the leg. <membrana interossea cruris> D

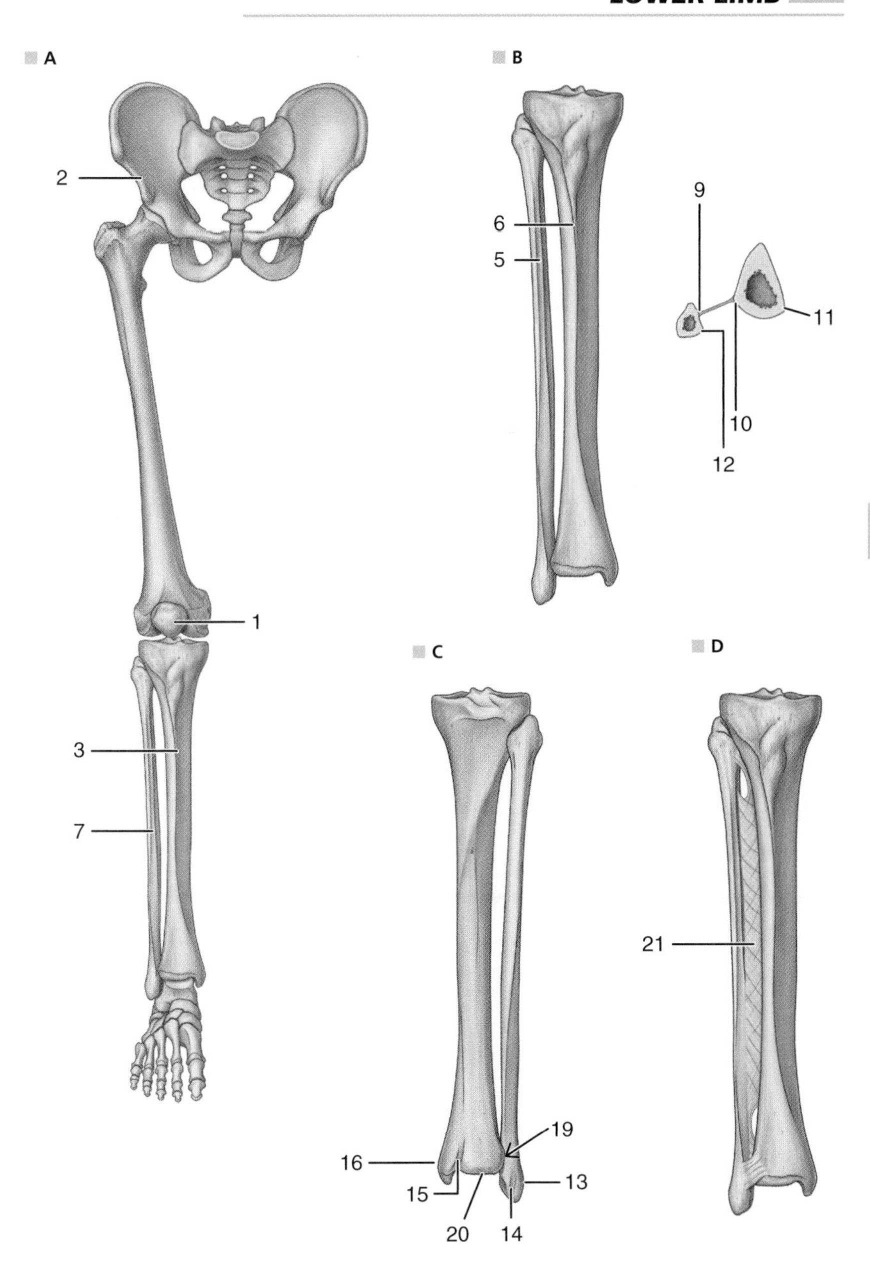
A
2
1
3
7
B
6
5
9
11
10
12
C
19
16
13
15
20
14
D
21

1 **base of metatarsal bone** the wedge-shaped proximal end of each metatarsal, which articulates with bone(s) of the tarsus and with adjacent metatarsals. <basis ossis metatarsi> A

2 **body of metatarsal** the long central part of a metatarsal bone. <corpus ossis metatarsi> A

3 **body of phalanx of foot** the long central part of a phalanx of the foot. <corpus phalangis pedis> A

4 **calcaneus** the irregular quadrangular bone at the back of the tarsus; called also *heel bone.* A

5 **cuboid bone** a bone on the lateral side of the tarsus between the calcaneus and the fourth and fifth metatarsal bones. <os cuboideum> A

6 **distal phalanx of toes** any one of the five terminal bones of the toes, articulating, except in the great toe, with the middle phalanx. <phalanx distalis digitorum pedis> A

7 **head of metatarsal bone** the distal extremity of a metatarsal bone, which articulates with the base of a toe. <caput ossis metatarsi> A

8 **head of phalanx of foot** the distal articular extremity of each of the proximal and middle phalanges of the toes. <caput phalangis pedis> A

9 **intermediate cuneiform bone** the intermediate and smallest of the three wedge-shaped tarsal bones located medial to the cuboid and between the navicular and the first three metatarsal bones. <os cuneiforme intermedium> A

10 **lateral cuneiform bone** the most lateral of the three wedge-shaped tarsal bones located medial to the cuboid and between the navicular and the first three metatarsal bones. <os cuneiforme laterale> A

11 **lateral process of talus** a large low process on the lateral surface of the talus, articulating with the lateral malleolus. <processus lateralis tali> A

12 **medial cuneiform bone** the medial and largest of the three wedge-shaped tarsal bones located medial to the cuboid and between the navicular and the first three metatarsal bones. <os cuneiforme mediale> A

13 **metatarsal bones** the five bones (metatarsals) extending from the tarsus to the phalanges of the toes, being numbered in the same sequence from the most medial to the most lateral. <ossa metatarsi> A

14 **middle phalanx of toes** any one of the four bones of the toes (excluding the great toe) situated between the proximal and distal phalanges. <phalanx media digitorum pedis> A

15 **navicular bone** the ovoid-shaped tarsal bone that is situated between the talus and the three cuneiform bones. <os naviculare> A

16 **phalanges of foot** the bones that compose the skeleton of the toes—two for the great toe and often the fifth toe and three for each of the other toes. <ossa digitorum pedis> A

17 **proximal phalanx of toes** any one of the five bones of the toes that articulate with the metatarsal bones and, except in the great toe, with the middle phalanx. <phalanx proximalis digitorum pedis> A

18 ***subcutaneous calcaneal bursa*** a bursa between the calcaneus and the skin on the sole of the foot; called also *subcalcaneal bursa.* <bursa subcutanea calcanea> A

19 **tarsal bones** the seven bones of the ankle (tarsus), including the calcaneus, talus, navicular, cuboid, and intermediate, lateral, and medial cuneiform bones. <ossa tarsi> A

20 ***tarsal ligaments*** the ligaments that connect the bones of the tarsus. <ligamenta tarsi> A

21 **trochlea of phalanx of foot** the pulleylike concavity of the head of a proximal or middle phalanx (except in the big toe, which has only two phalanges), to which the base of the articulating phalanx is adapted. <trochlea phalangis pedis> A

22 **tuberosity of navicular bone** a rough protuberance on the navicular bone of the foot, projecting downward and medially, and giving attachment to the tendon of the tibialis posterior muscle. <tuberositas ossis navicularis> B

23 **lateral malleolar facet of talus** the large triangular facet on the talus that articulates with the lateral malleolus. <facies malleolaris lateralis tali> C

24 **os trigonum** an external tubercle at the back of the talus, sometimes occurring as a separate bone. C

25 **talus** the highest of the tarsal bones and the one that articulates with the tibia and fibula to form the ankle joint. C

26 **body of talus** the roughly quadrilateral portion of the talus, which presents several surfaces for articulation with the calcaneus, tibia, and fibula. <corpus tali> D

27 **groove for tendon of flexor hallucis longus on talus** the sagittal groove on the posterior surface of the body of the talus that transmits the tendon of the flexor hallucis longus muscle. <sulcus tendinis musculi flexoris hallucis longi tali> D

28 **head of talus** the rounded anterior end of the talus. <caput tali> D

29 **medial malleolar facet of talus** the narrow facet on the talus continuous with the superior surface; it articulates with the medial malleolus. <facies malleolaris medialis tali> D

30 **navicular articular surface of talus** the surface of the head of the talus that articulates with the navicular bone. <facies articularis navicularis tali> D

31 **neck of talus** the constriction between the head and body of the talus. <collum tali> D

32 **superior facet of trochlea of talus** the smooth surface of the talus that articulates with the tibia. <facies superior trochleae tali> D

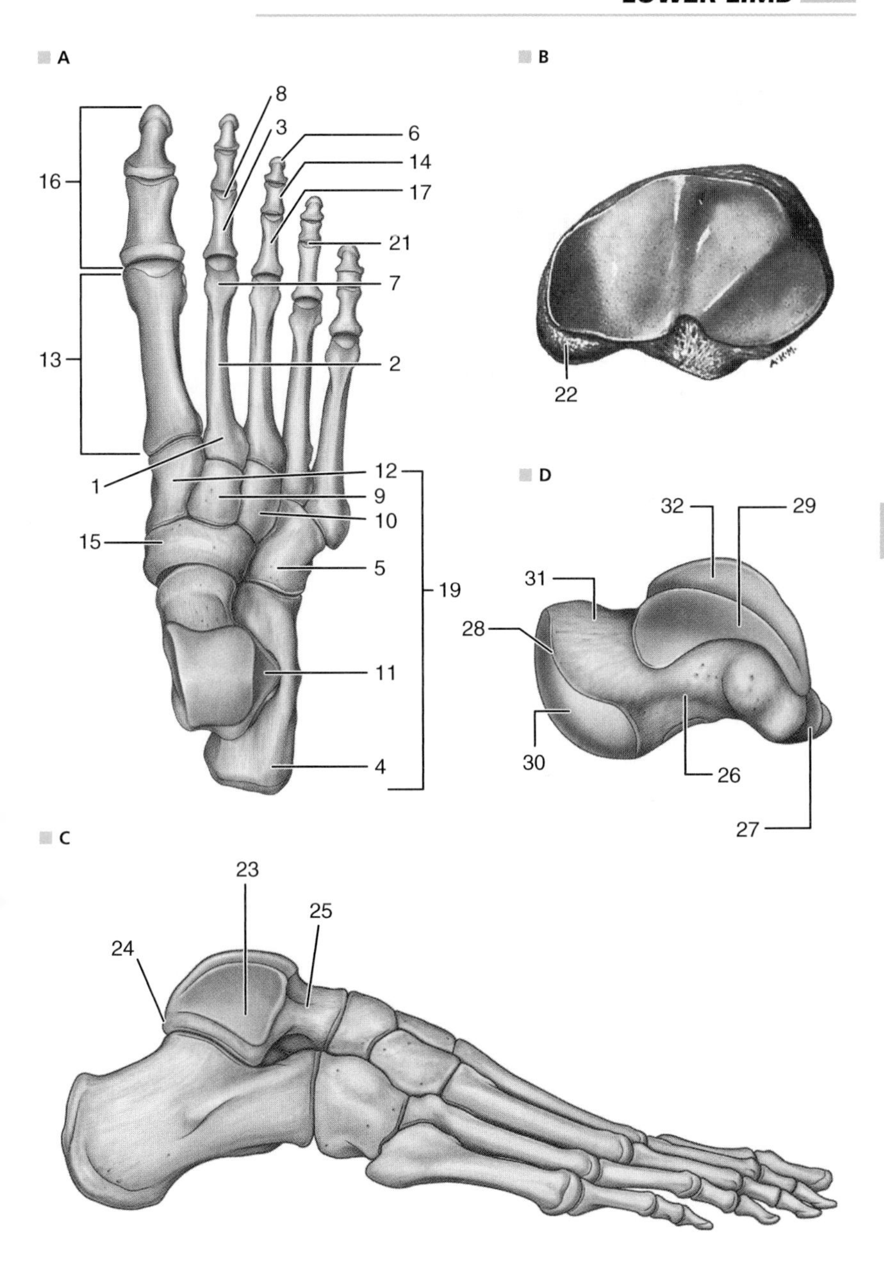
A
8
3
6
14
17
21
7
2
16
13
1
12
9
10
15
5
19
11
4
B
22
D
32
29
31
28
30
26
27
C
23
25
24

1 **anterior facet for calcaneus** the small surface on the head of the talus that rests upon the anterior articular surface of the calcaneus. <facies articularis calcanea anterior tali> A

2 **middle facet for calcaneus** the convex part of the head of the talus that articulates with the sustentaculum tali of the calcaneus. <facies articularis calcanea media tali> A

3 **posterior calcaneal articular facet of talus** a transverse concavity on the inferior surface of the talus, articulating with the calcaneus. <facies articularis calcanea posterior tali> A

4 **sulcus tali** a transverse groove on the inferior surface of the talus, between the middle and the posterior articular surface, which helps to form the tarsal sinus. A

5 **lateral tubercle** the lateral tubercle of the posterior process of the talus. <tuberculum laterale processus posterioris tali> B

6 **medial tubercle** the medial tubercle of the posterior process of the talus. <tuberculum mediale processus posterioris tali> B

7 **posterior process of talus** a backward projection from the posterior portion of the talus, divided into two unequal parts by the sulcus for tendon of flexoris hallucis longus of talus. <processus posterior tali> B

8 **anterior talar articular surface of calcaneus** the small area on the superior surface of the calcaneus just anterior to the middle articular surface, which articulates with the talus. <facies articularis talaris anterior calcanei> C

9 **calcaneal sulcus** a rough, deep groove on the upper surface of the calcaneus, between the medial and the posterior articular surfaces and giving attachment to the interosseous talocalcaneal ligament. <sulcus calcanei> C

10 **middle talar articular surface of calcaneus** the area on the superior surface of the calcaneus just in front of the calcaneal sulcus, which articulates with the talus. <facies articularis talaris media calcanei> C

11 **posterior talar articular surface of calcaneus** the area on the superior surface of the calcaneus just posterolateral to the calcaneal sulcus, which articulates with the talus. <facies articularis talaris posterior calcanei> C

12 **sustentaculum tali** a process of the calcaneus which supports the talus. C

13 **articular surface for cuboid** the saddle-shaped area on the anterior surface of the calcaneus where it articulates with the cuboid bone. <facies articularis cuboidea calcanei> D

14 **calcaneal tubercle** the eminence, often double, on the inferior surface of the calcaneus at the anterior extremity of the rough area for the attachment of the long plantar ligament. <tuberculum calcanei> D

15 **calcaneal tuberosity** the posteroinferior projection of the calcaneus that forms the heel. <tuber calcanei> D

16 **groove for tendon of flexor hallucis longus on calcaneus** a groove on the inferior surface of the sustentaculum tali of the calcaneus, lodging the tendon of the flexoris hallucis longus muscle. <sulcus tendinis musculi flexoris hallucis longi calcanei> D

17 **lateral process of calcaneal tuberosity** a rough process projecting downward from the lower lateral portion of the calcaneal tuberosity. <processus lateralis tuberis calcanei> D

18 **medial process of calcaneal tuberosity** a rough process projecting downward from the lower medial portion of the calcaneal tuberosity. <processus medialis tuberis calcanei> D

19 **groove for tendon of peroneus longus muscle** a deep groove on the inferior surface of the cuboid bone, which in certain foot positions lodges the tendon of the fibularis longus muscle. <sulcus tendinis musculi fibularis longi> E

20 **fibular trochlea** a small eminence on the lateral surface of the calcaneus, separating the tendons of the fibularis brevis and longus muscles. <trochlea fibularis calcanei> F

21 **tarsal sinus** the space between the calcaneus and talus, containing the interosseous ligament; called also *tarsal canal*. <sinus tarsi> E

22 **sesamoid bones of foot** sesamoid bones usually located in the metatarsal region, particularly in the tendon of the flexor hallucis brevis muscle. <ossa sesamoidea pedis> G

23 **tuberosity of fifth metatarsal** a large conical protuberance projecting backward and laterally from the base of the fifth metatarsal bone, to which the tendon of the fibularis brevis muscle is attached. <tuberositas ossis metatarsalis quinti> G

24 **tuberosity of first metatarsal** a blunt process projecting downward and laterally from the lower surface of the base of the first metatarsal bone, to which the tendon of the fibularis longus muscle is attached. <tuberositas ossis metatarsalis primi> G

25 **articular facet of lateral malleolus** the anterosuperior surface of the lateral malleolus, which articulates with the lateral side of the talus. <facies articularis malleoli lateralis> H

26 **articular facet of medial malleolus** the lateral aspect of the medial malleolus, which articulates with the talus. <facies articularis malleoli medialis> H

27 **trochlea of talus** the rounded part of the talus that articulates with the tibia and fibula. <trochlea tali> H

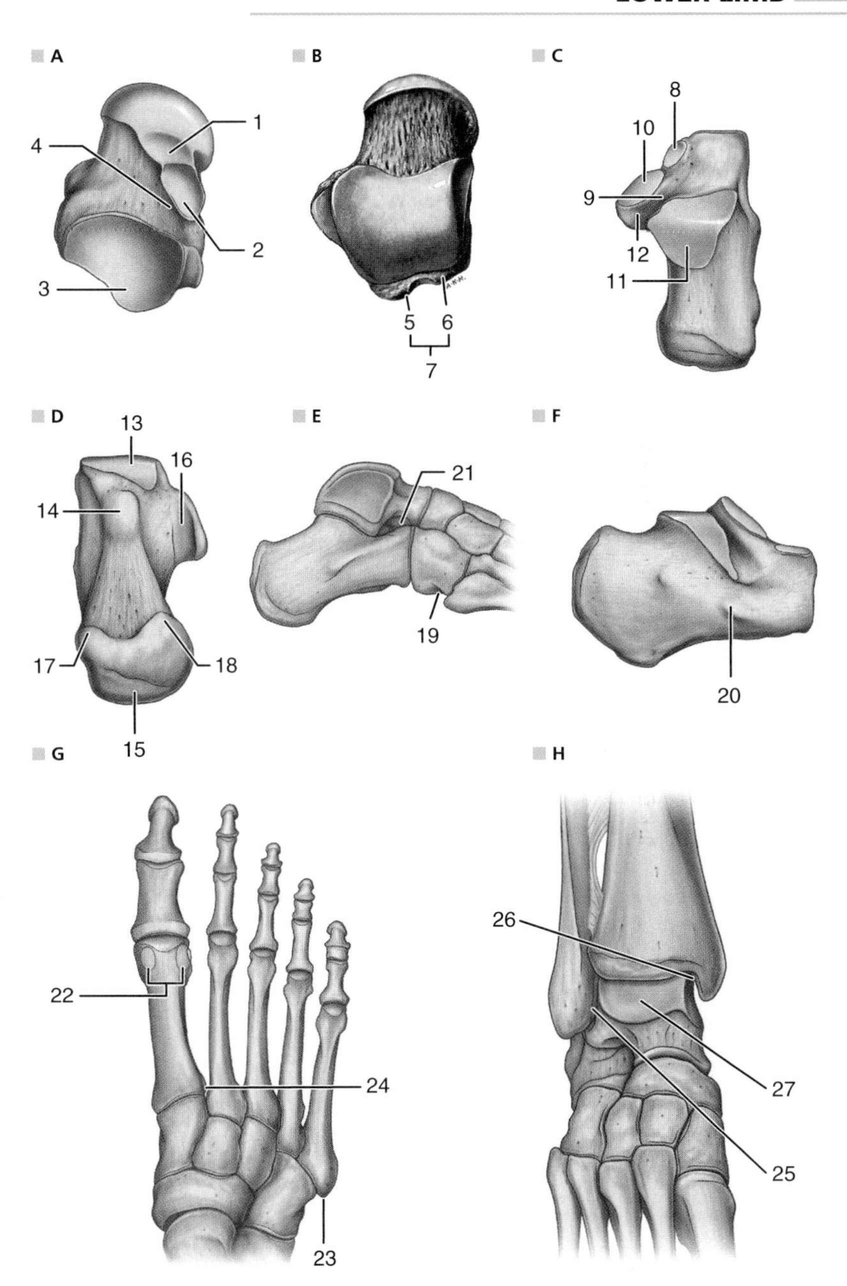
A
1
2
3
4
B
5
6
7
C
8
9
10
11
12
D
13
14
15
16
17
18
E
19
21
F
20
G
22
23
24
H
25
26
27

JOINTS

1 **hip joint** the joint formed between the head of the femur and the acetabulum of the pelvic bone <articulatio coxae> A

2 ***joints of lower limb*** the joints of the free lower limb and those of the pelvic girdle considered together. <juncturae membri inferioris> A

3 **knee joint** the compound joint formed between the articular surface of the patella, the condyles and patellar surface of the femur, and the superior articular surface of the tibia. <articulatio genus> A

4 **acetabular labrum** a ring of fibrocartilage attached to the rim of the acetabulum of the pelvic bone, increasing the depth of the cavity. <labrum acetabuli> B

5 **transverse acetabular ligament** a fibrous band continuous with the acetabular lip of the hip joint, which bridges the acetabular notch and converts it into a foramen. <ligamentum transversum acetabuli> B

6 **ligament of head of femur** a curved triangular or V-shaped fibrous band, attached by its apex to the anterosuperior part of the fovea of the head of the femur and by its base to the sides of the acetabular notch and the intervening transverse ligament of the acetabulum. <ligamentum capitis femoris> C

7 **iliofemoral ligament** a very strong triangular or inverted Y-shaped band that covers the anterior and superior portions of the hip joint. It arises by its apex from the lower part of the anterior inferior iliac spine and is inserted by its base into the intertrochanteric line of the femur. <ligamentum iliofemorale> D

8 **pubofemoral ligament** a band that arises from the entire length of the obturator crest of the pubic bone and passes laterally and inferiorly to merge into the capsule of the hip joint, some fibers reaching to the lower part of the neck of the femur. <ligamentum pubofemorale> D

9 **ischiofemoral ligament** a broad triangular band on the posterior surface of the hip joint. Its base is attached to the ischium posterior and inferior to the acetabulum; its fibers pass superiorly, laterally, and anteriorly across the capsule, bend over the neck, and in part are inserted into the inner side of the trochanteric fossa of the femur and in part blend into the zona orbicularis. <ligamentum ischiofemorale> E

10 **infrapatellar fat pad** a mass of fibrous fatty tissue inferior to the patella, in the angle between the deep surface of the patellar ligament and the tibia. <corpus adiposum infrapatellare> F

11 **lateral meniscus of knee joint** a crescent-shaped disk of fibrocartilage, but nearly circular in form, attached to the lateral margin of the superior articular surface of the tibia. <meniscus lateralis articulationis genus> G

12 **medial meniscus of knee joint** a crescent-shaped disk of fibrocartilage attached to the medial margin of the superior articular surface of the tibia. <meniscus medialis articulationis genus> G

13 **transverse ligament of knee** a more or less distinct bundle of fibers in the knee joint, joining together the anterior convex margin of the lateral meniscus and the anterior concave margin or anterior end of the medial meniscus. <ligamentum transversum genus> G

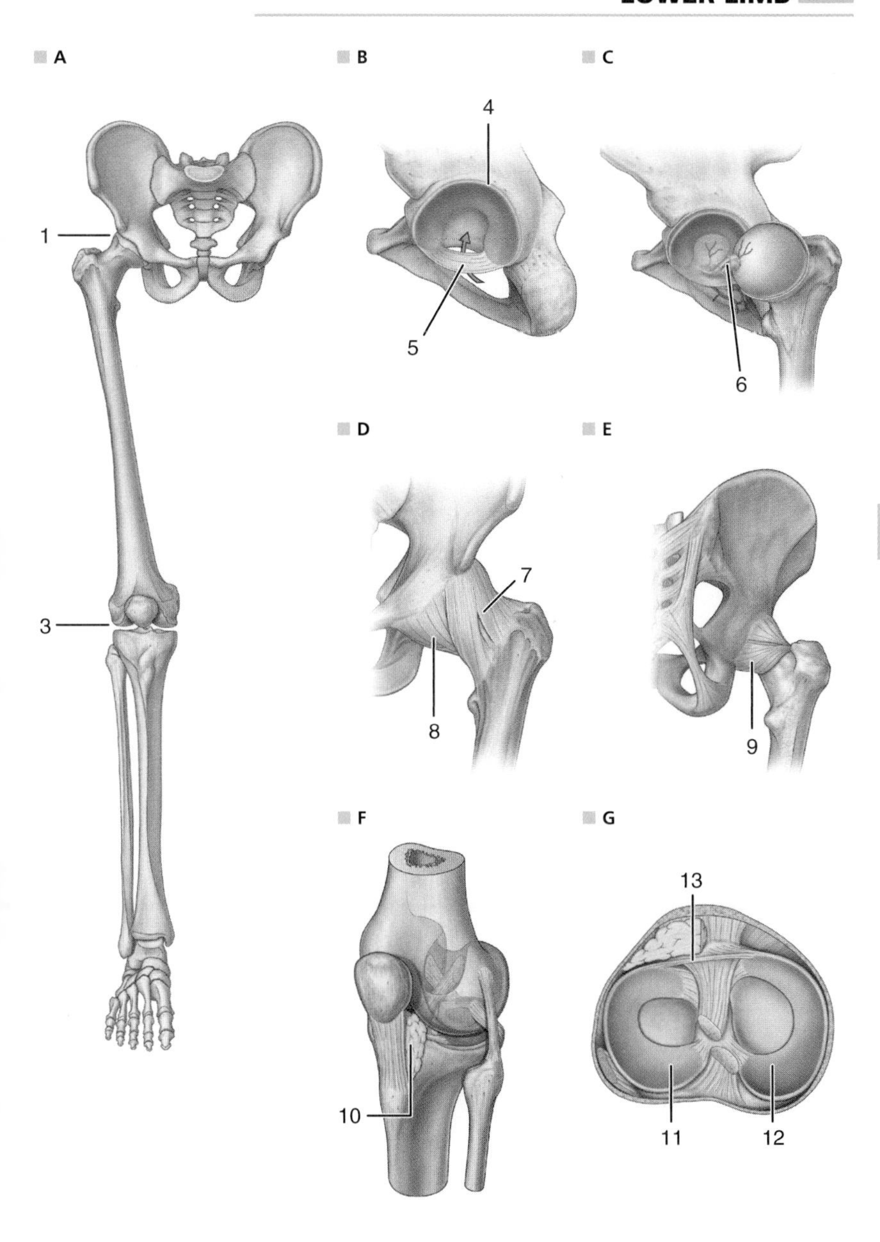
A
B
C
1
3
4
5
6
D
E
7
8
9
F
G
10
11
12
13

1 **anterior cruciate ligament of knee** a strong band that arises from the posteromedial portion of the lateral condyle of the femur, passes anteriorly and inferiorly between the condyles, and is attached to the depression in front of the intercondylar eminence of the tibia. <ligamentum cruciatum anterius genus> A

2 **anterior ligament of head of fibula** a band of fibers that passes obliquely superiorly from the anterior part of the head of the fibula to the lateral condyle of the tibia. <ligamentum capitis fibulae anterius> A

3 ***anterior meniscofemoral ligament*** a small fibrous band of the knee joint, attached to the posterior area of the lateral meniscus and passing superiorly and medially, anterior to the posterior cruciate ligament, to attach to the anterior cruciate ligament. <ligamentum meniscofemorale anterius> A

4 **posterior cruciate ligament of knee** a strong band that arises from the anterolateral surface of the medial condyle of the femur, passes posteriorly and inferiorly between the condyles, and is inserted into the posterior intercondylar area of the tibia. <ligamentum cruciatum posterius genus> A

5 **alar folds** a pair of folds of the synovial membrane of the knee joint; attached to the medial and lateral margins of the articular surface of the patella, they pass posteriorly, converge, and become continuous with the infrapatellar synovial fold. <plicae alares> B

6 **infrapatellar synovial fold** a large process of synovial membrane, containing some fat, which projects into the knee joint; attached to the infrapatellar adipose body, it passes posteriorly and superiorly to the intercondylar fossa of the femur. <plica synovialis infrapatellaris> B

7 **subpopliteal recess** a prolongation of the synovial tendon sheath of the popliteus muscle outside the knee joint into the popliteal space. <recessus subpopliteus> B

8 **arcuate popliteal ligament** a band of variable and ill-defined fibers at the posterolateral part of the knee joint; it is attached inferiorly to the apex of the head of the fibula, arches superiorly and medially over the popliteal tendon, and merges with the articular capsule. <ligamentum popliteum arcuatum> C

9 **posterior ligament of head of fibula** a band of fibers that passes obliquely superiorly from the posterior part of the head of the fibula to the lateral condyle of the tibia. <ligamentum capitis fibulae posterius> D

10 **posterior meniscofemoral ligament** a small fibrous band of the knee joint, attached to the posterior area of the lateral meniscus and passing superiorly and medially, posterior to the posterior cruciate ligament, to the medial condyle of the femur. <ligamentum meniscofemorale posterius> D

11 **deep infrapatellar bursa** a bursa between the patellar ligament and the tibia. <bursa infrapatellaris profunda> E

12 ***subcutaneous bursa of tibial tuberosity*** a bursa between the tibial tuberosity and the skin; called also *superficial infrapatellar bursa*. <bursa subcutanea tuberositatis tibiae> E

13 **subcutaneous infrapatellar bursa** a bursa between the upper end of the patellar ligament and the skin. Called also *superficial infrapatellar bursa.* <bursa subcutanea infrapatellaris> E

14 **subcutaneous prepatellar bursa** a bursa between the patella and the skin. <bursa subcutanea prepatellaris> E

15 ***subtendinous prepatellar bursa*** a bursa sometimes present between the quadriceps tendon and the patellar periosteum. <bursa subtendinea prepatellaris> E

16 **suprapatellar bursa** a bursa between the distal end of the femur and the quadriceps tendon. <bursa suprapatellaris> E

17 **subfascial prepatellar bursa** a bursa between the front of the patella and the investing fascia of the knee. <bursa subfascialis prepatellaris> G

18 **lateral patellar retinaculum** a fibrous membrane from the tendon of the vastus lateralis muscle, attached to the lateral margin of the patella and then along the side of the patellar ligament, and inserted into the tibia as far distal as the fibular collateral ligament; it also blends with the iliotibial tract of the fascia lata. Called also *lateral patellar ligament.* <retinaculum patellae laterale> F

19 **medial patellar retinaculum** a fibrous membrane from the tendon of the vastus medialis muscle, attached to the medial margin of the patella and then along the side of the patellar ligament, and inserted into the tibia as far distal as the tibial collateral ligament. <retinaculum patellae mediale> F

20 **patellar ligament** the continuation of the central portion of the tendon of the quadriceps femoris muscle distal to the patella; it extends from the patella to the tuberosity of the tibia. Called also *patellar tendon.* <ligamentum patellae> F

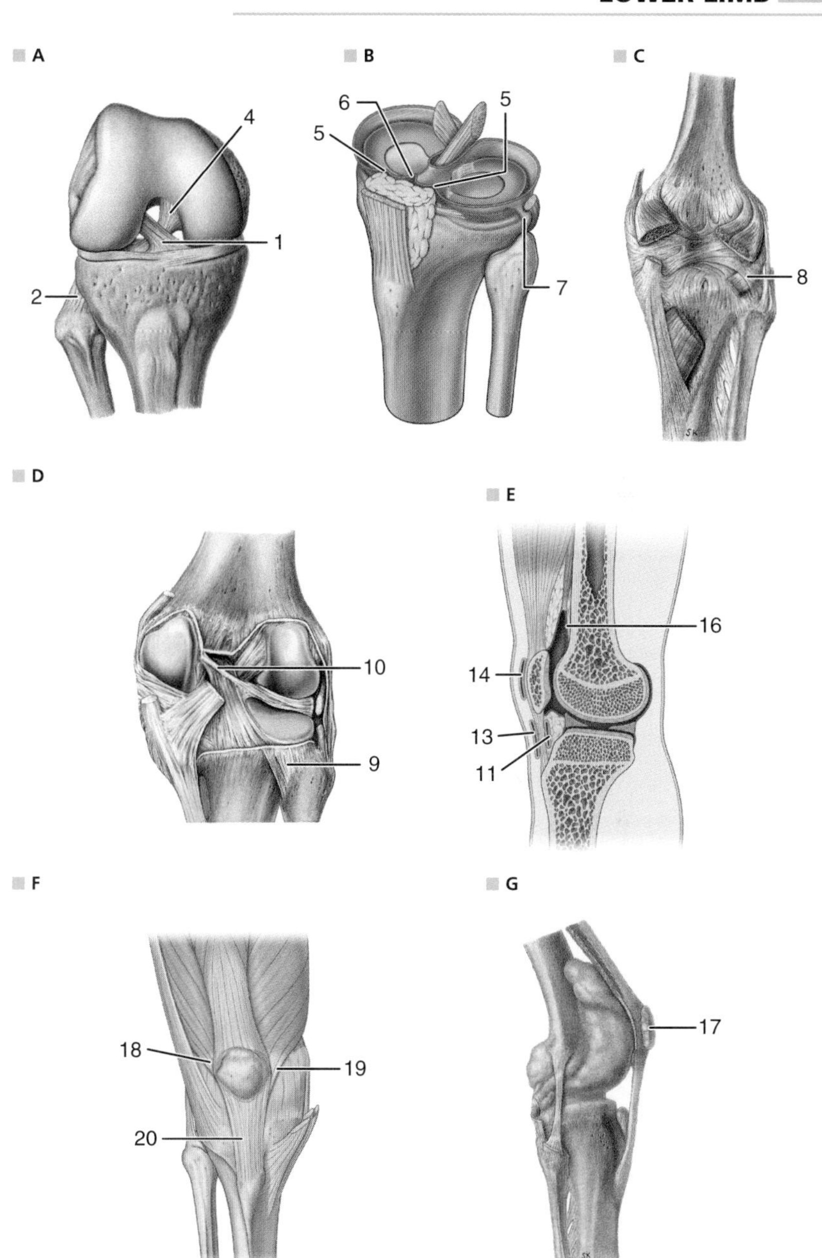
A
B
C
D
E
F
G
1
2
4
5
6
5
7
8
10
9
16
14
13
11
17
18
19
20

1 **oblique popliteal ligament** a broad band of fibers that arises from the medial condyle of the tibia, merges more or less with the tendon of the semimembranosus, and passes obliquely across the back of the knee joint to the lateral epicondyle of the femur. It contains large openings for the passage of vessels and nerves. <ligamentum popliteum obliquum> A

2 **fibular collateral ligament** a strong, round fibrous cord on the lateral side of the knee joint, entirely independent of the capsule of the knee joint; it is attached superiorly to the posterior part of the lateral epicondyle of the femur and inferiorly to the lateral side of the head of the fibula just in front of the styloid process. <ligamentum collaterale fibulare> B, D

3 **tibial collateral ligament** a broad, flat, longitudinal band on the medial side of the knee joint; it is attached superiorly to the medial epicondyle of the femur, inferiorly to the medial surface of the body of the tibia, and in between to the medial meniscus. <ligamentum collaterale tibiale> C, D

4 **tibiofibular joint** a plane joint between the lateral condyle of the tibia and the head of the fibula. <articulatio tibiofibularis> A

5 **calcaneocuboid joint** one formed between the cuboidal articular surface of the calcaneus and the cuboid bone. <articulatio calcaneocuboidea> E

6 **cuneonavicular joint** the joint between the anterior surface of the navicular bone and the proximal ends of the three cuneiform bones. <articulatio cuneonavicularis> E

7 **intercuneiform joints** the synovial joints between the cuneiform bones. <articulationes intercuneiformes> E

8 **intermetatarsal joints** the joints formed between the adjoining bases of the five metatarsal bones. <articulationes intermetatarsales> E

9 ***interosseous cuneocuboid ligament*** fibers connecting the central portions of the adjacent surfaces of the cuboid and lateral cuneiform bones, between the articular surfaces. <ligamentum cuneocuboideum interosseum> E

10 ***interosseous cuneometatarsal ligaments*** fibrous bands that join the adjacent surfaces of the cuneiform and the metatarsal bones. <ligamenta cuneometatarsalia interossea> E

11 ***interosseous intercuneiform ligaments*** short fibrous bands that join the adjacent surfaces of the medial and intermediate, and the intermediate and lateral, cuneiform bones. <ligamenta intercuneiformia interossea> E

12 ***interosseous metatarsal ligaments*** bands between the bases of the second to fifth metatarsal bones, similar to the corresponding ligaments of the hand. <ligamenta metatarsalia interossea> E

13 **interosseous metatarsal spaces** the four spaces between the metatarsal bones. <spatia interossea metatarsi> E

14 ***interosseous tarsal ligaments*** the interosseous cuneocuboid, intercuneiform, and talocalcaneal ligaments. <ligamenta tarsi interossea> E

15 **interphalangeal joints of foot** the hinge joints between the phalangeal articulations of the foot. <articulationes interphalangeae pedis> E

16 ***joints of foot*** the joints of the foot, including the talocrural, intertarsal, tarsometatarsal, metatarsophalangeal, and interphalangeal joints. <articulationes pedis> E

17 **metatarsophalangeal joints** the joints formed between the heads of the five metatarsal bones and the bases of the corresponding proximal phalanges. <articulationes metatarsophalangeae> E

18 **tarsometatarsal joints** joints formed by the cuneiform and cuboid bones together with the bases of the metatarsal bones; called also *Lisfranc joints*. <articulationes tarsometatarsales> E

19 **ankle joint** a joint formed by the inferior articular and malleolar articular surfaces of the tibia, the malleolar articular surface of the fibula, and the medial malleolar, lateral malleolar, and superior surfaces of the talus. <articulatio talocruralis> F

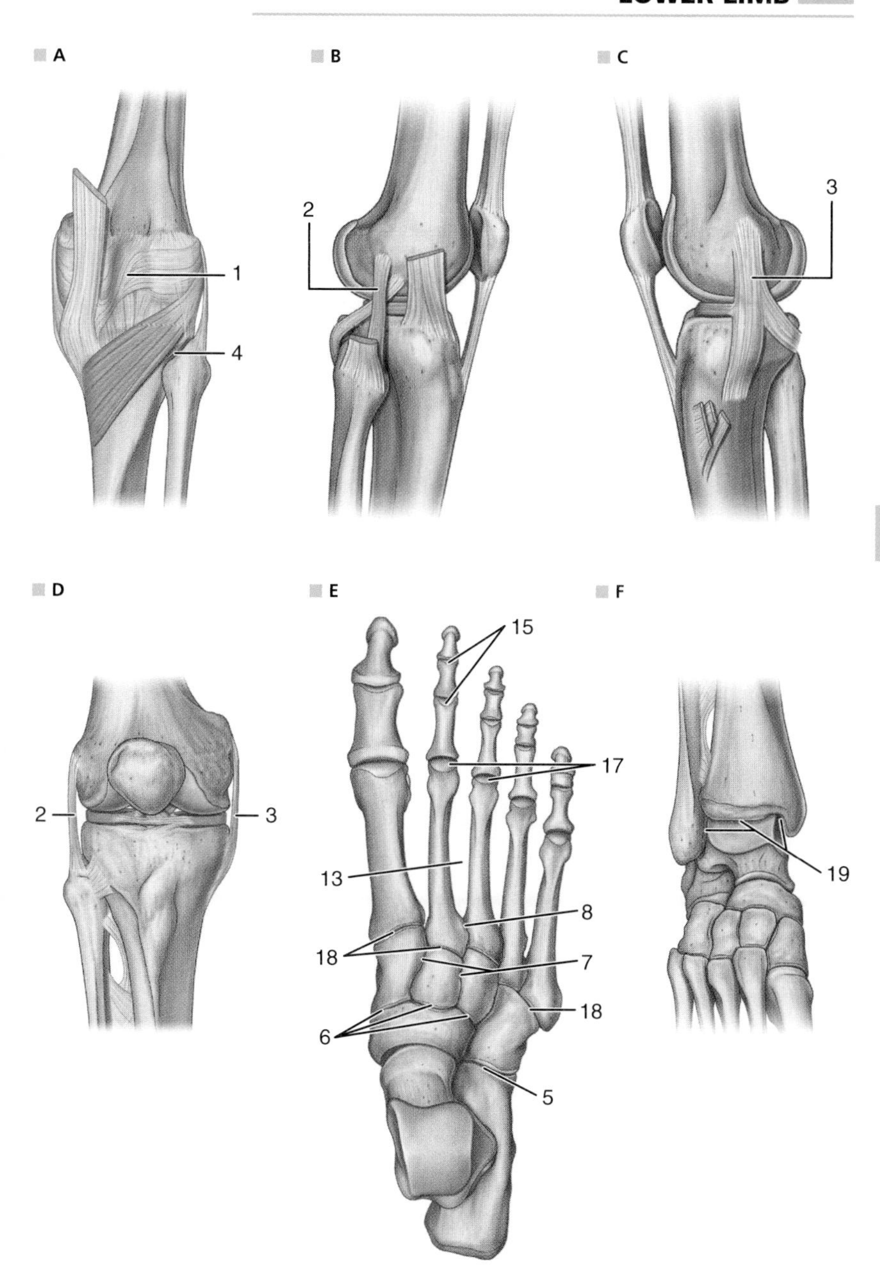
A
1
4
B
2
C
3
D
2
3
E
15
17
13
8
18
7
18
6
5
F
19

1 **anterior tibiotalar part of medial ligament of ankle joint** the deeper portion of the medial collateral ligament of the ankle joint, attached superiorly to the medial malleolus of the tibia and inferiorly to the medial surface of the talus. Called also *anterior talotibial ligament.* <pars tibiotalaris anterior ligamenti collateralis medialis> A

2 **medial ligament of ankle joint** a large fan-shaped ligament on the medial side of the ankle, passing from the medial malleolus of the tibia down onto the tarsal bones. It comprises four parts: *tibionavicular, tibiocalcaneal, anterior tibiotalar,* and *posterior tibiotalar parts.* Called also *deltoid ligament.* <ligamentum collaterale mediale articulationis talocruralis> A

3 **posterior tibiotalar part of medial ligament of ankle joint** the posterior portion of the superficial fibers of the medial collateral ligament of the ankle joint, attached superiorly to the posterior part of the medial malleolus of the tibia and inferiorly to the medial surface of the talus. <pars tibiotalaris posterior ligamenti collateralis medialis> A

4 **tibiocalcaneal part of medial ligament of ankle joint** the middle portion of the superficial fibers of the medial collateral ligament of the ankle joint, attached superiorly to the medial malleolus of the tibia and inferiorly into nearly the entire length of the sustentaculum tali of the calcaneus. Called also *tibiocalcaneal ligament.* <pars tibiocalcanea ligamenti collateralis medialis> A

5 **tibionavicular part of medial ligament of ankle joint** the anterior portion of the superficial fibers of the medial collateral ligament of the ankle joint, attached superiorly to the anterior surface of the medial malleolus of the tibia and inferiorly to the navicular bone and the margin of the calcaneonavicular ligament. Called also *tibionavicular ligament.* <pars tibionavicularis ligamenti collateralis medialis> A

6 **anterior talofibular ligament** one or more fibrous bands that pass from the anterior surface of the lateral malleolus of the fibula to the anterior margin of the lateral articular surface of the talus. <ligamentum talofibulare anterius> B

7 **calcaneofibular ligament** a band of fibers arising from the lateral surface of the lateral malleolus of the fibula just anterior to the apex and passing inferiorly and posteriorly to be attached to the lateral surface of the calcaneus. <ligamentum calcaneofibulare> B

8 **lateral ligament of ankle joint** the three ligamentous fascicles present on the lateral side of the ankle joint: the *calcaneofibular, anterior talofibular,* and *posterior talofibular ligaments* considered collectively. <ligamentum collaterale laterale articulationis talocruralis> B

9 **posterior talofibular ligament** a strong fibrous horizontal band passing from the posteromedial face of the lateral malleolus of the fibula to the area of the posterior process of the talus. <ligamentum talofibulare posterius> B

10 **transverse tarsal joint** a joint comprising the articulation of the calcaneus and the cuboid bone and the articulation of the talus and the navicular bone. <articulatio tarsi transversa> C

11 **bifurcate ligament** a Y-shaped ligament on the dorsum of the foot, comprising the calcaneonavicular and calcaneocuboid ligaments. <ligamentum bifurcatum> E, G

12 **calcaneocuboid ligament** the band of fibers connecting the superior surface of the calcaneus and the dorsal surface of the cuboid bone together with the calcaneonavicular ligament constituting the bifurcate ligament. <ligamentum calcaneocuboideum> E, G

13 **calcaneonavicular ligament** the band of fibers connecting the superior surface of the calcaneus and the lateral surface of the navicular bone, together with the calcaneocuboid ligament constituting the bifurcate ligament. <ligamentum calcaneonaviculare> E

14 **dorsal cuboideonavicular ligament** a fibrous bundle connecting the dorsal surfaces of the cuboid and navicular bones. <ligamentum cuboideonaviculare dorsale> E

15 **dorsal metatarsal ligaments** light transverse bands on the dorsal surfaces of the bases of the second to fifth metatarsal bones, similar to the corresponding ligaments on the metacarpal bones. <ligamenta metatarsalia dorsalia> E

16 **lateral talocalcaneal ligament** a fibrous band passing from the lateral surface of the talus to that of the calcaneus. <ligamentum talocalcaneum laterale> E

17 **medial talocalcaneal ligament** a fibrous band connecting the medial tubercle of the talus with the sustentaculum tali of the calcaneus. <ligamentum talocalcaneum mediale> D

18 **subtalar joint** the joint formed between the posterior calcaneal articular surface of the talus and the posterior articular surface of the calcaneus. <articulatio subtalaris> F

19 **interosseous talocalcaneal ligament** fibrous bands in the tarsal sinus, passing between the opposed surfaces of the calcaneus and the talus. <ligamentum talocalcaneum interosseum> G

20 **talonavicular ligament** a broad, thin fibrous band passing from the dorsal and lateral surfaces of the neck of the talus to the dorsal surface of the navicular bone. <ligamentum talonaviculare> G

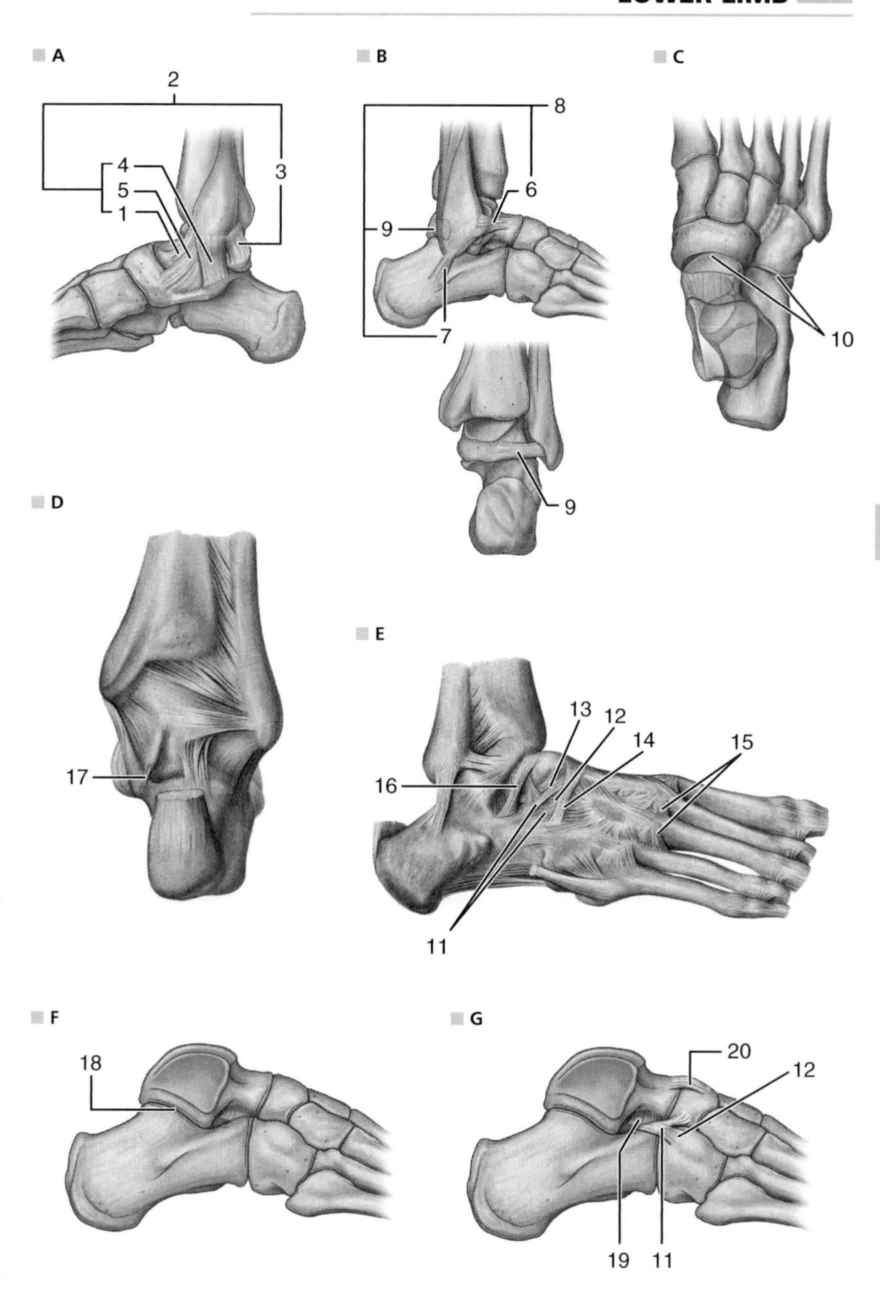
A
2
3
4
5
1
B
8
6
9
7
9
C
10
D
17
E
13
12
14
15
16
11
F
18
G
20
12
19
11

1 **collateral ligaments of interphalangeal joints of foot** fibrous bands, one on either side of each of the interphalangeal joints of the toes. <ligamenta collateralia articulationum interphalangealium pedis> A

2 **collateral ligaments of metatarsophalangeal joints** strong fibrous bands on either side of each metatarsophalangeal joint, holding the two bones involved in each joint firmly together. <ligamenta collateralia articulationum metatarsophalangealium> A

3 **deep transverse metatarsal ligament** a narrow fibrous band that extends across, is attached to the plantar surfaces of, and thus joins together the heads of all the metatarsal bones. <ligamentum metatarsale transversum profundum> A

4 **plantar ligaments of interphalangeal joints** thick, dense bands on the plantar surfaces of the interphalangeal articulations of the foot, between the collateral ligaments. <ligamenta plantaria articulationum interphalangealium pedis> A

5 **plantar ligaments of metatarsophalangeal joints** thick, dense bands on the plantar surface of the metatarsophalangeal articulations, between the collateral ligaments. <ligamenta plantaria articulationum metatarsophalangealium> A

6 **superficial transverse metatarsal ligament** fibers that lie in the superficial fascia of the sole of the foot beneath the heads of the metatarsal bones. <ligamentum metatarsale transversum superficiale> B

7 **talocalcaneonavicular joint** a joint formed by the head of the talus, the anterior and middle articular surfaces of the calcaneus, the plantar calcaneonavicular ligament, and the posterior surface of the navicular bone. <articulatio talocalcaneonavicularis> C

8 **anterior tibiofibular ligament** a flat triangular band that passes diagonally, inferiorly, and laterally from the anterior portion of the lateral surface of the distal end of the tibia to the anterior surface of the distal end of the fibula. <ligamentum tibiofibulare anterius> D

9 **tibiofibular syndesmosis** a firm fibrous union formed at the distal ends of the tibia and fibula between the fibular notch of the tibia and a roughened triangular surface on the fibula, which frequently contains a synovial prolongation of the cavity of the talocrural articulation. <syndesmosis tibiofibularis> D, F

10 **posterior tibiofibular ligament** a fibrous band that passes diagonally, inferiorly, and laterally from the posterior surface of the distal end of the tibia to the adjacent posterior surface of the distal end of the fibula. <ligamentum tibiofibulare posterius> F

11 **long plantar ligament** the longest ligament of the foot, arising from the lower surface of the calcaneus as far back as the lateral and the medial processes, passing forward over the tendon of the fibularis longus, and inserting into the bases of the second through fifth metatarsal bones. <ligamentum plantare longum> E

12 **plantar calcaneocuboid ligament** a short, wide, strong band connecting the plantar surfaces of the calcaneus and the cuboid bone; called also *short plantar ligament*. <ligamentum calcaneocuboideum plantare> E

13 **plantar cuboideonavicular ligament** a fibrous band connecting the plantar surfaces of the cuboid and navicular bones. <ligamentum cuboideonaviculare plantare> E

14 ***plantar cuneocuboid ligament*** a band of fibers connecting the plantar surfaces of the cuboid and lateral cuneiform bones. <ligamentum cuneocuboideum plantare> E

15 **plantar cuneonavicular ligaments** bands that join the plantar surface of the navicular bone to the adjacent plantar surfaces of the three cuneiform bones. <ligamenta cuneonavicularia plantaria> E

16 **plantar calcaneonavicular ligament** a broad, thick band passing from the anterior margin of the sustentaculum tali to the plantar surface of the navicular bone; it bears on its deep surface a fibrocartilage that helps to support the head of the talus. Called also *spring ligament*. <ligamentum calcaneonaviculare plantare> E

17 ***plantar intercuneiform ligaments*** fibrous bands that join the plantar surfaces of the cuneiform bones. <ligamenta intercuneiformia plantaria> G

18 **plantar metatarsal ligaments** strong transverse bands on the plantar surfaces of the bases of the second to fifth metatarsal bones. <ligamenta metatarsalia plantaria> G

19 ***plantar tarsal ligaments*** the inferior ligaments of the foot, comprising the long plantar and the plantar calcaneocuboid, calcaneonavicular, cuneonavicular, cuboideonavicular, intercuneiform, and cuneocuboid ligaments. <ligamenta tarsi plantaria> G

20 **plantar tarsometatarsal ligaments** fibrous bands passing from the plantar surfaces of the bases of the metatarsal bones to the plantar surfaces of the cuboid and the three cuneiform bones. <ligamenta tarsometatarsalia plantaria> G

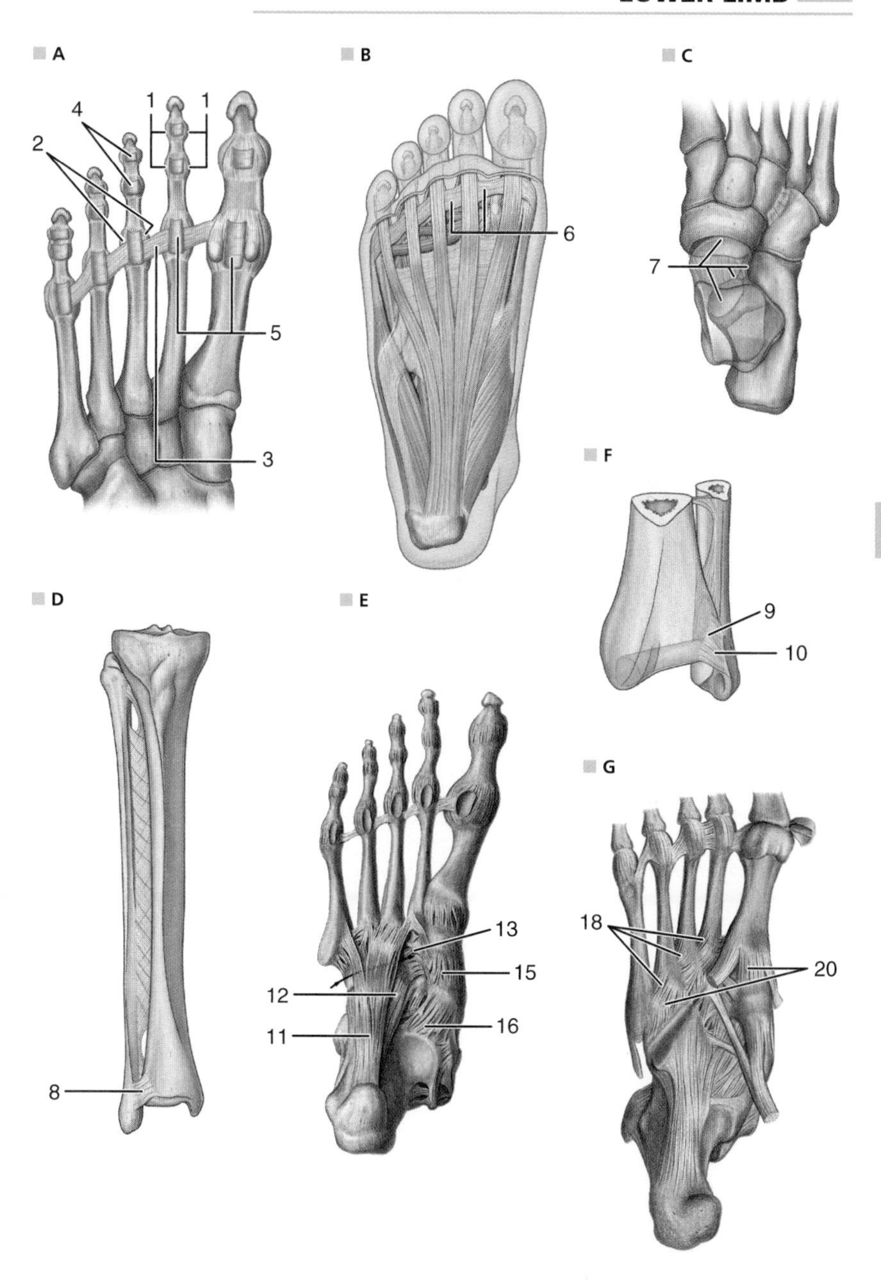
A
1
1
4
2
5
3
B
6
C
7
F
9
10
D
8
E
13
15
12
11
16
G
18
20

1 **dorsal cuneocuboid ligament** fibers connecting the dorsal surfaces of the cuboid and lateral cuneiform bones. <ligamentum cuneocuboideum dorsale> A

2 **dorsal cuneonavicular ligaments** bands that join the dorsal surface of the navicular bone to the dorsal surfaces of the three cuneiform bones. <ligamenta cuneonavicularia dorsalia> A

3 **dorsal intercuneiform ligaments** fibrous bands connecting the dorsal surfaces of the three cuneiform bones. <ligamenta intercuneiformia dorsalia> A

4 ***dorsal tarsal ligaments*** the bifurcate, dorsal cuboideonavicular, cuneocuboid, cuneonavicular, intercuneiform, and talonavicular ligaments; called also *dorsal intertarsal ligaments.* <ligamenta tarsi dorsalia> A

5 **dorsal tarsometatarsal ligaments** fibrous bands passing from the dorsal surfaces of the bases of the metatarsal bones to the dorsal surfaces of the cuboid and the three cuneiform bones. <ligamenta tarsometatarsalia dorsalia> A

6 **anserine bursa** a bursa between the tendons of the sartorius, gracilis, and semitendinosus muscles, and the tibial collateral ligament. <bursa anserina> B

GLUTEAL REGION

7 ***nerve to piriformis*** *origin,* posterior branches of anterior rami of S1-S2; *distribution,* anterior piriform muscle; *modality,* general sensory and motor. <nervus musculi piriformis> C

8 ***nerve to quadratus femoris*** *origin,* anterior branches of anterior rami of L4-L5; *distribution,* gemellus inferior, anterior quadratus femoris muscle, hip joint; *modality,* general sensory and motor. <nervus musculi quadrati femoris> C

9 **sacral plexus** a plexus that lies in front of the piriformis muscle, arising from the anterior branches of the last two lumbar nerves (which form the lumbosacral trunk) and the first four sacral nerves. It has twelve named branches, five supplying pelvic structures (the nerves to the piriformis, to levator ani and coccygeus, and to sphincter ani muscles, the pelvic splanchnic nerves, and the pudendal nerve) and seven helping supply the buttocks and lower limbs (superior and inferior gluteal, posterior femoral cutaneous, perforating cutaneous, and sciatic nerves, and nerves to the quadratus femoris and obturator internus muscles). <plexus sacralis> C, D

10 **gluteal fold** a curved transverse groove or fold on the posterior aspect of the upper thigh, separating the upper part of the thigh from the buttocks. <sulcus glutealis> E

11 ***bursa of piriformis muscle*** a bursa between the piriformis tendon, the superior gemellus muscle, and the femur. <bursa musculi piriformis> F

12 **gluteus medius muscle** *origin,* lateral surface of ilium between anterior and posterior gluteal lines; *insertion,* greater trochanter of femur; *innervation,* superior gluteal; *action,* abducts and rotates thigh medially. <musculus gluteus medius> F

13 **gluteus minimus muscle** *origin,* lateral surface of ilium between anterior and inferior gluteal lines; *insertion,* greater trochanter of femur; *innervation,* superior gluteal; *action,* abducts, rotates thigh medially. <musculus gluteus minimus> F

14 **inferior gemellus muscle** *origin,* tuberosity of ischium; *insertion,* greater trochanter of femur; *innervation,* nerve to quadrate muscle of thigh; *action,* rotates thigh laterally. <musculus gemellus inferior> F

15 ***piriformis fascia*** an extension of the parietal pelvic fascia that surrounds the piriformis muscle. <fascia musculi piriformis> F

16 **piriformis muscle** *origin,* ilium, second to fourth sacral vertebrae; *insertion,* upper border of greater trochanter of femur; *innervation,* first and second sacral; *action,* rotates thigh laterally. <musculus piriformis> F

17 **quadratus femoris muscle** *origin,* upper part of lateral border of tuberosity of ischium; *insertion,* quadrate tubercle of femur, intertrochanteric crest; *innervation,* fourth and fifth lumbar and first sacral; *action,* adducts, rotates thigh laterally. <musculus quadratus femoris> F

18 **superior gemellus muscle** *origin,* spine of ischium; *insertion,* greater trochanter of femur; *innervation,* nerve to internal obturator; *action,* rotates thigh laterally. <musculus gemellus superior> F

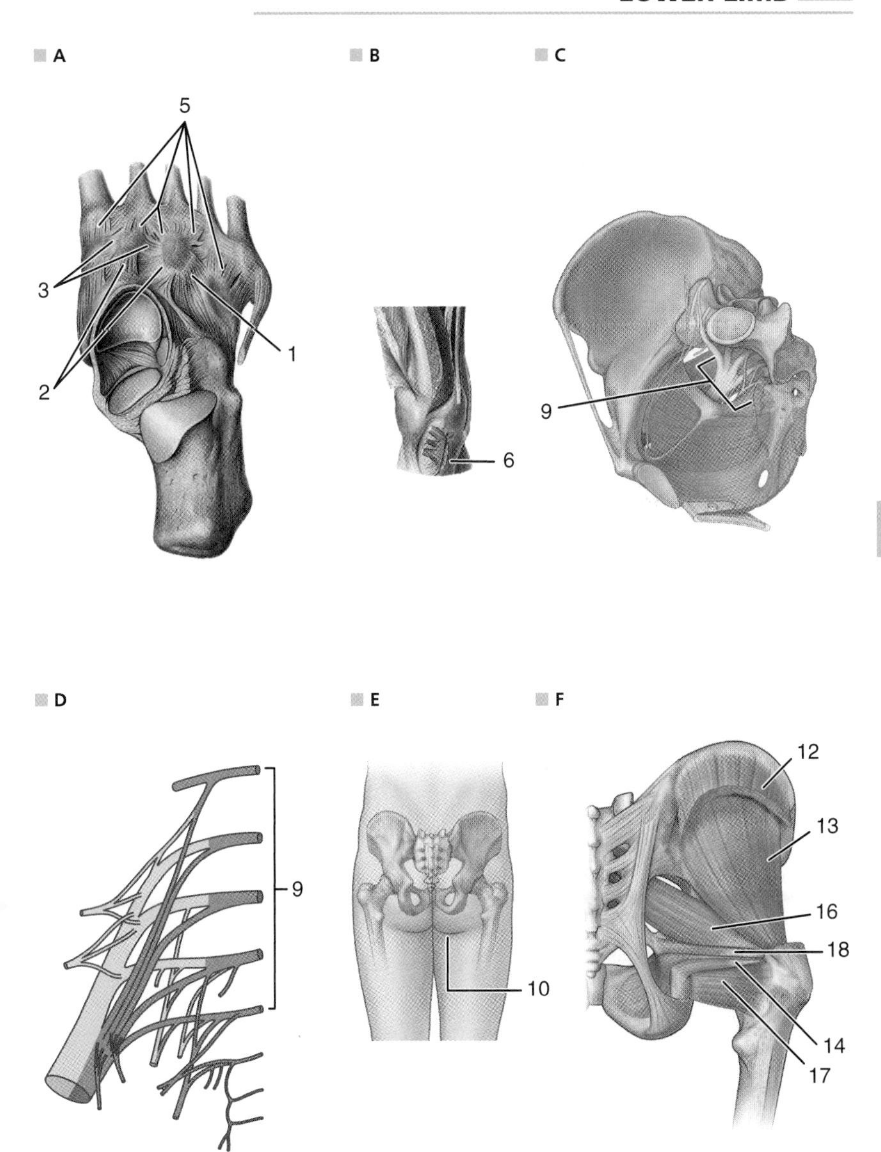
A
B
C
5
3
1
2
6
9
D
E
F
9
12
13
16
18
10
14
17

1 **gluteal aponeurosis** a dense sheet of fascia lying between the iliac crest and the superior border of the gluteus maximus; from it arises a part of the gluteus medius muscle. <aponeurosis glutealis> A

2 **gluteus maximus muscle** *origin,* posterior aspect of ilium, posterior surface of sacrum and coccyx, sacrotuberous ligament, fascia covering gluteus medius; *insertion,* iliotibial tract of fascia lata, gluteal tuberosity of femur; *innervation,* inferior gluteal; *action,* extends, abducts, and rotates thigh laterally. <musculus gluteus maximus> A

3 **tensor fasciae latae muscle** *origin,* iliac crest; *insertion,* iliotibial tract of fascia lata; *innervation,* superior gluteal; *action,* flexes, rotates thigh medially. <musculus tensor fasciae latae> B

4 **falciform process of sacrotuberous ligament** a prolongation of the sacrotuberous ligament, continuing forward along the inner border of the ramus of the ischium from the point of attachment of the ligament on the tuber of the ischium. <processus falciformis ligamenti sacrotuberalis> C

5 **intermuscular gluteal bursae** several sacs that surround the tendon attaching the gluteus maximus to the femur. <bursae intermusculares musculorum gluteorum> D

6 ***ischial bursa of gluteus maximus muscle*** a bursa between the ischial tuberosity and the gluteus maximus. <bursa ischiadica musculi glutei maximi> D

7 **sciatic bursa of obturator internus muscle** a bursa between the tendon of the obturator internus muscle and the lesser sciatic notch. <bursa ischiadica musculi obturatorii interni> E

8 ***subtendinous bursa of obturator internus muscle*** a bursa beneath the tendon of the obturator internus muscle. <bursa subtendinea musculi obturatorii interni> E

9 **trochanteric bursa of gluteus maximus muscle** a bursa between the fascial tendon of the gluteus maximus, the posterolateral surface of the greater trochanter, and the vastus lateralis muscle. <bursa trochanterica musculi glutei maximi> D, E

10 ***trochanteric bursa of gluteus minimus muscle*** a bursa between the edge of the gluteus minimus and the greater trochanter. <bursa trochanterica musculi glutei minimi> E

11 **trochanteric bursae of gluteus medius muscle** bursae between the gluteus medius and the lateral surface of the greater trochanter, and sometimes between the tendons of the gluteus medius and the piriformis. <bursae trochantericae musculi glutei medii> E

12 **lumbosacral plexus** a term applied to the lumbar and sacral nerve plexuses together, because of their continuous nature. <plexus lumbosacralis> F

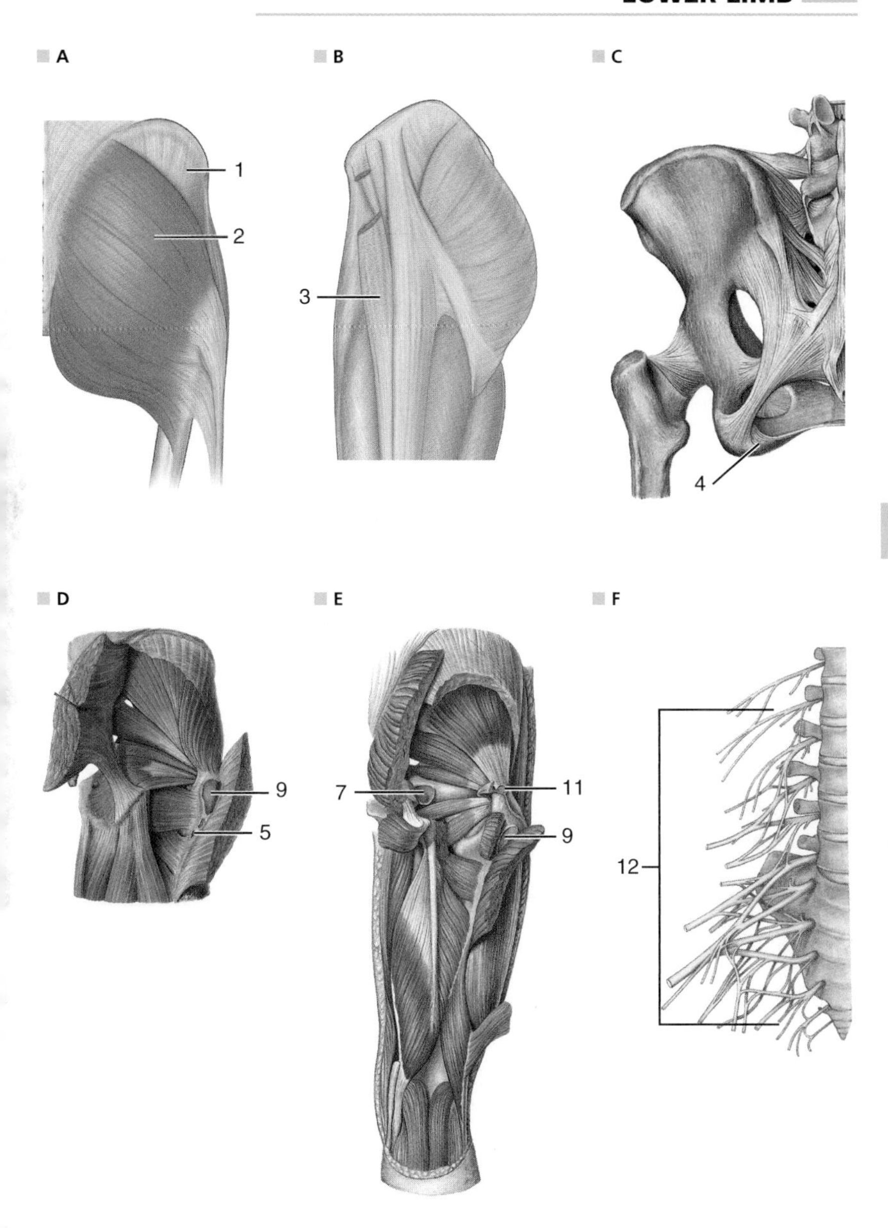
A
1
2
B
3
C
4
D
9
5
E
7
11
9
F
12

1 **inferior gluteal nerve** *origin,* sacral plexus (L5-S2); *distribution,* gluteus maximus muscle; *modality,* motor. <nervus gluteus inferior> A

2 **perforating cutaneous nerve** one arising from the posterior surface of the second and third sacral nerves, piercing the sacrotuberous ligament, and supplying the skin over the inferomedial gluteus maximus; it is absent in one-third of the population. <nervus cutaneus perforans> A

3 **posterior cutaneous nerve of thigh** *origin,* sacral plexus—S1-S3; *branches,* inferior cluneal nerves and perineal rami; *distribution,* skin of buttock, external genitalia, and back of thigh and calf; *modality,* general sensory. <nervus cutaneus femoris posterior> A

4 **sciatic nerve** the largest nerve of the body: *origin,* sacral plexus (L4-S3); it leaves the pelvis through the greater sciatic foramen; *branches,* divides into the tibial and common fibular nerves, usually in lower third of thigh. <nervus ischiadicus> A

5 **superior gluteal nerve** *origin,* sacral plexus (L4-S1); *distribution,* gluteus medius and minimus muscles, tensor fasciae latae, and hip joint; *modality,* motor and general sensory. <nervus gluteus superior> A

6 **inferior cluneal nerves** general sensory nerve branches of the posterior femoral cutaneous nerve, innervating the skin of the lower part of the buttocks. <nervi clunium inferiores> B

7 **middle cluneal nerves** general sensory nerve branches of the plexus formed by the lateral branches of posterior rami of the first four sacral nerves, innervating ligaments of the sacrum and the skin over the posterior buttocks. <nervi clunium medii> B

8 **superior cluneal nerves** general sensory nerve branches of the posterior rami of the upper lumbar nerves, innervating the skin of the upper part of the buttocks. <nervi clunium superiores> B

9 **deep branch of superior gluteal artery** a branch passing forward between the gluteus medius and minimus muscles, and dividing into superior and inferior branches. <ramus profundus arteriae gluteae superioris> C

10 **inferior branch of deep branch of superior gluteal artery** the lower division of the deep branch of the superior gluteal artery, accompanied by the superior gluteal nerve and helping supply the gluteus medius, gluteus minimus, and tensor fasciae latae muscles and the hip joint and ilium. <ramus inferior rami profundi arteriae gluteae superioris> C

11 ***inferior gluteal lymph nodes*** the internal iliac lymph nodes situated along the inferior gluteal artery. <nodi lymphoidei gluteales inferiores> C

12 **inferior gluteal veins** venae comitantes of the inferior gluteal artery; they drain the subcutaneous tissue of the back of the thigh and the muscles of the buttock, unite into a single vein after passing through the greater sciatic foramen, and empty into the internal iliac vein. <venae gluteae inferiores> C

13 **superficial branch of superior gluteal artery** a branch that ramifies to supply the gluteus maximus muscle. <ramus superficialis arteriae gluteae superioris> C

14 **superior branch of deep branch of superior gluteal artery** the upper division of the deep branch of the superior gluteal artery, extending as far as the anterior superior iliac spine and helping supply the gluteus medius, gluteus minimus, and tensor fasciae latae muscles. <ramus superior rami profundi arteriae gluteae superioris> C

15 ***superior gluteal lymph nodes*** the internal iliac lymph nodes situated along the superior gluteal artery. <nodi lymphoidei gluteales superiores> C

16 **superior gluteal veins** venae comitantes of the superior gluteal artery; they drain the muscles of the buttock, pass through the greater sciatic foramen, and empty into the internal iliac vein. <venae gluteae superiores> C

17 **superior gluteal artery** *origin,* internal iliac artery; *branches,* superficial and deep branches; *distribution,* buttocks. <arteria glutea superior> D

THIGH

Anterior Compartment of Thigh

18 **deep inguinal lymph nodes** nodes deep to the fascia lata along the femoral vein; they receive lymph from the deep structures of the lower limb and from the penis or clitoris, and superficial inguinal lymph nodes and drain into the external iliac lymph nodes. <nodi lymphoidei inguinales profundi> E

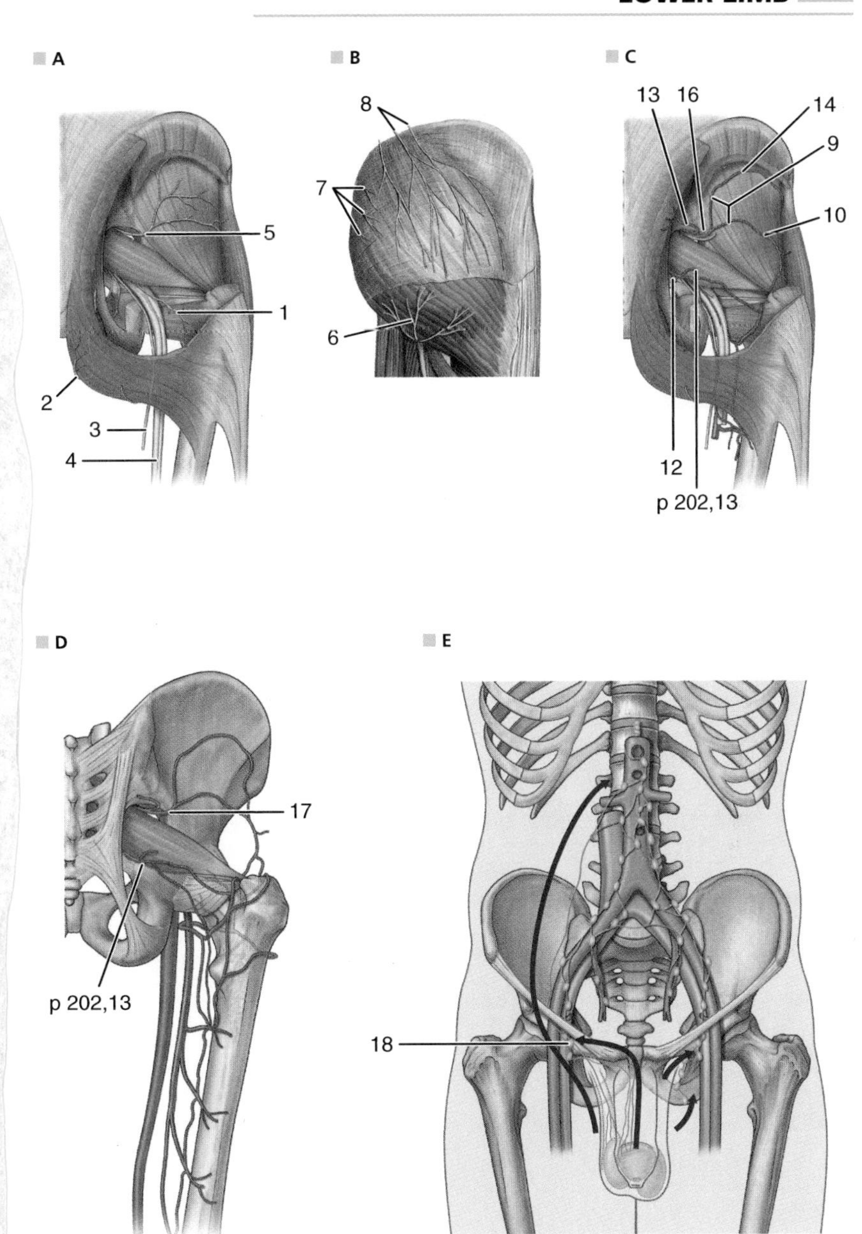
A
5
1
2
3
4
B
8
7
6
C
13 16
14
9
10
12
p 202,13
D
17
p 202,13
E
18

1 ***inferior superficial inguinal lymph nodes*** the lower superficial inguinal lymph nodes situated below the opening of the saphenous vein. <nodi lymphoidei inguinales superficiales inferiores> A

2 **superficial inguinal lymph nodes** lymph nodes situated in the subcutaneous tissue inferior to the inguinal ligament on either side of the proximal part of the greater saphenous vein, comprising two upper (supermedial and superolateral) groups and one lower (inferior) group; they drain the skin of the lower abdominal wall, penis, scrotum or labia majora, perineum, and buttocks. <nodi lymphoidei inguinales superficiales> A

3 ***superolateral superficial inguinal lymph nodes*** the upper superficial inguinal nodes situated on the lateral side of the opening of the saphenous vein. <nodi lymphoidei inguinales superficiales superolaterales> A

4 ***superomedial superficial inguinal lymph nodes*** the upper superficial inguinal lymph nodes situated on the medial side of the opening of the saphenous vein. <nodi lymphoidei inguinales superficiales superomediales> A

5 **falciform border of saphenous hiatus** the lateral margin of the saphenous hiatus. <margo falciformis hiatus sapheni> B

6 **inferior horn of falciform margin** the distal edge of the falciform margin of the saphenous hiatus, deep to the great saphenous vein. <cornu inferius marginis falciformis> B

7 **saphenous hiatus** the depression in the fascia lata that is bridged by the cribriform fascia and perforated by the great saphenous vein; called also *saphenous opening* or *ring*. <hiatus saphenus> B

8 **superior horn of falciform margin** the proximal end of the falciform margin of the saphenous hiatus. <cornu superius marginis falciformis> B

9 **cribriform fascia** the part of the superficial fascia of the thigh that covers the saphenous opening. <fascia cribrosa> C

10 **adductor canal** an intramuscular interval on the medial aspect of the middle third of the thigh, which contains the femoral vessels and the saphenous nerve. The lateral wall is formed by the vastus medialis, the posterior wall by the adductor longus and adductor magnus, the roof by a layer of fascia, and it is covered by the sartorius. <canalis adductorius> D

11 **medial femoral intermuscular septum** the fascial sheet in the thigh separating the vastus medialis from the adductor and the pectineus muscles. <septum intermusculare femoris mediale> F

12 **femoral triangle** a triangular subfascial area bounded superiorly by the inguinal ligament, laterally by the medial border of the sartorius muscle, and medially by the medial border of the adductor longus muscle. The term is also used to denote the surface area of the thigh overlying this region. <trigonum femorale> E

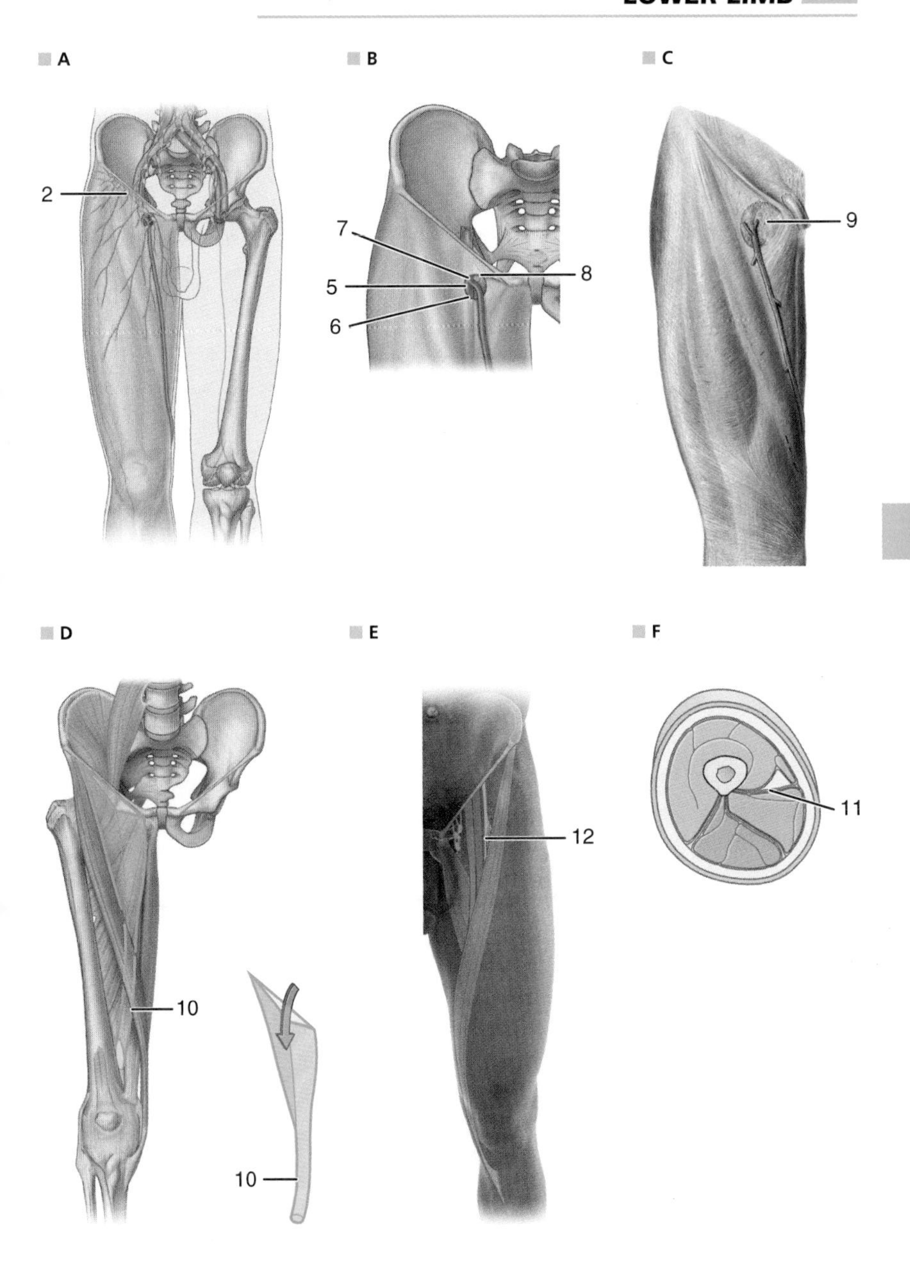
A
2
B
7
8
5
6
C
9
D
10
10
E
12
F
11

1 **femoral canal** the cone-shaped medial part of the femoral sheath lateral to the base of the lacunar ligament. <canalis femoralis> A

2 ***femoral ring*** the abdominal opening of the femoral canal, normally closed by the crural septum and peritoneum. <anulus femoralis> A

3 ***femoral septum*** the thin fibrous membrane that helps to close the femoral ring; it is derived from the transverse fascia, is perforated for the passage of lymphatic vessels, and is embedded in fat. <septum femorale> A

4 **muscular space** a compartment beneath the inguinal ligament for the passage of the iliopsoas muscle and femoral nerve, separated from the vascular space by the iliopectineal arch. <lacuna musculorum> A

5 **articularis genus muscle** *origin,* distal fourth of anterior surface of shaft of femur; *insertion,* synovial membrane of knee joint; *innervation,* femoral; *action,* lifts capsule of knee joint. <musculus articularis genus> B

6 **subtendinous bursae of sartorius muscle** bursae between the tendons of the sartorius, semitendinosus, and gracilis muscles. <bursae subtendineae musculi sartorii> C

7 **subtendinous iliac bursa** a bursa between the iliopsoas tendon and the lesser trochanter. <bursa subtendinea iliaca> C

Muscles of Anterior Compartment of Thigh

8 **psoas minor muscle** *origin,* last thoracic and first lumbar vertebrae; *insertion,* pectineal line, iliopectineal eminence, iliac fascia; *innervation,* first lumbar; *action,* flexes trunk. Absent in 40 to 50 per cent of persons. <musculus psoas minor> F

9 **lateral femoral intermuscular septum** the fascial sheet in the thigh separating the vastus lateralis muscle from the biceps femoris. <septum intermusculare femoris laterale> D

10 **iliacus muscle** *origin,* iliac fossa and base of sacrum; *insertion,* greater psoas tendon and lesser trochanter of femur; *innervation,* femoral; *action,* flexes thigh, trunk on limb. <musculus iliacus> E

11 **iliopsoas muscle** a compound muscle consisting of the iliacus and the psoas major. <musculus iliopsoas> E

12 **psoas major muscle** *origin,* lumbar vertebrae; *insertion,* lesser trochanter of femur; *innervation,* second and third lumbar; *action,* flexes thigh or trunk. <musculus psoas major> E

13 **patellar ligament** the continuation of the central portion of the tendon of the quadriceps femoris muscle distal to the patella; it extends from the patella to the tuberosity of the tibia. <ligamentum patellae> B

14 **quadriceps femoris muscle** a name applied collectively to the rectus femoris, vastus intermedius, vastus lateralis, and vastus medialis, inserting by a common tendon that surrounds the patella and ends on the tuberosity of the tibia, and acting to extend the leg upon the thigh. <musculus quadriceps femoris> B

15 **rectus femoris muscle** *origin,* anterior inferior iliac spine, rim of acetabulum; *insertion,* base of patella, tuberosity of tibia; *innervation,* femoral; *action,* extends knee, flexes thigh at hip. <musculus rectus femoris> B

16 **reflected head of rectus femoris muscle** the posterior head of the rectus femoris, it arises from a groove above the rim of the acetabulum, fuses with the straight or anterior head, and continues down into the belly of the muscle. <caput reflexum musculi recti femoris> B

17 **sartorius muscle** *origin,* anterior superior iliac spine; *insertion,* proximal part of medial surface of tibia; *innervation,* femoral; *action,* flexes leg at knee and thigh at pelvis. <musculus sartorius> B

18 **straight head of rectus femoris muscle** the anterior head of the rectus femoris, it arises from anteroinferior iliac spine, fuses with the reflected or posterior head, and continues down into the belly of the muscle. <caput rectum musculi recti femoris> B

19 **vastus intermedius muscle** *origin,* anterior and lateral surfaces of femur; *insertion,* patella, common tendon of quadriceps femoris; *innervation,* femoral; *action,* extends leg. <musculus vastus intermedius> B

20 **vastus lateralis muscle** *origin,* lateral aspect of femur; *insertion,* patella, common tendon of quadriceps femoris; *innervation,* femoral; *action,* extends leg. <musculus vastus lateralis> B

21 **vastus medialis muscle** *origin,* medial aspect of femur; *insertion,* patella, common tendon of quadriceps femoris; *innervation,* femoral; *action,* extends leg. <musculus vastus medialis> B

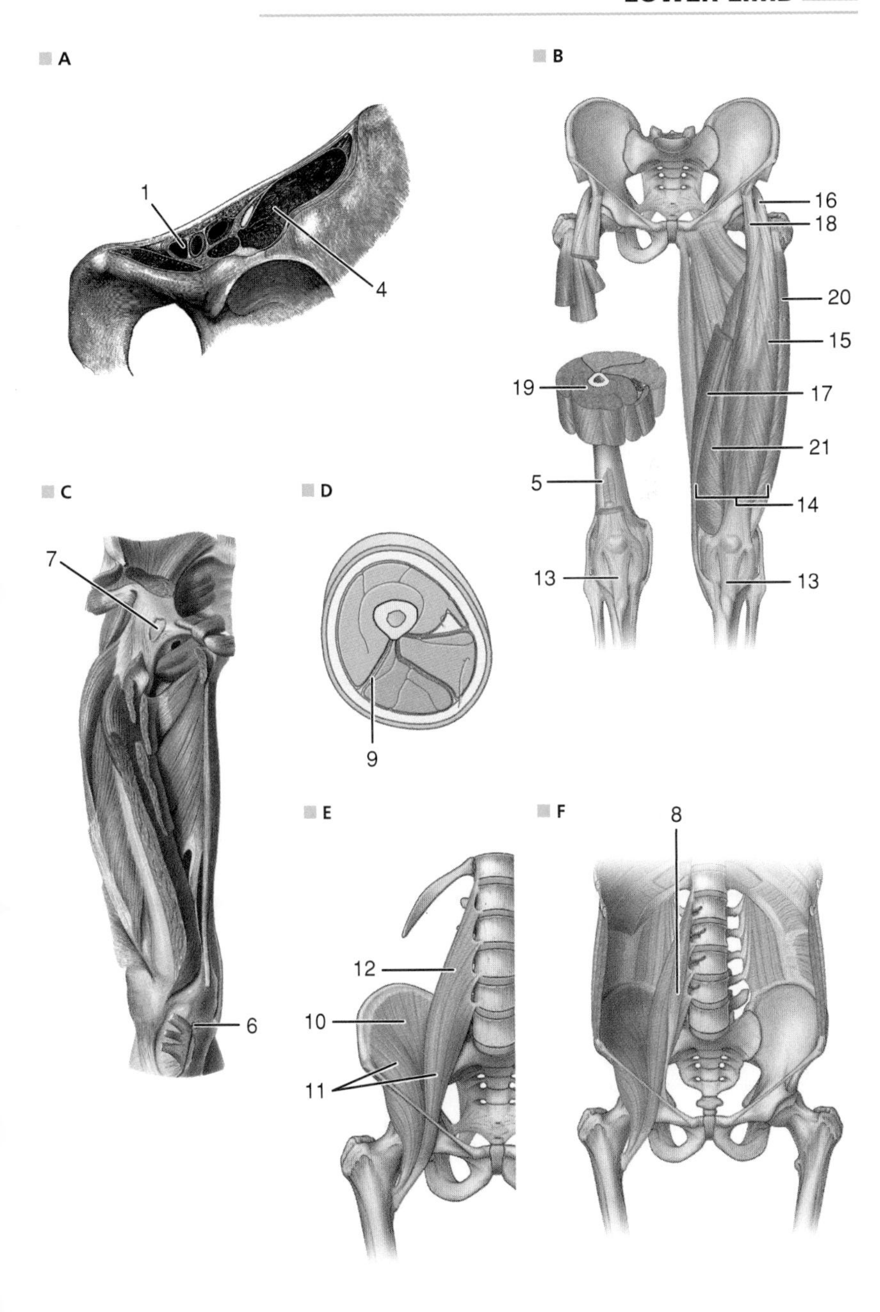
A
1
4
B
16
18
20
15
19
17
21
5
14
13
13
C
7
6
D
9
E
12
10
11
F
8

1 **pectineus muscle** *origin,* pectineal line of pubis; *insertion,* pectineal line of femur; *innervation,* obturator and femoral; *action,* flexes, adducts thigh. <musculus pectineus> A

Nerves of Anterior Compartment of Thigh

2 **femoral nerve** *origin,* lumbar plexus (L2-L4); descending behind the inguinal ligament to the femoral triangle; *branches,* saphenous nerve, muscular and anterior cutaneous branches; *distribution,* skin of thigh and leg, muscles of front of thigh, and hip and knee joints; *modality,* general sensory and motor. <nervus femoralis> B

3 **anterior cutaneous branches of femoral nerve** branches that innervate the skin on the front and medial aspect of the thigh and patella and contribute to the subsartorial and patellar plexuses; modality, general sensory. <rami cutanei anteriores nervi femoralis> C

4 **femoral branch of genitofemoral nerve** a branch arising by division of the genitofemoral nerve above the inguinal ligament; entering the femoral sheath, it turns forward and supplies the skin of the femoral triangle; modality, general sensory. <ramus femoralis nervi genitofemoralis> C

5 **lateral cutaneous nerve of thigh** *origin,* lumbar plexus—L2-L3; *distribution,* skin of lateral and frontal aspects of thigh; *modality,* general sensory. Called also *lateral femoral cutaneous nerve.* <nervus cutaneus femoris lateralis> C

6 **saphenous nerve** *origin,* termination of femoral nerve, descending first with femoral vessels and then on medial side of leg and foot; *branches,* infrapatellar and medial crural cutaneous branches; *distribution,* knee joint, subsartorial and patellar plexuses, skin on medial side of leg and foot; *modality,* general sensory. <nervus saphenus> C

7 **muscular branches of femoral nerve** branches that innervate the anterior thigh muscles; modality, motor. <rami musculares nervi femoralis> D

Vessels of Anterior Compartment of Thigh

8 **vascular space** a space for the passage of the femoral vessels into the thigh, separated from the lacuna musculorum by the iliopectineal arch. Called also *vascular compartment.* <lacuna vasorum> E

9 **accessory saphenous vein** a vein that, when present, drains the medial and posterior superficial parts of the thigh and opens into the great saphenous vein. <vena saphena accessoria> F

10 **patellar anastomosis** a network of arterial branches surrounding the patella and derived from the various arteries of the knee. <rete patellare> F

11 **superficial veins of lower limb** veins that drain the lower limb, found immediately beneath the skin, and anastomosing freely with the deep veins; the principal superficial veins are the great and small saphenous veins. <venae superficiales membri inferioris> F, G

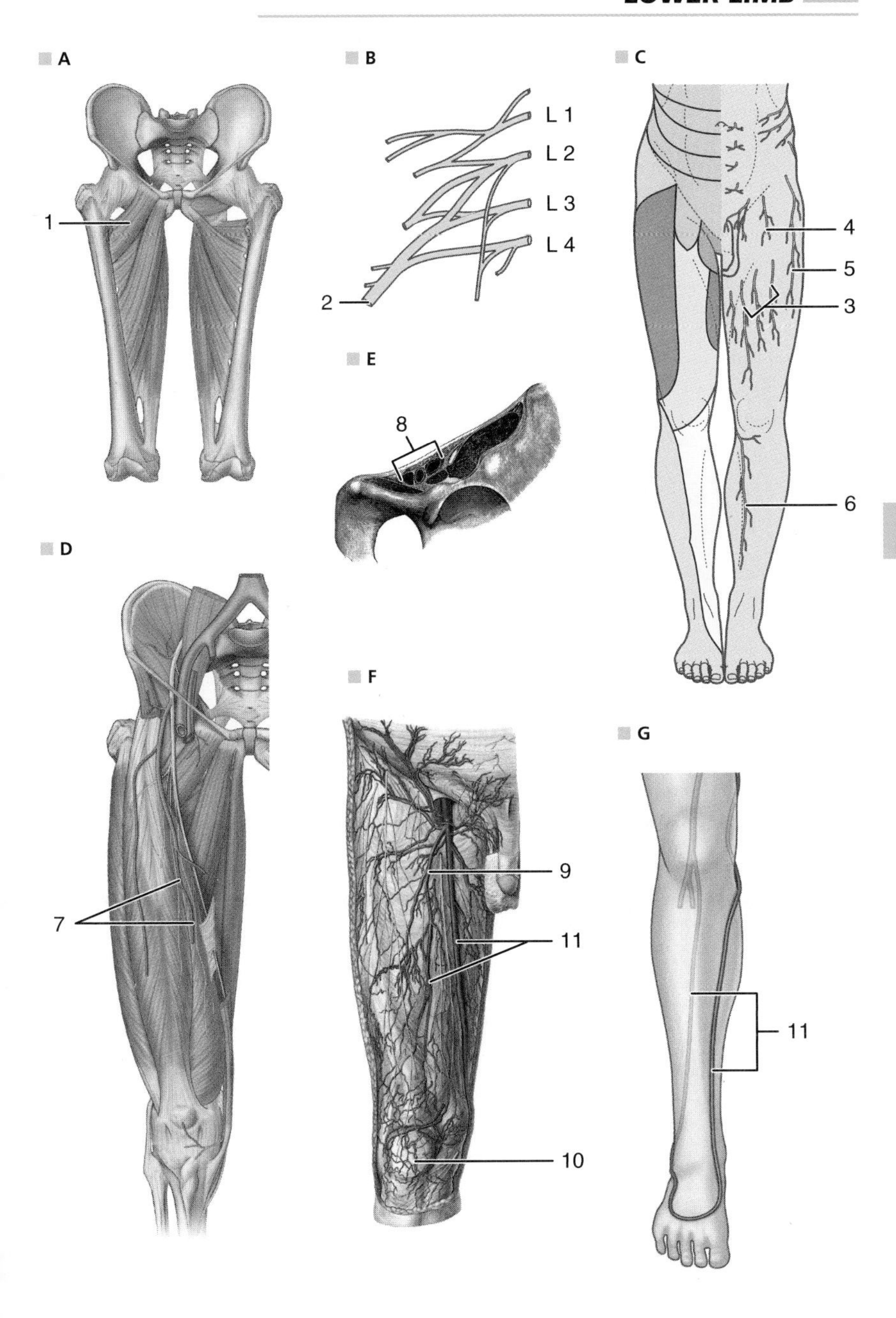
A
1
B
L 1
L 2
L 3
L 4
2
C
4
5
3
6
E
8
D
7
F
9
11
10
G
11

1 **femoral vein** a vein that lies in the proximal two-thirds of the thigh; it is a direct continuation of the popliteal vein, follows the course of the femoral artery, and at the inguinal ligament becomes the external iliac vein. NOTE: The portion of the femoral vein proximal to the branching of the deep femoral vein is sometimes known as the *common femoral vein,* and its continuation distal to the branching as the *superficial femoral vein.* <vena femoralis> A

2 **transverse branch of lateral circumflex femoral artery** a branch that pierces the vastus lateralis muscle, turning around the femur to anastomose with the transverse branch of the medial circumflex femoral artery and with other arteries, deep to the gluteus maximus muscle. <ramus transversus arteriae circumflexae femoris lateralis> B

3 **deep external pudendal artery** *origin,* femoral artery; *branches,* anterior scrotal or anterior labial branches, inguinal branches; *distribution,* external genitalia, upper medial thigh. <arteria pudenda externa profunda> C

4 ***external pudendal veins*** veins that follow the distribution of the external pudendal arteries, drain anterior parts of the labia or scrotum, and open into the great saphenous vein and femoral veins. <venae pudendae externae> C

5 **femoral artery** *origin,* continuation of external iliac; *branches,* superficial epigastric, superficial circumflex iliac, external pudendal, deep femoral, descending genicular; *distribution,* lower abdominal wall, external genitalia, lower extremity. NOTE: The portion of the femoral artery proximal to the branching of the deep femoral is sometimes known as the *common femoral artery,* and its continuation as the *superficial femoral artery,* with the descending genicular artery is then a branch of the superficial femoral artery. <arteria femoralis> C

6 ***femoral plexus*** a nerve plexus accompanying the femoral artery, derived chiefly from the aortic plexus by way of the common and external iliac plexuses. <plexus femoralis> C

7 ***inguinal branches of deep external pudendal artery*** branches arising from the deep external pudendal arteries and supplying the inguinal region. <rami inguinales arteriae pudendae externae profundae> C

8 **superficial epigastric artery** *origin,* femoral; *branches,* none; *distribution,* abdominal wall, groin. <arteria epigastrica superficialis> C

9 **superficial external pudendal artery** *origin,* femoral artery; *branches,* none; *distribution,* external genitalia. <arteria pudenda externa superficialis> C

10 ***acetabular branch of medial circumflex femoral artery*** a branch of the medial circumflex artery of the thigh, distributed to the head of the femur and to the acetabulum. Called also *acetabular artery.* <ramus acetabularis arteriae circumflexae femoris medialis> D

11 **ascending branch of lateral circumflex femoral artery** a branch that runs upward along the trochanteric line of the femur and between the gluteus medius and minimus muscles, and anastomoses with branches of the superior gluteal artery. It helps supply the upper thigh muscles. <ramus ascendens arteriae circumflexae femoris lateralis> D

12 **deep artery of thigh** *origin,* femoral artery; *branches,* medial and lateral circumflex arteries of thigh, perforating arteries; *distribution,* thigh muscles, hip joint, gluteal muscles, femur. Called also *profunda femoris artery.* <arteria profunda femoris> D

13 **deep branch of medial circumflex femoral artery** a branch ascending toward the trochanteric fossa, and anastomosing with gluteal branches. <ramus profundus arteriae circumflexae femoris medialis> D

14 ***deep femoral vein*** a vein that follows the distribution of the deep femoral artery and opens into the femoral vein. <vena profunda femoris> D

15 **descending branch of lateral circumflex femoral artery** a branch passing from the lateral circumflex artery (sometimes directly from the deep femoral) to the knee, and supplying the thigh muscles. <ramus descendens arteriae circumflexae femoris lateralis> D

16 ***femoral nutrient arteries*** *origin,* third perforating artery; *branches,* none; *distribution,* femur. <arteriae nutriciae femoris> D

17 **lateral circumflex femoral artery** *origin,* deep femoral artery; *branches,* ascending, descending, and transverse branches; *distribution,* hip joint, thigh muscles. <arteria circumflexa femoris lateralis> D

18 ***lateral circumflex femoral veins*** venae comitantes of the lateral circumflex femoral artery, emptying into the femoral or the deep femoral vein. <venae circumflexae femoris laterales> D

19 **medial circumflex femoral artery** *origin,* deep femoral artery; *branches,* deep, superficial, ascending, transverse, and acetabular branches; *distribution,* hip joint, thigh muscles. <arteria circumflexa femoris medialis> D

20 ***medial circumflex femoral veins*** venae comitantes of the medial circumflex femoral artery, emptying into the femoral or the deep femoral vein. <venae circumflexae femoris mediales> D

21 ***superficial branch of medial circumflex femoral artery*** a branch passing between the quadratus femoris and the proximal border of the adductor magnus and anastomosing with the inferior gluteal, lateral circumflex femoral, and first perforating arteries. <ramus superficialis arteriae circumflexae femoris medialis> D

22 **ascending branch of medial circumflex femoral artery** a branch that ascends anterior to the quadratus femoris muscle to the trochanteric fossa, and there anastomoses with gluteal arteries. <ramus ascendens arteriae circumflexae femoris medialis> E

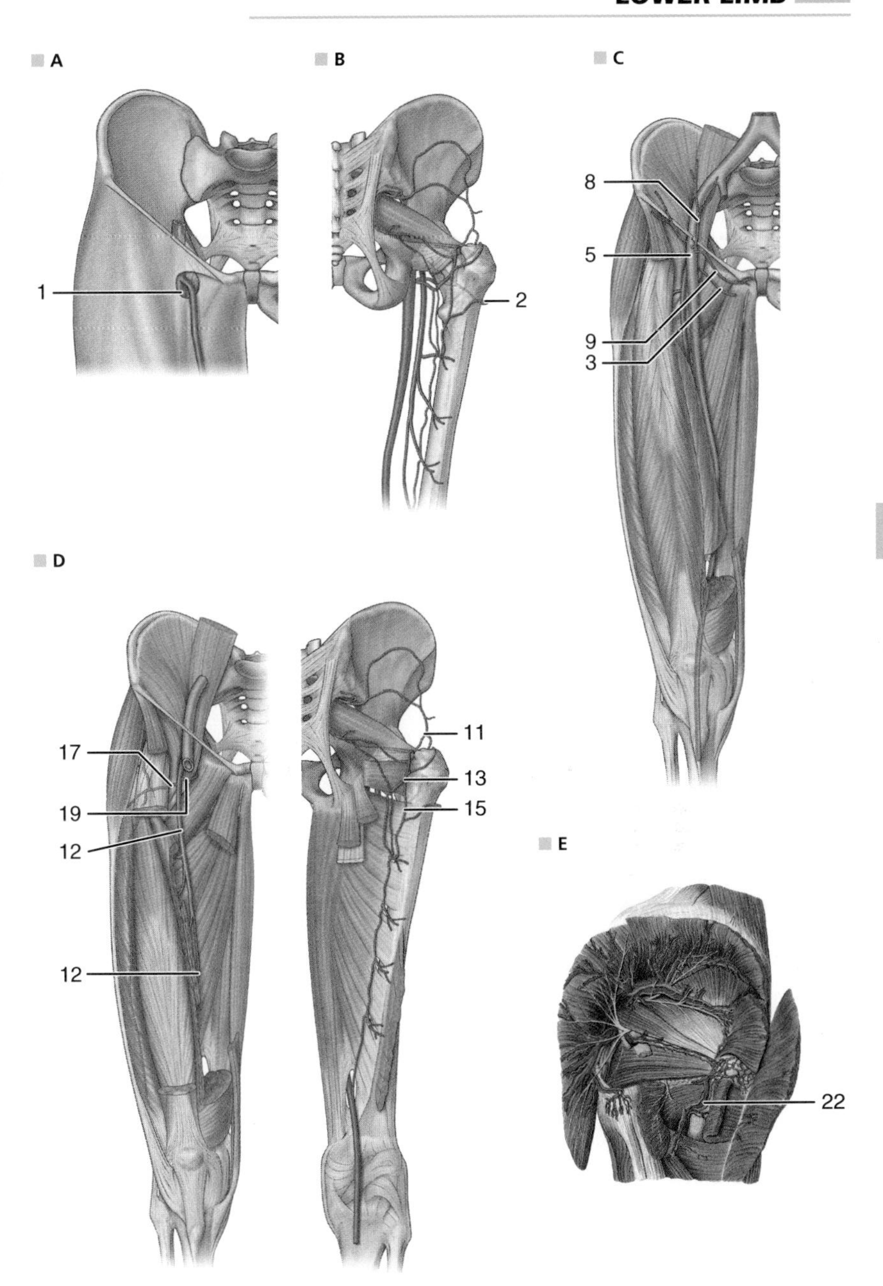
A
1
B
2
C
8
5
9
3
D
17
19
12
12
11
13
15
E
22

1 **descending genicular artery** *origin,* femoral artery; *branches,* saphenous, articular; *distribution,* knee joint, upper and medial leg. <arteria descendens genus> A

2 **articular branches of descending genicular artery** branches that pass downward in the vastus medialis muscle and help supply the knee joint. <rami articulares arteriae descendentis genicularis> B

Medial Compartment of Thigh

3 **obturator groove** a groove that obliquely crosses the inferior surface of the superior ramus of the pubis, giving passage to the obturator vessels and nerve. <sulcus obturatorius ossis pubis> C

Muscles of Medial Compartment of Thigh

4 **adductor brevis muscle** *origin,* outer surface of body and inferior ramus of pubis; *insertion,* upper part of linea aspera of femur; *innervation,* obturator; *action,* adducts, rotates, flexes thigh. <musculus adductor brevis> D

5 **adductor longus muscle** *origin,* crest and symphysis of pubis; *insertion,* linea aspera of femur; *innervation,* obturator; *action,* adducts, rotates, flexes thigh. <musculus adductor longus> D

6 **adductor magnus muscle** (2 parts): origin, inferior ramus of pubis, ramus of ischium, ischial tuberosity; insertion, linea aspera of femur, adductor tubercle of femur; innervation, obturator, sciatic; action, adducts thigh, extends thigh. <musculus adductor magnus> D

7 ***adductor minimus muscle*** a name given the anterior portion of the adductor magnus muscle; *insertion,* ischium, body and ramus of pubis; *innervation,* obturator and sciatic; *action,* adducts thigh. <musculus adductor minimus> D

8 **gracilis muscle** *origin,* lower half of body and entire inferior ramus of pubis; *insertion,* medial surface of shaft of tibia; *innervation,* obturator; *action,* adducts thigh, flexes knee joint. <musculus gracilis> D

9 **obturator externus muscle** *origin,* pubis, ischium, and external surface of obturator membrane; *insertion,* trochanteric fossa of femur; *innervation,* obturator; *action,* rotates thigh laterally. <musculus obturatorius externus> E

Nerves of Medial Compartment of Thigh

10 **obturator nerve** *origin,* lumbar plexus—L3-L4; *branches,* anterior, posterior, and muscular branches; *distribution,* adductor muscles and gracilis muscle, skin of medial part of thigh, and hip and knee joints; *modality,* general sensory and motor. <nervus obturatorius> E

11 **cutaneous branch of obturator nerve** a variable branch arising from the anterior branch of the obturator nerve, forming part of the subsartorial plexus, and supplying the skin of the medial aspect of the thigh and leg; *modality,* general sensory. <ramus cutaneus nervi obturatorii> F

12 ***accessory obturator nerve*** *origin,* anterior branches of anterior rami of L3-L4; *distribution,* pectineus muscle, hip joint, communicates with obturator nerve; *modality,* general sensory and motor. <nervus obturatorius accessorius> E

13 **anterior branch of obturator nerve** a branch that supplies the gracilis and the adductor longus and brevis muscles and the pectineus, and occasionally gives off a branch to the skin of the medial side of the thigh and leg; *modality,* motor and general sensory. <ramus anterior nervi obturatorii> E

14 ***muscular branches of anterior branch of obturator nerve*** branches arising from the anterior ramus of the obturator nerve and innervating the gracilis and adductor longus muscles, usually the adductor brevis, and often the pectineus; *modality,* motor. <rami musculares rami anterioris nervi obturatorii> E

15 ***muscular branches of posterior branch of obturator nerve*** branches arising from the posterior ramus of the obturator nerve and innervating the obturator externus and adductor brevis muscles, and sometimes the adductor magnus; *modality,* motor. <rami musculares rami posterioris nervi obturatorii> E

16 **posterior branch of obturator nerve** a branch that descends to innervate the knee joint, giving muscular branches to the obturator externus, adductor magnus, and sometimes the adductor brevis muscle; modality, general sensory and motor. <ramus posterior nervi obturatorii> E

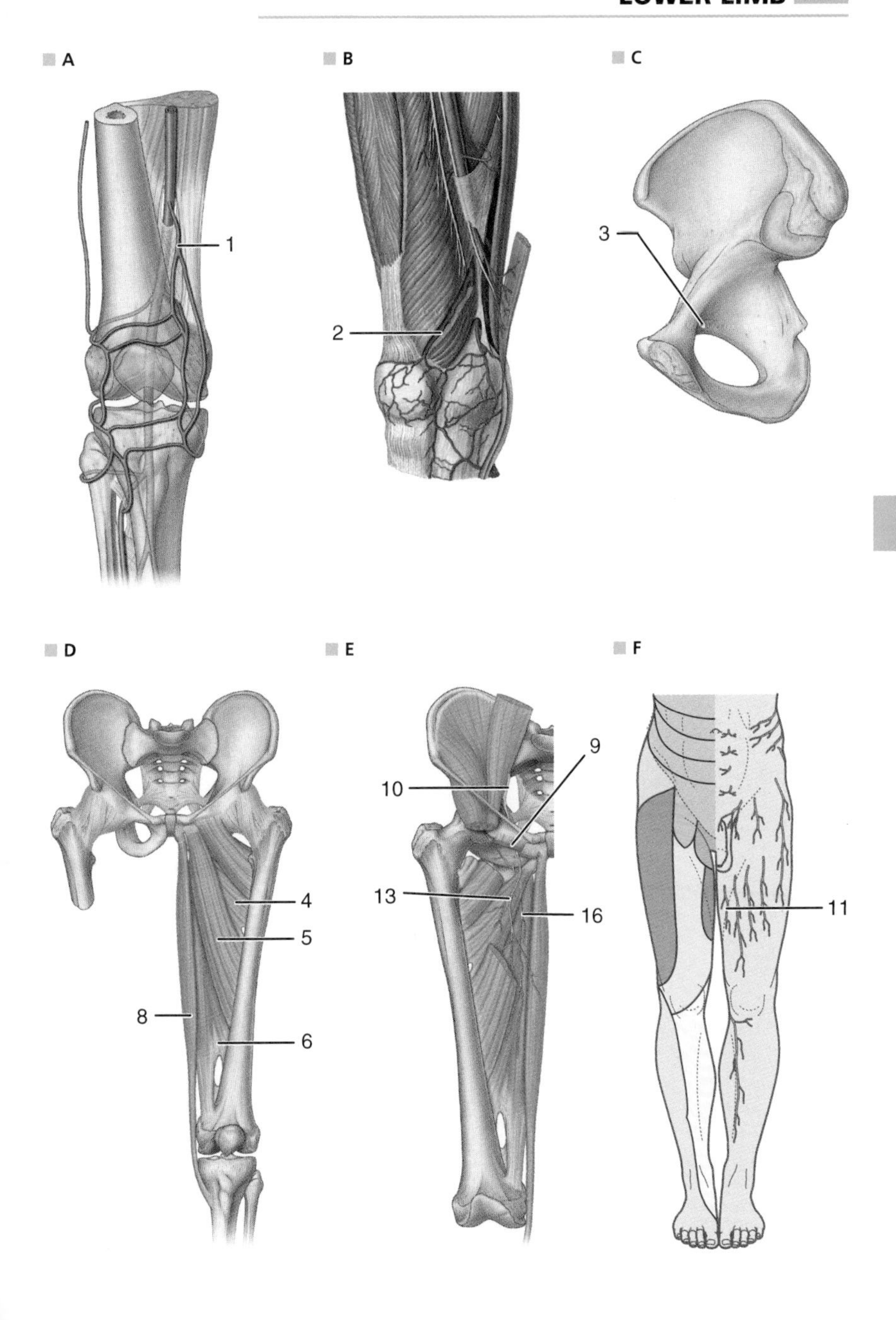
A
1
B
2
C
3
D
4
5
8
6
E
9
10
13
16
F
11

Vessels of Medial Compartment of Thigh

1 **acetabular branch of obturator artery** a branch distributed to the hip joint; called also *acetabular artery.* <ramus acetabularis arteriae obturatoriae> A

2 **anterior branch of obturator artery** a branch that passes forward around the medial margin of the obturator foramen, on the obturator membrane, and is distributed to the obturator and adductor muscles. <ramus anterior arteriae obturatoriae> B

3 **posterior branch of obturator artery** a branch passing backward around the lateral margin of the obturator foramen, on the obturator membrane, supplying muscles around the ischial tuberosity and giving off an acetabular branch. <ramus posterior arteriae obturatoriae> B

Posterior Compartment of Thigh

4 **iliotibial tract** a thickened longitudinal band of fascia lata extending from the tensor muscle downward along the lateral side of the thigh to the lateral condyle of the tibia. <tractus iliotibialis> C

5 **adductor hiatus** the opening between the long tendon of the adductor magnus and the femur, marking the distal end of the adductor canal. <hiatus adductorius> D

Muscles of Posterior Compartment of Thigh

6 **biceps femoris muscle** (2 heads): *origin,* ischial tuberosity, linea aspera of femur; *insertion,* head of fibula, lateral condyle of tibia; *innervation,* tibial, peroneal, popliteal; *action,* flexes leg, extends thigh. <musculus biceps femoris> E

7 **long head of the biceps femoris muscle** the head of the biceps femoris muscle arising from the ischial tuberosity. <caput longum musculi bicipitis femoris> E

8 **semimembranosus muscle** *origin,* tuberosity of ischium; *insertion,* medial condyle and border of tibia, lateral condyle of femur; *innervation,* tibial; *action,* flexes and rotates leg medially, extends thigh at hip. <musculus semimembranosus> E

9 **semitendinosus muscle** *origin,* tuberosity of ischium; *insertion,* upper part of medial surface of tibia; *innervation,* tibial; *action,* flexes and rotates leg medially, extends thigh. <musculus semitendinosus> E

10 **short head of biceps femoris muscle** the head of the biceps femoris arising from the linea aspera femoris. <caput breve musculi bicipitis femoris> E

11 **bursa of semimembranosus muscle** a bursa between the semimembranosus muscle and the medial head of the gastrocnemius. <bursa musculi semimembranosi> F

12 ***inferior subtendinous bursa of bicipitis femoris muscle*** a bursa between the tendon of the biceps femoris muscle and the fibular collateral ligament of the knee joint. <bursa subtendinea musculi bicipitis femoris inferior> F

13 **medial subtendinous bursa of gastrocnemius muscle** a bursa between the tendon of the medial head of the gastrocnemius, the condyle of the femur, and the joint capsule. Called also *medial supracondyloid bursa.* <bursa subtendinea musculi gastrocnemii medialis> F

14 ***superior bursa of biceps femoris muscle*** a bursa between the long head of the biceps, the semitendinosus, the tendon of the semimembranosus, and the ischial tuberosity. <bursa musculi bicipitis femoris superior> F

Nerves of Posterior Compartment of Thigh

15 **perineal branches of posterior femoral cutaneous nerve** branches arising from the posterior femoral cutaneous nerve at the lower margin of the gluteus maximus muscle and innervating the skin of the external genitalia; modality, general sensory. <rami perineales nervi cutanei femoris posterioris> G

16 **sciatic nerve** the largest nerve of the body: *origin,* sacral plexus (L4-S3); it leaves the pelvis through the greater sciatic foramen; *branches,* divides into the tibial and common peroneal nerves, usually in lower third of thigh; *modality,* general sensory and motor. <nervus ischiadicus> H

17 **tibial nerve** *origin,* sciatic nerve in lower part of thigh; *branches,* interosseous nerve of leg, medial cutaneous nerve of calf, sural nerve, and medial and lateral plantar nerves, and muscular and medial calcaneal branches; *distribution,* while still incorporated in the sciatic nerve, it supplies the semimembranosus and semitendinosus muscles, long head of biceps, and adductor magnus muscle; it supplies the knee joint as it descends in the popliteal fossa and, continuing into the leg, supplies the muscles and skin of the calf and sole of the foot, and the toes; *modality,* general sensory and motor. <nervus tibialis> H

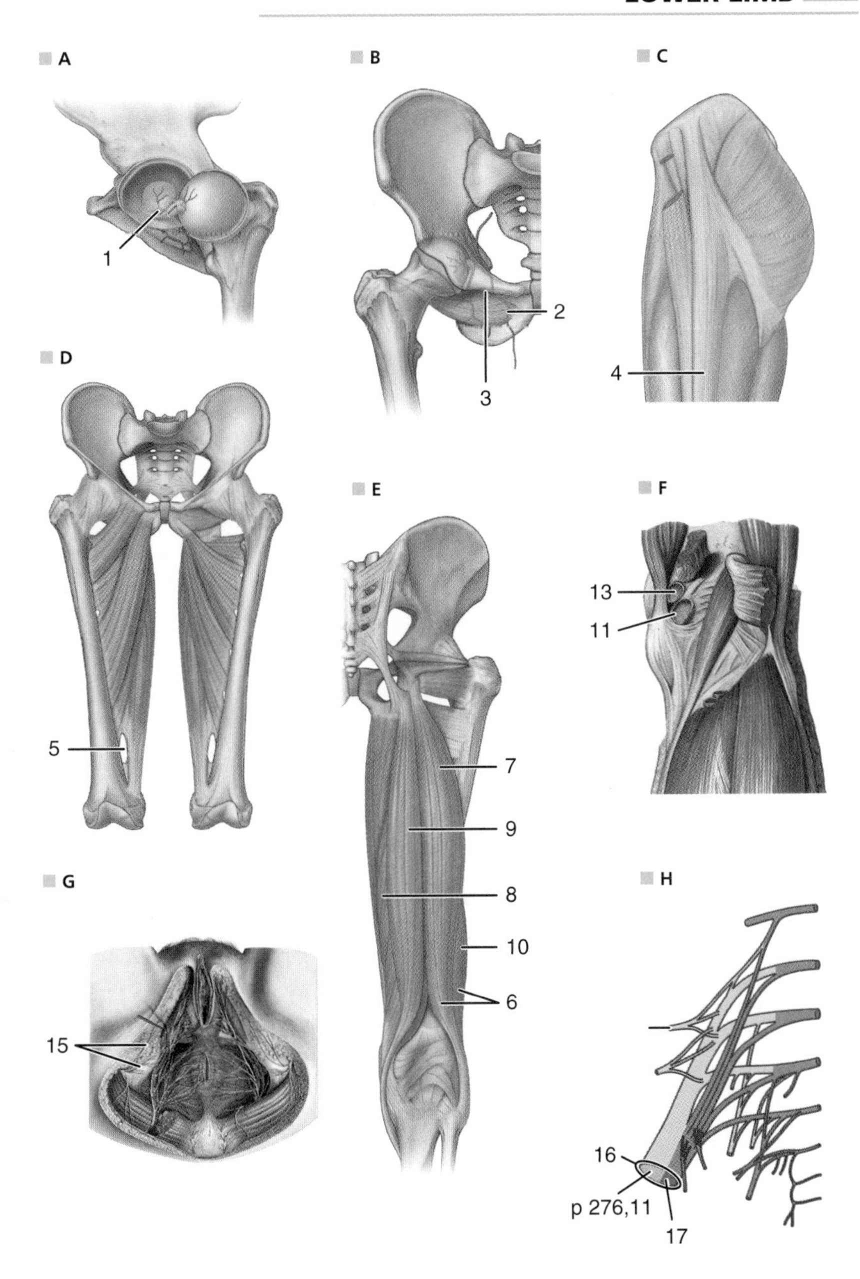
A
1
B
2
3
C
4
D
5
E
7
9
8
10
6
F
13
11
G
15
H
16
p 276,11
17

Vessels of Posterior Compartment of Thigh

1 **artery to sciatic nerve** *origin,* inferior gluteal artery; *branches,* none; *distribution,* accompanies sciatic nerve. Called also *sciatic artery.* <arteria comitans nervi ischiadici> A

2 **middle genicular artery** *origin,* popliteal artery; *branches,* none; *distribution,* knee joint, cruciate ligaments, patellar synovial and alar folds. Called also *middle artery of knee.* <arteria media genus> B

3 **perforating arteries** *origin,* branches (usually three) of the deep femoral artery that perforate the insertion of the adductor magnus to reach the back of the thigh; *branches,* nutrient arteries; *distribution,* adductor, hamstring, and gluteal muscles, and femur. <arteriae perforantes> C

4 ***genicular anastomosis*** an extensive arterial network on the capsule of the knee joint, supplying branches to the contiguous bones and joints. It is formed by the genicular arteries, the termination of the deep femoral artery, the descending branch of the lateral circumflex artery, and the tibial recurrent artery. <rete articulare genus> D

5 ***genicular veins*** veins accompanying the genicular arteries and draining into the popliteal vein. <venae geniculares> D

6 **inferior lateral genicular artery** *origin,* popliteal artery; *branches,* none; *distribution,* knee joint. Called also *lateral inferior genicular artery.* <arteria inferior lateralis genus> D

7 **inferior medial genicular artery** *origin,* popliteal artery; *branches,* none; *distribution,* knee joint. Called also *medial inferior genicular artery.* <arteria inferior medialis genus> D

8 **superior lateral genicular artery** *origin,* popliteal artery; *branches,* none; *distribution,* knee joint, femur, patella, contiguous muscles. Called also *lateral superior genicular artery.* <arteria superior lateralis genus> D

9 **superior medial genicular artery** *origin,* popliteal artery; *branches,* none; *distribution,* knee joint, femur, patella, contiguous muscles. Called also *medial superior genicular artery.* <arteria superior medialis genus> D

LEG

10 **leg** the part of the lower limb from the knee to the ankle. <crus> E

Anterior Compartment of Leg

11 **anterior tibial lymph node** a lymph node situated along the anterior tibial artery. <nodus lymphoideus tibialis anterior> F

12 **inferior extensor retinaculum** a Y-shaped band of fascia passing from the lateral side of the upper foot to attach by one arm to the medial malleolus and by the other arm to the medial side of the plantar aponeurosis. <retinaculum musculorum extensorum inferius pedis> G

13 **superior extensor retinaculum** the thickened lower portion of the fascia on the front of the leg, attached to the tibia on one side and the fibula on the other, and serving to hold in place the extensor tendons that pass beneath it. <retinaculum musculorum extensorum superius pedis> G

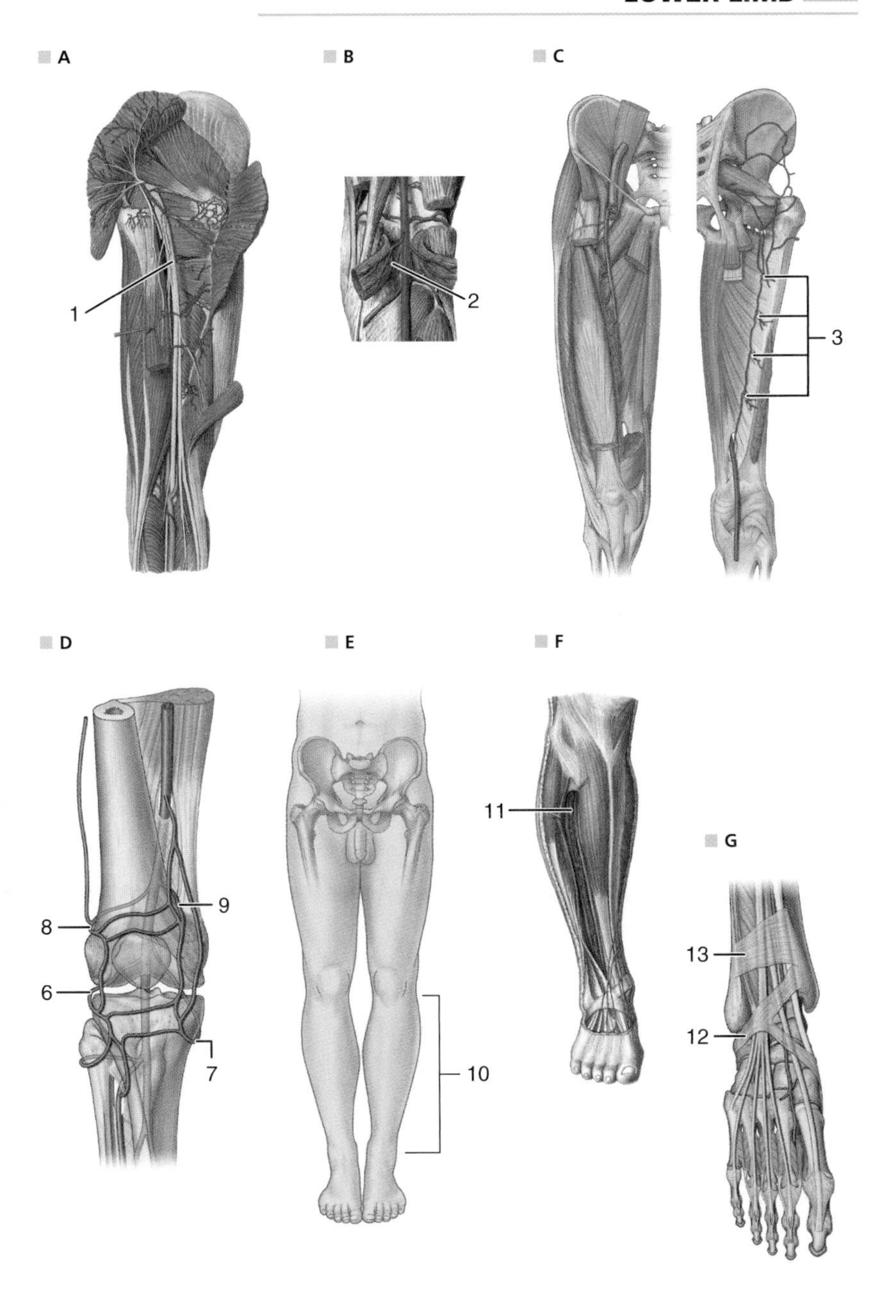
A
B
C
1
2
3
D
E
F
G
8
9
6
7
10
11
13
12

Muscles of Anterior Compartment of Leg

1 **anterior intermuscular septum of leg** a fascial sheet extending between the extensor digitorum longus and the fibularis muscles to the anterior fibular crest. <septum intermusculare cruris anterius> A

2 **extensor digitorum longus muscle** *origin,* anterior surface of fibula, lateral condyle of tibia, interosseous membrane; *insertion,* extensor expansion of each of the four lateral toes: *innervation,* deep fibular; *action,* extends toes. <musculus extensor digitorum longus> B

3 **extensor hallucis longus muscle** *origin,* front of fibula and interosseous membrane; *insertion,* dorsal surface of base of distal phalanx of great toe; *innervation,* deep fibular; *action,* dorsiflexes ankle joint, extends great toe. <musculus extensor hallucis longus> B

4 **fibularis tertius muscle** *origin,* anterior surface of fibula, interosseous membrane; *insertion,* base of fifth metatarsal; *innervation,* deep fibular; *action,* everts, dorsiflexes foot. Called also *peroneus tertius muscle* <musculus fibularis tertius> B

5 **tibialis anterior muscle** *origin,* lateral condyle and lateral surface of tibia, interosseous membrane; *insertion,* medial cuneiform and base of first metatarsal; *innervation,* deep fibular; *action,* dorsiflexes and inverts foot. <musculus tibialis anterior> B

Nerves of Anterior Compartment of Leg

6 **medial crural cutaneous branches of saphenous nerve** branches distributed by the saphenous nerve to the skin of the medial aspect of the leg; modality, general sensory. <rami cutanei cruris mediales nervi sapheni> C

7 **infrapatellar branch of saphenous nerve** a branch running inferolaterally from the saphenous nerve to the patellar plexus; modality, general sensory. <ramus infrapatellaris nervi sapheni> C, D

8 **saphenous nerve** *origin,* termination of femoral nerve, descending first with femoral vessels and then on medial side of leg and foot; *branches,* infrapatellar and medial crural cutaneous branches; *distribution,* knee joint, subsartorial and patellar plexuses, skin on medial side of leg and foot; *modality,* general sensory. <nervus saphenus> D

9 **deep fibular nerve** *origin,* a terminal branch of common fibular nerve; *branches and distribution,* winds around the neck of the fibula and descends on the interosseous membrane to the front of the ankle; gives off muscular branches, an articular branch, and lateral and medial terminal branches; *modality,* general sensory and motor. <nervus fibularis profundus> E

10 **muscular branches of deep fibular nerve** branches that innervate the tibialis anterior, extensor hallucis longus, extensor digitorum longus, and fibularis tertius muscles; modality, motor. <rami musculares nervi fibularis profundi> E

Vessels of Anterior Compartment of Leg

11 **anterior tibial artery** *origin,* popliteal artery; *branches,* posterior and anterior tibial recurrent, and lateral and medial anterior malleolar arteries, lateral and medial malleolar networks; *distribution,* leg, ankle, foot. <arteria tibialis anterior> E

12 **anterior tibial recurrent artery** *origin,* anterior tibial artery; *branches,* none; *distribution,* tibialis anterior, extensor digitorum longus, knee joint, contiguous fascia and skin. <arteria recurrens tibialis anterior> E

13 ***anterior tibial veins*** venae comitantes of the anterior tibial artery, which unite with the posterior tibial veins to form the popliteal vein. <venae tibiales anteriores> E

Posterior Compartment of Leg

14 ***posterior tibial lymph node*** a lymph node situated along the posterior tibial artery. <nodus lymphoideus tibialis posterior>

15 **bursa of tendo calcaneus** a bursa between the calcaneal tendon and the back of the calcaneus; called also *bursa of Achilles tendon* and *retrocalcaneal bursa.* <bursa tendinis calcanei> F

16 **popliteal fossa** the depression in the posterior region of the knee. <fossa poplitea> G

17 **posterior intermuscular septum of leg** a fascial sheet extending between the fibularis muscles and soleus to the lateral fibular crest. <septum intermusculare cruris posterius> A

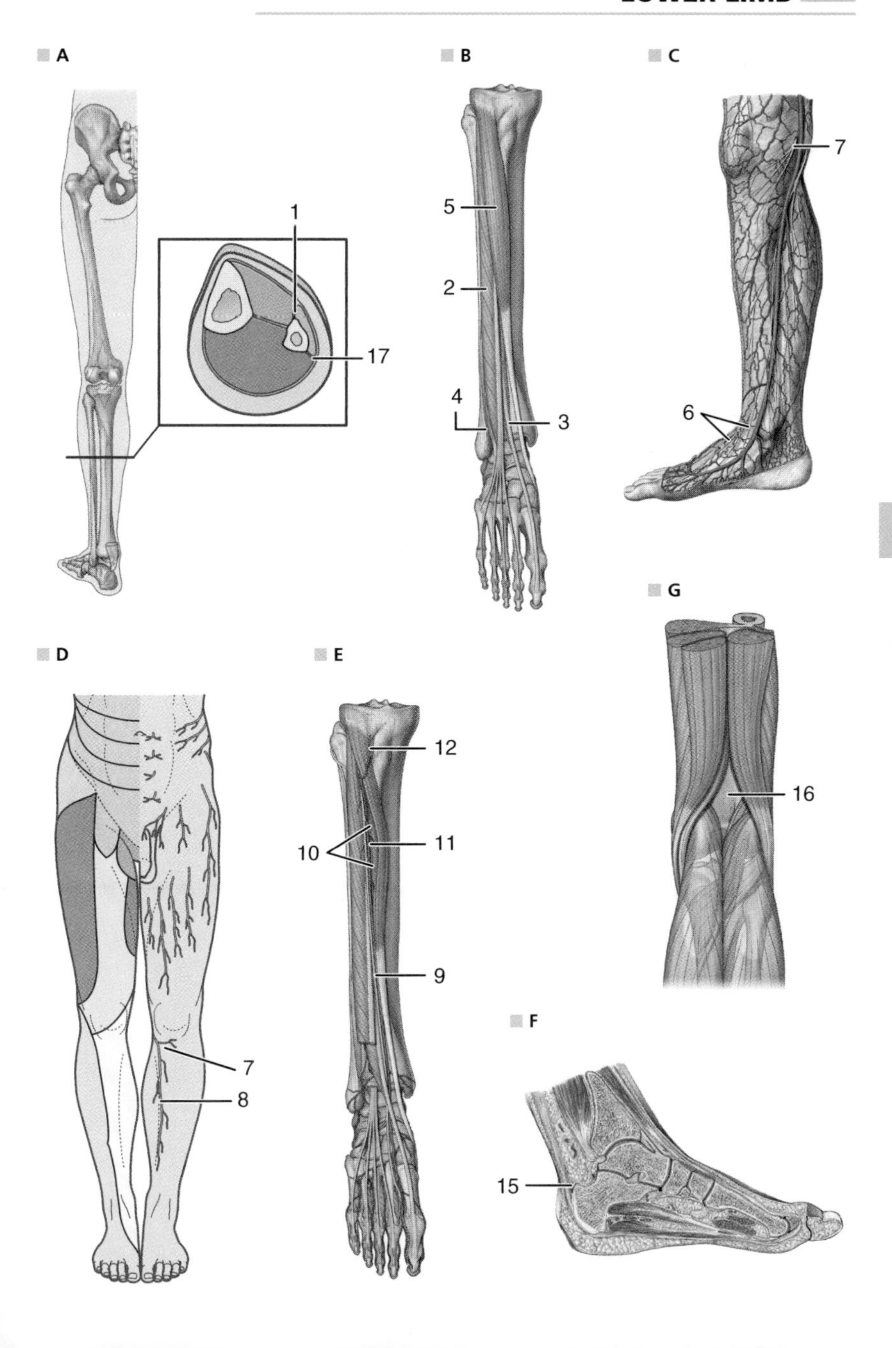
A
1
17
B
5
2
4
3
C
7
6
D
7
8
E
12
10
11
9
G
16
F
15

1 **calcaneal tendon** a powerful tendon at the back of the heel which attaches the triceps surae muscle to the tuberosity of the calcaneus; called also *Achilles tendon.* <tendo calcaneus> A

2 ***fibular lymph node*** a lymph node situated along the fibular artery. <nodus lymphoideus fibularis> B

3 **anterior lateral malleolar artery** *origin,* anterior tibial artery; *branches,* none; *distribution,* ankle joint. <arteria malleolaris anterior lateralis> G

4 **anterior medial malleolar artery** *origin,* anterior tibial artery; *branches,* none; *distribution,* ankle joint. <arteria malleolaris anterior medialis> G

5 **deep popliteal lymph nodes** the popliteal lymph nodes situated at the sides of the popliteal vessels. <nodi lymphoidei poplitei profundi> E

6 **popliteal lymph nodes** lymph nodes embedded in the fat of the popliteal fossa, comprising superficial and deep groups; their efferent vessels accompany the femoral vessels to the deep inguinal lymph nodes. <nodi lymphoidei poplitei> E

7 **superficial popliteal lymph nodes** the popliteal lymph nodes situated at the termination of the small saphenous vein. <nodi lymphoidei poplitei superficiales> E

Muscles of Posterior Compartment of Leg

8 **gastrocnemius muscle** (2 heads): *origin,* popliteal surface of femur, upper part of medial condyle, and capsule of knee, lateral condyle and capsule of knee; *insertion,* aponeurosis unites with tendon of soleus to form calcaneal (Achilles) tendon; *innervation,* tibial; *action,* plantar flexes ankle joint, flexes knee joint. <musculus gastrocnemius> A

9 **lateral head of gastrocnemius muscle** the head of the gastrocnemius muscle arising from the lateral condyle and posterior surface of the femur, and the capsule of the knee joint; called also *lateral gastrocnemius muscle.* <caput laterale musculi gastrocnemii> A

10 ***lateral subtendinous bursa of gastrocnemius muscle*** a bursa between the tendon of the lateral head of the gastrocnemius muscle and the joint capsule. <bursa subtendinea musculi gastrocnemii lateralis> A

11 **medial head of gastrocnemius muscle** the head of the gastrocnemius muscle arising from the medial condyle of the femur and the capsule of the knee joint; called also *medial gastrocnemius muscle.* <caput mediale musculi gastrocnemii> A

12 **plantaris muscle** *origin,* oblique popliteal ligament, lateral supracondylar line of femur; *insertion,* posterior part of calcaneus; *innervation,* tibial; *action,* plantar flexes foot; flexes knee. <musculus plantaris> A

13 **soleus muscle** *origin,* fibula, tibia, tendinous arch between tibia and fibula and passing over popliteal vessels; *insertion,* calcaneus by tendo calcaneus; *innervation,* tibial; *action,* plantar flexes foot. <musculus soleus> A

14 **tendinous arch of soleus muscle** an aponeurotic band in the front part of the soleus muscle, extending from a tubercle on the neck of the fibula to the soleal line of the tibia. <arcus tendineus musculi solei> A

15 ***triceps surae muscle*** the gastrocnemius and soleus considered together. <musculus triceps surae> A

16 **flexor digitorum longus muscle** *origin,* posterior surface of shaft of tibia; *insertion,* distal phalanges of four lateral toes; *innervation,* posterior tibial; *action,* flexes toes and plantar flexes foot. <musculus flexor digitorum longus> C

17 **flexor hallucis longus muscle** *origin,* posterior surface of fibula; *insertion,* base of distal phalanx of great toe; *innervation,* tibial; *action,* flexes great toe. <musculus flexor hallucis longus> C

18 **popliteus muscle** *origin,* lateral condyle of femur, lateral meniscus; *insertion,* posterior surface of tibia; *innervation,* tibial; *action,* flexes leg, rotates leg medially. <musculus popliteus> C

19 **tibialis posterior muscle** *origin,* tibia, fibula, interosseous membrane; *insertion,* bases of second to fourth metatarsals and tarsals, except talus; *innervation,* tibial; *action,* plantar flexes and inverts foot. <musculus tibialis posterior> C

Nerves of Posterior Compartment of Leg

20 ***interosseous nerve of leg*** *origin,* tibial nerve; *distribution,* interosseous membrane and tibiofibular syndesmosis; *modality,* general sensory. <nervus interosseus cruris> D

21 **muscular branches of tibial nerve** branches that supply muscles of the back of the leg; modality, motor. <rami musculares nervi tibialis> D

22 **tibial nerve** *origin,* sciatic nerve in lower part of thigh; *branches,* interosseous nerve of leg, medial cutaneous nerve of calf, sural nerve, and medial and lateral plantar nerves, and muscular and medial calcaneal branches; *distribution,* while still incorporated in the sciatic nerve, it supplies the semimembranosus and semitendinosus muscles, long head of biceps, and adductor magnus muscle; it supplies the knee joint as it descends in the popliteal fossa and, continuing into the leg, supplies the muscles and skin of the calf and sole of the foot, and the toes; *modality,* general sensory and motor. <nervus tibialis> D

23 **medial calcaneal branches of tibial nerve** branches supplying the medial side of the heel and of the posterior part of the sole; modality, general sensory. <rami calcanei mediales nervi tibialis> F

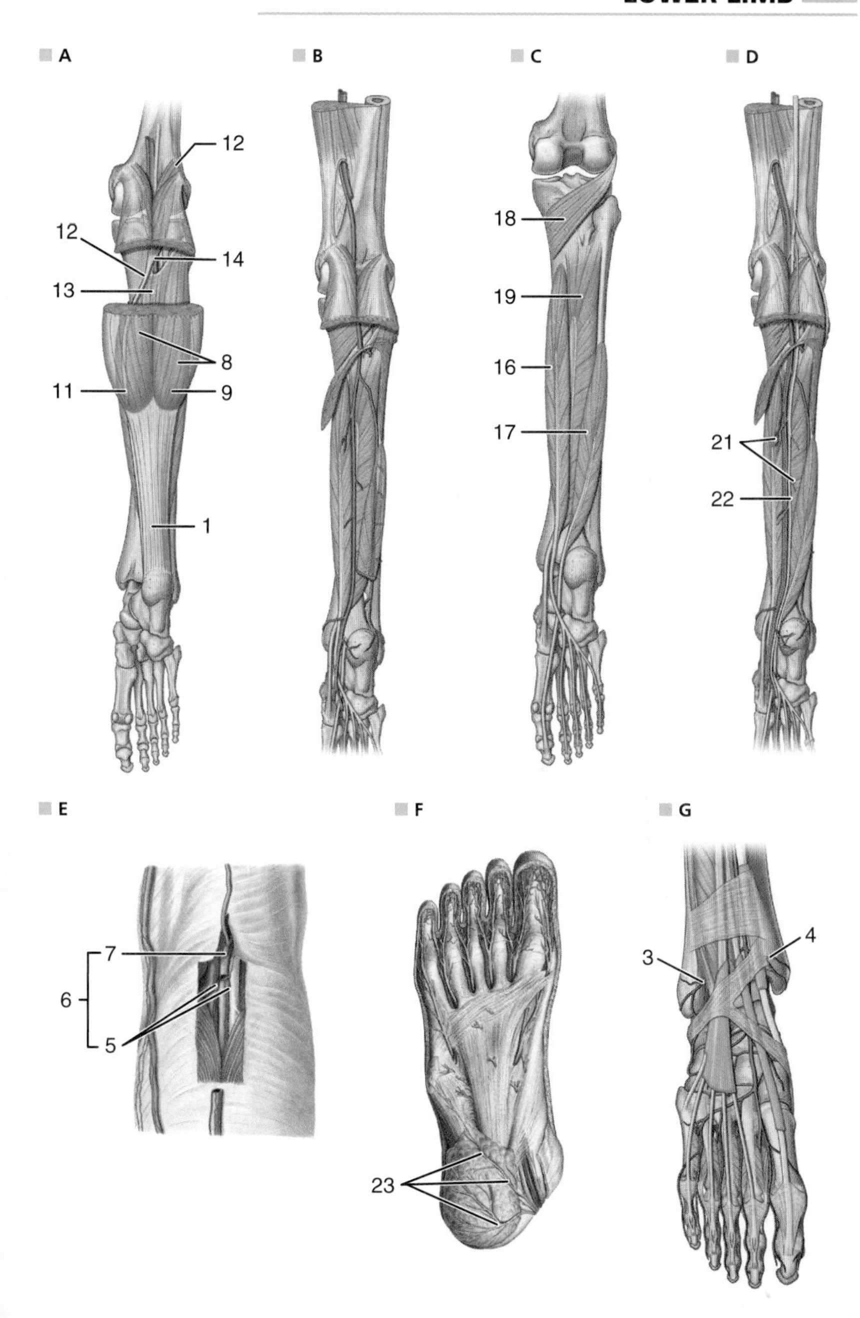
A
12
12
14
13
8
11
9
1
B
C
18
19
16
17
D
21
22
E
7
6
5
F
23
G
3
4

1 **lateral calcaneal branches of sural nerve** branches that innervate the skin on the back of the leg and the lateral side of the foot and heel; *modality,* general sensory. <rami calcanei laterales nervi suralis> A

2 **lateral sural cutaneous nerve** *origin,* common fibular nerve; *distribution,* skin of lateral side of back of leg, rarely may continue as the sural nerve; *modality,* general sensory. <nervus cutaneus surae lateralis> A

3 **medial sural cutaneous nerve** *origin,* tibial nerve; usually joins fibular communicating branch of common fibular nerve to form the sural nerve; *distribution,* may continue as the sural nerve; *modality,* general sensory. <nervus cutaneus surae medialis> A

4 **sural communicating branch of common fibular nerve** a small branch arising from the common fibular nerve, either with the lateral cutaneous sural nerve or separately; distally it joins the medial sural cutaneous nerve to form the sural nerve. <ramus communicans fibularis nervi fibularis communis> A

5 **sural nerve** *origin,* medial sural cutaneous nerve and communicating branch of common fibular nerve; *branches,* lateral dorsal cutaneous nerve and lateral calcaneal branches; *distribution,* skin on back of leg, and skin and joints on lateral side of heel and foot; *modality,* general sensory. <nervus suralis> A

Vessels of Posterior Compartment of Leg

6 **lateral malleolar network** a small arterial network on the lateral malleolus, formed by the lateral anterior malleolar artery, the perforating branch of the fibular artery, and the lateral tarsal artery. <rete malleolare laterale> B

7 **medial malleolar network** a small arterial network on the medial malleolus, formed by the medial anterior malleolar artery and branches from the posterior tibial artery. <rete malleolare mediale> B

8 **small saphenous vein** the vein that continues the lateral marginal vein from behind the malleolus and passes up the back of the leg to the knee joint, where it opens into the popliteal vein. <vena saphena parva> C

9 **fibular circumflex branch of posterior tibial artery** a branch that winds laterally around the neck of the fibula, helping supply the soleus muscle and contributing to the anastomosis around the knee joint. Called also *fibular circumflex artery.* <ramus circumflexus fibularis arteriae tibialis posterioris> D

10 **saphenous branch of descending genicular artery** a vessel that accompanies the saphenous nerve between the sartorius and gracilis muscles on the medial side of the knee, supplying the skin and anastomosing with the medial inferior genicular artery. <ramus saphenus arteriae descendentis genus> D

11 **fibular artery** *origin,* posterior tibial artery; *branches,* perforating, communicating, calcaneal, and lateral and medial malleolar branches, and calcaneal anastomosis; *distribution,* outside and back of ankle, deep calf muscles. <arteria fibularis> E

12 ***fibular veins*** the venae comitantes of the fibular artery, emptying into the posterior tibial vein. <venae fibulares> E

13 ***nutrient fibular artery*** *origin,* fibular artery; *branches,* none; *distribution,* fibula. <arteria nutricia fibulae> E

14 ***nutrient tibial artery*** *origin,* posterior tibial artery; *branches,* none; *distribution,* tibia. <arteria nutricia tibiae> E

15 ***perforating veins*** valved veins that drain blood from the superficial to the deep veins in the leg and foot. Called also *communicating veins.* <venae perforantes> E

16 **popliteal artery** *origin,* continuation of femoral artery; *branches,* superior lateral and medial genicular, middle genicular, sural, inferior lateral and medial genicular, anterior and posterior tibial arteries, and the genicular articular and the patellar anastomosis; *distribution,* knee, calf. <arteria poplitea> E

17 ***popliteal vein*** a vein following the popliteal artery, and formed by union of the venae comitantes of the anterior and posterior tibial arteries; at the adductor hiatus it becomes continuous with the femoral vein. <vena poplitea> E

18 **posterior tibial artery** *origin,* popliteal artery; *branches,* fibular circumflex branch, fibular, medial plantar, and lateral plantar arteries; *distribution,* leg, foot. <arteria tibialis posterior> E

19 ***posterior tibial veins*** accompanying veins of the posterior tibial artery, which unite with the anterior tibial veins to form the popliteal vein. <venae tibiales posteriores> E

20 **posterior tibial recurrent artery** *origin,* anterior tibial artery; *branches,* none; *distribution,* knee. <arteria recurrens tibialis posterior> F

21 **sural arteries** *origin,* popliteal artery; *branches,* none; *distribution,* popliteal space, calf. <arteriae surales> F

22 ***sural veins*** veins that ascend with the sural arteries and drain blood from the calf into the popliteal vein. <venae surales> F

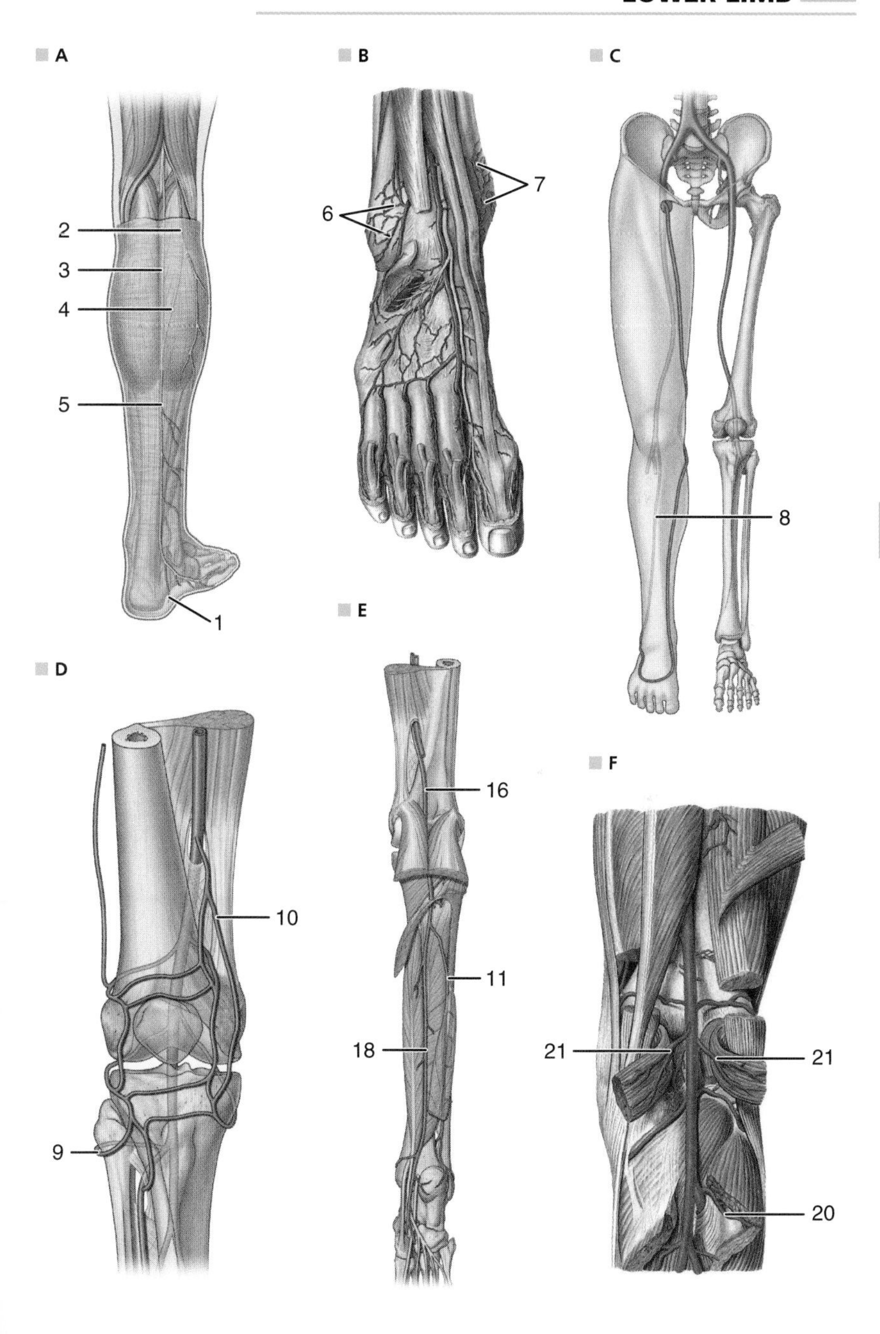
A
B
C
2
3
4
5
1
6
7
8
E
D
16
10
11
18
9
F
21
21
20

1 **calcaneal anastomosis** an arterial network on the posterior and lower surfaces of the calcaneus, receiving branches from the calcaneal branches of the fibular artery and the lateral malleolar branches of the fibular artery. <rete calcaneum> A

2 **calcaneal branches of lateral malleolar branches of fibular artery** branches distributed to the lateral aspect and back of the heel. <rami calcanei ramorum malleolarium lateralium arteriae fibularis> A

3 **calcaneal branches of posterior tibial artery** branches that arise from the posterior tibial artery and are distributed to the medial aspect and back of the heel. <rami calcanei arteriae tibialis posterioris> A

4 **communicating branch of fibular artery** a communicating branch between the fibular and the posterior tibial arteries, distributed to the interosseous membrane and supramalleolar region. <ramus communicans arteriae fibularis> A

5 **lateral malleolar branches of fibular artery** branches that supply the lateral aspect of the ankle and give off calcaneal branches to the lateral aspect and back of the heel. <rami malleolares laterales arteriae fibularis> A

6 **medial malleolar branches of posterior tibial artery** vessels supplying the area of the medial malleolus and giving off calcaneal branches to the medial aspect and back of the heel. <rami malleolares mediales arteriae tibiales posterioris> A

7 **perforating branch of fibular artery** a branch passing forward from the fibular artery where the interosseous membrane and the tibiofibular syndesmosis are continuous and descending to supply the syndesmosis and the ankle joint. <ramus perforans arteriae fibularis> B

Lateral Compartment of Leg

Muscles of Lateral Compartment

8 **fibularis brevis muscle** *origin,* lateral surface of fibula; *insertion,* tuberosity on base of fifth metatarsal bone; *innervation,* superficial fibular; *action,* everts, abducts, plantar flexes foot. Called also *peroneus brevis muscle.* <musculus fibularis brevis> C

9 **fibularis longus muscle** *origin,* lateral condyle of tibia, head and lateral surface of fibula; *insertion,* medial cuneiform, first metatarsal; *innervation,* superficial fibular; *action,* abducts, everts, plantar flexes foot. Called also *peroneus longus muscle.* <musculus fibularis longus> C

10 **common tendon sheath of fibulares** the tendon sheath for the fibularis longus and brevis muscles; it is single and double distally and extends about 4 cm proximally and distally from the tip of the lateral malleolus. <vagina communis tendinum musculorum fibularium> E

Nerves of Lateral Compartment

11 **common fibular nerve** *origin,* sciatic nerve in lower part of thigh; *branches and distribution,* supplies short head of biceps femoris muscle (while still incorporated in sciatic nerve), gives off lateral sural cutaneous nerve and communicating branch as it descends in popliteal fossa, supplies knee and superior tibiofibular joints and tibialis anterior muscle, and divides into superficial and deep fibular nerves; *modality,* general sensory and motor. Called also *common peroneal nerve.* <nervus fibularis communis> D, F

12 ***muscular branches of superficial fibular nerve*** branches that innervate the fibularis longus and brevis muscles; *modality,* motor. <rami musculares nervi fibularis superficialis> F

13 **superficial fibular nerve** *origin,* a terminal branch of common fibular nerve; *branches and distribution,* descends in front of the fibula, supplies fibularis longus and brevis muscles and, in the lower part of the leg, divides into muscular branches and medial and intermediate dorsal cutaneous nerves; *modality,* general sensory and motor. Called also *superficial peroneal nerve.* <nervus fibularis profundus> F

A

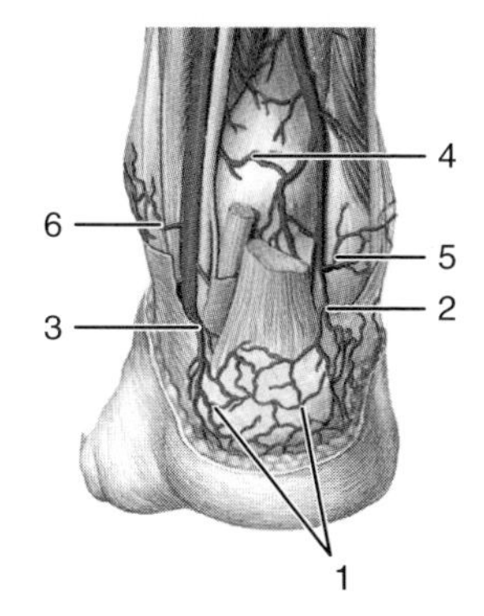

B

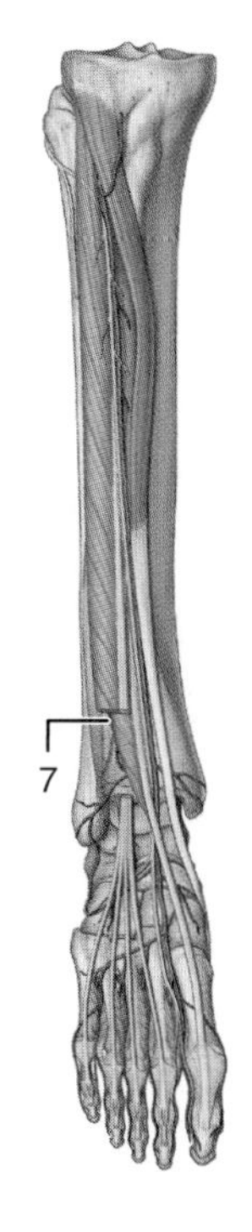

C

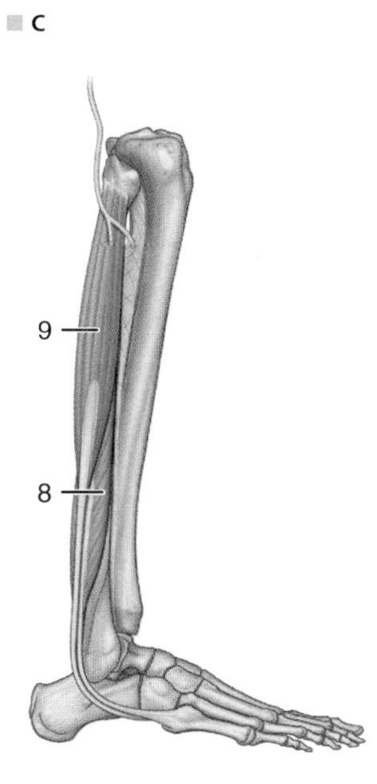

D

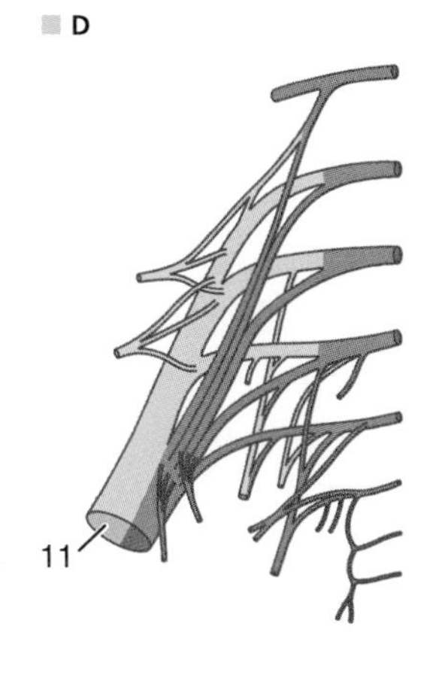

E

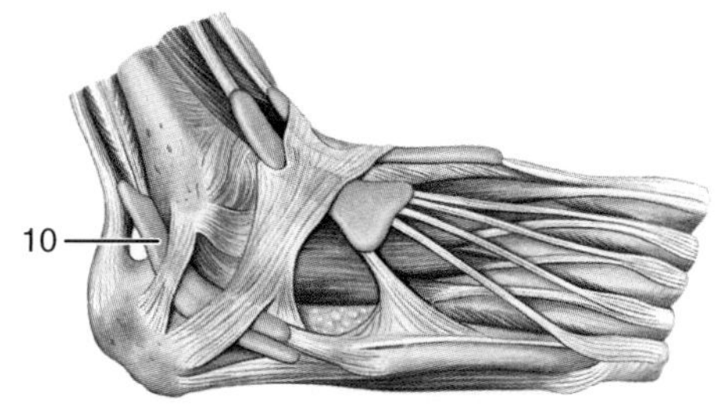

F

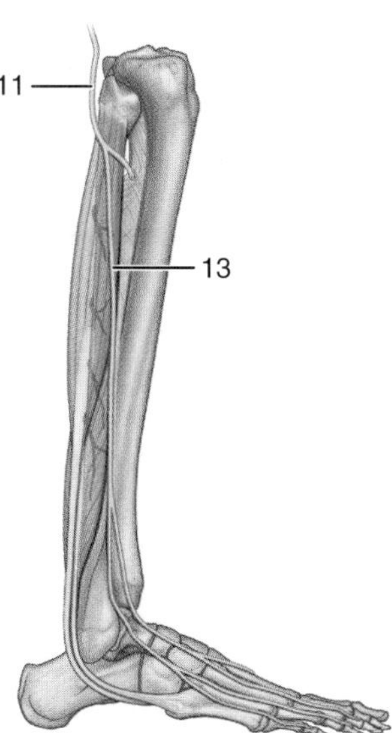

FOOT

1 **foot** the distal portion of the lower limb, upon which an individual stands and walks. In humans it consists of the tarsus, metatarsus, and phalanges and the tissues encompassing them. <pes> A

2 **fourth toe** the fourth digit of the foot. <digitus quartus (IV) pedis> A

3 **great toe** the first digit of the foot. <hallux> A

4 **heel** the hindmost projection of the foot. <calx> A

5 **lateral border of foot** the lateral, or fibular, border of the foot. <margo lateralis pedis> A

6 **little toe** the fifth, and smallest, digit of the foot. <digitus minimus pedis> A

7 **medial border of foot** the medial, or tibial, border of the foot. <margo medialis pedis> A

8 **metatarsal region** the region of the foot overlying the metatarsals. <regio metatarsalis> A

9 **metatarsus** the part of the foot between the tarsus and the toes, its skeleton being the five long bones (the metatarsals) extending from the tarsus to the phalanges. A

10 **nail** the horny cutaneous plate on the dorsal surface of the distal end of the terminal phalanx of a toe, made up of flattened epithelial scales developed from the stratum lucidum of the skin. <unguis> A

11 **second toe** the second digit of the foot. <digitus secundus (II) pedis> A

12 **third toe** the third digit of the foot. <digitus tertius (III) pedis> A

13 **toes** the digits of the foot. <digiti pedis> A

14 **base of phalanx of toes** the proximal end of each phalanx of the toes. <basis phalangis digitorum pedis> B

15 **flexor retinaculum of foot** a strong band of fascia that extends from the medial malleolus down onto the calcaneus. It holds in place the tendons of the tibialis posterior, flexor digitorum, and flexor hallucis muscles as they pass to the sole of the foot, and gives protection to the posterior tibial vessels and tibial nerve. <retinaculum musculorum flexorum pedis> E

16 **lateral longitudinal arch of foot** the part of the arch formed by the calcaneus, the cuboid bone, and the lateral two metatarsal bones. <pars lateralis arcus pedis longitudinalis> F

17 **longitudinal arch of foot** the arch running longitudinally along the sole of the foot, consisting of lateral and medial parts. <arcus pedis longitudinalis> F

18 **medial longitudinal arch of foot** the part of the longitudinal arch formed by the calcaneus, talus, navicular, cuneiform, and the first three metatarsal bones. <pars medialis arcus pedis longitudinalis> F

19 **plantar aponeurosis** bands of fibrous tissue radiating toward the bases of the toes from the medial process of the calcaneal tuberosity; called also plantar fascia. <aponeurosis plantaris> C

20 ***plantar*** a term designating relationship to the sole of the foot. <plantaris> C

21 ***sole*** the undersurface of the foot. <planta> C

22 **transverse fasciculi of plantar aponeurosis** transverse bundles in the plantar aponeurosis near the toes. <fasciculi transversi aponeurosis plantaris> C

23 **annular part of fibrous sheaths of toes** fibrous bands in the toes resembling those of similar name in the fingers; called also *annular ligaments of toes.* <pars anularis vaginae fibrosae digitorum pedis> D

24 **cruciate part of fibrous sheaths of toes** one of the bundles of fascial fibers in the toes resembling those of similar name in the fingers; called also *cruciate ligaments of toes.* <pars cruciformis vaginae fibrosae digitorum pedis> D

25 **fibrous sheaths of toes** more or less complete fascial sheaths surrounding the phalanges of the toes, for attachment of the tendons and their synovial membranes. <vaginae fibrosae digitorum pedis> D

26 **synovial sheaths of toes** the synovial sheaths surrounding the tendons of the toes. <vaginae synoviales digitorum pedis> D

27 **tendon sheaths of toes** the sheaths enclosing the tendons of the toes, each comprising a fibrous sheath surrounding a synovial sheath. <vaginae tendinum digitorum pedis> D

28 ***vincula tendinum*** bands connecting the tendons of the flexor digitorum longus and flexor digitorum brevis muscles to the phalanges and interphalangeal articulations of the foot. They are similar to the vincula found in the hand. <vincula tendinum digitorum pedis> D

29 **oblique head of adductor hallucis muscle** the head of the adductor hallucis muscle originating from the bases of the second, third, and fourth metatarsal bones, and the sheath of the fibularis longus muscle. <caput obliquum musculi adductoris hallucis> G

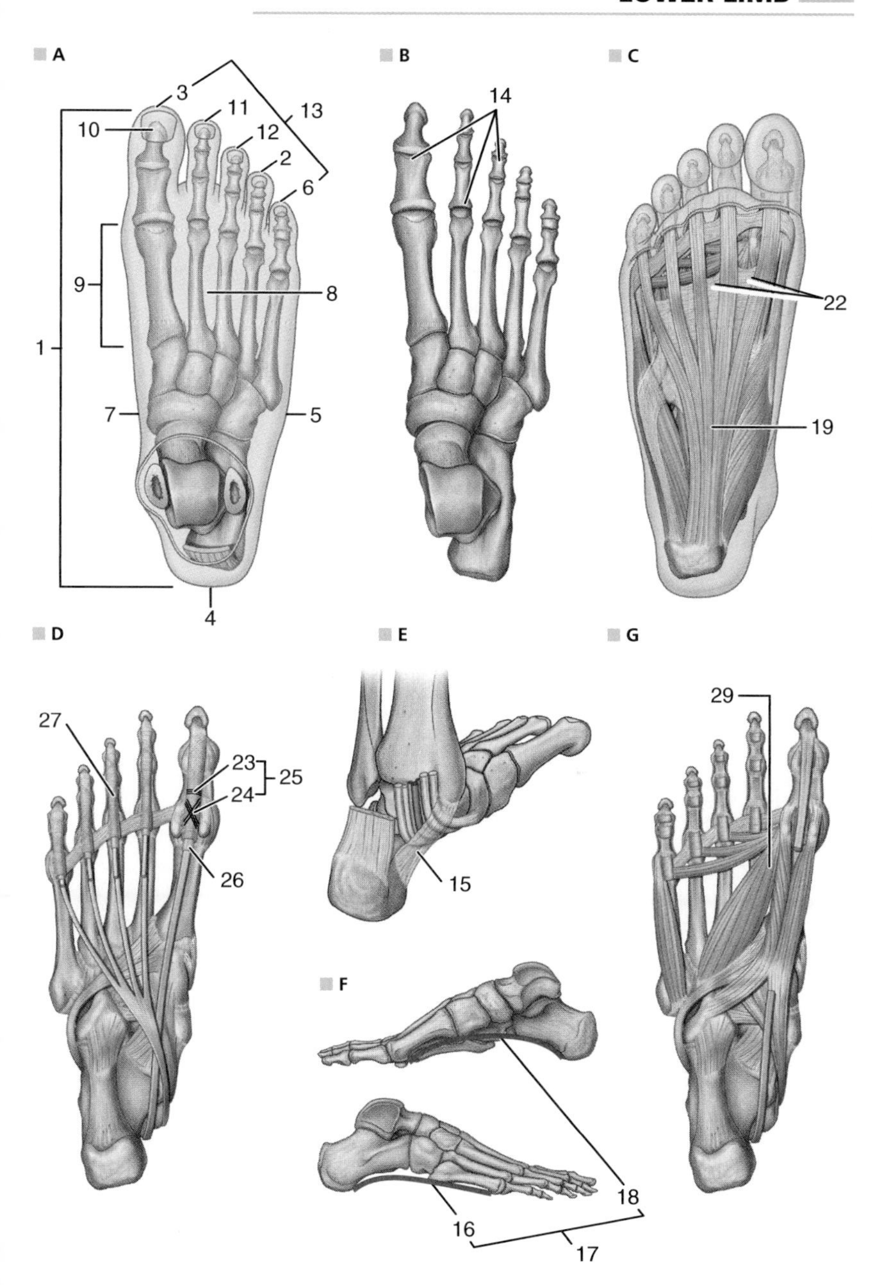
A
B
C
D
E
F
G
1
2
3
4
5
6
7
8
9
10
11
12
13
14
15
16
17
18
19
22
23
24
25
26
27
29

1 ***dorsal fascia of foot*** the investing fascia on the dorsum of the foot. <fascia dorsalis pedis> A

2 **dorsal surfaces of toes** the superior surfaces of the toes. <facies dorsales digitorum pedis> A

3 **dorsum of foot** the upper surface of the foot; the surface opposite the sole. <dorsum pedis> A

4 **plantar surfaces of toes** the inferior surfaces of the toes. <facies plantares digitorum pedis> C

Muscles of Foot

5 **extensor digitorum brevis muscle** *origin,* superior surface of calcaneus; *insertion,* tendons of extensor digitorum longus of first, second, third, fourth toes; *innervation,* deep fibular; *action,* extends toes. <musculus extensor digitorum brevis> B

6 **extensor hallucis brevis muscle** a name given the portion of the extensor digitorum brevis muscle that goes to the great toe. <musculus extensor hallucis brevis> B

7 ***subtendinous bursa of tibialis anterior muscle*** a bursa between the tendon of the tibialis anterior and the medial surface of the medial cuneiform bone. <bursa subtendinea musculi tibialis anterioris> B

8 **tendon sheath of extensor digitorum longus** the tendon sheath enclosing the tendons of the extensor digitorum longus muscle, extending from just above the malleoli to the base of the fifth metatarsal. <vaginae tendinum musculi extensoris digitorum longi> B

9 **tendon sheath of extensor hallucis longus** the tendon sheath of the extensor hallucis longus muscle, extending from the level of the malleoli and reaching the base of the first metatarsal dorsal fascia of the foot. <vagina tendinis musculi extensoris hallucis longi> B

10 **tendon sheath of tibialis anterior** the tendon sheath of the tibialis anterior muscle, extending from the proximal margin of the superior extensor retinaculum to the region between the diverging limbs of the inferior extensor retinaculum. <vagina tendinis musculi tibialis anterioris> B

11 **tendon sheath of flexor digitorum longus** the tendon sheath enclosing the tendons of the flexor digitorum longus muscle, extending from the medial malleolus to below the navicular bone. <vaginae tendinum musculi flexoris digitorum longi> D

12 **tendon sheath of flexor hallucis longus** the tendon sheath of the flexor hallucis longus muscle, extending from the medial malleolus to the base of the first metatarsal. <vagina tendinis musculi flexoris hallucis longi> D

13 **tendon sheath of tibialis posterior** the tendon sheath of the tibialis posterior muscle, beginning at the medial malleolus and extending into the foot. <vagina tendinis musculi tibialis posterioris> D

14 **plantar tendon sheath of fibularis longus** a tendon sheath beginning in the fibular groove of the cuboid bone. <vagina plantaris tendinis musculi fibularis longi> C

15 **abductor digiti minimi muscle of foot** *origin,* medial and lateral tubercles of calcaneus, plantar fascia; *insertion,* lateral surface of base of proximal phalanx of little toe; *innervation,* superficial branch of lateral plantar; *action,* abducts little toe. <musculus abductor digiti minimi pedis> E

16 **abductor hallucis muscle** *origin,* medial tubercle of calcaneus, plantar fascia; *insertion,* medial surface of base of proximal phalanx of great toe; *innervation,* medial plantar; *action,* abducts, flexes great toe. <musculus abductor hallucis> E

17 **flexor digitorum brevis muscle** *origin,* medial tuberosity of calcaneus, plantar fascia; *insertion,* middle phalanges of four lateral toes; *innervation,* medial plantar; *action,* flexes four lateral toes. <musculus flexor digitorum brevis> E

18 **lumbrical muscles of foot** *origin,* tendons of flexor digitorum longus; *insertion,* extensor expansions of four lateral toes; *innervation,* medial and lateral plantar; *action,* flex metatarsophalangeal joints, extend distal phalanges. <musculi lumbricales pedis> F

19 **quadratus plantae muscle** *origin,* calcaneus and plantar fascia; *insertion,* tendons of flexor digitorum longus; *innervation,* lateral plantar; *action,* aids in flexing toes. <musculus quadratus plantae> F

20 **adductor hallucis muscle** (2 heads): *origin,* oblique head, bases of second, third, and fourth metatarsals, and sheath of fibularis longus, transverse head, capsules of metatarsophalangeal joints of three lateral toes; *insertion,* lateral side of base of proximal phalanx of great toe; *innervation,* lateral plantar; *action,* adducts great toe. <musculus adductor hallucis> G

21 **flexor digiti minimi brevis muscle of foot** *origin,* base of fifth metatarsal, sheath of fibularis longus muscle; *insertion,* lateral surface of base of proximal phalanx of little toe; *innervation,* lateral plantar; *action,* flexes little toe. <musculus flexor digiti minimi brevis pedis> G

22 **flexor hallucis brevis muscle** *origin,* undersurface of cuboid, lateral cuneiform; *insertion,* both sides of base of proximal phalanx of great toe; *innervation,* medial plantar; *action,* flexes great toe. <musculus flexor hallucis brevis> G

23 **transverse head of adductor hallucis muscle** the head of the adductor hallucis muscle arising from the capsules of the metatarsophalangeal joints of the third, fourth, and fifth toes. <caput transversum musculi adductoris hallucis> G

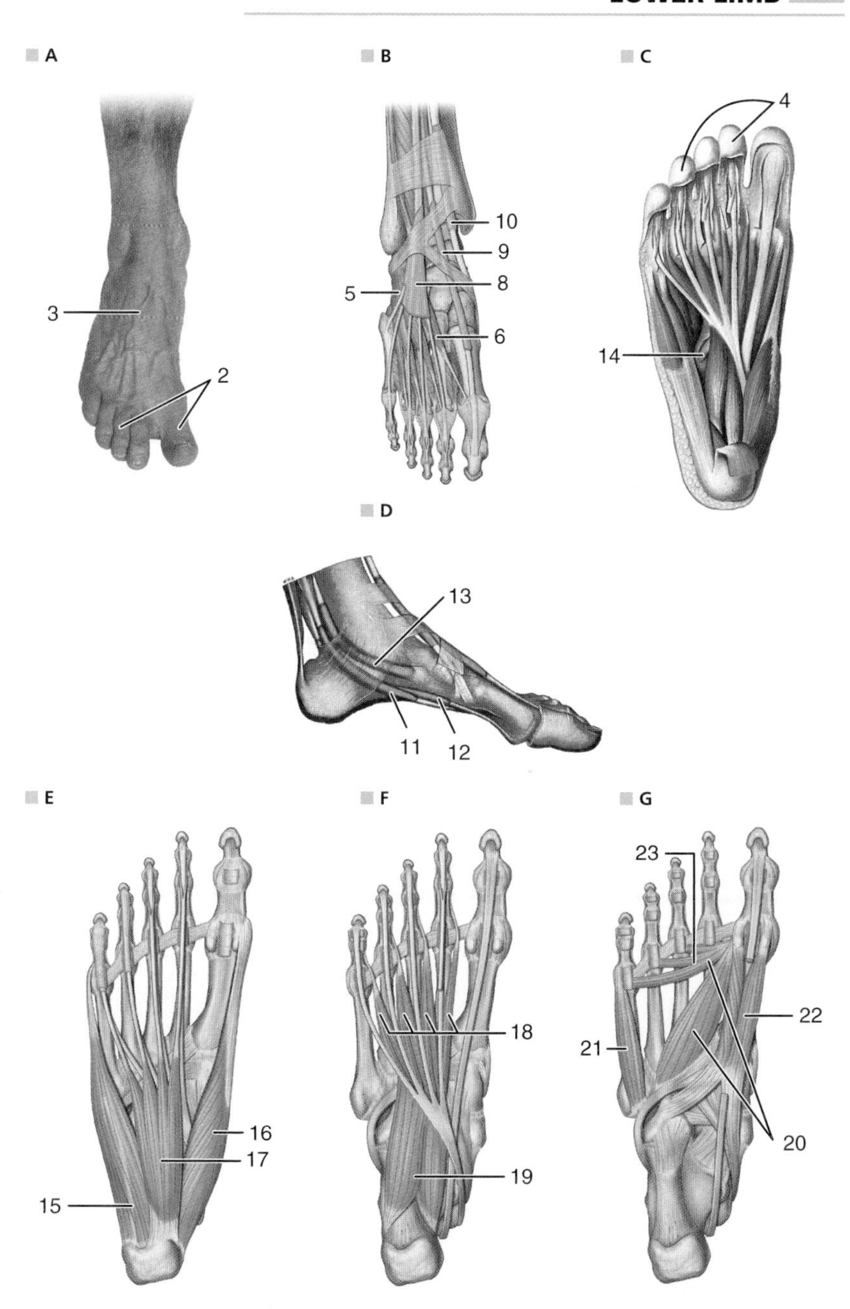
A
B
C
D
E
F
G
2
3
4
5
6
8
9
10
11
12
13
14
15
16
17
18
19
20
21
22
23

1 **dorsal interossei muscles of foot** (4): *origin,* adjacent surfaces of metatarsal bones; *insertion,* base of proximal phalanges of second, third, and fourth toes and their extensor expansions; *innervation,* lateral plantar; *action,* abduct, flex toes. <musculi interossei dorsales pedis> A

2 **plantar interossei muscles** (3): *origin,* medial surface of third, fourth, and fifth metatarsal bones; *insertion,* medial side of base of proximal phalanges of third, fourth, and fifth toes and their extensor expansions; *innervation,* lateral plantar; *action,* adduct, flex toes. <musculi interossei plantares> A

Nerves of Foot

3 **common plantar digital nerves of lateral plantar nerve** *number,* two; *origin,* superficial branch of lateral plantar nerve; *branches,* the medial nerve gives rise to two proper plantar digital nerves; *distribution,* the lateral one to the flexor digiti minimi brevis and to skin and joints of lateral side of sole and little toe; the medial one to adjacent sides of fourth and fifth toes; *modality,* motor and general sensory. <nervi digitales plantares communes nervi plantaris lateralis> B

4 **common plantar digital nerves of medial plantar nerve** *number,* four; *origin,* medial plantar nerve; *branches,* muscular branches and proper plantar digital nerves; *distribution,* flexor hallucis brevis muscle and first lumbrical muscles, skin and joints of medial side of foot and great toe, and adjacent sides of great and second, second and third, and third and fourth toes; *modality,* motor and general sensory. <nervi digitales plantares communes nervi plantaris medialis> B

5 **deep branch of lateral plantar nerve** a branch that accompanies the lateral plantar artery on its medial side and the plantar arch, innervating the interossei, the second, third, and fourth lumbrical, and the adductor hallucis muscles, and some articulations; modality, general sensory. <ramus profundus nervi plantaris lateralis> B

6 **lateral plantar nerve** *origin,* the smaller of the terminal branches of tibial nerve; *branches,* muscular, superficial, and deep branches; *distribution,* lying between first and second layers of muscles of sole, it supplies the quadratus plantae, abductor digiti minimi, flexor digiti minimi brevis, adductor hallucis, interossei, and second, third, and fourth lumbrical muscles, and gives off cutaneous and articular twigs to lateral side of sole and fourth and fifth toes; modality, general sensory and motor. <nervus plantaris lateralis> B

7 **medial plantar nerve** *origin,* the larger of the terminal branches of tibial nerve; *branches,* common plantar digital nerves and muscular branches; *distribution,* abductor hallucis, flexor digitorum brevis, flexor hallucis brevis, and first lumbrical muscles, and cutaneous and articular twigs to the medial side of the sole, and to the first to fourth toes; *modality,* general sensory and motor. <nervus plantaris medialis> B

8 **proper plantar digital nerves of lateral plantar nerve** *origin,* common plantar digital nerves; *distribution,* flexor digiti minimi brevis muscle, skin and joints of lateral side of sole and little toe, and adjacent sides of fourth and fifth toes; *modality,* motor and general sensory. <nervi digitales plantares proprii nervi plantaris lateralis> B

9 **proper plantar digital nerves of medial plantar nerve** *origin,* common plantar digital nerves; *distribution,* skin and joints of medial side of great toe, and adjacent sides of great and second, second and third, and third and fourth toes; the nerves extend to the dorsum to supply nail beds and tips of toes; *modality,* general sensory. <nervi digitales plantares proprii nervi plantaris medialis> B

10 **superficial branch of lateral plantar nerve** a branch that arises from the lateral plantar nerve at the lateral border of the quadratus plantae muscle and passes forward, dividing into a lateral part that innervates skin of the lateral side of the sole and little toe, joints of the toe, and the flexor digiti minimi brevis muscle, and a medial part, a common plantar digital nerve, that gives two proper plantar digital nerves to the adjacent sides of the fourth and fifth toes; modality, general sensory. <ramus superficialis nervi plantaris lateralis> B

11 **dorsal digital nerves of foot nerves supplying the dorsal aspect of the toes to the nails** 1. nerves supplying the third, fourth, and fifth toes; *origin,* intermediate dorsal cutaneous nerve; *distribution,* skin and joints of adjacent sides of third and fourth, and of fourth and fifth toes; *modality,* general sensory. 2. nerves supplying the first and second toes; *origin,* medial terminal division of deep fibular nerve; *distribution,* skin and joints of adjacent sides of great and second toes; *modality,* general sensory. <nervi digitales dorsales pedis> C

12 **intermediate dorsal cutaneous nerve** *origin,* superficial fibular nerve; *branches,* dorsal digital nerves of foot; *distribution,* skin of front of lower third of leg and dorsum of foot, and skin and joints of adjacent sides of third and fourth, and of fourth and fifth toes; *modality,* general sensory. <nervus cutaneus dorsalis intermedius> C

13 **lateral dorsal cutaneous nerve** *origin,* continuation of sural nerve; *distribution,* skin and joints of lateral side of foot and fifth toe; *modality,* general sensory. <nervus cutaneus dorsalis lateralis> C

14 **medial dorsal cutaneous nerve** *origin,* superficial fibular nerve; *distribution,* skin and joints of medial side of foot and big toe, and adjacent sides of second and third toes; *modality,* general sensory. <nervus cutaneus dorsalis medialis> C

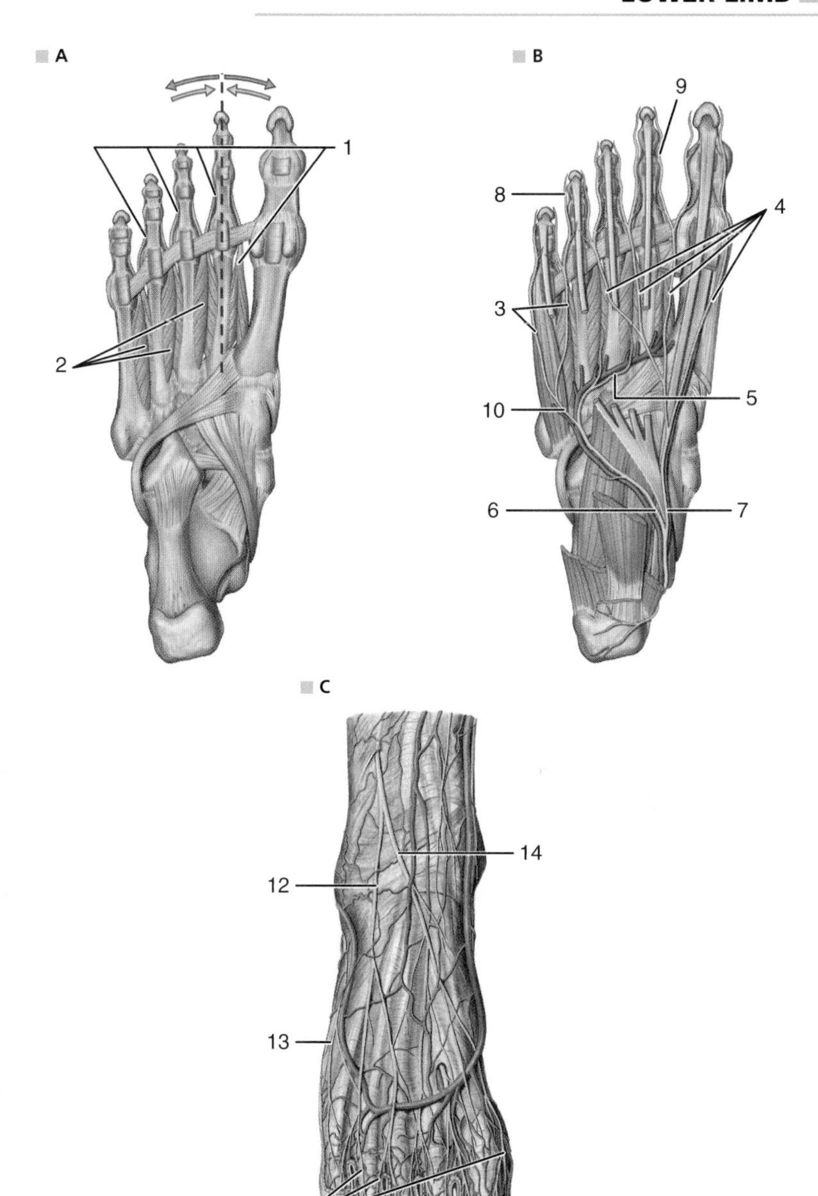
A
1
2
B
9
8
4
3
5
10
6
7
C
14
12
13
11

Vessels of Foot

1 **common plantar digital arteries** *origin,* plantar metatarsal arteries; *branches,* proper plantar digital arteries; *distribution,* toes. <arteriae digitales plantares communes> A

2 **deep plantar arch** the deep arterial arch in the foot, formed by the anastomosis of the lateral plantar artery with the deep plantar branch of the dorsalis pedis artery, and giving off the plantar metatarsal arteries. Called also *plantar arch.* <arcus plantaris profundus> A

3 **deep plantar artery** *origin,* dorsalis pedis artery; *branches,* none; *distribution,* sole of foot to help form deep plantar arch. <arteria plantaris profunda> A

4 **lateral plantar artery** *origin,* posterior tibial artery; *branches,* plantar arch and plantar metatarsal arteries; *distribution,* sole of foot and toes. <arteria plantaris lateralis> A

5 **medial plantar artery** *origin,* posterior tibial artery; *branches,* deep and superficial branches; *distribution,* sole of the foot and toes. <arteria plantaris medialis> A

6 **perforating branches of plantar metatarsal arteries** vessels connecting the plantar metatarsal arteries with the dorsal metatarsal arteries through the interosseous spaces. <rami perforantes arteriarum metatarsearum plantarium> A

7 **plantar metatarsal arteries** *origin,* plantar arch; *branches,* perforating branches, common and proper plantar digital arteries; *distribution,* toes. <arteriae metatarsales plantares> A

8 **proper plantar digital arteries** *origin,* common plantar digital arteries; *branches,* none; *distribution,* toes. <arteriae digitales plantares propriae> A

9 **superficial branch of medial plantar artery** a branch that supplies the medial side of the great toe. <ramus superficialis arteriae plantaris medialis> A

10 **deep branch of medial plantar artery** a branch that supplies the anteromedial aspect of the sole, anastomosing with the medial three plantar metatarsal arteries. <ramus profundus arteriae plantaris medialis> B

11 **arcuate artery of foot** *origin,* dorsalis pedis artery; *branches,* deep plantar branch and dorsal metatarsal artery; *distribution,* foot, toes. <arteria arcuata pedis> C

12 **dorsal digital arteries of foot** *origin,* dorsal metatarsal arteries; *branches,* none; *distribution,* dorsum of toes. <arteriae digitales dorsales pedis> C

13 **dorsal metatarsal arteries** *origin,* arcuate artery of foot; *branches,* dorsal digital arteries; *distribution,* dorsum of foot, including toes. <arteriae metatarsales dorsales> C

14 **dorsalis pedis artery** *origin,* continuation of anterior tibial; *branches,* lateral and medial tarsal, arcuate, and deep plantar arteries; *distribution,* foot, toes. <arteria dorsalis pedis> C

15 **lateral tarsal artery** *origin,* dorsal artery of foot; *branches,* none; *distribution,* tarsus. <arteria tarsalis lateralis> C

16 **medial tarsal arteries** *origin,* dorsalis pedis artery; *branches,* none; *distribution,* side of foot. <arteriae tarsales mediales> C

17 **dorsal digital veins of foot** the veins on the dorsal surfaces of the toes that unite in pairs around each cleft to form the dorsal metatarsal veins. <venae digitales dorsales pedis> D

18 **dorsal metatarsal veins** veins that are formed by the dorsal digital veins of the toes at the clefts of the toes, joining the dorsal venous arch. <venae metatarsales dorsales> D

19 **dorsal venous arch of foot** a transverse venous arch across the dorsum of the foot near the bases of the metatarsal bones. <arcus venosus dorsalis pedis> D

20 ***dorsal venous network of foot*** a superficial network of anastomosing veins on the dorsum of the foot proximal to the transverse venous arch, draining into the great and the small saphenous veins. <rete venosum dorsale pedis> D

21 ***intercapitular veins of foot*** veins at the clefts of the toes that pass between the heads of the metatarsal bones and establish communication between the dorsal and plantar venous systems of the foot. <venae intercapitulares pedis> D

22 **lateral marginal vein** a vein running along the lateral side of the dorsum of the foot, returning blood from the dorsal venous arch, dorsal venous network, and superficial veins of the sole, and draining into the small saphenous vein. <vena marginalis lateralis> D

23 **medial marginal vein** a vein running along the medial side of the dorsum of the foot, returning blood from the dorsal venous arch, the dorsal venous network, and superficial veins of the sole and draining into the great saphenous vein. <vena marginalis medialis> D

24 ***plantar digital veins*** veins from the plantar surfaces of the toes which unite at the clefts to form the plantar metatarsal veins of the foot. <venae digitales plantares> D

25 ***plantar metatarsal veins*** deep veins of the foot that arise from the plantar digital veins at the clefts of the toes and pass back to open into the plantar venous arch. <venae metatarsales plantares> D

26 ***plantar venous arch*** the deep venous arch that accompanies the plantar arterial arch. <arcus venosus plantaris> D

27 ***plantar venous network*** a thick venous network in the subcutaneous tissue of the sole of the foot. <rete venosum plantare> D

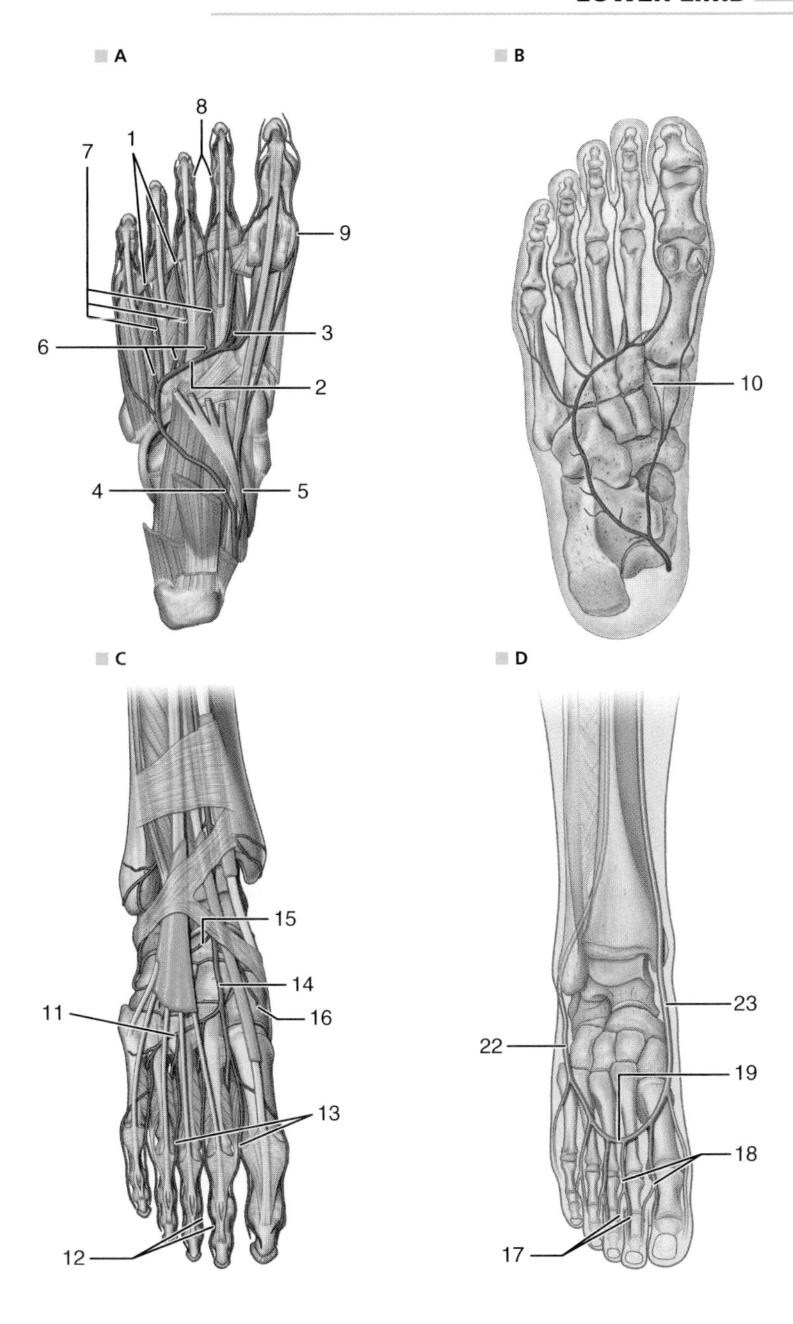
A
B
C
D
1
2
3
4
5
6
7
8
9
10
11
12
13
14
15
16
17
18
19
22
23

UPPER LIMB

GENERAL

1 **antebrachial region** the part of the upper limb between the elbow and wrist. <regio antebrachialis> C

2 **anterior region of wrist** the anterior aspect of the wrist; called also *anterior carpal region*. <regio carpalis anterior> C

3 **arm** the part of the upper limb from shoulder to elbow. <brachium> C

4 **brachial region** the part of the upper limb from shoulder to elbow; the arm. <regio brachialis> C

5 **forearm** the part of the upper limb of the body between the elbow and the wrist; it is divided into anterior and posterior compartments. <antebrachium> C

6 **hand region** the region of the upper limb distal to the antebrachial region and comprising the various regions of the hand and wrist. <regio manus> C

7 **pectoral girdle** the encircling bony structure supporting the upper limbs, comprising the clavicles and scapulae, articulating with each other and with the sternum and vertebral column, respectively; called also *shoulder girdle*. <cingulum pectorale> C

8 ***posterior region of wrist*** the posterior aspect of the wrist; called also *posterior carpal region*. <regio carpalis posterior> C

9 ***upper limb*** the limb of the body extending from the deltoid region to the hand. It is specialized for functions requiring great mobility, such as grasping and manipulating. <membrum superius> C

10 **wrist** the joint between the forearm and hand, made up of eight bones. <carpus> C

11 **axillary region** the region of the thorax around the fossa axillaris. <regio axillaris> A

12 **carpal region** the region of the wrist between the forearm and hand. <regio carpalis> A

13 **cubital fossa** the depression in the anterior region of the elbow. <fossa cubitalis> A

14 **cubital region** the area of the upper limb about the elbow; divided into anterior and posterior regions. <regio cubitalis> A

15 **elbow** the bend of the upper limb between the arm and forearm. <cubitus> C

16 **radial border of forearm** the lateral border of the forearm. <margo radialis antebrachii> A

17 **ulnar border of forearm** the ulnar, or medial, border of the forearm. <margo ulnaris antebrachii> A

18 **lateral bicipital groove** a longitudinal groove on the lateral side of the arm which marks the limit between the lateral border of the biceps muscle and the brachialis. <sulcus bicipitalis lateralis> B

19 **medial bicipital groove** a longitudinal groove on the medial side of the arm which marks the limit between the medial border of the biceps muscle and the brachialis. <sulcus bicipitalis medialis> B

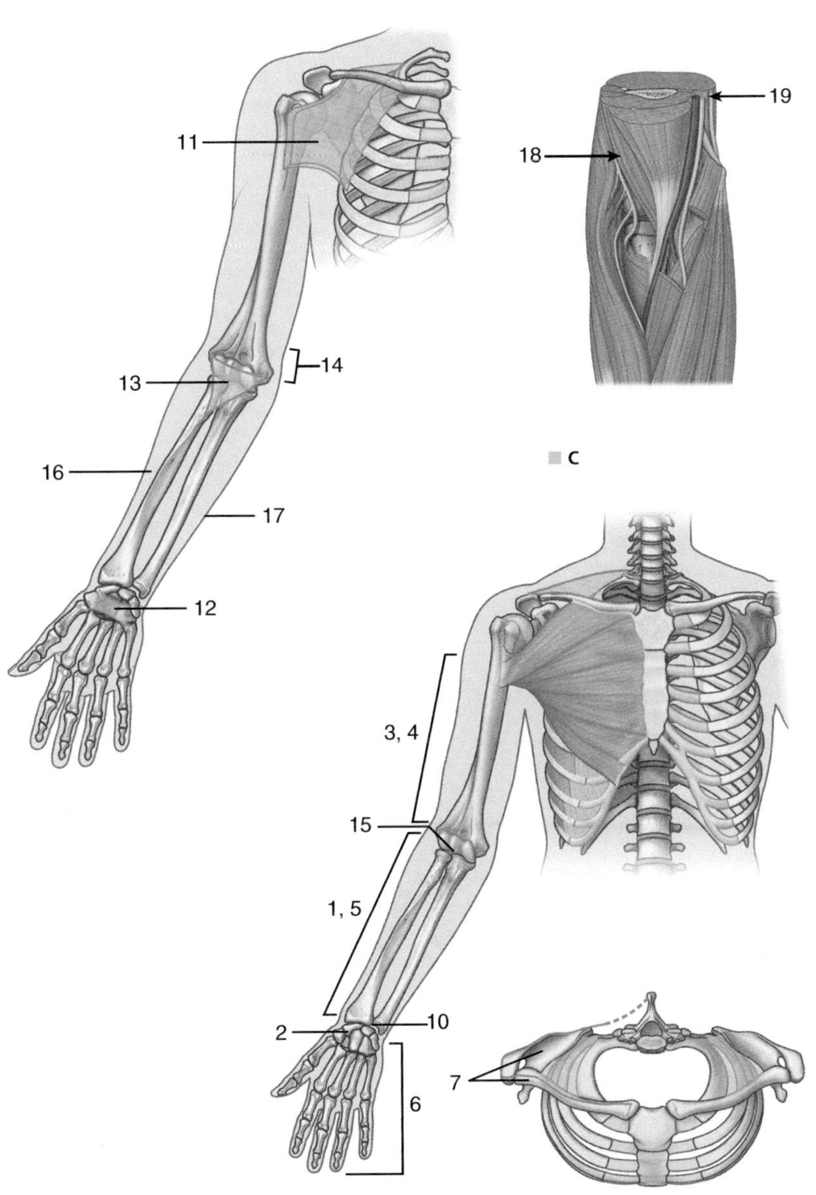
A
11
13
14
16
17
12
B
19
18
C
3, 4
15
1, 5
2
10
6
7

BONES

1 **anatomical neck of humerus** the somewhat constricted zone on the humerus just distal to the head, separating the articular surface from the tubercles. <collum anatomicum humeri> A

2 **anterolateral surface of humerus** the surface that provides attachment to the deltoid muscle and lateral part of the brachialis muscle. <facies anterolateralis humeri> A

3 **anteromedial surface of humerus** the surface of the humerus that begins above at the intertubercular groove and spreads out inferiorly to form the wide smooth area for origin of the brachialis muscle. <facies anteromedialis humeri> A

4 **body of humerus** the long central part of the humerus. <corpus humeri> A

5 **capitulum of humerus** an eminence on the distal end of the lateral epicondyle of the humerus for articulation with the head of the radius. <capitulum humeri> A

6 **condyle of humerus** the distal end of the humerus, including the various fossae as well as the trochlea and capitulum. <condylus humeri> A

7 **coronoid fossa** the cavity in the humerus that receives the coronoid process of the ulna when the elbow is flexed. Called also *ulnar fossa*. <fossa coronoidea humeri> A

8 **crest of greater tubercle** a projection on the greater tubercle of the humerus, forming one lip of the intertubercular groove. <crista tuberculi majoris> A

9 **crest of lesser tubercle** a projection on the lesser tubercle of the humerus, forming one lip of the intertubercular groove. <crista tuberculi minoris> A

10 **deltoid tuberosity** a rough, triangular elevation, about the middle of the anterolateral border of the shaft of the humerus, for attachment of the deltoid muscle. <tuberositas deltoidea humeri> A

11 **greater tubercle of humerus** a large flattened prominence at the upper end of the lateral surface of the humerus, just lateral to the highest part of the anatomical neck, giving attachment to the infraspinatus, supraspinatus, and teres minor muscles. <tuberculum majus humeri> A

12 **head of humerus** the proximal end of the humerus, which articulates with the glenoid cavity of the scapula. <caput humeri> A

13 **humerus** the bone that extends from the shoulder to the elbow articulating proximally with the scapula and distally with the radius and ulna. A

14 **intertubercular sulcus** a longitudinal groove on the anterior surface of the humerus, lying between the tubercles above and between the crests of the greater and lesser tubercles farther down, and lodging the tendon of the long head of the biceps muscle. <sulcus intertubercularis humeri> A

15 **lateral border of humerus** the edge of the humerus that extends from posteroinferior part of the greater tubercle to the lateral epicondyle. <margo lateralis humeri> A

16 **lateral epicondyle of humerus** a projection from the distal end of the humerus, giving attachment to a common tendon of origin of the extensor carpi radialis brevis, extensor digitorum communis, extensor digiti minimi, extensor carpi ulnaris, and supinator muscles. <epicondylus lateralis humeri> A

17 **lateral supraepicondylar crest of humerus** a prominent curved ridge on the lateral surface of the humerus, giving attachment in front to the brachioradialis and extensor carpi radialis longus muscles. <crista supraepicondylaris lateralis humeri> A

18 **lesser tubercle of humerus** a distinct prominence at the proximal end of the anterior surface of the humerus, just lateral to the anatomical neck; it gives insertion to the subscapularis muscle. <tuberculum minus humeri> A

19 **medial border of humerus** the edge of the humerus that begins at the lesser tubercle above and continues downward to the medial epicondyle. <margo medialis humeri> A

20 **medial epicondyle of humerus** a projection from the distal end of the humerus, giving attachment to the pronator teres above; a common tendon of origin of the flexor carpi radialis, palmaris longus, flexor digitorum superficialis, and flexor carpi ulnaris muscles in the middle, and the ulnar collateral ligament below. <epicondylus medialis humeri> A

21 **medial supraepicondylar crest of humerus** a prominent, curved ridge on the medial surface of the humerus, giving attachment to the brachialis muscle in front and to the medial head of the triceps behind; called also *medial supracondylar ridge of humerus*. <crista supraepicondylaris medialis humeri> A

22 **radial fossa** a depression on the anterior surface of the humerus just above the capitulum. <fossa radialis humeri> A

23 **surgical neck of humerus** the region on the humerus just below the tubercles, where the bone becomes constricted. <collum chirurgicum humeri> A

24 **trochlea of humerus** the pulleylike medial portion of the distal end of the humerus for articulation with the semilunar notch of the ulna. <trochlea humeri> A

A

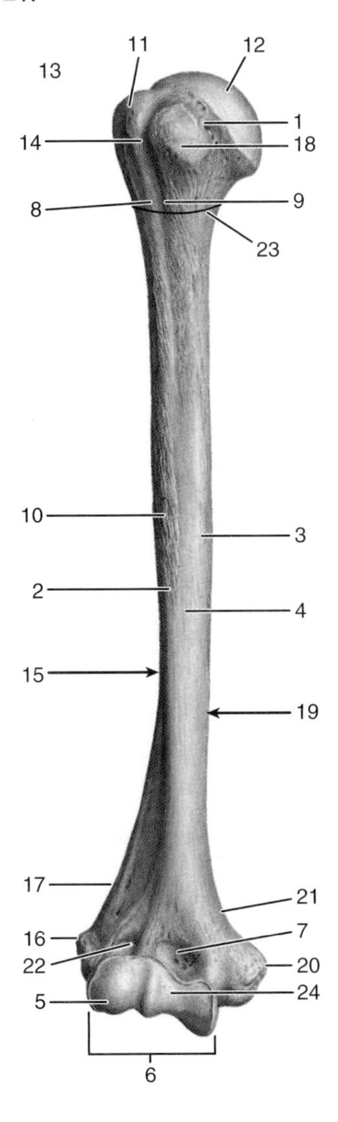

1 **groove for ulnar nerve** a shallow vertical groove on the posterior surface of the medial epicondyle of the humerus for the ulnar nerve. <sulcus nervi ulnaris> A

2 **olecranon fossa** a depression on the posterior surface of the humerus, above the trochlea, for lodging the olecranon of the ulna when the elbow is extended. <fossa olecrani> A

3 **posterior surface of humerus** the surface of the humerus that is subdivided obliquely by the radial groove to give attachment to the lateral and medial heads of the triceps muscle. <facies posterior humeri> A

4 **radial groove** a broad oblique groove on the posterior surface of the humerus for the radial nerve and the profunda brachii artery. <sulcus nervi radialis> A

5 **acromial extremity of clavicle** the lateral end of the clavicle, which articulates with the acromion of the scapula. <extremitas acromialis claviculae> B

6 **acromial facet of clavicle** the smooth area on the lateral end of the clavicle for articulation with the acromion of the scapula. <facies articularis acromialis claviculae> B

7 **body of clavicle** the long curved central part of the clavicle, extending between the acromial and sternal extremities. <corpus claviculae> B

8 **clavicle** a bone, curved like the letter *f*, that articulates with the sternum and scapula, forming the anterior portion of the pectoral girdle on either side; called also *collar bone*. <clavicula> B

9 **conoid tubercle** a prominent elevation on the inferior aspect of the lateral part of the clavicle, to which the conoid part of the coracoclavicular ligament is attached. <tuberculum conoideum> B

10 **groove for subclavius** a groove on the inferior surface of the clavicle into which the subclavius muscle is inserted by muscle fibers. <sulcus musculi subclavii> B

11 **sternal extremity of clavicle** the medial end of the clavicle, which articulates with the sternum. <extremitas sternalis claviculae> B

12 **sternal facet of clavicle** a triangular surface on the medial end of the clavicle for articulation with the sternum. <facies articularis sternalis claviculae> B

13 **trapezoid line** a ridge extending anterolaterally from the conoid tubercle on the inferior surface of the clavicle, giving attachment to the trapezoid portion of the coracoclavicular ligament. <linea trapezoidea> B

14 **tuberosity for coracoclavicular ligament** a protuberance on the inferior surface of the acromial extremity of the clavicle, giving attachment to the coracoclavicular ligament and including the surface for the acromial joint. <tuberositas ligamenti coracoclavicularis> B

15 **acromial angle** the easily palpable subcutaneous bony point where the lateral border of the acromion becomes continuous with the spine of the scapula. <angulus acromii> C

16 **acromion** the lateral extension of the spine of the scapula, projecting over the shoulder joint and forming the highest point of the shoulder. C

17 **inferior angle of scapula** the angle formed by the junction of the medial and lateral borders of the scapula. <angulus inferior scapulae> C

18 **infraspinous fossa** the large, slightly concave area below the spinous process on the dorsal surface of the scapula; it is the site of origin of the infraspinatus muscle. <fossa infraspinata> C

19 **posterior surface of scapula** the convex surface, which is divided into two unequal parts by the spine of the scapula. <facies posterior scapulae> C

20 **scapula** the flat, triangular bone in the back of the shoulder; the shoulder blade. C

21 **spine of scapula** a triangular plate of bone attached by one edge to the back of the scapula, its tip being at the vertebral border of the scapula; it passes laterally toward the shoulder joint and at its base bears the acromion. <spina scapulae> C

22 **suprascapular notch** a notch, converted into a foramen by a ligament, on the upper border of the scapula at the base of the coracoid process. <incisura scapulae> C

23 **supraspinous fossa** the deeply concave area above the spinous process on the dorsal surface of the scapula from which the supraspinous muscle takes origin. <fossa supraspinata> C

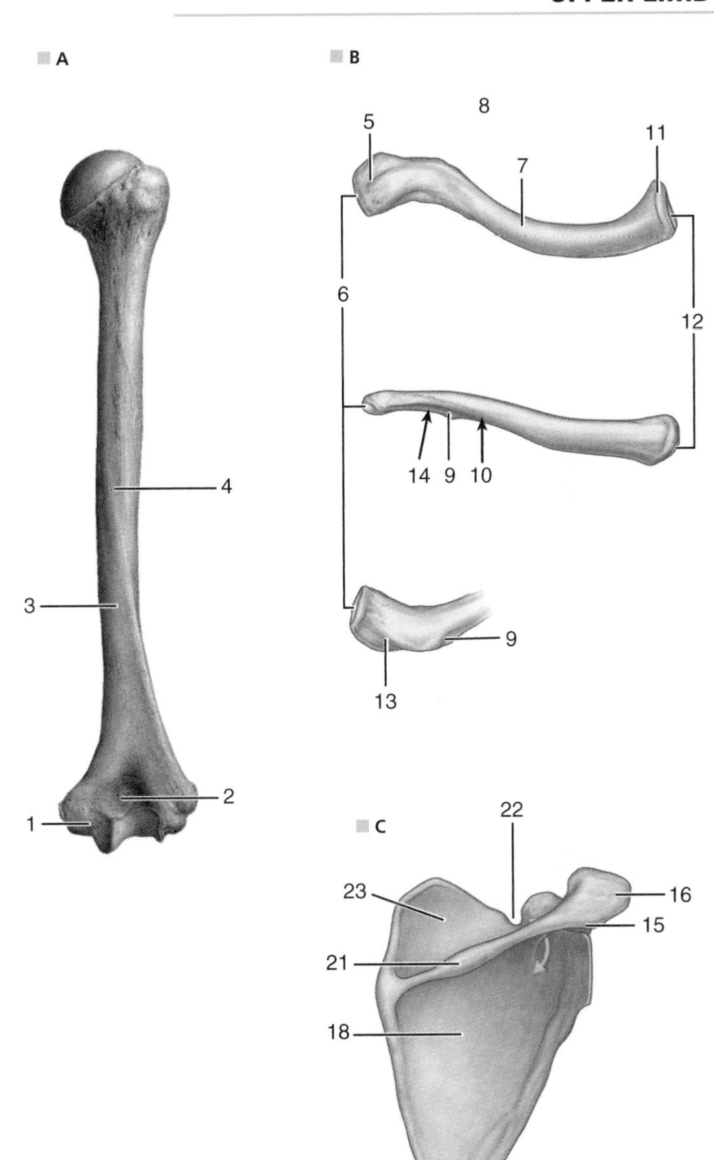
A
B
C
1
2
3
4
5
6
7
8
9
10
11
12
13
14
9
9
15
16
17
18
19, 20
21
22
23

1 **coracoid process of scapula** a strong curved process that arises from the upper part of the neck of the scapula and overhangs the shoulder joint. <processus coracoideus scapulae> A, B

2 **costal surface of scapula** the anteromedially facing, concave surface of the scapula. <facies costalis scapulae> B

3 **glenoid cavity** a depression in the lateral angle of the scapula for articulation with the humerus. <cavitas glenoidalis> A, B

4 **lateral angle of scapula** the head of the scapula, which bears the glenoid cavity and articulates with the head of the humerus. <angulus lateralis scapulae> A

5 **lateral border of scapula** the thick edge of the scapula, extending from the inferior margin of the glenoid cavity to the inferior angle. <margo lateralis scapulae> A

6 **medial border of scapula** the thin edge of the scapula extending from the superior to the inferior angle. <margo medialis scapulae> A, B

7 **neck of scapula** the somewhat constricted part of the scapula that surrounds the lateral angle. <collum scapulae> A

8 **subscapular fossa** the shallow concavity covering most of the costal surface of the body of the scapula. <fossa subscapularis> A

9 **superior angle of scapula** the angle made by the superior and medial borders of the scapula. <angulus superior scapulae> A

10 **superior border of scapula** the thin, short edge of the scapula, extending from the superior angle to the coracoid process. <margo superior scapulae> A

11 **supraglenoid tubercle** a raised roughened area, just superior to the glenoid cavity of the scapula, that gives attachment to the long head of the biceps brachii muscle. <tuberculum supraglenoidale> B

12 **anterior border of radius** the edge of the radius that runs obliquely between the radial tuberosity and the styloid process. <margo anterior radii> C

13 **anterior surface of radius** the surface of the radius that gives attachment to the flexor pollicis longus and pronator quadratus muscles. <facies anterior radii> C

14 **articular circumference of head of radius** the rounded surface of the head or capitulum of the radius, which articulates with the radial notch of the ulna. <circumferentia articularis capitis radii> C

15 **articular facet of head of radius** a depression on the proximal surface of the head of the radius for articulation with the capitulum of the humerus. <fovea articularis capitis radii> C

16 **body of radius** the long central part of the radius. <corpus radii> C

17 **carpal articular surface of radius** the convex surface of the distal end of the radius, which articulates with the lunate and scaphoid bones. <facies articularis carpalis radii> C

18 **dorsal tubercle of radius** an easily palpable prominence on the distal dorsal aspect of the radius; it is grooved by the tendon of the extensor pollicis longus muscle. <tuberculum dorsale radii> C

19 **head of radius** the disk on the proximal end of the radius that articulates with the capitulum of the humerus and the radial notch of the ulna. <caput radii> C

20 **interosseous border of radius** the prominent medial border of the radius, connected with a similar ridge on the ulna by a strong, wide fibrous sheet, the interosseous membrane. <margo interosseus radii> C

21 **lateral surface of radius** the surface of the radius that gives attachment to the supinator and pronator teres muscles proximally, and underlies the tendons of the extensor carpi radialis longus and brevis muscles distally. <facies lateralis radii> C

22 **neck of radius** the somewhat constricted portion of the radius just distal to the head. <collum radii> C

23 **posterior border of radius** the edge of the radius that extends from the posterior part of the radial tuberosity to the middle tubercle. <margo posterior radii> C

24 **posterior surface of radius** a surface giving attachment to the supinator, abductor pollicis longus, and extensor pollicis brevis muscles. <facies posterior radii> C

25 **pronator tuberosity** the apex of the lateral curve of the radius, where there is a roughened ridge for the insertion of the pronator teres muscle. <tuberositas pronatoria> C

26 **radial tuberosity** the tuberosity on the anterior inner surface of the neck of the radius, for the insertion of the tendon of the biceps muscle. <tuberositas radii> C

27 **radius** the bone on the outer or thumb side of the forearm, articulating proximally with the humerus and ulna and distally with the ulna and carpus. C

28 **styloid process of radius** a blunt projection from the lateral surface of the distal end of the radius. <processus styloideus radii> C

29 **ulnar notch of radius** a concavity on the medial side of the distal extremity of the radius, articulating with the head of the ulna. <incisura ulnaris radii> C

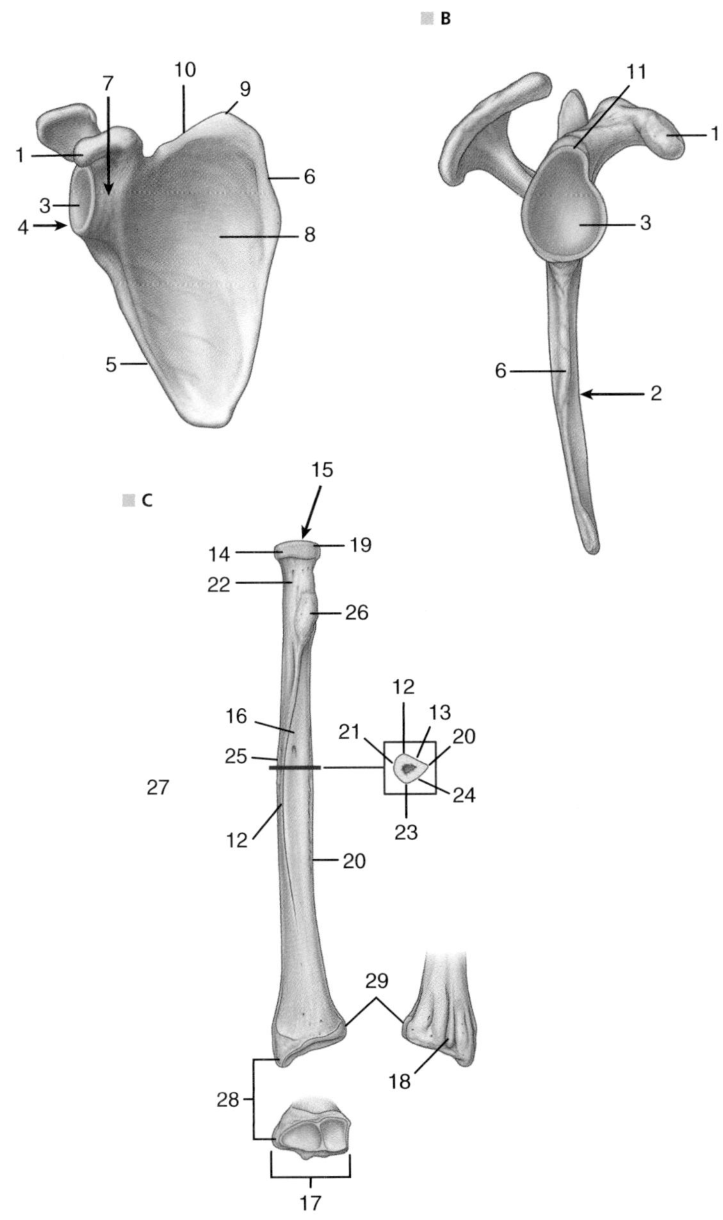
A
B
C
1
2
3
4
5
6
7
8
9
10
11
12
13
14
15
16
17
18
19
20
21
22
23
24
25
26
27
28
29

1 **anterior border of ulna** the smooth, rounded border of the ulna separating the medial and posterior surfaces. <margo anterior ulnae> A

2 **anterior surface of ulna** <facies anterior ulnae> A

3 **articular circumference of head of ulna** the semilunar surface of the head of the ulna, which articulates with the ulnar notch of the radius. <circumferentia articularis capitis ulnae> A

4 **body of ulna** the long central part of the ulna. <corpus ulnae> A

5 **coronoid process of ulna** a wide eminence at the proximal end of the ulna, forming the anterior and inferior part of the trochlear notch. <processus coronoideus ulnae> A

6 **head of ulna** the articular surface of the distal extremity of the ulna. <caput ulnae> A

7 **interosseous border of ulna** the prominent lateral border of the ulna, connected with a similar ridge on the radius by the interosseous membrane. <margo interosseus ulnae> A

8 **medial surface of ulna** the smooth, rounded, internal surface of the ulna. <facies medialis ulnae> A

9 **olecranon** the proximal bony projection of the ulna at the elbow, its anterior surface forming part of the trochlear notch. A

10 **posterior border of ulna** the sharp border of the ulna separating the posterior and medial surfaces. <margo posterior ulnae> A

11 **posterior surface of ulna** the posterolaterally directed surface of the ulna. <facies posterior ulnae> A

12 **radial notch of ulna** the cavity on the outer side of the coronoid process, articulating with the rim of the head of the radius. <incisura radialis ulnae> A

13 **styloid process of ulna** the medial, non-articular process on the distal extremity of the ulna. <processus styloideus ulnae> A

14 **supinator crest** a strong ridge forming the posterior margin of the supinator fossa below the radial notch of the ulna, and with it giving attachment to the supinator muscle. <crista musculi supinatoris> A

15 **trochlear notch of ulna** a large concavity on the anterior surface at the proximal end of the ulna, formed by the olecranon and coronoid processes, for articulation with the trochlea of the humerus. <incisura trochlearis ulnae> A

16 **tuberosity of ulna** a large roughened area on the palmar surface of the ulna, located just distal to the coronoid process, and giving attachment to the brachialis muscle. <tuberositas ulnae> A

17 **ulna** the inner and larger bone of the forearm, on the side opposite that of the thumb; it articulates with the humerus and with the head of the radius at its proximal end and with the radius and bones of the carpus at the distal end. A

18 **base of metacarpal bone** the proximal end of each metacarpal, which articulates with one or more carpals and with adjacent metacarpals. <basis ossis metacarpi> B

19 **base of phalanx of fingers** the proximal end of each phalanx of the fingers. <basis phalangis digitorum manus> B

20 **body of metacarpal** the long central part of a metacarpal bone. <corpus ossis metacarpi> B

21 **body of phalanx of hand** the long central part of a phalanx of the hand. <corpus phalangis manus> B

22 **capitate bone** the bone in the distal row of carpal bones lying between the trapezoid and hamate bones. <os capitatum> B

23 **carpal bones** the eight bones of the wrist (carpus). <ossa carpi> B

24 ***carpal groove*** a broad deep groove on the volar surface of the carpal bones, which transmits the flexor tendons and the median nerve into the palm of the hand. <sulcus carpi> B

25 ***central bone*** an accessory bone sometimes found on the back of the wrist. <os centrale> B

26 **distal phalanx of fingers** any one of the five terminal bones of the fingers, articulating, except in the thumb, with the middle phalanx. <phalanx distalis digitorum manus> B

27 **distal tuberosity of fingers** a roughened, raised bony mass on the palmar surface of the tip of a distal phalanx of the hand. <tuberositas phalangis distalis manus> B

28 **hamate bone** the medial bone in the distal row of carpal bones. <os hamatum> B

29 **head of metacarpal bone** the distal extremity of a metacarpal bone, which articulates with the base of a proximal digit. <caput ossis metacarpi> B

30 **head of phalanx of hand** the distal articular surface of each of the proximal and middle phalanges of the fingers. <caput phalangis manus> B

31 **hook of hamate bone** a hooklike process on the palmar surface of the hamate bone, to which numerous structures are attached. <hamulus ossis hamati> B

32 **lunate bone** the bone in the proximal row of carpal bones lying between the scaphoid and triquetral bones. <os lunatum> B

33 **metacarpal bones** the five cylindrical bones of the hand (metacarpals), which articulate proximally with the bones of the carpus and distally with the proximal phalanges of the fingers; numbered from that articulating with the proximal phalanx of the thumb to the most lateral one articulating with the proximal phalanx of the little finger. <ossa metacarpi> B

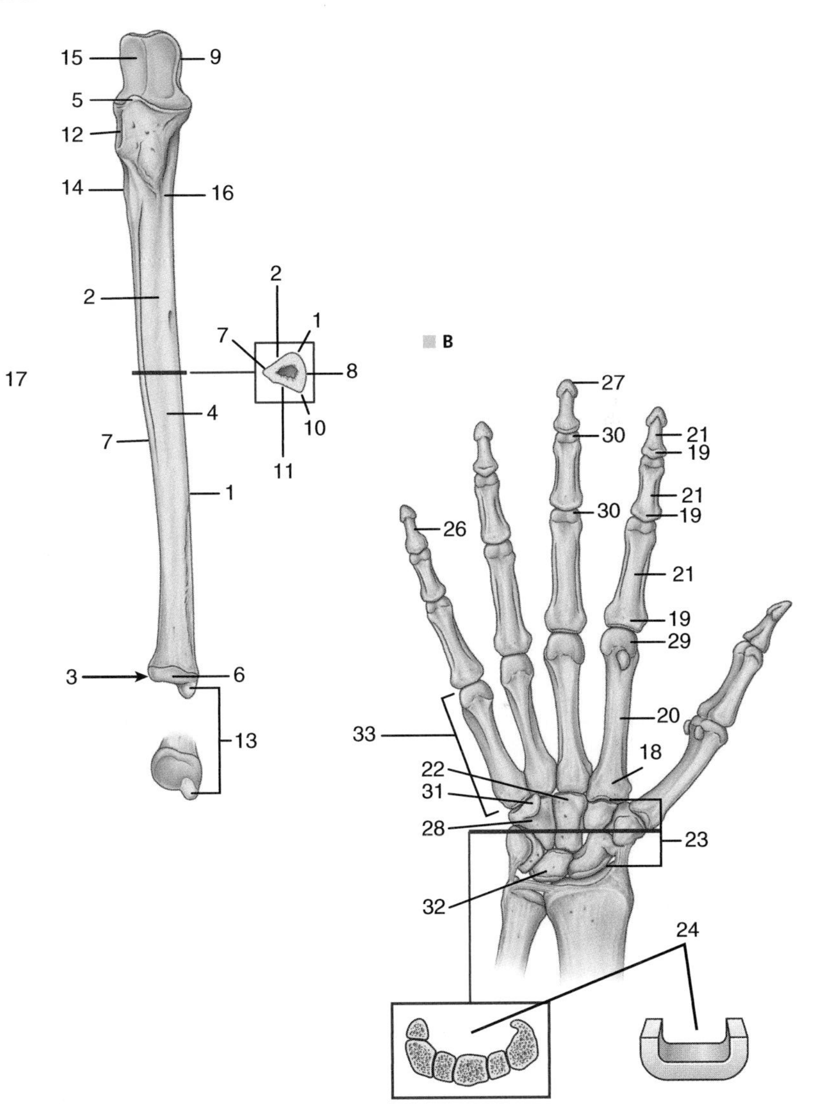
A
15
9
5
12
14
16
2
2
1
7
8
17
4
10
7
11
1
3
6
13
B
27
30
21
19
21
30
19
26
21
19
29
20
33
18
22
31
28
23
32
24

1 **middle phalanx of fingers** any one of the four bones of the fingers (excluding the thumb) situated between the proximal and distal phalanges. also *p. secunda digitorum manus*. <phalanx media digitorum manus> A

2 **phalanges of hand** the 14 bones that compose the skeleton of the fingers—two for the thumb and three for each finger. <ossa digitorum manus> A

3 **pisiform bone** the medial bone of the proximal row of carpal bones. <os pisiforme> A

4 **proximal phalanx of fingers** any one of the five bones of the fingers that articulate with the metacarpal bones and, except in the thumb, with the middle phalanx. <phalanx proximalis digitorum manus> A

5 **scaphoid bone** the most lateral bone of the proximal row of carpal bones. <os scaphoideum> A

6 **sesamoid bones of hand** sesamoid bones usually located in the palmar region in the tendons of the flexor pollicis brevis and adductor pollicis muscles. <ossa sesamoidea manus> A

7 ***styloid process of third metacarpal bone*** a prominent process projecting proximally from the base of the third metacarpal bone. <processus styloideus ossis metacarpi tertii> A

8 **third metacarpal** the middle metacarpal bone, which presents the styloid process on the dorsal surface of the radial side of its base. <os metacarpi tertium> A

9 **trapezium** the most lateral bone of the distal row of carpal bones. <os trapezium> A

10 **trapezoid** the bone in the distal row of carpal bones lying between the trapezium and capitate bones. <os trapezoideum> A

11 **triquetral bone** the bone in the proximal row of carpal bones lying between the lunate and pisiform bones. <os triquetrum> A

12 **trochlea of phalanx of hand** the pulleylike concavity of the head of a proximal or middle phalanx, to which the base of the articulating phalanx is adapted. <trochlea phalangis manus> A

13 **tubercle of scaphoid bone** a projection on the lateral palmar surface of the scaphoid bone of the wrist, giving attachment to the transverse carpal ligament. <tuberculum ossis scaphoidei> A

14 **tubercle of trapezium** a prominent ridge on the palmar surface of the trapezium bone, forming the lateral margin of the groove that transmits the tendon of the flexor carpi radialis muscle. <tuberculum ossis trapezii> A

JOINTS

15 **coracoacromial ligament** one of three intrinsic ligaments of the scapula, a strong broad triangular band that is attached by its base to the lateral border of the coracoid process and by its tip to the summit of the acromion just in front of the articular facet for the clavicle. <ligamentum coracoacromiale> B

16 **costoclavicular ligament** a short, powerful ligament that extends from the superior margin of the first costal cartilage to the inferior surface at the sternal end of the clavicle. <ligamentum costoclaviculare> C

17 ***impression of costoclavicular ligament*** the point on the inferior surface of the clavicle where the costoclavicular ligament is attached. <impressio ligamenti costoclavicularis> C

18 **interclavicular ligament** a flattened band that passes from the superior surface of the sternal end of one clavicle across the superior margin of the sternum to the same position on the other clavicle. <ligamentum interclaviculare> C

19 ***posterior sternoclavicular ligament*** a thick reinforcing band on the posterior portion of the articular capsule of the sternoclavicular joint. It is attached superiorly to the posterior and superior parts of the sternal extremity of the clavicle and inferiorly to the posterior surface of the manubrium of the sternum. <ligamentum sternoclaviculare posterius> C

20 **sternoclavicular joint** the joint formed by the sternal extremity of the clavicle, the clavicular notch of the manubrium of the sternum, and the first costal cartilage. <articulatio sternoclavicularis> C

21 **acromioclavicular joint** the joint formed by the acromion of the scapula and the acromial extremity of the clavicle. <articulatio acromioclavicularis> D

22 **acromioclavicular ligament** a dense band that joins the superior surface of the acromion and the acromial extremity of the clavicle together, and strengthens the superior part of the articular capsule. <ligamentum acromioclaviculare> D

23 ***articular disk of acromioclavicular joint*** a pad of fibrocartilage, sometimes present, commonly imperfect, within the articular cavity of the acromioclavicular joint. <discus articularis articulationis acromioclavicularis> D

24 **conoid ligament** the conical, posteromedial portion of the coracoclavicular ligament, attached inferiorly by its tip to the base of the coracoid process of the scapula and superiorly by its base to the inferior surface of the clavicle. <ligamentum conoideum> D

25 **coracoclavicular ligament** a strong band that joins the coracoid process of the scapula and the acromial extremity of the clavicle; it is divided into two parts, the trapezoid and conoid ligaments. <ligamentum coracoclaviculare> D

26 **trapezoid ligament** a broad, flat band forming the anterolateral portion of the coracoclavicular ligament; it is attached inferiorly to the superior surface of the coracoid process of the scapula and superiorly to the oblique ridge on the inferior surface of the clavicle. <ligamentum trapezoideum> D

A

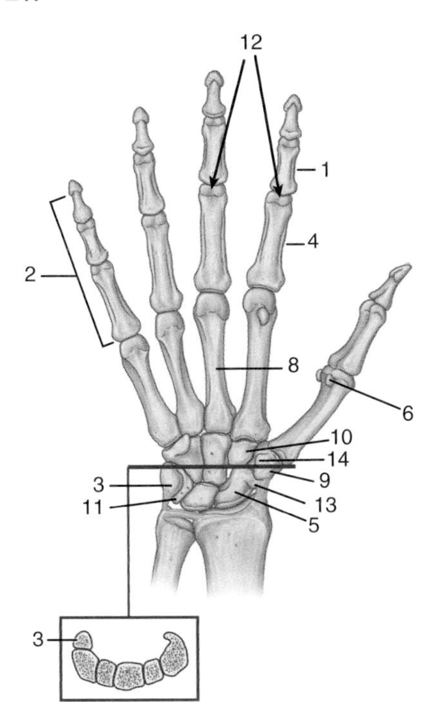

B

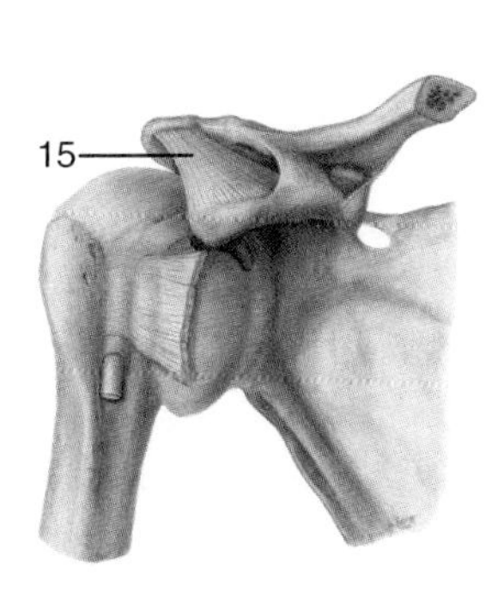

D

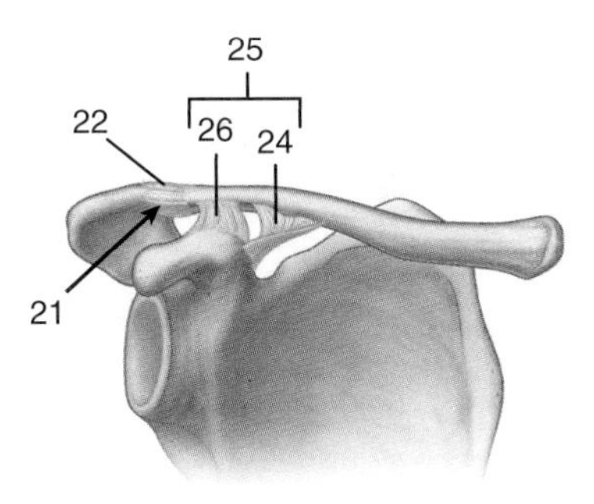

C

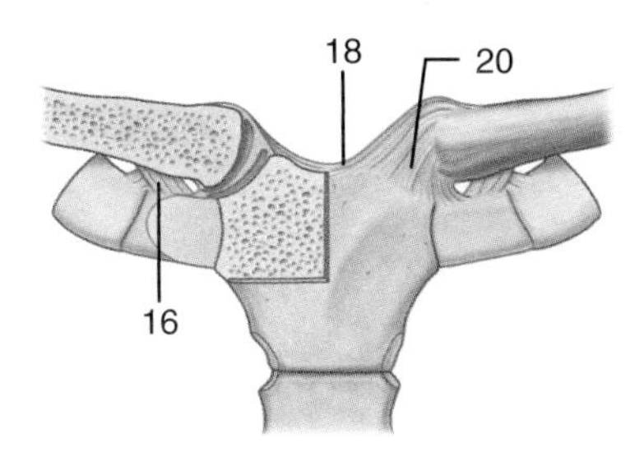

1 **coracohumeral ligament** a broad band that arises from the lateral border of the coracoid process of the scapula and passes downward and laterally to be attached to the greater tubercle of the humerus. <ligamentum coracohumerale> A

2 **glenohumeral ligaments** bands, usually three, on the inner surface of the articular capsule of the humerus, attached to the margin of the glenoid cavity and to the anatomical neck of the humerus. <ligamenta glenohumeralia> A

3 ***inferior transverse scapular ligament*** one of three intrinsic ligaments of the scapula, composed of more or less distinct fascial fibers that pass from the lateral border of the spine of the scapula to the adjacent margin of the glenoid cavity, thus converting the notch at the base of the spine into a foramen for the passage of the suprascapular vessels and nerves to the infraspinous fossa. <ligamentum transversum scapulae inferius> A

4 **shoulder joint** the joint formed by the head of the humerus and the glenoid cavity of the scapula; called also *glenohumeral joint*. <articulatio humeri> A

5 **glenoid labrum** a ring of fibrocartilage attached to the rim of the glenoid cavity of the scapula, increasing the depth of the cavity. <labrum glenoidale> B

6 **subacromial bursa** one between the acromion and the insertion of the supraspinatus muscle, extending between the deltoid and the greater tubercle of the humerus. <bursa subacromialis> B

7 ***subcutaneous acromial bursa*** a bursa between the acromion and the overlying skin. <bursa subcutanea acromialis> B

8 **subdeltoid bursa** a bursa between the deltoid and the shoulder joint capsule, usually connected to the subacromial bursa. <bursa subdeltoidea> B

9 **subtendinous bursa of infraspinatus muscle** a bursa between the tendon of the infraspinatus and the joint capsule or the greater tubercle. <bursa subtendinea musculi infraspinati> B

10 ***subtendinous bursa of latissimus dorsi muscle*** a bursa between the tendons of the latissimus dorsi and teres major muscles. <bursa subtendinea musculi latissimi dorsi> B

11 **subtendinous bursa of subscapularis muscle** a bursa between the tendon of the subscapularis muscle and the fibrous membrane of the capsule of the shoulder joint. <bursa subtendinea musculi subscapularis> B

12 ***subtendinous bursa of teres major muscle*** a bursa deep to the tendon of insertion of the teres major muscle. <bursa subtendinea musculi teretis majoris> B

13 ***subtendinous bursa of trapezius muscle*** a bursa between the trapezius and the medial end of the spine of the scapula. <bursa subtendinea musculi trapezii> B

14 **humeroradial joint** the joint between the humerus and the radius. <articulatio humeroradialis> C

15 **humeroulnar joint** the joint between the humerus and the ulna. <articulatio humeroulnaris> C

16 **annular ligament of radius** a strong fibrous band that encircles the head of the radius and holds it in position; it is attached to the anterior and posterior margins of the radial notch of the ulna, forming, with the notch, a complete ring. <ligamentum anulare radii> D

17 **elbow joint** the joint between the arm and forearm, comprising the humeroulnar, humeroradial, and proximal radioulnar joints; called also *cubital joint*. <articulatio cubiti> C

18 ***interosseous cubital bursa*** a bursa between the ulna, the biceps tendon, and nearby muscles. <bursa cubitalis interossea> D

19 ***intratendinous bursa of olecranon*** a bursa within the triceps tendon near its insertion. <bursa intratendinea olecrani> D

20 ***quadrate ligament*** a fibrous bundle connecting the distal margin of the radial notch of the ulna to the neck of the radius. <ligamentum quadratum> D

21 **radial collateral ligament** a large bundle of fibers arising from the lateral epicondyle of the humerus and fanning out to be attached to the lateral side of the annular ligament of the radius. <ligamentum collaterale radiale> D

22 **sacciform recess of articulation of elbow** the distal bulging of the articular capsule of the elbow joint, situated between the radial notch of ulna and the articular circumference of head of radius. <recessus sacciformis articulationis cubiti> D

23 ***subcutaneous bursa of olecranon*** a bursa between the olecranon process and the skin. <bursa subcutanea olecrani> D

24 ***subtendinous bursa of triceps brachii muscle*** an inconstant sac between the triceps tendon, the olecranon, and the dorsal ligament of the elbow. <bursa subtendinea musculi tricipitis brachii> D

25 **ulnar collateral ligament** a triangular bundle of fibers attached proximally to the medial epicondyle of the humerus, distally to the coronoid process of the ulna and the medial surface of the olecranon, and to a ridge running between the two. <ligamentum collaterale ulnare> D

A

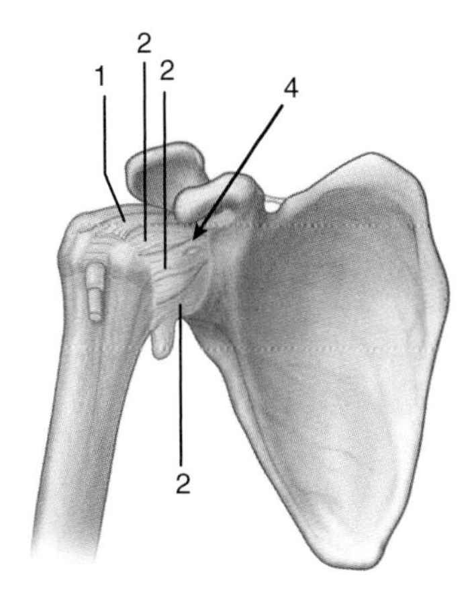

B

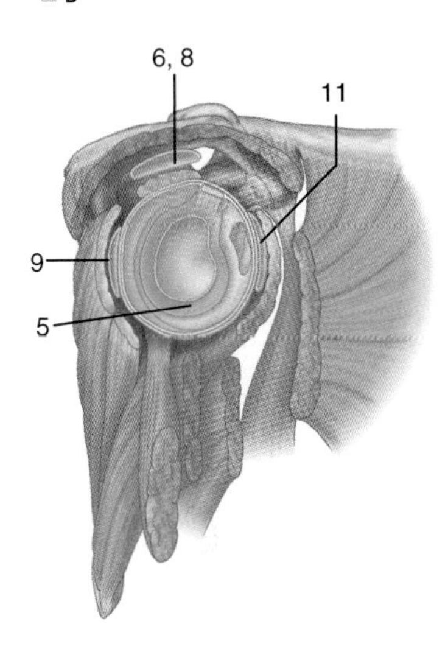

C

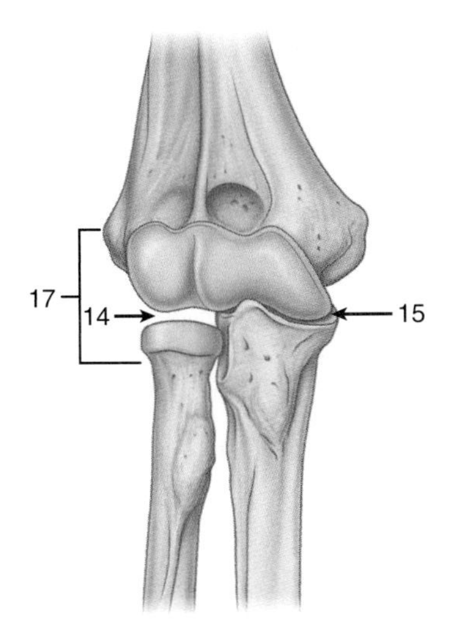

D

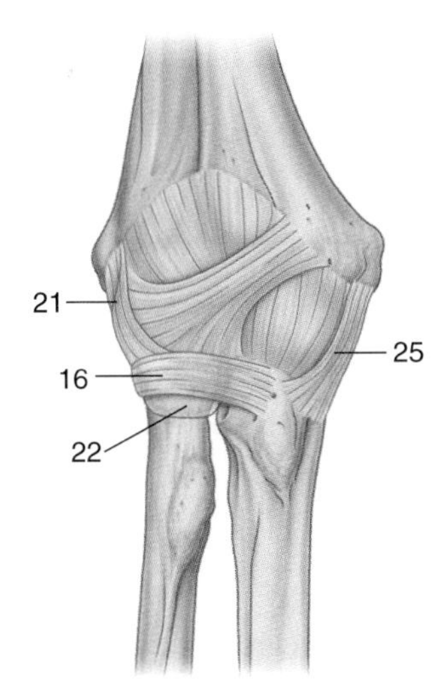

1 **distal radioulnar joint** the joint formed by the head of the ulna and the ulnar notch of the radius. <articulatio radioulnaris distalis> A

2 **interosseous membrane of forearm** a thin fibrous sheet that connects the bodies of the radius and ulna, passing from the interosseous margin of the radius to that of the ulna. <membrana interossea antebrachii> A

3 **oblique cord of elbow joint** a small ligamentous band extending from the lateral face of the tuberosity of the ulna to the radius a little distal to its tuberosity. <chorda obliqua membranae interosseae antebrachii> A

4 **proximal radioulnar joint** the proximal of the two joints between the radius and the ulna; it is involved in pronation and supination of the forearm. <articulatio radioulnaris proximalis> A

5 **radiocarpal joint** a condylar joint formed by the radius and the articular disk with the scaphoid, lunate, and triquetral bones; called also *brachiocarpal articulation, radiocarpal joint, wrist,* and *wrist joint.* <articulatio radiocarpalis> A

6 **radioulnar syndesmosis** the fibrous union of the radius and ulna, which consists of the interosseous membrane of the forearm and the oblique cord of the elbow; called also *articulatio radioulnaris* and *radioulnar articulation.* <syndesmosis radioulnaris> A

7 ***sacciform recess of distal radioulnar articulation*** a bulging of the synovial membrane of the articular capsule of the distal radioulnar joint, which extends proximally between the radius and ulna beyond the point of their articular surfaces. <recessus sacciformis articulationis radioulnaris distalis> A

8 **collateral ligaments of metacarpophalangeal joints** massive, strong fibrous bands on either side of each metacarpophalangeal joint, holding the two bones involved in each joint firmly together. <ligamenta collateralia articulationum metacarpophalangealium> B

9 **deep transverse metacarpal ligament** a narrow fibrous band that extends across and is attached to the palmar surfaces of the heads of the second to fifth metacarpal bones, joining them together. <ligamentum metacarpale transversum profundum> B

10 **interphalangeal joints of hand** the hinge joints between the phalangeal joints of the hand. <articulationes interphalangeae manus> B

11 **metacarpophalangeal joints** joints formed between the heads of the five metacarpal bones and the bases of the corresponding proximal phalanges. <articulationes metacarpophalangeae> B

12 **pisiform joint** the carpal joint formed by the pisiform and triquetral bones. <articulatio ossis pisiformis> B

13 **superficial transverse metacarpal ligament** transverse fibers occupying the intervals between the diverging longitudinal bands of the palmar aponeurosis. <ligamentum metacarpale transversum superficiale> C

14 **palmar carpometacarpal ligaments** a series of bands on the palmar surface of the carpometacarpal joints, joining the carpal bones to the second to fifth metacarpals. The second metacarpal bone is thus joined to the trapezium, the third to the trapezium, capitate, and hamate, the fourth to the hamate, and the fifth to the hamate. <ligamenta carpometacarpalia palmaria> D

15 ***palmar intercarpal ligaments*** several bands that extend transversely across the palmar surfaces of the carpal bones, connecting various ones together. <ligamenta intercarpalia palmaria> D

16 **palmar ligaments of metacarpophalangeal joints** thick dense fibrocartilaginous plates on the palmar surfaces of the metacarpophalangeal joints, between the collateral ligaments. Called also *palmar metacarpophalangeal ligaments.* <ligamenta palmaria articulationum metacarpophalangealium> D

17 **palmar metacarpal ligaments** bands that interconnect the bases of the second to fifth metacarpal bones by passing transversely from bone to bone on their palmar surfaces. <ligamenta metacarpalia palmaria> D

18 **palmar radiocarpal ligament** several bundles of fibers that pass obliquely from the styloid process and the distal anterior margin of the radius to the lunate, triquetral, capitate, and hamate bones. <ligamentum radiocarpale palmare> D

19 **palmar ulnocarpal ligament** several bundles of fibers that pass from the styloid process of the ulna to the carpal bones. <ligamentum ulnocarpale palmare> D

20 **pisohamate ligament** a fibrous band extending from the pisiform bone to the hook of the hamate bone. <ligamentum pisohamatum> D

21 **pisometacarpal ligament** a fibrous band extending from the pisiform bone to the bases of the fifth, usually the fourth, and sometimes the third metacarpal bones. <ligamentum pisometacarpeum> D

22 **radiate carpal ligament** a group of about seven fibrous bands which diverge in all directions on the palmar surface of the mediocarpal joint; the majority radiate from the capitate to the scaphoid, lunate, and triquetral bones. <ligamentum carpi radiatum> D

A

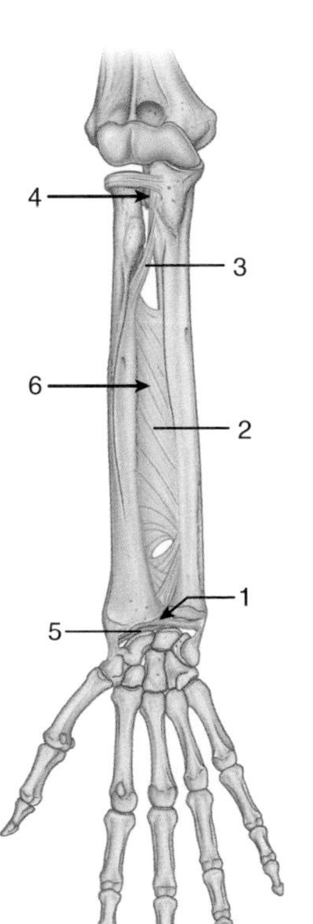

B

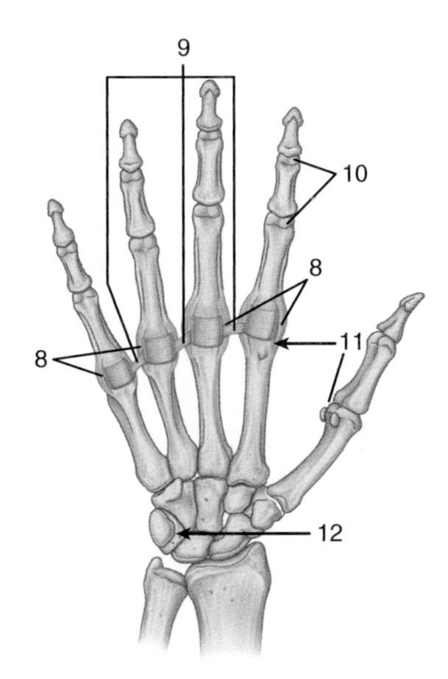

C

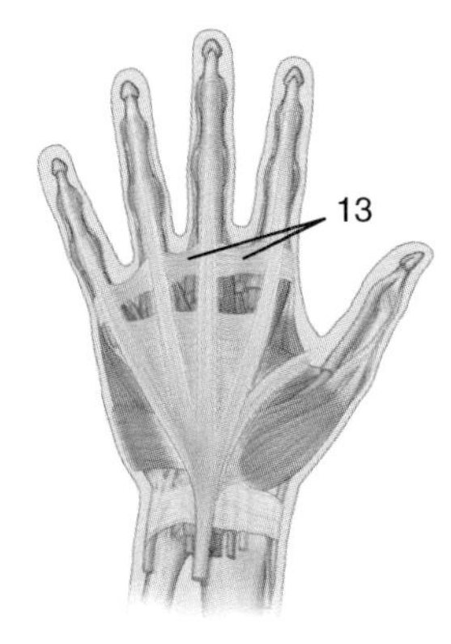

D

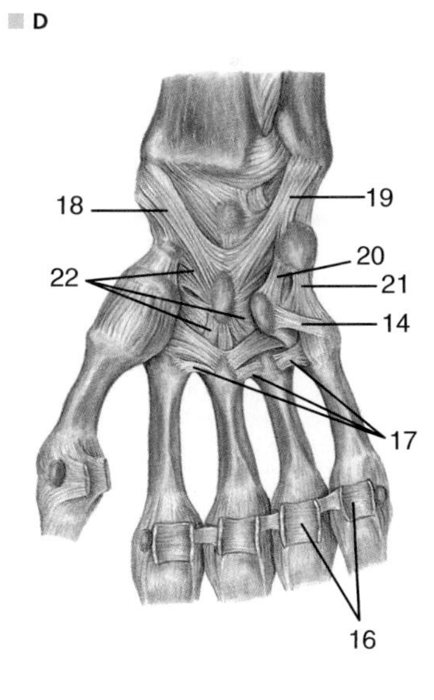

1 **dorsal carpometacarpal ligaments** a series of bands on the dorsal surface of the carpometacarpal joints, joining the carpal bones to the bases of the second to fifth metacarpals. The second metacarpal bone is thus joined to the trapezium, trapezoid, and capitate, the third to the capitate, the fourth to the capitate and hamate, and the fifth to the hamate. <ligamenta carpometacarpalia dorsalia> A

2 **dorsal intercarpal ligaments** several bands that extend transversely across the dorsal surfaces of the carpal bones, connecting various ones together. <ligamenta intercarpalia dorsalia> A

3 **dorsal metacarpal ligaments** bands that interconnect the bases of the second to fifth metacarpal bones by passing transversely from bone to bone on their dorsal surfaces. <ligamenta metacarpalia dorsalia> A

4 **dorsal radiocarpal ligament** a fibrous band that passes obliquely from the posterior border of the distal extremity of the radius to the dorsal surfaces of the proximal row of carpal bones, especially the triquetral and lunate, and to the dorsal intercarpal ligaments. <ligamentum radiocarpale dorsale> A

5 **articular disk of distal radioulnar joint** a triangular pad of fibrocartilage, attached at its base to the radius and at its apex to the base of the styloid process of the ulna; it usually separates the articular cavity of the distal radioulnar joint from that of the radiocarpal joint. <discus articularis articulationis radioulnaris distalis> B

6 **carpal joints** any of the joints that connect the carpal bones together, comprising: the joints between each row, distal and proximal; the joint between the distal and proximal rows; and the joint formed by the pisiform and triquetral bones. <articulationes carpi> B

7 **carpometacarpal joint of thumb** the joint formed by the first metacarpal bone and the trapezium. <articulatio carpometacarpalis pollicis> B

8 **carpometacarpal joints** joints formed by the trapezium, trapezoid, capitate, and hamate bones together with the bases of the four medial metacarpal bones. <articulationes carpometacarpales> B

9 **intercarpal joints** the joints between the bones, within each row, distal and proximal, of carpal bones. <articulationes intercarpales> B

10 **intermetacarpal joints** the joints formed between the adjoining bases of the second, third, fourth, and fifth metacarpal bones; called also *articulations of metacarpal bones.* <articulationes intermetacarpales> B

11 **interosseous intercarpal ligaments** short fibrous bands that join the adjacent surfaces of the various carpal bones. <ligamenta intercarpalia interossea> B

12 **interosseous metacarpal ligaments** short, strong fibrous bands situated between the adjacent surfaces of the bases of the second to fifth metacarpal bones, just distal to the articular surfaces. <ligamenta metacarpalia interossea> B

13 **midcarpal joint** the joint between the two rows, distal and proximal, of carpal bones. <articulatio mediocarpalis> B

14 **radial collateral ligament of wrist joint** a short, thick band that passes from the tip of the styloid process of the radius to attach to the scaphoid bone. <ligamentum collaterale carpi radiale> B

15 **ulnar collateral ligament of wrist joint** a strong fibrous band that passes from the tip of the styloid process of the ulna and is attached to the triquetral and pisiform bones. <ligamentum collaterale carpi ulnare> B

16 **collateral ligaments of interphalangeal joints of hand** massive fibrous bands on each side of the interphalangeal joints of the fingers; they are placed diagonally, the proximal ends being near the dorsal, and the distal ends near the palmar margins of the digits. <ligamenta collateralia articulationum interphalangealium manus> C

17 **palmar ligaments of interphalangeal joints** thick, dense fibrocartilaginous plates on the palmar surfaces of the interphalangeal joints of the hand, between the collateral ligaments. <ligamenta palmaria articulationum interphalangealium manus> C

18 **anterior sternoclavicular ligament** a thick reinforcing band on the anterior portion of the articular capsule of the sternoclavicular joint. It is attached superiorly to the anterior and superior parts of the sternal extremity of the clavicle and inferiorly to the anterior surface of the manubrium of the sternum. <ligamentum sternoclaviculare anterius> D

19 **articular disk of sternoclavicular joint** a pad of fibrocartilage, the circumference of which is connected to the articular capsule of the sternoclavicular joint; it is attached superiorly to the clavicle and inferiorly to the first costal cartilage near its union with the sternum, and it divides the joint cavity into two parts. <discus articularis articulationis sternoclavicularis> D

A

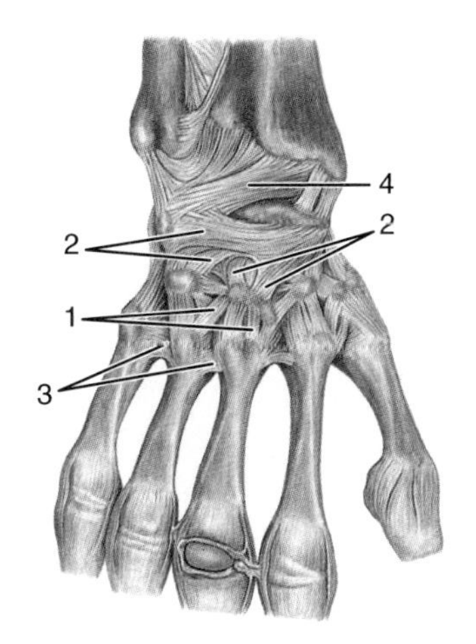

B

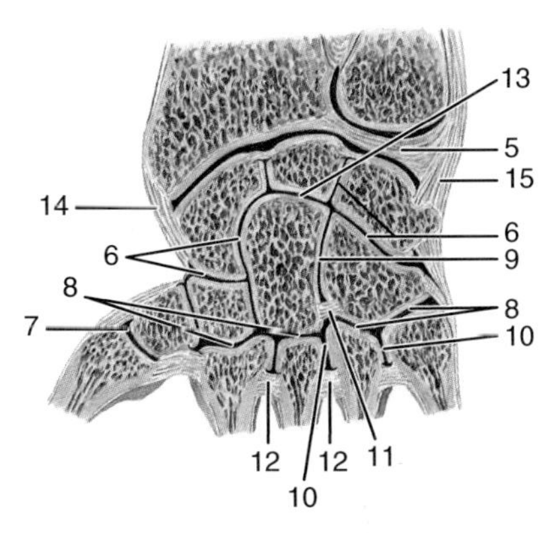

C

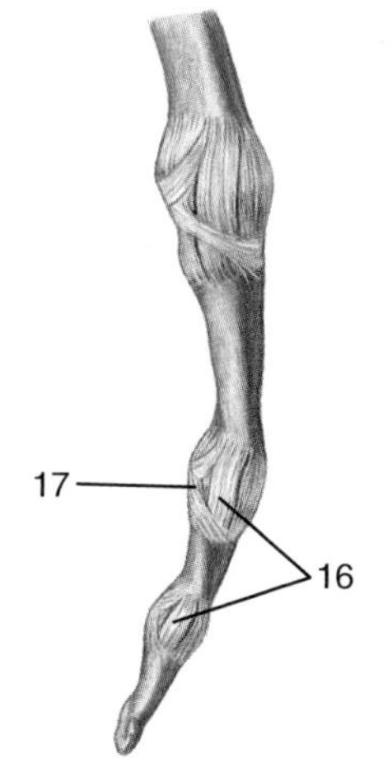

D

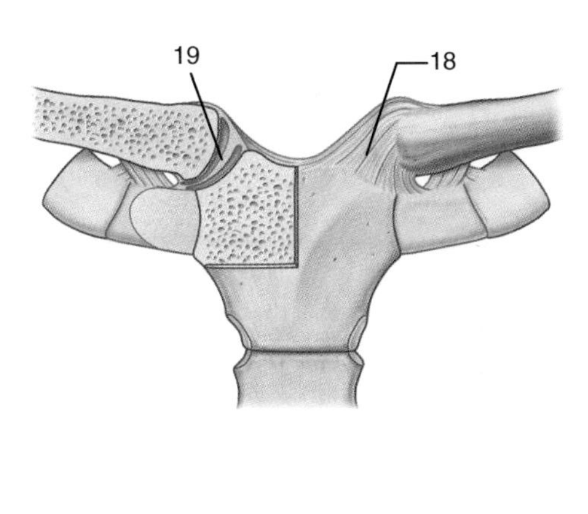

PECTORAL REGION

1 **clavipectoral fascia** a fascial sheet investing the subclavius muscle, attached to the clavicle above and continuing to the pectoralis minor muscle below. <fascia clavipectoralis> A

2 ***deltopectoral lymph nodes*** one or two lymph nodes in the groove between the pectoralis major and deltoid muscles, just inferior to the clavicle, which drain into the apical lymph nodes. Called also *infraclavicular nodes*. <nodi lymphoidei deltopectorales> A

3 **pectoral branches of thoracoacromial artery** branches that descend between the pectoralis major and minor muscles, supplying these muscles and the mammary gland. <rami pectorales arteriae thoracoacromialis> A

4 **serratus anterior muscle** *origin,* eight or nine upper ribs; *insertion,* medial border of scapula; *innervation,* long thoracic; *action,* draws scapula forward; rotates scapula to raise shoulder in abduction of arm. <musculus serratus anterior> A

5 **subclavius muscle** *origin,* first rib and its cartilage; *insertion,* lower surface of clavicle; *innervation,* fifth and sixth cervical; *action,* depresses lateral end of clavicle. <musculus subclavius> A, B

6 ***acromial anastomosis*** a network formed by ramification of the acromial branch of the thoracoacromial artery on the acromion process. <rete acromiale> A

7 **acromial branch of thoracoacromial artery** a branch distributed to the deltoid muscle and acromion process. <ramus acromialis arteriae thoracoacromialis> B

8 **clavicular branch of thoracoacromial artery** a vessel that passes medially to supply the subclavius muscle. <ramus clavicularis arteriae thoracoacromialis> B

9 **deltoid branch of thoracoacromial artery** a branch of the thoracoacromial artery descending with the cephalic vein and helping to supply the deltoid and pectoralis major muscles. <ramus deltoideus arteriae thoracoacromialis> B

10 **thoracoacromial artery** *origin,* axillary artery; *branches,* clavicular, pectoral, deltoid, acromial rami; *distribution,* deltoid, clavicular, and thoracic regions. Called also *acromiothoracic artery, thoracicoacromial artery,* and *thoracic axis.* <arteria thoracoacromialis> B

11 ***thoracoacromial vein*** the vein that follows the homonymous artery and opens into the subclavian vein. <vena thoracoacromialis> B

POSTERIOR SCAPULAR REGION

12 ***deltoid fascia*** the deep fascia covering the deltoid muscle of the shoulder. <fascia deltoidea> C

13 **deltoid muscle** *origin,* clavicle, acromion, spine of scapula; *insertion,* deltoid tuberosity of humerus; *innervation,* axillary; *action,* abducts, flexes, extends arm. <musculus deltoideus> C

14 ***deltoid region*** the surface region overlying the deltoid muscle. <regio deltoidea> C

15 **infraspinatus muscle** *origin,* infraspinous fossa of scapula; *insertion,* greater tubercle of humerus; *innervation,* suprascapular; *action,* rotates humerus laterally. <musculus infraspinatus> D

16 **supraspinatus muscle** *origin,* supraspinous fossa of scapula; *insertion,* greater tubercle of humerus; *innervation,* suprascapular; *action,* abducts humerus. <musculus supraspinatus> D

17 **teres major muscle** *origin,* inferior angle of scapula; *insertion,* lip of intertubercular sulcus of humerus; *innervation,* lower subscapular; *action,* adducts, extends, rotates arm medially. <musculus teres major> D

18 **teres minor muscle** *origin,* lateral margin of scapula; *insertion,* greater tubercle of humerus; *innervation,* axillary; *action,* rotates arm laterally. <musculus teres minor> D

A

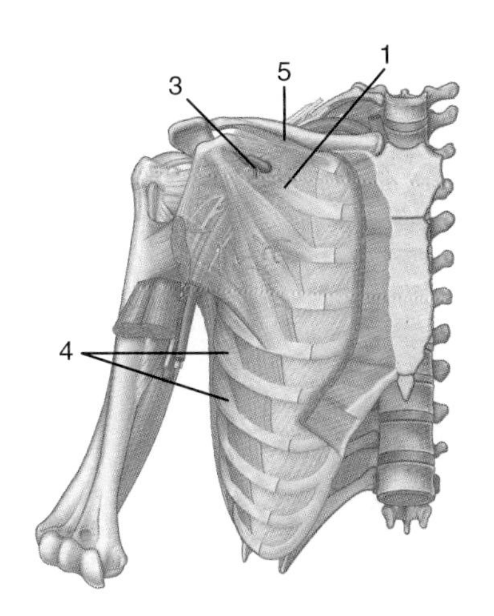

B

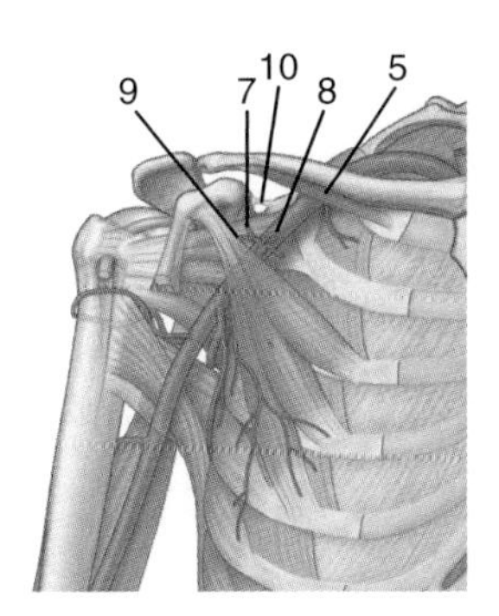

C

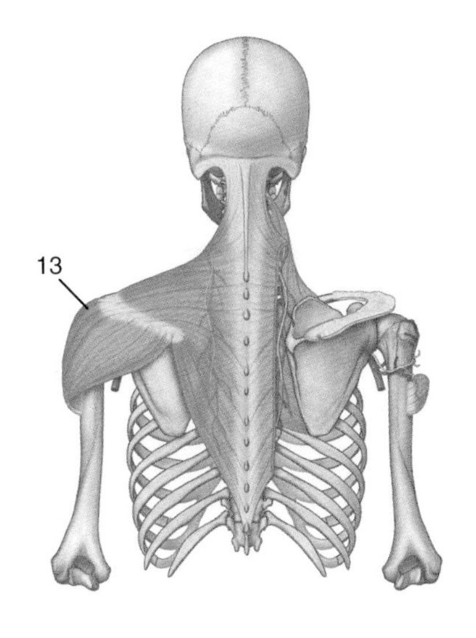

D

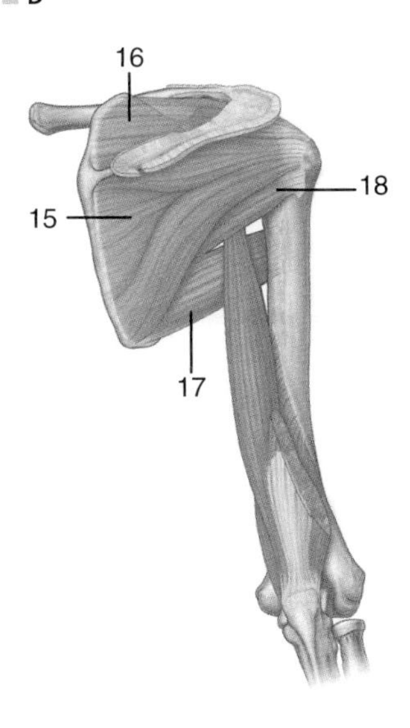

1 ***acromial branch of suprascapular artery*** a branch distributed to the acromion process. <ramus acromialis arteriae suprascapularis> A

2 **circumflex scapular artery** *origin,* subscapular artery; *branches,* none; *distribution,* inferolateral muscles of the scapula. <arteria circumflexa scapulae> A

3 **muscular branches of axillary nerve** branches that innervate the deltoid and teres minor muscles; *modality,* motor. <rami musculares nervi axillaris> A

4 **superior transverse scapular ligament** one of three intrinsic ligaments of the scapula, a band of fibers that bridges the scapular notch, thus forming a foramen for the passage of the suprascapular nerve. One end is attached to the base of the coracoid process, the other end to the medial border of the scapular notch. <ligamentum transversum scapulae superius> A

5 **suprascapular artery** *origin,* thyrocervical trunk; *branches,* acromial branch; *distribution,* clavicular, deltoid, and scapular regions. Called also *transverse scapular artery.* <arteria suprascapularis> A

6 **suprascapular nerve** *origin,* brachial plexus (C5-C6); *distribution,* descends through suprascapular and spinoglenoid notches and supplies acromioclavicular and shoulder joints, and supraspinatus and infraspinatus muscles; *modality,* motor and general sensory. <nervus suprascapularis> A

7 ***suprascapular vein*** the vein that accompanies the homonymous artery (sometimes as two veins that unite), opening usually into the external jugular, or occasionally into the subclavian vein; called also *transverse scapular vein.* <vena suprascapularis> A

8 **deltoid branch of profunda brachii artery** a branch distributed to the brachialis and deltoid muscles, and anastomosing with the posterior circumflex humeral artery. <ramus deltoideus arteriae profundae brachii> B

AXILLA

9 **axilla** the pyramidal region between the upper thoracic wall and the upper limb, its base formed by the skin and apex bounded by the approximation of the clavicle, coracoid process, and first rib; it contains axillary vessels, the brachial plexus of nerves, many lymph nodes and vessels, and loose areolar tissue. C

10 **axillary fascia** the investing fascia of the armpit, which passes between the lateral borders of the pectoralis major and latissimus dorsi muscles. <fascia axillaris> C

11 **axillary fossa** the small hollow at the axilla where the upper limb joins the body at the shoulder. <fossa axillaris> C

12 **lateral pectoral nerve** *origin,* lateral cord of brachial plexus or anterior divisions of upper and middle trunks (C5-C7); *distribution,* usually several nerves supplying the pectoralis minor muscle and acromioclavicular and shoulder joints; *modality,* motor and general sensory. <nervus pectoralis lateralis> D

A

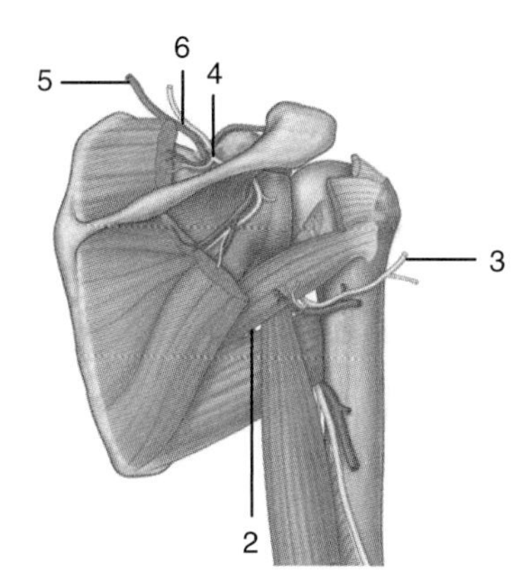

B

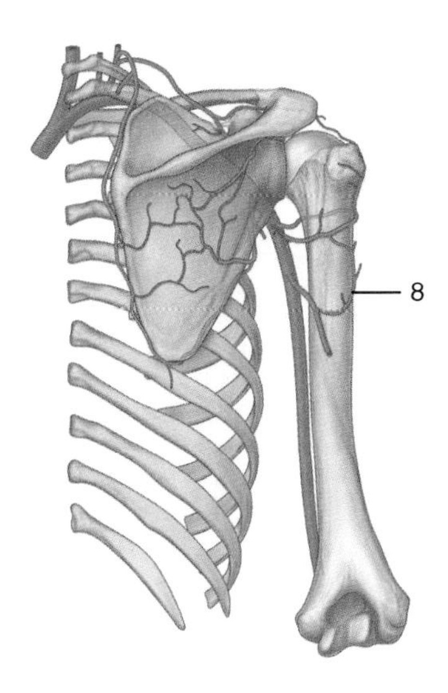

C

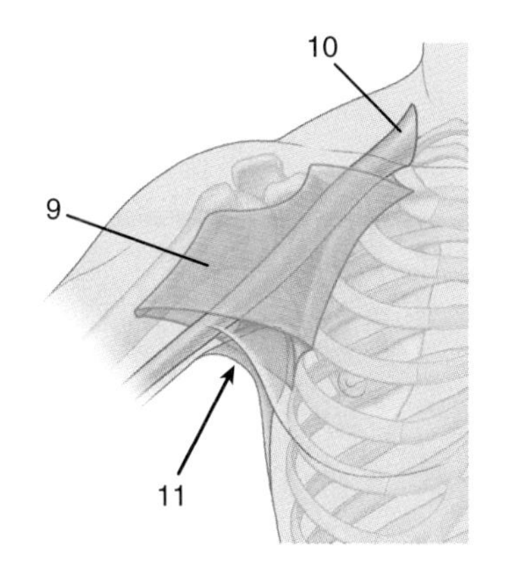

D

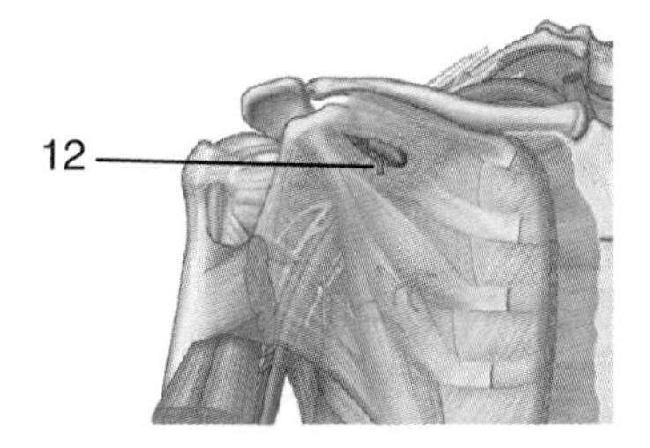

1 **subscapularis muscle** *origin,* subscapular fossa of scapula; *insertion,* lesser tubercle of humerus; *innervation,* subscapular; *action,* rotates humerus medially. <musculus subscapularis> A

2 **dorsal scapular nerve** *origin,* brachial plexus-anterior ramus of C5; *distribution,* rhomboid muscles and occasionally the levator scapulae muscle; *modality,* motor. <nervus dorsalis scapulae> B

3 **intercostobrachial nerves** two nerves arising from the intercostal nerves and supplying the skin of the upper limb. The first is constant: *origin,* second intercostal nerve; *distribution,* skin on back and medial aspect of arm; *modality,* general sensory. A second intercostobrachial nerve is often present; *origin,* third intercostal nerve; *distribution,* skin of axilla and medial aspect of arm; *modality,* general sensory. <nervi intercostobrachiales> B

4 **lateral root of median nerve** the fibers contributed to the median nerve by the lateral cord of the brachial plexus. <radix lateralis nervi mediani> B

5 **long thoracic nerve** *origin,* brachial plexus (anterior rami of C5-C7); *distribution,* descends behind brachial plexus to anterior serratus; *modality,* motor. <nervus thoracicus longus> B

6 **medial cutaneous nerve of arm** *origin,* medial cord of brachial plexus (T1); *distribution,* skin on medial and posterior aspects of arm; *modality,* general sensory. Called also *medial brachial cutaneous nerve.* <nervus cutaneus brachii medialis> B

7 **medial cutaneous nerve of forearm** *origin,* medial cord of brachial plexus (C8, T1); *branches,* anterior and ulnar; *distribution,* skin of front, medial, and posteromedial aspects of forearm; *modality,* general sensory. Called also *medial antebrachial cutaneous nerve.* <nervus cutaneus antebrachii medialis> B

8 **medial pectoral nerve** *origin,* medial cord or lower trunk of brachial plexus (C8, T1); *distribution,* usually several nerves supplying the pectoralis major and pectoralis minor muscles; *modality,* motor. <nervus pectoralis medialis> B

9 **medial root of median nerve** the fibers contributed to the median nerve by the medial cord of the brachial plexus. <radix medialis nervi mediani> B

10 **median nerve** *origin,* lateral and medial cords of brachial plexus—C6-T1; *branches,* anterior interosseous nerve of forearm, common palmar digital nerves, muscular and palmar branches, and a communicating branch with the ulnar nerve; *distribution,* the elbow, wrist, and intercarpal joints, anterior muscles of the forearm, muscles of the digits, skin of the palm, thenar eminence, and digits; *modality,* general sensory. <nervus medianus> B

11 **radial nerve** *origin,* posterior cord of brachial plexus (C6-C8, and sometimes C5 and T1); *branches,* posterior cutaneous and inferior lateral cutaneous nerves of arm, posterior cutaneous nerve of forearm, muscular, deep, and superficial branches; *distribution,* descending in the back of arm and forearm, it is ultimately distributed to skin on back of arm, forearm, and hand, extensor muscles on back of arm and forearm, and elbow joint and many joints of hand; *modality,* general sensory and motor. <nervus radialis> B

12 **subclavian nerve** *origin,* upper trunk of brachial plexus (C5); *distribution,* subclavius muscle and sternoclavicular joint; *modality,* motor and general sensory. <nervus subclavius> B

13 **subscapular nerves** *origin,* posterior cord of brachial plexus (C5); *distribution,* usually two or more nerves, upper and lower, supplying subscapularis and teres major muscles; *modality,* motor. <nervi subscapulares> B

14 **suprascapular nerve** *origin,* brachial plexus (C5-C6); *distribution,* descends through suprascapular and spinoglenoid notches and supplies acromioclavicular and shoulder joints, and supraspinatus and infraspinatus muscles; *modality,* motor and general sensory. <nervus suprascapularis> B

15 **thoracodorsal nerve** *origin,* posterior cord of brachial plexus (C7-C8); *distribution,* latissimus dorsi muscle; *modality,* motor. <nervus thoracodorsalis> B

16 **ulnar nerve** *origin,* medial and lateral cords of brachial plexus (C7-T1); *branches,* muscular, dorsal, palmar, superficial, and deep branches; *distribution,* ultimately to skin on front and back of medial part of hand, some flexor muscles on front of forearm, many short muscles of hand, elbow joint, many joints of hand; *modality,* general sensory and motor. <nervus ulnaris> B

Brachial Plexus

17 **brachial plexus** a plexus originating from the anterior branches of the last four cervical spinal nerves and most of the anterior branch of the first thoracic spinal nerves. Situated partly in the neck and partly in the axilla; it successively progresses medial to lateral, from roots to trunks to divisions to cords to terminal nerves, and is divided into supraclavicular and infraclavicular parts. <plexus brachialis> B

18 **anterior divisions of brachial plexus** the three anterior divisions into which each of the three trunks (superior, middle, and inferior) of the brachial plexus splits. The anterior divisions of the superior and middle trunks unite to form the lateral cord; and the anterior division of the inferior trunk forms the medial cord of the plexus. <divisiones anteriores plexus brachialis> B

19 **axillary nerve** *origin,* posterior cord of brachial plexus (C5-C6); *branches,* lateral superior brachial cutaneous nerve and muscular branches; *distribution,* deltoid and teres minor muscles, skin on back of arm; *modality,* motor and general sensory. <nervus axillaris> B

A

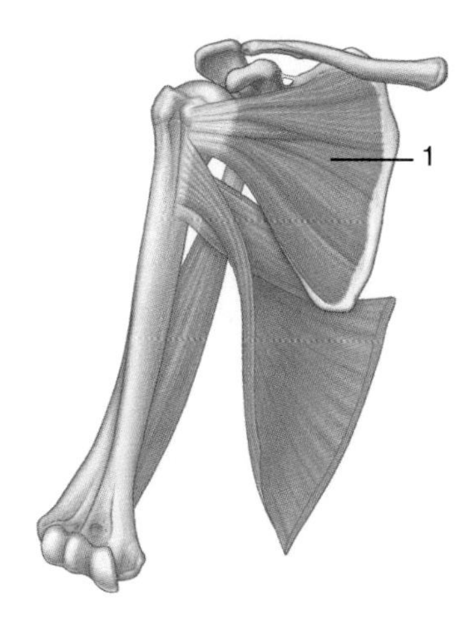
1

B

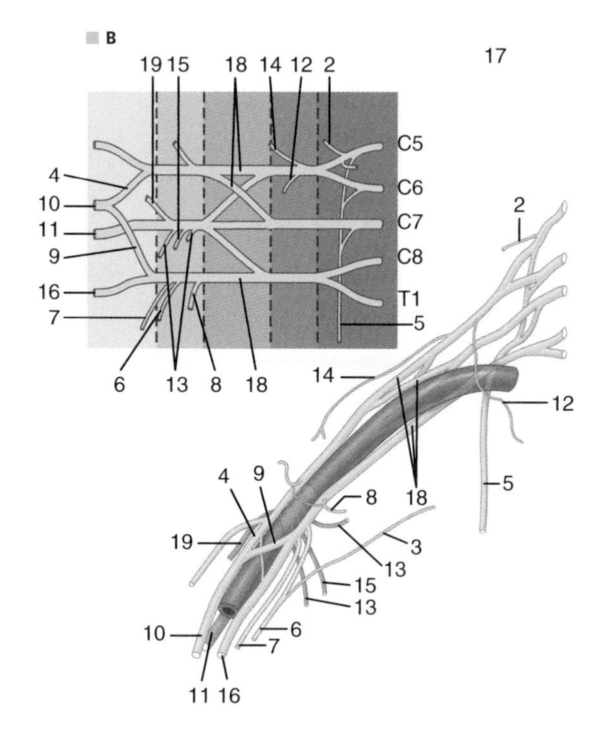
17
19 15
18 14 12 2
C5
C6
C7
C8
T1
4
10
11
9
16
7
5
6
13 8 18
2
14
12
18
5
4 9
8
19
3
13
15
13
10
6
7
11 16

1 **inferior trunk of brachial plexus** the trunk of the brachial plexus that is formed by the anterior branches of the eighth cervical and first thoracic nerves; medial pectoral nerves may arise from it. Its anterior division becomes the medial cord of the plexus, and its posterior division helps form the posterior cord; *modality,* general sensory and motor. <truncus inferior plexus brachialis> A

2 ***infraclavicular part of brachial plexus*** the part of the brachial plexus that lies in the axilla, below the level of the clavicle. In it arise the medial and lateral pectoral, musculocutaneous, medial cutaneous of arm, medial cutaneous of forearm, median, ulnar, radial, subscapular, thoracodorsal, and axillary nerves. <pars infraclavicularis plexus brachialis> A

3 **lateral cord of brachial plexus** the lateral bundle of fibers of the brachial plexus, formed by the union of the anterior divisions of the superior and middle trunks, C5 through C7, and from which arise the lateral pectoral and musculocutaneous nerves and the lateral root of the median and the ulnar nerves. <fasciculus lateralis plexus brachialis> A

4 **medial cord of brachial plexus** the medial bundle of fibers of the brachial plexus, formed by the anterior division of the inferior trunk, C8 through T1, and from which arise the medial pectoral nerve, medial cutaneous nerves of arm and of forearm, and the medial root of the ulnar and the median nerves. <fasciculus medialis plexus brachialis> A

5 **middle trunk of brachial plexus** the trunk of the brachial plexus that is formed by the anterior branch of the seventh cervical nerve. Its anterior division, from which lateral pectoral nerves may arise, helps form the lateral cord of the plexus, and its posterior division helps form the posterior cord; *modality,* general sensory and motor. <truncus medius plexus brachialis> A

6 **posterior cord of brachial plexus** the posterior bundle of fibers of the brachial plexus, formed by the union of the posterior divisions of the superior, middle, and inferior trunks, C5 through C8 and sometimes T1, and from which arise the subscapular, thoracodorsal, radial, and axillary nerves. <fasciculus posterior plexus brachialis> A

7 **posterior divisions of brachial plexus** the three posterior divisions into which each of the three trunks (superior, middle, and inferior) of the brachial plexus splits. All three posterior divisions unite to form the posterior cord of the plexus. <divisiones posteriores plexus brachialis> A

8 **roots of brachial plexus** the anterior branches of the four lower cervical spinal nerves and the first thoracic nerve, which combine to form the trunks of the brachial plexus. <radices plexus brachialis> A

9 **superior trunk of brachial plexus** the trunk of the brachial plexus that is formed by the anterior branches of the fifth and sixth cervical nerves. Its anterior division helps form the lateral cord of the plexus, its posterior division helps form the posterior cord, and it gives rise directly to the suprascapular and subclavian nerves; *modality,* general sensory and motor. <truncus superior plexus brachialis> A

10 ***supraclavicular part of brachial plexus*** the part of the brachial plexus lying in the cervical region above the level of the clavicle, in which arise the dorsal scapular, long thoracic, and suprascapular nerves, and the nerve to the subclavius muscle. <pars supraclavicularis plexus brachialis> A

11 ***trunks of brachial plexus*** the three trunks (superior, middle, and inferior) of the brachial plexus, arising from the anterior branches of the lower four cervical nerves and the first thoracic nerve near the lateral border of the scalenus anterior muscle; they continue laterally and downward, above and behind the subclavian artery, and near the clavicle each splits into an anterior and a posterior division. The anterior divisions of the superior and middle trunks unite to form the lateral cord and that of the inferior trunk forms the medial cord of the plexus; and the posterior divisions of the three trunks form the posterior cord of the plexus. <trunci plexus brachialis> A

A

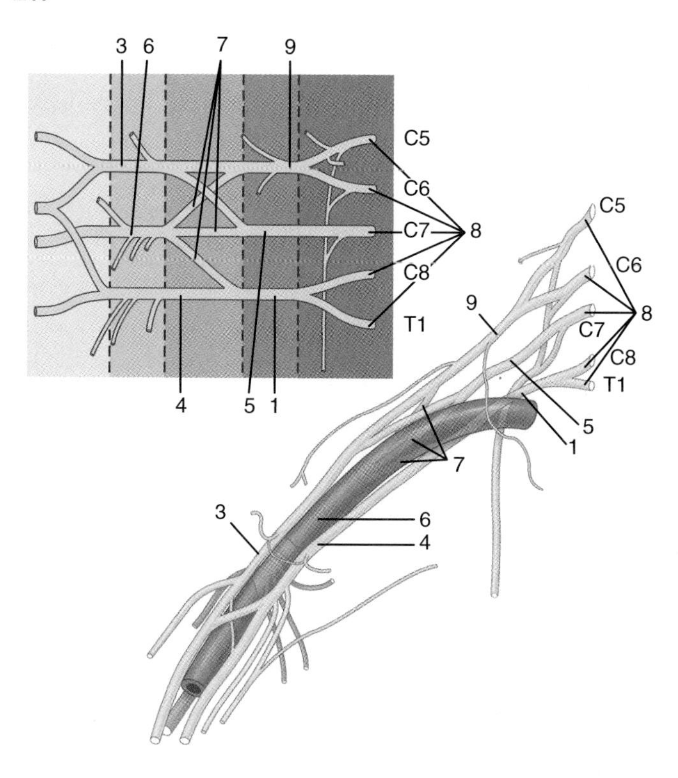

Vessels of Axilla

1 **anterior circumflex humeral artery** *origin,* axillary artery; *branches,* none; *distribution,* shoulder joint and head of humerus, long tendon of biceps, tendon of pectoralis major muscle. <arteria circumflexa humeri anterior> A

2 **axillary artery** *origin,* continuation of subclavian artery; *branches,* subscapular branches, and superior thoracic, thoracoacromial, lateral thoracic, subscapular, and anterior and posterior circumflex humeral arteries; *distribution,* upper limb, axilla, chest, shoulder. <arteria axillaris> A

3 ***dorsal scapular artery*** *origin,* second or third part of subclavian artery; *branches,* none; *distribution,* rhomboid, latissimus dorsi, and trapezius muscles. <arteria dorsalis scapulae> A

4 **posterior circumflex humeral artery** *origin,* axillary artery; *branches;* none; *distribution,* deltoideus, shoulder joint, teres minor and triceps muscles. <arteria circumflexa humeri posterior> A

5 **subscapular artery** *origin,* axillary artery; *branches,* thoracodorsal and circumflex scapular arteries; *distribution,* scapular and shoulder region. <arteria subscapularis> A

6 ***subscapular branches of axillary artery*** branches that supply the subscapularis muscle. <rami subscapulares arteriae axillaris> A

7 **superior thoracic artery** *origin,* axillary artery; *branches,* none; *distribution,* axillary aspect of chest wall. Called also *highest thoracic artery.* <arteria thoracica superior> A

8 ***suprascapular artery*** *origin,* thyrocervical trunk; *branches,* acromial branch; *distribution,* clavicular, deltoid, and scapular regions. Called also *transverse scapular artery.* <arteria suprascapularis> A

9 **thoracodorsal artery** *origin,* subscapular artery; *branches,* none; *distribution,* subscapularis and teres muscles. <arteria thoracodorsalis> A

10 **axillary vein** the venous trunk of the upper limb; it begins at the lower border of the teres major muscle by junction of the basilic and brachial veins, and at the lateral border of the first rib is continuous with the subclavian vein. <vena axillaris> B

11 ***dorsal scapular vein*** an occasional branch that contributes to the subclavian vein. <vena scapularis dorsalis> B

12 ***suprascapular vein*** the vein that accompanies the homonymous artery (sometimes as two veins that unite), opening usually into the external jugular, or occasionally into the subclavian vein; called also *transverse scapular vein.* <vena suprascapularis> B

LYMPHATICS OF UPPER LIMB

13 **apical lymph nodes** six to twelve axillary lymph nodes partly posterior to the superior part of the pectoralis minor muscle and partly in the apex of the axilla, receiving afferent vessels that accompany the cephalic vein and draining all other axillary nodes; their efferent vessels unite to form the subclavian trunk. <nodi lymphoidei axillares apicales> C

14 ***axillary lymph nodes*** the 20 to 30 lymph nodes of the axilla, which receive lymph from all the lymph vessels of the upper limb, most of those of the breast, and the cutaneous vessels from the trunk above the level of the umbilicus. They are divided into groups: apical, central, brachial, pectoral, and subscapular. <nodi lymphoidei axillares> C

15 ***axillary lymphatic plexus*** a plexus of lymph vessels and nodes in the axillary fossa. <plexus lymphaticus axillaris> C

16 ***brachial lymph nodes*** four to six axillary lymph nodes lying medial to, and behind, the axillary vein, which drain most of the upper limb. <nodi lymphoidei brachiales> C

17 **central lymph nodes** three or four axillary lymph nodes embedded in adipose tissue near the base of the axilla; they receive lymph from the brachial, pectoral, and subscapular nodes and drain into the apical nodes. <nodi lymphoidei axillares centrales> C

18 ***deep lymph nodes of upper limb*** the lymph nodes situated internal to the deep fascia of the upper limb, most of which are grouped in the axilla; they accompany the radial, ulnar, interosseous, and brachial arteries and end in the brachial axillary lymph nodes. <nodi lymphoidei profundi membri superioris> C

19 **subscapular axillary lymph nodes** six or seven axillary lymph nodes along the inferior margins of the posterior axillary wall along the course of the subscapular artery; they receive lymph from the skin and superficial muscles of the posterior part of the neck and the posterior thoracic wall and drain into the apical and central nodes. <nodi lymphoidei axillares subscapulares> C

20 ***superficial lymph nodes of upper limb*** the lymph nodes of the upper limb that are superficially placed, such as the cubital lymph nodes; all except those in the hand and on the back of the forearm converge toward and accompany the superficial veins. <nodi lymphoidei superficiales membri superioris> C

ARM

Anterior Compartment of Arm

Muscles of Anterior Compartment of Arm

21 **lateral intermuscular septum of arm** the fascial sheet extending from the lateral border of the humerus to the undersurface of the fascia investing the arm. <septum intermusculare brachii laterale> D

22 **medial intermuscular septum of arm** the fascial sheet extending from the medial border of the humerus to the undersurface of the fascia investing the arm. <septum intermusculare brachii mediale> D

A

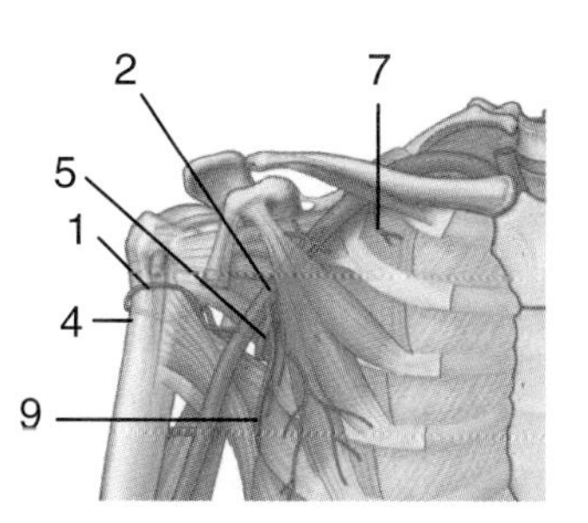

B

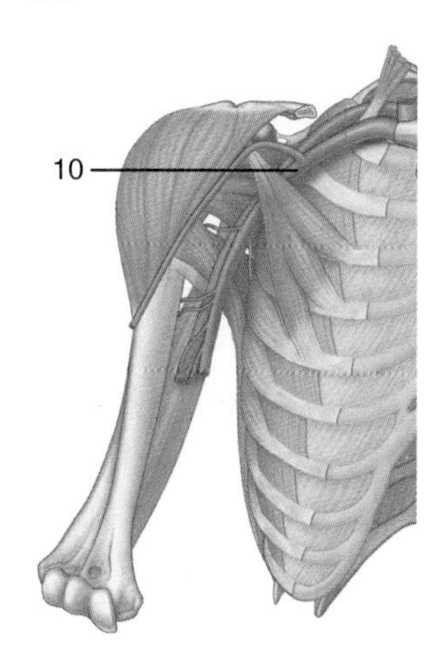

C

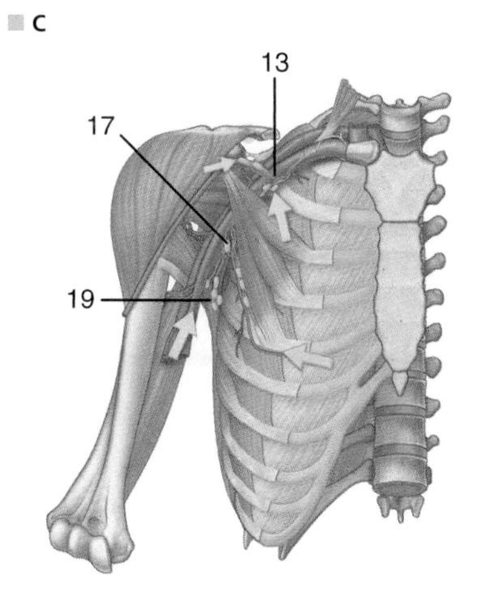

D

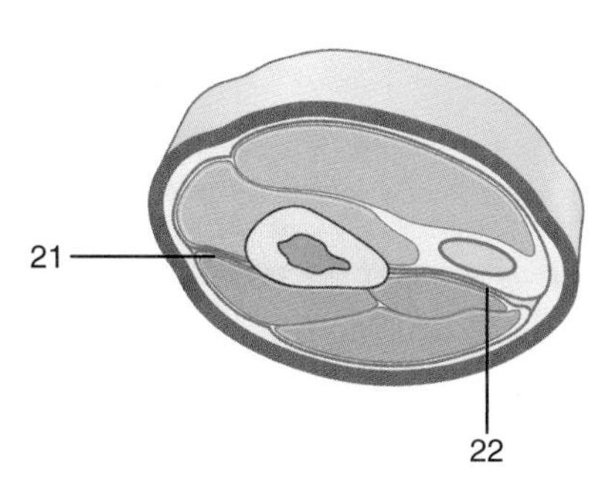

1 **biceps brachii muscle** *two heads—long and short; insertion,* radial tuberosity and fascia of forearm; *innervation,* musculocutaneous; *action,* flexes forearm, supinates hand. <musculus biceps brachii> A

2 **bicipital aponeurosis** an expansion of the tendon of the biceps brachii muscle by which it is attached to the fascia of the forearm and to the ulna. <aponeurosis musculi bicipitis brachii> A

3 ***brachial fascia*** the investing fascia of the upper limb. <fascia brachii> A

4 **brachialis muscle** *origin,* anterior surface of humerus; *insertion,* coronoid process of ulna; *innervation,* radial, musculocutaneous; *action,* flexes forearm. <musculus brachialis> A

5 ***bursa of coracobrachialis muscle*** a bursa between the coracobrachialis and subscapularis muscles and the coracoid process. <bursa musculi coracobrachialis> A

6 **coracobrachialis muscle** *origin,* coracoid process of scapula; *insertion,* medial surface of shaft of humerus; *innervation,* musculocutaneous; *action,* flexes, adducts arm. <musculus coracobrachialis> A

7 ***intertubercular tendon sheath*** the tendon sheath that surrounds the long head of the biceps brachii muscle as it passes through the intertubercular sulcus. <vagina tendinis intertubercularis> A

8 **long head of biceps brachii muscle** the head of the biceps brachii arising from the upper border of the glenoid cavity. <caput longum musculi bicipitis brachii> A

9 **short head of biceps brachii muscle** the head of the biceps brachii arising from the apex of the coracoid process. <caput breve musculi bicipitis brachii> A

Nerves of Anterior Compartment of Arm

10 **lateral cutaneous nerve of forearm** *origin,* continuation of musculocutaneous nerve; *distribution,* skin over radial side of forearm and sometimes an area of skin of dorsum of hand; *modality,* general sensory. Called also *lateral antebrachial cutaneous nerve.* <nervus cutaneus antebrachii lateralis> B

11 **median nerve** *origin,* lateral and medial cords of brachial plexus—C6-T1; *branches,* anterior interosseous nerve of forearm, common palmar digital nerves, muscular and palmar rami, and a communicating branch with the ulnar nerve; *distribution,* the elbow, wrist, and intercarpal joints, anterior muscles of the forearm, muscles of the digits, skin of the palm, thenar eminence, and digits; *modality,* general sensory. <nervus medianus> B

12 **muscular branches of musculocutaneous nerve** branches that innervate the coracobrachialis, biceps, and brachialis muscles; *modality,* motor and general sensory. <rami musculares nervi musculocutanei> B

13 **musculocutaneous nerve** *origin,* lateral cord of brachial plexus—C5-C7; *branches,* lateral cutaneous nerve of forearm, and muscular branches; *distribution,* coracobrachialis, biceps brachii, and brachialis muscles, the elbow joint, and skin of radial side of forearm; *modality,* general sensory and motor. <nervus musculocutaneus> B

14 **ulnar nerve** *origin,* medial and lateral cords of brachial plexus (C7-T1); *branches,* muscular, dorsal, palmar, superficial, and deep branches; *distribution,* ultimately to skin on front and back of medial part of hand, some flexor muscles on front of forearm, many short muscles of hand, elbow joint, many joints of hand; see individual branches; *modality,* general sensory and motor. <nervus ulnaris> B

Vessels of Anterior Compartment of Arm

15 **brachial artery** *origin,* continuation of axillary artery; *branches,* superficial brachial, profunda brachii, humeral nutrient, superior ulnar and inferior ulnar collateral, radial, and ulnar arteries; *distribution,* shoulder, arm, forearm, hand. <arteria brachialis> C

16 **humeral nutrient arteries** *origin,* brachial and profunda brachii arteries; *branches,* none; *distribution,* humerus. <arteriae nutriciae humeri> C

17 **inferior ulnar collateral artery** *origin,* brachial artery; *branches,* none; *distribution,* arm muscles at back of elbow. <arteria collateralis ulnaris inferior> C

18 ***superficial brachial artery*** an occasional vessel that arises from high bifurcation of the brachial artery and assumes a more superficial course than usual. <arteria brachialis superficialis> C

19 **superior ulnar collateral artery** *origin,* brachial artery; *branches,* none; *distribution,* elbow joint, triceps muscle. <arteria collateralis ulnaris superior> C

20 **brachial veins** the venae comitantes of the brachial artery, which join with the basilic vein to form the axillary vein. <venae brachiales> D

A

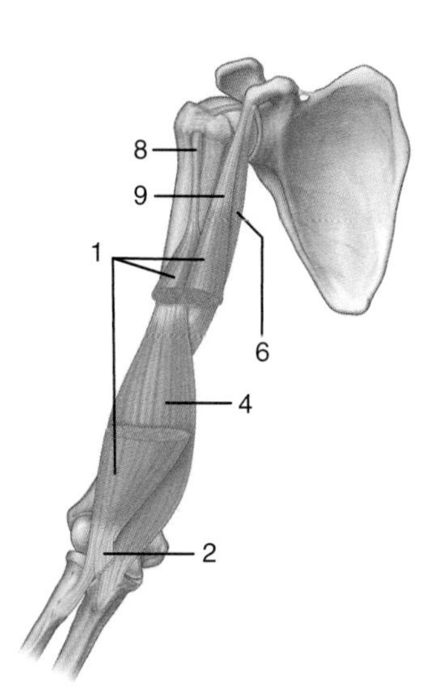

B

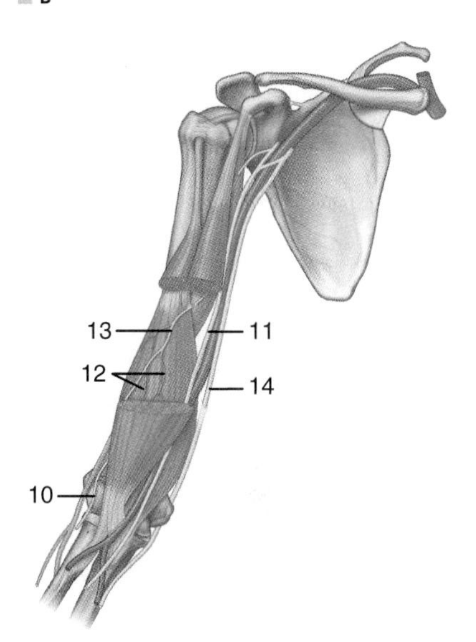

C

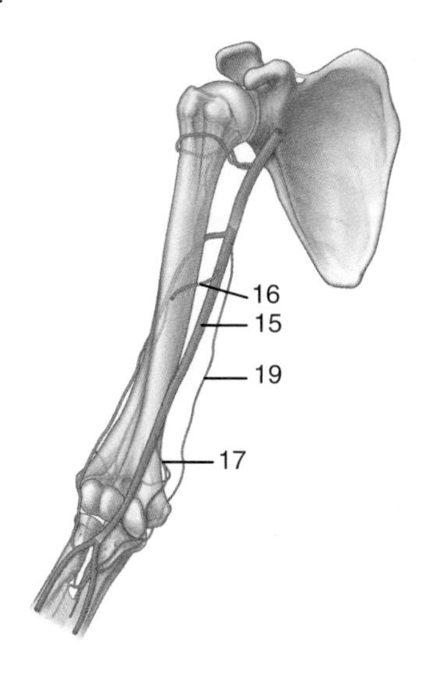

D

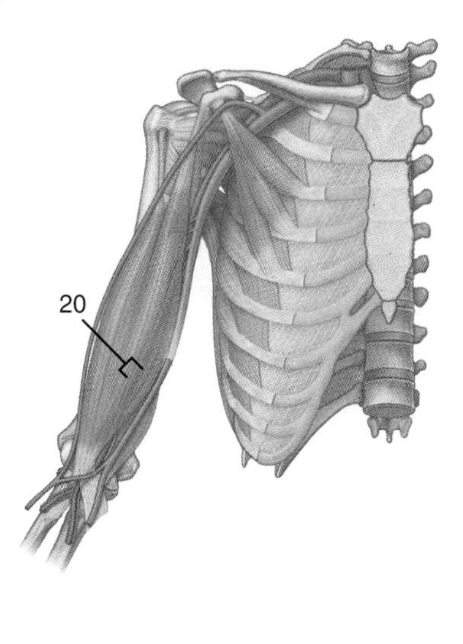

Posterior Compartment of Arm

Muscles of Posterior Compartment of Arm

1 ***articularis cubiti muscle*** a few fibers of the deep surface of the triceps brachii that insert into the posterior ligament and synovial membrane of the elbow joint. <musculus articularis cubiti> A

2 **lateral head of triceps brachii muscle** the head of the triceps brachii muscle arising from the posterior surface of the humerus, the lateral border of the humerus, and the lateral intermuscular septum. <caput laterale musculi tricipitis brachii> A

3 **long head of triceps brachii muscle** the head of the triceps brachii muscle arising from the infraglenoid tubercle of the scapula. <caput longum musculi tricipitis brachii> A

4 **medial head of triceps brachii muscle** the head of the triceps brachii muscle arising from the posterior surface of the humerus below the radial groove, the medial border of the humerus, and the medial intermuscular septum. <caput mediale musculi tricipitis brachii> A

5 **triceps brachii muscle** *three heads—long, lateral, and medial; insertion,* olecranon of ulna; *innervation,* radial; *action,* extends forearm, long head adducts and extends arm. <musculus triceps brachii> A

Nerves of Posterior Compartment of Arm

6 **inferior lateral cutaneous nerve of arm** *origin,* radial nerve; *distribution,* skin of lateral surface of lower part of arm; *modality,* general sensory. Called also *inferior lateral brachial cutaneous nerve.* <nervus cutaneus brachii lateralis inferior> A

7 **muscular branches of radial nerve** branches that innervate the triceps, anconeus, brachioradialis, and extensor carpi radialis muscles; a branch to the brachialis muscle is probably sensory; modality, motor and sensory. <rami musculares nervi radialis> A

8 ***posterior cutaneous nerve of arm*** *origin,* radial nerve in the axilla; *distribution,* skin on back of arm; *modality,* general sensory. Called also *posterior brachial cutaneous nerve.* <nervus cutaneus brachii posterior> A

9 **radial nerve** *origin,* posterior cord of brachial plexus (C6-C8, and sometimes C5 and T1); *branches,* posterior cutaneous and inferior lateral cutaneous nerves of arm, posterior cutaneous nerve of forearm, muscular, deep, and superficial branches; *distribution,* descending in the back of arm and forearm, it is ultimately distributed to skin on back of arm, forearm, and hand, extensor muscles on back of arm and forearm, and elbow joint and many joints of hand; *modality,* general sensory and motor. <nervus radialis> A

10 ***superior lateral cutaneous nerve of arm*** *origin,* axillary nerve; *distribution,* skin of back of arm; *modality,* general sensory. Called also *superior lateral brachial cutaneous nerve.* <nervus cutaneus brachii lateralis superior> A

Vessels of Posterior Compartment of Arm

11 **cubital anastomosis** an arterial network formed on the posterior aspect of the elbow by the posterior ulnar recurrent, inferior and superior ulnar collateral, and interosseous recurrent arteries. <rete articulare cubiti> B

12 **middle collateral artery** *origin,* profunda brachii artery; *branches,* none; *distribution,* triceps muscle, elbow joint. <arteria collateralis media> B

13 **profunda brachii artery** *origin,* brachial artery; *branches,* deltoid branch, nutrient artery, medial and radial collateral arteries; *distribution,* humerus, muscles and skin of arm. Called also *deep artery of arm.* <arteria profunda brachii> B

14 **radial collateral artery** *origin,* profunda brachii artery; *branches,* none; *distribution,* brachioradialis and brachialis muscles. <arteria collateralis radialis> B

15 **cephalic vein** the superficial vein that arises from the radial side of the dorsal rete of the hand, and winds anteriorly to pass along the anterior border of the brachioradialis muscle; above the elbow it ascends along the lateral border of the biceps muscle and the pectoral border of the deltoid muscle, and opens into the axillary vein. <vena cephalica> C

FOREARM

Anterior Compartment of Forearm

16 **bicipitoradial bursa** a bursa between the radial tuberosity and the biceps tendon. <bursa bicipitoradialis> D

A

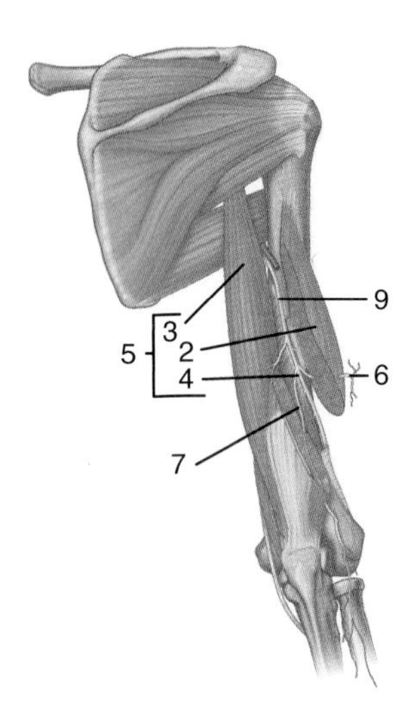

B

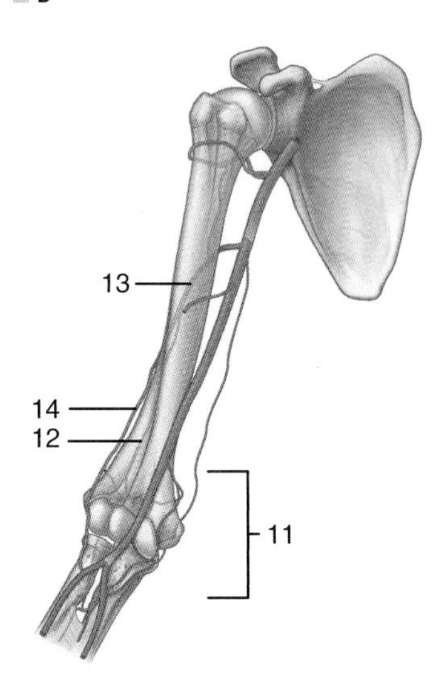

C

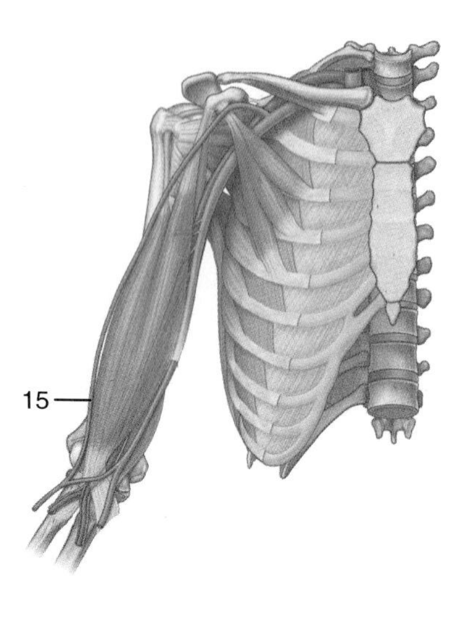

D

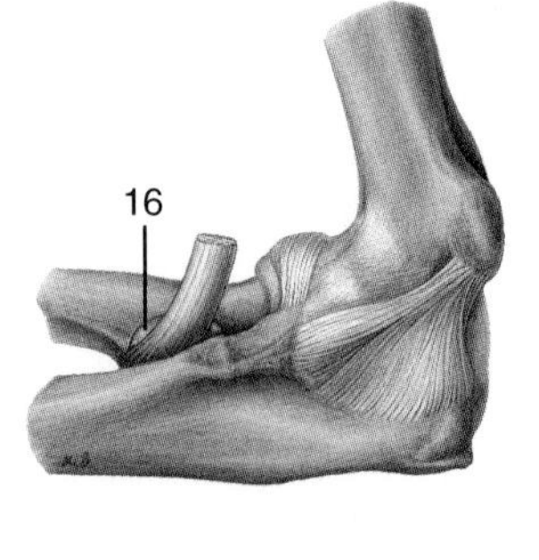

1 ***cubital lymph nodes*** one or two superficially placed lymph nodes situated above the medial epicondyle, medial to the basilic vein, the efferent vessels of which accompany the basilic vein and join the deep lymph vessels. <nodi lymphoidei cubitales> A

2 **supratrochlear lymph nodes** one or two lymph nodes superficial to the deep fascia proximal to the medial epicondyle and medial to the basilic vein and draining into the deep lymph vessels. <nodi lymphoidei supratrochleares> A

3 **carpal tunnel** an osseofibrous tunnel for passage of the tendons of the flexor muscles of the hand and digits, formed by the flexor retinaculum as it roofs over the concavity of the carpus on the palmar surface. <canalis carpi> B

Muscles of Anterior Compartment of Forearm

4 ***antebrachial fascia*** the investing fascia of the forearm. <fascia antebrachii> C

5 **flexor carpi radialis muscle** *origin,* medial epicondyle of humerus; *insertion,* base of second metacarpal; *innervation,* median; *action,* flexes and abducts wrist joint. <musculus flexor carpi radialis> C

6 **flexor carpi ulnaris muscle** two heads—humeral, ulnar; *insertion,* pisiform, hook of hamate, proximal end of fifth metacarpal; *innervation,* ulnar; *action,* flexes and adducts wrist joint. <musculus flexor carpi ulnaris> C

7 **humeral head of pronator teres muscle** the head of the pronator teres muscle arising from the medial epicondyle of the humerus. <caput humerale musculi pronatoris teretis> C

8 **palmaris longus muscle** *origin,* medial epicondyle of humerus; *insertion,* flexor retinaculum, palmar aponeurosis; *innervation,* median; *action,* flexes wrist joint, anchors skin and fascia of hand. <musculus palmaris longus> C

9 **pronator teres muscle** two heads—humeral, ulna; *insertion,* lateral surface of radius; *innervation,* median; *action,* flexes elbow and pronates forearm. <musculus pronator teres> C

10 **ulnar head of pronator teres muscle** the head of the pronator teres muscle arising from the coronoid process of the ulna. <caput ulnare musculi pronatoris teretis> C

11 **humeral head of flexor carpi ulnaris muscle** the head of the flexor carpi ulnaris muscle arising from the medial epicondyle of the humerus. <caput humerale musculi flexoris carpi ulnaris> D

12 **ulnar head of flexor carpi ulnaris muscle** the head of the flexor carpi ulnaris muscle arising from the olecranon and the adjacent part of the ulna. <caput ulnare musculi flexoris carpi ulnaris> D

13 **flexor digitorum superficialis muscle** two heads—humeroulnar, radial; *insertion,* sides of middle phalanges of four (nonthumb) fingers; *innervation,* median; *action,* primarily flexes middle phalanges. <musculus flexor digitorum superficialis> E

14 **flexor retinaculum of muscles of hand** a heavy fibrous band continuous with the distal part of the antebrachial fascia, completing the carpal tunnel through which pass the tendons of the flexor muscles of the hand and fingers. <retinaculum musculorum flexorum manus> E

15 **humeroulnar head of flexor digitorum superficialis muscle** the head of the flexor digitorum superficialis muscle arising from the medial epicondyle of the humerus and coronoid process of ulna. <caput humeroulnare musculi flexoris digitorum superficialis> E

16 **radial head of flexor digitorum superficialis muscle** the head of the flexor digitorum superficialis muscle arising from the oblique line and anterior border of the radius. <caput radiale musculi flexoris digitorum superficialis> E

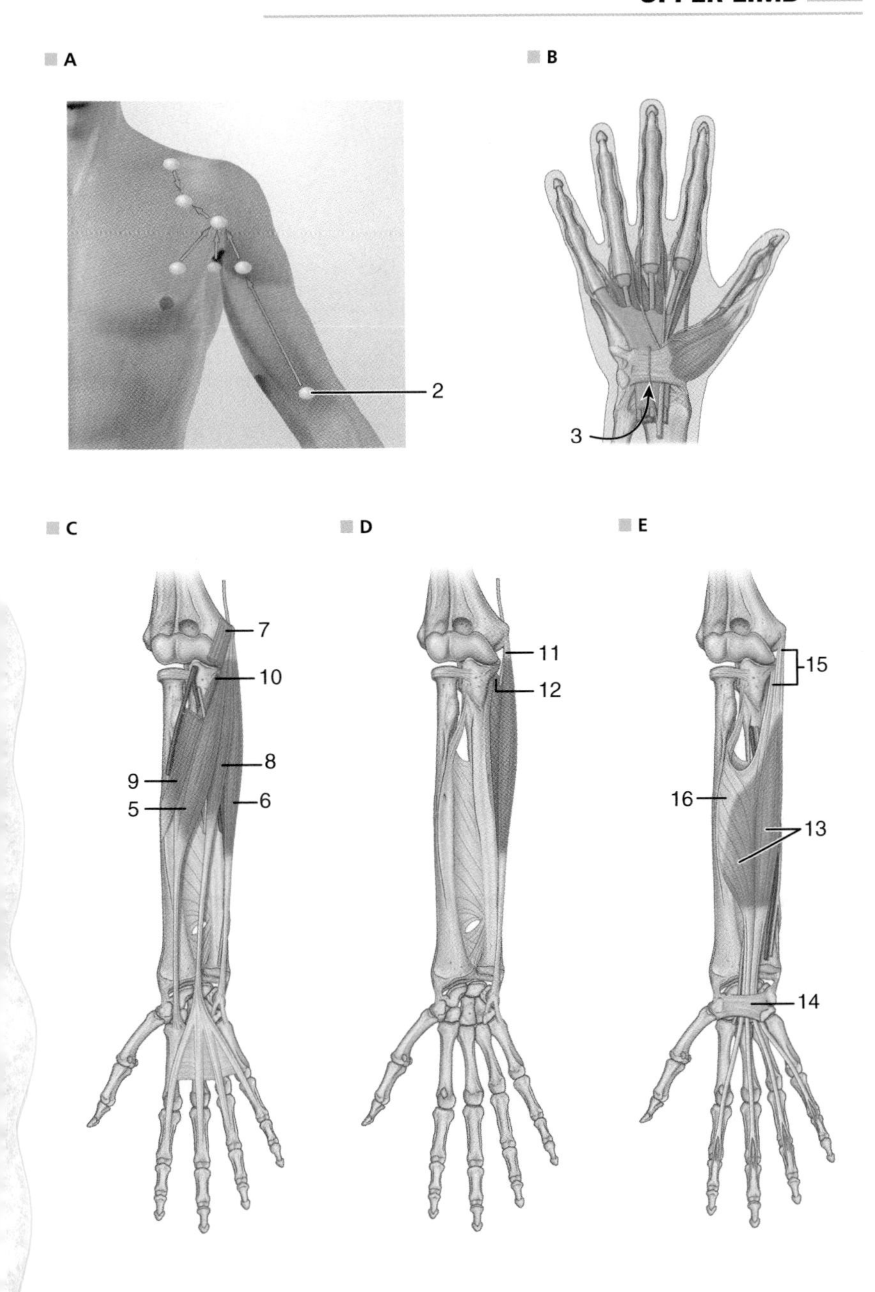
A
2
B
3
C
7
10
8
9
5
6
D
11
12
E
15
16
13
14

1 **flexor digitorum profundus muscle** *origin,* shaft of ulna, coronoid process, interosseous membrane; *insertion,* bases of distal phalanges of fingers; *innervation,* ulnar and anterior interosseous; *action,* flexes distal phalanges. <musculus flexor digitorum profundus> A

2 **flexor pollicis longus muscle** *origin,* anterior surface of radius, interosseous membrane, and medial epicondyle of humerus or coronoid process of ulna; *insertion,* base of distal phalanx of thumb; *innervation,* anterior interosseous; *action,* flexes thumb. <musculus flexor pollicis longus> A

3 **pronator quadratus muscle** *origin,* anterior surface and border of distal third or fourth of shaft of ulna; *insertion,* anterior surface and border of distal fourth of shaft of radius; *innervation,* anterior interosseous; *action,* pronates forearm. <musculus pronator quadratus> A

Nerves of Anterior Compartment of Forearm

4 ***anterior branch of medial cutaneous nerve of forearm*** a branch that innervates the skin of the front and medial aspect of the forearm; modality, general sensory. <ramus anterior nervi cutanei antebrachii medialis> B

5 **anterior interosseous nerve of forearm** *origin,* median nerve; *distribution,* flexor pollicis longus, flexor digitorum profundus, and pronator quadratus muscles, wrist and intercarpal joints; *modality,* motor and general sensory. <nervus interosseus antebrachii anterior> B

6 ***communicating branch of median nerve with ulnar nerve*** a small branch across the flexor digitorum profundus muscle, connecting the median with the ulnar nerve. <ramus communicans nervi mediani cum nervo ulnari> B

7 **median nerve** *origin,* lateral and medial cords of brachial plexus—C6-T1; *branches,* anterior interosseous nerve of forearm, common palmar digital nerves, muscular and palmar branches, and a communicating branch with the ulnar nerve; *distribution,* the elbow, wrist, and intercarpal joints, anterior muscles of the forearm, muscles of the digits, skin of the palm, thenar eminence, and digits; *modality,* general sensory. <nervus medianus> B

8 **muscular branches of median nerve** branches that innervate most of the flexor muscles on the front of the forearm and most of the short muscles of the thumb; *modality,* motor. <rami musculares nervi mediani> B

9 **muscular branches of ulnar nerve** branches that innervate the flexor carpi ulnaris muscle and the ulnar half of flexor digitorum profundus; *modality,* motor. <rami musculares nervi ulnaris> B

10 **palmar branch of median nerve** a branch arising from the median nerve in the lower part of the forearm and supplying the skin of the central palm and of the thenar eminence; *modality,* general sensory. <ramus palmaris nervi mediani> B

11 **palmar branch of ulnar nerve** a branch arising from the ulnar nerve in the lower part of the forearm, supplying the cutaneous structures of the medial part of the palm; *modality,* general sensory. <ramus palmaris nervi ulnaris> B

12 **posterior branch of ulnar nerve** a large cutaneous branch that arises from the ulnar nerve and passes down the distal portion of the forearm to the medial side of the back of the hand, where it divides usually into three, sometimes four, dorsal digital nerves; *modality,* general sensory. <ramus dorsalis nervi ulnaris> B

13 **superficial branch of radial nerve** the continuation of the radial nerve that accompanies the radial artery in the forearm, winds dorsalward, supplies the lateral side of the back of the hand, and divides into dorsal digital nerves that supply the skin of the dorsal surface and adjacent surfaces of the thumb, index, and middle fingers, and sometimes the radial side of the ring finger; *modality,* general sensory. <ramus superficialis nervi radialis> B

14 **ulnar nerve** *origin,* medial and lateral cords of brachial plexus (C7-T1); *branches,* muscular, dorsal, palmar, superficial, and deep branches; *distribution,* ultimately to skin on front and back of medial part of hand, some flexor muscles on front of forearm, many short muscles of hand, elbow joint, many joints of hand; *modality,* general sensory and motor. <nervus ulnaris> B

Vessels of Anterior Compartment of Forearm

15 ***median artery*** *origin,* anterior interosseous artery; *branches,* none; *distribution,* median nerve, muscles of front of forearm. <arteria comitans nervi mediani> C

A

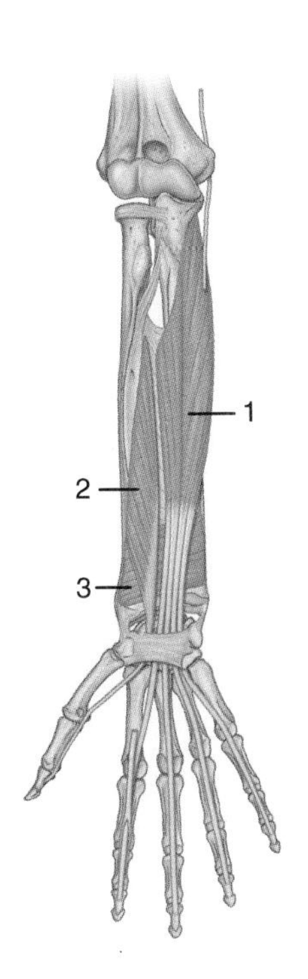

B

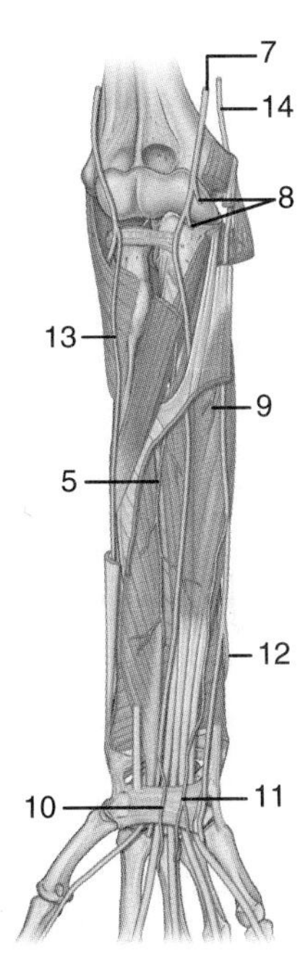

C

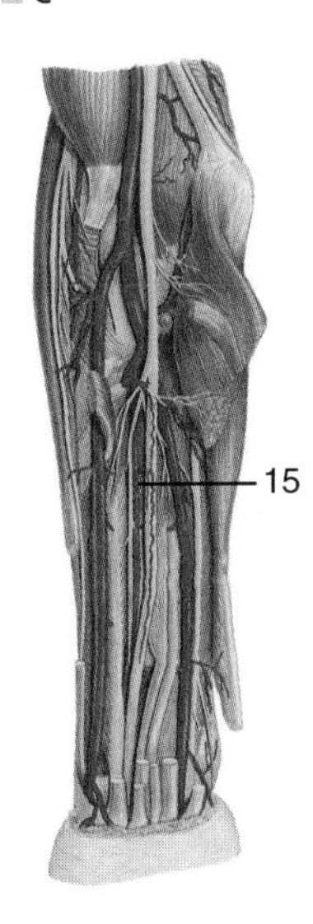

1 **basilic vein** the superficial vein that arises from the ulnar side of the dorsal rete of the hand, passes up the forearm, and joins with the brachial veins to form the axillary vein. <vena basilica> A

2 **median antebrachial vein** a vein that arises from a palmar venous plexus and passes up the forearm between the cephalic and the basilic veins to the elbow, where it either joins one of these, bifurcates to join both, or joins the median cubital vein; called also *median vein of forearm.* <vena mediana antebrachii> A

3 ***median basilic vein*** a vein sometimes present as the medial branch, ending in the basilic vein, of a bifurcation of the median antebrachial vein. <vena basilica antebrachii> A

4 **median cubital vein** the large connecting branch that arises from the cephalic vein below the elbow and passes obliquely upward over the cubital fossa to join the basilic vein; called also *median vein of elbow.* <vena mediana cubiti> A

5 **anterior branch of ulnar recurrent artery** a branch that helps supply the pronator teres and brachialis muscles and runs to the front of the medial epicondyle, supplying the elbow joint and adjacent structures. <ramus anterior arteriae recurrentis ulnaris> B

6 **posterior branch of ulnar recurrent artery** a branch running to the back of the medial epicondyle, supplying the elbow joint and neighboring muscles. <ramus posterior arteriae recurrentis ulnaris> B

7 **radial artery** *origin,* brachial artery; *branches,* palmar carpal, superficial palmar and dorsal carpal branches, recurrent radial and princeps pollicis arteries, deep palmar arch; *distribution,* forearm, wrist, hand. <arteria radialis> B, C

8 **radial recurrent artery** *origin,* radial artery; *branches,* none; *distribution,* brachioradialis, brachialis, elbow region. <arteria recurrens radialis> B

9 ***radial veins*** the venae comitantes of the radial artery, which open into the brachial veins. <venae radiales> B

10 **recurrent interosseous artery** *origin,* posterior interosseous or common interosseous artery; *branches,* none; *distribution,* back of elbow joint. <arteria interossea recurrens> B

11 **ulnar recurrent artery** *origin,* ulnar artery; *branches,* anterior and posterior; *distribution,* elbow joint region. <arteria recurrens ulnaris> B

12 **anterior interosseous artery** *origin,* posterior or common interosseous artery; *branches,* median artery; *distribution,* deep parts of front of forearm. <arteria interossea anterior> C

13 ***anterior interosseous veins*** the veins accompanying the anterior interosseous artery, which join the ulnar veins near the elbow. <venae interosseae anteriores> C

14 **common interosseous artery** *origin,* ulnar artery; *branches,* anterior and posterior interosseous arteries; *distribution,* antecubital fossa. <arteria interossea communis> C

15 **ulnar artery** *origin,* brachial artery; *branches,* palmar carpal, dorsal carpal, and deep palmar branches, ulnar recurrent and common interosseous arteries, superficial palmar arch; *distribution,* forearm, wrist, hand. <arteria ulnaris> C

16 ***ulnar veins*** the venae comitantes of the ulnar artery, which unite with the radial veins at the elbow to form the brachial veins. <venae ulnares> C

Posterior Compartment of Forearm

17 **abductor pollicis longus muscle** *origin,* posterior surfaces of radius and ulna; *insertion,* radial side of base of first metacarpal bone; *innervation,* posterior interosseous; *action,* abducts, extends thumb. <musculus abductor pollicis longus> D

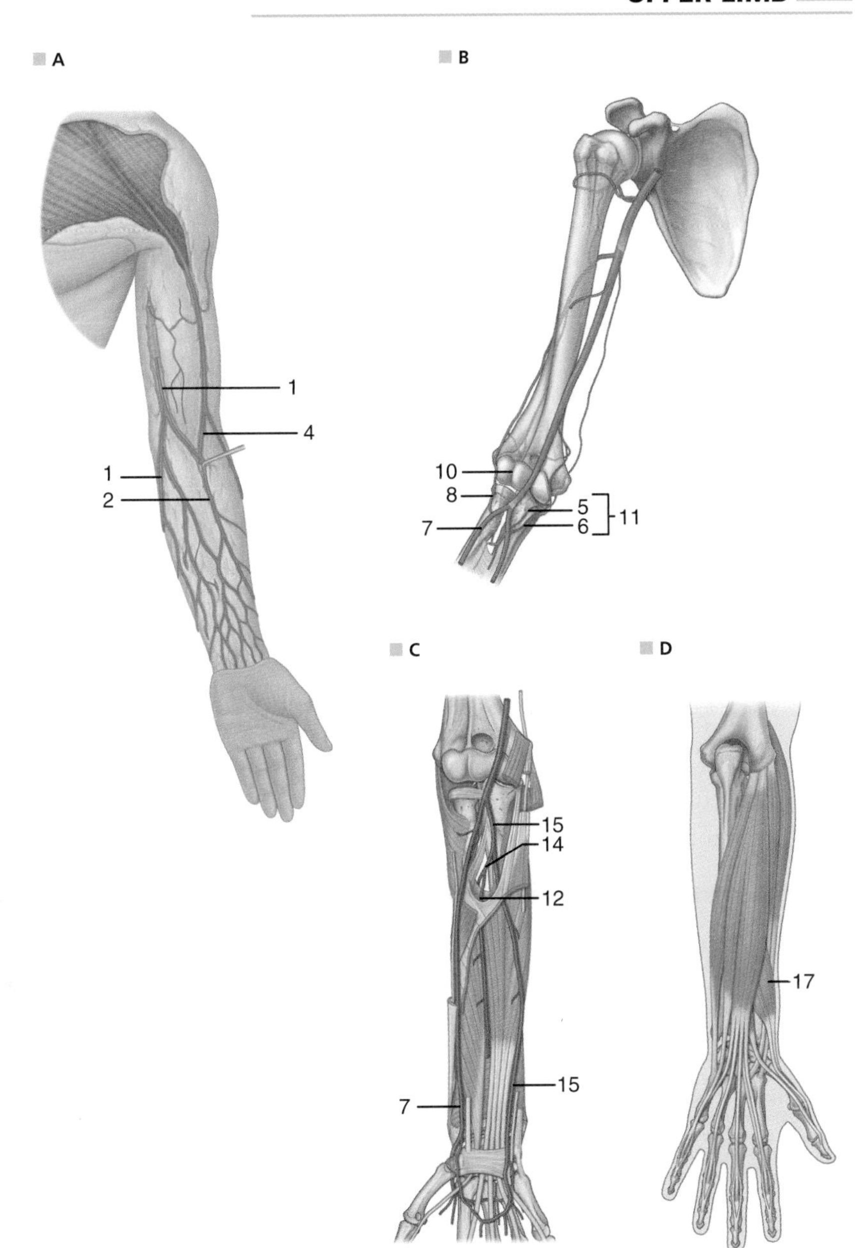
A
1
4
1
2
B
10
8
7
5
6
11
C
15
14
12
15
7
D
17

Muscles of Posterior Compartment of Forearm

1 **brachioradialis muscle** *origin,* lateral supracondylar ridge of humerus; *insertion,* lower end of radius; *innervation,* radial; *action,* flexes forearm. <musculus brachioradialis> A

2 **anconeus muscle** *origin,* back of lateral epicondyle of humerus; *insertion,* olecranon and posterior surface of ulna; *innervation,* radial; *action,* extends forearm. <musculus anconeus> B

3 **extensor carpi radialis brevis muscle** *origin,* lateral epicondyle of humerus, *insertion,* base of third metacarpal bone; *innervation,* deep branch of radial; *action,* extends and abducts wrist joint. <musculus extensor carpi radialis brevis> B

4 **extensor carpi radialis longus muscle** *origin,* lateral supracondylar ridge of humerus; *insertion,* base of second metacarpal bone; *innervation,* radial; *action,* extends and abducts wrist joint. <musculus extensor carpi radialis longus> B

5 **extensor carpi ulnaris muscle** (2 heads): *origin,* lateral epicondyle of humerus, posterior border of ulna; *insertion,* base of fifth metacarpal bone; *innervation,* deep branch of radial; *action,* extends and adducts wrist joint. <musculus extensor carpi ulnaris> B

6 **extensor digiti minimi muscle** *origin,* common extensor tendon and adjacent intermuscular septa; *insertion,* extensor expansion of little finger; *innervation,* deep branch of radial; *action,* extends little finger. <musculus extensor digiti minimi> B

7 **extensor digitorum muscle** *origin,* lateral epicondyle of humerus; *insertion,* extensor expansion of each (nonthumb) finger; *innervation,* posterior interosseus; *action,* extends wrist joint and phalanges. <musculus extensor digitorum> B

8 **extensor retinaculum of muscles of hand** the distal part of the antebrachial fascia, overlying the extensor tendons. <retinaculum musculorum extensorum manus> B

9 **extensor indicis muscle** *origin,* posterior surface of body of ulna, interosseous membrane; *insertion,* extensor expansion of index finger; *innervation,* posterior interosseous; *action,* extends index finger. <musculus extensor indicis> C

10 **extensor pollicis brevis muscle** *origin,* posterior surface of radius and interosseous membrane; *insertion,* dorsolateral base of proximal phalanx of thumb; *innervation,* posterior interosseous; *action,* extends thumb. <musculus extensor pollicis brevis> C

11 **extensor pollicis longus muscle** *origin,* posterior surface of ulna and interosseous membrane; *insertion,* base of distal phalanx of thumb; *innervation,* posterior interosseous; *action,* extends thumb, adducts and rotates thumb laterally. <musculus extensor pollicis longus> C

12 **supinator muscle** *origin,* lateral epicondyle of humerus, ulna, elbow joint fascia; *insertion,* radius; *innervation,* deep radial; *action,* supinates forearm. <musculus supinator> D

Nerves of Posterior Compartment of Forearm

13 **posterior branch of medial cutaneous nerve of forearm** a branch that innervates the skin of the posteromedial and medial aspects of the forearm; *modality,* general sensory. <ramus posterior nervi cutanei antebrachii medialis> E

14 **posterior cutaneous nerve of forearm** *origin,* radial nerve; *distribution,* skin of dorsal aspect of forearm; *modality,* general sensory. Called also *posterior antebrachial cutaneous nerve.* <nervus cutaneus antebrachii posterior> E

15 **radial nerve** *origin,* posterior cord of brachial plexus (C6-C8, and sometimes C5 and T1); *branches,* posterior cutaneous and inferior lateral cutaneous nerves of arm, posterior cutaneous nerve of forearm, muscular, deep, and superficial branches; *distribution,* descending in the back of arm and forearm, it is ultimately distributed to skin on back of arm, forearm, and hand, extensor muscles on back of arm and forearm, and elbow joint and many joints of hand; see individual branches; *modality,* general sensory and motor. <nervus radialis> F

16 **muscular branches of radial nerve** branches that innervate the triceps, anconeus, brachioradialis, and extensor carpi radialis muscles; a branch to the brachialis muscle is probably sensory; *modality,* motor and sensory. <rami musculares nervi radialis> F

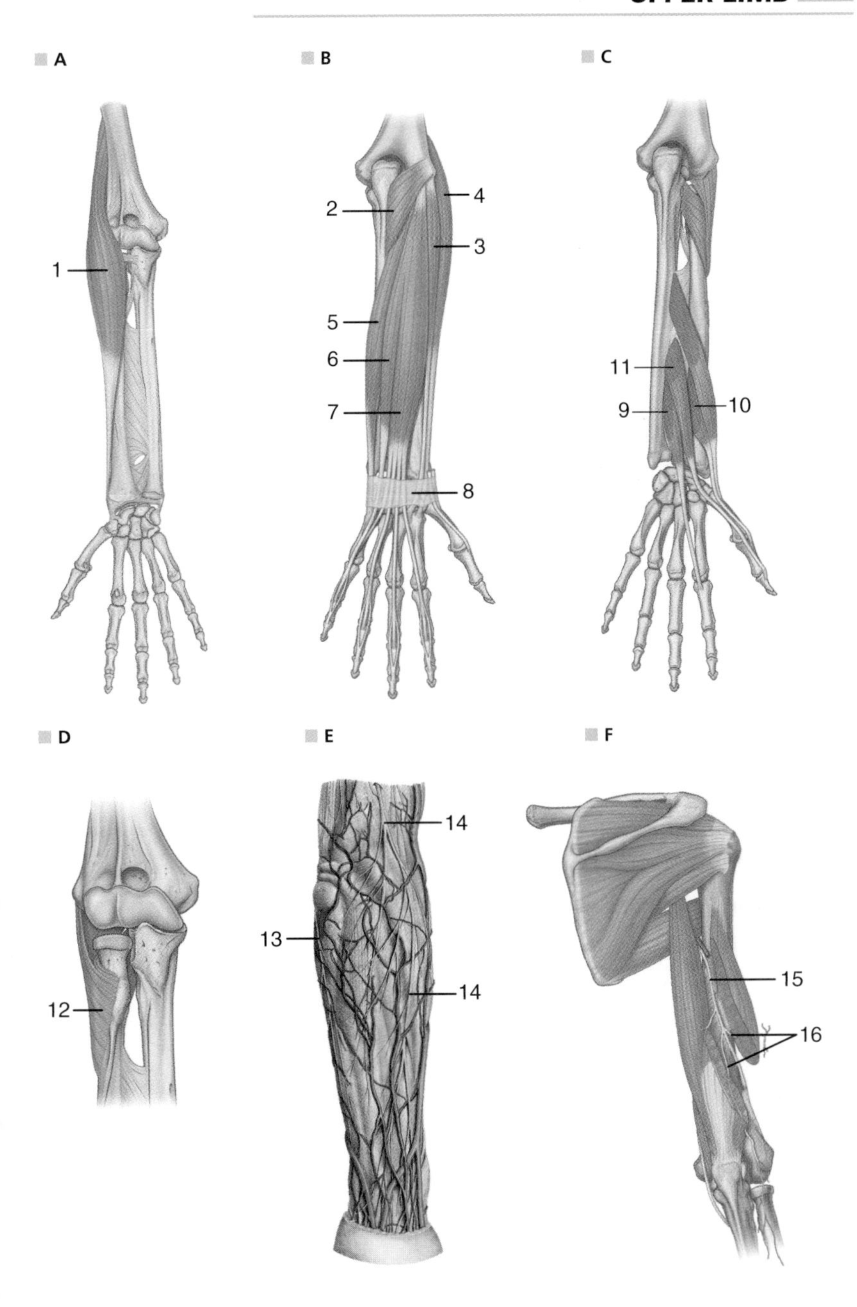
A
1
B
2
4
3
5
6
7
8
C
11
9
10
D
12
E
14
13
14
F
15
16

1 **deep branch of radial nerve** a branch arising from the radial nerve and winding laterally around the radius to the back of the forearm, supplying the supinator, extensor digitorum, extensor digiti minimi, and extensor carpi ulnaris muscles, and often the extensor carpi radialis brevis muscle. Its continuation is the posterior interosseous nerve; *modality,* motor. <ramus profundus nervi radialis> A

2 **posterior interosseous nerve of forearm** *origin,* continuation of deep branch of radial nerve; *distribution,* abductor pollicis longus, extensors of the thumb and second finger, and wrist and intercarpal joints; *modality,* motor and general sensory. <nervus interosseus antebrachii posterior> B

3 **dorsal branch of ulnar nerve** a large cutaneous branch that arises from the ulnar nerve and passes down the distal portion of the forearm to the medial side of the back of the hand, where it divides usually into three, sometimes four, dorsal digital nerves; *modality,* general sensory. <ramus dorsalis nervi ulnaris> C

Vessels of Posterior Compartment of Forearm

4 **median cephalic vein** a vein sometimes present as the lateral branch, ending in the cephalic vein, of a bifurcation of the median antebrachial vein. <vena cephalica antebrachii> D

5 **posterior interosseous artery** *origin,* common interosseous artery; *branches,* recurrent interosseous; *distribution,* deep parts of back of forearm. Called also *dorsal interosseous artery.* <arteria interossea posterior> E

6 ***posterior interosseous veins*** the veins accompanying the posterior interosseous artery, which join the ulnar veins near the elbow. <venae interosseae posteriores> E

7 **accessory cephalic vein** a vein arising from the dorsal rete of the hand, passing up the forearm to join the cephalic vein just above the elbow. <vena cephalica accessoria> F

8 **cephalic vein** the superficial vein that arises from the radial side of the dorsal rete of the hand, and winds anteriorly to pass along the anterior border of the brachioradialis muscle; above the elbow it ascends along the lateral border of the biceps muscle and the pectoral border of the deltoid muscle, and opens into the axillary vein. <vena cephalica> F

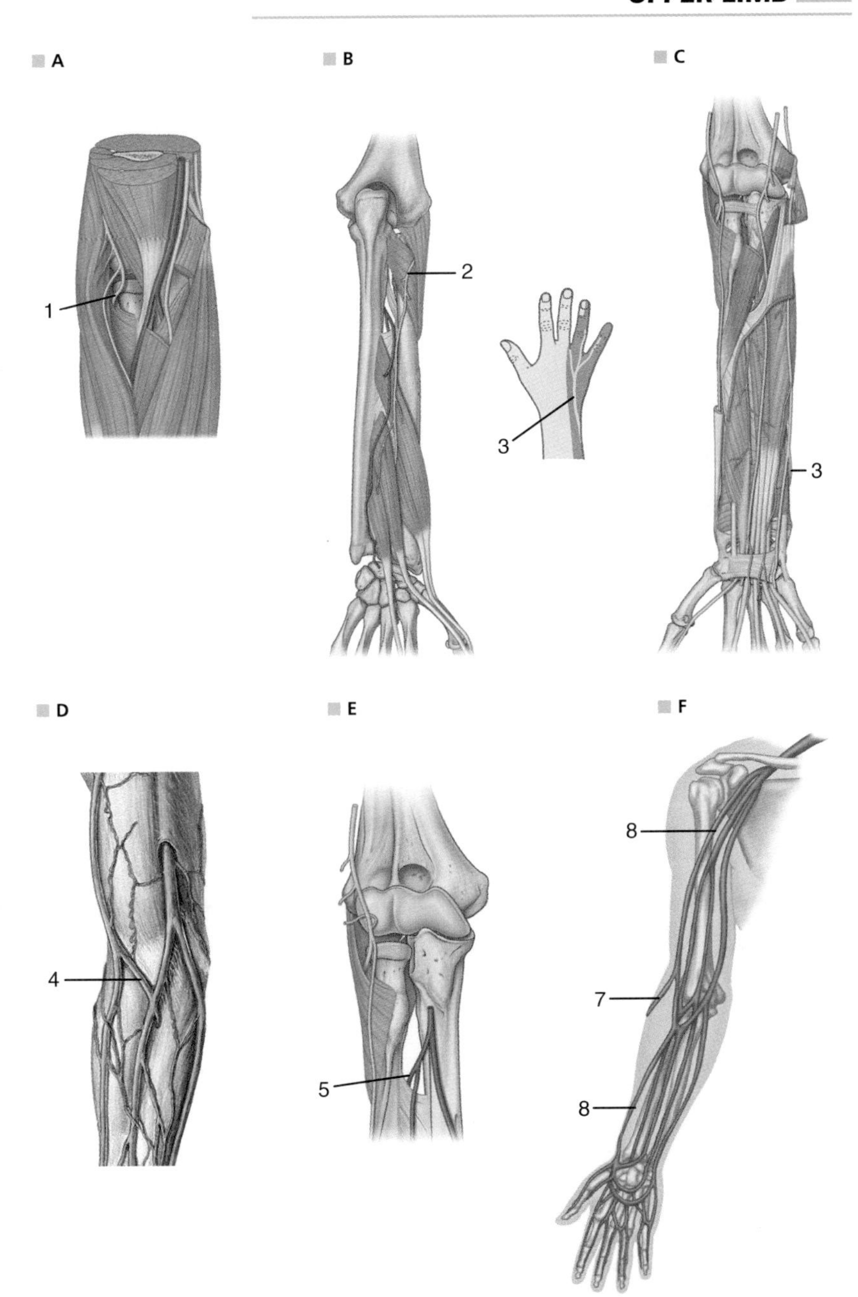
A
B
C
1
2
3
3
D
E
F
4
5
8
7
8

1 **intertendinous connections** narrow bands extending obliquely between the tendons of insertion of the extensor digitorum muscles on the dorsum of the hand. <connexus intertendinei> A

2 **tendon sheath of abductor pollicis longus and extensor pollicis brevis** the tendon sheath enclosing the tendons of the abductor pollicis longus and extensor pollicis brevis muscles, passing into the hand on the lateral side of the wrist. <vagina tendinum musculorum abductoris longi et extensoris pollicis brevis> A

3 **tendon sheath of extensor carpi ulnaris** the tendon sheath enclosing the tendon of the extensor carpi ulnaris muscle, passing into the hand along the medial side of the wrist. <vagina tendinis musculi extensoris carpi ulnaris> A

4 **tendon sheath of extensor digiti minimi muscle** the tendon sheath enclosing the tendon of the extensor digiti minimi muscle, passing into the hand along the medial side of the wrist. <vagina tendinis musculi extensoris digiti minimi> A

5 **tendon sheath of extensor pollicis longus** the tendon sheath enclosing the tendon of the extensor pollicis longus muscle, passing into the hand on the lateral side of the wrist. <vagina tendinis musculi extensoris pollicis longi> A

6 **tendon sheath of extensores carpi radiales** the tendon sheath enclosing the tendons of the extensor carpi radialis brevis and longus muscles, passing into the hand along the posterior side of the wrist. <vagina tendinum musculorum extensorum carpi radialium> A

7 ***tendon sheath of flexor carpi radialis*** the tendon sheath enclosing the tendon of the flexor carpi radialis muscle, which passes through a tubular compartment formed by attachment of the lateral aspect of the flexor retinaculum to the margins of a groove on the medial side of the tubercle of trapezium, on the palmar aspect of the wrist. <vagina tendinis musculi flexoris carpi radialis> A

8 **tendon sheath of the extensor digitorum and extensor indicis** the tendon sheath enclosing the tendons of the extensor digitorum and extensor indicis muscles, passing into the hand along the posterior side of the wrist. <vagina tendinum musculorum extensoris digitorum et extensoris indicis> A

HAND

9 **fingers** the digits of the hand, sometimes excluding the thumb. <digiti manus> B

10 **hand** the distal region of the upper limb, including the carpus, metacarpus, and digits. <manus> B

11 **index finger** the second digit of the hand, the finger adjacent to the thumb; called also *forefinger*. <index> B

12 **interosseous metacarpal spaces** the four spaces between the metacarpal bones. <spatia interossea metacarpi> B

13 **little finger** the fifth, and smallest, digit of the hand. <digitus minimus manus> B

14 **metacarpal region** the region of the surface of the hand overlying the metacarpals. <regio metacarpalis> B

15 **metacarpus** the part of the hand between the wrist and the fingers, its skeleton being five cylindric bones (metacarpals) extending from the carpus to the phalanges. B

16 **middle finger** the third digit of the hand. <digitus medius> B

17 **ring finger** the fourth digit of the hand. <digitus anularis> B

18 **thumb** the first digit of the hand; it is the most preaxial of the five fingers, having only the two phalanges and being opposable to the other four fingers. <pollex> B

19 ***dorsal fascia of hand*** the investing fascia of the back of the hand. <fascia dorsalis manus> C

20 **dorsal surfaces of fingers** the posterior surfaces of the fingers. <facies dorsales digitorum manus> C

21 **dorsum of hand** the hand surface opposite the palm. <dorsum manus> C

22 **nail** the horny cutaneous plate on the dorsal surface of the distal end of the terminal phalanx of a finger, made up of flattened epithelial scales developed from the stratum lucidum of the skin. <unguis> C

23 **palmar surfaces of fingers** the anterior surfaces of the fingers. <facies palmares digitorum manus> C

24 **hypothenar eminence** the fleshy eminence on the palm along the ulnar margin. <hypothenar> D

25 **palm** the flexor surface of the hand. Called also *vola* [TA alternative] and *regio palmaris* [TA alternative]. <palma> D

26 ***palmar*** a general term designating relationship to the palm of the hand. <palmaris> D

27 **thenar eminence** the mound on the palm at the base of the thumb. <thenar> D

28 **fibrous sheaths of fingers** strong fibrous, semicylindrical sheaths investing the grooved palmar surface of the proximal and middle phalanges of the fingers. <vaginae fibrosae digitorum manus> E

29 **synovial sheaths of fingers** the synovial sheaths surrounding the tendons of the fingers. <vaginae synoviales digitorum manus> E

A

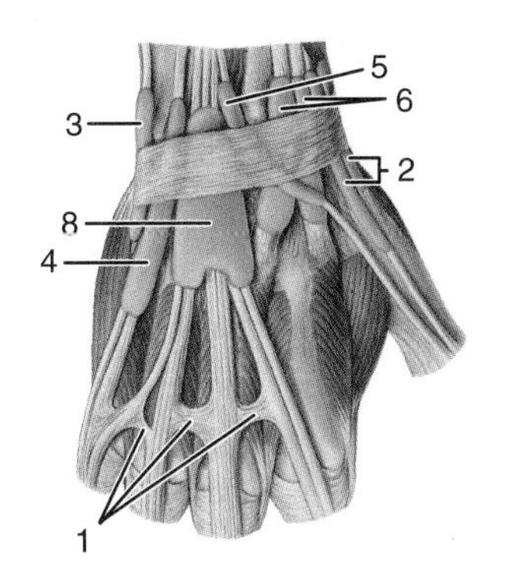

B

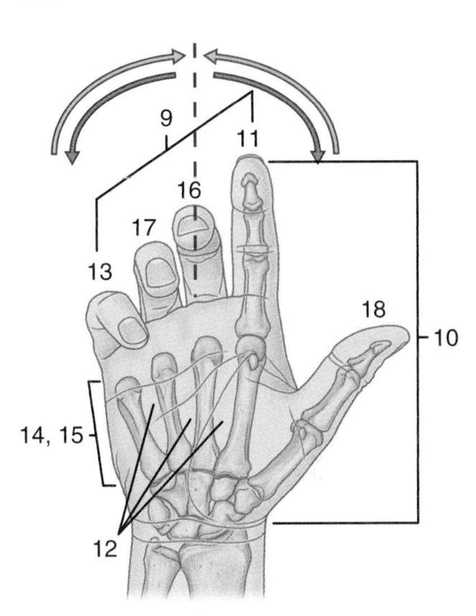

C

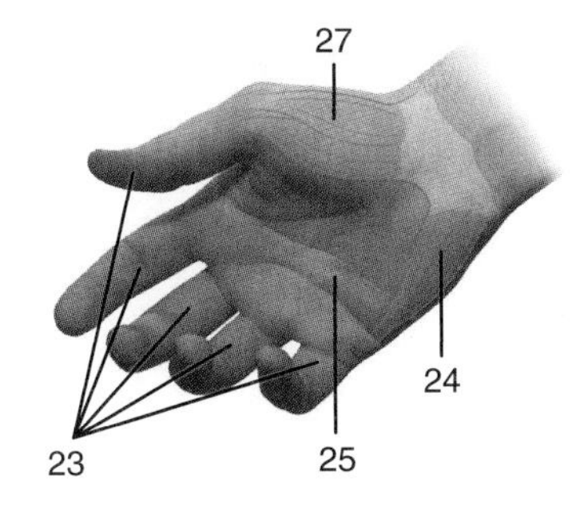

D

27
24
23
25

E

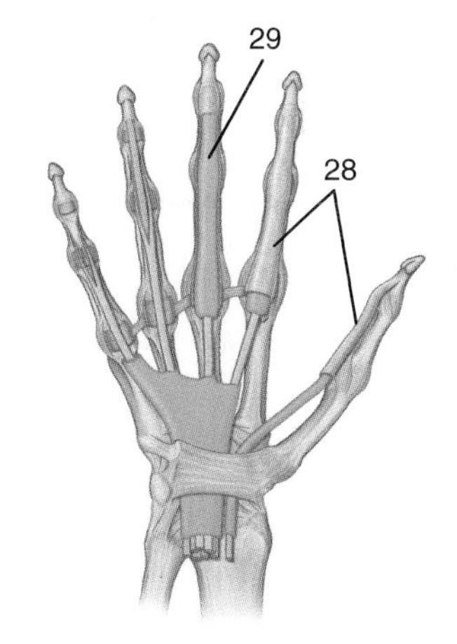

1 **annular part of fibrous sheaths of fingers** strong transverse bands of fibrous tissue, one in the fibrous sheath of each finger, crossing the flexor tendons at the level of the upper half of the proximal phalanx; called also *annular ligaments of fingers.* <pars anularis vaginae fibrosae digitorum manus> A

2 **cruciate part of fibrous sheaths of fingers** one of the diagonal bundles of the fascia of the fingers that cross each other on the dorsal surface of each digit at the level of the distal end of the proximal phalanx; called also *ligamenta cruciata digitorum manus* and *cruciate ligaments of fingers.* <pars cruciformis vaginae fibrosae digitorum manus> A

3 ***vincula of tendons of fingers*** small vascular bands that connect the tendons of the flexor digitorum profundus and flexor digitorum superficialis muscles to the phalanges and interphalangeal joints of the hand. The tendon blood supply is also carried in them. <vincula tendinum digitorum manus> B

4 **vinculum breve of fingers** either of two fan-shaped expansions near the ends of the flexor tendons of a finger, one connecting the superficial tendon to the proximal interphalangeal joint and the other connecting the deep tendon to the intermediate interphalangeal joint. <vinculum breve digitorum manus> B

5 **vinculum longum of fingers** either of two independent pairs of slender bands in each finger, one connecting the deep flexor tendon to the superficial tendon after the latter becomes subjacent, and the other connecting the superficial tendon to the proximal phalanx. <vinculum longum digitorum manus> B

Muscles of Hand

6 **palmar aponeurosis** bundles of fibrous tissue radiating toward the bases of the fingers from the tendon of the palmaris longus muscle. <aponeurosis palmaris> C

7 **palmaris brevis muscle** *origin,* palmar aponeurosis; *insertion,* skin of medial border of hand; *innervation,* ulnar; *action,* assists in deepening hollow of palm. <musculus palmaris brevis> C

8 **common flexor sheath** the common sheath for the flexor tendons as they pass through osteofibrous canals of the fingers. <vagina communis musculorum flexorum manus> D

9 **tendon sheath of flexor pollicis longus** the tendon sheath enclosing the tendon of the flexor pollicis longus muscle passing into the hand on the lateral side of the wrist. <vagina tendinis musculi flexoris pollicis longi> D

10 **dorsal interossei muscles of hand** (4): *origin,* by two heads from adjacent sides of metacarpal bones; *insertion,* bases of proximal phalanges and corresponding extensor expansions of second, third, and fourth fingers; *innervation,* ulnar; *action,* abduct fingers, flex proximal phalanges, extend middle and distal phalanges. <musculi interossei dorsales manus> E

11 **palmar interossei muscles** (3): *origin,* sides of second, fourth, and fifth metacarpal bones; *insertion,* bases of proximal phalanges and corresponding extensor expansions of second, fourth, and fifth fingers; *innervation,* ulnar; *action,* adduct fingers, flex proximal phalanges, extend middle and distal phalanges. <musculi interossei palmares> F

A

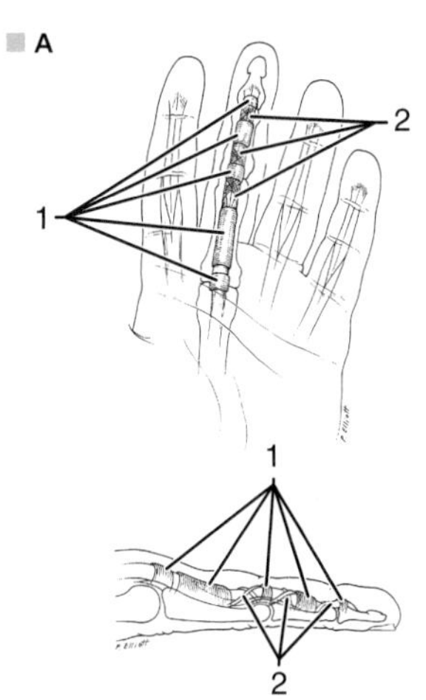

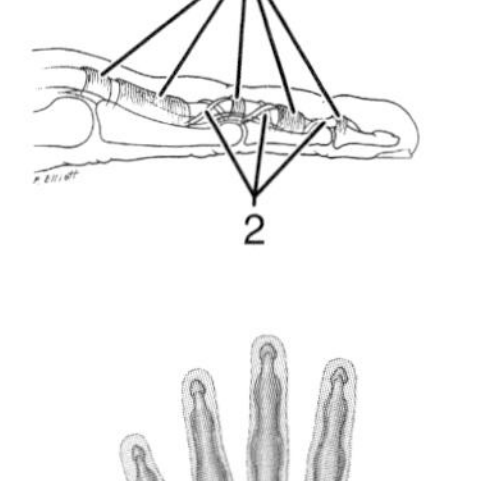

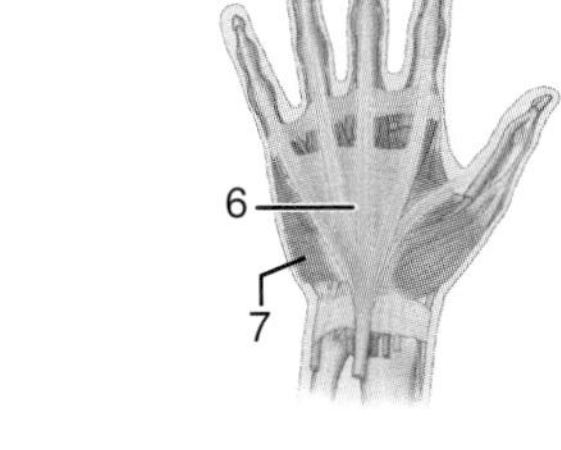

B

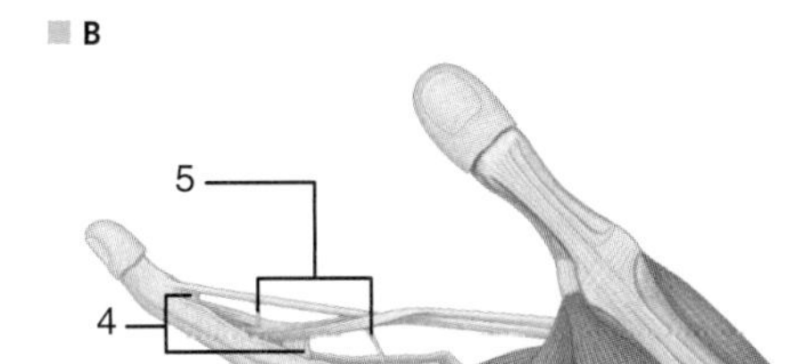

C

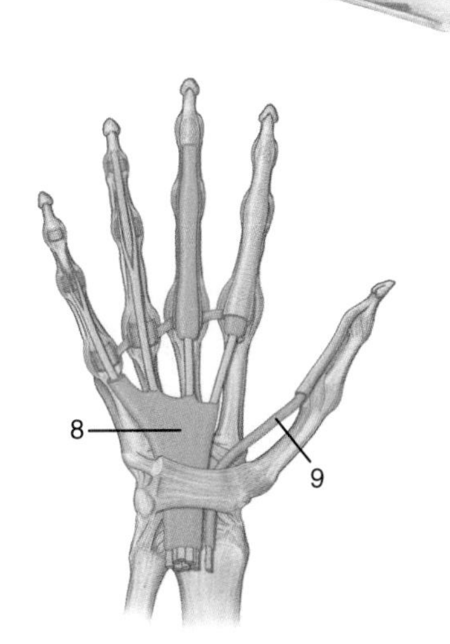

D

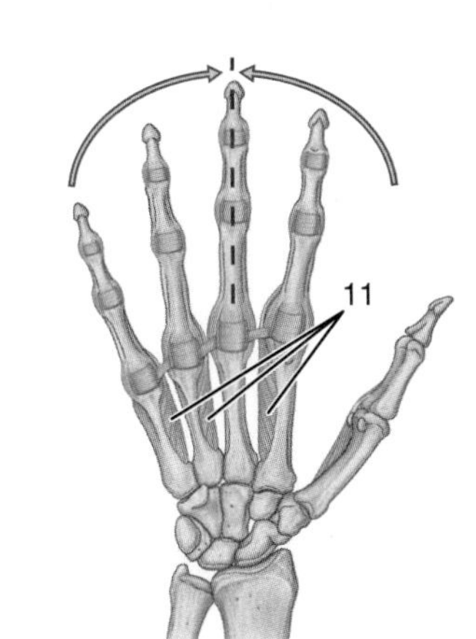

E

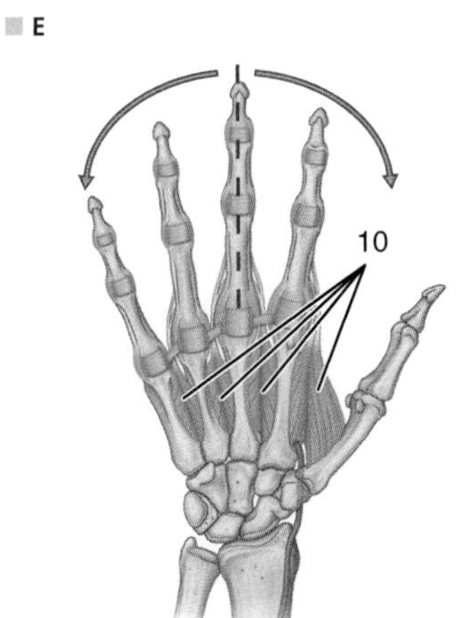

F

1 **adductor pollicis muscle** two heads—oblique and transverse; *insertion,* medial surface of base of proximal phalanx of thumb; *innervation,* ulnar; *action,* adducts, opposes thumb. <musculus adductor pollicis> A

2 **oblique head of adductor pollicis muscle** the head of the adductor pollicis muscle arising from the capitate bone and the bases of the second and third metacarpals. <caput obliquum musculi adductoris pollicis> A

3 **transverse head of adductor pollicis muscle** the head of the adductor pollicis muscle arising from the lower anterior surface of the third metacarpal. <caput transversum musculi adductoris pollicis> A

4 **abductor digiti minimi muscle of hand** *origin,* pisiform bone, flexor carpi ulnaris tendon; *insertion,* medial surface of base of proximal phalanx of little finger; *innervation,* ulnar; *action,* abducts little finger. <musculus abductor digiti minimi manus> B

5 **abductor pollicis brevis muscle** *origin,* scaphoid, ridge of trapezium, flexor retinaculum of hand; *insertion,* lateral surface of base of proximal phalanx of thumb; *innervation,* median; *action,* abducts thumb. <musculus abductor pollicis brevis> B

6 **flexor digiti minimi brevis muscle of hand** *origin,* hook of hamate bone, transverse carpal ligament; *insertion,* medial side of proximal phalanx of little finger; *innervation,* ulnar; *action,* flexes little finger. <musculus flexor digiti minimi brevis manus> B

7 **flexor pollicis brevis muscle** *origin,* flexor retinaculum, distal part of tubercle of trapezium; *insertion,* radial side of base of proximal phalanx of thumb; *innervation,* median, ulnar; *action,* flexes and adducts thumb. <musculus flexor pollicis brevis> B

8 **opponens digiti minimi muscle** *origin,* hook of hamate bone, flexor retinaculum; *insertion,* ulnar margin of fifth metacarpal; *innervation,* eighth cervical through ulnar; *action,* rotates, abducts, and flexes fifth metacarpal. <musculus opponens digiti minimi> B

9 **opponens pollicis muscle** *origin,* tubercle of trapezium, flexor retinaculum; *insertion,* radial side of first metacarpal; *innervation,* sixth and seventh cervical through median; *action,* flexes and opposes thumb. <musculus opponens pollicis> B

10 **lumbrical muscles of hand** *origin,* tendons of flexor digitorum profundus; *insertion,* extensor expansions of four nonthumb fingers; *innervation,* median and ulnar; *action,* flex metacarpophalangeal joint and extend middle and distal phalanges. <musculi lumbricales manus> C

11 **tendinous chiasm of fingers** the crossing of the tendons of the flexor digitorum profundus through the tendons of the flexor digitorum superficialis. <chiasma tendinum digitorum manus> C

Nerves of Hand

12 **common palmar digital nerves of ulnar nerve** (2): *origin,* superficial branch of ulnar nerve; *branches,* proper palmar digital nerves; *distribution,* little and ring fingers; *modality,* general sensory. <nervi digitales palmares communes nervi ulnaris> D

13 **palmar branch of median nerve** a branch arising from the median nerve in the lower part of the forearm and supplying the skin of the central palm and of the thenar eminence; *modality,* general sensory. <ramus palmaris nervi mediani> D

14 **palmar branch of ulnar nerve** a branch arising from the ulnar nerve in the lower part of the forearm, supplying the cutaneous structures of the medial part of the palm; *modality,* general sensory. <ramus palmaris nervi ulnaris> D

15 **proper palmar digital nerves of median nerve** *origin,* common palmar digital nerves; *distribution,* first two lumbrical muscles, skin and joints of both sides and palmar aspect of thumb, index, and middle fingers, radial side of ring finger, and back of distal aspect of these digits; *modality,* general sensory and motor. <nervi digitales palmares proprii nervi mediani> D

16 **proper palmar digital nerves of ulnar nerve** *origin,* the lateral of the two common palmar digital nerves from the superficial branch of the ulnar nerve; *distribution,* skin and joints of adjacent sides of fourth and fifth fingers; *modality,* general sensory. <nervi digitales palmares proprii nervi ulnaris> D

17 **superficial branch of ulnar nerve** the branch of the ulnar nerve in the hand that supplies the palmaris brevis muscle and divides into a proper palmar digital nerve for the medial side of the little finger, a common palmar digital nerve giving off two proper nerves to supply adjacent sides of the little and fourth fingers, and sometimes palmar digital nerves also for the adjacent sides of the third and fourth fingers; *modality,* general sensory and motor. <ramus superficialis nervi ulnaris> D

18 **common palmar digital nerves of median nerve** (4): *origin,* lateral and medial divisions of median nerve; *branches,* proper palmar digital nerves; *distribution,* thumb, index, middle, and ring fingers, and first two lumbrical muscles; *modality,* motor and general sensory. <nervi digitales palmares communes nervi mediani> D

A

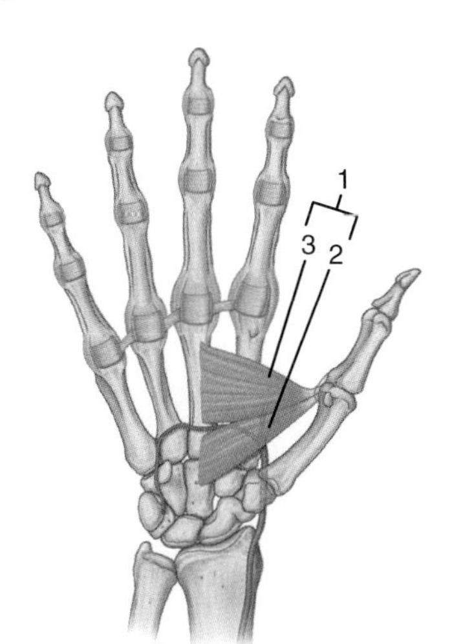

B

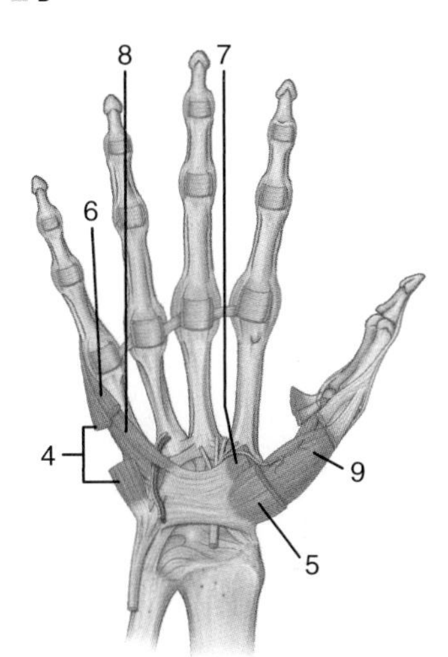

C

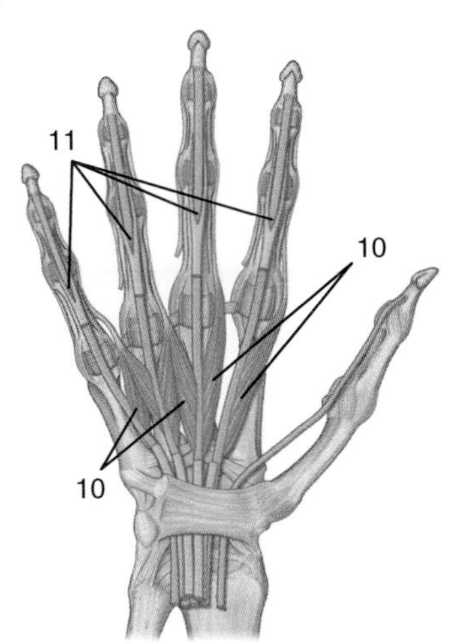

D

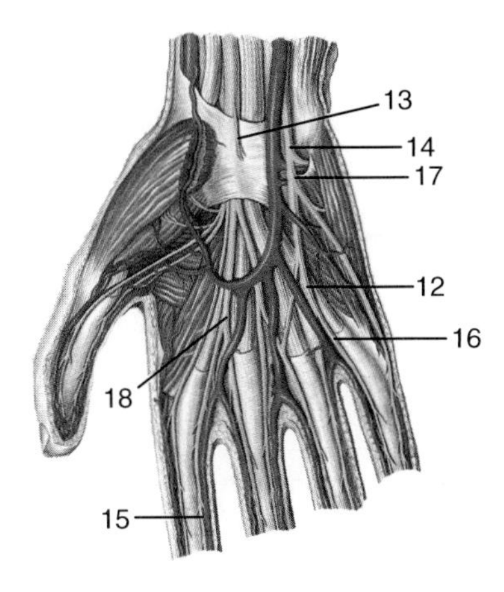

1 **deep branch of ulnar nerve** the deep branch that is accompanied by the deep palmar branch of the ulnar artery, rounds the hook of the hamate bone, and follows the deep palmar arch beneath the flexor tendons, supplying the wrist joint, the interossei, third and fourth lumbrical, and adductor pollicis muscles, and usually the deep head of the flexor pollicis brevis muscle; *modality,* general sensory and motor. <ramus profundus nervi ulnaris> A

2 **dorsal digital nerves of radial nerve** *origin,* superficial branch of radial nerve; *distribution,* skin and joints of back of thumb, index finger, and part of middle finger, as far distally as the distal phalanx; *modality,* general sensory. <nervi digitales dorsales nervi radialis> B

3 **dorsal digital nerves of ulnar nerve** *origin,* dorsal branch of ulnar nerve; *distribution,* skin and joints of medial side of little finger, dorsal aspects of adjacent sides of little and ring fingers and of ring and middle fingers; *modality,* general sensory. <nervi digitales dorsales nervi ulnaris> B

4 **superficial branch of radial nerve** the continuation of the radial nerve that accompanies the radial artery in the forearm, winds dorsalward, supplies the lateral side of the back of the hand, and divides into dorsal digital nerves that supply the skin of the dorsal surface and adjacent surfaces of the thumb, index, and middle fingers, and sometimes the radial side of the ring finger; *modality,* general sensory. <ramus superficialis nervi radialis> B

5 **ulnar communicating branch of radial nerve** a small branch in the hand that interconnects the most medial dorsal digital nerve from the superficial branch of the radial nerve with the adjacent most lateral dorsal digital nerve from the dorsal branch of the ulnar nerve. <ramus communicans ulnaris nervi radialis> B

Vessels of Hand

6 **palmar carpal branch of radial artery** a branch that passes medially behind the flexor tendons on the palmar aspect of the wrist and forms a network with a corresponding branch of the ulnar artery. <ramus carpalis palmaris arteriae radialis> A

7 **proper palmar digital arteries** *origin,* common palmar digital arteries; *branches,* none; *distribution,* fingers. <arteriae digitales palmares propriae> C

8 **palmar carpal branch of ulnar artery** a branch that passes laterally behind the flexor tendons on the palmar aspect of the wrist and forms a network with a corresponding branch of the radial artery. <ramus carpalis palmaris arteriae ulnaris> A

9 **palmar metacarpal arteries** *origin,* deep palmar arch; *branches,* none; *distribution,* deep parts of metacarpus. <arteriae metacarpales palmares> A

10 ***palmar metacarpal veins*** the venae comitantes of the palmar metacarpal arteries, which open into the deep palmar venous arch. <venae metacarpales palmares> A

11 **dorsal metacarpal veins** veins that arise from the union of dorsal veins of adjacent fingers and pass proximally to join in forming the dorsal venous rete of the hand. <venae metacarpales dorsales> B

12 **dorsal venous rete of hand** a venous network on the back of the hand, formed by the dorsal metacarpal veins. <rete venosum dorsale manus> B

13 **intercapitular veins of hand** veins at the clefts of the finger that pass between the heads of the metacarpal bones and establish communication between the dorsal and palmar venous systems of the hand. <venae intercapitulares manus> B

14 **dorsal carpal arch** an arterial rete formed by the dorsal radial carpal and dorsal ulnar carpal arteries and giving off the second, third, and fourth dorsal metacarpal arteries to the dorsum of the hand and the second, third, and fourth fingers. <rete carpale dorsale> D

15 **dorsal carpal branch of radial artery** a branch running medially deep to the extensor tendons, and helping form the dorsal carpal arch. <ramus carpalis dorsalis arteriae radialis> D

16 **dorsal carpal branch of ulnar artery** a variable branch of the ulnar artery that runs laterally deep to the tendons of the ulnar muscles of the wrist, helping to form the dorsal carpal arch. <ramus carpalis dorsalis arteriae ulnaris> D

17 **dorsal metacarpal arteries** *origin,* dorsal carpal arch and radial artery; *branches,* dorsal digital arteries; *distribution,* dorsum of fingers. <arteriae metacarpales dorsales> D

A

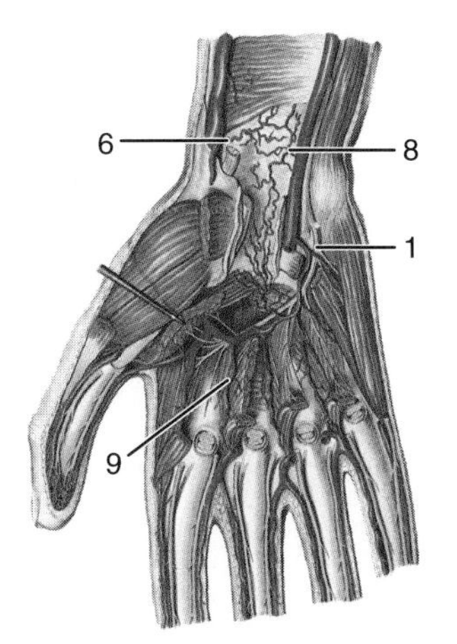

B

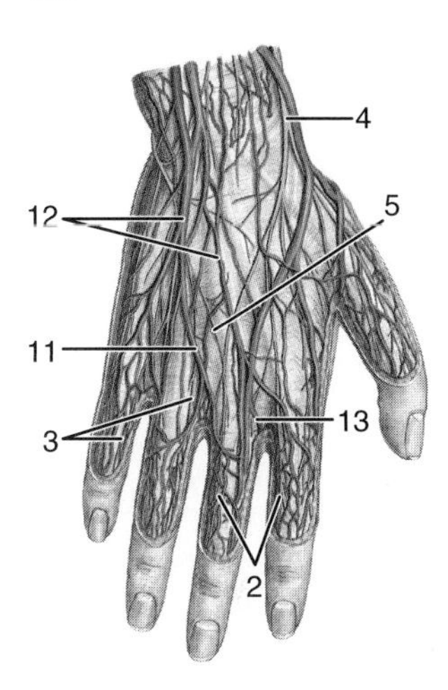

C

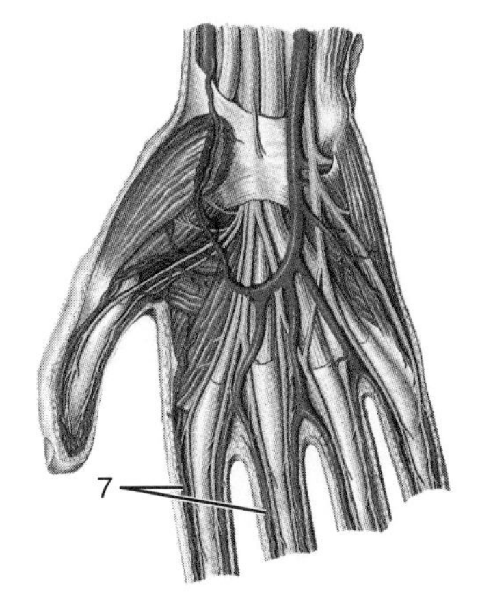

D

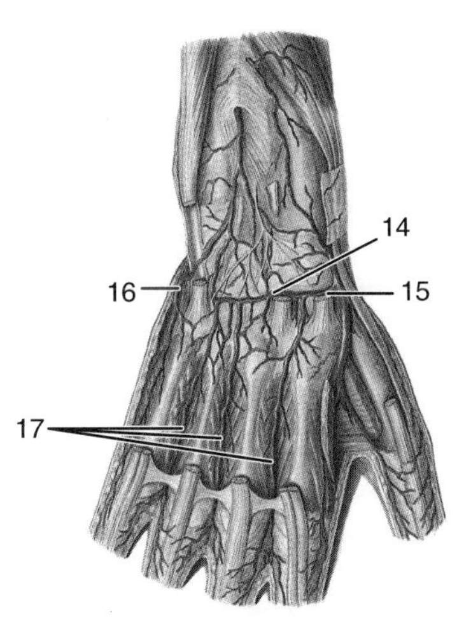

1 **dorsal digital arteries of hand** *origin,* dorsal metacarpal arteries; *branches,* none; *distribution,* dorsum of fingers. <arteriae digitales dorsales manus> A

2 **perforating branches of deep palmar arch** vessels connecting the palmar metacarpal arteries and deep palmar arch with the dorsal metacarpal arteries, between the bases of the metacarpal bones and in the interosseous spaces. <rami perforantes arcus palmaris profundus> A

3 **common palmar digital arteries** common palmar digital arteries: *origin,* superficial palmar arch; *branches,* proper palmar digital arteries; *distribution,* fingers. <arteriae digitales palmares communes> A, B

4 ***palmar digital veins*** the venae comitantes of the proper and common palmar digital arteries, which join the superficial palmar venous arch. <venae digitales palmares> A, B

5 **superficial palmar arch** an arterial arch formed by the terminal part of the ulnar artery and its anastomosis with the superficial palmar branch of the radial, giving rise to the palmar digital arteries and supplying blood to the palmar aspect of the hands and fingers. <arcus palmaris superficialis> B

6 **superficial palmar branch of radial artery** a branch arising from the radial artery in the lower part of the forearm and supplying the thenar eminence. <ramus palmaris superficialis arteriae radialis> B

7 **deep palmar arch** an arterial arch formed by the terminal part of the radial artery and its anastomosis with the deep branch of the ulnar, and extending from the base of the metacarpal bone of the little finger to the proximal end of the first interosseous space; it gives off palmar metacarpal arteries and perforating branches. <arcus palmaris profundus> C

8 **deep palmar branch of ulnar artery** a branch that accompanies the deep palmar branch of the ulnar nerve and joins the radial artery to form the deep palmar arch. <ramus palmaris profundus arteriae ulnaris> C

9 ***deep palmar venous arch*** a venous arch accompanying the deep palmar arterial arch. <arcus venosus palmaris profundus> C

10 **princeps pollicis artery** *origin,* radial artery; *branches,* radial of index finger; *distribution,* each side and palmar aspect of thumb. <arteria princeps pollicis> C

11 **radialis indicis artery** *origin,* princeps pollices artery; *branches,* none; *distribution,* index finger. <arteria radialis indicis> C

12 ***superficial palmar venous arch*** a venous arch accompanying the superficial palmar arterial arch. <arcus venosus palmaris superficialis> C

A

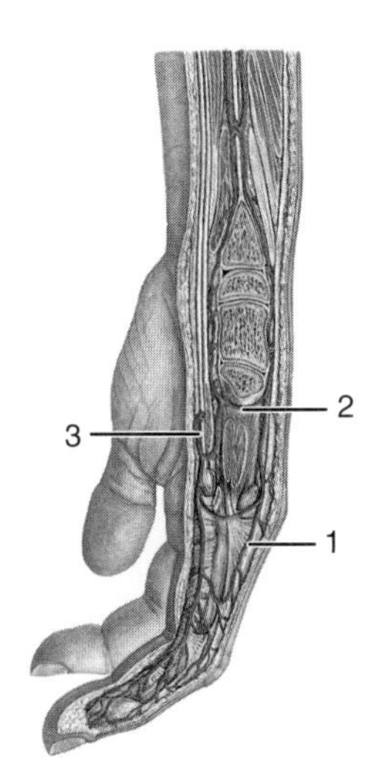

B

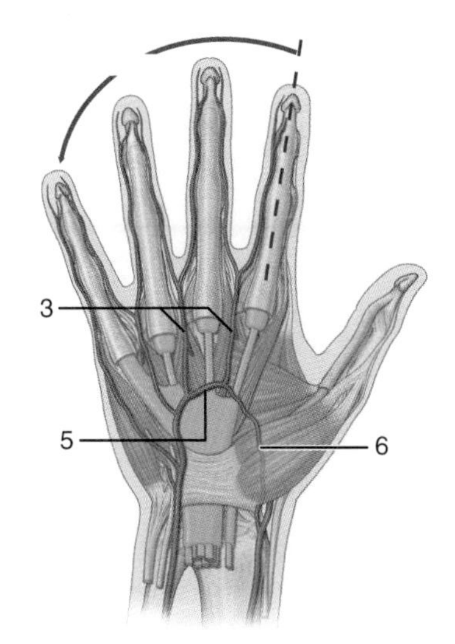

C

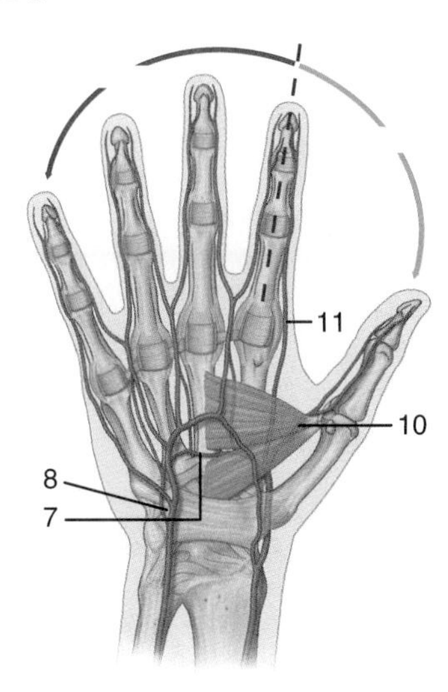

HEAD

General

1 ***head*** the superior extremity of the body, comprising the cranium and face, and containing the brain, the organs of special sense, and the first organs of the digestive system. <caput> A

2 ***sense organs*** organs that receive stimuli that give rise to sensations; they are characterized by highly specialized neuroreceptors and relationships, and include the visual, vestibulocochlear, olfactory, and gustatory organs. <organa sensuum>

3 **mastoid region** the region of the head on either side roughly corresponding to the outline of the mastoid process of the temporal bone. <regio mastoidea> A

4 ***occipital*** pertaining to the occiput or to the occipital bone. <occipitalis> A

5 **occipital region** the surface region of the head overlying the occipital bone. <regio occipitalis> A

6 **occiput** the posterior part of the head. A

7 **parietal region** the surface region of the head on either side roughly corresponding to the outline of the parietal bone. <regio parietalis> A

8 **parotideomasseteric region** the region of the face on either side, about the parotid gland and masseter muscle. <regio parotideomasseterica> A

9 ***regions of the head*** the various anatomical regions of the head, including the frontal, parietal, occipital, temporal, auricular, mastoid, and facial regions. <regiones capitis> A

10 **sinciput** the anterior and superior part of the head superior to the supercilliary arches. A

11 **temples** the regions on each side of the head superior to the zygomatic arches. <tempora> A

12 **temporal region** the surface region of the head on either side roughly corresponding to the outline of the temporal bone. <regio temporalis> A

13 **zygomatic region** the region of the face on either side, about the zygomatic bone. <regio zygomatica> A

14 **greater supraclavicular fossa** a depression on the surface of the body, located superior and posterior to the clavicle, lateral to the tendon of the sternocleidomastoid muscle. <fossa supraclavicularis major> B, C

15 **forehead** the anterior region of the head, superior to the supercilliary arches. Called also *brow*. <frons> B, C

16 **frontal region** the surface region of the head overlying the frontal bone. <regio frontalis> B, C

17 **oral region** the region of the face about the mouth. <regio oralis> B, C

18 **orbital region** the region of the face about the eye. <regio orbitalis> B, C

A

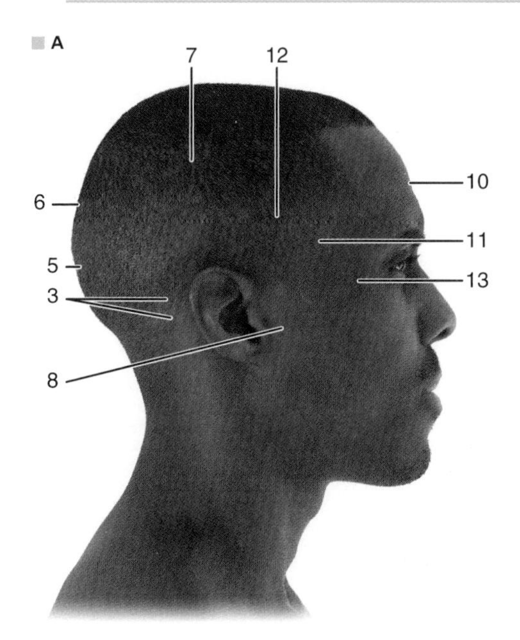

B

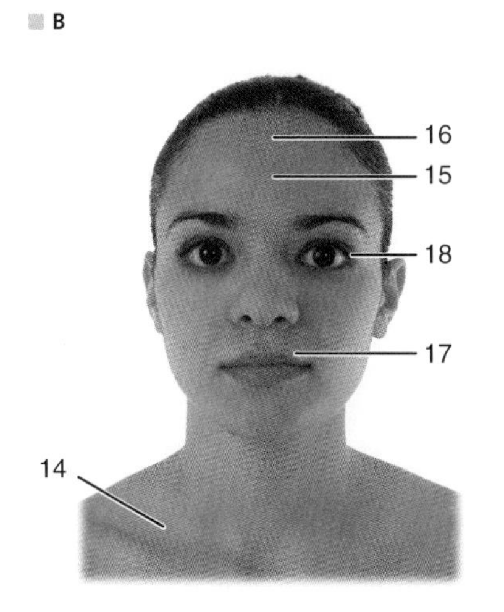

C

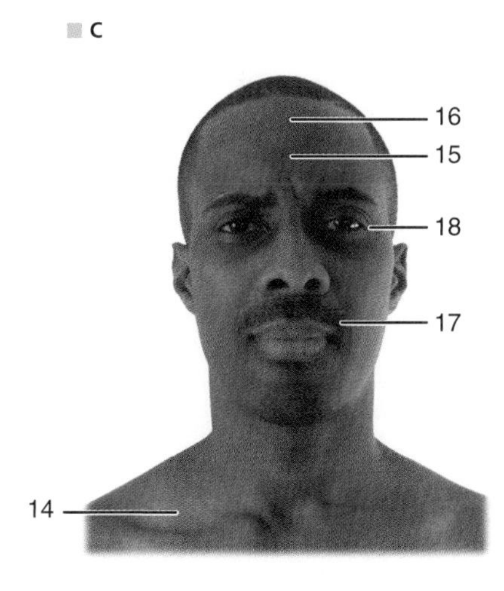

Bones

1 ***chondrocranium*** that part of the neurocranium formed by endochondral ossification and comprising the bones of the base of the skull; called also *cartilaginous neurocranium.*

2 ***cranial bones*** the bones of the cranium, including the occipital, temporal, parietal, frontal, ethmoid, sphenoid, lacrimal, and nasal bones; the nasal conchae; and the vomer. Some authorities also include the maxilla, the palatine bone, and the zygomatic bone. <ossa cranii>

3 ***cranial cartilaginous*** joints the cranial synchondroses. <juncturae cartilagineae cranii>

4 ***cranial fibrous joints*** the sutures and ligaments connecting the bones of the skull to each other, as well as the syndesmoses holding the teeth in their sockets. <juncturae fibrosae cranii>

5 ***cranial sutures*** the sutures between the various bones of the skull, named generally for the specific components participating in their formation. <suturae cranii>

6 ***cranial synovial joints*** the temporomandibular joint and the atlanto-occipital joint. <articulationes cranii>

7 ***cranium*** the large round superior part of the skull, enclosing the brain and made up of the cranial bones.

8 ***joints of skull*** junctions between bones of the skull or between skull bones and other bones, comprising fibrous, cartilaginous, and synovial joints. <juncturae cranii>

9 ***neurocranium*** brain box; the portion of the cranium which encloses the brain.

10 **anterior fontanelle** the unossified area of the skull situated at the junction of the frontal, coronal, and sagittal sutures. <fonticulus anterior> A

11 **fontanelles** the membrane-covered spaces, or soft spots, remaining at the incomplete angles of the parietal and adjacent bones, until ossification of the skull is completed. <fonticuli cranii> A

12 **posterior fontanelle** the unossified area of the skull at the junction of the sagittal and lambdoidal sutures; called also *occipital or triangular fontanelle.* <fonticulus posterior> A

13 **frontal angle of parietal bone** the anterosuperior angle of the parietal bone, which is membranous at birth and forms part of the anterior fontanelle. <angulus frontalis ossis parietalis> A

14 **mastoid fontanelle** the unossified area of the skull at the junction of the lambdoid, parietomastoid, and occipitomastoid sutures. Called also *posterolateral fontanelle.* <fonticulus mastoideus> B

15 **sphenoidal fontanelle** the unossified area at the junction of the parietal and frontal bones, the greater wing of the sphenoidal, and the squamous part of the temporal bones. Called also *anterolateral fontanelle.* <fonticulus sphenoidalis> B

16 **anterior nasal spine of maxilla** the sharp anterosuperior projection at the anterior extremity of the nasal crest of the maxilla. <spina nasalis anterior maxillae> C

17 **anterior surface of body of maxilla** the surface of the body of the maxilla that is directed forward and somewhat laterally; it is bounded roughly by the infraorbital margin, root of the frontal process, nasal notch, alveolar process, and zygomatic process. <facies anterior corporis maxillae> C

18 **frontal bone** a single bone that closes the anterior part of the cranial cavity and forms the skeleton of the forehead; it is developed from two halves, the line of separation (the frontal or metopic suture) sometimes persisting in adult life. <os frontale> C

19 ***frontal notch*** a notch located in the supraorbital margin of the frontal bone medial to the supraorbital notch or foramen, for transmission of branches of the supraorbital nerve and vessels; frequently converted into a foramen *(frontal foramen)* by a bridge of osseous tissue. <incisura frontalis> C

20 **frontomaxillary suture** the line of junction between the frontal bone and the frontal process of the maxilla. <sutura frontomaxillaris> C

21 **frontonasal suture** the line of junction between the frontal and the two nasal bones; called also *nasofrontal suture.* <sutura frontonasalis> C

22 **frontozygomatic suture** the line of junction between the zygomatic bone and the zygomatic process of the frontal bone; called also *zygomaticofrontal suture.* <sutura frontozygomatica> C

23 **glabella** the most prominent point in the median plane between the eyebrows; used as an anthropometric landmark. C

24 **gonion** an anthropometric landmark located at the most inferior, posterior, and lateral point on the external angle of the mandible, being the apex of the maximum curvature of the mandible, where the ascending ramus becomes confluent with the body. C

25 **infraorbital foramen** the opening of the infraorbital canal on the anterior surface of the maxilla giving passage to the infraorbital nerve and vessels. <foramen infraorbitale> C

26 **intermaxillary suture** the line of junction between the maxillary bones of either side, just inferior to the anterior nasal spine. <sutura intermaxillaris> C

27 **internasal suture** the line of junction between the two nasal bones. <sutura internasalis> C

28 **mandibular symphysis** the line of fusion in the median plane of the mandible that marks the union of the two halves of the mandible. <symphysis mandibulae> C

29 **maxilla** the irregularly shaped bone that with its fellow forms the upper jaw; it assists in the formation of the orbit, the nasal cavity, and the palate, and lodges the upper teeth. C

30 **mental foramen** an opening on the lateral part of the body of the mandible, opposite the second bicuspid tooth, for passage of the mental nerve and vessels. <foramen mentale> C

A

B

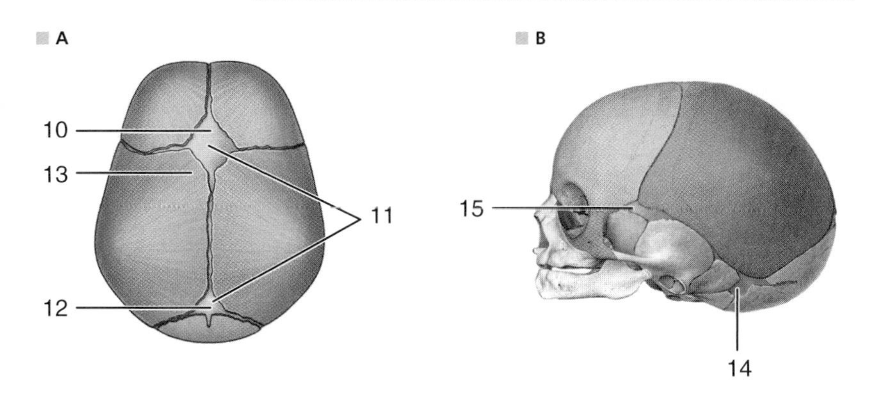

C

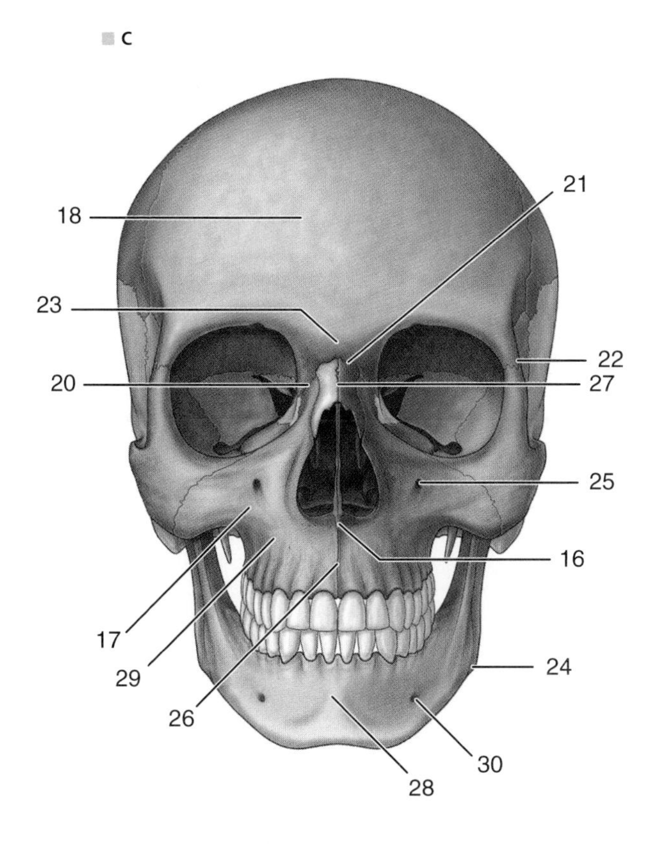

1 **mental protuberance** a more or less distinct and triangular prominence on the anterior surface of the body of the mandible, on or near the median line. <protuberantia mentalis> A

2 **mental tubercle** a more or less distinct prominence on the inferior border of either side of the mental protuberance of the mandible. <tuberculum mentale mandibulae> A

3 ***metopic suture*** the name given to the inferior part of the frontal suture when it persists in the adult. <sutura frontalis persistens> A

4 **nasal bone** either of the two small, oblong bones that together form the bridge of the nose. <os nasale> A

5 ***nasal foramina*** openings on the outer surface of each nasal bone for the transmission of blood vessels. <foramina nasalia> A

6 **nasal margin of frontal bone** the articular surface, on each nasal part of the frontal bone, that articulates with the nasal bones and with the frontal processes of the maxilla. <margo nasalis ossis frontalis> A

7 **nasal notch of maxilla** the large notch in the anterior border of the maxilla that forms the lateral and inferior margins of the anterior nasal aperture. <incisura nasalis maxillae> A

8 **nasal part of frontal bone** the small, irregularly shaped process that projects inferiorly from the medial part of the squama of the frontal bone to articulate with the nasal bones and the frontal processes of the maxillae. Called also *prefrontal bone.* <pars nasalis ossis frontalis> A

9 ***nasal spine of frontal bone*** a rough and somewhat irregular process of bone projecting downward and forward from the front art of the inferior surface of the nasal part of the frontal bone and fitting between the nasal bones and the ethmoid bone. <spina nasalis ossis frontalis> A

10 **nasion** a cephalometric landmark located where the internasal and nasofrontal sutures meet; it corresponds roughly to the depression at the root of the nose just inferior to the level of the eyebrows. A

11 **nasomaxillary suture** the line of junction between the lateral edge of the nasal bone and the frontal process of the maxilla. <sutura nasomaxillaris> A

12 **piriform aperture** the anterior end of the bony nasal opening, connecting the external nose with the skull; called also *anterior nasal aperture and base of nose.* <apertura piriformis> A

13 **superciliary arch** a smooth elevation arching superolaterally from the glabella, slightly superior to the margin of the orbit. <arcus superciliaris> A

14 **supraorbital border of frontal bone** the anteroinferior edge of the frontal bone, bending down laterally to the zygomatic bone and medially to the frontal process of the maxilla; it marks the junction between the squama and the orbital portion of the bone. <margo supraorbitalis ossis frontalis> A

15 **supraorbital foramen** an opening in the frontal bone in the supraorbital margin, giving passage to the supraorbital artery and nerve; it is often present as a notch *(supraorbital notch)* bridged only by fibrous tissue. <foramen supraorbitale> A

16 **supraorbital margin of orbit** the superior edge of the entrance to the orbit, formed by the supraorbital margin of the frontal bone. <margo supraorbitalis orbitae> A

17 **supraorbital notch** a palpable notch in the frontal bone at the junction of the medial one-third and lateral two-thirds of the supraorbital margin, for transmission of the supraorbital nerve and vessels to the forehead. In life it is bridged by fibrous tissue, which is sometimes ossified, forming a bony aperture *(supraorbital foramen).* <incisura supraorbitalis> A

18 **zygomatic process of maxilla** the rough triangular eminence that articulates with the zygomatic bone and marks the separation of the anterior, infratemporal, and orbital surfaces. <processus zygomaticus maxillae> A

19 **zygomaticomaxillary suture** the line of junction between the zygomatic bone and the zygomatic process of the maxilla. <sutura zygomaticomaxillaris> A

20 **articular tubercle of temporal bone** an enlargement of the inferior border of the zygomatic process of the temporal bone, forming the anterior boundary of the mandibular fossa and marking the termination of the anterior root of the zygomatic arch; it gives attachment to the lateral ligament of the temporomandibular joint. <tuberculum articulare ossis temporalis> B

21 **asterion** the point on the surface of the skull where the lambdoid, parietomastoid, and occipitomastoid sutures meet. B

22 **base of mandible** the lower margin of the body of the mandible. <basis mandibulae> B

23 **body of mandible** the horizontal horseshoe-shaped portion of the mandible. <corpus mandibulae> B

24 **body of maxilla** the large central portion of the maxilla, roughly pyramidal in shape, to which four major processes are connected; it contains the maxillary sinus. <corpus maxillae> B

25 **calvaria** the domelike superior portion of the cranium, derived from the membranous neurocranium and consisting of the frontal and parietal bones and the squamous parts of the occipital and temporal bones. Called also *skull cap.* B

26 **condylar process of mandible** the posterior process on the ramus of the mandible that articulates with the mandibular fossa of the temporal bone. <processus condylaris mandibulae> B

27 **coronoid process of mandible** the anterior part of the upper end of the ramus of the mandible, to which the temporalis muscle is attached. <processus coronoideus mandibulae> B

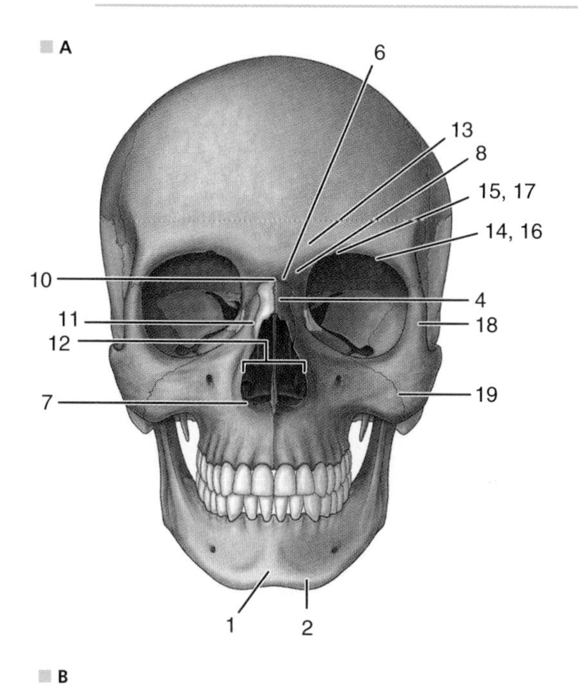
A
6
13
8
15, 17
14, 16
10
4
11
18
12
7
19
1
2

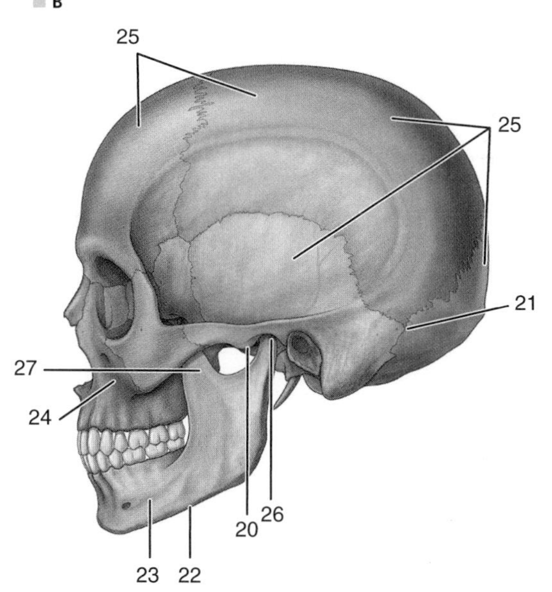
B
25
25
21
27
24
20
26
23
22

1 **external surface of frontal bone** the external surface of the squama of the frontal bone; called also *outer table of frontal bone.* <facies externa ossis frontalis> A

2 **external surface of parietal bone** the externally directed surface of the parietal bone. <facies externa ossis parietalis> A

3 **frontal border of parietal bone** the edge of the parietal bone that articulates with the frontal bone along the coronal suture. <margo frontalis ossis parietalis> A

4 **frontal process of maxilla** a large, strong, irregular process of bone that projects upward from the body of the maxilla, its medial surface forming part of the lateral wall of the nasal cavity. <processus frontalis maxillae> A

5 **frontal process of zygomatic bone** the strong, superiorly projecting triangular process of the zygomatic bone lying posterior to the malar surface and between the orbital and temporal surfaces; it unites superiorly with the zygomatic process of the frontal bone and posteriorly with the greater wing of the sphenoid bone. <processus frontalis ossis zygomatici> A

6 **frontal squama** the broad, curved portion of the frontal bone, situated superior to the supraorbital margin and forming the forehead. <squama frontalis> A

7 **frontal tuber** one of the slight rounded prominences on the frontal bone on either side superior to the eyes, forming the most prominent portions of the forehead; called also *frontal eminence.* <tuber frontale> A

8 **frontolacrimal suture** the line of junction between the upper edge of the lacrimal bone and the orbital part of the frontal bone. <sutura frontolacrimalis> A

9 **groove for middle temporal artery** a nearly vertical groove running just superior to the external acoustic meatus on the external surface of the squamous part of the temporal bone; it lodges the middle temporal artery. <sulcus arteriae temporalis mediae> A

10 **inferior temporal line of parietal bone** a curved line on the external surface of the parietal bone, marking the limit of attachment of the temporalis muscle. <linea temporalis inferior ossis parietalis> A

11 **lateral surface of zygomatic bone** the more anterior convex surface of the zygomatic bone. <facies lateralis ossis zygomatici> A

12 **marginal tubercle** a process on the superior part of the temporal border of the zygomatic bone to which a strong slip of the temporal fascia is attached. <tuberculum marginale ossis zygomatici> A

13 **mastoid angle of parietal bone** the posteroinferior angle of the parietal bone, which articulates with the posterior part of the temporal bone and the occipital bone. <angulus mastoideus ossis parietalis> A

14 **mastoid border of occipital bone** the edge of the occipital bone that extends from the jugular process to the lateral angle, articulating with the part of the temporal bone that bears the mastoid process. <margo mastoideus ossis occipitalis> A

15 **mastoid process of temporal bone** a conical process projecting forward and downward from the external surface of the petrous part of the temporal bone just posterior to the external acoustic meatus. <processus mastoideus ossis temporalis> A

16 ***norma facialis*** the outline of the skull as viewed from the front; called also anterior, *facial,* or *frontal aspect of cranium.* A

17 ***norma inferior*** the outline of the inferior aspect of the skull, viewed from above. A

18 ***norma lateralis*** the outline of the skull as viewed from either side; called also *temporal aspect of cranium.* A

19 ***norma occipitalis*** the outline of the skull as viewed from behind; called also *occipital aspect of cranium.* A

20 ***norma superior*** the outline of the superior surface of the skull; called also *superior* or *vertical aspect of cranium.* A

21 **parietal bone** either of the two quadrilateral bones forming part of the superior and lateral surfaces of the skull, and joining each other in the midline at the sagittal suture. <os parietale> A

22 **parietal border of frontal bone** the posterior border of the frontal bone, semicircular in shape, which articulates with the parietal bones. <margo parietalis ossis frontalis> A

23 **parietal border of squamous part of temporal bone** the superior border of the squamous part of the temporal bone where it articulates with the parietal bone. <margo parietalis partis squamosae ossis temporalis> A

24 **parietal margin of greater wing of sphenoid bone** the superior extremity of the squamous portion of the greater wing, where it articulates with the parietal bone. Called also *parietal angle of sphenoid bone.* <margo parietalis alae majoris> A

25 **parietal notch of temporal bone** the notch found on the upper margin of the temporal bone where the squamous and parietomastoid sutures meet. <incisura parietalis ossis temporalis> A

26 **parietomastoid suture** the line of junction between the posterior inferior angle of the parietal bone and the mastoid process of the temporal bone. <sutura parietomastoidea> A

27 **pterion** a point at the junction of the frontal, parietal, temporal, and greater wing of the sphenoid bones. A

28 ***skull*** the skeleton of the head, including the cranium and the mandible. A

29 **sphenoid angle of parietal bone** the anteroinferior angle of the parietal bone, which articulates with the greater wing of the sphenoid bone and the frontal bone; called also *anterior inferior angle of parietal bone.* <angulus sphenoidalis ossis parietalis> A

A

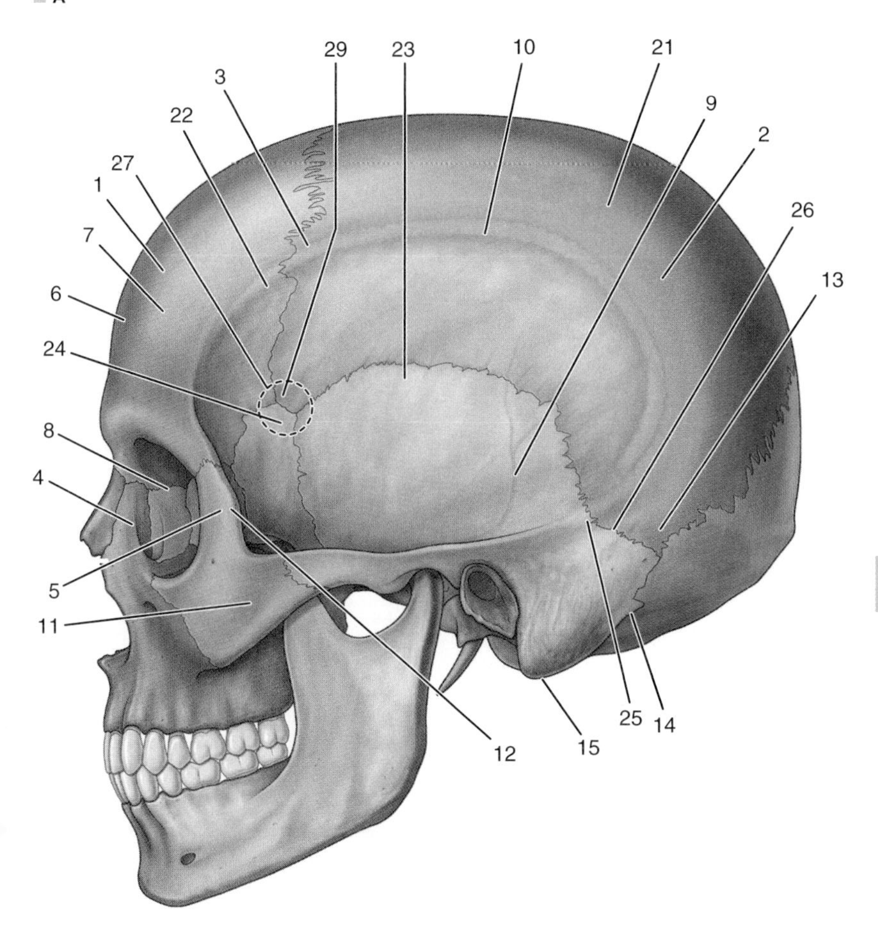

1 **sphenoidal border of squamous part of temporal bone** the anterior border of the temporal bone, articulating with the greater wing of the sphenoid bone. <margo sphenoidalis partis squamosae ossis temporalis> A

2 **sphenoparietal suture** the line of junction between the greater wing of the sphenoid bone and the parietal bone. <sutura sphenoparietalis> A

3 **sphenosquamous suture** the line of junction between the greater wing of the sphenoid bone and the squamous part of the temporal bone. <sutura sphenosquamosa> A

4 **sphenozygomatic suture** the line of junction between the greater wing of the sphenoid bone and the zygomatic bone. <sutura sphenozygomatica> A

5 ***squamomastoid suture*** a suture existing early in life between the squamous and mastoid portions of the temporal bone. <sutura squamomastoidea> A

6 **squamous border of parietal bone** the inferior edge of the parietal bone, which articulates with the sphenoid and temporal bones along the squamous suture. <margo squamosus ossis parietalis> A

7 **squamous margin of great wing of sphenoid bone** the border of the greater wing of the sphenoid bone that articulates with the squama of the temporal bone. <margo squamosus alae majoris> A

8 **squamous part of temporal bone** the flat, scale-like, anterior and superior portion of the temporal bone. <pars squamosa ossis temporalis> A

9 **squamous suture of cranium** the suture between the squamous part of the temporal bone and the parietal bone. <sutura squamosa cranii> A

10 **superior temporal line of parietal bone** a curved line on the external surface of the parietal bone, superior and parallel to the inferior temporal line, giving attachment to the temporal fascia. <linea temporalis superior ossis parietalis> A

11 **supramastoid crest** a ridge on the temporal bone that is a continuation of the superior border of the posterior root of the zygomatic process of the temporal bone. <crista supramastoidea> A

12 ***suprameatal spine*** a pointed process that sometimes projects from the temporal bone, just above and at the back of the external acoustic meatus. Called also *Henle's spine.* <spina suprameatica> A

13 ***suprameatal triangle*** a small triangular depression at the junction of the posterior and superior borders of the external acoustic meatus, posterior to the suprameatal spine; called also *mastoid* or *supramastoid fossa,* and *Macewen's triangle.* <foveola suprameatica> A

14 **temporal bone** one of the two irregular bones forming part of the lateral surfaces and base of the skull, and containing the organs of hearing. It is divided anatomically into four parts: the *mastoid, petrous, squamous,* and *tympanic parts.* <os temporale> A

15 ***temporal crest of mandible*** a ridge on the medial aspect of the coronoid process, extending from near the apex of the coronoid process to the level of the last molar, which gives attachment to the temporalis muscle. <crista temporalis mandibulae> A

16 **temporal line of frontal bone** a ridge extending superiorly and posteriorly from the zygomatic process of the frontal bone, dividing into superior and inferior parts that are continuous with corresponding lines on the parietal bone, and giving attachment to the temporal fascia. <linea temporalis ossis frontalis> A

17 **temporal plane** the depressed area on the side of the skull inferior to the inferior temporal line. <planum temporale> A

18 **temporal process of zygomatic bone** the posterior blunt process of the zygomatic bone that articulates with the zygomatic process of the temporal bone to form the zygomatic arch. <processus temporalis ossis zygomatici> A

19 **temporal surface of frontal bone** the slightly concave surface of the frontal bone that forms the superior part of the wall of the temporal fossa and gives attachment to the anterosuperior part of the temporalis muscle. <facies temporalis ossis frontalis> A

20 **temporal surface of greater wing** the lateral and inferior surface of the greater wing of the sphenoid bone, divided by the infratemporal crest into a superior part that forms a portion of the wall of the temporal fossa, and an inferior part that forms part of the wall of the infratemporal fossa. <facies temporalis alae majoris> A

21 **temporal surface of squamous part of temporal bone** the external surface of the squamous part, the anterior part of which forms a portion of the temporal fossa. <facies temporalis partis squamosae ossis temporalis> A

22 **temporal surface of zygomatic bone** the internal, concave surface of the bone, facing the temporal and infratemporal fossae. <facies temporalis ossis zygomatici> A

23 **temporozygomatic suture** the line of junction between the zygomatic process of the temporal bone and the temporal process of the zygomatic bone. <sutura temporozygomatica> A

24 ***tympanic ring*** the bony ring forming part of the temporal bone at the time of birth and developing into the tympanic part of the bone. <anulus tympanicus> A

25 **tympanomastoid fissure** an external fissure on the inferior and lateral aspect of the skull between the tympanic portion and the mastoid process of the temporal bone; the auricular branch of the vagus nerve often passes through it. Called also *petromastoid fissure.* <fissura tympanomastoidea> A

26 ***tympanosquamous fissure*** a line seen on the posterior wall of the external acoustic meatus at the junction between the tympanic and squamous parts of the temporal bone. Called also *squamotympanic fissure.* <fissura tympanosquamosa> A

A

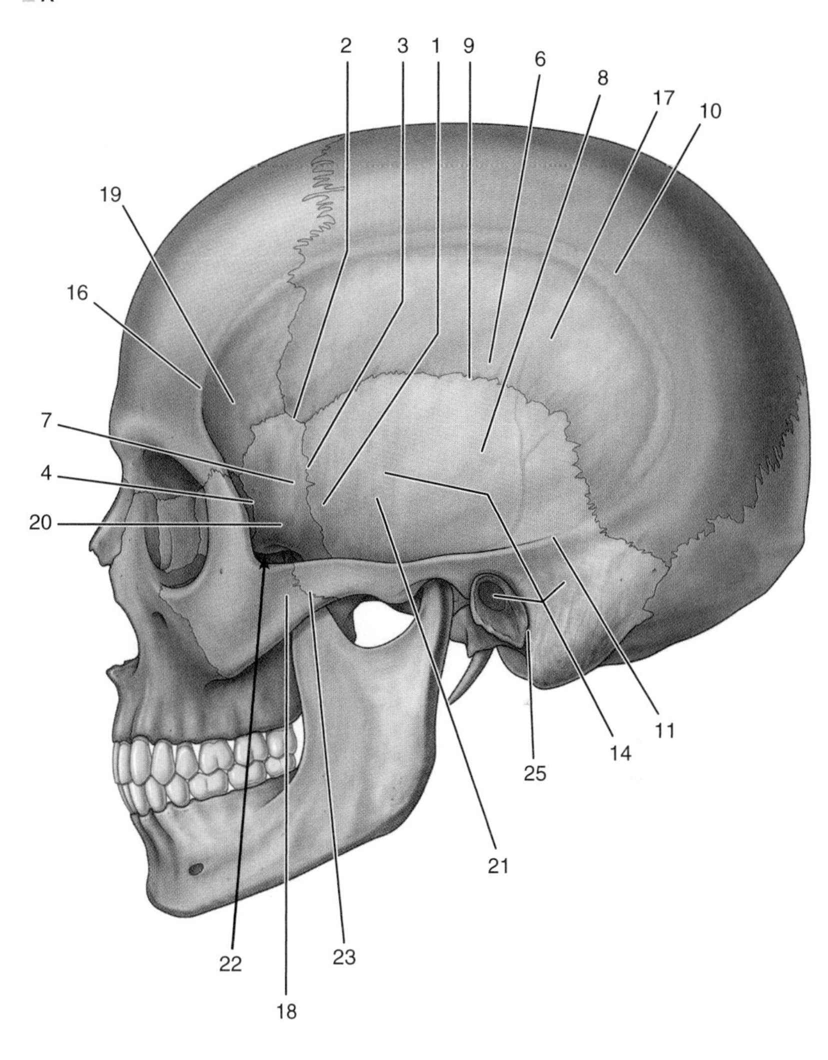

1 **zygomatic arch** the arch formed by the articulation of the broad temporal process of the zygomatic bone and the slender zygomatic process of the temporal bone, giving attachment to the masseter muscle and serving as a line of demarcation between the temporal and infratemporal fossae. <arcus zygomaticus> A

2 **zygomatic bone** the quadrangular bone of the cheek, articulating with the frontal bone, the maxilla, the zygomatic process of the temporal bone, and the greater wing of the sphenoid bone. <os zygomaticum> A

3 **zygomatic margin of greater wing of sphenoid bone** the border on the greater wing separating its temporal and orbital surfaces and articulating with the zygomatic bone. <margo zygomaticus alae majoris> A

4 **zygomatic process of frontal bone** a thick, strong process of the frontal bone, situated at the lateral end of the supraorbital margin and articulating with the zygomatic bone, and from which the temporal line starts. <processus zygomaticus ossis frontalis> A

5 **zygomatic process of temporal bone** a long, strong process arising from the inferior portion of the squamous part of the temporal bone, passing anteriorly from just superior to the entrance of the external acoustic meatus to join the zygomatic bone and thus forming the zygomatic arch. It has an anterior root and a posterior root extending along the temporal bone. <processus zygomaticus ossis temporalis> A

6 ***zygomatico-orbital foramen*** either of the two openings on the orbital surface of each zygomatic bone, which transmit branches of the zygomatic branch of the trigeminal nerve and branches of the lacrimal artery. <foramen zygomaticoorbitale> A

7 ***anterior intraoccipital synchondrosis*** the cartilaginous union of the basilar part with the lateral parts of the occipital bone in the newborn. <synchondrosis intraoccipitalis anterior> B

8 ***highest nuchal line*** a sometimes indistinct line arching superiorly from the external occipital protuberance and running toward the lateral angle of the occipital bone: the epicranial aponeurosis attaches to it. <linea nuchalis suprema> B

9 **inferior nuchal line** the most inferior of the three nuchal lines found on the outer surface of the occipital bone, extending laterally from the middle of the external occipital crest to the jugular process. <linea nuchalis inferior> B

10 **inion** the most prominent point of the external occipital protuberance. B

11 ***interparietal bone*** the portion of the squamous part of the occipital bone that lies superior to the highest nuchal line when this portion remains separate throughout life. <os interparietale> B

12 **lambdoid border of occipital bone** the edge of the occipital bone that extends from the lateral angle to the superior angle, articulating with the parietal bone to help form the lambdoid suture. <margo lambdoideus ossis occipitalis> B

13 **lambdoid suture** the line of junction between the occipital and parietal bones, shaped like the Greek letter lambda. <sutura lambdoidea> B

14 **occipital angle of parietal bone** the posterosuperior angle of the parietal bone, which during fetal life participates in the formation of the posterior fontanelle. <angulus occipitalis ossis parietalis> B

15 **occipital bone** a single trapezoid-shaped bone situated at the posterior and inferior part of the cranium, articulating with the two parietal and two temporal bones, the sphenoid bone, and the atlas; it contains a large opening, the foramen magnum. <os occipitale> B

16 **occipital border of temporal bone** the border of the petrous part of the temporal bone that articulates with the occipital bone along the occipitomastoid suture. <margo occipitalis ossis temporalis> B

17 **occipital margin of parietal bone** the edge of the parietal bone that articulates with the occipital bone at the lambdoid suture. <margo occipitalis ossis parietalis> B

18 **occipital plane** the outer surface of the occipital bone superior to the superior nuchal line. <planum occipitale> B

19 **occipitomastoid suture** an extension of the lambdoid suture between the occipital bone and the posterior edge of the mastoid portion of the temporal bone. <sutura occipitomastoidea> B

20 **parietal tuber** the somewhat laterally bulging prominence just superior to the superior temporal line on the external surface of the parietal bone. Called also *parietal eminence.* <tuber parietale> B

21 ***posterior intraoccipital synchondrosis*** the cartilaginous union of the squama with the lateral parts of the occipital bone in the newborn. <synchondrosis intraoccipitalis posterior> B

22 **squamous part of occipital bone** the largest of the four parts of the occipital bone, extending from the posterior edge of the foramen magnum to the lambdoid suture, its external surface bearing the external occipital protuberance and nuchal lines. <squama occipitalis> B

23 **superior nuchal line** a curved line on the outer surface of the occipital bone, extending from the external occipital protuberance toward the lateral angle and giving attachment medially to the trapezius muscle and laterally to the sternocleidomastoid muscle. <linea nuchalis superior> B

24 **sutural bone** any of the small irregular bones in the sutures between the bones of the skull, most frequently in the course of the lambdoid suture and often at the fontanelles; called also *epactal bone* and *wormian bone.* <os suturale> B

A

B

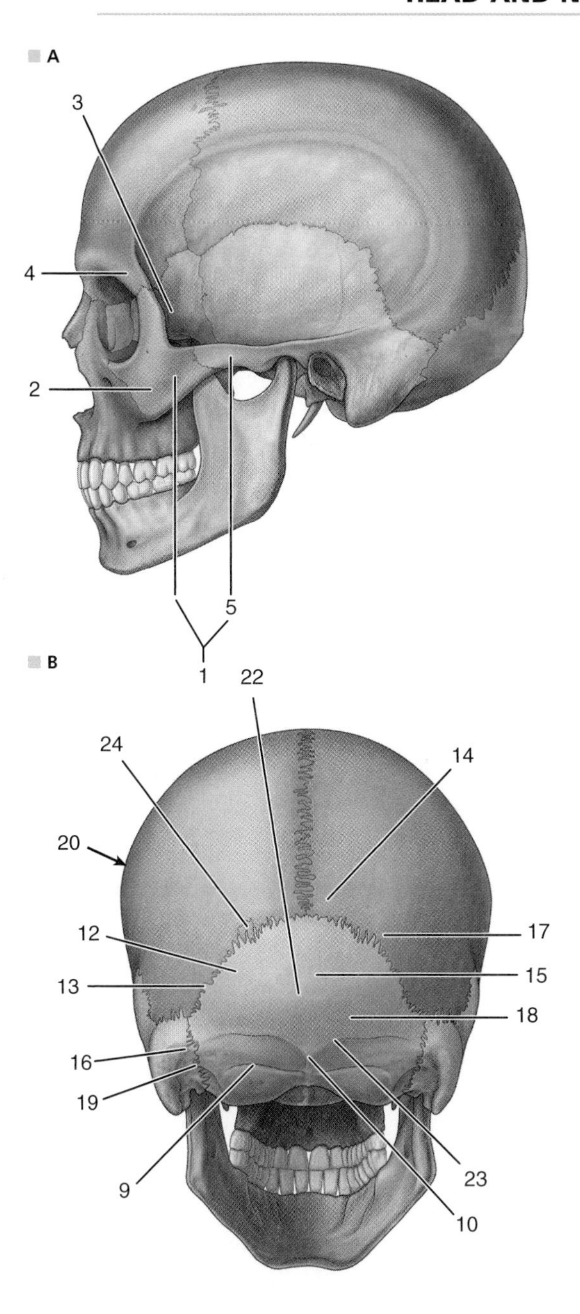

1 **bregma** the point on the surface of the skull at the junction of the coronal and sagittal sutures; used as a craniometric landmark. A

2 **coronal suture** the line of junction of the frontal bone with the two parietal bones. <sutura coronalis> A

3 **lambda** the point at the site of the posterior fontanelle where the lambdoid and sagittal sutures meet; used as a craniometric landmark. A

4 **parietal foramen** an opening on the posterior part of the superior portion of the parietal bone near the sagittal suture, for the passage of a vein and arteriole. <foramen parietale> A

5 **sagittal border of parietal bone** the edge of the parietal bone that articulates with the other parietal bone along the sagittal suture; called also *superior margin of parietal bone.* <margo sagittalis ossis parietalis> A

6 **sagittal suture** the line of junction between the two parietal bones. <sutura sagittalis> A

7 **basilar part of occipital bone** a quadrilateral plate of the occipital bone that projects superiorly and anteriorly from the foramen magnum. <pars basilaris ossis occipitalis> B

8 ***basion*** a craniometric landmark located at the midpoint of the anterior border of the foramen magnum in the median plane. Called also *point Ba.* B

9 **canal for auditory tube** a groove on the medial part of the base of the spine of the sphenoid bone; it lodges a portion of the cartilaginous part of the auditory tube. <semicanalis tubae auditoriae> B

10 **carotid canal** a passage in the petrous portion of the temporal bone, beginning on the inferior surface just anterior to the jugular foramen, and running anteromedially for about 2 cm; it is seen interiorly in the floor of the middle cranial fossa, where it meets the carotid sulcus on the body of the sphenoid bone. It houses the internal carotid artery. <canalis caroticus> B

11 **external occipital crest** a variable crest of bone that sometimes extends from the external occipital protuberance toward the foramen magnum; called also *middle nuchal line.* <crista occipitalis externa> B

12 **external occipital protuberance** a prominence at the center of the outer surface of the squama of the occipital bone which gives attachment to the ligamentum nuchae. <protuberantia occipitalis externa> B

13 ***external surface of cranial base*** the outer surface of the inferior aspect of the skull; called also *norma ventralis, external base of skull,* and *scaphion.* <basis cranii externa> B

14 **foramen lacerum** an irregular gap formed at the junction of the base of the greater wing of the sphenoid bone, the tip of the petrous part of the temporal bone, and the basilar part of the occipital bone; in life, it does not exist, being occupied by an unossified part of the petrous part of the temporal bone. B

15 **foramen magnum** the large opening in the anterior and inferior part of the occipital bone, interconnecting the vertebral canal and the cranial cavity. B

16 **greater palatine foramen** the inferior opening of the great palatine canal, found laterally on the horizontal plate of each palatine bone opposite the root of each third molar tooth; it transmits a palatine nerve and artery. Called also *pterygopalatine foramen.* <foramen palatinum majus> B

17 ***groove of pterygoid hamulus*** a smooth groove on the lateral surface of the medial pterygoid plate of the sphenoid bone, in the angle at the base of the pterygoid hamulus; it lodges the tendon of the tensor veli palatini muscle. <sulcus hamuli pterygoidei> B

18 **inferior surface of petrous part of temporal bone** that surface of the petrous part of the temporal bone which appears on the external surface of the base of the cranium. <facies inferior partis petrosae ossis temporalis> B

19 **infratemporal surface of body of maxilla** the posterior convex surface of the body of the maxilla, bounded roughly by the inferior orbital fissure, the zygomatic process and associated ridge, maxillary tuberosity, and posterior margin of the nasal surface. <facies infratemporalis corporis maxillae> B

20 ***intrajugular process of occipital bone*** a small process that subdivides the jugular notch of the occipital bone into a lateral and a medial part. <processus intrajugularis ossis occipitalis> B

21 ***intrajugular process of temporal bone*** a small ridge on the petrous part of the temporal bone that separates the jugular notch into a medial and a lateral part, corresponding to similar parts of the jugular notch of the facing occipital bone. <processus intrajugularis ossis temporalis> B

22 **jugular foramen** the opening formed by the jugular notches on the temporal and occipital bones, for the transmission of various veins, arteries, and nerves. <foramen jugulare> B

23 **jugular fossa** a prominent depression on the inferior surface of the petrous part of the temporal bone, forming the major part of the jugular notch; it forms the anterior and lateral wall of the jugular foramen and lodges the superior bulb of the internal jugular vein. <fossa jugularis ossis temporalis> B

24 **jugular notch of occipital bone** a notch on the anterior surface of the jugular process of the occipital bone, forming the posterior wall of the jugular foramen. <incisura jugularis ossis occipitalis> B

25 **jugular notch of temporal bone** a prominent depression on the inferior surface of the petrous part of the temporal bone. It forms the anterior and lateral wall of the jugular foramen and lodges the superior bulb of the internal jugular vein in its lateral part and the glossopharyngeal, vagus, and accessory nerves in its medial part. <incisura jugularis ossis temporalis> B

A

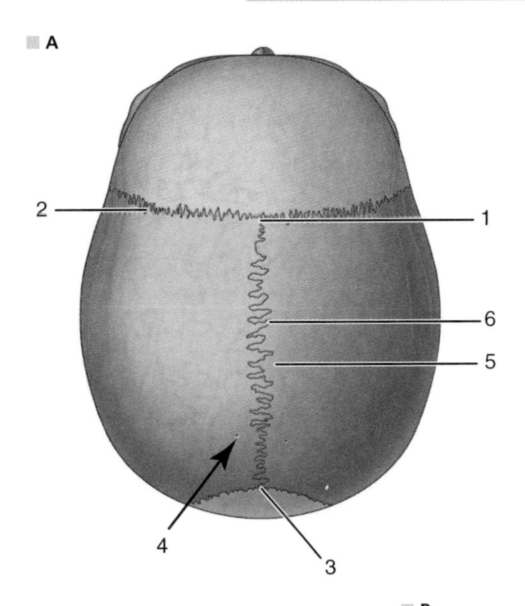

B

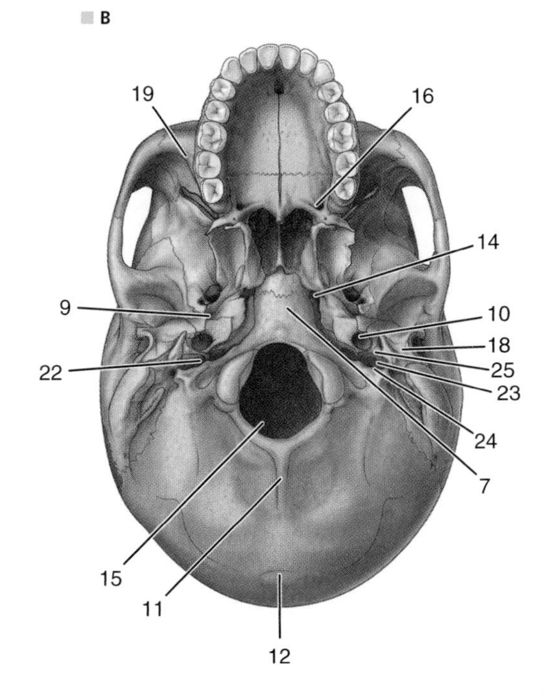

1 **jugular process of occipital bone** either of two processes on the occipital bone that project laterally from the occipital condyles and form the posterior boundary of the jugular foramen. <processus jugularis ossis occipitalis> A

2 **lateral part of occipital bone** one of the paired parts of the occipital bone that form the lateral boundaries of the foramen magnum, each being prominently characterized by the presence of one of the occipital condyles. <pars lateralis ossis occipitalis> A

3 **lateral plate of pterygoid process** either of a pair of bony plates projecting downward from the roots of the greater wings of the sphenoid bone and forming the medial wall of the ipsilateral infratemporal fossa; called also *lateral pterygoid plate.* <lamina lateralis processus pterygoidei> A

4 **lesser palatine foramina** the openings of the lesser palatine canals behind the palatine crest and the greater palatine foramina. <foramina palatina minora> A

5 **mandibular fossa** a prominent depression in the inferior surface of the squamous part of the temporal bone at the base of the zygomatic process, in which the condyloid process of the mandible rests. <fossa mandibularis> A

6 ***mastoid canaliculus*** a minute passage beginning in the lateral wall of the jugular fossa of the temporal bone and passing into the temporal bone. The auricular branch of the vagus nerve passes through it to exit via the tympanomastoid fissure. <canaliculus mastoideus> A

7 ***mastoid emissary vein*** a small vein passing through the mastoid foramen of the skull and connecting the sigmoid sinus with the occipital or the posterior auricular vein. <vena emissaria mastoidea> A

8 **mastoid foramen** a prominent opening in the temporal bone posterior to the mastoid process and near its occipital articulation; an artery and vein usually pass through it. <foramen mastoideum> B

9 **mastoid notch** a deep groove on the medial surface of the mastoid process of the temporal bone, which gives origin to the posterior belly of the digastric muscle. <incisura mastoidea ossis temporalis> A

10 **medial plate of pterygoid process** either of a pair of bony plates projecting inferiorly from the roots of the greater wings of the sphenoid bone and forming the lateral boundary of the ipsilateral posterior aperture of the nasal cavity and the most posterior part of the lateral wall of the nasal cavity. Called also *medial pterygoid plate.* <lamina medialis processus pterygoidei> A

11 **median palatine suture** the line of junction between the horizontal part of the palatine bones of either side. <sutura palatina mediana> A

12 **occipital condyle** one of two oval processes on the lateral portions of the occipital bone, on either side of the foramen magnum, for articulation with the atlas. <condylus occipitalis> A

13 **occipital groove** the groove just medial to the mastoid notch on the temporal bone, lodging the occipital artery. <sulcus arteriae occipitalis> A

14 ***opisthion*** a craniometric landmark located at the midpoint of the posterior border of the foramen magnum. A

15 ***paramastoid process of occipital bone*** a tubercle on the inferior surface of the jugular process. <processus paramastoideus ossis occipitalis> A

16 **petrotympanic fissure** a narrow transversely running slit just posterior to the articular surface of the mandibular fossa of the temporal bone; an arteriole and the chorda tympani nerve pass through it, and it lodges a portion of the malleus. Called also *glaserian fissure.* <fissura petrotympanica> A

17 **pharyngeal tubercle** a midline eminence on the inferior surface of the basilar part of the occipital bone, for attachment of the pharynx (superior constrictor and pharyngeal raphe). <tuberculum pharyngeum> A

18 **pterygoid fossa** the posteriorly facing fossa which is formed by the divergence of the medial and lateral pterygoid plates of the sphenoid bone, and lodges the origins of the medial pterygoid muscle and tensor veli palatini muscle. <fossa pterygoidea ossis sphenoidalis> A

19 **pterygoid hamulus** a hooklike process on the inferior extremity of the medial pterygoid plate of the sphenoid bone, around which the tendon of the tensor veli palatini muscle passes. <hamulus pterygoideus> A

20 **spheno-occipital synchondrosis** the cartilaginous union of the anterior end of the basilar portion of the occipital bone with the posterior surface of the body of the sphenoid bone. <synchondrosis sphenooccipitalis> A

21 **stylomastoid foramen** a foramen on the inferior part of the temporal bone between the styloid and mastoid processes, for the facial nerve and the stylomastoid artery. <foramen stylomastoideum> A

22 **condylar canal** an opening sometimes present in the floor of the condylar fossa for the transmission of a vein from the transverse sinus. <canalis condylaris> B

23 ***condylar emissary vein*** a small vein running through the condylar canal of the skull, connecting the sigmoid sinus with the vertebral or the internal jugular vein. <vena emissaria condylaris> B

24 **condylar fossa** either of two pits on the lateral portions of the occipital bone, one on either side of the foramen magnum, posterior to the occipital condyle. <fossa condylaris> B

A

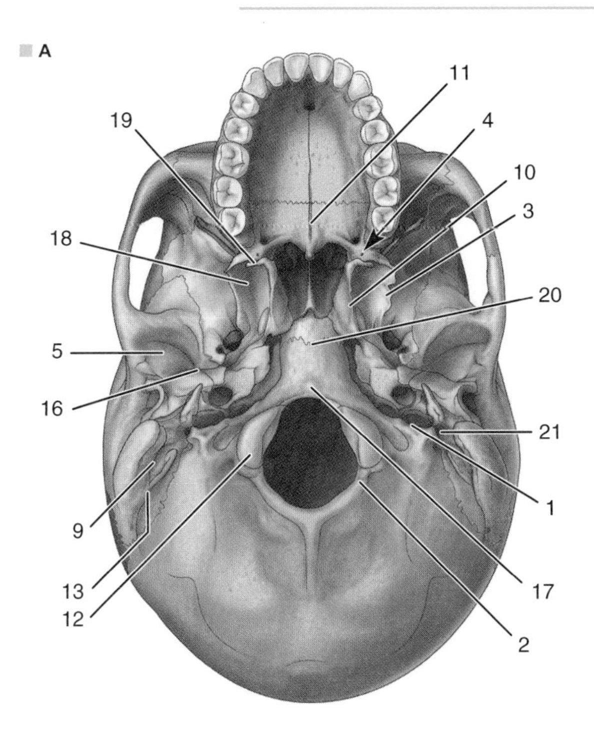

B

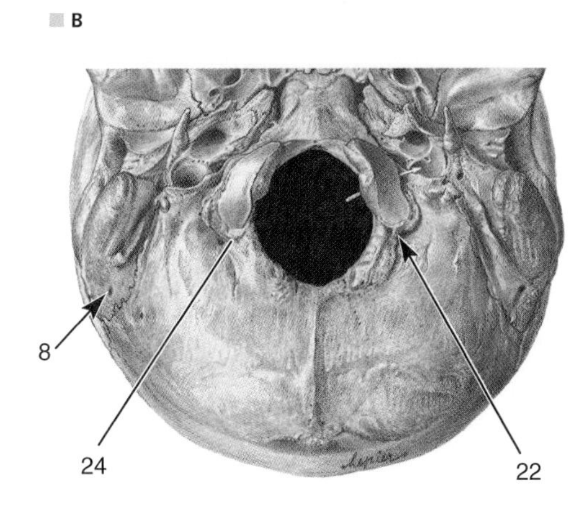

1 ***petrosal foramen*** a small opening sometimes present posterior to the foramen ovale for transmission of the lesser petrosal nerve; called also *innominate canaliculus.* <foramen petrosum> A

2 **petrous part of temporal bone** a pyramid of dense bone located at the base of the cranium; one of the three parts of the temporal bone, it houses the organ of hearing. Some anatomists divide it into petrous and mastoid subparts and call it the *petromastoid part of temporal bone.* <pars petrosa ossis temporalis> B

3 **posterior clinoid process** either of two tubercles found on the superior angle of either side of the dorsum sellae of the sphenoid bone, and giving attachment to the tentorium of the cerebellum. <processus clinoideus posterior> A

4 **sella turcica** a transverse depression crossing the midline on the superior surface of the body of the sphenoid bone, and containing the hypophysis. A

5 **sphenopetrosal fissure** a fissure in the floor of the middle cranial fossa between the posterior edge of the greater wing of the sphenoid bone and the petrous part of the temporal bone; called also *angular* or *petrosphenoidal fissure.* <fissura sphenopetrosa> A

6 ***sphenopetrosal synchondrosis*** the cartilaginous union of the inferior border of the greater wing of the sphenoid bone with the petrous portion of the temporal bone in the sphenopetrosal fissure. <synchondrosis sphenopetrosa> A

7 **tegmen tympani** the thin layer of translucent bone, on the petrous part of the temporal bone in the floor of the middle cranial fossa, separating the epitympanic recess from the cranial cavity. A

8 **trigeminal impression of temporal bone** the shallow impression in the floor of the middle cranial fossa on the petrous part of the temporal bone, lodging the semilunar ganglion of the trigeminal nerve. <impressio trigeminalis ossis temporalis> A

9 **tubercle of sella turcica** a transverse ridge on the superior surface of the body of the sphenoid bone; it is anterior to the sella turcica, posterior to the prechiasmatic sulcus, and between the anterior clinoid processes. <tuberculum sellae turcicae> A

10 ***petrosquamous fissure*** a slight fissure of varying distinctness in the floor of the middle cranial fossa, marking the line of fusion between the squamous and petrous portions of the temporal bone. <fissura petrosquamosa> B

11 **jugular tubercle of occipital bone** a smooth eminence overlying the hypoglossal canal on the superior surface of the lateral part of the occipital bone. <tuberculum jugulare ossis occipitalis> C

12 **posterior margin of petrous part of temporal bone** the border of the petrous part extending from the apex to the jugular notch and articulating with part of the occipital bone. Called also *posterior border of petrous part of temporal bone.* <margo posterior partis petrosae ossis temporalis> C

13 **posterior surface of petrous part of temporal bone** the surface of the petrous part of the temporal bone that forms part of the anterior portion of the floor of the posterior cranial fossa. <facies posterior partis petrosae ossis temporalis> C

14 **external table of calvarium** the outer compact layer of bone of the flat bones of the skull. <lamina externa calvariae> D

A

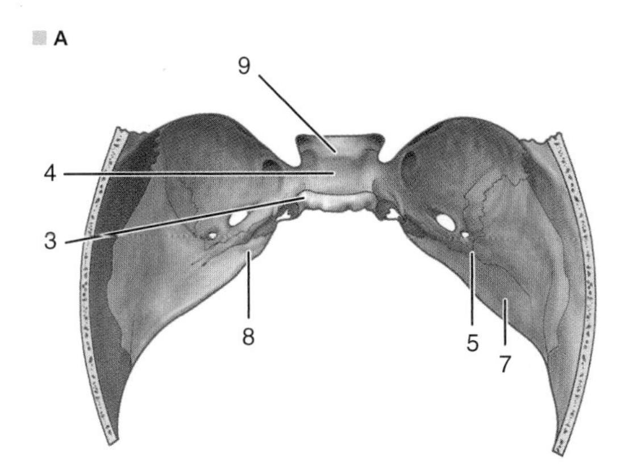

B

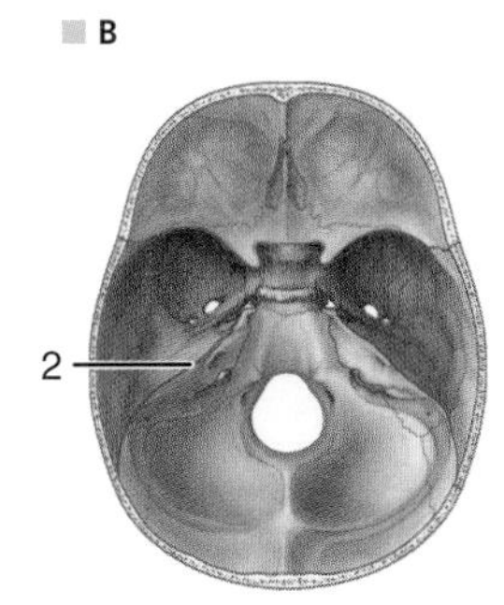

C

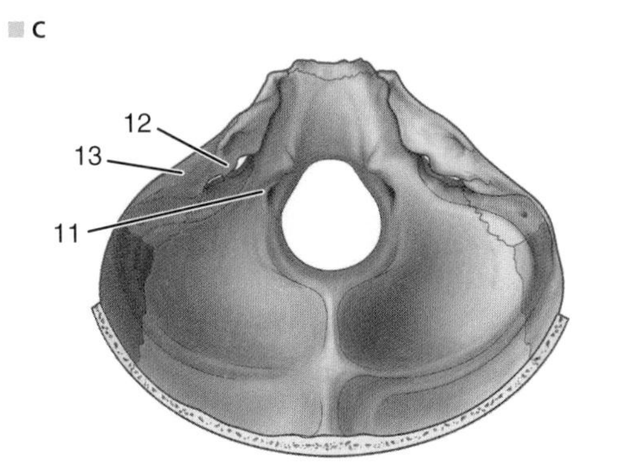

D

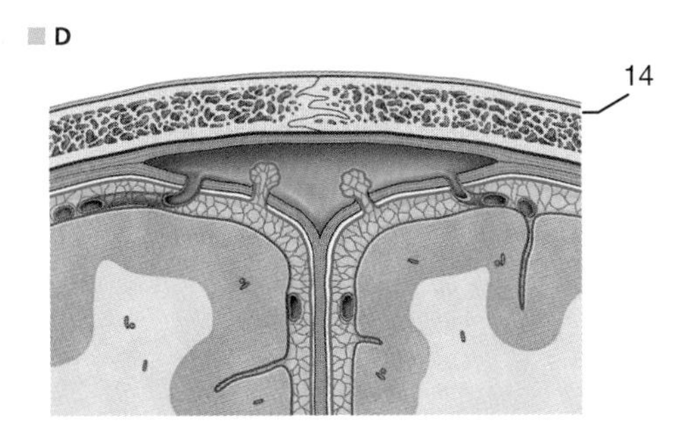

1 **ethmoidolacrimal suture** the vertical line of junction, on the medial wall of the orbit, between the lacrimal bone and the orbital plate of the ethmoid bone. <sutura ethmoidolacrimalis> A

2 **ethmoidomaxillary suture** the line of junction between the orbital plate of the ethmoid bone and the orbital surface of the maxilla. <sutura ethmoidomaxillaris> A

3 **optic canal** one of the paired openings in the sphenoid bone where the small wings are attached to the body of the bone at the apex of the orbit; each canal transmits one of the optic nerves and the ophthalmic artery of that side. <canalis opticus> A

4 **orbit** the bony cavity that contains the eyeball and its associated muscles, vessels, and nerves; the ethmoid, frontal, lacrimal, nasal, palatine, sphenoid, and zygomatic bones, and the maxilla contribute to its formation. <orbita> A

5 **orbital margin** the edge of the entrance to the orbit, formed mainly by the frontal and zygomatic bones and the maxilla. <margo orbitalis> A

6 **orbital opening** the opening to the orbit in the cranium; called also *orbital aperture* and *anterior opening of orbital cavity.* <aditus orbitalis> A

7 **orbital part of frontal bone** the horizontal part of the bone; it forms the greater part of the roof of the orbit and of the floor of the anterior cranial fossa and is separated from its fellow of the other side by the ethmoid incisure. Called also *orbital plate of frontal bone.* <pars orbitalis ossis frontalis> A

8 **orbital plate of ethmoid bone** a thin plate of bone laterally bounding the ethmoid labyrinth on either side and forming part of the medial wall of the orbit. <lamina orbitalis ossis ethmoidalis> A

9 **orbital process of palatine bone** a pyramidal process on the uppermost part of the palatine bone, one surface of it forming the posterior angle of the floor of the orbit. <processus orbitalis ossis palatini> A

10 **orbital surface of body of maxilla** a triangular surface on the body of the maxilla that forms the greater part of the floor of the orbit. <facies orbitalis corporis maxillae> A

11 **orbital surface of frontal bone** the triangular plates of the frontal bone that form most of the roof of each orbit and the floor of the anterior cranial fossa; they are separated by the ethmoid notch. <facies orbitalis ossis frontalis> A

12 **orbital surface of greater wing of sphenoid bone** the quadrilateral surface on the greater wing of the sphenoid bone that forms the major part of the lateral wall of the orbit; called also *orbital border of sphenoid bone.* <facies orbitalis alae majoris ossis sphenoidalis> A

13 **orbital surface of zygomatic bone** the part of the zygomatic bone that helps form the lateral wall of the orbit. <facies orbitalis ossis zygomatici> A

14 **palatoethmoidal suture** the line of junction between the orbital process of the palatine bone and the orbital lamina of the ethmoid bone. <sutura palatoethmoidalis> A

15 **palatomaxillary suture** the suture in the floor of the orbit, between the orbital processes of the palatine bone and the orbital portion of the maxilla. <sutura palatomaxillaris> A

16 **posterior lacrimal crest** a vertical ridge dividing the lateral or orbital surface of the lacrimal bone into two parts, and forming one margin of the fossa for the lacrimal sac. <crista lacrimalis posterior> A

17 **sphenoethmoidal suture** the line of junction between the body of the sphenoid bone and the orbital lamina of the ethmoid bone. <sutura sphenoethmoidalis> A

18 ***sphenoethmoidal synchondrosis*** the cartilaginous union between the body of the sphenoid and the labyrinth of the ethmoid bone. <synchondrosis sphenoethmoidalis> A

19 **superior orbital fissure** an elongated cleft between the lesser and greater wings of the sphenoid bone, which transmits various nerves and vessels. <fissura orbitalis superior> A

20 **trochlear fovea** a depression on the anteromedial part of the orbital surface of the frontal bone for the attachment of the trochlea of the superior oblique muscle; it is often replaced by the trochlear spine. <fovea trochlearis> A

21 ***trochlear spine*** a spicule of bone on the anteromedial part of the orbital surface of the frontal bone for attachment of the trochlea of the superior oblique muscle; when absent, it is represented by the trochlear fovea. <spina trochlearis> A

22 **mylohyoid groove** a groove on the medial surface of the ramus of the mandible, passing downward and forward from the mandibular foramen and lodging the mylohyoid artery and nerve. <sulcus mylohyoideus mandibulae> B

23 **mylohyoid line of mandible** a ridge on the inner surface of the mandible from the base of the symphysis to the ascending ramus behind the last molar tooth; it affords attachment to the mylohyoid muscle and superior constrictor of the pharynx. <linea mylohyoidea mandibulae> B

24 **pterygoid fovea of mandible** a depression on the inner side of the neck of the condyloid process of the mandible, for attachment of the lateral pterygoid muscle; called also *pterygoid depression* or *pit.* <fovea pterygoidea mandibulae> B

25 **pterygoid tuberosity** a roughened area on the inner side of the angle of the mandible for the insertion of the medial pterygoid muscle; called also *pterygoid tubercle.* <tuberositas pterygoidea mandibulae> B

■ A

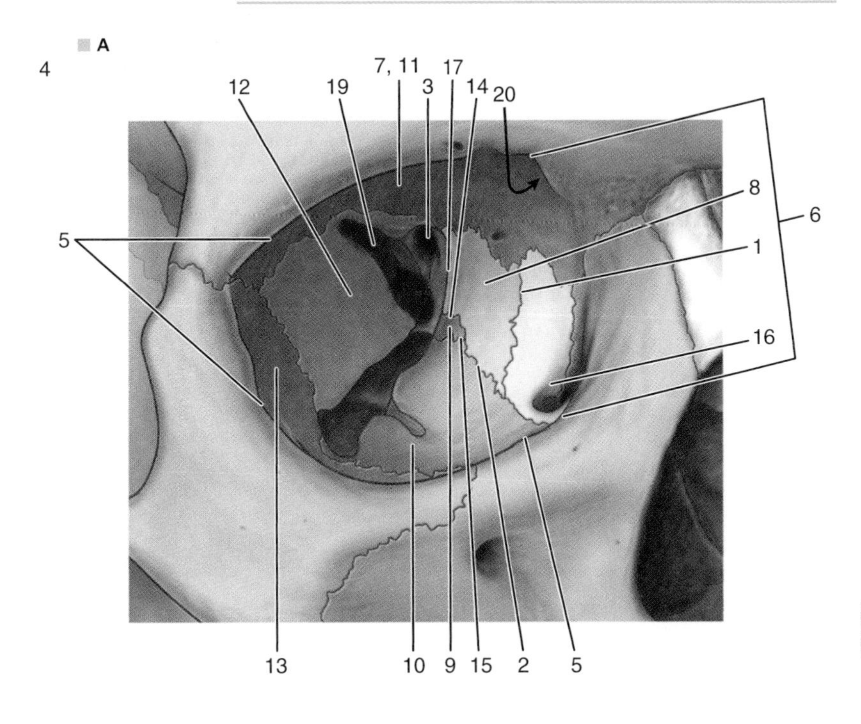

■ B

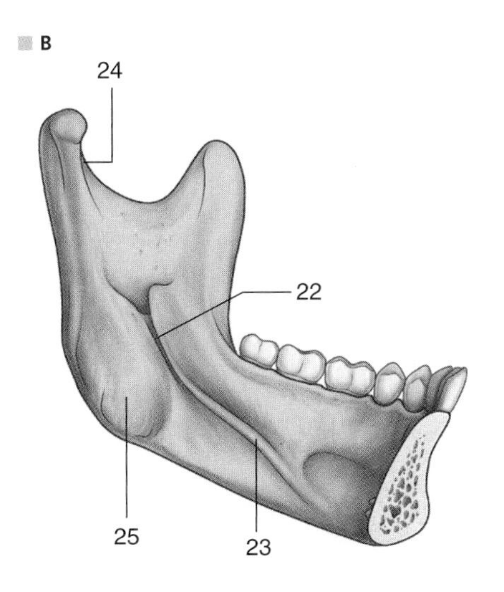

1 ***superior synovial membrane of temporomandibular joint*** the synovial membrane that lines the articular capsule of the joint above the articular disk. <membrana synovialis superior articulationis temporomandibularis> A

2 ***synovial membrane of temporomandibular joint*** the synovial membrane that lines the articular capsule of the joint below the articular disk. <membrana synovialis inferior articulationis inferior temporomandibularis> A

3 **temporomandibular joint** a bicondylar joint formed by the head of the mandible and the mandibular fossa, and the articular tubercle of the temporal bone. <articulatio temporomandibularis> A

4 **sphenomandibular ligament** a thin aponeurotic band that extends from the angular spine of the sphenoid bone downward medial to the temporomandibular joint and attaches to the lingula of the mandible. <ligamentum sphenomandibulare> B

5 **stylomandibular ligament** an aponeurotic band attached superiorly to the tip of the styloid process of the temporal bone and inferiorly to the angle and posterior margin of the ramus of the mandible. <ligamentum stylomandibulare> B

6 **zygomaticotemporal foramen** the opening on the temporal surface of the zygomatic bone for passage of the zygomaticotemporal nerve. <foramen zygomaticotemporale> C

7 **maxillary tuberosity** a rounded eminence at the posteroinferior angle of the infratemporal surface of the maxilla. <tuber maxillae> D

8 **maxillary surface of perpendicular plate of palatine bone** the lateral surface of the perpendicular plate of the palatine bone, which is in relation to the maxilla. Posteriorly and inferiorly it contains the greater palatine sulcus, which forms the greater palatine canal with a corresponding groove on the maxilla. <facies maxillaris laminae perpendicularis ossis palatini> D

9 **pterygomaxillary fissure** a cleft just posterior to the inferior orbital fissure between the lateral pterygoid plate and the maxilla. <fissura pterygomaxillaris> D

10 ***sphenomaxillary suture*** a suture occasionally seen between the pterygoid process of the sphenoid bone and the maxilla. <sutura sphenomaxillaris> D

11 **palatovaginal canal** a narrow canal located in the roof of the nasal cavity between the inferior surface of the body of the sphenoid bone and the sphenoidal process of the palatine bone; it opens posteriorly into the nasal cavity and anteriorly into the pterygopalatine fossa. Called also *pharyngeal* or *pterygopalatine canal.* <canalis palatovaginalis> E

12 **body of sphenoid** the central, cuboidal part of the sphenoid bone to which the greater wings, lesser wings, and pterygoid processes are attached; it contains the sphenoidal sinuses. <corpus ossis sphenoidalis> E

13 **maxillary surface of greater wing of sphenoid bone** a small surface on the inferior part of the greater wing superior to the pterygoid processes; it is perforated by the foramen rotundum. <facies maxillaris alae majoris> E

14 **palatovaginal groove** the groove on the vaginal process of the sphenoid bone that participates in formation of the palatovaginal canal. <sulcus palatovaginalis> E

15 **pterygoid notch** a notch on the inferior portion of the pterygoid processes of the sphenoid bone, where the pyramidal process of the palatine bone is inserted between the diverging medial and lateral pterygoid plates; called also *palatine notch* and *pterygoid fissure.* <incisura pterygoidea> E

16 **sphenoid bone** a single irregular, wedge-shaped bone at the base of the skull, forming a part of the floor of the anterior, middle, and posterior cranial fossae. <os sphenoidale> E

17 **sphenoidal concha** a thin curved plate of bone at the anterior and lower part of the body of the sphenoid bone, on either side, forming part of the roof of the nasal cavity; called also *sphenoturbinal bone* or *ossicle.* <concha sphenoidalis> E

18 **sphenoidal crest** a median ridge on the anterior surface of the body of the sphenoid bone, articulating with the perpendicular plate of the ethmoid. <crista sphenoidalis> E

19 **sphenoidal lingula** a slender ridge of bone on the lateral margin of the carotid sulcus, projecting posteriorly between the body and greater wing of the sphenoid bone. <lingula sphenoidalis> F

20 **sphenoidal yoke** the portion of the body of the sphenoid bone that connects its lesser wings. <jugum sphenoidale> F

A

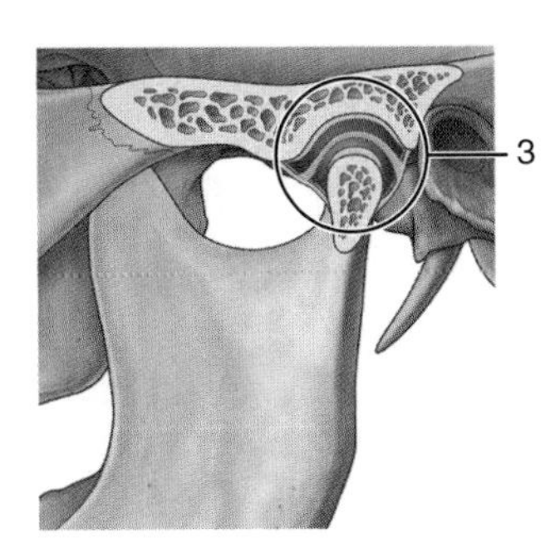

B

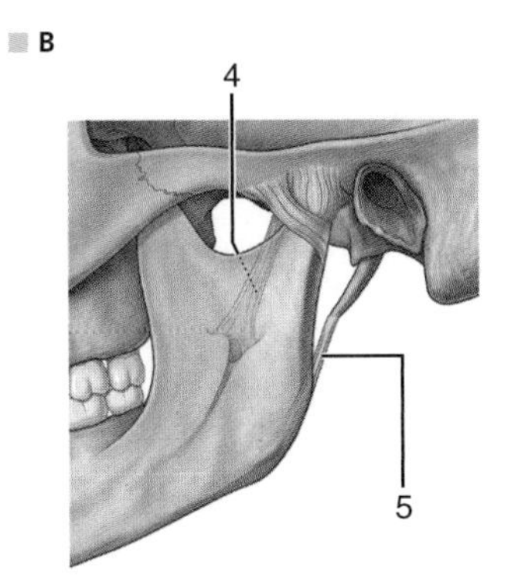

C

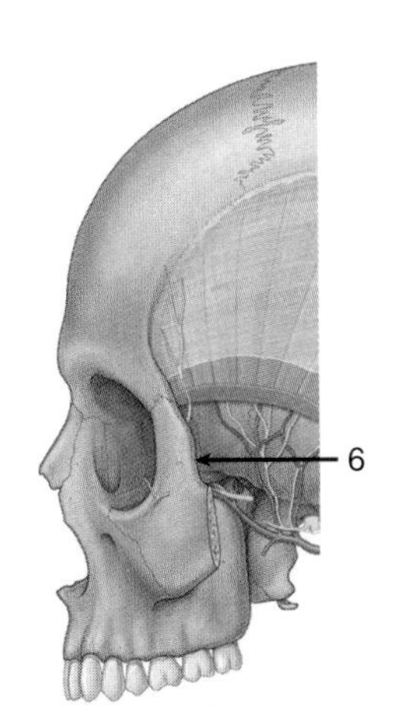

E

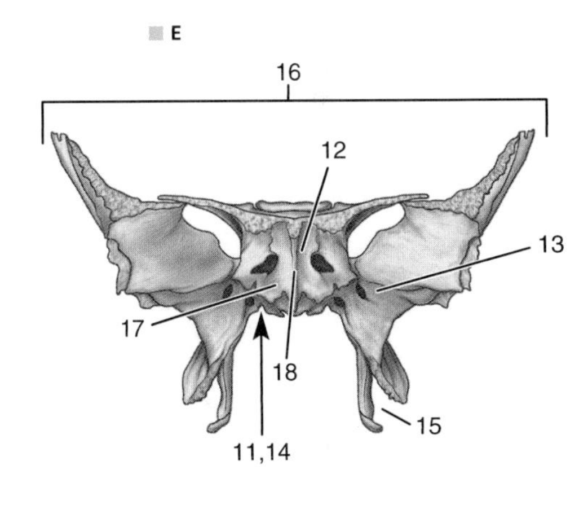

D

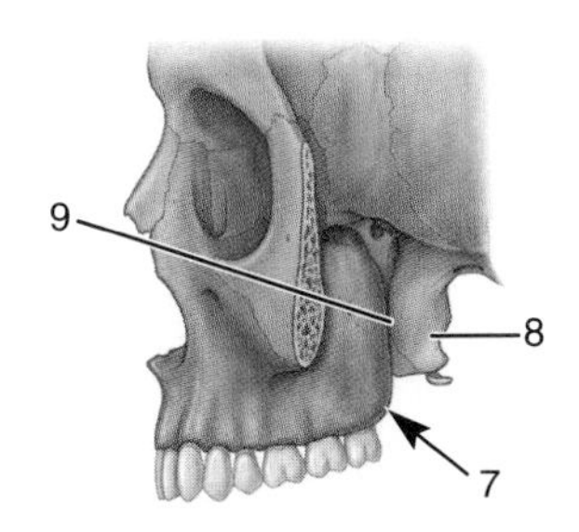

F

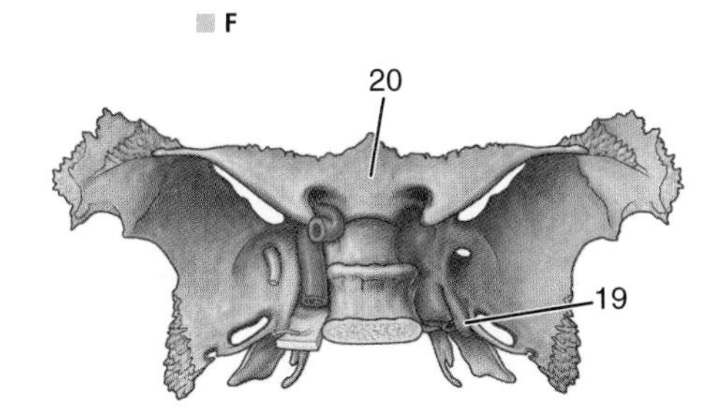

1 **greater palatine canal** a passage in the sphenoid and palatine bones for the greater palatine vessels and nerve; it ends at the greater palatine foramen. Called also *pterygopalatine canal.* <canalis palatinus major> A

2 **greater palatine groove of maxilla** the sulcus on the nasal surface of the maxilla which, along with the corresponding one on the perpendicular plate of the palatine bone, forms the canal for the greater palatine nerve. <sulcus palatinus major maxillae> A

3 **greater palatine groove of palatine bone** a vertical groove on the maxillary surface of the perpendicular plate of the palatine bone; it articulates with the maxilla to form the canal for the greater palatine nerve. <sulcus palatinus major ossis palatini> A

4 **ethmoid bone** the cubical bone located between the orbits and consisting of the cribriform plate, the perpendicular plate, and the paired ethmoid labyrinths. <os ethmoidale> B

5 **ethmoid labyrinth** either of the paired lateral masses of the ethmoid bone, consisting of numerous thin-walled cellular cavities, including the ethmoidal cells and the nasal conchae. <labyrinthus ethmoidalis> B

6 **ethmoid notch of frontal bone** a space between the orbital parts of the frontal bone, in which the ethmoid bone is lodged. <incisura ethmoidalis ossis frontalis> B

7 **perpendicular plate of ethmoid bone** a thin bony plate that descends from the inferior surface of the cribriform plate of the ethmoid bone and participates in forming the nasal septum. <lamina perpendicularis ossis ethmoidalis> B

8 **nasal crest of maxilla** a ridge, raised along the medial border of the palatine process of the maxilla, with which the vomer articulates. <crista nasalis maxillae> C

9 **nasal crest of palatine bone** a thick ridge projecting superiorly from the medial part of the horizontal plate of the palatine bone and articulating with the posterior part of the vomer. <crista nasalis ossis palatini> C

10 **vomer** the unpaired flat bone that forms the inferior and posterior part of the nasal septum. C

11 **vomerine groove** the cleft in the inferior half on the anterior border of the vomer that receives the inferior border of the septal cartilage of the nose. <sulcus vomeris> C

12 ***vomerorostral canal*** a canal located between the vomer and sphenoidal rostrum. <canalis vomerorostralis> C

13 ***vomerovaginal canal*** an inconstant opening formed by the articulating margins of the ala of the vomer and the body of the sphenoid bone; called also *vomerine canal.* <canalis vomerovaginalis> C

14 ***vomerovaginal groove*** the groove on the vaginal process of the pterygoid process of the sphenoid bone that helps form the vomerovaginal canal. <sulcus vomerovaginalis> C

15 **nasal surface of body of maxilla** the surface of the body of the maxilla that helps form the lateral wall of the nasal cavity; it is bounded roughly by the following: medial margin of the orbital surface, medial margin of the infratemporal surface, the palatine process, and the nasal notch. <facies nasalis corporis maxillae> D

16 **nasal surface of horizontal plate of palatine bone** the superior surface of the horizontal plate, which forms the posterior part of the floor of the nasal cavity. <facies nasalis laminae horizontalis ossis palatini> D

17 **nasal surface of perpendicular plate of palatine bone** the medial surface of the perpendicular plate; it articulates with the middle and inferior nasal conchae. <facies nasalis laminae perpendicularis ossis palatini> D

18 **palatine bone** the irregularly shaped bone forming the posterior part of the hard palate, the lateral wall of the nasal fossa between the medial pterygoid plate and the maxilla, and the posterior part of the floor of the orbit. <os palatinum> D

19 **sphenopalatine notch of palatine bone** a notch between the orbital and sphenoid processes of the palatine bone; it is converted into a foramen by the inferior surface of the sphenoid bone; called also *palatine notch.* <incisura sphenopalatina ossis palatini> D

20 **uncinate process of ethmoid bone** a curved plate of bone that extends inferiorly and posteriorly from the anterior part of the ethmoid labyrinth. <processus uncinatus ossis ethmoidalis> D

21 **ethmoid crest of maxilla** a low, oblique ridge on the medial surface of the frontal process of the maxilla, which articulates with the middle nasal concha. <crista ethmoidalis maxillae> D

22 ***ethmoid crest of palatine bone*** a ridge near the upper end of the medial surface of the palatine bone, which articulates with the middle nasal concha. <crista ethmoidalis ossis palatini> D

23 **perpendicular plate of palatine bone** the flat, vertical, bony plate that extends superiorly on either side from the palatine bone; it is surmounted by the orbital and sphenoidal processes. <lamina perpendicularis ossis palatini> D

24 **superior nasal concha** the upper of two bony plates projecting from the inner wall of the ethmoid labyrinth and forming the upper boundary of the superior meatus of the nose, and the mucous membrane covering the plate. Called also *superior ethmoidal concha, superior turbinate bone,* and *superior turbinate.* <concha nasi superior> D

A

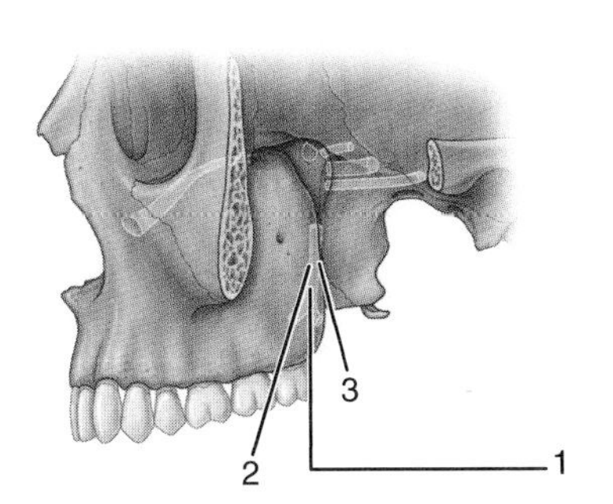

B

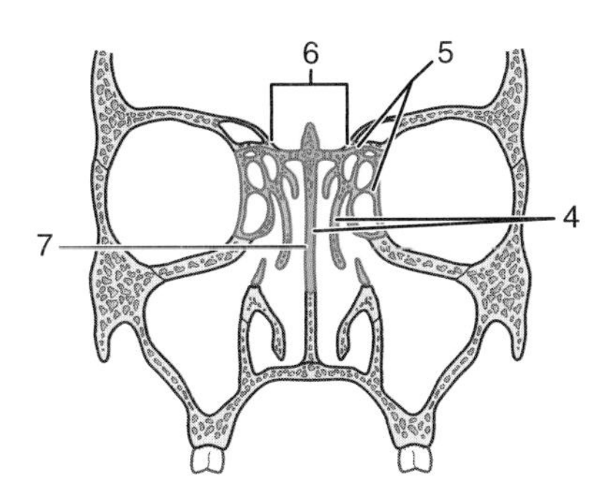

C

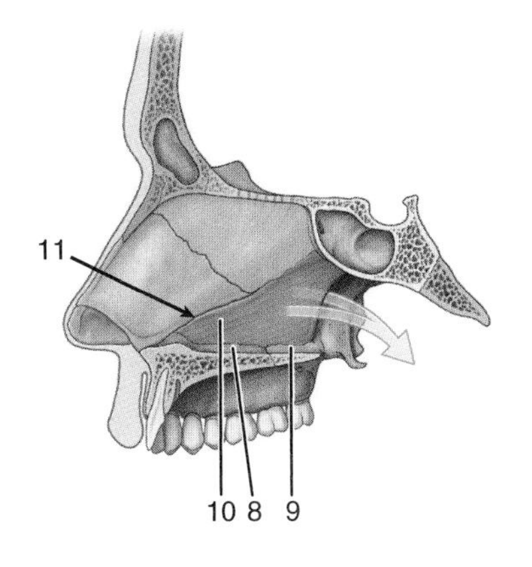

D

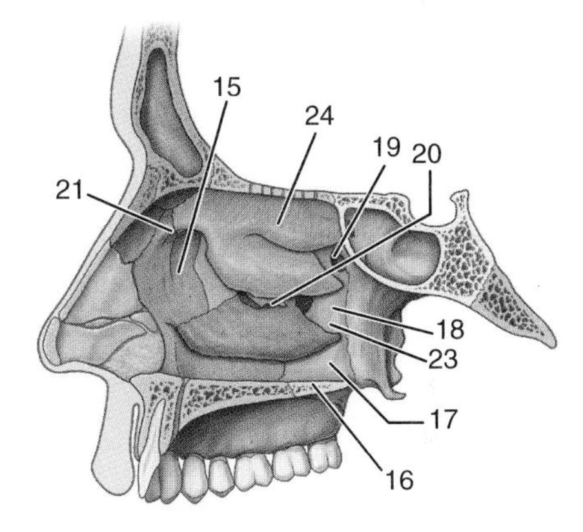

1 **ethmoid bulla of ethmoid bone** a rounded projection of the ethmoid bone into the lateral wall of the middle nasal meatus just below the middle nasal concha, enclosing a large ethmoid air cell; called also *ethmoid antrum.* <bulla ethmoidalis ossis ethmoidalis> A

2 **sphenoid process of palatine bone** an irregular mass of bone that projects superiorly and medially from the posterior portion of the superior margin of the perpendicular portion of the palatine bone, and articulates with the body of the sphenoid bone and with the ala of vomer. <processus sphenoidalis ossis palatini> B

3 **sphenovomerine suture** the line of junction between the vaginal processes of the medial pterygoid plates of the sphenoid and the ala of the vomer. <sutura sphenovomeralis> B

4 **vaginal process of sphenoid bone** a small plate on the inferior surface of the body of the sphenoid bone on either side, running medially from the medial pterygoid plate to articulate with the ala of the vomer and with the sphenoid process of the palatine bone. <processus vaginalis ossis sphenoidalis> B

5 **horizontal plate of palatine bone** the horizontal part of the palatine bone, forming the posterior part of the hard palate. <lamina horizontalis ossis palatini> C

6 **palatine crest of palatine bone** a transverse crest often seen on the inferior surface of the horizontal plate of the palatine bone a short distance anterior to the posterior border. <crista palatina ossis palatini> C

7 **palatine process of maxilla** a horizontally arched plate of bone that helps to form the lower part of the maxilla and with its fellow of the opposite side the anterior two-thirds of the hard palate. <processus palatinus maxillae> C

8 **palatine surface of horizontal plate of palatine bone** the inferior surface of the horizontal plate, forming the posterior part of the hard palate. <facies palatina laminae horizontalis ossis palatini> C

9 **palatine torus** a bony protuberance sometimes found on the hard palate at the junction of the intermaxillary suture and the transverse palatine suture. <torus palatinus> C

10 **petro-occipital fissure** a fissure extending posteriorly from the foramen lacerum to the jugular foramen, between the basioccipital area and the posterior and inner border of the petrous portion of the temporal bone; called also *petrobasilar fissure.* <fissura petrooccipitalis> C

11 ***petro-occipital synchondrosis*** the plate of cartilage in the petro-occipital fissure which helps to unite the basilar portion of the occipital bone and the petrous portion of the temporal bone. <synchondrosis petrooccipitalis> C

12 **petrosal fossula** a small depression on the inferior surface of the petrous portion of the temporal bone, on a small ridge separating the jugular fossa from the external opening of the carotid canal. <fossula petrosa> C

13 **posterior nasal spine of palatine bone** a small, sharp, backward-projecting bony spine forming the medial posterior angle of the horizontal plate of the palatine bone; called also *nasal spine of palatine bone.* <spina nasalis posterior ossis palatini> C

14 **pterygoid process of sphenoid bone** either of two processes on the sphenoid bone descending from the points of junction of the greater wings and body of the bone, and each consisting of a lateral and a medial plate. <processus pterygoideus ossis sphenoidalis> B, C

15 ***pterygospinal ligament*** a band of fibers extending from the superior part of the superior border of the lateral pterygoid plate to the spine of the sphenoid bone. <ligamentum pterygospinale> C

16 ***pterygospinous process*** a small spine on the posterior edge of the lateral pterygoid plate of the sphenoid bone, giving attachment to the pterygospinal ligament. <processus pterygospinosus> C

17 **pyramidal process of palatine bone** a strong process projecting downward, backward, and laterally from the lateral part of the posterior margin of the palatine bone and helping to form the pterygoid fossa. <processus pyramidalis ossis palatini> B, C

18 **scaphoid fossa** a depression on the superior part of the posterior portion of the medial plate of the pterygoid process of the sphenoid bone, giving attachment to the tensor veli palatini muscle. Called also *navicular f. of sphenoid bone.* <fossa scaphoidea ossis sphenoidalis> C

19 **spine of sphenoid bone** a small bony process projecting inferiorly from the inferior aspect of the greater wing of the sphenoid bone where the wing projects into the angle between the petrous and squamous portions of the temporal bone; it is just posterior to the foramen spinosum and serves for attachment of the sphenomandibular and pterygospinous ligaments. <spina ossis sphenoidalis> C

20 **sulcus for auditory tube** a groove on the medial part of the base of the spine of the sphenoid bone; it lodges a portion of the cartilaginous part of the auditory tube. Called also *groove for auditory tube.* <sulcus tubae auditoriae> C

21 **transverse palatine suture** the line of junction between the palatine processes of the maxillae and the horizontal parts of the palatine bones. <sutura palatina transversa> C

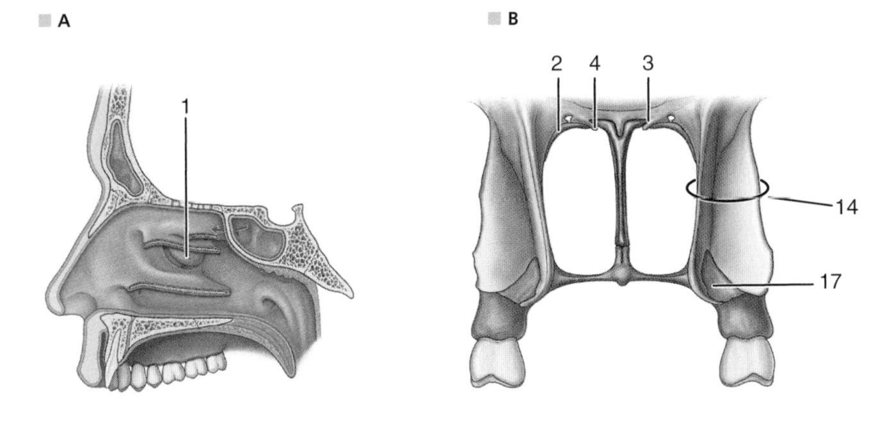
A
1
B
2
4
3
14
17

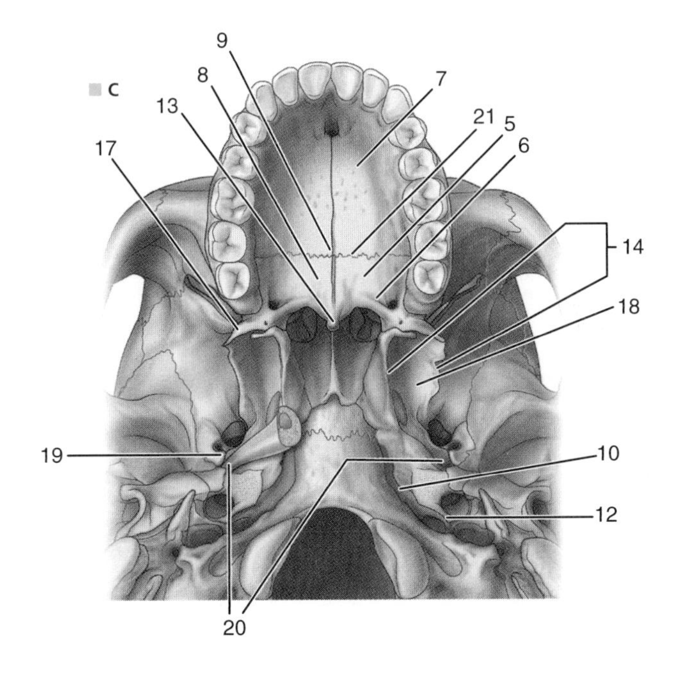
C
9
8
7
13
21
5
17
6
14
18
19
10
12
20

1 **sheath of styloid process** a ridge on the inferior surface of the temporal bone, partly enclosing the base of the styloid process. <vagina processus styloidei> A

2 **stylohyoid ligament** a vertical fibroelastic aponeurotic cord attached superiorly to the tip of the styloid process of the temporal bone and inferiorly to the lesser horn of the hyoid bone. <ligamentum stylohyoideum> A

3 **styloid process of temporal bone** a long spine projecting inferiorly from the inferior surface of the temporal bone just anterior to the stylomastoid foramen, giving attachment to three muscles and two ligaments. <processus styloideus ossis temporalis> A

4 **palatine grooves of maxilla** the laterally placed furrows, between the palatine spines on the inferior surface of the hard palate, that lodge the palatine vessels and nerves. <sulci palatini maxillae> B

5 **palatine spines** ridges that are laterally placed on the inferior surface of the maxillary part of the hard palate, separating the palatine sulci. <spinae palatinae> B

6 **head of mandible** the articular surface of the condyloid process of the mandible. <caput mandibulae> C

7 **mandible** the horseshoe-shaped bone forming the lower jaw; the largest and strongest bone of the face, consisting of a body and a pair of rami, which articulate with the skull at the temporomandibular joints. <mandibula> C

8 **mandibular notch** a deep notch on the upper edge of the ramus of the mandible between the condyle and the coronoid process. <incisura mandibulae> C

9 **masseteric tuberosity** an elongated, raised and roughened area on the lateral side of the angle of the mandible, for the insertion of tendinous bundles of the masseter muscles. <tuberositas masseterica> C

10 **neck of mandible** the narrow portion supporting the condyle of the mandible. <collum mandibulae> C

11 **oblique line of mandible** a ridge on the external surface of the body of the mandible extending from the mental tubercle to the anterior border of the ascending ramus on either side. <linea obliqua mandibulae> C

12 **ramus of mandible** a quadrilateral process projecting superiorly from the posterior part of either side of the mandible. <ramus mandibulae> C

13 **lingula of mandible** the sharp medial boundary of the mandibular foramen, to which is attached the sphenomandibular ligament. <lingula mandibulae> D

14 **mandibular foramen** the opening on the medial surface of the ramus of the mandible, leading into the mandibular canal. <foramen mandibulae> D

15 ***mandibular torus*** a prominence sometimes seen on the lingual aspect of the mandible at the base of its alveolar part, adjacent to the postcanine teeth. <torus mandibularis> D

16 **sublingual fossa** a depression on the inner surface of the body of the mandible, lodging a portion of the sublingual gland. <fovea sublingualis> D

17 **submandibular fossa** a depression on the medial aspect of the body of the mandible, lodging a small portion of the submandibular gland; called also *submaxillary fossa.* <fovea submandibularis> D

18 **digastric fossa** a depression on the internal surface of the body of the mandible on each side of the symphysis to which is attached the anterior belly of the digastric muscle. <fossa digastrica> E

19 **inferior mental spine** the lower part of a small bony projection located on the internal surface of the mandible, near the lower end of the midline and above the anterior end of the mylohyoid line, serving for attachment of the geniohyoid muscle. <spina mentalis inferior> F

20 **superior mental spine** the upper part of a small bony projection located on the internal surface of the mandible, near the lower end of the midline and above the anterior end of the mylohyoid line, serving for attachment of the genioglossus muscle. Called also *superior genial tubercle.* <spina mentalis superior> F

A

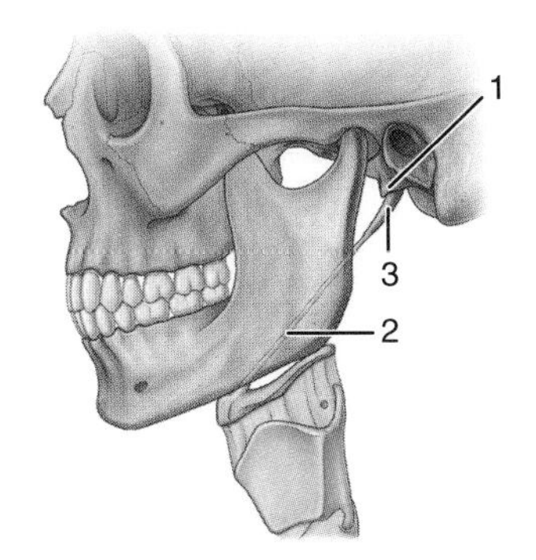

B

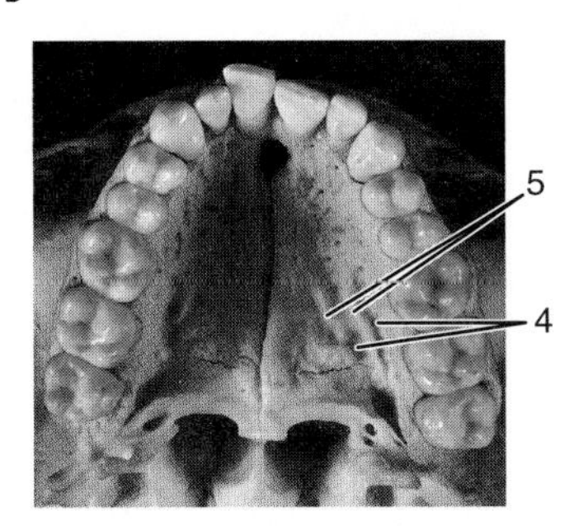

C

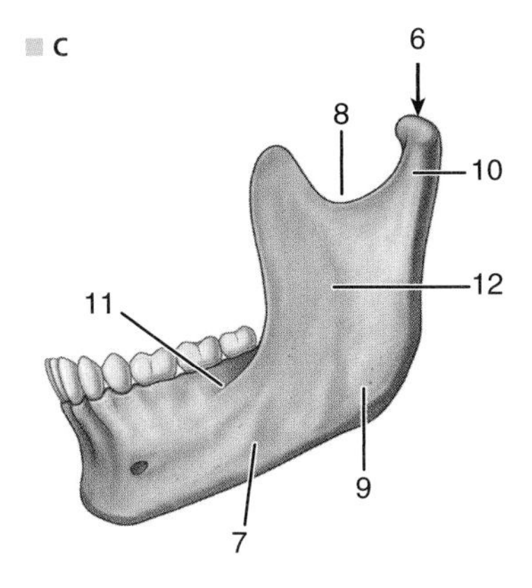

D

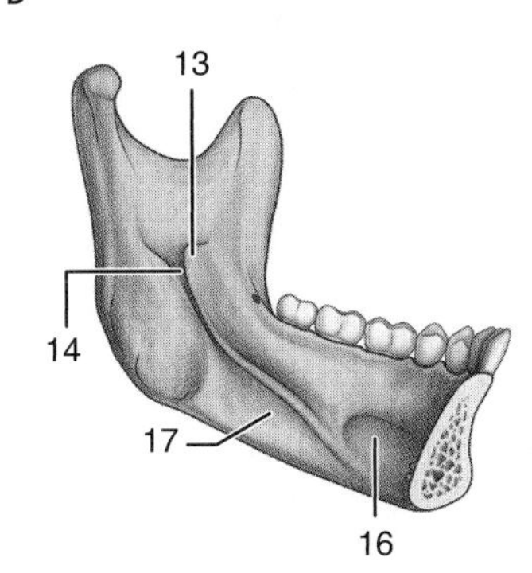

E

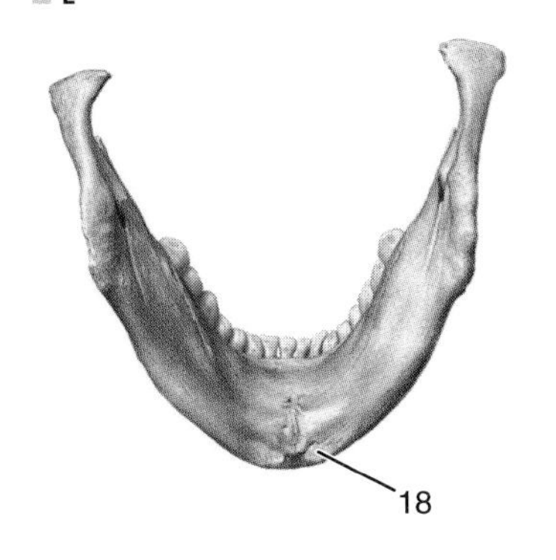

F

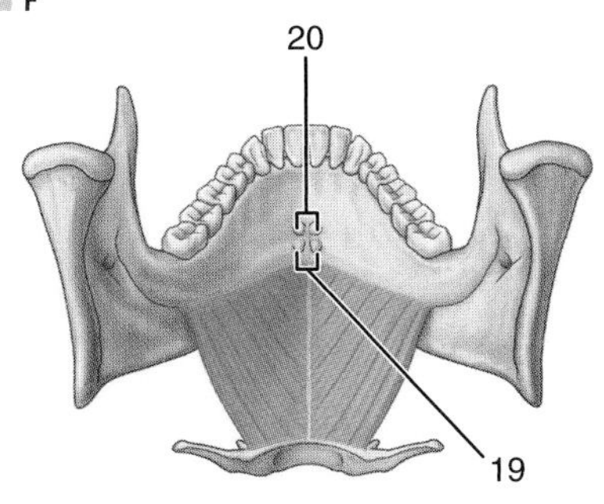

1 **alveolar arch of mandible** the superior free border of the alveolar process of the mandible. <arcus alveolaris mandibulae> A

2 **alveolar arch of maxilla** the inferior free border of the alveolar process of the maxilla. <arcus alveolaris maxillae> A

3 **alveolar canals of maxilla** several canals in the maxilla for the passage of the posterior superior alveolar vessels and nerves, each canal beginning on the infratemporal surface of the maxilla at an alveolar foramen. <canales alveolares maxillae> A

4 **alveolar foramina** the openings of the alveolar canals at the deepest portion of the tooth sockets in the maxilla. <foramina alveolaria maxillae> A

5 **alveolar part of mandible** the superior portion of the body of the mandible, which contains sockets for the teeth. <pars alveolaris mandibulae> A

6 **alveolar process of maxilla** the thick parabolically curved ridge that projects downward and forms the free lower border of the maxilla; it is in front of and lateral to the palatine process and it bears the teeth. <processus alveolaris maxillae> A

7 **alveolar yokes of mandible** depressions on the anterior surface of the alveolar process of the mandible, between the ridges caused by the roots of the incisor teeth. <juga alveolaria mandibulae> B

8 **alveolar yokes of maxilla** the depressions on the anterior surface of the alveolar process of the maxilla, between the ridges caused by the roots of the incisor teeth. <juga alveolaria maxillae> B

9 **angle of mandible** the angle created at the junction of the posterior edge of the ramus and the lower edge of the mandible. <angulus mandibulae> A

10 **dental alveoli of mandible** the cavities or sockets in the alveolar process of the mandible in which the roots of the teeth are held by the periodontal ligament. <alveoli dentales mandibulae> B

11 **dental alveoli of maxilla** the cavities or sockets in the alveolar process of the maxilla in which the roots of the teeth are held by the periodontal ligament. <alveoli dentales maxillae> B

12 **dental alveolus** one of the cavities or sockets in the alveolar process of the mandible or maxilla, in which the roots of the teeth are held by fibers of the periodontal ligament. Called also *tooth sockets.* <alveolus dentalis> B

13 **interalveolar septa of mandible** the partitions between the tooth sockets in the alveolar part of the mandible. <septa interalveolaria mandibulae> B

14 **interalveolar septa of maxilla** the partitions between the tooth sockets in the alveolar process of the maxilla. <septa interalveolaria maxillae> B

15 **interradicular septa of mandible** the thin bony partitions separating the crypts of a mandibular dental alveolus occupied by the separate roots of a multirooted tooth. <septa interradicularia mandibulae> B

16 **interradicular septa of maxilla** the thin bony partitions separating the crypts of a maxillary alveolus occupied by the separate roots of a multirooted tooth. <septa interradicularia maxillae> B

17 **mandibular canal** a canal that traverses the ramus and body of the mandible between the mandibular and mental foramina, transmitting the inferior alveolar vessels and nerve; beneath the first or second premolars it splits into the mental canal and the incisive canal. <canalis mandibulae> A

18 **mandibular dental arch** the portion of the dental arch formed by the teeth of the mandible. <arcus dentalis mandibularis> A

Cranial Cavity

19 ***cranial cavity*** the space enclosed by the bones of the cranium. <cavitas cranii>

20 ***venous grooves*** grooves on the internal surfaces of the cranial bones for the meningeal veins. <sulci venosi>

21 **granular foveolae** small pits on the internal surface of the cranial bones on either side of the sagittal sulcus; they are occupied by the arachnoidal granulations. <foveolae granulares> C

22 **groove for superior sagittal sinus** a groove on the internal surface of the frontal, parietal, and occipital bones that lodges the superior sagittal sinus; called also *sagittal groove* and *sagittal sulcus.* <sulcus sinus sagittalis superioris> C

23 **internal surface of frontal bone** the vertically situated, concave cerebral surface of the frontal bone; in its midline the sagittal sulcus is seen superiorly and the frontal crest inferiorly. Called also *inner table of frontal bone.* <facies interna ossis frontalis> C

24 **internal surface of parietal bone** the internal, or cerebral, surface of the parietal bone; called also *cerebral surface of parietal bone.* <facies interna ossis parietalis> C

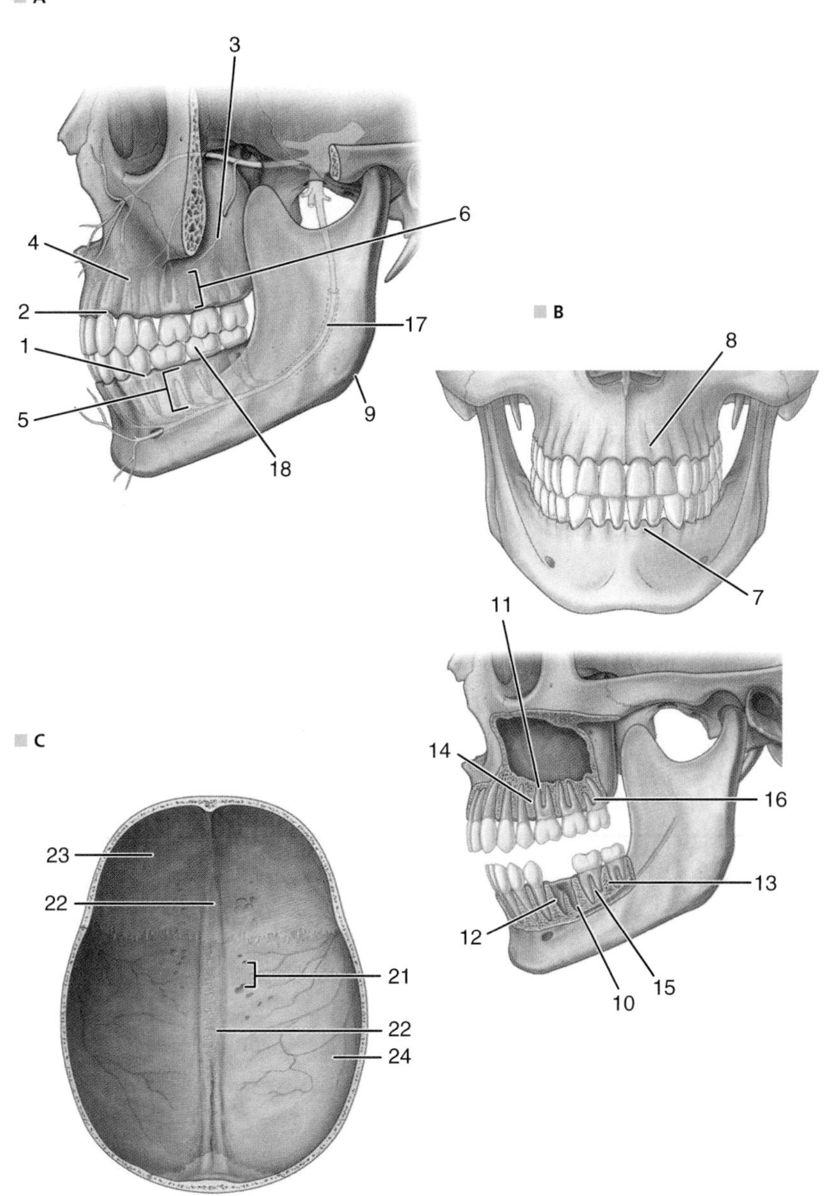
A
3
6
4
2
17
1
5
9
18
B
8
7
11
14
16
13
12
10
15
C
23
22
21
22
24

1 **anterior cranial fossa** the anterior subdivision of the floor of the cranial cavity, supporting the frontal lobes of the brain, and composed of portions of three bones: the ethmoid, frontal, and sphenoid. <fossa cranii anterior> A, B

2 ***frontal margin of greater wing of sphenoid bone*** a roughened area on the greater wing where it articulates with the frontal bone; it is at the superolateral margin of the orbital surface of the greater wing at its junction with the cerebral and temporal surfaces. <margo frontalis alae majoris> B

3 **greater wing of sphenoid bone** a large wing-shaped process arising from either side of the body of the sphenoid bone; its cerebral surface forms the anterior part of the floor of the middle cranial fossa, and its orbital surface forms the chief part of the lateral wall of the orbit. Called also *major* or *temporal wing of sphenoid bone.* <ala major ossis sphenoidalis> B

4 **lesser wing of sphenoid bone** the thin triangular plate of bone that extends horizontally and laterally from either side of the anterior part of the body of the sphenoid bone; it articulates with the frontal bone and helps form the roof of the orbit and the floor of the anterior cranial fossa. Called also *minor* or *small wing of sphenoid bone.* <ala minor ossis sphenoidalis> B

5 **middle clinoid process** either of two small inconstant eminences on the internal surface of the sphenoid bone, one on either side of the anterior part of the hypophysial fossa. <processus clinoideus medius> B

6 **middle cranial fossa** the middle subdivision of the floor of the cranial cavity, supporting the temporal lobes of the brain and the pituitary gland; it is composed of the body and greater wings of the sphenoid bone and the squamous and petrous portions of the temporal bone. <fossa cranii media> A

7 **posterior cranial fossa** the posterior subdivision of the floor of the cranial cavity, lodging the cerebellum, pons, and medulla oblongata; it is formed by portions of the sphenoid, temporal, parietal, and occipital bones. <fossa cranii posterior> A

8 **ala of crista galli** a small winglike process on the anterior part of the crista galli of the ethmoid bone. <ala cristae galli> B

9 **anterior clinoid process** the bony process found on the medial extremity of the posterior border of the lesser wing of the sphenoid bone. <processus clinoideus anterior> B

10 **anterior surface of petrous part of temporal bone** the surface of the petrous part of the temporal bone that forms the posterior portion of the floor of the middle cranial fossa. <facies anterior partis petrosae ossis temporalis> B

11 **apex of petrous part of temporal bone** the truncated portion of the petrous part of the temporal bone that is directed anteriorly and medially and ends at the medial opening of the carotid canal. <apex partis petrosae ossis temporalis> B

12 **arcuate eminence** an arched prominence on the internal surface of the petrous part of the temporal bone in the floor of the middle cranial fossa, marking the position of the superior semicircular canal. It is particularly prominent in young skulls. Called also *eminence of superior semicircular canal.* <eminentia arcuata> B

13 **arterial grooves** grooves on the internal surfaces of the cranial bones for the meningeal arteries. <sulci arteriosi> B

14 **carotid groove** the groove on the side of the body of the sphenoid bone that lodges the internal carotid artery and the cavernous sinus. <sulcus caroticus ossis sphenoidalis> B

15 **cerebral surface of greater wing** the smooth, concave part of the greater wing of the sphenoid bone that forms the anterior part of the floor of the middle cranial fossa, lying anterior to the petrous and squamous parts of the temporal bone. <facies cerebralis alae majoris> B

16 **cerebral surface of squamous part of temporal bone** the inner surface of the squamous part forming the lateral wall of the middle cranial fossa. <facies cerebralis partis squamosae ossis temporalis> B

17 **chiasmatic sulcus** a furrow on the superior surface of the sphenoid bone, located just anterior to the tuberculum sellae; it lodges the optic chiasm. Called also *prechiasmatic sulcus, optic groove,* and *optic sulcus.* <sulcus prechiasmaticus> B

18 **clivus of occipital bone** the lower part of the clivus, formed by the basilar portion of the occipital bone; called also *basilar clivus* and *basilar groove of occipital bone.* <clivus ossis occipitalis> B

19 **clivus** a bony surface in the posterior cranial fossa, sloping superiorly from the foramen magnum to the dorsum sellae, the inferior part being formed by a portion of the basilar part of the occipital bone and the superior part by a surface of the body of the sphenoid bone. B

20 **cribriform foramina** the openings in the cribriform plate of the ethmoid bone for passage of the olfactory nerves. Called also *olfactory foramina.* <foramina cribrosa ossis ethmoidalis> B

21 **cribriform plate of ethmoid bone** the horizontal plate of the ethmoid bone that forms the roof of the nasal cavity; it is perforated by many foramina for the passage of the olfactory nerves. On its superior surface is a projection called the *crista galli.* <lamina cribrosa ossis ethmoidalis> B

22 **crista galli** a thick triangular process projecting superiorly from the cribriform plate of the ethmoid bone; the falx cerebri attaches to it. B

23 **cruciform eminence of occipital bone** the cross-shaped bony prominence on the internal surface of the squama of the occipital bone, at the intersection of the ridges associated with the sulci of the superior sagittal sinus and the transverse sinuses. Called also *cruciate line.* <eminentia cruciformis> B

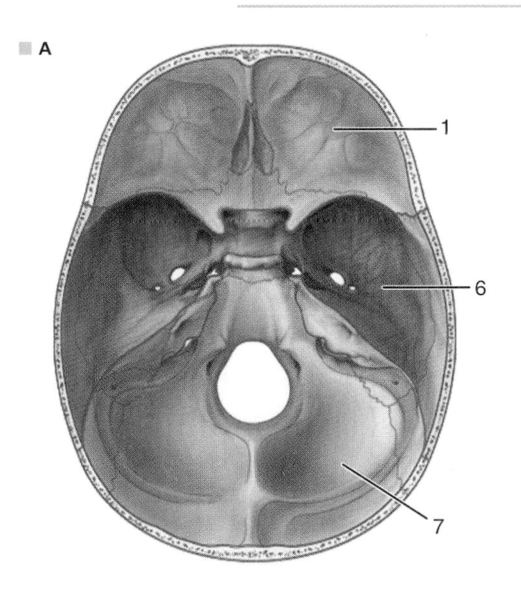
A
1
6
7

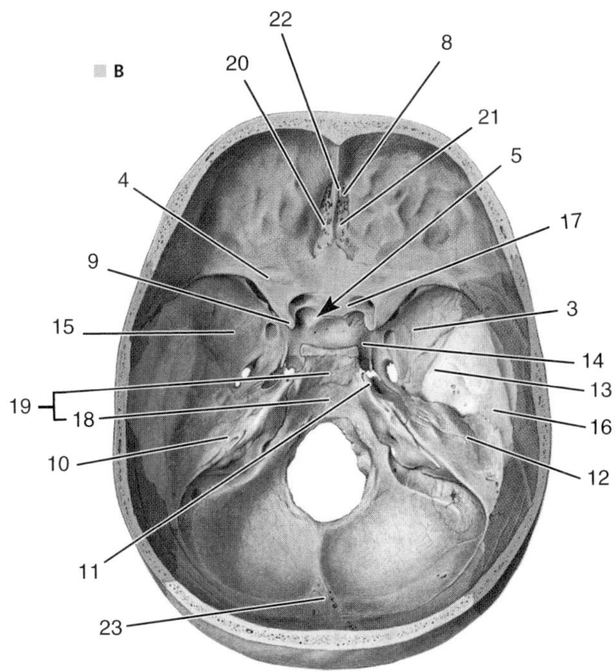
B
22
8
20
21
5
4
17
9
3
15
14
13
19
18
16
10
12
11
23

1 **dorsum sellae** the quadrilateral plate on the sphenoid bone that forms the posterior boundary of the sella turcica; the posterior clinoid processes project from its superior extremity, and it is continuous inferiorly with the clivus. A

2 **foramen cecum of frontal bone** a blind opening formed between the frontal crest and the crista galli, which sometimes transmits a vein from the nasal cavity to the superior sagittal sinus; called also *cecal foramen.* <foramen caecum ossis frontalis> A

3 **foramen ovale of sphenoid** an opening in the posterior part of the medial portion of the greater wing of the sphenoid bone; it transmits the mandibular branch of the trigeminal nerve and some vessels. <foramen ovale ossis sphenoidalis> A

4 **foramen rotundum of sphenoid** a round opening in the medial part of the greater wing of the sphenoid bone that transmits the maxillary branch of the trigeminal nerve; called also *superior maxillary foramen* or *canal.* <foramen rotundum ossis sphenoidalis> A

5 **foramen spinosum** an opening in the greater wing of the sphenoid bone, near its posterior angle, for the middle meningeal artery. A

6 **frontal crest** a median ridge on the internal surface of the frontal bone, extending superiorly from the foramen cecum to unite with the sulcus for the superior sagittal sinus. <crista frontalis> A

7 **frontoethmoidal suture** the line of junction in the anterior cranial fossa between the frontal bone and the cribriform plate of the ethmoid bone. <sutura frontoethmoidalis> A

8 **groove for inferior petrosal sinus on occipital bone** the groove in the floor of the posterior cranial fossa at the line of junction between the basilar part of the occipital and the petrous portion of the temporal bone; it lodges the inferior petrosal sinus. <sulcus sinus petrosi inferioris ossis occipitalis> A

9 **groove for inferior petrosal sinus on temporal bone** a groove on the posteromedial edge of the internal surface of the petrous portion of the temporal bone, which, with a corresponding groove on the adjacent basilar part of the occipital bone, lodges the inferior petrosal sinus. <sulcus sinus petrosi inferioris ossis temporalis> A

10 **groove for middle meningeal artery** either of two grooves in the parietal bone for branches of the meningeal artery; one, posterior to the coronal suture, lodges the frontal branch, and the other, anterior to the mastoid angle, lodges the parietal branch. <sulcus arteriae meningeae mediae> A

11 **groove for superior petrosal sinus** a small posterolaterally directed sulcus that runs along the internal surface of the petrous part of the temporal bone on the angle separating the posterior and middle cranial fossae; it lodges the superior petrosal sinus. <sulcus sinus petrosi superioris> A

12 **hypophysial fossa** a deep depression in the middle of the sella turcica of the sphenoid bone, lodging the pituitary gland; called also *pituitary* or *sellar fossa.* <fossa hypophysialis> A

13 ***internal surface of cranial base*** the inner surface of the inferior region of the skull, constituting the floor of the cranial cavity. <basis cranii interna> A

14 **sphenofrontal suture** a long suture joining the orbital part of the frontal bone to the greater and lesser wings of the sphenoid bone on either side of the skull. <sutura sphenofrontalis> A

15 ***subarcuate fossa*** a small fossa on the internal surface of the petrous part of the temporal bone just inferior to the arcuate eminence, most prominent in the fetus. In the adult it lodges a piece of dura and transmits a small vein. <fossa subarcuata ossis temporalis> A

16 **groove for sigmoid sinus on occipital bone** the portion of the groove for the sigmoid sinus found on the occipital bone. <sulcus sinus sigmoidei ossis occipitalis> B

17 **groove for sigmoid sinus on parietal bone** a short groove on the internal surface of the posteroinferior angle of the parietal bone, continuous with both the sulcus for the sigmoid sinus on the temporal bone and the sulcus for the transverse sinus on the occipital bone; it lodges the superior part of the sigmoid sinus. <sulcus sinus sigmoidei ossis parietalis> B

18 **groove for sigmoid sinus on temporal bone** the portion of the groove for the sigmoid sinus found on the temporal bone. <sulcus sinus sigmoidei ossis temporalis> B

19 **groove for transverse sinus** a wide groove that passes horizontally, lateralward, and anteriorly from the internal occipital protuberance to the parietal bone, where it becomes continuous with the sulcus of the sigmoid sinus; it lodges the transverse sinus. <sulcus sinus transversi> B

20 **hypoglossal canal** an opening in the lateral part of the occipital bone at the base of the condyle, which transmits the hypoglossal nerve and a branch of the posterior meningeal artery; called also *anterior condyloid canal* or *foramen.* <canalis nervi hypoglossi> B

21 **internal occipital crest** a median ridge on the internal surface of the occipital bone extending from the midpoint of the cruciform eminence toward the foramen magnum. <crista occipitalis interna> B

22 **internal occipital protuberance** the projection of bone at the midpoint of the cruciform eminence, on the internal surface of the squama of the occipital bone, sometimes presenting as a ridge *(internal occipital crest).* <protuberantia occipitalis interna> B

23 **frontal branch of middle meningeal artery** a branch lodged in grooves on the sphenoid and parietal bones and supplying the dura mater of the front of the brain. A part of it is sometimes enclosed in a bony canal. <ramus frontalis arteriae meningeae mediae> C

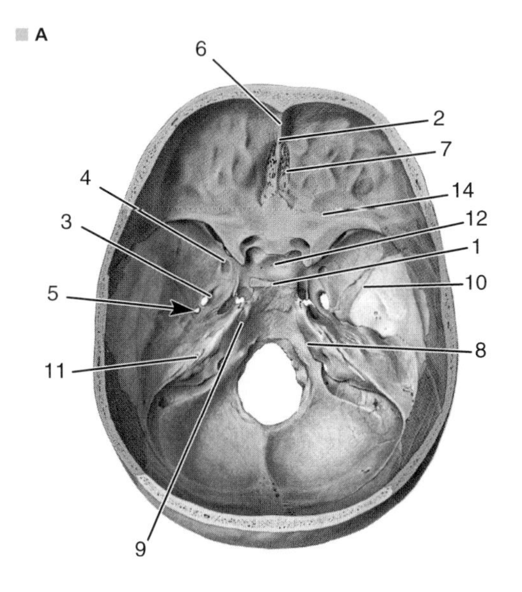
A
6
2
7
4
14
3
12
1
10
5
11
8
9

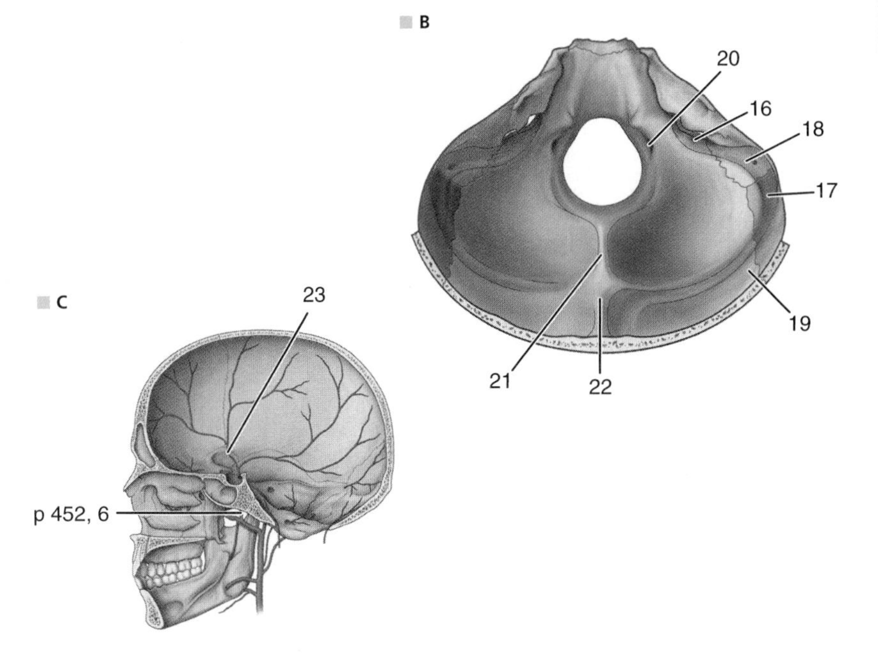
B
20
16
18
17
19
21
22
C
23
p 452, 6

1 **diploë** the loose osseous tissue between the two tables of the cranial bones. A

2 **internal table of calvaria** the inner compact layer of bone of the flat bones of the skull. Called also *inner table of skull.* <lamina interna calvariae> A

3 ***anterior temporal diploic vein*** a vein that drains the lateral portion of the frontal and the anterior part of the parietal bone, opening internally into the sphenoparietal sinus and externally into a deep temporal vein. <vena diploica temporalis anterior> B

4 **diploic canals** bony canals in the cranial bones, located in the spongy bone between the compact tables and providing for passage of the veins of the diploë; called also *Breschet's canals.* <canales diploici> B

5 **diploic veins** veins of the skull, including the frontal, occipital, anterior temporal, and posterior temporal diploic veins, which form sinuses in the cancellous tissue between the laminae of the cranial bones. They send branches to the external and the internal lamina, the periosteum, and the dura mater, and empty in part inside and in part outside the skull, communicating with meningeal veins, dural sinuses, pericranial veins, and each other. <venae diploicae> B

6 **emissary vein** one of the small, valveless veins that pass through foramina of the skull, connecting the dural venous sinuses with scalp veins or with deep veins below the base of the skull. <vena emissaria> B

7 ***frontal diploic vein*** a vein that drains the frontal bone, emptying externally into the supraorbital vein and internally into the superior sagittal sinus. <vena diploica frontalis> B

8 ***occipital diploic vein*** the largest of the diploic veins, which drains blood from the occipital bone and empties into the occipital vein or the transverse sinus. <vena diploica occipitalis> B

9 ***occipital emissary vein*** an occasional small vein running through a minute foramen in the occipital protuberance of the skull and connecting the confluence of the sinuses with the occipital vein. <vena emissaria occipitalis> B

10 ***posterior temporal diploic vein*** a vein that drains the parietal bone and empties into the transverse sinus. <vena diploica temporalis posterior> B

11 **groove for greater petrosal nerve** a small groove sometimes present on the anterior surface of the petrous part of the temporal bone (the floor of the middle cranial fossa), running anteromedially from the hiatus of the facial canal to the foramen lacerum, and lodging the greater petrosal nerve. <sulcus nervi petrosi majoris> C

12 **groove for lesser petrosal nerve** a small groove on the anterior surface of the petrous part of the temporal bone (the floor of the middle cranial fossa), running anteromedially just lateral to the groove for the greater petrosal nerve, and lodging the lesser petrosal nerve. <sulcus nervi petrosi minoris> C

13 **hiatus for greater petrosal nerve** an opening in the petrous part of the temporal bone in the floor of the middle cranial fossa that transmits the greater petrosal nerve and a branch of the middle meningeal artery. <hiatus canalis nervi petrosi majoris> C

14 **hiatus for lesser petrosal nerve** the small, laterally placed opening on the anterior surface of the pyramid of the temporal bone that transmits the lesser petrosal nerve. <hiatus canalis nervi petrosi minoris> C

15 **internal spiral sulcus** the C-shaped concavity within the cochlear duct formed by the spiral limbus and its tympanic and vestibular lips along the edge of the osseous spiral lamina. <sulcus spiralis internus> D

16 **infratemporal crest** a crest separating the temporal surface of the greater wing of the sphenoid bone into a superior temporal portion and an inferior infratemporal portion. <crista infratemporalis> E

17 **greater petrosal nerve** *origin*, intermediate nerve via geniculate ganglion; *distribution*, running forward from the geniculate ganglion, it joins the deep petrosal nerve of the pterygoid canal, and reaches lacrimal, nasal, and palatine glands and nasopharynx, via pterygopalatine ganglion and its branches; *modality*, parasympathetic and special sensory (taste). <nervus petrosus major> F

A

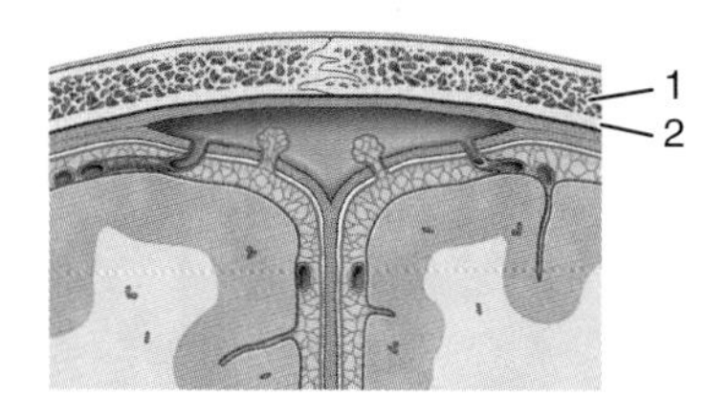

B

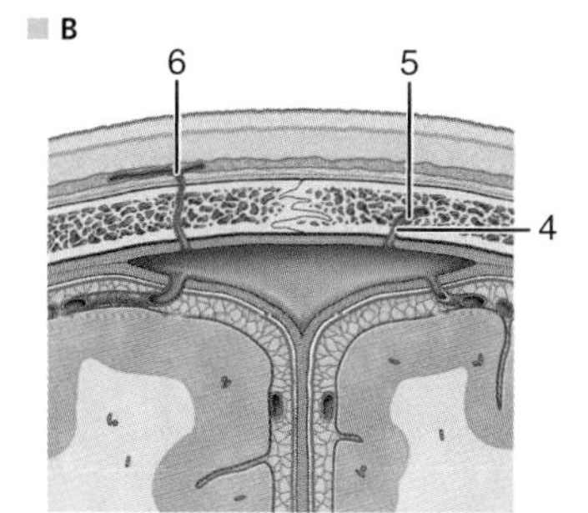

C

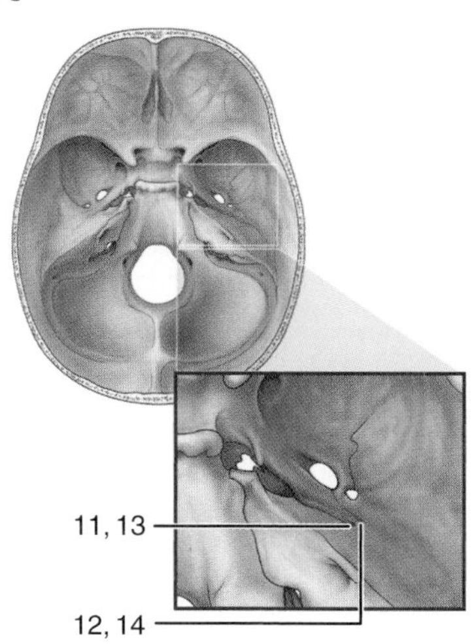

D

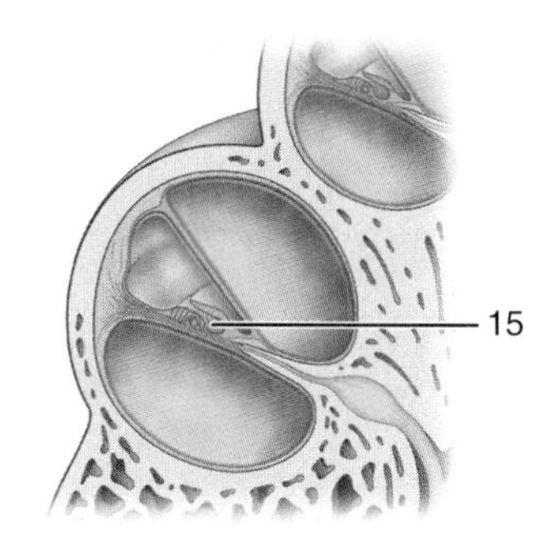

E

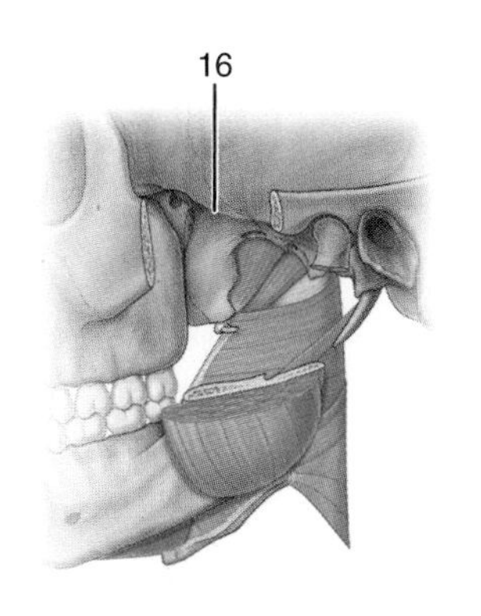

F

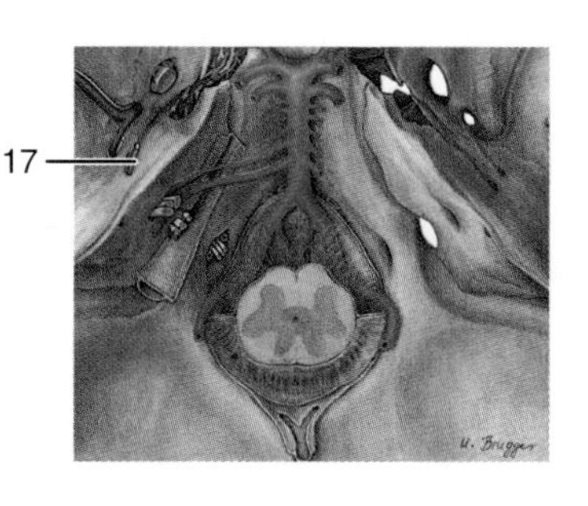

Meninges

1 ***meningeal branch of maxillary nerve*** a branch arising from the maxillary nerve in the middle cranial fossa, accompanying the middle meningeal artery, and supplying the dura mater; *modality*, general sensory. Called also *meningeal nerve*. <ramus meningeus nervi maxillaris>

2 ***meningeal branch of spinal nerve*** the small branch of each spinal nerve that re-enters the intervertebral foramen to supply the dura mater, vertebral column, and associated ligaments. <ramus meningeus nervi spinalis>

3 ***meningeal branch of vagus nerve*** a branch that arises in the jugular foramen from the superior ganglion of the vagus nerve, innervating dura mater of the posterior cranial fossa. <ramus meningeus nervi vagi>

4 **diaphragma sellae** a ring-shaped fold of dura mater covering the sella turcica and containing an aperture for passage of the infundibulum of the hypophysis. A

5 **falx cerebelli** the small fold of dura mater in the midline of the posterior cranial fossa, projecting forward toward the vermis of the cerebellum. A

6 **falx cerebri** the fold of dura mater, sickle-shaped when viewed in sagittal section, that extends downward in the longitudinal cerebral fissure and separates the two cerebral hemispheres. A

7 ***tentorial branch of ophthalmic nerve*** a branch that arises from the ophthalmic nerve close to its origin from the trigeminal ganglion, turning back to innervate the dura mater of the tentorium cerebelli and falx cerebri; *modality*, general sensory. <ramus meningeus recurrens nervi ophthalmici> A

8 **tentorial notch** an opening at the anterior part of the cerebellum, formed by the free, internal border of the tentorium and the dorsum sellae of the sphenoid, and occupied chiefly by the mesencephalon. <incisura tentorii cerebelli> A

9 **tentorium cerebelli** the process of dura mater that supports the occipital lobes and covers the cerebellum. Its internal border is free and bounds the tentorial notch; its external border is attached to the skull and encloses the transverse sinus behind. A

10 ***trigeminal cave*** the small outpocketing of the dura mater surrounding the ganglion and divisions of the trigeminal nerve at the end of the petrous portion of the temporal bone; it contains the trigeminal ganglion. Called also *Meckel's space* and *trigeminal cavity*. <cavum trigeminale> A

11 ***anastomotic branch of middle meningeal artery with lacrimal artery*** a branch of the middle meningeal artery that is distributed to the orbit and anastomoses with the anastomotic branch of the lacrimal artery. <ramus anastomoticus arteriae meningeae mediae cum arteria lacrimali> B

12 **anterior meningeal branch of anterior ethmoidal artery** a branch that supplies the dura mater. Called also *anterior meningeal artery*. <ramus meningeus anterior arteriae ethmoidalis anterioris> B

13 **meningeal branches of vertebral artery** branches, anterior and posterior, arising from the vertebral artery in the foramen magnum, and ramifying in the posterior cranial fossa to supply the dura mater, including the falx cerebelli and bone. <rami meningei arteriae vertebralis> B

14 ***meningeal veins*** the venae comitantes of the meningeal arteries, which drain the dura mater, communicate with the lateral lacunae, and empty into the regional sinuses and veins. <venae meningeae> B

15 **parietal branch of middle meningeal artery** a vessel that arises in the middle cranial fossa, grooves the temporal and parietal bones, and supplies the posterior dura mater. <ramus parietalis arteriae meningeae mediae> B

16 **posterior meningeal artery** *origin*, ascending pharyngeal; *branches*, none; *distribution*, bones and dura mater of posterior cranial fossa. <arteria meningea posterior> B

17 **arachnoid granulations** small elevations, visible to the naked eye, thought by some to be enlargements of arachnoid villi, which project into the superior sagittal sinus and associated venous lacunae and create slight depressions on the inner surface of the cranium; these granulations are the structures through which cerebrospinal fluid is reabsorbed into the blood in the venous system. <granulationes arachnoideae> C

18 **arachnoid mater** a delicate membrane interposed between the dura mater and the pia mater, separated from the pia mater by the subarachnoid space. <arachnoidea mater> C

19 **arachnoid trabeculae** delicate fibrous threads connecting the inner surface of the arachnoid to the pia mater. <trabeculae arachnoideae> C

20 **cranial arachnoid mater** the arachnoid covering the brain. <arachnoidea mater cranialis> C

21 **cranial dura mater** the dura mater covering the brain, composed of two mostly fused layers: an endosteal outer layer (endocranium) adherent to the inner aspect of the cranial bones, analogous to the periosteum of the bony skeleton, and an inner meningeal layer. Venous sinuses and the trigeminal ganglion are located between the layers. <dura mater cranialis> C

22 **cranial pia mater** the pia mater covering the brain, very thin over the cerebral cortex, and thicker over the brainstem; the blood vessels for the brain ramify within it and, as they enter the brain, are accompanied for a short distance by a pial sheath. <pia mater cranialis> C

A

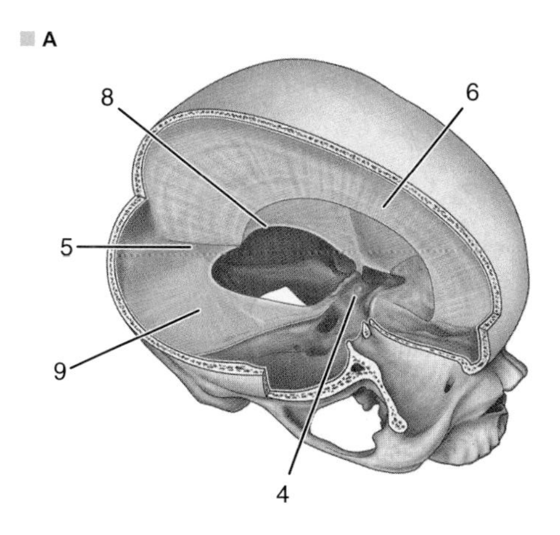

B

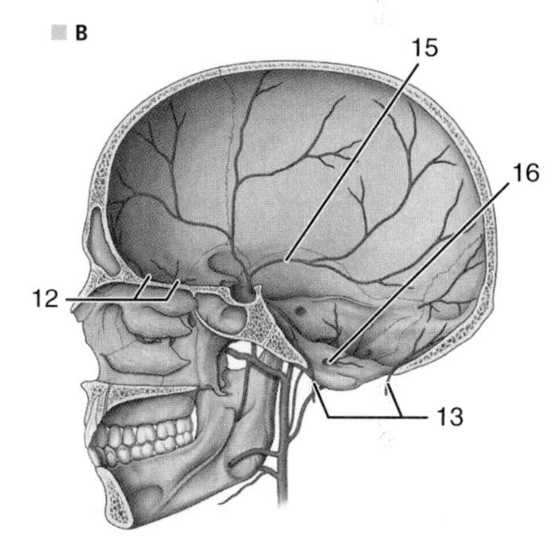

C

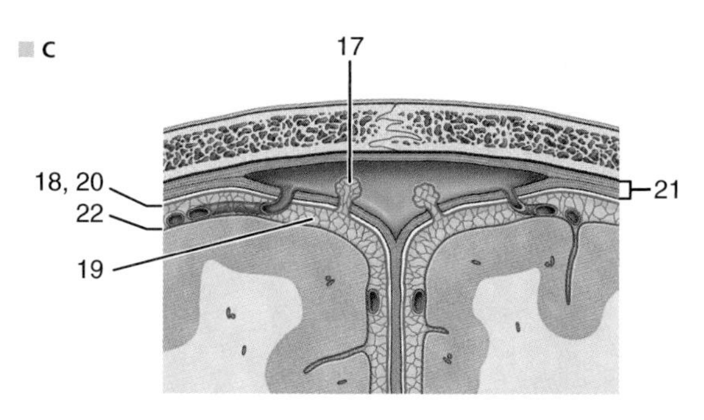

1 **dural venous sinuses** large venous channels forming an anastomosing system between the layers of the cranial dura mater. They are devoid of valves, do not collapse when drained, and in some parts contain numerous trabeculae. They drain the cerebral veins and some diploic and meningeal veins into the veins of the neck. Those at the base of the skull also drain most of the blood from the orbit. In some places they communicate with superficial veins by small emissary vessels. Called also *cranial sinuses.* <sinus durae matris> A

2 **extradural space** the space between the dura mater and the walls of the vertebral canal, containing venous plexuses and fibrous and alveolar tissue. In the head, the dura mater is normally attached to the inside of the skull, and the extradural space is only a potential space. Called also *epidural space.* <spatium epidurale> A

3 **lateral lacunae** venous meshworks within the dura mater on either side of the superior sagittal sinus; arachnoidal granulations project into them. <lacunae laterales> A

4 **pia mater** the innermost of the three membranes (meninges) covering the brain and spinal cord, investing them closely and extending into the depths of the fissures and sulci; it consists of reticular, elastic, and collagenous fibers. A

5 ***posterior cerebellomedullary cistern*** the enlarged subarachnoid space between the inferior surface of the cerebellum and the posterior surface of the medulla oblongata, and continuous below with the spinal subarachnoid space. <cisterna cerebellomedullaris posterior> A

6 ***quadrigeminal cistern*** the superior confluent of the subarachnoid space, lying in the angle between the splenium of the corpus callosum and the superior surfaces of the cerebellum and mesencephalon, and containing the great cerebral vein. <cisterna venae magnae cerebri> A

7 ***subarachnoid cisterns*** localized enlargements of the subarachnoid space, occurring in areas where the dura mater and arachnoid do not closely follow the contour of the brain with its covering pia mater, and serving as reservoirs of cerebrospinal fluid. <cisternae subarachnoideae> A

8 **subarachnoid space** the space between the arachnoid mater and the pia mater, containing cerebrospinal fluid and bridged by delicate trabeculae. <spatium subarachnoideum> A

9 **subdural space** a potential space between the dura mater and the arachnoid. <spatium subdurale> A

A

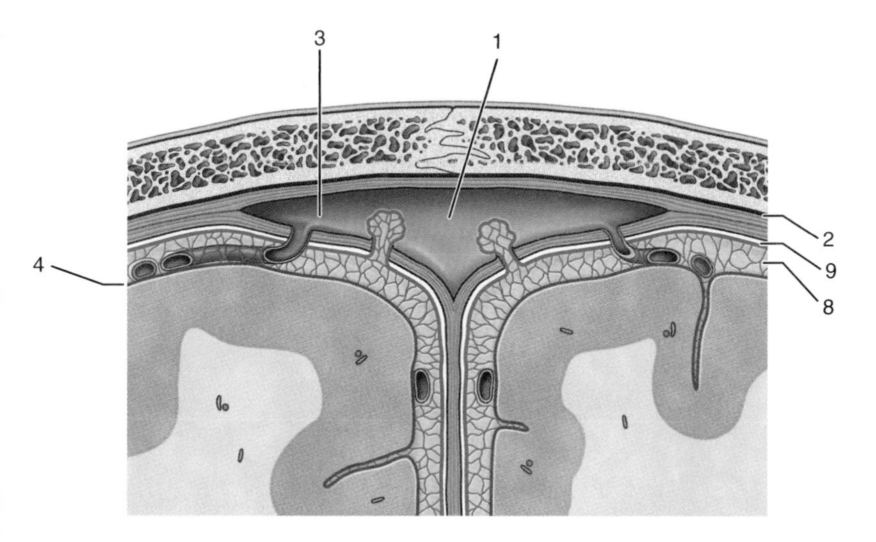

1 **anterior intercavernous sinus** the anterior of the two sinuses of the dura mater connecting the two cavernous sinuses, passing anterior to the infundibulum of the hypophysis. <sinus intercavernosus anterior> A

2 ***cavernous plexus*** a plexus of sympathetic nerve fibers related to the cavernous sinus of the dura mater. <plexus cavernosus> A

3 **cavernous sinus** either of two sinuses of the dura mater, irregularly shaped and located at either side of the body of the sphenoid bone, extending from the medial end of the superior orbital fissure in front to the apex of the petrous part of the temporal bone behind. Each sinus receives the corresponding superior ophthalmic vein, superficial middle cerebral vein, and the sphenoparietal sinus, and communicates with the opposite cavernous sinus and with the transverse sinuses and internal jugular vein by way of the petrosal sinuses. Each cavernous sinus commonly comprises one or more main venous channels and contains the internal carotid artery and abducent nerve. <sinus cavernosus> A

4 **confluence of sinuses** the dilated point of confluence of the superior sagittal, straight, occipital, and two transverse sinuses of the dura mater, lodged in a depression at one side of the internal occipital protuberance. <confluens sinuum> A

5 **great cerebral vein** a short median trunk formed by union of the two internal cerebral veins, which curves around the splenium of the corpus callosum and empties into, or is continued as, the straight sinus. <vena magna cerebri> A

6 **inferior petrosal sinus** one of the sinuses of the dura mater, arising from the cavernous sinus and running along the line of the petro-occipital synchondrosis to the superior bulb of the internal jugular vein. <sinus petrosus inferior> A

7 **inferior sagittal sinus** one of the sinuses of the dura mater; it is small and situated in the posterior half of the lower concave border of the falx cerebri, and it opens into the upper end of the straight sinus. Called also *inferior longitudinal sinus.* <sinus sagittalis inferior> A

8 ***marginal sinus*** any of the variably developed veins that connect the occipital sinus with the vertebral venous plexuses. <sinus marginalis> A

9 ***occipital sinus*** one of the sinuses of the dura mater; it begins in right and left branches, the marginal sinuses, and passes upward along the attached margin of the falx cerebelli to end in the confluence of the sinuses. <sinus occipitalis> A

10 ***petrosquamous sinus*** an inconstant sinus, one of the sinuses of the dura mater; it runs along the petrosquamous fissure, connecting posteriorly with the transverse sinus and anteriorly with the retromandibular vein. <sinus petrosquamosus> A

11 **posterior intercavernous sinus** the posterior of the two sinuses of the dura mater connecting the two cavernous sinuses, passing posterior to the infundibulum of the hypophysis. <sinus intercavernosus posterior> A

12 **sigmoid sinus** either of two sinuses of the dura mater, continuations of the transverse sinuses; each curves downward from the tentorium cerebelli to become continuous with the superior bulb of the internal jugular vein. <sinus sigmoideus> A

13 **sphenoparietal sinus** either of two sinuses of the dura mater, each beginning at a meningeal vein next to the apex of the lesser wing of the sphenoid bone and draining into the anterior part of the cavernous sinus. <sinus sphenoparietalis> A

14 **straight sinus** one of the sinuses of the dura mater, situated in the line of union of the falx cerebri and the tentorium cerebelli, formed by the junction of the great cerebral vein and the inferior sagittal sinus, and commonly ending in the opposite transverse sinus at the confluence of the sinuses. <sinus rectus> A

15 **superior petrosal sinus** one of the sinuses of the dura mater, arising at the cavernous sinus, passing along the attached margin of the tentorium cerebelli, and draining into the transverse sinus. <sinus petrosus superior> A

16 **superior sagittal sinus** one of the sinuses of the dura mater; it begins in front of the crista galli and extends backward in the convex border of the falx cerebri. Near the internal occipital protuberance it ends in a variable way in the confluence of the sinuses. It receives the superior cerebral veins, communicates with the lateral lacunae of adjacent dura mater, and is partially invaginated by arachnoid granulations. Called also *superior longitudinal sinus.* <sinus sagittalis superior> A

17 **transverse sinus** either of two large sinuses of the dura mater that begin in a variable fashion at the confluence of the sinuses near the internal occipital protuberance. Each follows the attached margin of the tentorium cerebelli to the petrous part of the temporal bone, where it becomes the sigmoid sinus. At their origin in the confluence, the right and left sinuses communicate with each other, and with the superior sagittal sinus and the straight sinus. <sinus transversus durae matris> A

18 ***venous plexus of foramen ovale*** a venous plexus that connects the cavernous sinus through the foramen ovale with the pterygoid plexus and the pharyngeal plexus; called also *rete of foramen ovale.* <plexus venosus foraminis ovalis> A

19 ***venous plexus of hypoglossal canal*** a venous plexus surrounding the hypoglossal nerve in its canal, and connecting the occipital sinus with the vertebral vein and with the longitudinal vertebral venous sinuses. Called also *rete of hypoglaossal canal.* <plexus venosus canalis hypoglossi> A

A

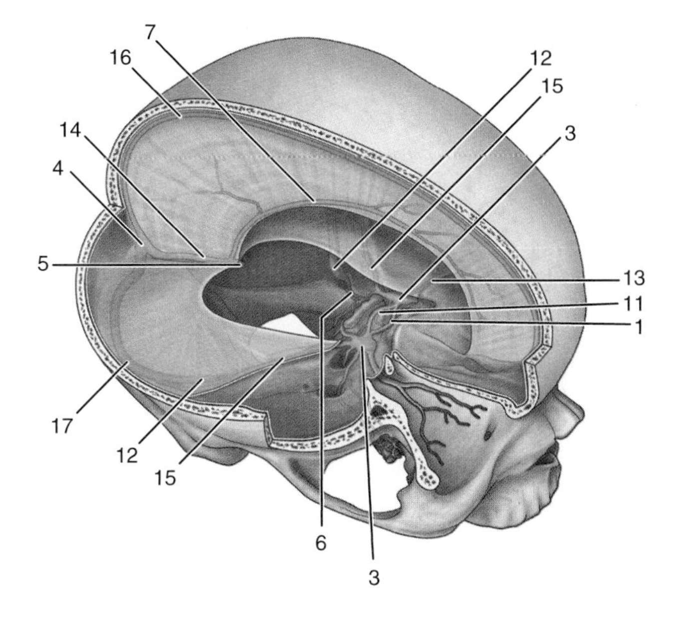

Brain and Blood Supply

1 **brain** that part of the central nervous system contained within the cranium, comprising the prosencephalon, mesencephalon, and rhombencephalon; it is derived (developed) from the anterior part of the embryonic neural tube. <encephalon> A

2 ***cerebellar cortex*** the superficial gray matter of the cerebellum; it consists of three layers, the molecular, granular, and Purkinje cell layers. <cortex cerebelli> A

3 **cerebellum** the part of the metencephalon that occupies the posterior cranial fossa behind the brainstem and is concerned in the coordination of movements. It is a fissured mass that is functionally subdivided into anterior, posterior, and flocculonodular lobes. A

4 ***cerebral cortex*** the thin (about 3 mm) layer or mantle of gray substance covering the surface of each cerebral hemisphere, folded into gyri that are separated by sulci. It is responsible for the higher mental functions, for general movement, for visceral functions, perception, and behavioral reactions, and for the association and integration of these functions. <cortex cerebri> A

5 **cerebral hemisphere** either of the pair of structures, formed by evagination of the embryonic telencephalon, lying on either side of the midline, partly separated by the longitudinal cerebral fissure, containing a central cavity, the lateral ventricle, and covered by a layer of gray substance, *(cerebral cortex)*; together they constitute the largest part of the brain in humans. <hemispherium cerebri> A

6 **cerebral sulci** the furrows on the surface of the brain between the cerebral gyri. <sulci cerebri> A

7 **cerebrum** the main portion of the brain, occupying the upper part of the cranial cavity, divided into two *cerebral hemispheres.* A

A

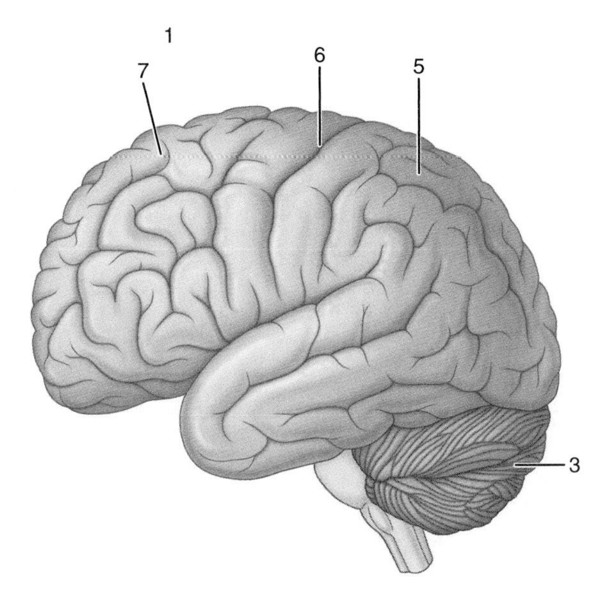

1 **medulla oblongata** the truncated cone of nerve tissue continuous above with the pons and below with the spinal cord. It lies anterior to the cerebellum, and the upper part of its posterior surface forms the floor of the lower part of the fourth ventricle; it contains ascending and descending tracts, and important collections of nerve cells that deal with vital functions, such as respiration, circulation, and special senses. Called also *bulb* and *myelencephalon.* A

2 ***neural lobe of pituitary gland*** the major portion of the neurohypophysis; called also *neural lobe of neurohypophysis.* <lobus nervosus neurohypophyseos> A

3 ***pharyngeal pituitary gland*** a small median residual collection of adenohypophysial glandular tissue in the mucoperiosteum of the roof of the nasopharynx; it develops from Rathke's pouch in the embryo. Called also *pharyngeal hypophysis.* <hypophysis pharyngealis> A

4 **pineal gland** a small flattened cone-shaped body in the epithalamus, lying above the superior colliculi and below the splenium of the corpus callosum. Its hormonal function in human physiology is not firmly established; in response to norepinephrine it synthesizes and releases melatonin, whose rate of release declines when light activates retinal photoreceptors. Called also *pineal body.* <glandula pinealis> A

5 **pituitary gland** an epithelial body located at the base of the brain in the sella turcica, attached by a stalk to the hypothalamus, from which it receives an important neural and vascular outflow. It consists of two lobes of differing embryonic origin, the *anterior lobe (adenohypophysis)*, which secretes most of the hormones, and the *posterior lobe (neurohypophysis)*, which stores and releases neurohormones that it receives from the hypothalamus. Called also *pituitary body.* <hypophysis> A

6 **anterior cerebral artery** divided into two parts: the first or *precommunical part* and the second or *postcommunical part*; *origin,* internal carotid artery; *branches,* (precommunical part) anteromedial central arteries; (postcommunical part) distal medial striate, medial frontobasal, polar frontal, callosomarginal (and its branches), and pericallosal (and it branches) arteries; *distribution,* orbital, frontal, and parietal cortex, corpus callosum, diencephalon, corpus striatum, internal capsule, and choroid plexus of lateral ventricle. <arteria cerebri anterior> B

7 **anterior communicating artery** *origin,* precommunical part of anterior cerebral artery; *branches,* anteromedial central arteries; *distribution,* establishes connection between the right and left anterior cerebral arteries. <arteria communicans anterior> B

8 **anterior inferior cerebellar artery** *origin,* basilar artery; *branches,* posterior, spinal (usually) and labyrinthine (usually) arteries; *distribution,* antero-inferior part of cerebellum, lower and lateral parts of pons and sometimes upper part of medulla oblongata. <arteria inferior anterior cerebelli> B

9 **basilar artery** *origin,* from junction of right and left vertebral arteries; *branches,* pontine, anterior inferior cerebellar, mesencephalic, superior cerebellar, and posterior cerebral arteries; *distribution,* brainstem, internal ear, cerebellum, posterior cerebrum. <arteria basilaris> B

10 ***branches of internal carotid artery to nerves*** twigs from the cavernous part of the internal carotid artery that supply the nerves of the walls of the cavernous and inferior petrosal sinuses. <rami nervorum arteriae carotidis internae> B

11 ***branches to trigeminal ganglion*** short branches from the cavernous part of the internal carotid artery that supply the trigeminal ganglion. <rami ganglionares trigeminales> B

12 ***cavernous part of internal carotid artery*** the part located in the cavernous sinus; it has numerous branches, including tentorial basal, tentorial marginal, meningeal, and cavernous branches, branches to trigeminal ganglion, and the inferior hypophysial artery. <pars cavernosa arteriae carotidis internae> B

13 **cerebral arterial circle** the important polygonal anastomosis formed by the internal carotid, the anterior, middle, and posterior cerebral arteries, the anterior communicating artery, and the posterior communicating arteries; called also *circle of Willis.* <circulus arteriosus cerebri> B

14 ***cerebral part of internal carotid artery*** the terminal part of the artery, which gives off branches comprising the ophthalmic, superior hypophysial, posterior communicating, anterior choroidal, and uncal arteries, and clival and meningeal branches, and then divides into the anterior and middle cerebral arteries in the middle cranial fossa. <pars cerebralis arteriae carotidis internae> B

15 ***inferior hypophysial artery*** a small branch from the cavernous part of the internal carotid artery that supplies the pituitary gland. <arteria hypophysialis inferior> B

16 **intracranial part of vertebral artery** the fourth part of the artery; it pierces the dura and arachnoid mater and ascends anterior to the hypoglossal roots and then medially in front of the medulla oblongata where, at about the lower border of the pons, it joins the opposite artery to form the basilar artery. <pars intracranialis arteriae vertebralis> B

A

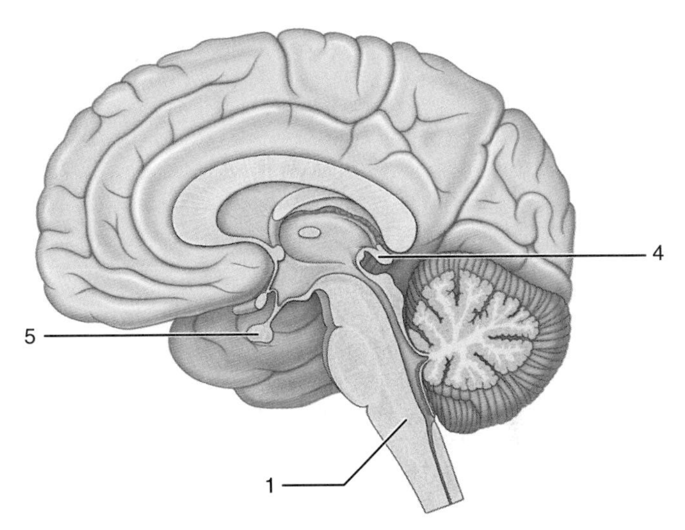

B

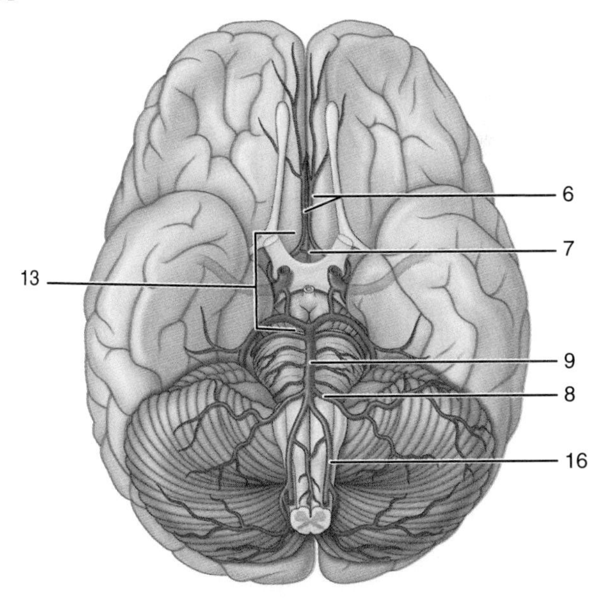

1 **middle cerebral artery** divided into four parts: *first or sphenoidal part, second* or *insular part, third part* or *inferior terminal branches,* and *fourth part* or *superior terminal branches; origin,* internal carotid; *branches,* (sphenoidal part) anterolateral central arteries; (insular part) insular arteries; (inferior terminal branches) anterior temporal, middle temporal, posterior temporal, and temporooccipital branches, and branch to angular gyrus; (superior terminal branches) lateral frontobasal, prefrontal, anterior parietal, and posterior parietal arteries, and arteries of precentral, central, and postcentral sulci; *distribution*, orbital, frontal, parietal, and temporal cortex, corpus striatum, internal capsule. Called also sylvian artery. <arteria cerebri media> A

2 **olfactory tract** a narrow triangular band in the olfactory sulcus of the frontal lobe, which arises from the olfactory bulb and extends posteriorly, to end by dividing into medial and lateral olfactory striae, the latter ending in the primary olfactory cortex. <tractus olfactorius> A

3 **optic chiasm** the part of the hypothalamus formed by the decussation, or crossing, of the fibers of the optic nerve from the medial half of each retina. <chiasma opticum> A

4 ***petrous part of internal carotid artery*** the portion of the artery located in the carotid canal; its branches include the caroticotympanic arteries. <pars petrosa arteriae carotidis internae> A

5 **pontine arteries** *origin,* basilar artery; *branches,* none; *distribution,* pons and adjacent areas of brain. There are shorter (medial branches) and longer (lateral branches) pontine arteries. <arteriae pontis> A

6 **posterior cerebral artery** divided into four segments: *first* or *precommunical part, second* or *postcommunical part, third segment* or *lateral occipital artery,* and *fourth segment* or *medial occipital artery; origin,* terminal bifurcation of basilar artery; *branches,* (precommunical part) posteromedial central, short circumferential, thalamoperforating, and collicular arteries; (postcommunical part) posterolateral central and thalamogeniculate arteries, and medial and lateral posterior choroidal, and peduncular branches; (lateral occipital artery) anterior, intermediate, and posterior temporal branches; (medial occipital artery) parietal, parieto-occipital, calcarine, and occipitotemporal branches, and dorsal branch to corpus callosum; *distribution,* occipital and temporal cortex, diencephalon, midbrain, choroid plexus of lateral and third ventricles, and visual area of cerebral cortex and other structures associated with the visual pathway. <arteria cerebri posterior> A

7 **posterior communicating artery** establishes connection between internal carotid and posterior cerebral arteries; *branches*, posteromedial central, thalamotuberal, and mammillary arteries, artery of tuber cinereum, and branches to the optic chiasm, oculomotor nerve, and hypothalamus. <arteria communicans posterior> A

8 **posterior inferior cerebellar artery** *origin,* vertebral artery; *branches,* medial, lateral and cerebellar tonsillar branches, choroidal branch to fourth ventricle, posterior spinal artery; *distribution,* lower cerebellum, medulla. <arteria inferior posterior cerebelli> A

9 **superior cerebellar artery** *origin,* basilar artery; *branches,* lateral and medial branches; *distribution,* upper cerebellum, midbrain, pineal gland, choroid plexus of third ventricle. <arteria superior cerebelli> A

10 ***superior hypophysial artery*** a small branch from the cerebral part of the internal carotid artery that supplies the pituitary gland. <arteria hypophysialis superior> A

A

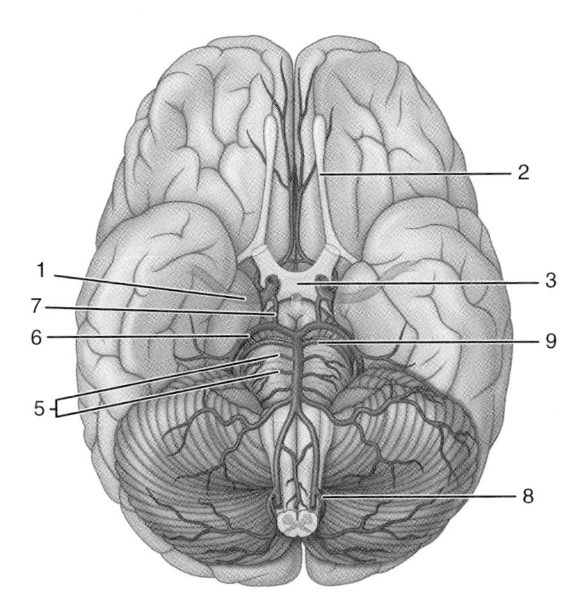

1 **labyrinthine artery** *origin*, anterior inferior cerebellar or basilar artery; *branches*, anterior vestibular and common cochlear arteries; *distribution*, through the internal acoustic meatus to the internal ear. Called also *internal auditory artery*. <arteria labyrinthi> A

Cranial Nerves

2 ***cranial nerves*** the twelve pairs of nerves that are connected with the brain: the olfactory (I), optic (II), oculomotor (III), trochlear (IV), trigeminal (V), abducent (VI), facial (VII), vestibulocochlear (VIII), glossopharyngeal (IX), vagus (X), accessory (XI), and hypoglossal (XII) nerves. <nervi craniales> B, D

3 ***inner sheath of optic nerve*** the inner sheath of the optic nerve, continuous with the pia mater and arachnoid mater. <vagina interna nervi optici> B, D

4 **abducent nerve** sixth cranial nerve; origin, a nucleus in the pons, beneath the floor of the fourth ventricle, emerging from the brain stem anteriorly between the pons and medulla oblongata; distribution, lateral rectus muscle of eye; modality, motor. <nervus abducens> B, D

5 **accessory nerve** eleventh cranial nerve; *origin*, by cranial roots from the side of the medulla oblongata, and by spinal roots from the side of the spinal cord (from the upper three or more cervical segments); the roots unite to form the trunk of the accessory nerve, which divides into an internal branch (cranial portion) and an external branch (spinal portion); *distribution*, the internal branch to the vagus and thereby to the palate, pharynx, larynx, and thoracic viscera; the external branch branches to the sternocleidomastoid and trapezius muscles; *modality*, parasympathetic and motor. <nervus accessorius> B, C, D

6 **cranial root of accessory nerves** any of the nerve roots originating from the nucleus ambiguus and emerging from the side of the medulla oblongata below the roots of the vagus nerve; considered to part of the vagus nerve. The roots then unite with the spinal portion in the jugular foramen. Their constituent fibers then form the internal branch, which joins the vagus nerve and is distributed to the soft palate, constrictors of the pharynx, and the larynx. <radix cranialis nervi accessorii> C, D

7 ***internal branch of accessory nerve*** the branch that continues from the cranial roots of the nerve, carrying motor fibers that are distributed by branches of the vagus to the soft palate, pharyngeal constrictors, and larynx. <ramus internus nervi accessorii> B, D

8 **inferior ganglion of glossopharyngeal nerve** the lower of two ganglia on the glossopharyngeal nerve as it passes through the jugular foramen; it contains cell bodies for some of the afferent fibers of the nerve. Called also *inferior petrosal ganglion* and *petrosal ganglion*. <ganglion inferius nervi glossopharyngei> C

9 **inferior ganglion of vagus nerve** a ganglion of the vagus nerve just below the jugular foramen, in front of the transverse processes of the first and second cervical vertebrae; it contains cell bodies for some of the afferent fibers of the nerve. <ganglion inferius nervi vagi> C

10 **superior ganglion of glossopharyngeal nerve** the upper of two ganglia on the glossopharyngeal nerve as it passes through the jugular foramen; it contains cell bodies for some of the afferent fibers of the nerve. <ganglion superius nervi glossopharyngei> C

11 **superior ganglion of vagus nerve** a small ganglion on the vagus nerve in the jugular foramen, giving off a meningeal and an auricular branch and containing cell bodies for some of the afferent fibers of the nerve. <ganglion superius nervi vagi> C

12 **oculomotor nerve** third cranial nerve; *origin*, brainstem, emerging medial to cerebral peduncles and running forward in the cavernous sinus; branches, superior and inferior branches; *distribution*, entering the orbit through the superior orbital fissure, the branches supply the levator palpebrae superioris, all extrinsic eye muscles except the lateral rectus and superior oblique, and carry parasympathetic fibers for the ciliary and sphincter pupillae muscles; *modality*, motor and parasympathetic. <nervus oculomotorius> B, D

13 **optic nerve** second cranial nerve, the so-called nerve of sight, actually part of the central nervous system throughout its course, misnamed as a nerve because of its cordlike appearance; it consists chiefly of axons and central processes of cells of the ganglionic layer of the retina, which leave each orbit through the optic canal, joining with those of the opposite side to form the optic chiasm (the medial fibers of each nerve crossing over to the opposite side), then continuing on each side as the optic tract. <nervus opticus> B, D

A

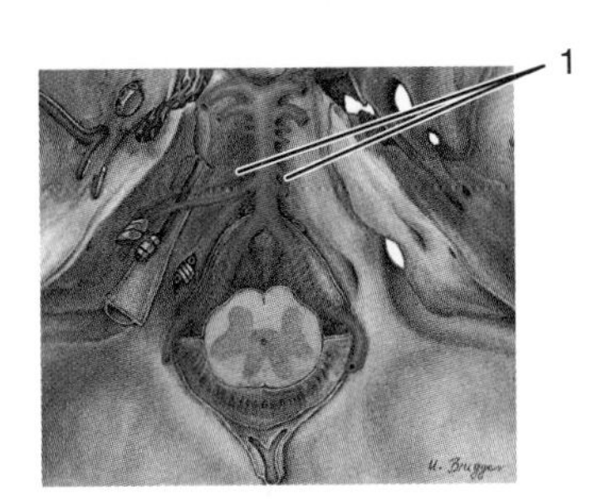

B

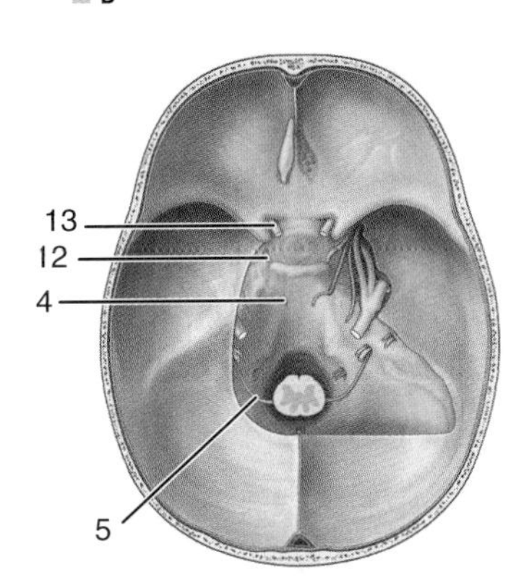

C

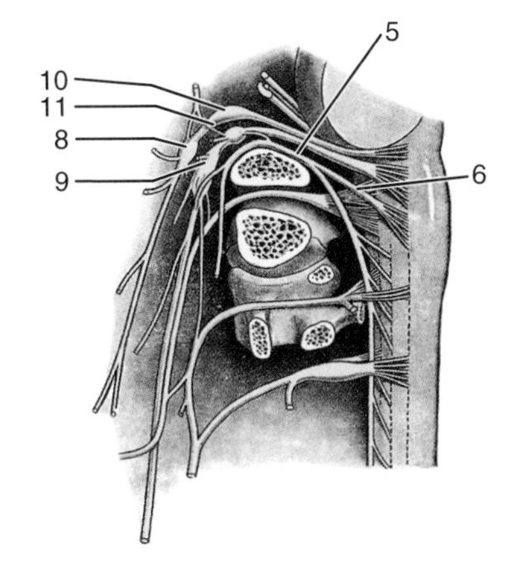

D

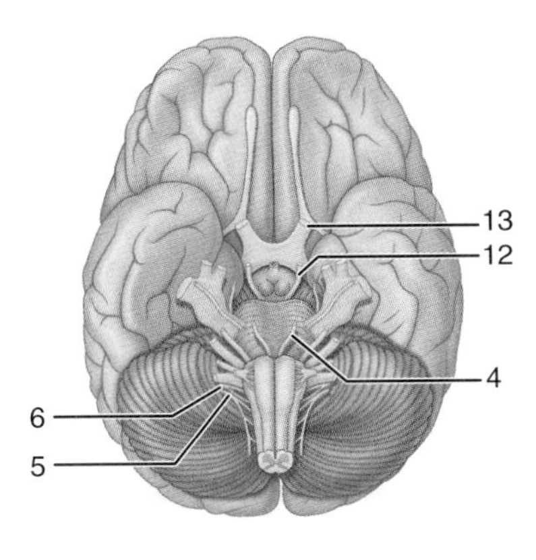

1 ***outer sheath of optic nerve*** the thick outer sheath of the optic nerve, continuous with the dura mater and connecting it with the sclera; called also *fibrous sheath of optic nerve* and *dural sheath.* <vagina externa nervi optici> A, B

2 **trochlear nerve** fourth cranial nerve; *origin,* the fibers of each trochlear nerve (one on either side) decussate across the median plane and emerge from the back of the brainstem below the corresponding inferior colliculus; *distribution,* runs forward in lateral wall of cavernous sinus, traverses the superior orbital fissure, and supplies superior oblique muscle of eyeball; *modality,* motor. <nervus trochlearis> A, B

3 **vagus nerve** tenth cranial nerve; *origin,* by numerous rootlets from lateral side of medulla oblongata in the groove between the olive and the inferior cerebellar peduncle; *branches,* superior and recurrent laryngeal nerves, meningeal, auricular, pharyngeal, cardiac, bronchial, gastric, hepatic, celiac, and renal branches, pharyngeal, pulmonary, and esophageal plexuses, and anterior and posterior trunks; *distribution,* descending through the jugular foramen, it presents as a superior and an inferior ganglion, and continues through the neck and thorax into the abdomen. It supplies sensory fibers to the ear, tongue, pharynx, and larynx, motor fibers to the pharynx, larynx, and esophagus, and parasympathetic and visceral afferent fibers to thoracic and abdominal viscera; *modality,* parasympathetic, visceral afferent, motor, general sensory. <nervus vagus> A, B

4 ***communicating branch of facial nerve with glossopharyngeal nerve*** a branch that interconnects the glossopharyngeal nerve and the facial nerve after emergence of the latter from the stylomastoid foramen. <ramus communicans nervi facialis cum nervo glossopharyngeo>

5 ***communicating branch of glossopharyngeal nerve with auricular branch of vagus nerve*** a small branch connecting the glossopharyngeal nerve with the auricular branch of the vagus nerve. <ramus communicans nervi glossopharyngei cum ramo auriculari nervi vagi>

6 ***communicating branch of glossopharyngeal nerve with auriculotemporal nerve*** a branch carrying postganglionic parasympathetic fibers from the glossopharyngeal nerve to the auriculotemporal nerve for distribution to the parotid gland. <ramus communicans nervi glossopharyngei cum nervo auriculotemporali>

7 ***communicating branch of glossopharyngeal nerve with chorda tympani*** a small branch that interconnects the glossopharyngeal nerve and the chorda tympani. <ramus communicans nervi glossopharyngei cum chorda tympani>

8 ***communicating branch of glossopharyngeal nerve with meningeal branch of vagal nerve*** a branch that carries autonomic fibers destined for the meninges from the glossopharyngeal nerve to the meningeal branch of the vagal nerve. <ramus communicans nervi glossopharyngei ramo meningeo nervi vagi>

9 ***communicating branch of intermediate nerve with vagus nerve*** a branch of the intermediate nerve that communicates with the vagus nerve. <ramus communicans nervi intermedii cum nervo vago>

10 ***communicating branch of superior laryngeal nerve with recurrent laryngeal nerve*** a small branch interconnecting the internal branch of the superior laryngeal nerve with the recurrent laryngeal nerve, behind or in the posterior cricoarytenoid muscle. <ramus communicans nervi laryngei superioris cum nervo laryngeo recurrente>

11 ***communicating branches of auriculotemporal nerve with facial nerve*** branches containing sensory fibers from the auriculotemporal nerve that join the facial nerve within the parotid gland, to be distributed with branches of the latter. <rami communicantes nervi auriculotemporalis cum nervo faciali>

12 ***communicating branches of lingual nerve with hypoglossal nerve*** plexiform terminal branches interconnecting the lingual and hypoglossal nerves just in front of the hyoglossus muscle. <rami communicantes nervi lingualis cum nervo hypoglosso>

13 **medial pterygoid nerve** *origin,* mandibular nerve; *branches,* nerve of tensor tympani, nerve of tensor veli palatini; *distribution,* medial pterygoid, tensor tympani, and tensor veli palatini muscles; *modality,* motor. <nervus pterygoideus medialis> C

14 **mandibular nerve** one of three terminal divisions of the trigeminal nerve, passing through the foramen ovale to the infratemporal fossa. *Origin,* trigeminal ganglion; branches, meningeal branch, masseteric, deep temporal, lateral and medial pterygoid, buccal, auriculotemporal, lingual, and inferior alveolar nerves; *distribution,* extensive distribution to muscles of mastication, skin of face, mucous membrane of mouth, and teeth; *modality,* general sensory and motor. <nervus mandibularis> A, B

15 **maxillary nerve** one of the three terminal divisions of the trigeminal nerve, passing through the foramen rotundum, and entering the pterygopalatine fossa. *Origin,* trigeminal ganglion; *branches,* meningeal branch, zygomatic, superior alveolar, and infraorbital nerves, ganglionic branches to pterygopalatine ganglion, and, indirectly, the branches of the pterygopalatine ganglion; *distribution,* extensive distribution to skin of face and scalp, mucous membrane of maxillary sinus and nasal cavity, and teeth; *modality,* general sensory. <nervus maxillaris> A

A

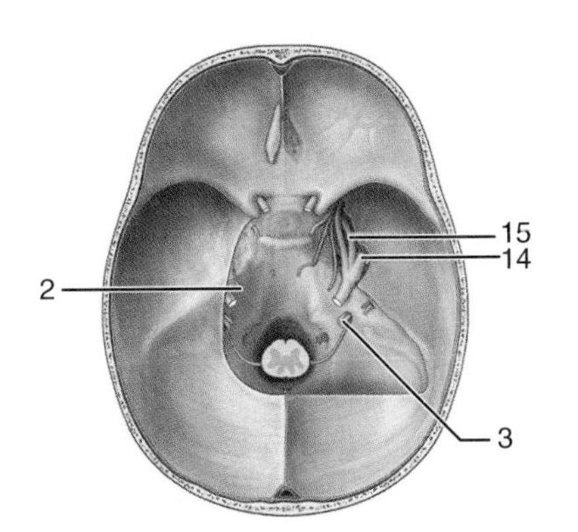

B

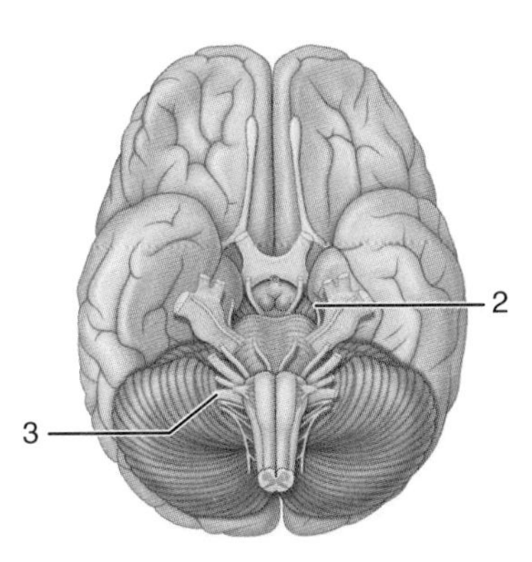

C

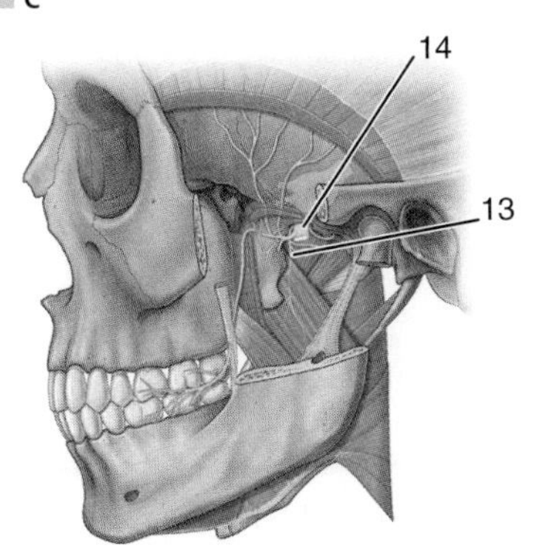

1 **ophthalmic nerve** one of the three terminal divisions of the trigeminal nerve. *Origin,* trigeminal ganglion; *branches,* recurrent meningeal (tentorial) branch, frontal, lacrimal, and nasociliary nerves; *distribution,* eyeball and conjunctiva, lacrimal gland and sac, nasal mucosa and frontal sinus, external nose, upper eyelid, forehead, and scalp; *modality,* general sensory. <nervus ophthalmicus> A

2 **trigeminal ganglion** a ganglion on the sensory root of the fifth cranial nerve, situated in a cleft within the dura mater (trigeminal cave) on the anterior surface of the petrous portion of the temporal bone, and giving off the ophthalmic and maxillary and part of the mandibular nerve; it contains the cells of origin of most of the sensory fibers of the trigeminal nerve. Called also *gasserian ganglion.* <ganglion trigeminale> A

3 **digastric branch of facial nerve** a branch that innervates the posterior belly of the digastric muscle; *modality,* motor; called also *digastric nerve.* <ramus digastricus nervi facialis> B

4 **internal branch of superior laryngeal nerve** the larger of the two branches of the superior laryngeal nerve, which innervates the mucosa of the epiglottis, base of the tongue, and larynx; *modality,* general sensory. <ramus internus nervi laryngei superioris> C

5 **hypoglossal nerve** twelfth cranial nerve; *origin,* several rootlets in the anterolateral sulcus between the olive and the pyramid of the medulla oblongata; it passes through the hypoglossal canal to the tongue; branches, lingual rami; *distribution,* styloglossus, hyoglossus, and genioglossus muscles and intrinsic muscles of the tongue; *modality,* motor. <nervus hypoglossus> A, D

6 **trigeminal nerve** fifth cranial nerve, which emerges from the lateral surface of the pons as a motor and a sensory root, together with some intermediate fibers. The sensory root expands into the *trigeminal ganglion,* from which the three divisions of the nerve arise *(mandibular, maxillary, and ophthalmic nerves).* It is sensory in supplying the face, teeth, mouth, and nasal cavity, and motor in supplying the muscles of mastication. <nervus trigeminus> A, D

7 **motor root of trigeminal nerve** the smaller of the two roots by which the trigeminal nerve is attached to the side of the pons; it contains proprioceptive as well as motor fibers, and continues deep to the trigeminal ganglion to join the mandibular nerve. <radix motoria nervi trigemini> D

8 **optic tract** the tract arising from the optic chiasm, proceeding backward, around the cerebral peduncle, and dividing into a lateral and a medial root; the roots end in the cranial colliculus and lateral geniculate body, respectively. <tractus opticus> D

9 **sensory root of trigeminal nerve** the larger of the two roots by which the trigeminal nerve is attached to the side of the pons. It contains sensory fibers and expands into a large flat ganglion (the trigeminal ganglion) which gives rise to the ophthalmic, maxillary, and mandibular nerves. <radix sensoria nervi trigemini> D

10 **spinal root of accessory nerve** any of the nerve roots originating from the gray matter of the spinal cord and emerging from the side of the cord as far down as a level between the third and seventh cervical nerves; they then form a trunk that ascends in the vertebral canal, passes through the foramen magnum, and unites with the cranial portion in the jugular foramen. The constituent fibers then form the external branch, which supplies the sternocleidomastoid and trapezius muscles. <radix spinalis nervi accessorii> D

11 ***trunk of accessory nerve*** the nerve trunk formed by the cranial and spinal roots of the accessory nerve, which separates into an internal and external terminal branch. <truncus nervi accessorii> D

12 **olfactory nerve** first cranial nerve, the central processes of the olfactory receptor cells, or fila olfactoria, considered collectively. <nervus olfactorius> A

13 **olfactory nerves** first cranial nerve; the nerves of smell, consisting of about 20 bundles which arise in the olfactory epithelium and pass through the cribriform plate of the ethmoid bone to the olfactory bulb. <fila olfactoria> A

A

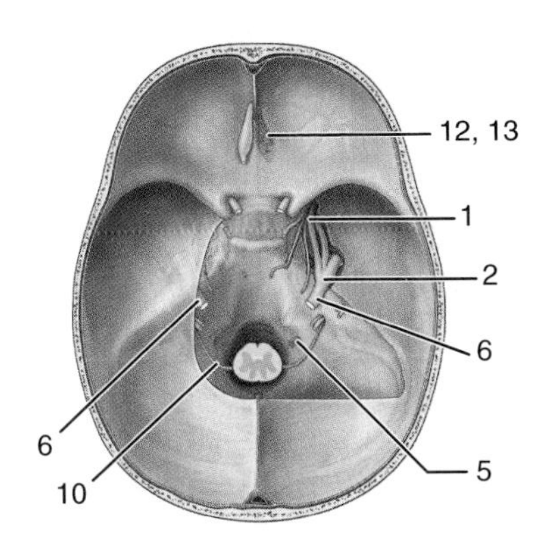

B

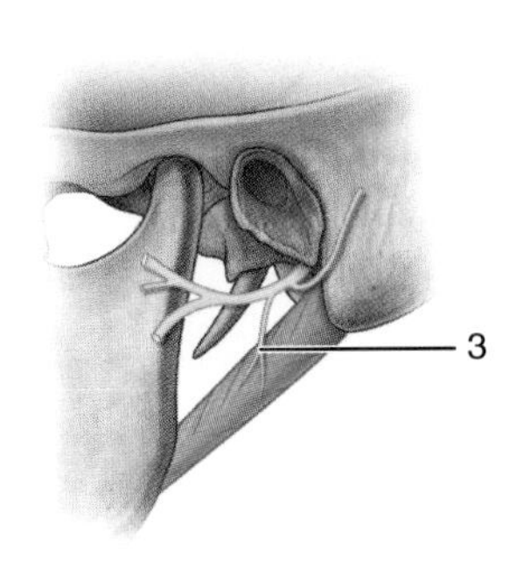

C

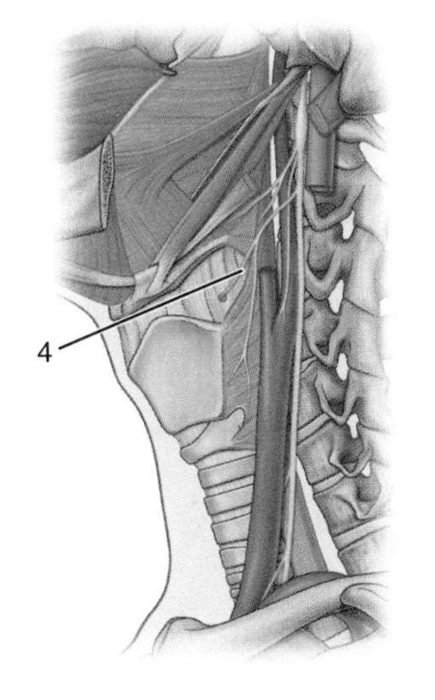

D

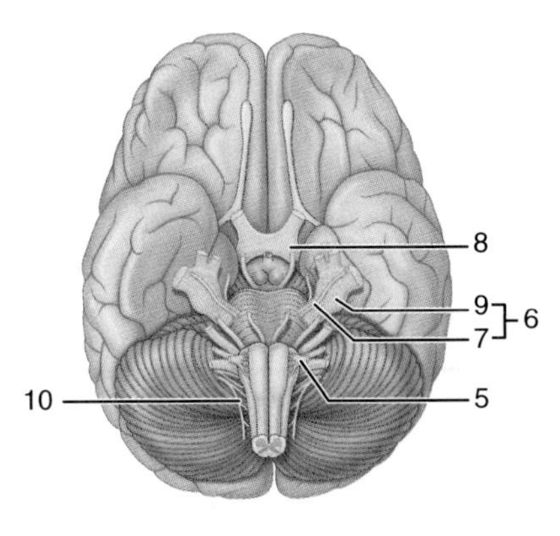

Scalp

1 **scalp** that part of the skin of the head, exclusive of the face and ears, which normally is covered with hair. A

2 **occipital vein** a vein in the scalp that follows the distribution of the occipital artery and opens under the trapezius muscle into the suboccipital venous plexus; it may continue with the occipital artery and end in the internal jugular vein. <vena occipitalis> C

3 **pericranium** the external periosteum of the skull. A

4 **frontal belly of occipitofrontal muscle** a part that originates from the epicranial aponeurosis and inserts into the skin of the eyebrows and the root of the nose. <venter frontalis musculi occipitofrontalis> B

5 **epicranial aponeurosis** the aponeurotic structure of the scalp, connecting the frontal and occipital bellies of the occipitofrontalis muscle. <galea aponeurotica> B

6 **epicranius muscle** a name given the muscular covering of the scalp, including the occipitofrontalis and temporoparietalis muscles, and the epicranial aponeurosis. <musculus epicranius> A

7 ***branch to tympanic membrane of auriculotemporal nerve*** a branch given to the tympanic membrane by the nerve of the external acoustic meatus, a branch of the auriculotemporal nerve; *modality,* general sensory. <ramus membranae tympani nervi auriculotemporalis> D

8 **greater occipital nerve** *origin,* medial branch of posterior ramus of C2; *distribution,* semispinalis capitis muscle and skin of scalp as far forward as the vertex; *modality,* general sensory and motor. <nervus occipitalis major> D

9 **lateral branch of supraorbital nerve** a branch of the supraorbital nerve that supplies the frontal sinus, upper eyelid, and skin and subcutaneous tissue of the forehead and scalp laterally as far as the temporal region; *modality,* general sensory. <ramus lateralis nervi supraorbitalis> D

10 ***occipital branch of posterior auricular nerve*** a branch supplying the occipital belly of the occipitofrontalis muscle; *modality,* motor. <ramus occipitalis nervi auricularis posterioris> D

11 **superficial temporal branches of auriculotemporal nerve** branches to the skin of the scalp in the temporal region; *modality,* general sensory. <rami temporales superficiales nervi auriculotemporalis> D

12 **frontal branch of superficial temporal artery** a tortuous terminal branch that supplies the forehead and frontal scalp. <ramus frontalis arteriae temporalis superficialis> C

13 ***mastoid branches of posterior auricular artery*** branches that supply the mastoid cells. <rami mastoidei arteriae auricularis posterioris> C

14 ***middle temporal vein*** the vein that arises in the substance of the temporalis muscle and passes down under the fascia to the zygoma, where it breaks through to join the superficial temporal vein. <vena temporalis media> C

15 **occipital branch of posterior auricular artery** a branch distributed to the epicranius muscle. <ramus occipitalis arteriae auricularis posterioris> C

16 **occipital branches of occipital artery** a medial and a lateral branch of the occipital artery, distributed to the scalp and, through the meningeal branch, to the dura mater. <rami occipitales arteriae occipitalis> C

17 **parietal branch of superficial temporal artery** the posterior terminal branch of the superficial temporal artery, supplying the scalp in the parietal region. <ramus parietalis arteriae temporalis superficialis> C

18 ***parietal emissary vein*** a small vein passing through the parietal foramen of the skull and connecting the superior sagittal sinus with the superficial temporal veins. <vena emissaria parietalis> C

19 **superficial temporal artery** *origin,* external carotid; *branches,* parotid, auricular, and occipital branches, transverse facial, zygomatico-orbital, and middle temporal arteries; *distribution,* parotid and temporal regions. <arteria temporalis superficialis> C

20 **superficial temporal veins** veins that drain the lateral part of the scalp in the frontal and parietal regions, the tributaries forming a single superficial temporal vein in front of the ear, just above the zygoma. This descending vein receives the middle temporal and transverse facial veins and, entering the parotid gland, unites with the maxillary vein deep to the neck of the mandible to form the retromandibular vein. <venae temporales superficiales> C

A

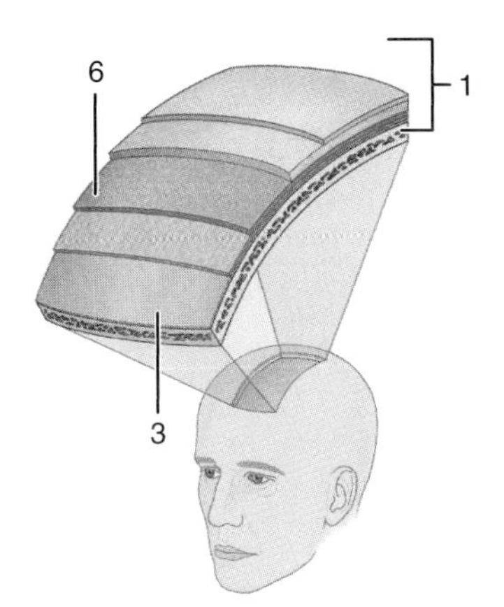

B

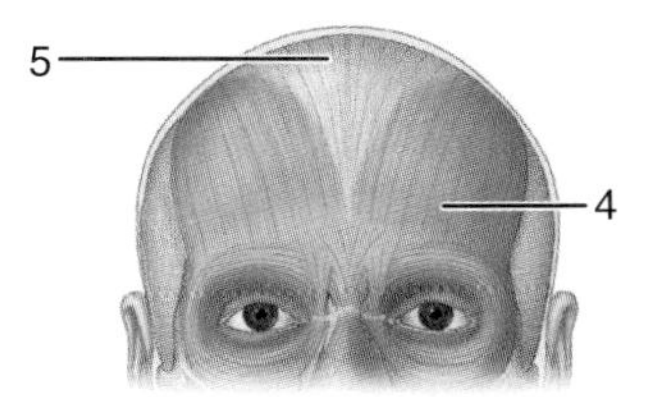

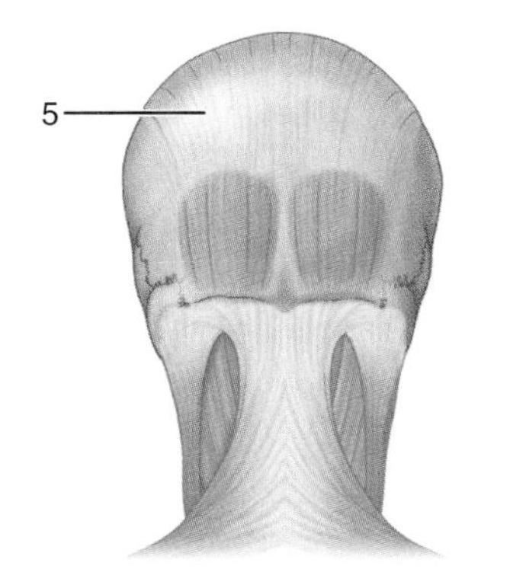

C

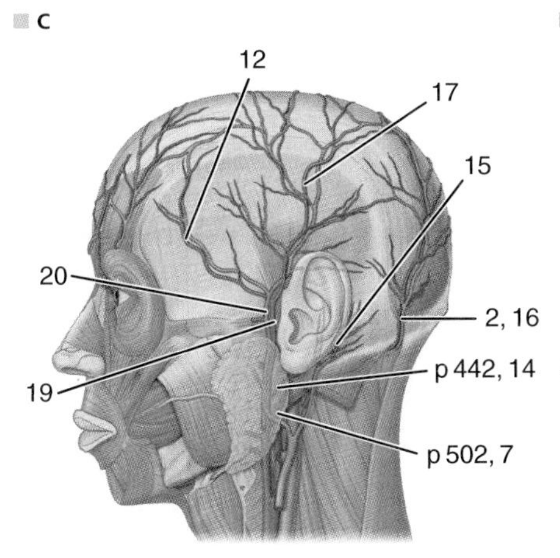

D

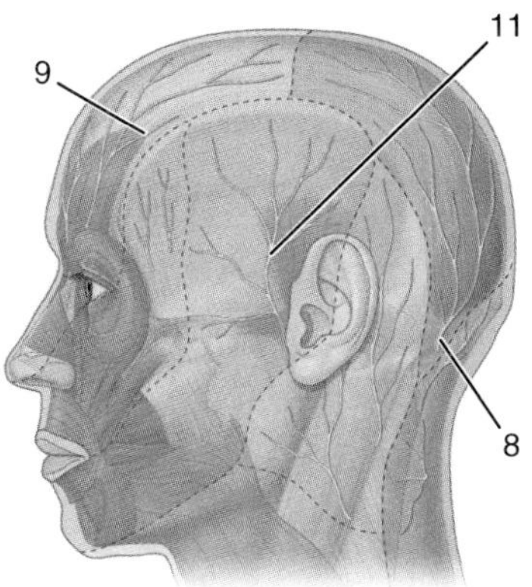

Face

1 **face** the anterior, or ventral, aspect of the head from the supercilliary arches to the chin, including the eyes, nose, mouth, cheeks, and chin but excluding the auricles. <facies> A

2 **angle of mouth** the angle formed at either side of the mouth by junction of the upper and the lower lip. <angulus oris> A

3 **apex of nose** the most distal portion of the nose; called also *tip of nose.* <apex nasi> A

4 ***buccal fat pad*** an encapsulated mass of fat in the cheek, separated from the subcutaneous fascia by a facial cleft, and situated between the masseter and the external surface of the buccinator muscles; especially well developed in infants and said to aid in sucking. Called also *sucking pad.* <corpus adiposum buccae> A

5 **buccal region** the region of the cheek. <regio buccalis> A

6 ***buccinator lymph node*** one of a variable number of facial lymph nodes lying on a line between the angle of the mandible and the mouth, receiving the afferent vessels draining the temporal and infratemporal fossae and nasopharynx; their efferent vessels drain into the superior deep cervical nodes. Called also *buccal lymph node.* <nodus lymphoideus buccinatorius> A

7 **cheek** the fleshy portion of the side of the face, constituting the lateral wall of the oral cavity. <bucca> A

8 **dorsum of nose** that part of the external surface of the nose formed by junction of the lateral surfaces. <dorsum nasi> A

9 **eyebrow** 1. the transverse elevation at the junction of the forehead and the upper eyelid, consisting of five layers: skin, subcutaneous tissue, a layer of interwoven fibers of the orbicularis oculi and occipitofrontalis muscles, a submuscular areolar layer, and pericranium. 2. the hairs growing on this elevation. <supercilium> A

10 ***frenulum of lower lip*** the fold of mucous membrane on the inside of the middle of the lower lip, connecting the lip with the gums. <frenulum labii inferioris> A

11 ***frenulum of upper lip*** the fold of mucous membrane on the inside of the middle of the upper lip, connecting the lip with the gums. <frenulum labii superioris> A

12 **labial commissure** the junction of the upper and lower lips at either side of the mouth. <commissura labiorum oris> A

13 **lateral angle of eye** the angle formed by the lateral junction of the superior and inferior eyelids. <angulus oculi lateralis> A

14 **lateral commissure of eyelids** the junction of the upper and lower eyelids on the lateral side. Called also *lateral palpebral commissure.* <commissura lateralis palpebrarum> A

15 **lips** the fleshy upper and lower margins of the mouth. <labia oris> A

16 **lower eyelid** the inferior of the paired movable folds that protect the surface of the eyeball. <palpebra inferior> A

17 **lower lip** the fleshy margin of the inferior border of the mouth. <labium inferius oris> A

18 **medial angle of eye** the angle formed by the medial junction of the superior and inferior eyelids. <angulus oculi medialis> A

19 ***mental region*** the region of the chin. <regio mentalis> A

20 **mentolabial sulcus** the depression between the lower lip and the chin. Called also *mentolabial furrow.* <sulcus mentolabialis> A

21 **mentum** the chin. A

22 ***nasal region*** the region of the face about the nose. <region nasalis> A

23 **nasolabial sulcus** the depression between the nose and the upper lip. <sulcus nasolabialis> A

24 **nose** the specialized structure of the face that serves as the organ of the sense of smell and as part of the respiratory system. <nasus> A

25 **philtrum** the vertical groove in the median portion of the upper lip. A

26 **root of nose** the upper portion of the nose, which is attached to the frontal bone. <radix nasi> A

27 **suprapalpebral sulcus** the furrow above the upper eyelid. <sulcus suprapalpebralis> A

28 **tubercle of upper lip** the central prominence of the superior border between the skin and the mucous membrane of the upper lip, marking the distal termination of the philtrum. Called also *procheilon* and *labial tubercle.* <tuberculum labii superioris> A

29 **upper lip** the fleshy margin of the superior border of the mouth. <labium superius oris> A

30 ***facial lymph nodes*** lymph nodes situated along the course of the facial artery and vein, which receive afferent vessels draining the eyelids, conjunctiva, nose, cheeks, lips, and gums, and send efferent vessels to the submandibular nodes. <nodi lymphoidei faciales>

31 ***facial muscles*** a group of cutaneous muscles of the facial structures, which includes the muscles of the scalp, ear, eyelids, nose, and mouth, and the platysma; called also *muscles of expression.* <musculi faciei>

32 ***infraorbital region*** the region inferior to the eye, adjacent to the nasal region. <regio infraorbitalis>

33 ***labial glands*** the serous and mucous glands on the inner part of the lips. <glandulae labiales>

A

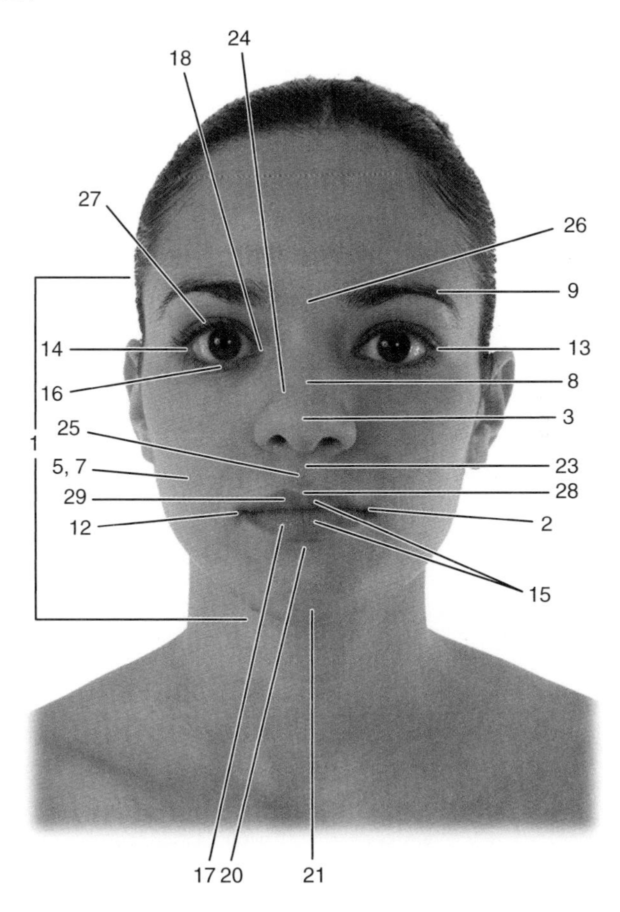

1 **zygomaticofacial foramen** the opening on the anterior surface of the zygomatic bone for the zygomaticofacial nerves and vessels. <foramen zygomaticofaciale> A

2 **platysma** a platelike muscle that originates from the fascia of the cervical region and inserts in the mandible and the skin around the mouth. It is innervated by the cervical branch of the facial nerve, and acts to wrinkle the skin of the neck and to depress the jaw. B

3 **corrugator supercilii muscle** *origin,* medial end of superciliary arch; *insertion,* skin of eyebrow; *innervation,* facial; *action,* draws eyebrow downward and medially. Called also *Koyter's muscle.* <musculus corrugator supercilii> C

4 ***depressor supercilii muscle*** a name given a few fibers of the orbital part of the orbicularis oculi muscle that are inserted in the eyebrow, which they depress. <musculus depressor supercilii> C

5 **orbicularis oculi muscle** orbicular muscle of eye: the oval sphincter muscle surrounding the eyelids, consisting of three parts: *origin,* ORBITAL PART—medial margin of orbit, including frontal process of maxilla, PALPEBRAL PART—medial palpebral ligament, LACRIMAL PART–posterior lacrimal crest; *insertion,* ORBITAL PART—near origin after encircling orbit, PALPEBRAL PART—fibers intertwine to form lateral palpebral raphe, LACRIMAL PART—lateral palpebral raphe, superior and inferior tarsi; *innervation,* facial; *action,* closes eyelids, wrinkles forehead, compresses lacrimal sac. <musculus orbicularis oculi> C

6 **orbital part of orbicularis oculi muscle** the part of the orbicularis oculi muscle that arises from the medial margin of the orbit and surrounds it and the palpebral part of the muscle, inserting near the site of origin. <pars orbitalis musculi orbicularis oculi> C

7 **palpebral part of orbicularis oculi muscle** the part of the orbicularis oculi muscle that is contained in the eyelids, originating from the medial palpebral ligament and inserting in the lateral canthus. <pars palpebralis musculi orbicularis oculi> C

8 **alar part of nasalis muscle** the part of the nasalis muscle arising from the maxilla on either side above the lateral incisor tooth and attaching to the alar cartilage of the nose; it assists in opening the nasal aperture. Called also *dilator muscle of naris* and *dilator naris.* <pars alaris musculi nasalis> D

9 **depressor septi nasi muscle** *origin,* incisor fossa of maxilla; *insertion,* ala and septum of nose; *innervation,* facial; *action,* contracts nostril and depresses ala. <musculus depressor septi nasi> D

10 **nasalis muscle** *origin,* maxilla; *insertion,* ALAR PART—ala nasi, TRANSVERSE PART—by aponeurotic expansion with fellow of opposite side; *innervation,* facial; *action,* ALAR PART—aids in widening nostril, TRANSVERSE PART—depresses cartilage of nose. <musculus nasalis> D

11 **procerus muscle** *origin,* fascia over nasal bone; *insertion,* skin of forehead; *innervation,* facial; *action,* draws medial angle of eyebrows down. <musculus procerus> D

12 **transverse part of nasalis muscle** the part of the nasalis muscle arising from the maxilla on either side just lateral to the nasal notch and inserting on the bridge of the nose; it compresses the nasal opening. Called also *compressor muscle of naris* and *compressor naris.* <pars transversa musculi nasalis> D

13 **depressor anguli oris muscle** *origin,* lower border of mandible; *insertion,* angle of mouth; *innervation,* facial; *action,* pulls down angle of mouth. <musculus depressor anguli oris> E

14 **depressor labii inferioris muscle** *origin,* anterior portion of lower border of mandible; *insertion,* orbicularis oris and skin of lower lip; *innervation,* facial; *action,* depresses lower lip. <musculus depressor labii inferioris> E

15 **labial part of orbicularis oris muscle** the part of the muscle whose fibers are restricted to the lips. <pars labialis musculi orbicularis oris> E

16 **levator anguli oris muscle** *origin,* canine fossa of maxilla; *insertion,* orbicularis oris and skin at angle of mouth; *innervation,* facial; *action,* raises angle of mouth. <musculus levator anguli oris> E

17 **levator labii superioris alaeque nasi muscle** *origin,* upper part of frontal process of maxilla; *insertion,* cartilage and skin of ala nasi, and upper lip; *innervation,* infraorbital branch of facial; *action,* raises upper lip and dilates nostril. <musculus levator labii superioris alaeque nasi> E

18 **levator labii superioris muscle** *origin,* lower orbital margin; *insertion,* muscle of upper lip; *innervation,* facial nerve; *action,* raises upper lip. <musculus levator labii superioris> E

19 **marginal part of orbicularis oris muscle** the part of the muscle whose fibers blend with those of adjacent muscles. <pars marginalis musculi orbicularis oris> E

20 **mentalis muscle** *origin,* incisive fossa of mandible; *insertion,* skin of chin; *innervation,* facial; *action,* wrinkles skin of chin. <musculus mentalis> E

21 **orbicularis oris muscle** orbicular muscle of mouth, comprising a *labial part,* fibers restricted to the lips, and a *marginal part,* fibers blending with those of adjacent muscles; *innervation,* facial; *action,* closes and protrudes lips. <musculus orbicularis oris> E

22 **risorius muscle** *origin,* fascia over masseter; *insertion,* skin at angle of mouth; *innervation,* buccal branch of facial; *action,* draws angle of mouth laterally. <musculus risorius> E

23 ***transversus menti muscle*** superficial fibers of the depressor anguli oris which turn back and cross to the opposite side. <musculus transversus menti> E

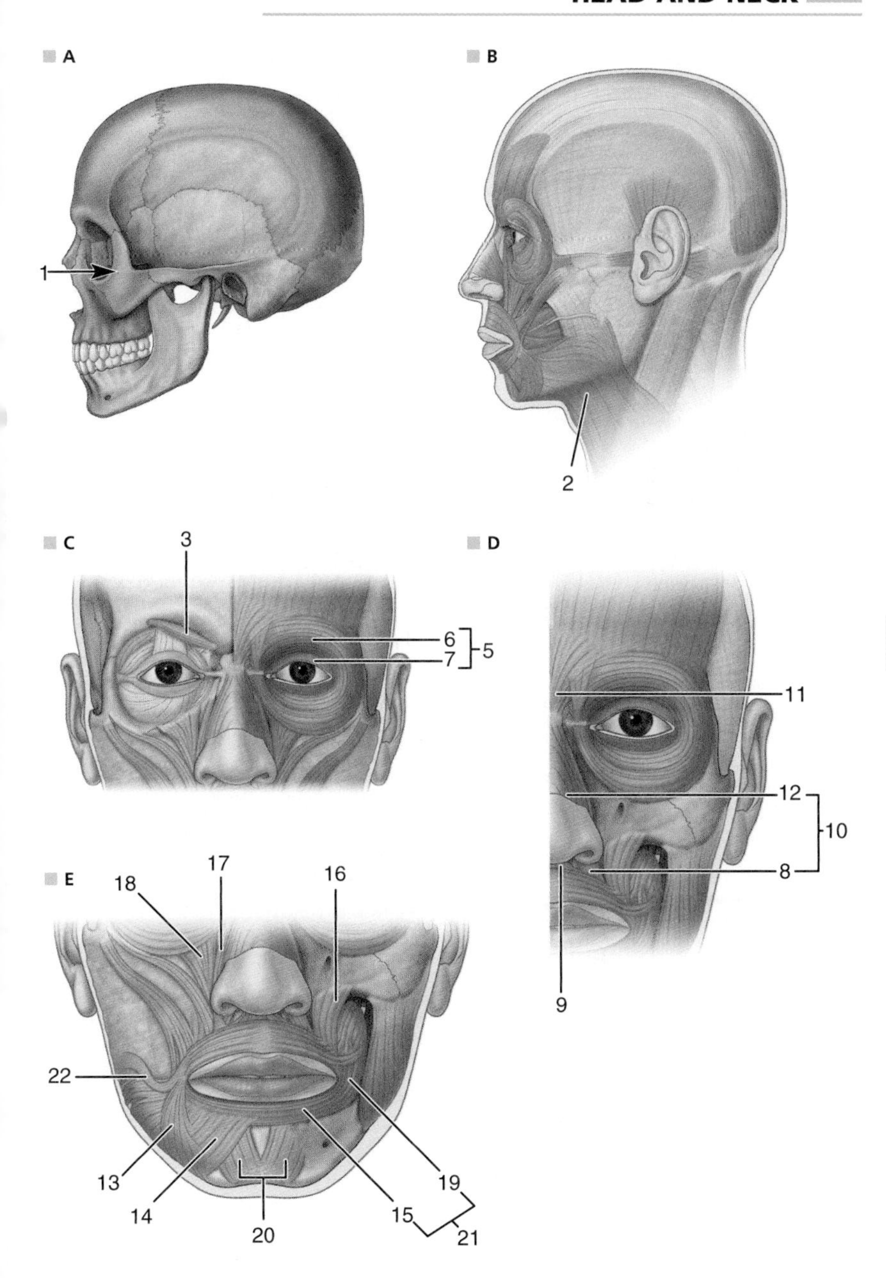
A
1
B
2
C
3
6
5
7
D
11
12
10
8
9
E
18
17
16
22
13
14
20
15
19
21

1 **zygomaticus major muscle** *origin,* zygomatic bone in front of temporal process; *insertion,* angle of mouth; *innervation,* facial; *action,* draws angle of mouth backward and upward. <musculus zygomaticus major> A

2 **zygomaticus minor muscle** *origin,* zygomatic bone near maxillary suture; *insertion,* orbicularis oris and levator labii superioris; *innervation,* facial; *action,* draws upper lip upward and laterally. <musculus zygomaticus minor> A

3 **buccinator muscle** *origin,* buccinator ridge of mandible, alveolar process of maxilla, pterygomandibular ligament; *insertion,* orbicularis oris at angle of mouth; *innervation,* buccal branch of facial; *action,* compresses cheek and retracts angle of the mouth. <musculus buccinator> A, B

4 ***anterior auricular ligament*** the auricular ligament that passes from the eminence of the concha to the mastoid part of the temporal bone. <ligamentum auriculare anterius> C

5 **anterior auricular muscle** *origin,* superficial temporal fascia; *insertion,* cartilage of ear; *innervation,* facial; *action,* draws the auricle forward. <musculus auricularis anterior> C

6 **posterior auricular muscle** *origin,* mastoid process; *insertion,* cartilage of ear; *innervation,* facial; *action,* draws auricle backward. <musculus auricularis posterior> C

7 **superior auricular muscle** *origin,* epicranial aponeurosis; *insertion,* cartilage of ear; *innervation,* facial; *action,* raises auricle. <musculus auricularis superior> C

8 ***temporoparietalis muscle*** *origin,* temporal fascia above ear; *insertion,* epicranial aponeurosis; *innervation,* temporal branches of facial; *action,* tightens scalp. <musculus temporoparietalis> C

9 ***accessory parotid gland*** a frequently present, more or less detached portion of the parotid gland that extends along the parotid duct. <glandula parotidea accessoria> D

10 ***articular veins*** small vessels that drain the plexus around the temporomandibular joint into the retromandibular vein; called also *temporomandibular articular veins.* <venae articulares> D

11 **buccal branches of facial nerve** branches that innervate the zygomatic, levator labii superioris, buccinator, and orbicularis oris muscles; *modality,* motor and general sensory. <rami buccales nervi facialis> D

12 **cervical branch of facial nerve** a branch lying deep to and innervating the platysma muscle; *modality,* motor. <ramus colli nervi facialis> D

13 **facial nerve** seventh cranial nerve, consisting of two roots: a large motor root, which supplies the muscles of facial expression, and a smaller root, the intermediate nerve. *Origin,* inferior border of pons, between olive and inferior cerebellar peduncle; *branches* (of motor root), stapedius and posterior auricular nerves, parotid plexus, digastric, stylohyoid, temporal, zygomatic, buccal, lingual, marginal mandibular, and cervical branches, and a communicating branch with the tympanic plexus; *modality,* motor, parasympathetic, general sensory, special sensory. <nervus facialis> D

14 ***lingual branch of facial nerve*** an inconstant branch of the facial nerve sometimes arising together with the stylohyoid branch, and helping to supply the styloglossus and glossopalatinus muscles; *modality,* motor. <ramus lingualis nervi facialis> D

15 **marginal mandibular branch of facial nerve** a branch of the facial nerve that runs forward from the front of the parotid gland along the border of the mandible, deep to the platysma and depressor anguli oris muscles, supplying the latter and the risorius, depressor labii inferioris, and mentalis muscles; *modality,* motor. <ramus marginalis mandibularis nervi facialis> D

16 **parotid duct** the duct that drains the parotid gland and empties into the oral cavity opposite the second superior molar; called also *Stensen's canal* or *duct.* <ductus parotideus> B, D

17 ***parotid fascia*** an extension of the deep cervical fascia that splits to enclose the parotid gland and sends extensions into the gland that become continuous with its stroma. <fascia parotidea> D

18 **parotid gland** the largest of the three glands occurring in pairs, which together with numerous small glands in the mouth constitute the salivary glands; it is located below the zygomatic arch, below and in front of the external acoustic meatus. <glandula parotidea> D

19 ***parotid plexus*** a plexus formed by anastomosis of the terminal branches of the temporal, zygomatic, buccal, marginal mandibular, and cervical branches of the facial nerve, arising in the parotid gland. <plexus intraparotideus> D

20 ***stylohyoid branch of facial nerve*** a branch that arises from the facial nerve just below the base of the skull to innervate the stylohyoid muscle; *modality,* motor. <ramus stylohyoideus nervi facialis> D

21 **temporal branches of facial nerve** terminal branches of the facial nerve that innervate the anterior and superior auricular muscles, the frontal belly of the occipitofrontalis muscle, and the orbicularis oculi and corrugator muscles; *modality,* motor. <rami temporales nervi facialis> D

22 **zygomatic branches of facial nerve** branches that cross the zygomatic bone and innervate the orbicularis oculi muscle; *modality,* motor. <rami zygomatici nervi facialis> D

A

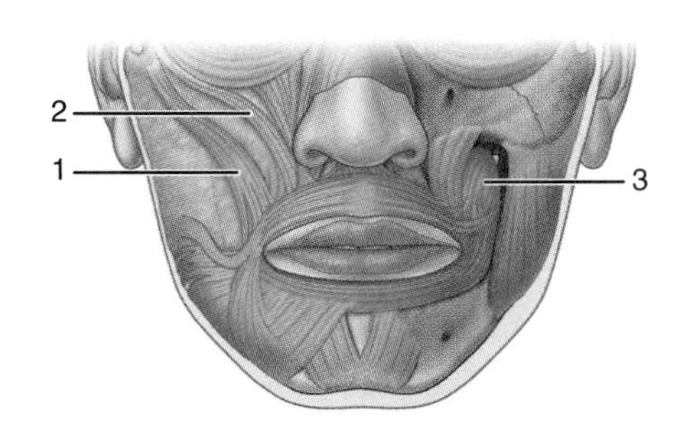

B

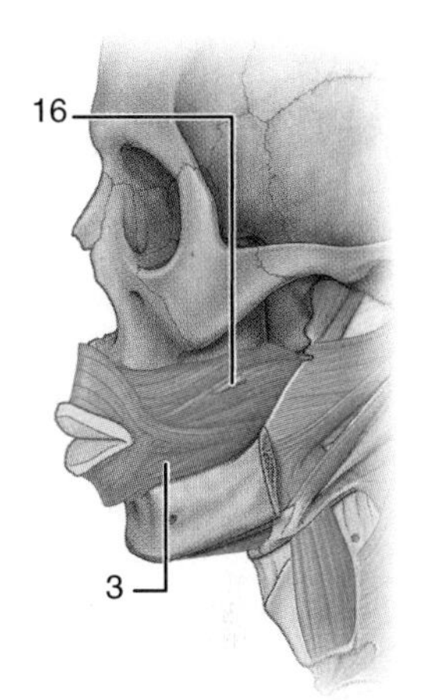

C

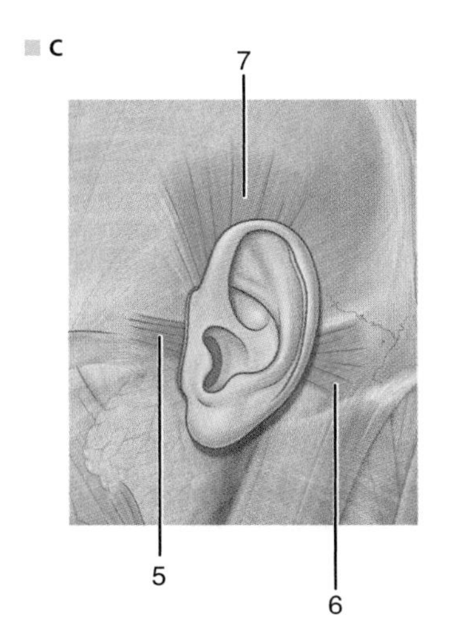

D

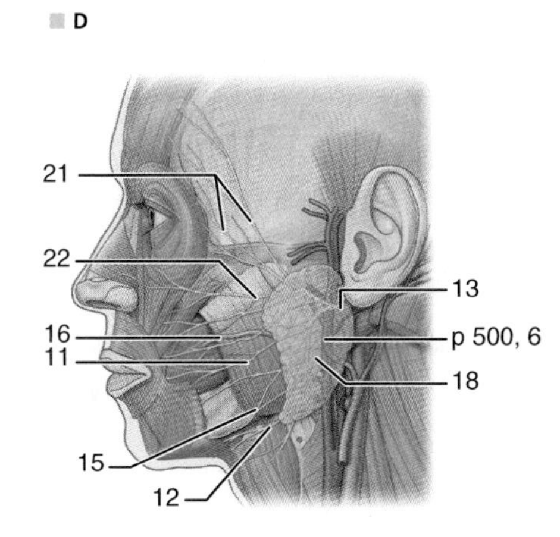

1 **deep part of parotid gland** that part of the gland located deep to the facial nerve. <pars profunda glandulae parotideae> A

2 **superficial part of parotid gland** that part of the parotid gland located superficial to the facial nerve. <pars superficialis glandulae parotideae> A

3 **anterior auricular nerves** *origin,* auriculotemporal nerve; *distribution,* skin of anterosuperior part of external ear; *modality,* general sensory. <nervi auriculares anteriores> B

4 **buccal nerve** *origin,* mandibular nerve; *distribution,* skin and mucous membrane of cheeks, gums, and perhaps the first two molars and the premolars; *modality,* general sensory. <nervus buccalis> B

5 **external nasal branch of anterior ethmoidal nerve** a branch that is essentially a continuation, or terminal branch, of the anterior ethmoidal nerve, innervating the skin of the dorsal part of the nose; *modality,* general sensory. <ramus nasalis externus nervi ethmoidalis anterioris> B

6 **external nasal branches of infraorbital nerve** they innervate the skin of the side of the nose; *modality,* general sensory. <rami nasales externi nervi infraorbitalis> B

7 **inferior palpebral branches of infraorbital nerve** branches that supply the skin and conjunctiva of the lower eyelid; *modality,* general sensory. <rami palpebrales inferiores nervi infraorbitalis> B

8 **infraorbital nerve** *origin,* continuation of the maxillary nerve, entering the orbit through the inferior orbital fissure, and occupying in succession the infraorbital groove, canal, and foramen; *branches,* middle and anterior superior alveolar, inferior palpebral, internal and external nasal, and superior labial branches; *modality,* general sensory. <nervus infraorbitalis> B

9 **infratrochlear nerve** *origin,* nasociliary nerve from ophthalmic nerve; *branches,* palpebral branches; *distribution,* skin of root and upper bridge of nose and lower eyelid, conjunctiva, lacrimal duct; *modality,* general sensory. <nervus infratrochlearis> B

10 **labial branches of mental nerve** branches of the mental nerve that innervate the lower lip; *modality,* general sensory. <rami labiales nervi mentalis> B

11 **medial branch of supraorbital nerve** a nerve branch that supplies the frontal sinus, upper eyelid, and skin and subcutaneous tissue of the forehead and adjacent scalp as far back as the parietal bone; *modality,* general sensory. <ramus medialis nervi supraorbitalis> B

12 **mental branches of mental nerve** branches that innervate the skin of the chin; *modality,* general sensory. <rami mentales nervi mentalis> B

13 **mental nerve** *origin,* inferior alveolar nerve; *branches,* mental, gingival, and inferior labial branches; *distribution,* skin of chin, and lower lip; *modality,* general sensory. <nervus mentalis> B

14 ***parotid branches of auriculotemporal nerve*** branches that bear postganglionic fibers from the otic ganglion to the parotid gland; *modality,* parasympathetic. <rami parotidei nervi auriculotemporalis> B

15 **superior labial branches of infraorbital nerve** branches of the infraorbital nerve that are distributed to mucous membranes of the mouth and skin of the upper lip; *modality,* general sensory. <rami labiales superiores nervi infraorbitalis> B

16 **zygomaticofacial branch of zygomatic nerve** a branch that passes from the lateral wall of the orbit, piercing the zygomatic bone to supply overlying skin; *modality,* general sensory. <ramus zygomaticofacialis nervi zygomatici> B

17 **zygomaticotemporal branch of zygomatic nerve** a branch that passes from the lateral wall of the orbit, piercing the zygomatic bone to innervate skin of the anterior temporal region; *modality,* general sensory. <ramus zygomaticotemporalis nervi zygomatici> B

18 **posterior auricular nerve** *origin,* facial nerve; *branches,* occipital branch; *distribution,* posterior auricularis and occipitofrontalis muscles and skin of external acoustic meatus; *modality,* motor and general sensory. <nervus auricularis posterior> C

A

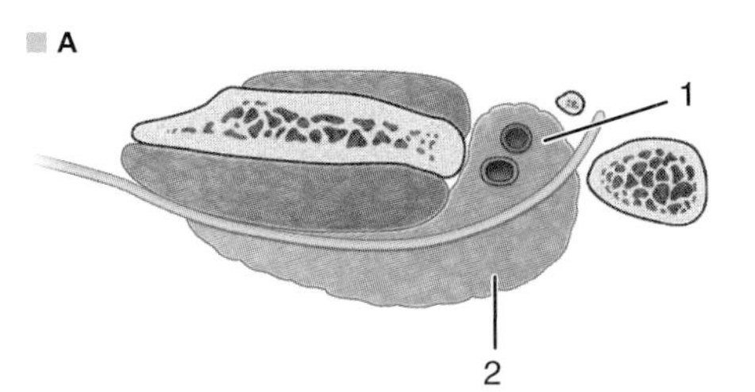

B

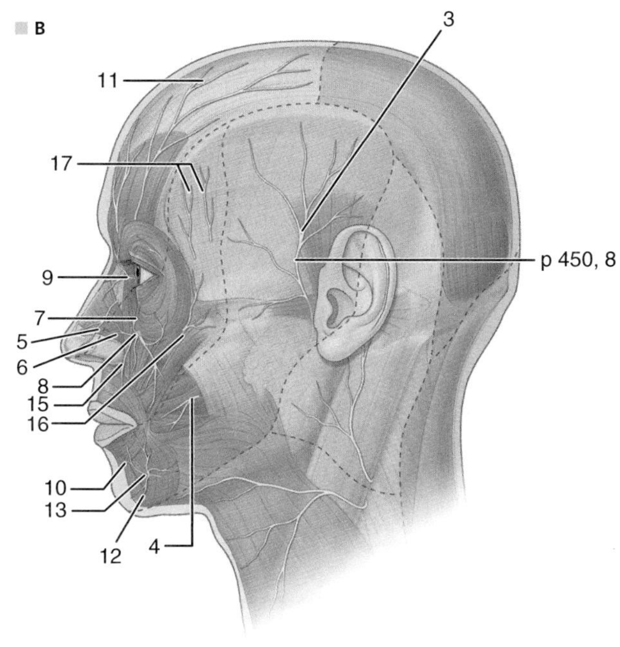

C

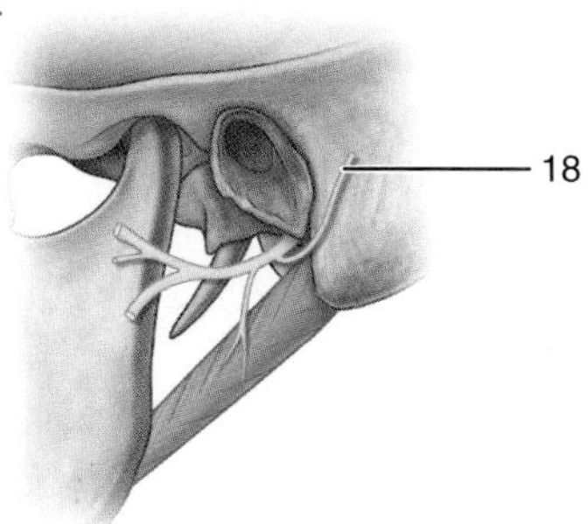

1 ***zygomatico-orbital artery*** *origin,* superficial temporal; *branches,* none; *distribution,* lateral side of orbit. <arteria zygomaticoorbitalis> A

2 **angular artery** *origin,* facial artery; *branches,* none; *distribution,* lacrimal sac, lower eyelid, nose. <arteria angularis> A

3 **angular vein** a short vein between the eye and the root of the nose; it is formed by union of the supratrochlear and supraorbital veins and continues inferiorly as the facial vein. <vena angularis> A

4 ***anterior auricular branches of superficial temporal artery*** branches that supply the lateral aspect of the pinna and the external acoustic meatus. Called also *anterior auricular arteries.* <rami auriculares anteriores arteriae temporalis superficialis> A

5 ***anterior auricular veins*** branches from the anterior part of the pinna that enter the superficial temporal vein. <venae auriculares anteriores> A

6 ***common cochlear artery*** *origin,* labyrinthine artery; *branches,* vestibulocochlear, proper cochlear, and spiral modiolar arteries; *distribution,* cochlea and vestibule. <arteria cochlearis communis> A

7 ***dorsal nasal artery*** *origin,* ophthalmic artery; *branches,* branch to nasolacrimal sac and branch anastomosing with terminal part of facial artery; *distribution,* skin of external nose. Called also *external nasal artery.* <arteria dorsalis nasi> A

8 ***external palatine vein*** the vein that drains blood from the tonsils and the soft palate into the facial vein. <vena palatina externa> A

9 **facial vein** the vein that begins at the medial angle of the eye as the angular vein, descends behind the facial artery, and usually ends in the internal jugular vein; this vessel sometimes joins the retromandibular vein to form a common trunk. <vena facialis> A

10 **inferior labial artery** *origin,* facial artery; *branches,* none; *distribution,* lower lip. <arteria labialis inferior> A

11 **inferior labial veins** veins that drain the region of the lower lip into the facial vein. <venae labiales inferiores> A

12 ***inferior palpebral veins*** branches that drain the blood from the lower eyelid into the facial vein. <venae palpebrales inferiores> A

13 **lateral nasal branch of the facial artery** a branch supplying the ala and dorsum of the nose. <ramus lateralis nasi arteriae facialis> A

14 ***nasal septal branch of superior labial artery*** a branch that ramifies on the lower and front part of the nasal septum. <ramus septi nasi arteriae labialis superioris> A

15 ***parotid branch of posterior auricular artery*** a branch supplying the parotid gland. <ramus parotideus arteriae auricularis posterioris> A

16 ***parotid branch of superficial temporal artery*** a branch supplying the parotid gland and the temporomandibular joint. <ramus parotideus arteriae temporalis superficialis> A

17 ***parotid branches of facial vein*** small veins from the substance of the parotid gland, which follow the parotid duct and open into the facial vein; called also *anterior parotid veins.* <rami parotidei venae facialis> A

18 ***parotid veins*** small veins from the parotid gland that open into the facial vein or into the retromandibular vein. Those opening into the facial vein are called also *parotid branches of facial vein.* <venae parotideae> A

19 **posterior auricular vein** a vein that begins in a plexus on the side of the head, passes down behind the pinna, and joins with the retromandibular vein to form the external jugular vein. <vena auricularis posterior> A

20 ***proper cochlear artery*** *origin,* common cochlear artery; *branches,* none; *distribution,* cochlea. Called also *cochlear artery.* <arteria cochlearis propria> A

21 **submental artery** *origin,* facial artery; *branches,* none; *distribution,* tissues under chin. <arteria submentalis> A

22 **submental vein** a vein that follows the submental artery and opens into the facial vein. <vena submentalis> A

23 **superior labial artery** *origin,* facial artery; *branches,* septal and alar; *distribution,* upper lip, nose. <arteria labialis superior> A

24 **superior labial vein** the vein that drains blood from the region of the upper lip into the facial vein. <vena labialis superior> A

25 **transverse facial artery** *origin,* superficial temporal artery; *branches,* none; *distribution,* parotid region. <arteria transversa faciei> A

26 **transverse facial vein** a vein that passes backward with the transverse facial artery just below the zygomatic arch to join the retromandibular vein. <vena transversa faciei> A

27 **facial artery** *origin,* external carotid; *branches,* ascending palatine, tonsillar, submental, inferior labial, superior labial, septal, lateral nasal, angular, glandular; *distribution,* face, tonsil, palate, submandibular gland. Called also *external maxillary artery.* <arteria facialis> A, B

28 ***glandular branches of facial artery*** branches given off to the submandibular gland by the facial artery as it passes over the lateral surface of the gland. <rami glandulares arteriae facialis> B

29 **infraorbital artery** *origin,* maxillary artery; *branches,* anterior superior alveolar; *distribution,* maxilla, maxillary sinus, upper teeth, lower eyelid, cheek, nose. <arteria infraorbitalis> B

A

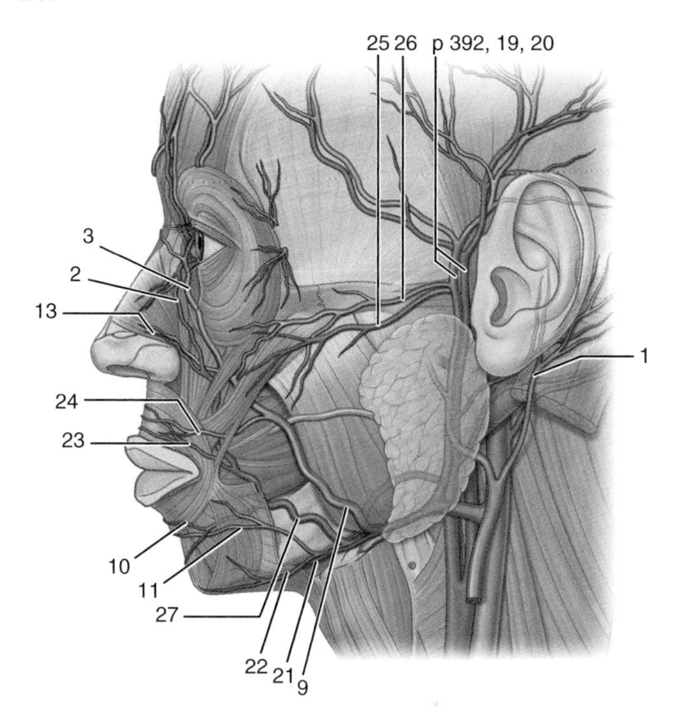

B

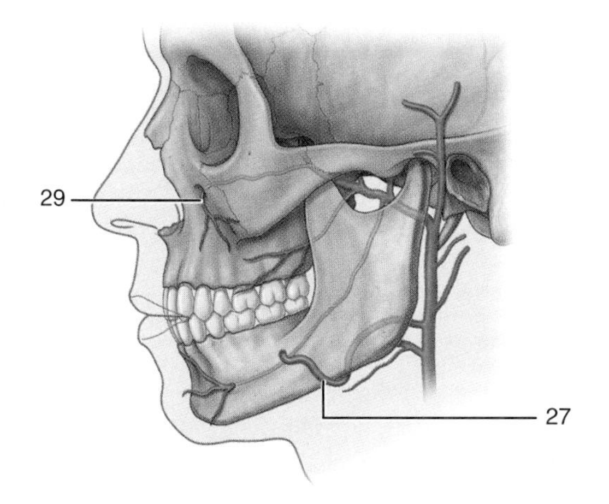

1 **mental branch of inferior alveolar artery** a branch arising from the inferior alveolar artery in the mandibular canal, which leaves the canal at the mental foramen, supplies the chin, and anastomoses with its fellow of the opposite side and with the submental and inferior labial arteries. Called also *mental artery.* <ramus mentalis arteriae alveolaris inferioris> A

2 **deep facial vein** a vein draining from the pterygoid plexus to the facial vein. <vena profunda faciei> B

3 **external nasal veins** small ascending branches from the nose that open into the angular and facial veins. <venae nasales externae> B

4 ***palpebral veins*** small branches from the eyelids that open into the superior ophthalmic vein. <venae palpebrales> B

5 **occipitofrontalis muscle** *origin,* FRONTAL BELLY—epicranial aponeurosis, OCCIPITAL BELLY—highest nuchal line of occipital bone; *insertion,* FRONTAL BELLY—skin of eyebrows and root of nose, OCCIPITAL BELLY—epicranial aponeurosis; *innervation,* FRONTAL BELLY—temporal branch of facial, OCCIPITAL BELLY—posterior auricular branch of facial; *action,* FRONTAL BELLY—raises eyebrows, OCCIPITAL BELLY—draws scalp posteriorly. <musculus occipitofrontalis> C

6 **occipital belly of occipitofrontalis muscle** a part that originates from the highest nuchal line of the occipital bone and inserts into the epicranial aponeurosis. <venter occipitalis musculi occipitofrontalis> C

7 **lacrimal part of orbicularis oris muscle** the part of the orbicularis oculi muscle that arises from the posterior lacrimal ridge of the lacrimal bone, to become continuous with the palpebral portion. <pars lacrimalis musculi orbicularis oculi> D

A

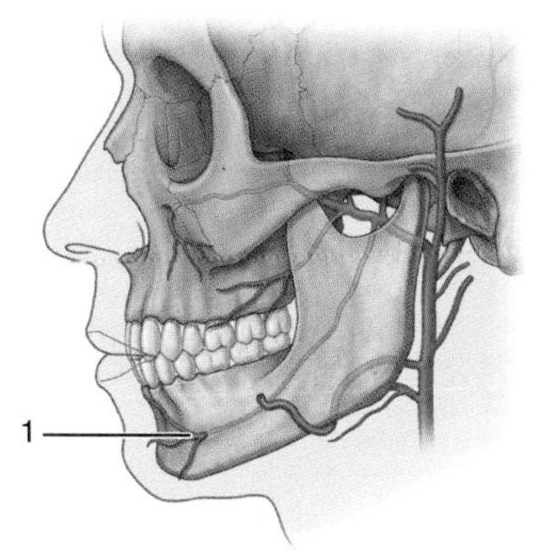

B

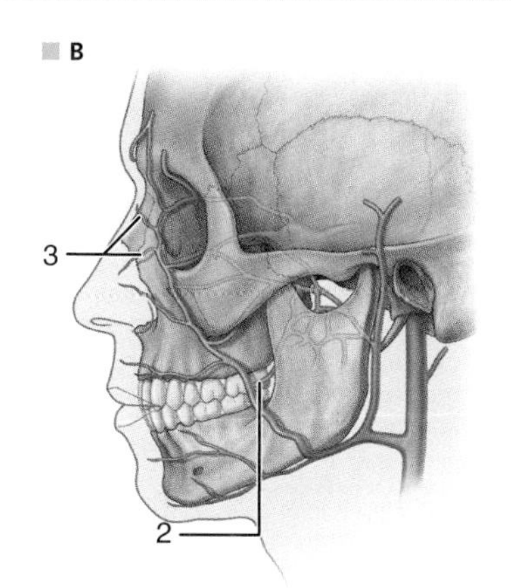

C

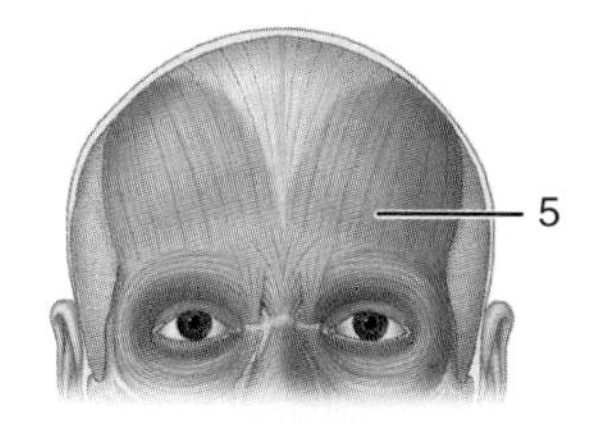

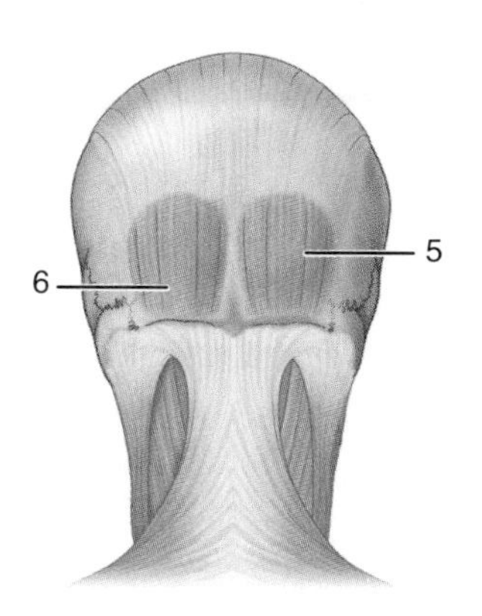

D

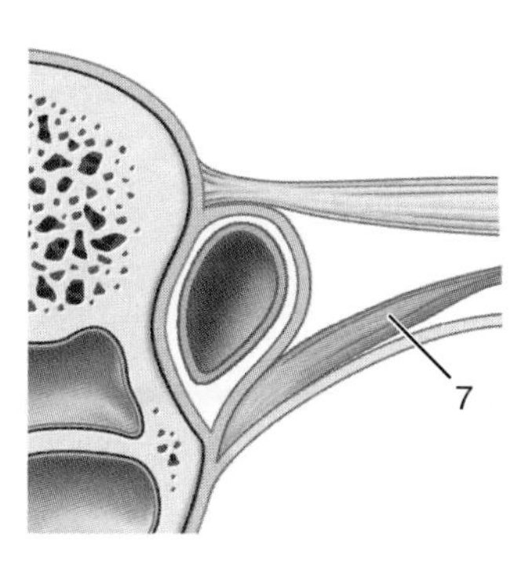

Orbit

1 ***anastomotic branch of lacrimal artery with middle meningeal artery*** a recurrent meningeal branch of the lacrimal artery that passes back via the lateral part of the orbital fissure and anastomoses with the anastomotic branch of the middle meningeal artery. <ramus anastomoticus arteriae lacrimalis cum arteria meningea media> A

2 **anterior ethmoidal artery** *origin,* ophthalmic artery; *branches,* anterior meningeal, anterior septal, and anterior lateral nasal branches; *distribution,* dura mater, nose, frontal sinus, anterior ethmoidal cells. <arteria ethmoidalis anterior> A

3 **central artery of retina** *origin,* ophthalmic artery; *branches,* none; *distribution,* retina. Called also *artery of Zinn.* <arteria centralis retinae> A

4 **central vein of retina** the vein that is formed by union of the retinal veins; it passes out of the eyeball in the optic nerve to empty into the superior ophthalmic vein. <vena centralis retinae> B

5 ***diploic branch of supraorbital artery*** a small branch arising as the artery passes through the supraorbital notch and supplying the diploë of the frontal bone and the lining of the frontal sinus. <ramus diploicus arteriae supraorbitalis> A

6 ***ethmoidal veins*** veins that follow the anterior and posterior ethmoidal arteries, emerge from the ethmoidal foramina, and empty into the superior ophthalmic vein. <venae ethmoidales> A

7 ***extraocular part of central artery of retina*** the portion of the artery that courses within the dural sheath with the optic nerve, then pierces the nerve approximately 1.2 cm behind the eyeball, and travels forward within the center of the nerve to the retina. <pars extraocularis arteriae centralis retinae> A

8 **inferior macular arteriole** the inferior arteriole supplying the macula lutea. <arteriola macularis inferior> B

9 **inferior nasal arteriole of retina** a small branch of the central artery of the retina supplying the inferior nasal region of the retina. <arteriola nasalis retinae inferior> B

10 **inferior temporal arteriole of retina** a branch of the central artery of the retina, supplying the inferior temporal region of the retina. <arteriola temporalis retinae inferior> B

11 **lacrimal artery** *origin,* ophthalmic artery; *branches,* lateral palpebral arteries and anastomotic branch with middle meningeal artery; *distribution,* lacrimal gland, upper and lower eyelids, conjunctiva. <arteria lacrimalis> A

12 ***medial arteriole of retina*** the small branch supplying blood to the central region of the retina. <arteriola macularis media> A

13 ***muscular arteries*** branches of the ophthalmic artery consisting of a superior group and an inferior group; the inferior group gives origin to the anterior ciliary arteries. <arteriae musculares> A

14 **ophthalmic artery** *origin,* cerebral part of internal carotid artery; *branches,* lacrimal, supraorbital, central artery of retina, ciliary, muscular, posterior and anterior ethmoidal, palpebral, supratrochlear, dorsal nasal, recurrent meningeal; *distribution,* eye, orbit, adjacent facial structures. <arteria ophthalmica> A

15 **posterior ethmoidal artery** *origin,* ophthalmic artery; *branches,* meningeal and septal and lateral nasal branches; *distribution,* posterior ethmoidal cells, dura mater, nose. <arteria ethmoidalis posterior> A

16 **superior macular arteriole** the superior arteriole supplying the macula lutea. <arteriola macularis superior> B

17 **superior nasal arteriole of retina** a small branch of the central artery of the retina, supplying the superior nasal region of the retina. <arteriola nasalis retinae superior> B

18 **superior temporal arteriole of retina** a branch of the central artery of the retina, supplying the superior temporal region of the retina. <arteriola temporalis retinae superior> B

19 **supraorbital artery** *origin,* ophthalmic artery; *branches,* superficial, deep, diploic; *distribution,* forehead, upper muscles of orbit, upper eyelid, frontal sinus. <arteria supraorbitalis> A

20 **supratrochlear artery** *origin,* ophthalmic artery; *branches,* none; *distribution,* anterior scalp. Called also *frontal artery.* <arteria supratrochlearis> A

21 ***vascular circle of optic nerve*** a circle of arteries in the sclera surrounding the site of entrance of the optic nerve. <circulus vasculosus nervi optici> A

22 ***conjunctival veins*** small veins that drain blood from the conjunctiva to the superior ophthalmic vein. <venae conjunctivales> C

23 **inferior ophthalmic vein** a vein formed by confluence of muscular and ciliary branches, and running backward either to join the superior ophthalmic vein or to open directly into the cavernous sinus; it sends a communicating branch through the inferior orbital fissure to join the pterygoid venous plexus. <vena ophthalmica inferior> C

24 ***lacrimal vein*** the vein that drains blood from the lacrimal gland into the superior ophthalmic vein. <vena lacrimalis> C

25 **nasofrontal vein** a vein that begins at the supraorbital vein, enters the orbit, and joins the superior ophthalmic vein. <vena nasofrontalis> C

26 **orbital veins** the veins that drain the orbit and its structures, including the superior ophthalmic vein and its tributaries and the inferior ophthalmic vein. <venae orbitae> C

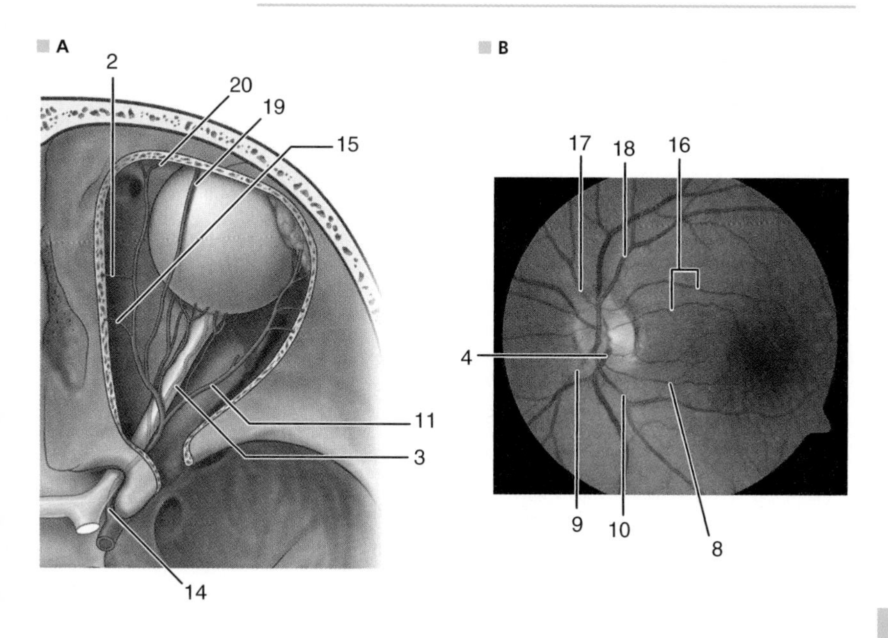
A
2
20
19
15
11
3
14
B
17
18
16
4
9
10
8

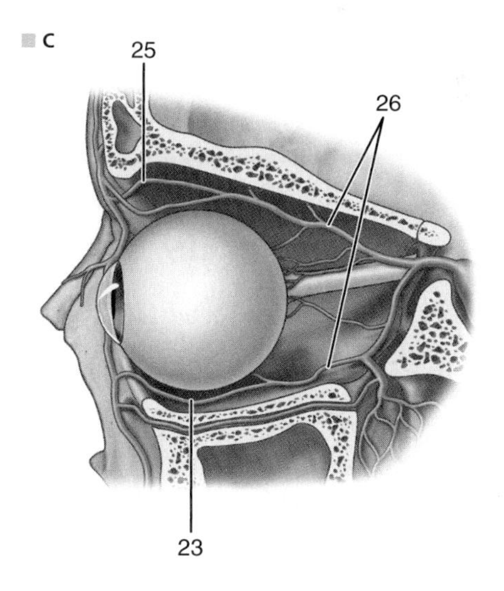
C
25
26
23

1 **superior ophthalmic vein** the vein that begins at the medial angle of the eye, where it communicates with the frontal, supraorbital, and angular veins; it follows the distribution of the ophthalmic artery, and may be joined by the inferior ophthalmic vein at the superior orbital fissure before opening into the cavernous sinus. <vena ophthalmica superior> A

2 **supraorbital vein** the vein that passes down the forehead lateral to the supratrochlear vein, joining it at the root of the nose to form the angular vein. <vena supraorbitalis> A

3 **vorticose veins** four veins that pierce the sclera and carry blood from the choroid to the superior ophthalmic vein; called also *posterior ciliary veins.* <venae vorticosae> A

4 **anterior ethmoidal nerve** *origin*, continuation of nasociliary nerve, from ophthalmic nerve; *branches*, internal, external, lateral, and medial nasal branches; *distribution*, mucosa of upper and anterior nasal septum, lateral wall of nasal cavity, skin of lower bridge and tip of nose; *modality*, general sensory. <nervus ethmoidalis anterior> B

5 ***communicating branch of lacrimal nerve with zygomatic nerve*** a branch that carries parasympathetic postganglionic fibers originating in the pterygopalatine ganglion and destined for the lacrimal gland. <ramus communicans nervi lacrimalis cum nervo zygomatico> B

6 **frontal nerve** *origin*, ophthalmic division of trigeminal nerve; enters the orbit through the superior orbital fissure; *branches*, supraorbital and supratrochlear nerves; *distribution*, chiefly to the forehead and scalp; *modality*, general sensory. <nervus frontalis> B

7 **lacrimal nerve** *origin*, ophthalmic division of trigeminal nerve, entering the orbit through the superior orbital fissure; *distribution*, lacrimal gland, conjunctiva, lateral commissure of eye, and skin of upper eyelid; *modality*, general sensory. <nervus lacrimalis> B

8 **long ciliary nerves** *origin*, nasociliary nerve, from ophthalmic nerve; *distribution*, dilator pupillae, uvea, cornea; *modality*, sympathetic and general sensory. <nervi ciliares longi> B, C

9 **ophthalmic nerve** one of the three terminal divisions of the trigeminal nerve. *Origin*, trigeminal ganglion; *branches*, recurrent meningeal (tentorial) branch, frontal, lacrimal, and nasociliary nerves; *distribution*, eyeball and conjunctiva, lacrimal gland and sac, nasal mucosa and frontal sinus, external nose, upper eyelid, forehead, and scalp; *modality*, general sensory. <nervus ophthalmicus> B

10 **posterior ethmoidal nerve** *origin*, nasociliary nerve, from ophthalmic nerve; *distribution*, mucosa of posterior ethmoidal cells and of sphenoidal sinus; *modality*, general sensory. <nervus ethmoidalis posterior> B

11 **supraorbital nerve** *origin*, continuation of frontal nerve, from ophthalmic nerve; *branches*, lateral and medial branches; *distribution*, leaves orbit through supraorbital notch or foramen, and supplies the skin of upper eyelid, forehead, anterior scalp (to vertex), mucosa of frontal sinus; *modality*, general sensory. <nervus supraorbitalis> B

12 **supratrochlear nerve** *origin*, frontal nerve, from ophthalmic nerve; *distribution*, leaves orbit at medial end of supraorbital margin and supplies the forehead and upper eyelid; *modality*, general sensory. <nervus supratrochlearis> B

13 **ciliary ganglion** a parasympathetic ganglion in the posterior part of the orbit; it receives preganglionic fibers from the oculomotor nerve, and its postganglionic fibers supply the ciliary muscle and the sphincter pupillae. Sensory and postganglionic sympathetic fibers pass through the ganglion. <ganglion ciliare> C

14 **inferior branch of oculomotor nerve** the branch of the oculomotor nerve that innervates the medial and inferior rectus and inferior oblique muscles of the eyeball and, via the motor root of the ciliary ganglion and then the short ciliary nerves, supplies the sphincter pupillae and ciliary muscles; modality, motor and parasympathetic. <ramus inferior nervi oculomotorii> C

15 **nasociliary nerve** *origin*, ophthalmic division of trigeminal nerve; *branches*, long ciliary, posterior ethmoidal, anterior ethmoidal, and infratrochlear nerves, and a communicating branch to the ciliary ganglion; *modality*, general sensory. <nervus nasociliaris> C

16 ***parasympathetic root of ciliary ganglion*** a short collection of fibers passing from the inferior branch of the oculomotor nerve to the posterior inferior portion of the ciliary ganglion; it contains preganglionic parasympathetic fibers for the sphincter papillae and the ciliary muscle. Called also *motor, short,* or *oculomotor root of ciliary ganglion.* <radix parasympathica ganglii ciliaris> C

17 **sensory root of ciliary ganglion** sensory fibers from the cornea, iris, ciliary body, and choroid that pass through the ciliary ganglion to the nasociliary nerve. Called also *long* or *nasociliary root of ciliary ganglion.* <radix sensoria ganglii ciliaris> C

18 **short ciliary nerves** *origin,* ciliary ganglion; *distribution,* smooth muscle and tunics of eye; *modality,* parasympathetic, sympathetic, and general sensory. <nervi ciliares breves> C

19 **superior branch of oculomotor nerve** the upper and smaller of the two branches of the oculomotor nerve, which supplies the superior rectus muscle and, terminally, the levator palpebrae superioris; *modality,* motor. <ramus superior nervi oculomotorii> C

20 ***sympathetic root of ciliary ganglion*** postganglionic fibers from the superior cervical ganglion, derived from the internal carotid plexus, to the ciliary ganglion, for distribution by the short ciliary nerves to the dilator pupillae, orbitalis, and tarsal muscles, and blood vessels of the eyeball. <radix sympathica ganglii ciliaris> C

A

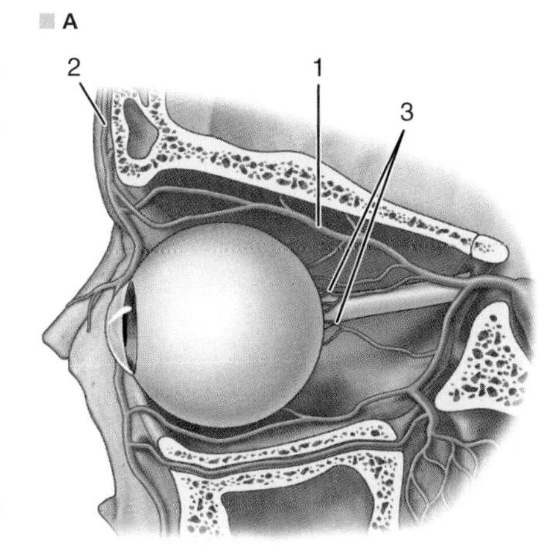

C

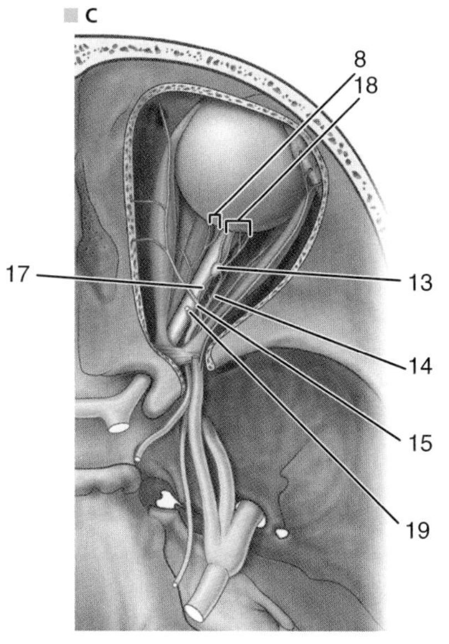

B

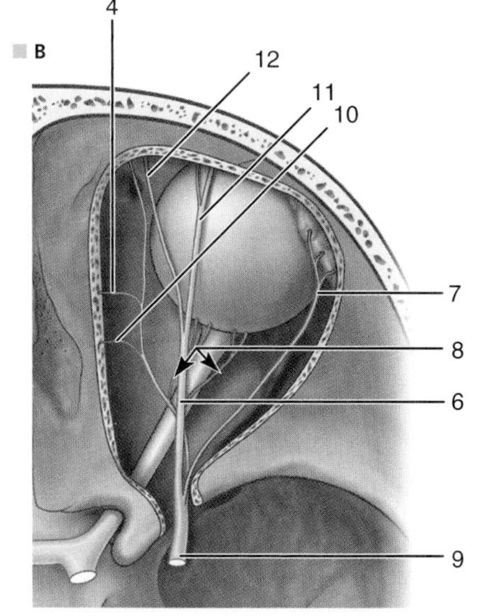

1 **anterior ciliary arteries** *origin,* muscular arteries; *branches,* episcleral and anterior conjunctival arteries; *distribution,* iris, conjunctiva. <arteriae ciliares anteriores> A

2 **anterior surface of lens** the surface of the lens directed toward the anterior surface of the eye. <facies anterior lentis> A

3 **anterior surface of the iris** the surface of the iris directed toward the anterior chamber of the eye. <facies anterior iridis> A

4 **aqueous humor** the fluid produced in the eye, occupying the anterior and posterior chambers, and diffusing out of the eye into the blood; regarded as the lymph of the eye, its composition varies from that of lymph in the body generally. <humor aquosus> A

5 ***axis of lens*** an imaginary line joining the anterior and posterior poles of the lens of the eye. <axis lentis> A

6 **basal lamina of choroid** the transparent inner layer of the choroid, which is in contact with the pigmented layer of the retina. Called also *vitreous lamina.* <lamina basalis choroideae> A

7 **capsule of lens** the elastic envelope covering the lens of the eye and fusing with the fibers of the ciliary zonule; called also *crystalline capsule.* <capsula lentis> A

8 **choroid** the thin, pigmented, vascular coat of the eye extending from the ora serrata to the optic nerve; it furnishes blood supply to the retina and conducts arteries and nerves to the anterior structures. <choroidea> A

9 ***ciliary folds*** low ridges in the furrows between the ciliary processes. <plicae ciliares> A

10 **ciliary part of retina** the two layers of epithelium lining the basal lamina of the ciliary body. <pars ciliaris retinae> A

11 ***ciliary veins*** veins that arise inside the eyeball by branches from the ciliary muscle and drain into the superior ophthalmic vein. The *anterior ciliary veins* follow the anterior ciliary arteries, and receive branches from the sinus venosus, sclerae, the episcleral veins, and the conjunctiva. The *posterior ciliary veins* follow the posterior ciliary arteries and empty also into the inferior ophthalmic vein. <venae ciliares> A

12 **cornea** the transparent structure forming the anterior part of the sclera of the eye. It consists of five layers: (1) the anterior corneal epithelium, continuous with that of the conjunctiva; (2) the anterior limiting layer (Bowman's membrane); (3) the substantia propria, or stroma; (4) the posterior limiting layer (Descemet's membrane); and (5) the endothelium of the anterior chamber. A

13 **corneal vertex** the central, thinner portion of the cornea. <vertex corneae> A

14 **corneoscleral junction** the junctional region between the cornea and the sclera, marked on the outer surface of the eyeball by a slight furrow, the scleral sulcus; called also *corneal limbus* and *sclerocorneal junction.* <limbus corneae> A

15 **cortex of lens** the softer, external part of the lens of the eye. <cortex lentis> A

16 ***episcleral arteries*** *origin,* anterior ciliary arteries; *branches,* none; *distribution,* iris, ciliary processes. <arteriae episclerales> A

17 ***episcleral lamina*** loose connective and elastic tissue covering the sclera and anteriorly connecting it with the conjunctiva. <lamina episcleralis> A

18 **episcleral space** the space between the bulbar fascia and the eyeball. <spatium episclerale> A

19 ***episcleral veins*** the veins that ring the cornea and drain into the vorticose and ciliary veins. <venae episclerales> A

20 **epithelium of lens** the cuboidal epithelium on the front of the lens. <epithelium lentis> A

21 ***equator of eyeball*** an imaginary line encircling the eyeball equidistant from the anterior and the posterior poles, dividing the eye into anterior and posterior halves. <equator bulbi oculi> A

22 **equator of lens** the rounded peripheral margin of the lens at which the anterior and posterior surfaces meet. <equator lentis> A

23 ***external axis of eye*** an imaginary line that passes from the anterior to the posterior pole of the eyeball. <axis bulbi externus> A

24 **eye** the organ of vision. It is composed of three coats: the external *fibrous layer,* the middle *vascular layer* (or *uvea*), and the *inner layer.* Within the three coats are the refracting media: the *aqueous humor,* the *crystalline lens,* and the *vitreous humor.* Posteriorly, fibers of the optic nerve enter the ganglionic layer and receive sensations from the visual cells of the retina (the *retinal rods* and *retinal cones*). Called also *oculus.* A

25 **eyeball** the spherical bulb of the eye. Called also *globe.* <bulbus oculi> A

26 ***fibers of lens*** long bands, derived from the epithelium, that make up the substance of the lens. <fibrae lentis> A

27 **fibrous layer of eyeball** the outer of the three tunics of the eye, comprising the cornea and the sclera. <tunica fibrosa bulbi> A

28 ***greater arterial circle of iris*** a circle of anastomosing arteries situated in the ciliary body along the ciliary margin of the iris. <circulus arteriosus iridis major> A

29 ***hyaloid artery*** a fetal vessel that continues forward from the central artery of retina through the vitreous body to supply the lens; it normally is not present after birth. <arteria hyaloidea> A

30 **hyaloid canal** a passage running from in front of the optic disk to the lens of the eye; in the fetus it transmits the hyaloid artery. Called also *central canal of Stilling, central canal of vitreous,* and *Cloquet's canal.* <canalis hyaloideus> A

A

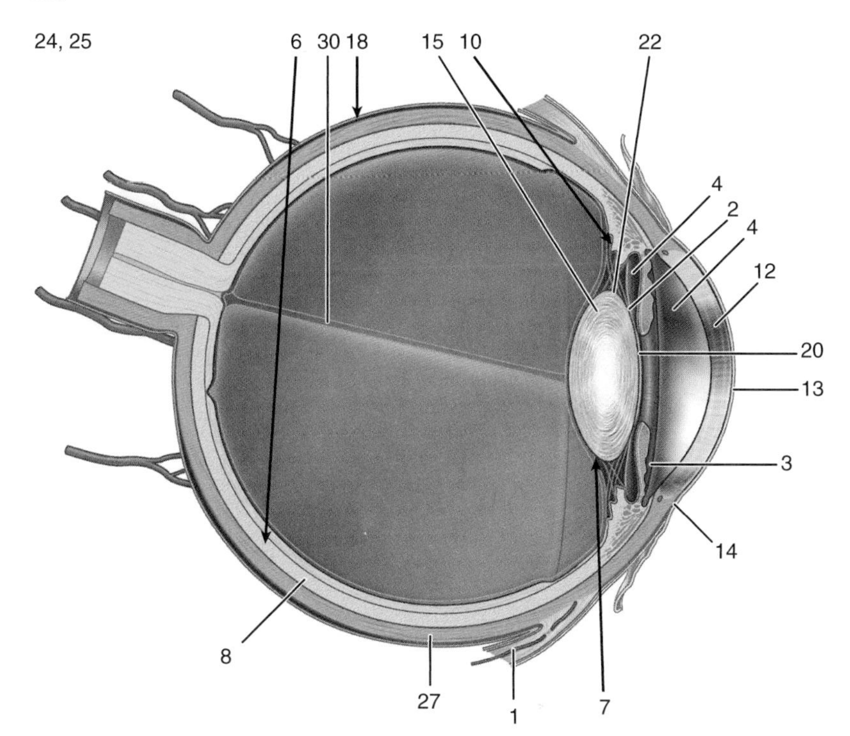

1 **hyaloid fossa** a depression on the anterior surface of the vitreous body, in which the lens is lodged; called also *lenticular fossa of vitreous body.* <fossa hyaloidea> A

2 **inner layer of eyeball** the innermost of the three tunics of the eye; it is nervous and sensory and consists primarily of the retina and its blood vessels. <tunica interna bulbi> A

3 ***internal axis of eye*** an imaginary line in the eyeball, passing from the anterior pole to a point on the anterior surface of the retina just deep to the posterior pole. <axis bulbi internus> A

4 ***intracanalicular part of optic nerve*** the part of the nerve that runs through the optic canal. <pars canalis nervi optici> A

5 ***intracranial part of optic nerve*** the part of the nerve that lies between the optic canal and the optic chiasm. <pars intracranialis nervi optici> A

6 **intralaminar part of intraocular optic nerve** the portion of the intraocular part of the optic nerve that runs through the lamina cribrosa of the sclera. <pars intralaminaris nervi optici intraocularis> A

7 **intraocular part of optic nerve** the part of the nerve that is within the eyeball, separated into postlaminar, intralaminar, and prelaminar parts. <pars intraocularis nervi optici> A

8 **lamina cribrosa of sclera** the perforated portion of the sclera through which pass the axons of the ganglion cells of the retina; called also *optic foramen of sclera.* <lamina cribrosa sclerae> A

9 ***lens substance*** the fibrous material making up the bulk of the lens of the eye. <substantia lentis> A

10 **lens** the transparent biconvex body of the eye situated between the posterior chamber and the vitreous body, constituting part of the refracting mechanism of the eye. Called also *crystalline lens.* A

11 **long posterior ciliary arteries** *origin,* ophthalmic artery; *branches,* none; *distribution,* iris, ciliary processes. Called also *long ciliary arteries.* <arteriae ciliares posteriores longae> A

12 ***meridians of eyeball*** imaginary lines encircling the eyeball, marking the intersection with its surface of planes passing through its anteroposterior axis. <meridiani bulbi oculi> A

13 **nucleus of lens** the harder internal part of the lens of the eye. <nucleus lentis> A

14 ***optic axis*** a line connecting the center of anterior and posterior poles. <axis opticus> A

15 **optic part of retina** the part of the retina that contains receptors sensitive to light, extending posteriorly from the ora serrata on the inner surface of the choroid and continuous at the optic disk with the optic nerve; it consists of an outer pigmented layer and an inner, multilayered nervous layer. <pars optica retinae> A

16 **ora serrata retinae** the irregular anterior margin of the optic part of retina, lying internal to the junction of the choroid and the ciliary body. A

17 ***orbital part of optic nerve*** the part of the nerve located between the optic canal and the eyeball. <pars orbitalis nervi optici> A

18 **posterior chamber of eye** that portion of the aqueous humor-containing space between the cornea and the lens that is bounded in front by the iris, and behind by the lens and ciliary zonule. <camera posterior bulbi> A

19 ***posterior epithelium of cornea*** the mesothelial layer covering the posterior surface of the posterior limiting lamina of the cornea; it was once believed to extend to the anterior surface of the stroma of the iris. Called also *anterior endothelium of cornea,* and *corneal endothelium.* <epithelium posterius corneae> A

20 ***posterior limiting lamina corneae*** a thin hyaline membrane between the substantia propria and the endothelial layer of the cornea. <lamina limitans posterior> A

21 **posterior pole of eyeball** the center of the posterior curvature of the eyeball. <polus posterior bulbi oculi> A

22 **posterior pole of lens** the central point of the posterior surface of the lens. <polus posterior lentis> A

23 **posterior segment of eyeball** the vitreous, retina, and optic nerve. <segmentum posterius bulbi oculi> A

24 **posterior surface of cornea** the posterior surface of the cornea, which forms the anterior boundary of the anterior chamber. <facies posterior corneae> A

25 **posterior surface of iris** the surface of the iris directed toward the posterior chamber of the eye. <facies posterior iridis> A

26 **posterior surface of lens** the surface of the lens directed toward the vitreous body of the eye. <facies posterior lentis> A

27 ***postlaminar part of intraocular optic nerve*** that portion of the intraocular part of the optic nerve that is posterior to the lamina cribrosa of the sclera. <pars postlaminaris nervi optici intraocularis> A

28 ***prelaminar part of intraocular optic nerve*** that portion of the intraocular part of the optic nerve that is anterior to the lamina cribrosa of the sclera. <pars prelaminaris nervi optici intraocularis> A

29 **retina** the innermost of the three tunics of the eyeball, surrounding the vitreous body and continuous posteriorly with the optic nerve. A

30 **sclera** the tough white outer coat of the eyeball, covering approximately the posterior five-sixths of its surface, and continuous anteriorly with the cornea and posteriorly with the external sheath of the optic nerve. A

31 ***scleral veins*** tributaries of the anterior ciliary veins that drain the sclera. <venae sclerales> A

32 **short posterior ciliary arteries** *origin,* ophthalmic artery; *branches,* none; *distribution,* choroid. Called also *short ciliary arteries.* <arteriae ciliares posteriores breves> A

A

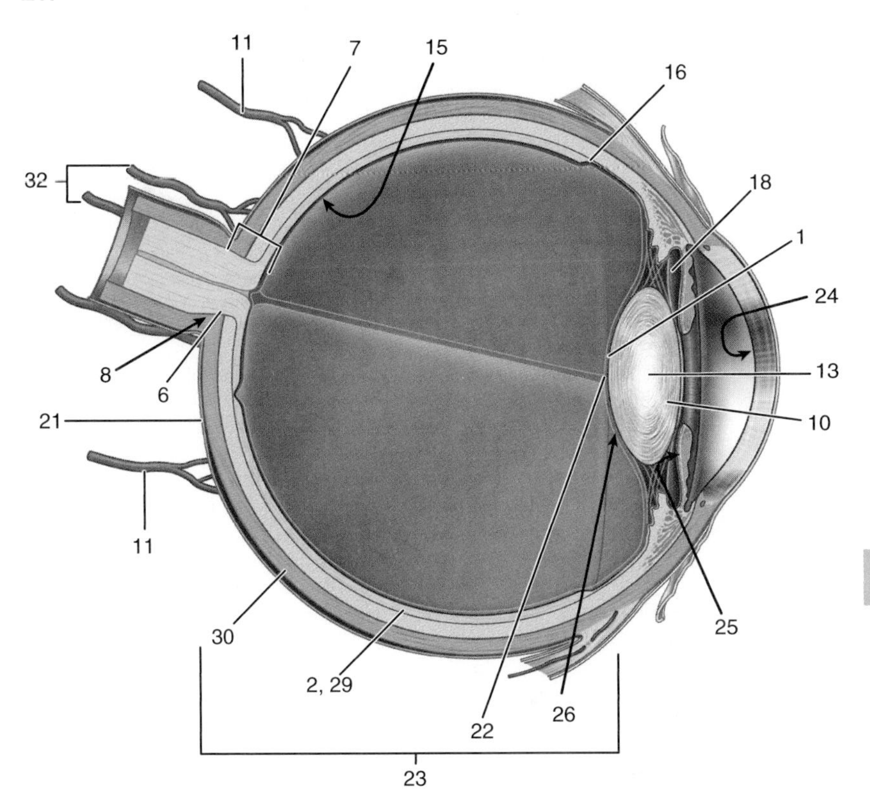

1 **substantia propria of cornea** the fibrous, tough, and transparent main part of the cornea, between the anterior and posterior limiting laminae. <substantia propria corneae> A

2 **substantia propria of sclera** the chief part of the sclera, lying between the suprachoroid and episcleral laminae, composed of dense bands of fibrous tissue, mostly parallel with the surface, and crossing each other in all directions. It is structurally continuous with the substantia propria of the cornea. <substantia propria sclerae> A

3 **suprachoroid lamina** a thin layer of loose, pigmented connective tissue on the inner surface of the sclera, connecting it with the choroid. <lamina fusca sclerae> A

4 ***trabecular meshwork*** a trabecula of loose fibers found at the iridocorneal angle between the anterior chamber of the eye and the scleral venous sinus; the aqueous humor filters through the spaces between the fibers into the sinus and passes into the bloodstream. It is divided into a corneoscleral part and a uveal part. <reticulum trabeculare> A

5 ***vascular lamina of choroid*** the layer of the choroid between the suprachoroid and choriocapillary layers, containing the largest blood vessels; called also *Haller's layer.* <lamina vasculosa choroideae> A

6 **vascular layer of eyeball** the middle, pigmented, vascular coat of the eye, comprising the choroid, the ciliary body, and the iris; called also *uvea* and *uveal tract.* <tunica vasculosa bulbi> A

7 **vitreous body** the transparent gel that fills the inner portion of the eyeball between the lens and the retina. <corpus vitreum> A

8 **vitreous chamber of eye** the space in the eyeball enclosing the vitreous body, bounded anteriorly by the lens and ciliary body and posteriorly by the posterior wall of the eyeball. <camera vitrea bulbi> A

9 ***vitreous humor*** the watery substance, resembling aqueous humor, contained within the interstices of the stroma in the vitreous body. <humor vitreus> A

10 **vitreous membrane** a delicate boundary layer investing the vitreous body of the eye. <membrana vitrea> A

11 ***vitreous stroma*** the framework of firmer material making up the vitreous body of the eye, and enclosing within its meshes the more fluid portion *(vitreous humor).* <stroma vitreum> A

12 **depression of optic disk** a depression in the center of the optic disk; called also *optic cup.* <excavatio disci> A

13 **fovea centralis** a tiny pit, about 1 degree wide, in the center of the macula lutea, which in turn presents an extremely small depression (foveola) containing rodlike elongated cones; it is the area of most acute vision, because here the layers of the retina are spread aside, permitting light to fall directly on the cones. <fovea centralis retinae> A

14 **foveola of retina** an extremely small depression in the floor of the fovea centralis, which is devoid of rod cells but contains rodlike elongated cones. <foveola retinae> A

15 **macula lutea** an irregular yellowish depression on the retina, about 3 degrees wide, lateral to and slightly below the optic disk; it is the site of absorption of short wavelengths of light, and it is thought that its variation in size, shape, and coloring may be related to variant types of color vision. Called also *macula retinae.* B

16 ***optic disk*** the intraocular portion of the optic nerve formed by fibers converging from the retina and appearing as a pink to white disk; there are no sensory receptors in this region and hence no response to stimuli restricted to it. Called also *blind spot* and *optic papilla.* <discus nervi optici> A, B

17 ***retinal blood vessels*** the blood vessels of the retina, including all the arterioles, derived from the central artery of the retina, and the venules, which return blood to the central vein. <vasa sanguinea retinae> B

18 **basal lamina of ciliary body** the innermost layer of the ciliary body, continuous with the basal lamina of the choroid. <lamina basalis corporis ciliaris> C

19 **ciliary body** the thickened part of the vascular layer of the eye anterior to the ora serrata, connecting the choroid with the iris; it is composed of the corona ciliaris, ciliary processes and folds, ciliary orbiculus, the ciliary muscle, and a basal lamina. <corpus ciliare> C

20 **ciliary disk** the thin part of the ciliary body extending between its crown and the ora serrata retinae. <orbiculus ciliaris> C

21 **ciliary margin of iris** the outer border of the iris, where it is continuous with the ciliary body. <margo ciliaris iridis> A, C

22 **ciliary muscle** *origin,* scleral spur; *insertion,* outer layers of choroid and ciliary processes; *innervation,* oculomotor, parasympathetic; *action,* affects shape of lens in visual accommodation. <musculus ciliaris> C

23 **ciliary processes** about 70 meridionally arranged ridges or folds projecting from the crown of the ciliary body; they secrete the aqueous humor into the posterior chamber of the eye. <processus ciliares> C

24 **ciliary zonule** a system of fibers extending between the ciliary body and the equator of the lens, holding the lens in place; called also *zonule of Zinn.* <zonula ciliaris> C

25 **circular fibers of ciliary muscle** the most internal fibers of the ciliary muscle that form a discrete portion of the ciliary muscle and extend around the apex of the ciliary body close to the root of the iris. Called also *Müller's muscle* and *sphincter fibers of ciliary muscle.* <fibrae circulares musculi ciliaris> C

26 ***corona ciliaris*** the region on the anterior inner surface of the ciliary body of the eye from which radiate the ciliary processes. C

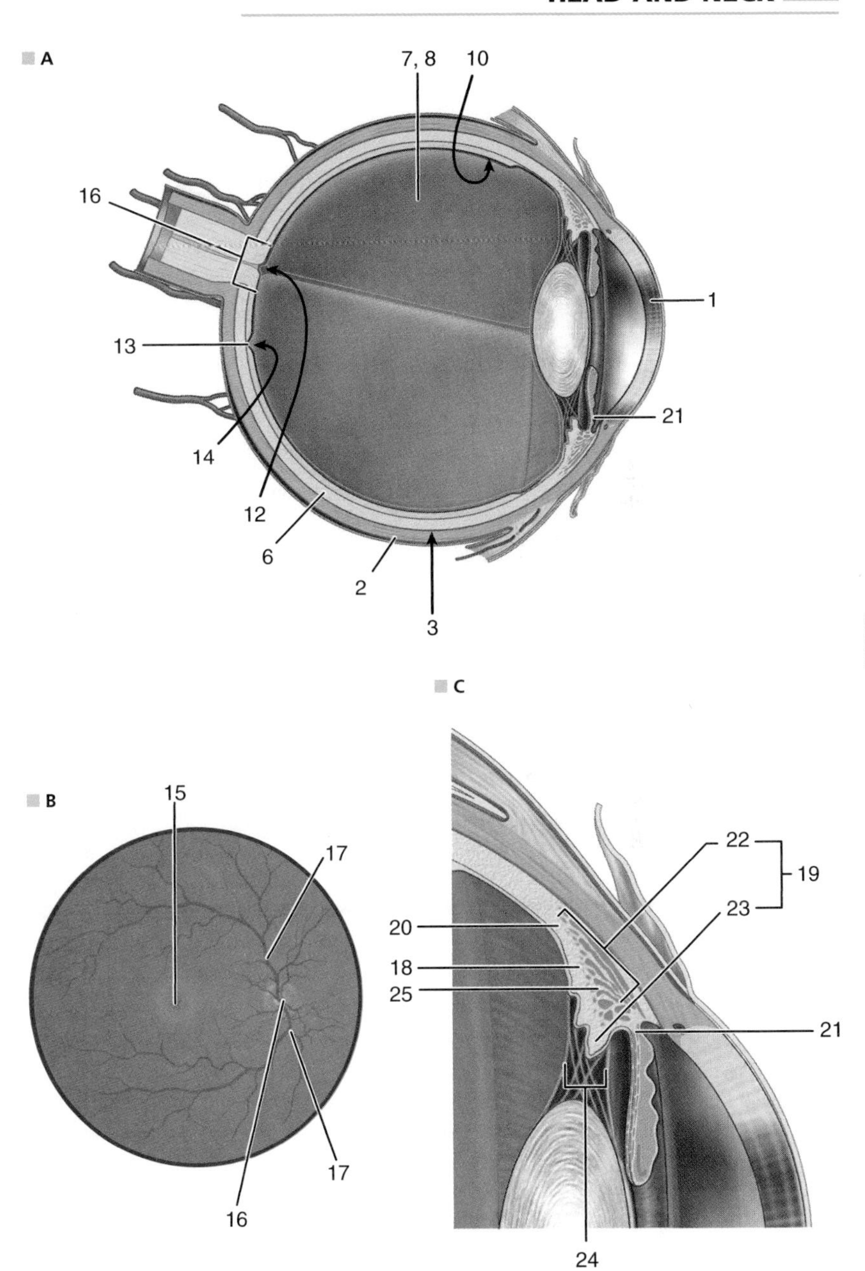
A
7, 8
10
16
1
13
21
14
12
6
2
3
B
15
17
17
16
C
22
19
23
20
18
25
21
24

1 **dilator pupillae muscle** a name given to fibers extending radially from the sphincter pupillae to the ciliary margin; *innervation,* sympathetic; *action,* dilates iris. <musculus dilatator pupillae> A

2 **iridocorneal angle** a narrow recess between the sclerocorneal junction and the attached margin of the iris, marking the periphery of the anterior chamber of the eye; it is the principal exit site for the aqueous humor. Called also *iridial angle.* <angulus iridocornealis> A

3 **meridional fibers of ciliary muscle** the most external fibers of the ciliary muscle that run meridionally or longitudinally from the trabecular meshwork toward the ciliary processes. Called also *Brücke's fibers* and *longitudinal fibers of ciliary muscle.* <fibrae meridionales musculi ciliaris> A

4 **radial fibers of ciliary muscle** the fibers of the ciliary muscle lying between the meridional (external) fibers and the circular (internal) fibers; they run in a radial or oblique direction from one to another and may form a fibrous network. Called also *oblique fibers of ciliary muscle.* <fibrae radiales musculi ciliaris> A

5 **scleral venous sinus** a circular channel at the junction of the sclera and cornea, which is the main pathway for elimination of aqueous humor from the eye. Called also *Schlemm's canal.* <sinus venosus sclerae> A

6 ***spaces of iridocorneal angle*** the spaces between the fibers of the pectinate ligament through which communication is effected between the anterior chamber and the canal of Schlemm. Called also *spaces of Fontana.* <spatia anguli iridocornealis> A

7 **sphincter pupillae** circular fibers of the iris, innervated by the ciliary nerves (parasympathetic), and acting to contract the pupil. <musculus sphincter pupillae> A

8 **sulcus sclerae** the groove at the junction of the sclera and cornea. A

9 **zonular fibers** the fibers that anchor the lens capsule to the ciliary body and the retina. <fibrae zonulares> A

10 **zonular spaces** the lymph-filled interstices between the fibers of the ciliary zonule, communicating with the posterior chamber of the eye; called also *Petit's canal.* <spatia zonularia> A

11 **zygomatic nerve** *origin,* maxillary nerve, entering the orbit through the inferior orbital fissure; *branches,* zygomaticofacial and zygomaticotemporal branches; *distribution,* communicates with the lacrimal nerve and supplies the skin of the temple and adjacent part of the face; *modality,* general sensory. <nervus zygomaticus> B

12 **anterior palpebral margin** the rounded anterior edges of the free margin of the eyelids, from which the eyelashes arise. <limbus anterior palpebrae> C

13 **anterior surface of eyelid** the exterior surface of the eyelid. <facies anterior palpebrae> C

14 **eyelashes** the hairs growing on the edges of the eyelids. <cilia> C

15 **infrapalpebral sulcus** the furrow below the lower eyelid. <sulcus infrapalpebralis> C

16 ***inner border of iris*** the more coarsely striated inner concentric circle on the anterior surface of the iris; called also *lesser circle* or *ring of iris.* <anulus iridis minor> C

17 ***iridial folds*** the numerous minute folds on the posterior surface of the iris. <plicae iridis> C

18 ***iridial part of retina*** the two layers of pigmented epithelium lining the posterior part of the iris. <pars iridica retinae> C

19 **iris** the circular pigmented membrane behind the cornea, perforated by the pupil; the most anterior portion of the vascular tunic of the eye, it is made up of a flat bar of circular muscular fibers surrounding the pupil, a thin layer of smooth muscle fibers by which the pupil is dilated, thus regulating the amount of light entering the eye, and posteriorly two layers of pigmented epithelial cells. C

20 **lacrimal caruncle** the red eminence at the medial angle of the eye. <caruncula lacrimalis> C

21 **lacrimal lake** the triangular space at the medial angle of the eye, where the tears collect. <lacus lacrimalis> C

22 ***lesser arterial circle of iris*** a circle of anastomosing arteries in the iris near the pupillary margin. <circulus arteriosus iridis minor> C

23 **medial commissure of eyelids** the junction of the upper and lower eyelids on the medial side; called also *medial palpebral commissure.* <commissura medialis palpebrarum> C

24 ***outer border of iris*** the less coarsely striated outer concentric circle on the anterior surface of the iris; called also *greater circle* or *ring of iris.* <anulus iridis major> C

25 **palpebral fissure** the longitudinal opening between the eyelids. <rima palpebrarum> C

26 ***pigmented epithelium of iris*** the anterior epithelium of the iris, situated just posterior to the stroma, which contains pigment cells. <epithelium pigmentosum iridis> C

27 **plica semilunaris conjunctivae** a fold of mucous membrane at the medial angle of the eye. C

28 **pupil** the opening at the center of the iris of the eye, through which light enters the eye. <pupilla> C

29 **pupillary margin of iris** the inner edge of the iris, surrounding the pupil. <margo pupillaris iridis> C

30 ***pupillary membrane*** that portion of the vascular envelope surrounding the developing lens of the fetus that is in front of the pupil; it is a mesodermal layer attached to the rim or front of the iris during embryonic development, sometimes persisting in the adult. <membrana pupillaris> C

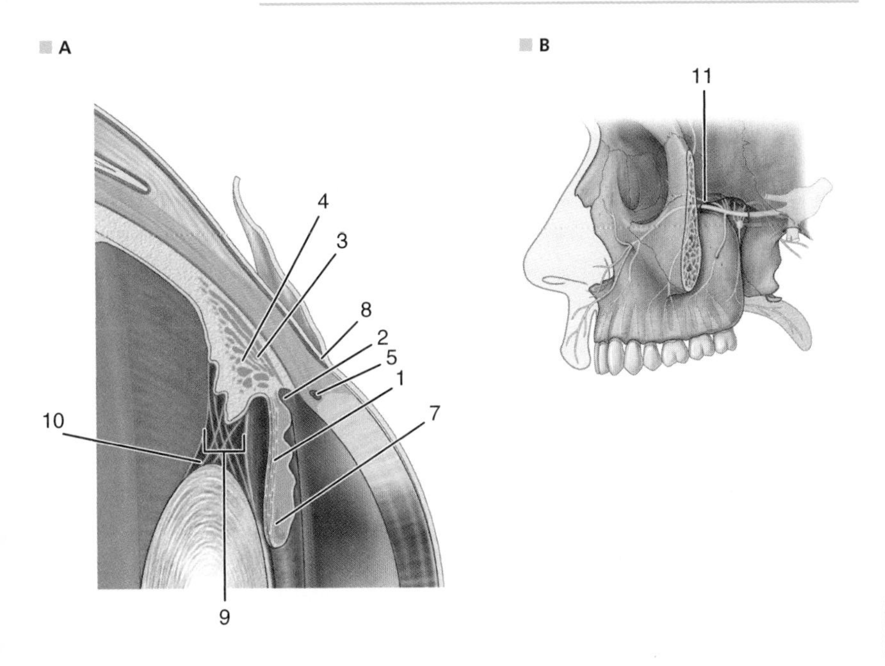
A
4
3
8
2
5
1
10
7
9
B
11

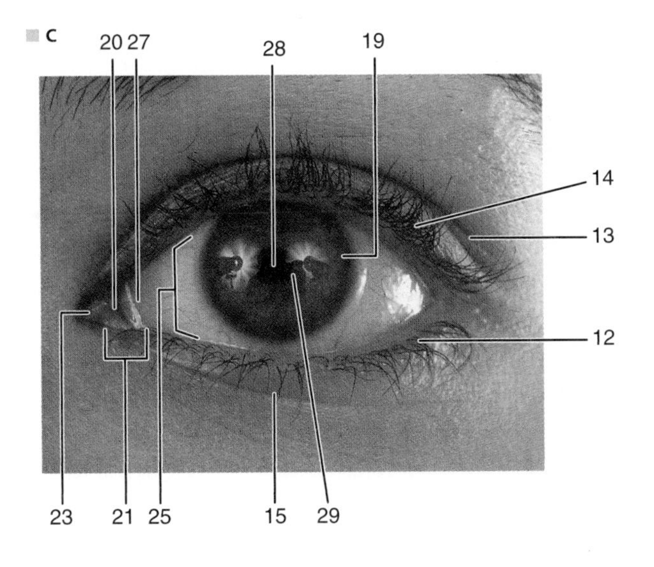
C
20 27
28
19
14
13
12
23
21
25
15
29

1 ***stroma of iris*** the soft mass of connective tissue fibers that make up the major portion of the iris. <stroma iridis> A

2 **upper eyelid** the superior of the paired movable folds that protect the surface of the eyeball. <palpebra superior> A

3 **anterior ethmoidal foramen** the anterior of two small grooves found as a pair crossing the superior surface on both sides of the ethmoid labyrinth, at its junction with the roof of each orbit; it transmits the nasal branch of the ophthalmic nerve and the anterior ethmoid artery and vein. <foramen ethmoidale anterius> B

4 **anterior lacrimal crest** the lateral margin of the groove on the posterior border of the frontal process of the maxilla. <crista lacrimalis anterior> C

5 **floor of orbit** an inner orbital surface formed by surfaces of the maxilla, the zygomatic bone, and the palatine bone. <paries inferior orbitae> B

6 **fossa for lacrimal sac** the fossa that lodges the lacrimal sac, formed by the lacrimal sulcus of the lacrimal bone and the frontal process of the maxilla; called also *lacrimal groove.* <fossa sacci lacrimalis> C

7 **inferior orbital fissure** a cleft in the inferolateral wall of the orbit bounded by the greater wing of the sphenoid and the orbital process of the maxilla; it transmits the infraorbital and zygomatic nerves and the infraorbital vessels. <fissura orbitalis inferior> B

8 **infraorbital canal** a passage beneath the orbital surface of the maxilla, continuous posteriorly with the infraorbital sulcus, and opening anteriorly on the anterior surface of the body of the maxilla in the infraorbital foramen. It contains the infraorbital vessels and nerve. <canalis infraorbitalis> B

9 **infraorbital groove** a groove in the orbital surface of the maxilla, commencing near the middle of the posterior edge of the surface and running anteriorly for a short distance to become continuous with the infraorbital canal. <sulcus infraorbitalis maxillae> B

10 **infraorbital margin of orbit** the inferior edge of the entrance to the orbit, formed by the infraorbital process of the zygomatic bone and the infraorbital margin of the maxilla. <margo infraorbitalis orbitae> B

11 **lacrimal bone** a thin scalelike bone at the anterior part of the medial wall of the orbit, articulating with the frontal and ethmoid bones and the maxilla and inferior nasal concha. <os lacrimale> B

12 **lacrimal fossa** a shallow depression in the lateral part of the roof of the orbit, lodging the lacrimal gland. <fossa glandulae lacrimalis> B

13 **lacrimal groove of lacrimal bone** a deep vertical groove on the anterior part of the lateral surface of the lacrimal bone, which with the maxilla forms the fossa for the lacrimal sac. <sulcus lacrimalis ossis lacrimalis> C

14 **lacrimal groove of maxilla** a groove directed inferiorly and somewhat posteriorly on the nasal surface of the body of the maxilla, just anterior to the large opening into the maxillary sinus; it is converted into the nasolacrimal canal by the lacrimal bone and inferior nasal concha. <sulcus lacrimalis maxillae> C

15 ***lacrimal hamulus*** the hooklike process on the anterior part of the inferolateral border of the lacrimal bone, articulating with the maxilla. <hamulus lacrimalis> B

16 **lacrimal margin of maxilla** the posterior border of the frontal process of the maxilla where it articulates with the lacrimal bone. <margo lacrimalis maxillae> C

17 **lacrimal notch of maxilla** an indentation on the posterior border of the frontal process of the maxilla, which lodges the lacrimal sac. <incisura lacrimalis maxillae> C

18 **lacrimomaxillary suture** a suture on the inner wall of the orbit, between the lacrimal bone and the maxilla. <sutura lacrimomaxillaris> B

19 **lateral margin of orbit** the orbital border formed by the zygomatic process of the frontal bone and the frontal process of the zygomatic bone. <margo lateralis orbitae> B

20 **lateral wall of orbit** an inner orbital surface formed by the orbital surfaces of the great wing of the sphenoid bone, the zygomatic bone, and the zygomatic process of the frontal bone. <paries lateralis orbitae> B

21 **medial margin of orbit** the orbital border formed above by the bone and below by the lacrimal crest of the frontal process of the maxilla. <margo medialis orbitae> B

22 **medial wall of orbit** an inner orbital surface formed by parts of the maxillary, lacrimal, ethmoid, and sphenoid bones. <paries medialis orbitae> B

23 **nasolacrimal canal** a canal formed by the lacrimal sulcus of the maxilla, lacrimal bone, and inferior nasal concha; it contains the nasolacrimal duct. Called also *lacrimal canal* and *nasal canal.* <canalis nasolacrimalis> B

24 ***orbitalis muscle*** a thin layer of nonstriated muscle that bridges the inferior orbital fissure; *innervation,* sympathetic branches. <musculus orbitalis> B

25 **posterior ethmoidal foramen** the posterior of two small grooves found as a pair crossing the superior surface on both sides of the ethmoid labyrinth, at its junction with the roof of each orbit; it transmits the posterior ethmoid artery and vein. <foramen ethmoidale posterius> B

26 **roof of orbit** an inner orbital surface formed chiefly by the orbital plate of the frontal bone and the orbital surface of the lesser wing of the sphenoid bone; called also *superior wall of orbit.* <paries superior orbitae> B

A

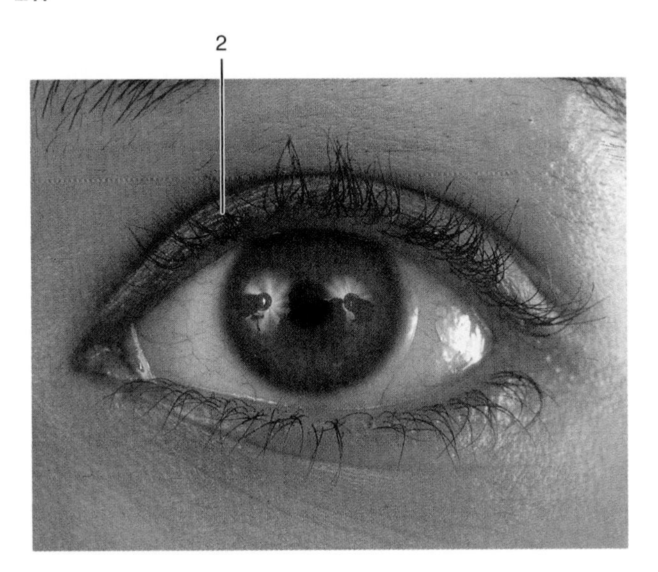

B

C

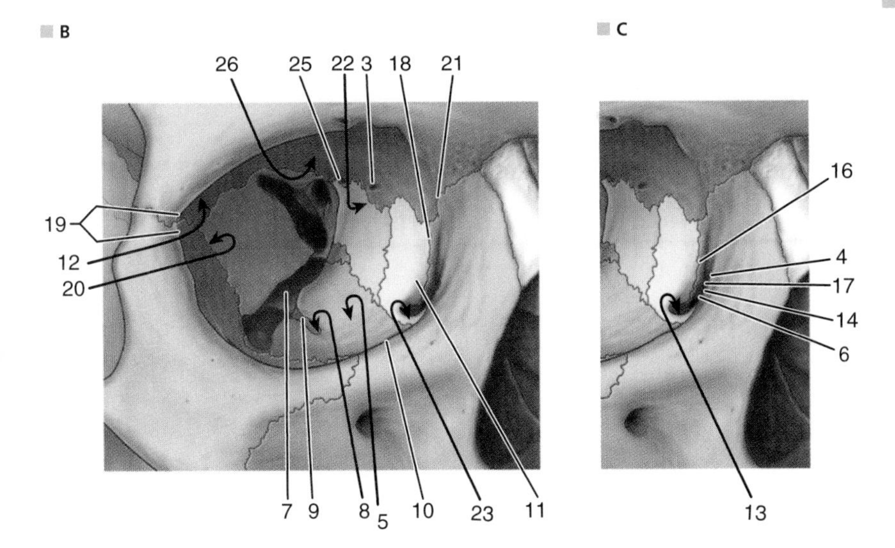

1 **anterior surface of cornea** the surface of the cornea directed toward the anterior surface of the eye. <facies anterior corneae> A

2 **bulbar conjunctiva** the portion of the conjunctiva covering the cornea and front part of the sclera, appearing white because of the sclera behind it. <tunica conjunctiva bulbi> A

3 ***ciliary glands*** sweat glands that have become arrested in their development, situated obliquely in contact with and parallel to the bulbs of the eyelashes; called also *Moll's glands.* <glandulae ciliares> A

4 **conjunctiva** the delicate membrane that lines the eyelids and covers the exposed surface of the sclera, divided into the *palpebral conjunctiva* and the *bulbar* or *ocular conjunctiva.*. <tunica conjunctiva> A

5 ***conjunctival glands*** accessory lacrimal glands situated deep in the subconjunctival connective tissue, mainly in the upper fornix; called also *Krause's glands.* <glandulae conjunctivales> A

6 ***conjunctival ring*** a ring at the junction of the conjunctiva and cornea. <anulus conjunctivae> A

7 **conjunctival sac** the potential space, lined by conjunctiva, between the eyelids and the eyeball. <saccus conjunctivalis> A

8 **orbital septum** a fibrous membrane anchored to the periorbita along the entire margin of the orbit, extending to the levator palpebrae superioris muscle in the upper lid and to the tarsal plate in the lower lid; called also *tarsal membrane.* <septum orbitale> A

9 **palpebral conjunctiva** the portion of the conjunctiva lining the eyelids, appearing red because of its great vascularity. <tunica conjunctiva palpebrarum> A

10 **posterior palpebral margin** the sharp posterior edges of the free margin of the eyelids, closely applied to the eyeball. <limbus posterior palpebrae> A

11 **posterior surface of eyelid** the inner surface of the eyelid, which is covered with conjunctiva and in contact with the eyeball. <facies posterior palpebrae> A

12 **sebaceous glands of eyelids** modified rudimentary sebaceous glands attached directly to the follicles of the eyelashes; called also *glands of Zeis.* <glandulae sebaceae palpebrarum> A

13 **superior tarsal muscle** *origin,* levator palpebrae superioris muscle; *insertion,* tarsal plate of upper eyelid; *innervation,* sympathetic; *action,* widens palpebral fissure. <musculus tarsalis superior> A

14 **superior tarsus** the firm framework of connective tissue that gives shape to the upper eyelids. <tarsus superior palpebrae> A

15 **tarsal glands** sebaceous follicles between the tarsi and the conjunctiva of the eyelids; called also *palpebral* or *meibomian glands.* <glandulae tarsales> A

16 **lateral palpebral ligament** a ligament that anchors the lateral end of the superior and inferior tarsal plates to the margin of the orbit. <ligamentum palpebrale laterale> B

17 **medial palpebral ligament** fibrous bands that connect the medial ends of the tarsi to the bones of the orbit, an anterior bundle passing in front of the lacrimal sac and being attached to the frontal process of the maxilla, and a posterior bundle passing behind the lacrimal sac and being attached to the posterior crest of the lacrimal bone. <ligamentum palpebrale mediale> B

18 **inferior palpebral arch** an arch derived from the inferior medial palpebral artery, supplying the lower lid of the eye. <arcus palpebralis inferior> C

19 **lateral palpebral arteries** *origin,* lacrimal artery; *branches,* none; *distribution,* eyelids, conjunctiva. <arteriae palpebrales laterales> C

20 **medial palpebral arteries** two arteries, superior and inferior medial palpebral arteries: *origin,* ophthalmic artery; *branches,* posterior conjunctival arteries, superior and inferior palpebral arches; *distribution,* eyelids. <arteriae palpebrales mediales> C

21 ***posterior conjunctival arteries*** *origin,* medial palpebral arteries; *branches,* none; *distribution,* lacrimal caruncle, conjunctiva. <arteriae conjunctivales posteriores> C

22 **superior palpebral arch** an arch derived from the superior medial palpebral artery, supplying the upper lid of the eye. <arcus palpebralis superior> C

23 ***superior palpebral veins*** branches that drain the blood from the upper eyelid to the angular vein. <venae palpebrales superiores> C

24 **supratrochlear veins** two veins, each beginning in a venous plexus high up on the forehead and descending to the root of the nose, where it joins with the supraorbital vein to form the angular vein. Called also *frontal veins.* <venae supratrochleares> C

25 ***palpebral branches of infratrochlear nerve*** branches that help supply the eyelids; *modality,* general sensory. <rami palpebrales nervi infratrochlearis> D

26 ***accessory lacrimal glands*** portions of the lacrimal gland sometimes found near the superior fornix of the conjunctiva. <glandulae lacrimales accessoriae> E

27 **excretory ductules of lacrimal gland** numerous ductules that traverse the palpebral part of the lacrimal gland and open into the superior fornix of the conjunctiva. <ductuli excretorii glandulae lacrimalis> E

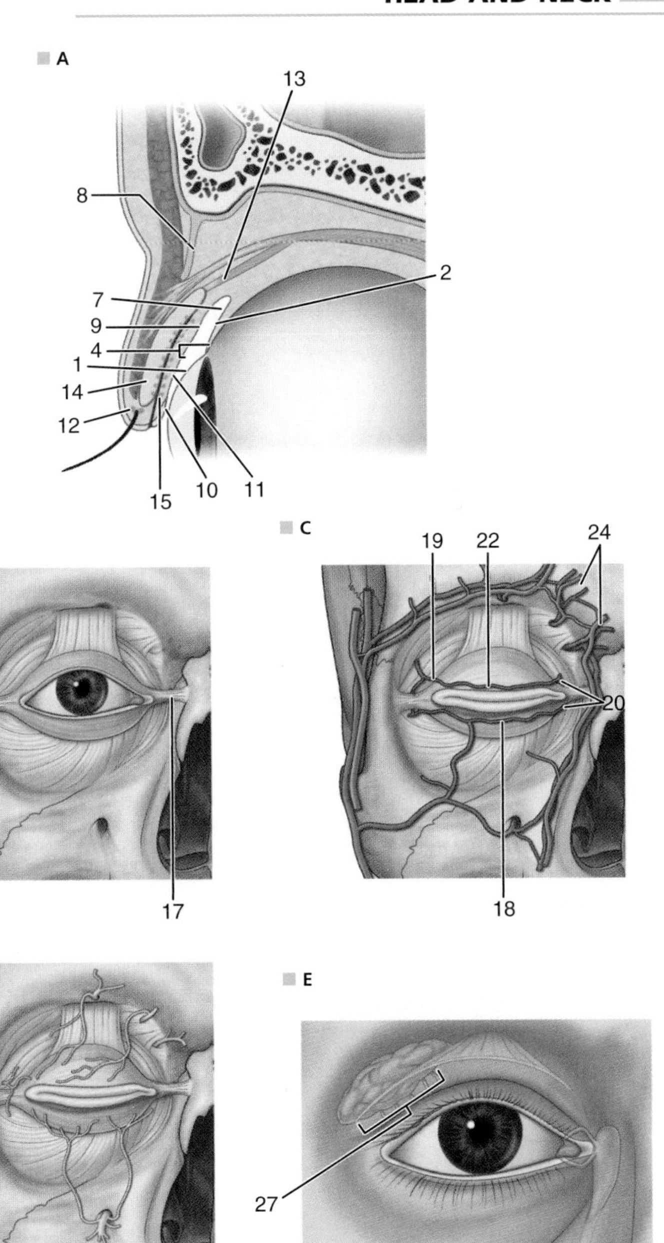
A
13
8
2
7
9
4
1
14
12
15
10
11
B
16
17
C
19
22
24
20
18
D
25
E
27

1 ***lacrimal apparatus*** the system concerned with the secretion and circulation of the tears and the normal fluid of the conjunctival sac; it consists of the lacrimal gland and ducts, and associated structures. <apparatus lacrimalis> A

2 **lacrimal canaliculus** the short passage in an eyelid, beginning at the lacrimal punctum, that leads from the lacrimal lake to the lacrimal sac. <canaliculus lacrimalis> A

3 ***lacrimal fold*** a fold of mucous membrane at the lower opening of the nasolacrimal duct. <plica lacrimalis> A

4 **lacrimal gland** either of a pair of glands, one at the upper outer angle of each orbit, secreting the tears; they are divided into two portions, orbital and palpebral, by the orbital fascia. <glandula lacrimalis> A

5 ***lacrimal papilla*** a papilla in the conjunctiva near the medial angle of the eye. <papilla lacrimalis> A

6 ***lacrimal pathway*** the pathway by which the tears reach the lacrimal lake from the excretory ductules of the lacrimal gland. <rivus lacrimalis> A

7 **lacrimal punctum** the opening on the lacrimal papilla of an eyelid, near the medial angle of the eye, into which tears from the lacrimal lake drain to enter the lacrimal canaliculi. <punctum lacrimale> A

8 **lacrimal sac** the dilated upper end of the nasolacrimal duct. <saccus lacrimalis> A

9 **nasolacrimal duct** the passage that conveys the tears from the lacrimal sac into the inferior nasal meatus; called also *lacrimonasal* or *nasal duct.* <ductus nasolacrimalis> A

10 **orbital part of lacrimal gland** the main part of the lacrimal gland, limited in front by the orbicularis muscle and the orbital septum. <pars orbitalis glandulae lacrimalis> A

11 **palpebral part of lacrimal gland** the part of the gland that projects laterally into the upper eyelid; called also *Rosenmüller's gland* or *node.* <pars palpebralis glandulae lacrimalis> A

12 **eyelid** either of the two movable folds that protect the anterior surface of the eyeball. <palpebra> A

13 **fornix of lacrimal sac** the upper, blind extremity of the lacrimal sac. <fornix sacci lacrimalis> A

14 **fascial sheath of eyeball** connective tissue that forms the capsule enclosing the posterior part of the eyeball, extending anteriorly to the conjunctival fornix, and continuous with the muscular fascia of the eye; called also *bulbar fascia* and *Tenon's capsule.* <vagina bulbi> B

15 **inferior conjunctival fornix** the inferior line of reflection of the conjunctiva from the eyelid to the eyeball. <fornix conjunctivae inferior> B

16 **inferior tarsus** the firm framework of connective tissue that gives shape to the lower eyelids. <tarsus inferior palpebrae> B

17 **muscular fasciae of eyeball** the sheets of fascia investing the extraocular muscles, continuous with the fascial sheath of eyeball. <fasciae musculares bulbi> B

18 **orbital fat body** a mass of fatty tissue in the posterior part of the orbit, around the optic nerve, extraocular muscles, and vessels. Called also *retrobulbar fat.* <corpus adiposum orbitae> B

19 **periorbita** the periosteal covering of the bones forming the orbit, or eye socket. B

20 **superior conjunctival fornix** the superior line of reflection of the conjunctiva from the eyelid to the eyeball; it receives the openings of the lacrimal duct. <fornix conjunctivae superior> B

21 **check ligament of lateral rectus muscle** an extension from the fascia covering the lateral rectus muscle, attaching to the orbital tubercle of the zygomatic bone and limiting the action of the muscle. <lacertus musculi recti lateralis bulbi> C

22 **check ligament of medial rectus muscle** an extension from the fascia covering the medial rectus muscle, attaching to the anterior medial wall of the orbit and limiting the action of the muscle. <lacertus musculi recti lateralis bulbi> C

23 **extraocular muscles** the six voluntary muscles that move the eyeball, including the superior, inferior, middle, and lateral recti, and the superior and inferior oblique muscles. Called also *muscles of eye* and *ocular muscles.* <musculi externi bulbi oculi> D

24 **lateral rectus muscle** *origin,* common tendinous ring; *insertion,* lateral side of sclera; *innervation,* abducent; *action,* abducts eyeball. <musculus rectus lateralis bulbi> D

25 **medial rectus muscle** *origin,* common tendinous ring; *insertion,* medial side of sclera; *innervation,* oculomotor; *action,* adducts eyeball. <musculus rectus medialis bulbi> D

26 **superior oblique muscle of eyeball** *origin,* lesser wing of sphenoid above optic canal; *insertion,* sclera; *innervation,* trochlear; *action,* rotates eyeball downward and outward. <musculus obliquus superior bulbi> D

27 **superior rectus muscle** *origin,* common tendinous ring; *insertion,* upper aspect of sclera; *innervation,* oculomotor; *action,* adducts, rotates eyeball upward and medially. <musculus rectus superior bulbi> D

28 ***tendon sheath of superior oblique muscle*** the synovial sheath of the superior oblique muscle, particularly where its tendon passes through the trochlea; called also *trochlear synovial bursa.* <vagina tendinis musculi obliqui superioris> D

29 **trochlea of superior oblique muscle** the fibrocartilaginous pulley near the internal angular process of the frontal bone, through which the tendon of the superior oblique muscle of the eyeball passes. <trochlea musculi obliqui superioris bulbi> D

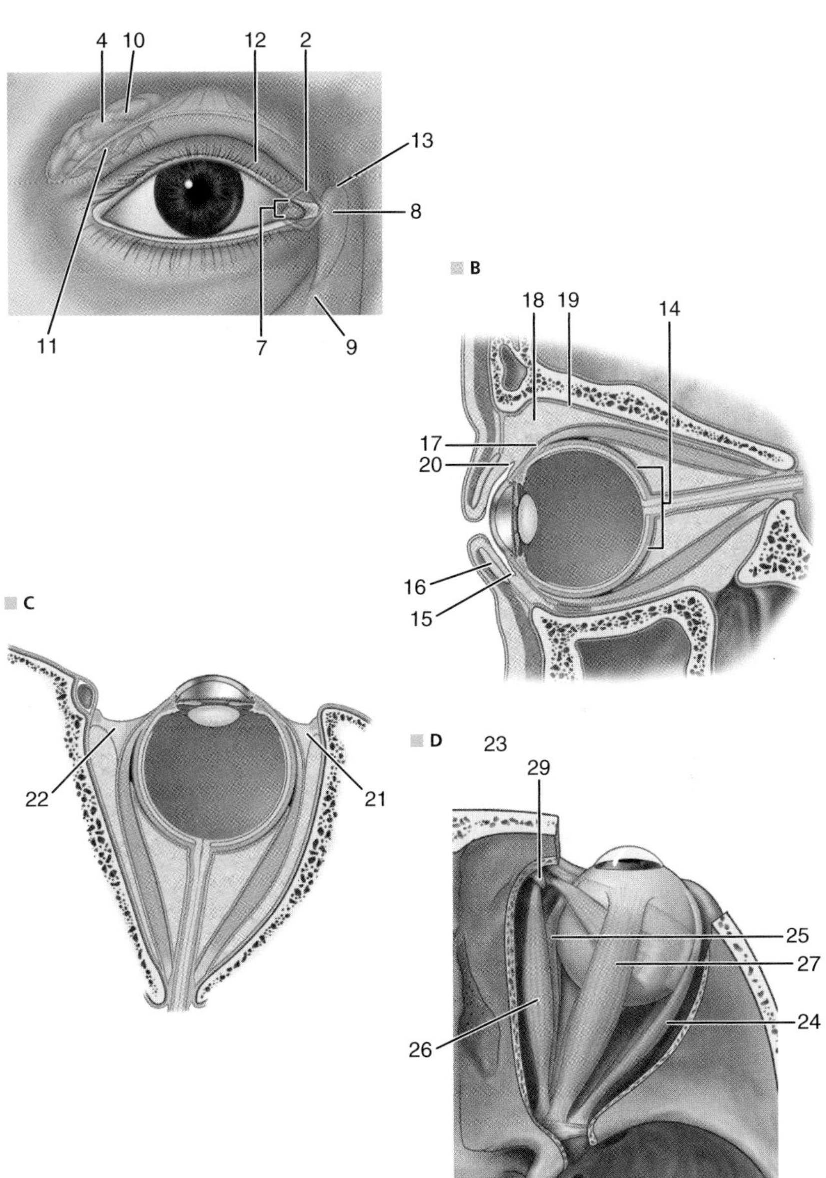
A
4
10
12
2
13
8
11
7
9
B
18
19
14
17
20
16
15
C
22
21
D
23
29
25
27
24
26

1 ***deep layer of levator palpebrae superioris*** the deeper of the two layers of the levator palpebrae superioris muscle, the fibers of which are attached to the superior tarsus. <lamina profunda musculi levatoris palpebrae superioris> A

2 **inferior oblique muscle of eyeball** *origin,* orbital surface of maxilla; *insertion,* sclera; *innervation,* oculomotor; *action,* rotates eyeball upward and outward. <musculus obliquus inferior bulbi> A

3 **inferior rectus muscle** *origin,* annulus tendineus communis; *insertion,* underside of sclera; *innervation,* oculomotor; *action,* adducts, rotates eyeball downward and medially. <musculus rectus inferior bulbi> A

4 ***inferior tarsal muscle*** *origin,* inferior rectus muscle; *insertion,* tarsal plate of lower eyelid; *innervation,* sympathetic; *action,* widens palpebral fissure. <musculus tarsalis inferior> A

5 **levator palpebrae superioris muscle** *origin,* sphenoid bone above optic canal; *insertion,* tarsal plate and skin of upper eyelid; *innervation,* oculomotor; *action,* raises upper lid. <musculus levator palpebrae superioris> A

6 ***superficial layer of levator palpebrae superioris*** the superficial of the two layers of the levator palpebrae superioris muscle. <lamina superficialis musculi levatoris palpebrae superioris> A

7 **common tendinous ring** the annular ligament of origin common to the recti muscles of the eye, attached to the edge of the optic canal and the inner part of the superior orbital fissure; called also *annulus of Zinn.* <anulus tendineus communis> B

8 **epicanthus** a vertical fold of skin on either side of the nose, sometimes covering the inner canthus. It is present as a normal characteristic in persons of certain races and as a congenital anomaly in others. Called also *epicanthal* or *palpebronasal fold.* <plica palpebronasalis> C

Ear

9 **auricular region** the surface region of the head about the ear. <regio auricularis> D

10 **tympanic part of temporal bone** the curved bony plate, developed from the tympanic ring of the fetus, forming the anterior and inferior walls and part of the posterior wall of the external auditory meatus in the adult; called also *tympanic plate.* <pars tympanica ossis temporalis> E

11 **cartilage of acoustic meatus** the trough-shaped cartilage of the cartilaginous part of the external acoustic meatus; called also *meatal cartilage.* <cartilago meatus acustici> F

12 **cartilage of auditory tube** the cartilage on the inferomedial surface of the temporal bone that supports the walls of the cartilaginous part of the auditory tube; called also *tubal* or *eustachian cartilage.* <cartilago tubae auditivae> F

13 **cartilaginous external acoustic meatus** the cartilaginous part of the external acoustic meatus, found lateral to the bony part. <meatus acusticus externus cartilagineus> F

14 **cartilaginous part of auditory tube** the part of the tube that is chiefly supported by the tubal cartilage, extending from the bony part to the pharyngeal opening of the auditory tube. <pars cartilaginea tubae auditivae> F

15 **ear** the organ of hearing and of equilibrium, consisting of the external ear, middle ear, and inner ear. <auris> D

16 **external acoustic meatus** the passage of the external ear leading to the tympanic membrane, divided into an outer cartilaginous meatus and an inner bony meatus. Called also *external auditory canal.* <meatus acusticus externus> F

17 **external acoustic pore** the outer end of the external acoustic meatus. <porus acusticus externus> F

18 **external ear** the portion of the auditory organ comprising the auricle and the external acoustic meatus. <auris externa> F

19 **internal ear** the labyrinth, comprising the vestibule, cochlea, and semicircular canals; called also *inner ear.* <auris interna> F

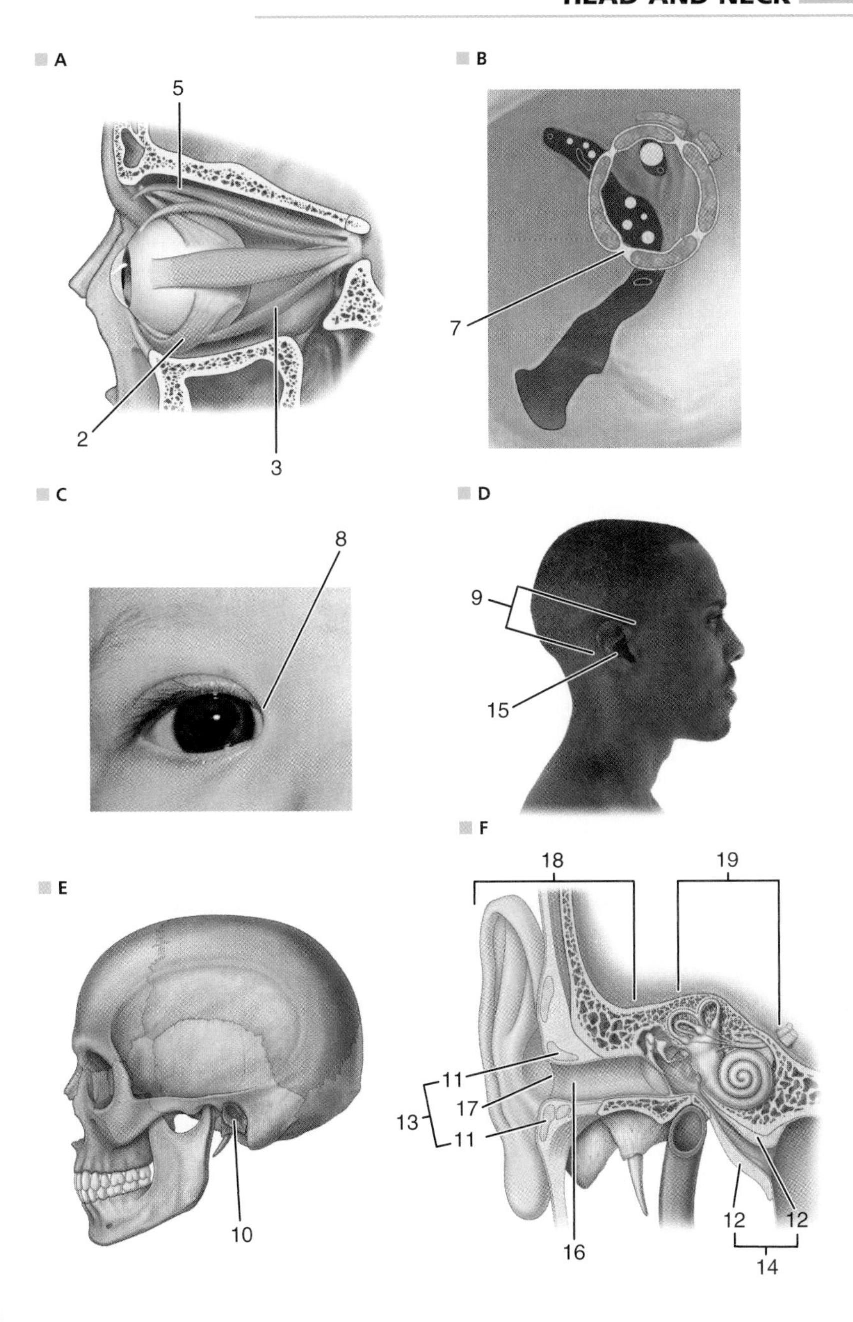
A
5
2
3
B
7
C
8
D
9
15
E
10
F
18
19
11
13
17
11
16
12
12
14

1 **air cells of auditory tube** air cells in the floor of the auditory tube close to the carotid canal, being similar to the air cells of the mastoid part of the temporal bone; called also *tubal air cells.* <cellulae pneumaticae tubae auditivae> A

2 **auditory ossicles** the malleus, incus, and stapes, the small bones of the middle ear, which transmit vibrations from the tympanic membrane to the oval window. <ossicula auditus> A

3 **auditory tube** a channel about 3.6 cm long, lined with mucous membrane, that establishes communication between the tympanic cavity and the nasopharynx and serves to adjust the pressure of gas in the cavity to the external pressure, as well as for mucociliary clearance of the middle ear. It comprises an *bony part* and a *cartilaginous part.* Called also *eustachian canal* or *tube* and *pharyngotympanic tube.* <tuba auditiva> A

4 **bony part of auditory tube** the part of the tube that lies within the temporal bone, extending from the tympanic orifice to the cartilaginous part of the auditory tube. <pars ossea tubae auditivae> A

5 ***carotid wall of tympanic cavity*** the anterior wall of the cavity, related to the carotid canal, in which is lodged the internal carotid artery. <paries caroticus cavitatis tympani> A

6 **cupular space** the part of the epitympanic recess above the head of the malleus. <pars cupularis recessus epitympanici> A

7 **epitympanic recess** the portion of the tympanic cavity above the level of the tympanic membrane, containing the greater part of the incus and the upper half of the malleus. Called also *attic.* <recessus epitympanicus> A

8 **fossa of incus** a groove in the posterior wall of the tympanic cavity, lodging the short limb of the incus and the posterior ligament of the incus. <fossa incudis> A

9 **fossa of oval window** a depression on the medial wall of the tympanic cavity, at the bottom of which is the oval window. <fossula fenestrae vestibuli> A

10 **fossa of round window** a depression on the medial wall of the tympanic cavity, at the bottom of which is the cochlear window. <fossula fenestrae cochleae> A

11 ***greater tympanic spine*** a spine of the temporal bone forming the anterior edge of the tympanic notch. <spina tympanica major> A

12 **incudomalleolar joint** the junction of the incus and the malleus. <articulatio incudomallearis> A

13 **incudostapedial joint** the junction of the incus and the stapes. <articulatio incudostapedialis> A

14 **incus** the middle of the three ossicles of the ear, which, with the stapes and malleus, serves to conduct vibrations from the tympanic membrane to the inner ear. Called also *anvil.* A

15 **isthmus of auditory tube** the narrowest part of the auditory tube, at the junction of the bony and cartilaginous parts of the tube. <isthmus tubae auditivae> A

16 ***joints of auditory ossicles*** junctions between two of the auditory ossicles, including the *incudomalleolar* and *incudostapedial joints.* <articulationes ossiculorum auditus> A

17 **jugular wall of tympanic cavity** the floor of the tympanic cavity, which is in intimate relation with the jugular fossa, which lodges the bulb of the internal jugular vein. <paries jugularis cavitatis tympani> A

18 ***lateral lamina of cartilage of auditory tube*** the smaller of the two laminae that compose the tubal cartilage; it lies in the lateral wall of the auditory tube. <lamina lateralis cartilaginis tubae auditivae> A

19 ***lesser tympanic spine*** a spine of the temporal bone forming the posterior edge of the tympanic notch. <spina tympanica minor> A

20 **malleus** the outermost of the auditory ossicles, and the one attached to the tympanic membrane; its club-shaped head articulates with the incus. Called also *hammer.* A

21 ***medial lamina of cartilage of auditory tube*** the larger of the two laminae that compose the tubal cartilage; it lies in the medial wall of the auditory tube. <lamina medialis cartilaginis tubae auditivae> A

22 ***membranous lamina of auditory tube*** the connective tissue lamina that supports the medial and lateral parts of the auditory tube. <lamina membranacea tubae auditivae> A

23 **membranous wall of tympanic cavity** the outer, or lateral, wall of the tympanic cavity, formed mainly by the tympanic membrane. <paries membranaceus cavitatis tympani> A

24 **middle ear** the cavity in the temporal bone comprising the tympanic cavity, auditory ossicles, and auditory tube. <auris media> A

25 ***mucosa of auditory tube*** the mucous membrane (tunica mucosa) lining the auditory tube. <tunica mucosa tubae auditivae> A

26 ***mucosa of tympanic cavity*** the mucous membrane (tunica mucosa) covering the walls and much of the contents of the tympanic cavity. <tunica mucosa cavitatis tympanicae> A

27 **oval window** an oval opening in the inner ear, which is closed by the base of the stapes. <fenestra vestibuli> A

28 **round window** a round opening in the inner wall of the middle ear inferior to and a little posterior to the oval window; it is covered by the secondary tympanic membrane. <fenestra cochleae> A

29 **stapes** the innermost of the auditory ossicles, shaped somewhat like the stirrup used in horse riding; it articulates by its head with the incus, and its base is inserted into the oval window. Called also *stirrup.* A

A

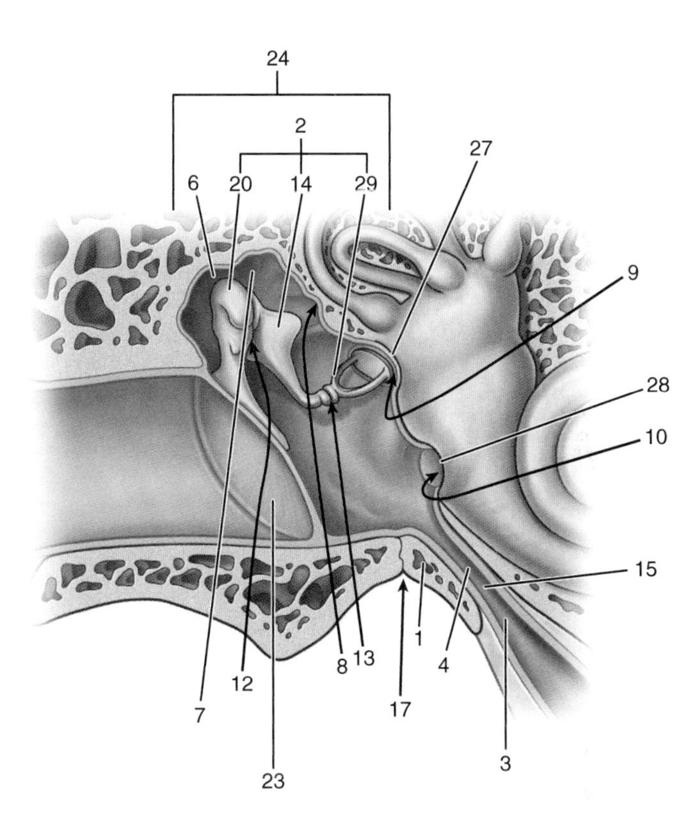

1 ***styloid prominence*** an irregular nodule on the posterior portion of the floor of the tympanic cavity, corresponding to the base of the styloid process. <prominentia styloidea> A

2 ***superior tympanic artery*** *origin,* middle meningeal artery; *branches,* none; *distribution,* tympanic cavity. <arteria tympanica superior> A

3 **tegmental wall of tympanic cavity** the upper surface of the cavity, formed by part of the petrous portion of the temporal bone. Called also *roof of tympanic cavity.* <paries tegmentalis cavitatis tympani> A

4 ***tubal glands*** mucous glands within the mucosa of the auditory tube, especially near its nasopharyngeal end; called also *mucous glands of auditory* or *eustachian tube.* <glandulae tubariae> A

5 **tympanic cavity** the major portion of the middle ear, consisting of a narrow air-filled cavity in the temporal bone that contains the auditory ossicles. It communicates with the mastoid air cells and the mastoid antrum via the aditus and with the nasopharynx via the auditory tube. Called also *tympanum.* <cavitas tympani> A

6 ***tympanic cells*** spaces in the tympanic cavity between the bony projections from the floor, or jugular wall; they sometimes communicate with the tubal air cells. Called also *tympanic air cells.* <cellulae tympanicae> A

7 **tympanic membrane** the obliquely placed, thin membranous partition between the external acoustic meatus and the tympanic cavity. The greater portion, the pars tensa, is attached by a fibrocartilaginous ring to the tympanic plate of the temporal bone; the much smaller, triangular portion, the pars flaccida, is situated anterosuperiorly between the two malleolar folds. Called also *drum, eardrum,* and *tympanum.* <membrana tympanica> A

8 ***tympanic notch*** a defect in the upper portion of the tympanic part of the temporal bone, between the greater and lesser tympanic spines, which is filled in by the pars flaccida of the tympanic membrane. Called also *Rivinus' notch.* <incisura tympanica> A

9 **tympanic opening of auditory tube** the opening of the auditory tube on the carotid wall of the tympanic cavity. <ostium tympanicum tubae auditivae> A

10 ***tympanic sulcus*** a narrow groove in the medial part of the external acoustic meatus of the temporal bone, into which the tympanic membrane fits; it is deficient above. <sulcus tympanicus ossis temporalis> A

11 ***tympanic veins*** small veins from the tympanic cavity that pass through the petrotympanic fissure, open into the plexus around the temporomandibular joint, and finally drain into the retromandibular vein. <venae tympanicae> A

12 **tympanostapedial syndesmosis** the connection of the base of the stapes with the secondary membrane in the oval window. <syndesmosis tympanostapedialis> A

13 **anterior malleolar fold of tympanic membrane** the line in the tympanic membrane that extends anteriorly from the malleolar prominence and separates the pars tensa from the pars flaccida. <plica mallearis anterior membranae tympanicae> B

14 **anterior recess of tympanic membrane** a pocket in the tympanic membrane formed by the tunica mucosa between the anterior malleolar fold and the anterior superior part of the pars tensa of the membrane, ending blindly above. <recessus anterior membranae tympanicae> B

15 **fibrocartilaginous ring of tympanic membrane** the margin of the pars tensa of the tympanic membrane, which attaches to the tympanicu sulcus. <anulus fibrocartilagineus membranae tympani> B

16 **malleolar prominence of tympanic membrane** a small projection at the upper extremity of the malleolar stria, formed by the lateral process of the malleus. Called also *short process of malleus.* <prominentia mallearis membranae tympanicae> B

17 **malleolar stria of tympanic membrane** a nearly vertical radial band seen on the outer surface of the tympanic membrane; it extends from the umbo upward to the malleolar prominence and is a bulge caused by the manubrium of malleus. <stria mallearis membranae tympanicae> B

18 **pars flaccida of tympanic membrane** a small portion of the tympanic membrane between the anterior and posterior malleolar folds; it is lax and thin. Called also *Shrapnell's membrane.* <pars flaccida membranae tympanicae> B

19 **pars tensa of tympanic membrane** the larger portion of the tympanic membrane; it is tense and firm. <pars tensa membranae tympanicae> B

20 **posterior malleolar fold of tympanic membrane** the line in the tympanic membrane that extends posteriorly from the malleolar prominence and marks the line of separation of the pars tensa and pars flaccida. <plica mallearis posterior membranae tympanicae> B

21 **posterior recess of tympanic membrane** a pocket in the tympanic membrane formed by the tunica mucosa between the posterior malleolar fold and the posterior superior part of the pars tensa of the membrane, ending blindly above. <recessus posterior membranae tympanicae> B

22 **superior recess of tympanic membrane** a recess in the tympanic membrane formed by the tunica mucosa between the neck of the malleus and the pars flaccida of the membrane, and ending blindly below. <recessus superior membranae tympanicae> B

23 **umbo of tympanic membrane** the slight projection at the center of the outer surface of the tympanic membrane, corresponding to the point of attachment of the tip of the manubrium of malleus. <umbo membranae tympanicae> B

A

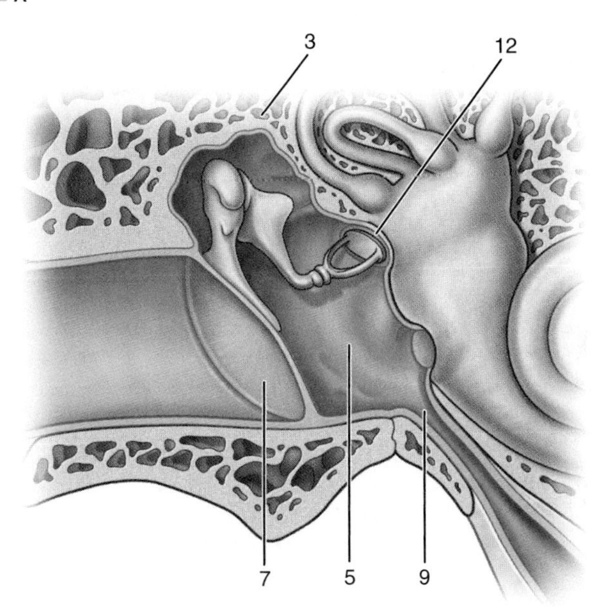

B

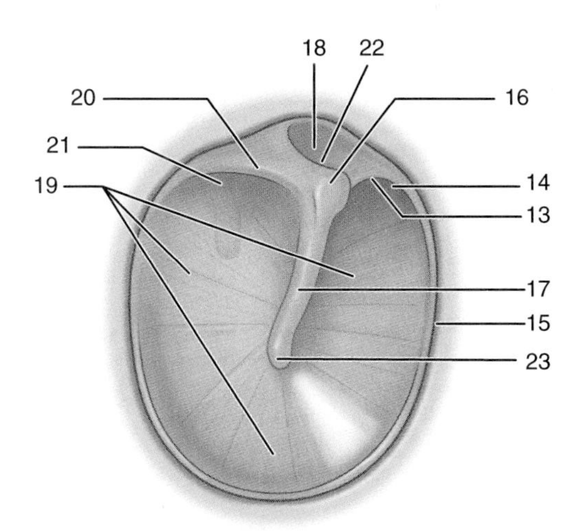

1 **aditus to mastoid antrum** an opening between the epitympanic recess and the mastoid antrum. <aditus ad antrum mastoideum> A

2 **mastoid antrum** an air space in the mastoid portion of the temporal bone, communicating with the tympanic cavity and the mastoid cells. <antrum mastoideum> A

3 **mastoid cells** the air cells in the mastoid process of the temporal bone. <cellulae mastoideae> A

4 **mastoid wall of tympanic cavity** the posterior wall of the cavity, related to the mastoid portion of the temporal bone. <paries mastoideus cavitatis tympani> A

5 **anterior process of malleus** a slender bony process that arises from the anterior aspect of the neck of the malleus, passes anteriorly and inferiorly to the petrotympanic fissure, and is attached to the petrous portion of the temporal bone by ligamentous fibers. Called also *process of Folius* or *of Rau* and *long process of malleus.* <processus anterior mallei> B

6 **head of malleus** the upper portion of the malleus, which includes the surface *(facet for incus)* that articulates with the incus; called also *capitulum of malleus.* <caput mallei> B

7 **lateral process of malleus** a small tapered process that projects laterally from the base of the manubrium of malleus and produces the malleolar prominence. <processus lateralis mallei> B

8 **manubrium of malleus** the largest process of the malleus; it is attached to the middle layer of the tympanic membrane and has the tendon of the tensor tympani muscle attached to it. Called also *handle of malleus.* <manubrium mallei> B

9 **neck of malleus** the constricted portion of the malleus below its head. <collum mallei> B

10 **body of incus** the central part of the incus, which contains an excavation (facet for malleus) in which the head of the malleus articulates. <corpus incudis> C

11 **lenticular process of incus** a small process on the medial side of the tip of the long limb of the incus, which articulates with the head of the stapes. <processus lenticularis incudis> C

12 **long limb of incus** a process on the incus directed downward and inward, parallel with the manubrium of malleus. <crus longum incudis> C

13 **short limb of incus** the backward-projecting process on the incus that is connected to the posterior wall of the tympanic cavity by the posterior incudal ligament. <crus breve incudis> C

14 **anterior crus of stapes** the anterior of the two bony limbs that connect the footplate and head of the stapes; called also *anterior limb of stapes.* <crus anterius stapedis> F

15 **base of stapes** the flat oval plate of bone on the stapes that fits into the oval window on the medial wall of the middle ear. Called also *footplate* and *stapedial footplate.* <basis stapedis> F

16 **head of stapes** the part that articulates with the incus; called also *capitulum of stapes.* <caput stapedis> F

17 **posterior crus of stapes** the posterior of the two bony limbs that connect the footplate and head of the stapes; called also *posterior limb of stapes.* <crus posterius stapedis> F

18 **muscles of auditory ossicles** the two muscles of the middle ear, the tensor tympani and the stapedius. <musculi ossiculorum auditoriorum> E

19 **musculotubal canal** the combined semicanals of the auditory tube and the tensor tympani muscle in the temporal bone. <canalis musculotubarius> E

20 **posterior sinus of tympanic cavity** a groove in the posterior wall of the tympanic cavity inferior to the pyramidal eminence. <sinus posterior cavitatis tympanicae> E

21 **tympanic aperture of canaliculus for chorda tympani** the opening in the posterior part of the middle ear through which the chorda tympani nerve enters the tympanic cavity. <apertura tympanica canaliculi chordae tympani> E

22 **caroticotympanic canaliculi** tiny passages in the temporal bone interconnecting the carotid canal and the tympanic cavity, and carrying communicating twigs between the internal carotid and tympanic nerve plexuses; called also *caroticotympanic foramina.* <canaliculi caroticotympanici> D

23 ***caroticotympanic nerves*** *origin,* internal carotid plexus; inferior and superior nerves can be distinguished; together with tympanic nerve, they form the tympanic plexus; *distribution,* tympanic region and parotid gland; *modality,* sympathetic. <nervi caroticotympanici> D

24 **groove of promontory of tympanic cavity** a groove in the surface of the promontory of the tympanic cavity, lodging the tympanic nerve. <sulcus promontorii cavitatis tympani> D

25 **labyrinthine wall of tympanic cavity** the wall of the cavity facing medially toward the inner ear; it contains the round window, the oval window, and the promontory. Called also *medial wall of tympanic cavity.* <paries labyrinthicus cavitatis tympani> D

26 **lesser petrosal nerve** *origin,* tympanic plexus; *distribution,* parotid gland via otic ganglion and auriculotemporal nerve; *modality,* parasympathetic. <nervus petrosus minor> D

27 **promontory of tympanic cavity** the prominence on the medial wall of the tympanic cavity, formed by the first turn of the cochlea. <promontorium tympani> D

28 ***tubal branch of tympanic plexus*** a branch given to the auditory tube from the tympanic plexus; *modality,* general sensory. <ramus tubarius plexus tympanici> D

A

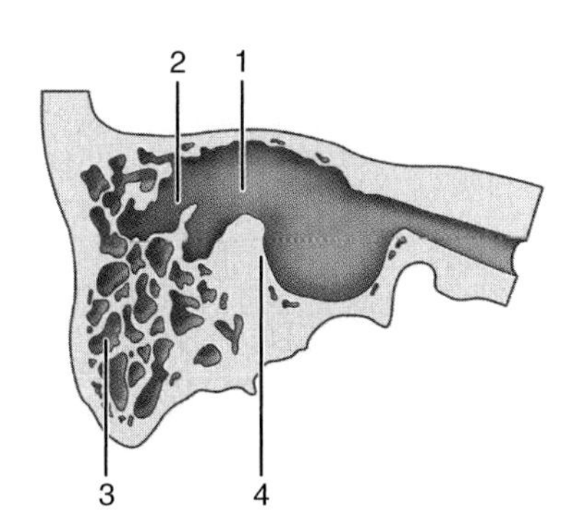

B

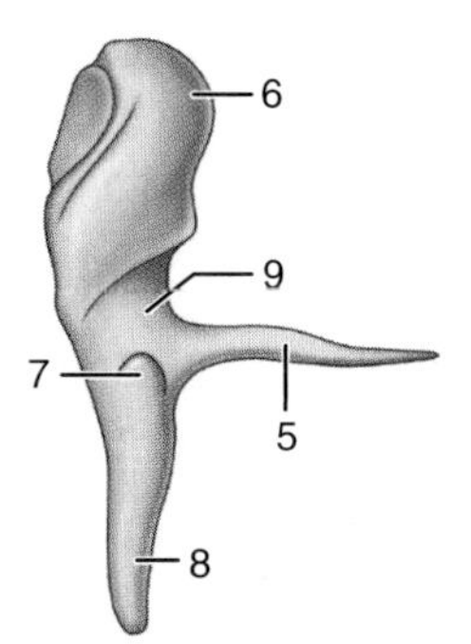

C

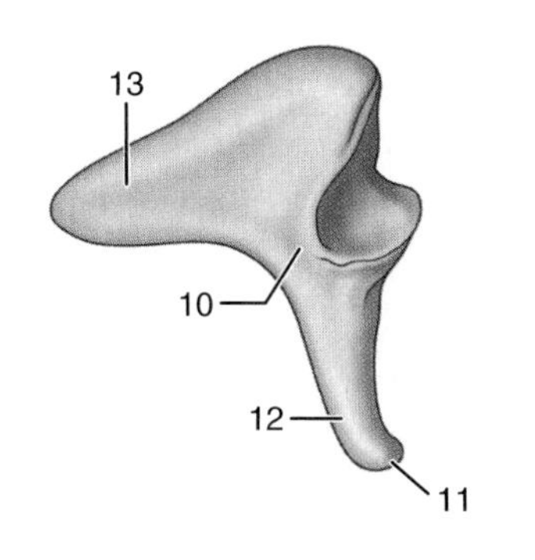

D

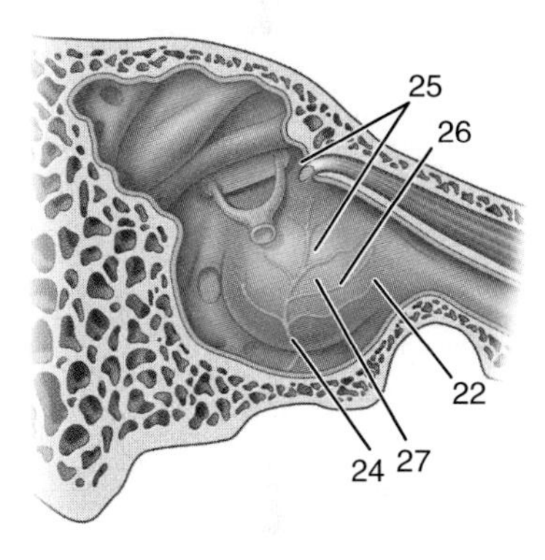

E

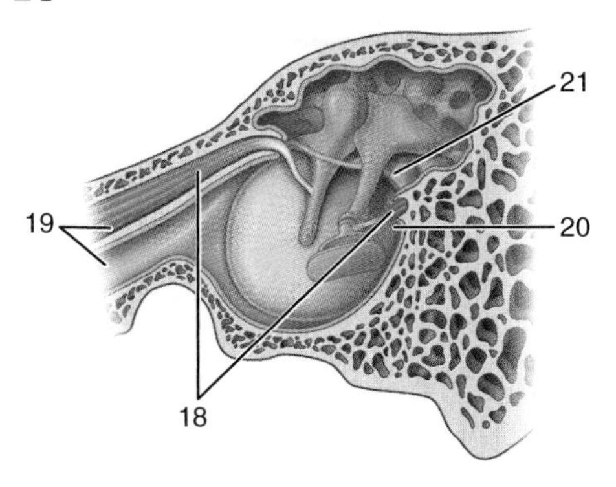

F

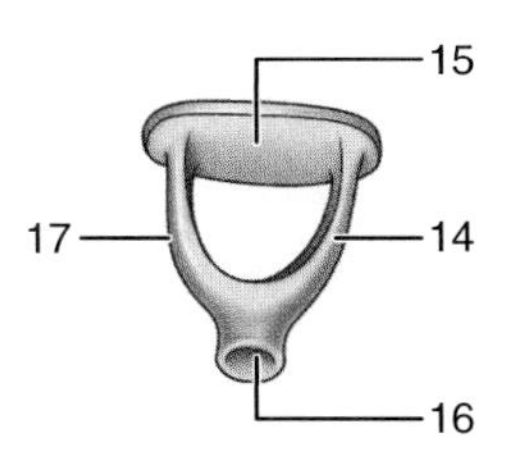

1 **tympanic canaliculus** a small opening on the inferior surface of the petrous part of the temporal bone in the floor of the petrosal fossa; it transmits the tympanic branch of the glossopharyngeal nerve and a small artery. <canaliculus tympanicus> A

2 **tympanic nerve** *origin,* inferior ganglion of glossopharyngeal nerve; *branches,* helps form tympanic plexus; *distribution,* mucous membrane of tympanic cavity, mastoid air cells, auditory tube, and, via lesser petrosal nerve and otic ganglion, the parotid gland; *modality,* general sensory and parasympathetic. <nervus tympanicus> A

3 **tympanic plexus** a nerve plexus on the promontory of the middle ear, formed by the tympanic and caroticotympanic nerves. It gives off the lesser petrosal nerve and a branch of the greater petrosal nerve and sends sensory fibers to the mucous membrane of the tympanic cavity, the auditory tube, and the mastoid air cells. <plexus tympanicus> A

4 **anterior ampullary nerve** the branch of the vestibular nerve that innervates the ampulla of the anterior semicircular duct, ending around the hair cells of the ampullary crest. <nervus ampullaris anterior> B

5 ***area of facial nerve*** the part of the fundus of the internal acoustic meatus where the facial nerve enters the facial canal. <area nervi facialis> B

6 **base of cochlea** the posterior of the cochlea, which rests upon the internal acoustic meatus. <basis cochleae> B

7 **cochlear nerve** the part of the vestibulocochlear nerve concerned with hearing, consisting of fibers that arise from the bipolar cells in the spiral ganglion and have their receptors in the spiral organ of the cochlea. <nervus cochlearis> B

8 **cupula of cochlea** the rounded or dome-shaped apex of the cochlea. <cupula cochleae> B

9 ***foramen singulare*** the opening in the inferior vestibular area of the fundus of the internal acoustic meatus that gives passage to the nerves of the ampulla of the posterior semicircular duct. B

10 ***foramina nervosa*** numerous small openings in the tympanic lip of the spiral limbus for the passage of the cochlear nerves. B

11 ***fundus of internal acoustic meatus*** the laterally placed end or bottom of the internal acoustic meatus. <fundus meatus acustici interni> B

12 ***inferior vestibular area of internal acoustic meatus*** the lower portion of the fundus of the internal acoustic meatus, transmitting fibers of the saccular nerve. <area vestibularis inferior meatus acustici interni> B

13 **internal acoustic meatus** the passage in the petrous portion of the temporal bone through which the facial, intermediate, and vestibulocochlear nerves and the labyrinthine artery pass. Called also *internal auditory canal.* <meatus acusticus internus> B

14 **internal acoustic pore** the opening of the internal acoustic meatus. <porus acusticus internus> B

15 ***labyrinthine veins*** several small veins that pass through the internal acoustic meatus from the cochlea into the inferior petrosal or the transverse sinus. <venae labyrinthi> B

16 **lateral ampullary nerve** the branch of the vestibular nerve that innervates the ampulla of the lateral semicircular duct, ending around the hair cells of the ampullary crest. <nervus ampullaris lateralis> B

17 **posterior ampullary nerve** the branch of the vestibular nerve that innervates the ampulla of the posterior semicircular duct, ending around the hair cells of the ampullary crest. <nervus ampullaris posterior> B

18 **saccular nerve** the branch of the vestibular nerve that innervates the macula of the saccule. <nervus saccularis> B

19 ***spiral modiolar artery*** *origin,* common cochlear artery; *branches,* none; *distribution,* internal auditory meatus, running a spiral course around the auditory nerve. <arteria spiralis modioli> B

20 ***superior vestibular area of internal acoustic meatus*** the upper portion of the fundus of the internal acoustic meatus, transmitting fibers of the utricular and superior ampullary nerves. <area vestibularis superior meatus acustici interni> B

21 ***transverse crest of internal acoustic meatus*** a ridge of bone that divides the fundus of the internal acoustic meatus into a superior and an inferior fossa. <crista transversa meati acustici interni> B

22 ***utricular nerve*** the branch of the vestibular nerve that innervates the macula of the utricle. <nervus utricularis> B

23 ***utriculoampullary nerve*** a nerve that arises by peripheral division of the vestibular nerve, and supplies the utricle and ampullae of the semicircular ducts. <nervus utriculoampullaris> B

24 ***vestibular branches of labyrinthine artery*** vessels supplying the vestibule of the ear. Called also *vestibular arteries.* <rami vestibulares arteriae labyrinthinae> B

25 ***vestibular ganglion*** the sensory ganglion located in the upper part of the lateral end of the internal acoustic meatus, the bipolar nerve cells of which give rise to the fibers of the vestibular nerve. <ganglion vestibulare> B

26 **vestibular nerve** the posterior part of the vestibulocochlear nerve, which is concerned with the sense of equilibrium. It consists of fibers arising from bipolar cells in the vestibular ganglion, and divides peripherally into a superior and an inferior part, with receptors in the ampullae of the semicircular canals, the utricle, and the saccule. <nervus vestibularis> B

A

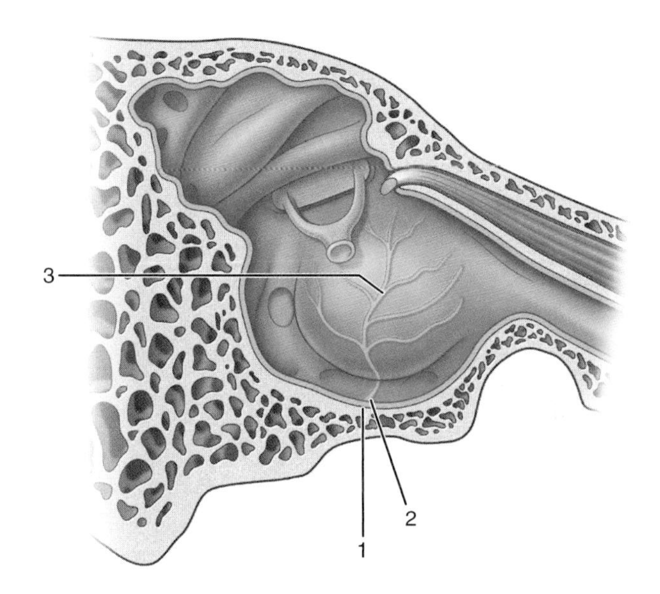

B

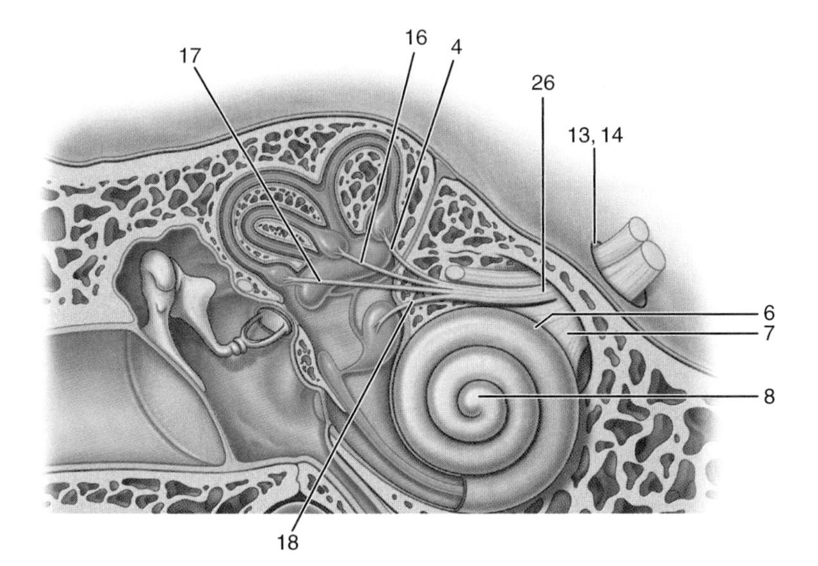

1 ***vestibular veins*** branches draining blood from the vestibule into the labyrinthine veins. <venae vestibulares> A

2 ***vestibulocochlear artery*** *origin,* common cochlear artery; *branches,* cochlear and posterior vestibular branches; *distribution,* cochlea, saccule, semicircular canals. <arteria vestibulocochlearis> A

3 **vestibulocochlear nerve** eighth cranial nerve; it emerges from the brain between the pons and the medulla oblongata, at the cerebellopontine angle and behind the facial nerve. It divides near the lateral end of the internal acoustic meatus into two functionally distinct and incompletely united components, the vestibular nerve and the cochlear nerve, and is connected with the brain by corresponding roots, the vestibular and the cochlear roots. Called also *acoustic nerve.* <nervus vestibulocochlearis> A

4 **vestibule of ear** an oval cavity in the middle of the bony labyrinth, communicating anteriorly with the cochlea and posteriorly with the semicircular canals, and containing perilymph surrounding the sacculus and utriculus. <vestibulum auris> A, B

5 **ampullary bony crura** the parts of the bony semicircular canals of the ear that lodge the membranous crura of the semicircular duct. <crura ossea ampullaria> A

6 ***ampullary crest*** the most prominent part of a localized thickening of the membrane that lines the ampullae of the semicircular ducts, covered with neuroepithelium containing endings of the vestibular nerve; called also *acoustic crest.* <crista ampullaris> B

7 **anterior bony ampulla** the dilatation at one end of the anterior semicircular canal. <ampulla ossea anterior> B

8 **anterior membranous ampulla** the dilatation at the end of the anterior membranous semicircular duct. <ampulla membranacea anterior> B

9 **anterior semicircular canal** the anterior of the bony semicircular canals, lodging the anterior semicircular duct of the membranous labyrinth. Called also *superior semicircular canal.* <canalis semicircularis anterior> B

10 **anterior semicircular duct** the semicircular duct occupying the anterior semicircular canal; called also *superior semicircular duct.* <ductus semicircularis anterior> B

11 **bony labyrinth** a layer of dense bone in the petrous portion of the temporal bone, in which the membranous labyrinth, the vestibular aqueduct, and the cochlear aqueduct are lodged; it consists of three parts: the vestibule, the semicircular canals, and the cochlea. <labyrinthus osseus> B

12 **bony semicircular canals** three long canals of the bony labyrinth of the ear, forming loops and opening into the vestibule by five openings; they lodge the semicircular ducts. See *anterior semicircular canal, lateral semicircular canal,* and *posterior semicircular canal.* Called also *semicircular canals.* <canales semicirculares ossei> B

13 **cochlea** a spirally wound tube, resembling a snail shell, which forms part of the inner ear. Its base lies against the lateral end of the internal acoustic meatus and its apex is directed anterolaterally. It consists of the modiolus and the bony spiral lamina, which partially divides the cochlea into the essential organs of hearing, the scala vestibuli and scala tympani; they communicate through the helicotrema. B

14 ***cochlear aqueduct*** a small channel that connects the scala tympani with the subarachnoid space; called also *perilymphatic duct.* <aqueductus cochleae> B

15 ***cochlear area*** the anterior part of the inferior portion of the fundus of the internal acoustic meatus, near the base of the cochlea. <area cochleae> B

16 ***cochlear canaliculus*** a small canal in the petrous part of the temporal bone that interconnects the scala tympani of the inner ear with the subarachnoid cavity; it houses the perilymphatic duct and a small vein. Called also *aqueduct of Cotunnius.* <canaliculus cochleae> B

17 **cochlear duct** a spirally arranged membranous tube in the bony canal of the cochlea along its outer wall, lying between the scala tympani below and the scala vestibuli above; called also *cochlear canal* and *membranous cochlea.* <ductus cochlearis> B

18 **cochlear labyrinth** the part of the membranous labyrinth that includes the perilymphatic space and the cochlear duct. <labyrinthus cochlearis> B

19 ***cochlear recess of vestibule*** a small depressed area on the medial wall of the vestibule of the ear, situated just below the posterior end of the vestibular crest, and perforated with foramina through which nerve fibers pass to the posterior portion of the cochlear duct. <recessus cochlearis vestibuli> B

20 ***common bony crus*** the part of a bony semicircular canal of the ear that lodges the common membranous crus of semicircular duct. <crus osseum commune> B

21 **common membranous crus of semicircular duct** an area consisting of the joined nonampullary ends of the anterior and posterior semicircular ducts of the ear. <crus membranaceum commune ductus semicircularis> B

22 **crest of round window** the ledge of bone that overhangs the round window of the middle ear. <crista fenestrae cochleae> B

23 **cupular cecum of cochlear duct** the closed blind apical end of the cochlear duct. <caecum cupulare ductus cochlearis> B

24 **ductus reuniens** a small canal leading from the saccule to the cochlear duct. B

A

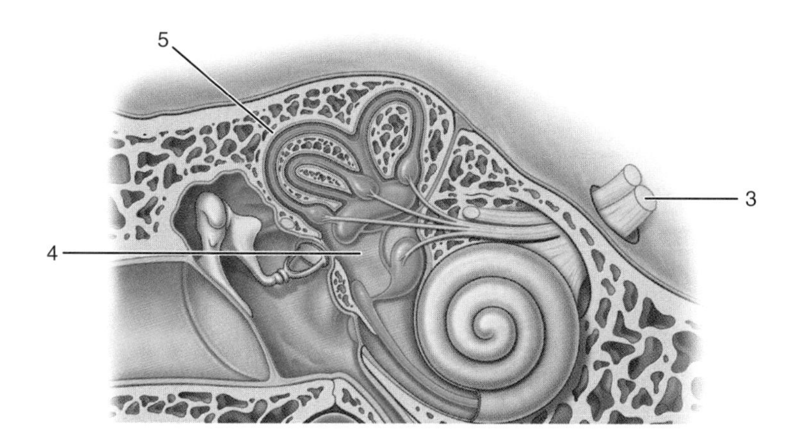

B

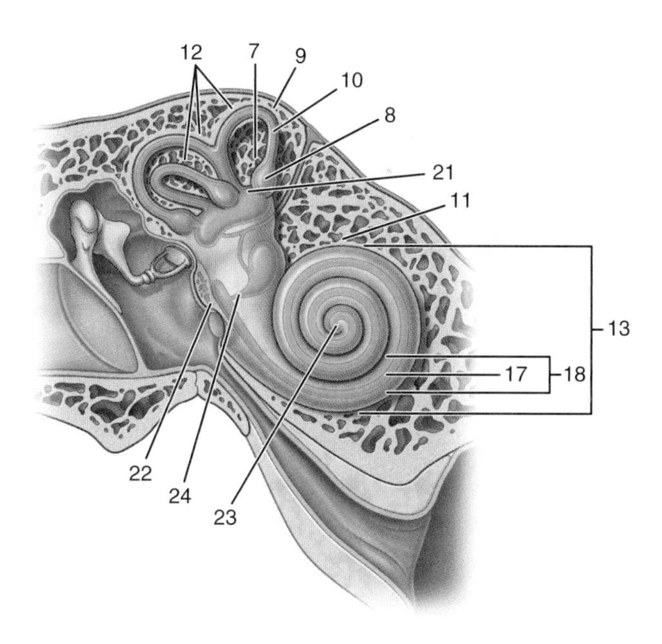

1 ***elliptical recess of vestibule*** an oval depressed area in the roof and medial wall of the vestibule of the inner ear, situated above and behind the crista and pierced by 25 to 30 small foramina through which nerves come from the internal acoustic meatus to the utricle, which occupies the depression. <recessus ellipticus vestibuli> A

2 ***external spiral sulcus*** a concavity within the cochlear duct immediately above the basilar crest. <sulcus spiralis externus> A

3 **helicotrema** the passage of the ear that connects the scala tympani and scala vestibuli at the apex of the cochlea; called also *Scarpa's hiatus.* A

4 **lateral bony ampulla** the dilatation at one end of the lateral semicircular canal. <ampulla ossea lateralis> A

5 **lateral membranous ampulla** the dilatation at the end of the lateral membranous semicircular duct. <ampulla membranacea lateralis> A

6 **lateral semicircular canal** the lateral of the bony semicircular canals, lodging the lateral semicircular duct of the membranous labyrinth; called also *horizontal semicircular canal.* <canalis semicircularis lateralis> A

7 **lateral semicircular duct** the semicircular duct occupying the lateral semicircular canal. <ductus semicircularis lateralis> A

8 ***macula cribrosa inferior*** the perforated area on the wall of the vestibule through which branches of the vestibulocochlear nerve pass to the ampullary crest and the posterior semicircular canal. A

9 ***macula cribrosa media*** the perforated area on the vestibular wall through which branches of the vestibulocochlear nerve pass to the macula of the saccule. A

10 ***macula cribrosa superior*** the perforated area on the vestibular wall through which branches of the vestibulocochlear nerve pass to the macula of the utricle and to the ampullary crest of the anterior and lateral semicircular canals. A

11 ***macula of saccule*** a thickening in the wall of the saccule where the epithelium contains hair cells that are stimulated by linear acceleration and deceleration and gravity. This and the macula of utricle together are called the *acoustic maculae.* <macula sacculi> A

12 ***macula of utricle*** a thickening in the wall of the utricle where the epithelium contains hair cells that are stimulated by linear acceleration and deceleration and gravity. This and the macula of saccule together are called the *acoustic maculae.* <macula utriculi> A

13 **membranous crura of semicircular duct** the ends of the semicircular ducts of the ear, in which the membranous ampullae are situated. <crura membranacea ampullaria ductus ampullary semicircularis> A

14 **membranous labyrinth** a system of communicating epithelial sacs and ducts, including the endolymphatic duct, cochlear duct, utricle, saccule, and semicircular ducts, lodged within and attached at certain points to the wall of the osseous labyrinth but separated from the major portion of the bony labyrinth by the perilymphatic space, and containing endolymph; it is divided into vestibular and cochlear parts. <labyrinthus membranaceus> A

15 ***opening of cochlear canaliculus*** the external opening of the cochlear canaliculus on the margin of the jugular foramen in the temporal bone. <apertura canaliculus cochlearis> A

16 ***opening of vestibular canaliculus*** the external opening for the vestibular canaliculus, located on the posterior surface of the petrous part of the temporal bone, lateral to the opening for the internal acoustic meatus. <apertura canaliculi vestibuli> A

17 ***otolithic membrane*** the gelatinous membrane surmounting the maculae of the saccule and utricle, containing the statoconia, and having special sensory hairs projecting into it. <membrana statoconiorum macularum> A

18 **posterior bony ampulla** the dilatation at one end of the posterior semicircular canal. <ampulla ossea posterior> A

19 **posterior membranous ampulla** the dilatation at the end of the posterior membranous semicircular duct. <ampulla membranacea posterior> A

20 **posterior semicircular canal** the posterior of the semicircular canals, lodging the posterior semicircular duct of the membranous labyrinth. <canalis semicircularis posterior> A

21 **posterior semicircular duct** the semicircular duct occupying the posterior semicircular canal. <ductus semicircularis posterior> A

22 **saccule** the smaller of the two divisions of the membranous labyrinth within the vestibule; it communicates with the cochlear duct by way of the ductus reuniens. <sacculus> A

23 **secondary tympanic membrane** the membrane that closes in the round window; called also *membrane of round window.* <membrana tympanica secundaria> A

24 **semicircular ducts** the long ducts of the membranous labyrinth of the ear, corresponding to the semicircular canals of the bony labyrinth and designated anterior, posterior, and lateral, according to the canal they occupy. Their diameter is only one-fourth that of the bony canals containing them, and each is affixed by one wall to the endosteal lining of the canal. They give information about angular acceleration and deceleration. Called also *membranous semicircular canals.* <ductus semicirculares> A

25 **simple bony crus** that part of the lateral bony semicircular canal of the ear that lodges the simple membranous crus of semicircular duct. <crus osseum simplex> A

A

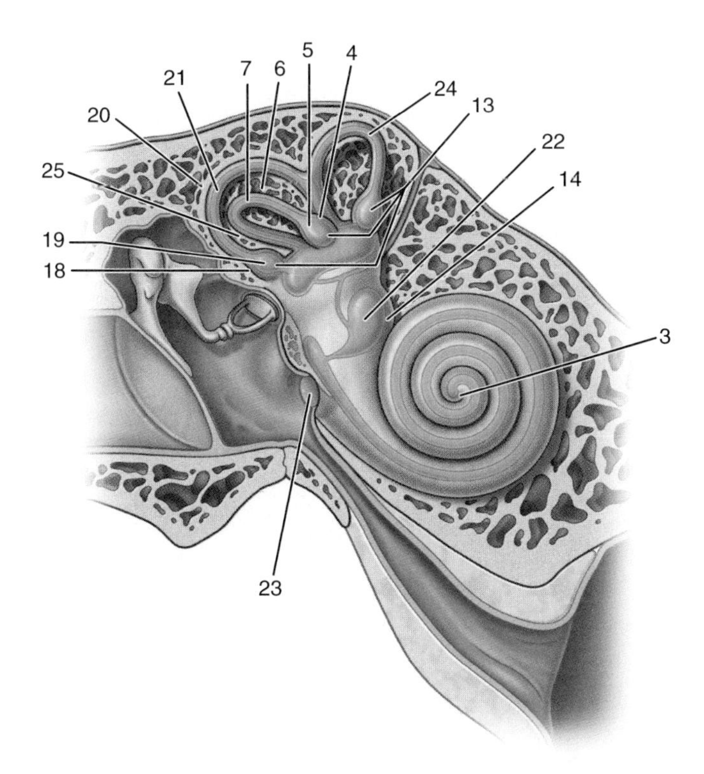

1 **simple membranous crus of semicircular duct** the nonampullary end of the lateral semicircular duct of the ear, opening into the utricle. <crus membranaceum simplex ductus semicircularis> A

2 ***spherical recess of vestibule*** a circular depressed area in the anteroinferior portion of the medial wall of the vestibule of the inner ear. It is pierced by 12 to 15 small foramina through which nerves come from the internal acoustic meatus to the saccule, which occupies the depression. <recessus sphericus vestibuli> A

3 ***statoconium*** one of the minute calciferous granules within the gelatinous statoconic membrane surmounting the acoustic maculae. Called also *otolith, otoconium,* and *ear crystal.* A

4 **utricle** the larger of the two divisions of the membranous labyrinth, located in the posterosuperior region of the vestibule. It is the major organ of the vestibular system, which gives information about position and movements of the head. <utriculus> A

5 **utriculosaccular duct** a tiny Y-shaped duct in the membranous labyrinth with one branch to the utricle, one branch to the saccule, and one branch to the endolymphatic duct. Called also *sacculoutricular duct* or *canal* and *utriculosaccular canal.* <ductus utriculosaccularis> A

6 ***vein of cochlear aqueduct*** a vein along the cochlea that empties into the superior bulb of the internal jugular vein; called also *vein of cochlear canaliculus.* <vena aqueductus cochleae> A

7 ***vein of vestibular aqueduct*** a small vein from the internal ear that passes through the aqueduct of the vestibule and empties into the superior petrosal sinus. <vena aqueductus vestibuli> A

8 ***vestibular aqueduct*** a small canal extending from the vestibule of the inner ear to open onto the posterior surface of the petrous part of the temporal bone. It lodges the endolymphatic duct and an arteriole and a venule. <aqueductus vestibuli> A

9 ***vestibular canaliculus*** a small canal extending from the vestibule of the inner ear to the posterior surface of the petrous part of the temporal bone; it houses the vestibular aqueduct. <canaliculus ventriculi> A

10 **vestibular cecum of cochlear duct** a small blind outpouching at the vestibular end of the cochlear duct. <caecum vestibulare ductus cochlearis> A

11 ***vestibular crest*** a ridge between the spherical and elliptical recesses of the vestibule, dividing posteriorly to bound the cochlear recess. <crista vestibuli> A

12 **vestibular labyrinth** the part of the membranous labyrinth that includes the utricle, saccule, and semicircular ducts. <labyrinthus vestibularis> A

13 **base of modiolus** the broad part of the modiolus situated near the lateral part of the internal acoustic meatus. <basis modioli> B

14 **modiolus** the central pillar of the cochlea. B

15 **spiral canal of modiolus** a canal following the course of the bony spiral lamina of the cochlea and containing the spiral ganglion of the cochlear division of the vestibulocochlear nerve. <canalis spiralis modioli> B

16 **endolymph** the fluid contained in the membranous labyrinth of the ear; it is entirely separate from the perilymph. <endolympha> C

17 **endolymphatic duct** the membranous tube connecting the utriculosaccular duct with the endolymphatic sac, located within the bony vestibular aqueduct. <ductus endolymphaticus> A, C

18 **endolymphatic sac** the blind, flattened cerebral end of the endolymphatic duct. <saccus endolymphaticus> A, C

19 **perilymph** the fluid contained within the space separating the membranous labyrinth from the bony labyrinth; it is entirely separate from the endolymph. <perilympha> C

20 **perilymphatic space** the fluid-filled space separating the membranous from the bony labyrinth. <spatium perilymphaticum> C

21 **scala tympani** the perilymph-filled part of the cochlea that is continuous with the scala vestibuli at the helicotrema, is separated from other cochlear structures by the spiral lamina and the basilar membrane of the cochlear duct, and ends blindly near the round window. C

22 **scala vestibuli** the perilymph-filled part of the cochlea that begins in the vestibule, is separated from other cochlear structures by the spiral lamina and the vestibular surface of the cochlear duct, and becomes continuous with the scala tympani at the helicotrema. C

23 **spiral canal of cochlea** a winding tube that makes two and one-half turns about the modiolus of the cochlea; it is divided into two compartments, scala tympani and scala vestibuli, by the spiral lamina. <canalis spiralis cochleae> C

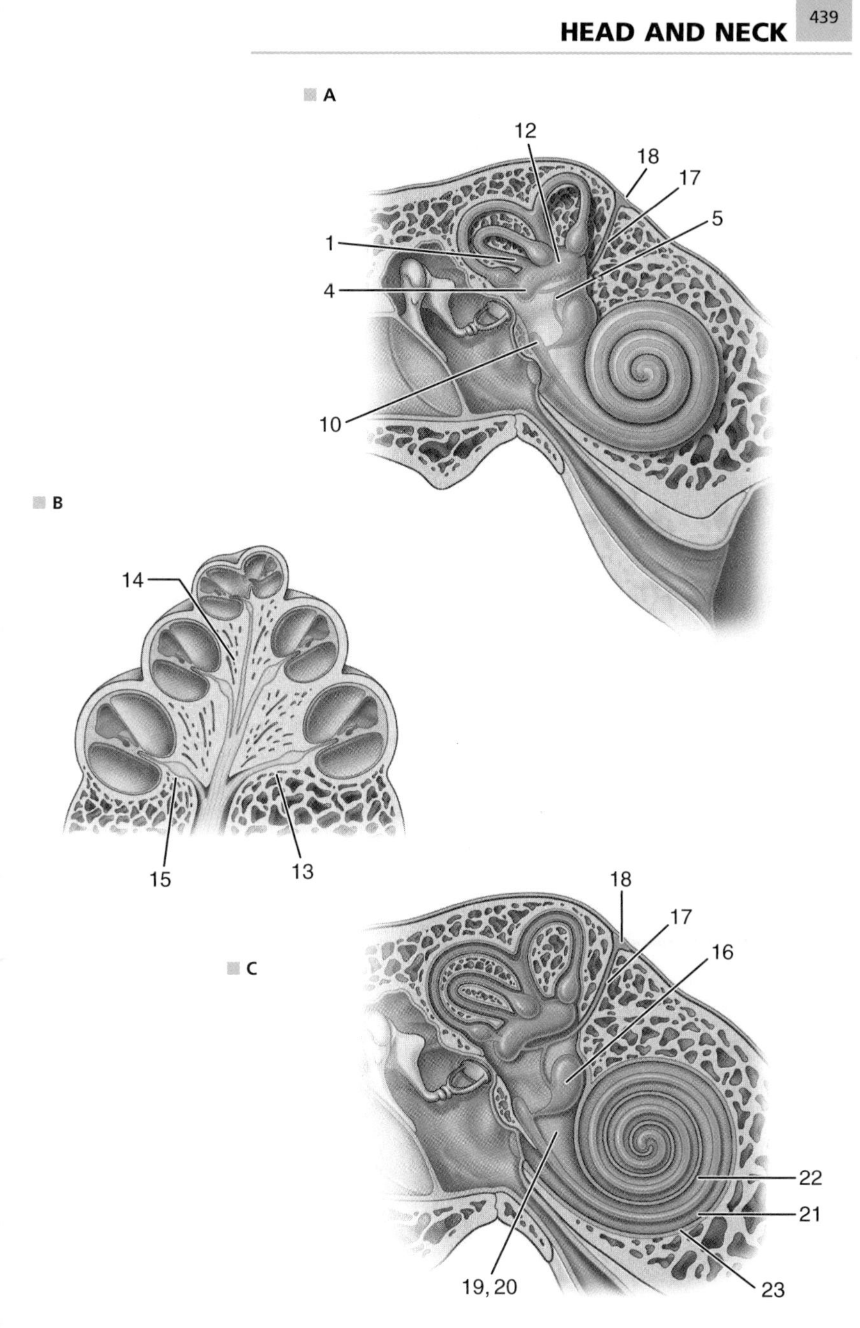
A
12
18
17
5
1
4
10
B
14
15
13
C
18
17
16
22
21
19, 20
23

1 **basilar crest of cochlear duct** the triangular eminence on the spiral ligament of the cochlea, providing a site of attachment for the basilar membrane. <crista basilaris ductus cochlearis> A

2 **basilar lamina of cochlear duct** the wall of the cochlear duct, which separates it from the scala tympani; the spiral organ lies against it. <lamina basilaris ductus cochlearis> A

3 **bony spiral lamina** a double plate of bone winding spirally around the modiolus and dividing the spiral canal of the cochlea incompletely into two parts, the scala tympani and the scala vestibuli. <lamina spiralis ossea> A

4 **cochlear ganglion** the sensory ganglion located within the spiral canal of the modiolus. It consists of bipolar cells that send fibers peripherally through the foramina nervosa to the spiral organ and centrally through the internal acoustic meatus to the cochlear nuclei of the brainstem. Called also *Corti's ganglion* and *spiral ganglion.* <ganglion cochleare> A

5 **external wall of cochlear duct** the part of the ductal wall adjacent to the outer wall of the cochlea. <paries externus ductus cochlearis> A

6 ***hamulus of spiral lamina*** the hooklike upper end of the bony spiral lamina. <hamulus laminae spiralis> A

7 ***lamina of modiolus*** a bony plate extending upward toward the cupula as a continuation of the modiolus and of the bony spiral lamina of the cochlea. <lamina modioli> A

8 ***longitudinal canals of modiolus*** short tunnels in the modiolus that transmit blood vessels and nerves. <canales longitudinales modioli> A

9 **reticular membrane of spiral organ** a netlike membrane over the spiral organ; the free ends of the outer hair cells pass through its apertures. Called also *reticular lamina.* <membrana reticularis organi spiralis> A

10 ***secondary spiral lamina*** a bony projection on the outer wall of the bony spiral lamina in the lower part of the first turn of the cochlea. <lamina spiralis secundaria> A

11 ***spiral ligament of cochlea*** a band of thickened periosteum in the bony cochlea. <ligamentum spirale ductus cochlearis> A

12 **spiral limbus** the thickened periosteum of the bony spiral lamina at the attachment of the vestibular membrane. Called also *limbus of bony spiral lamina.* <limbus spiralis> A

13 **spiral organ** the organ, resting on the basilar membrane in the cochlear duct, that contains the special sensory receptors for hearing; it consists of neuroepithelial hair cells and several types of supporting cells, including the inner and outer pillar cells, inner and outer phalangeal cells, border cells, and Hensen's cells. Called also *organ of Corti.* <organum spirale> A

14 ***spiral prominence*** a prominence on the external wall of the cochlear duct, separating the stria vascularis from the external spiral sulcus. <prominentia spiralis> A

15 **tectorial membrane of cochlear duct** a delicate gelatinous mass extending from the limbus and resting on the spiral organ of the ear and connected with the hairs of the hair cells. <membrana tectoria ductus cochlearis> A

16 **tympanic surface of cochlear duct** the wall of the cochlear duct that separates it from the scala tympani, composed of the bony spiral laminae and the basilar membrane. Called also *spiral membrane of cochlear duct.* <paries tympanicus ductus cochlearis> A

17 **vestibular surface of cochlear duct** the thin anterior wall of the cochlear duct, which separates it from the scala vestibuli; called also *vestibular membrane of cochlear duct.* <paries vestibularis ductus cochlearis> A

18 **facial canal** a canal in the temporal bone for the facial nerve, beginning in the internal acoustic meatus and passing anterolaterally dorsal to the vestibule of the inner ear for about 2 mm. Turning sharply backward at the geniculum of facial canal, it runs along the medial wall of the tympanic cavity, then turns inferiorly and reaches the exterior of the petrous part of the bone at the stylomastoid foramen. Called also *canal for facial nerve* and *fallopian canal.* <canalis nervi facialis> B

19 **geniculate ganglion** the sensory ganglion of the facial nerve, situated on the geniculum of facial nerve. <ganglion geniculi nervi facialis> B

20 **geniculum of facial canal** the bend in the facial canal lodging the geniculum of facial nerve; called also *genu of facial canal.* <geniculum canalis nervi facialis> B

21 **geniculum of facial nerve** the part of the facial nerve at the lateral end of the internal acoustic meatus, where the fibers turn sharply posteroinferiorly, and where the geniculate ganglion is found; called also *external genu of facial nerve.* <geniculum nervi facialis> B

22 **prominence of facial canal** an elongated elevation on the medial wall of the tympanic cavity, just inferior to the prominence of the lateral semicircular canal and superior and posterior to the vestibular window. <prominentia canalis facialis> B

23 **prominence of lateral semicircular canal** a large rounded prominence on the upper portion of the medial wall of the tympanic cavity, between the vestibular window and the mastoid antrum. <prominentia canalis semicircularis lateralis> B

24 **pyramidal eminence** an elevation in the posterior wall of the middle ear, which contains the stapedius muscle. <eminentia pyramidalis> B

25 **stapedius muscle** *origin,* interior of pyramidal eminence of tympanic cavity; *insertion,* posterior surface of neck of stapes; *innervation,* stapedial branch of facial; *action,* dampens stapedial movement. <musculus stapedius> B

A

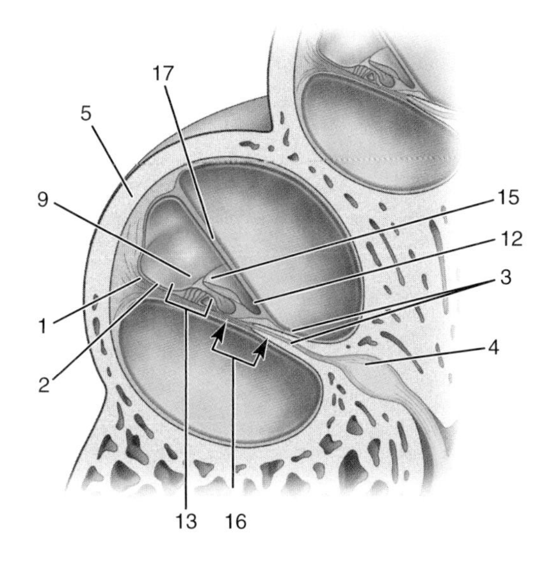

B

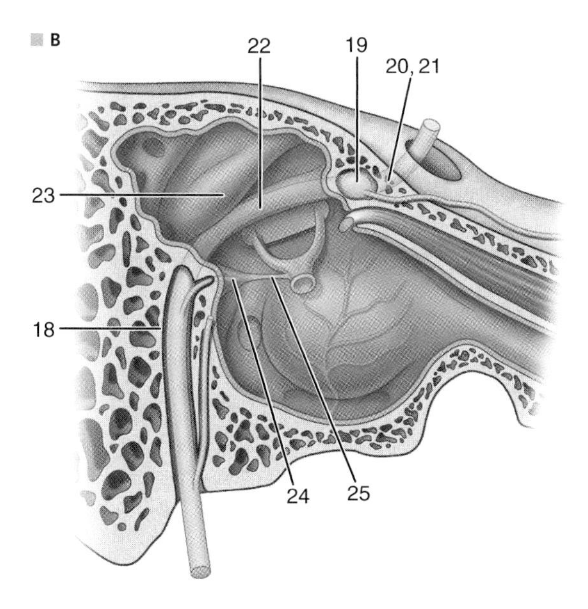

1. **stapedius nerve** *origin,* facial nerve; *distribution,* stapedius muscle; *modality,* motor. <nervus stapedius> A
2. **subiculum of promontory of tympanic cavity** a ridge of bone bounding the tympanic sinus inferiorly. <subiculum promontorii cavitatis tympani> A
3. **tympanic sinus** a deep fossa in the posterior part of the tympanic cavity; it is bounded superiorly by the pyramidal eminence and inferiorly by the subiculum of promontory of tympanic cavity, and it opens anteriorly into the fossa of round window. <sinus tympani> A
4. **canal for tensor tympani muscle** a small canal hidden in the temporal bone, constituting the superior part of the musculotubal canal, and lodging the tensor tympani muscle. <semicanalis musculi tensoris tympani> B
5. **canaliculus for chorda tympani** a small canal that opens off the facial canal just before its termination, transmitting the chorda tympani nerve into the tympanic cavity; called also *canal of chorda tympani* and *Civinini's canal.* <canaliculus chordae tympani> B
6. **chorda tympani** a nerve originating from the facial nerve and distributed to the submandibular, sublingual, and lingual glands and the anterior two-thirds of the tongue; *modality;* parasympathetic and special sensory. B
7. **cochleariform process** a small hollow cone of bone at the end of the septum of the musculotubular canal, just anterior to the vestibular window, with an opening through which the tendon of the tensor tympani passes. <processus cochleariformis> B
8. **fold of chorda tympani** a fold in the mucous membrane of the tympanic cavity overlying the chorda tympani nerve. <plica chordae tympani> B
9. ***nerve to tensor tympani*** *origin,* medial pterygoid nerve; *distribution,* tensor tympani muscle; *modality,* motor. <nervus musculi tensoris tympani> B
10. **septum of musculotubal canal** the thin lamella of bone that divides the musculotubal canal into the semicanals for the tensor tympani muscle and the auditory tube. <septum canalis musculotubarii> B
11. **tensor tympani muscle** *origin,* cartilaginous portion of auditory tube; *insertion,* manubrium of malleus; *innervation,* mandibular; *action,* tenses tympanic membrane. <musculus tensor tympani> B
12. ***nerve to external acoustic meatus*** *origin,* auriculotemporal nerve; *distribution,* skin lining external acoustic meatus, and tympanic membrane; *modality,* general sensory. <nervus meatus acustici externi> C
13. ***auricular branch of posterior auricular artery*** a branch supplying the pinna and adjacent skin. <ramus auricularis arteriae auricularis posterioris> D
14. **posterior auricular artery** *origin,* external carotid; *branches,* auricular and occipital branches, stylomastoid artery; *distribution,* middle ear, mastoid cells, auricle, parotid gland, digastric and other muscles. <arteria auricularis posterior> D
15. ***posterior tympanic artery*** *origin,* stylomastoid branch of posterior auricular artery; *branches,* none; *distribution,* tympanic cavity. <arteria tympanica posterior> D
16. ***stapedial branch of posterior auricular artery*** a variable branch supplying the stapedius muscle and tendon. <ramus stapedius arteriae auricularis posterioris> D
17. ***stylomastoid artery*** *origin,* posterior auricular; *branches,* mastoid and stapedial branches, posterior tympanic artery; *distribution,* tympanic cavity walls, mastoid cells, stapedius muscle. <arteria stylomastoidea> D

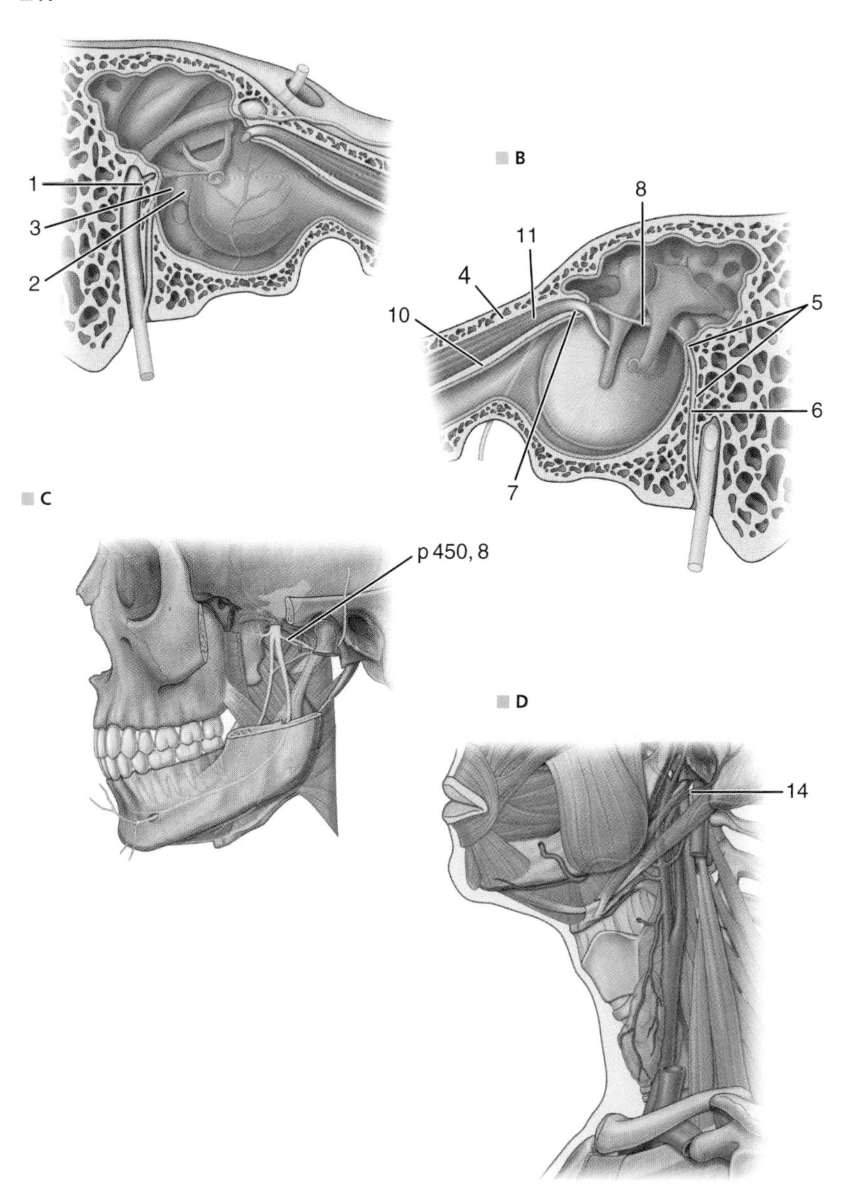
A
1
3
2
B
8
11
4
10
5
6
7
C
p 450, 8
D
14

1 **anterior notch of auricle** a depression between the crus of the helix and the tragus; called also *auricular notch.* <incisura anterior auriculae> A

2 ***antihelical fossa*** the depression on the medial surface of the auricle of the ear that corresponds to the antihelix on the lateral surface. <fossa antihelica> A

3 **antihelix** the prominent semicircular ridge seen on the lateral aspect of the auricle of the external ear, anteroinferior to the helix; called also *anthelix.* A

4 **antitragus** a projection opposite the tragus, bounding the cavity of the concha posteroinferiorly and continuous above with the antihelix. A

5 ***apex of auricle*** a point sometimes present on the posterior superior part of the helix of the ear; called also *tip of ear.* <apex auriculae> A

6 ***auricle*** the portion of the external ear not contained within the head; the flap of the ear. Called also *pinna.* <auricula> A

7 ***auricular tubercle*** a small projection sometimes found on the edge of the helix, and conjectured by some to be a relic of a simian ancestry. Called also *darwinian tubercle.* <tuberculum auriculare> A

8 **cavity of concha** the inferior part of the concha of the auricle, which leads into the external acoustic meatus. <cavitas conchae> A

9 **concha of auricle** the hollow of the auricle of the external ear, bounded anteriorly by the tragus and posteriorly by the antihelix. <concha auriculae> A

10 **crura of antihelix** the two ridges on the external ear marking the superior termination of the antihelix and bounding the triangular fossa; called also *limbs of antihelix.* <crura anthelicis> A

11 **crus of helix** the anterior termination of the helix of the external ear, located above the entrance to the external acoustic meatus. <crus helicis> A

12 **cymba conchae** the upper part of the concha of the auricle. A

13 ***eminence of concha*** the projection on the medial surface of the auricle that corresponds to the concha on the lateral surface. <eminentia conchae> A

14 ***eminence of scapha*** the prominence on the medial side of the auricle of the external ear that corresponds to the scapha on the lateral side. <eminentia scaphae> A

15 ***eminence of triangular fossa of auricle*** the protuberance on the medial surface of the auricle of the ear that corresponds to the triangular fossa on the lateral surface. Called also *triangular eminence.* <eminentia fossae triangularis auriculae> A

16 ***groove of crus of helix*** a transverse sulcus on the medial surface of the pinna, corresponding to the crus of helix on the lateral surface. <sulcus cruris helicis> A

17 **helix** the superior and posterior free margin of the pinna of the ear. A

18 **intertragic notch** the notch at the lower part of the pinna of the ear between the tragus and the antitragus. <incisura intertragica> A

19 ***ligaments of auricle*** the three ligaments (anterior, superior, and posterior) that help attach the auricle to the side of the head. <ligamenta auricularia> A

20 **lobule of auricle** the inferior, dependent part of the auricle below the antitragus, which contains fibrous and fatty tissue but no cartilage. Called also *earlobe* or *ear lobe.* <lobulus auriculae> A

21 **posterior sulcus of auricle** the slight depression on the pinna that separates the antihelix from the antitragus. <sulcus posterior auriculae> A

22 **scapha** the long curved depression that separates the helix from the antihelix; called also *scaphoid fossa* and *fossa of helix.* A

23 ***spine of helix*** a small, forward-projecting cartilaginous process on the anterior portion of the helix at about the junction of the helix and its crus, just above the tragus. <spina helicis> A

24 ***supratragic tubercle*** a small tubercle sometimes seen on the pinna just superior to the tragus. <tuberculum supratragicum> A

25 **tragus** 1. the cartilaginous projection anterior to the external opening of the ear. 2. *(in the pl.)* hairs growing on the pinna of the external ear, especially on the cartilaginous projection anterior to the external opening. A

26 **triangular fossa of auricle** the cavity just above the concha of the ear between the crura of the antihelix. <fossa triangularis auriculae> A

27 ***auricular branch of vagus nerve*** a branch arising from the superior ganglion of the vagus, innervating the cranial surface of the auricle, the floor of the external acoustic meatus, and the adjacent part of the tympanic membrane; *modality,* general sensory. <ramus auricularis nervi vagi>

A

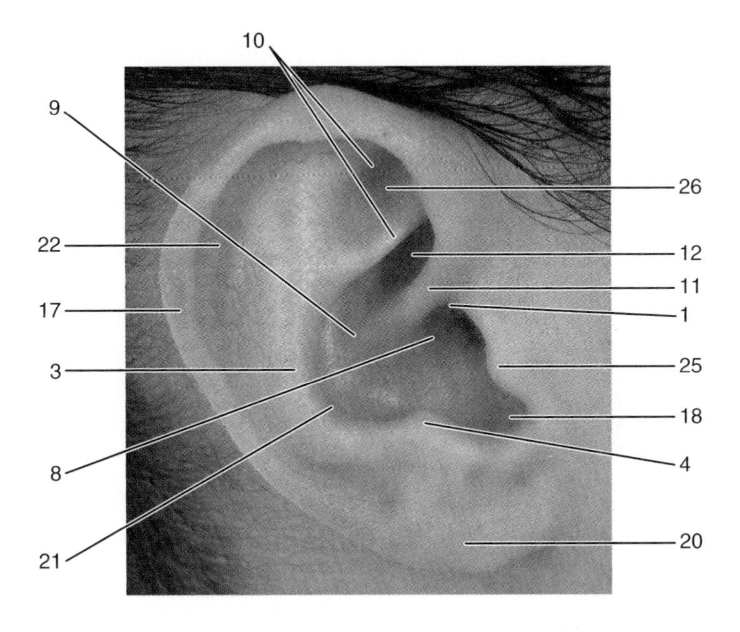

1 **auricular cartilage** the internal plate of elastic cartilage which is found in the external ear. <cartilago auriculae> A

2 **fissura antitragohelicina** a fissure in the auricular cartilage between the tail of the helix and the antitragus; called also *posterior fissure of auricle.* A

3 **isthmus of cartilaginous auricle** a bridge of cartilage connecting the cartilage of the external acoustic meatus with the main part of the cartilage of the auricle of the external ear. <isthmus cartilaginis auricularis> A

4 **notch in cartilage of acoustic meatus** two vertical fissures in the anterior part of the cartilage of the external acoustic meatus. <incisura cartilaginis meatus acustici> A

5 **tail of helix** the termination of the posterior margin of the cartilage of the helix. <cauda helicis> A

6 ***terminal notch of auricle*** a deep notch separating the tragal lamina and cartilage of the external acoustic meatus from the main auricular cartilage. <incisura terminalis auricularis> A

7 **tragal lamina** the longitudinal curved lamina of cartilage in the tragus of the auricle, at the beginning of the cartilaginous part of the external acoustic meatus. <lamina tragi> A

8 **antitragicus muscle** *origin,* outer part of antitragus; *insertion,* caudate process of helix and antihelix; *innervation,* temporal and posterior auricular. <musculus antitragicus> B

9 **helicis major muscle** *origin,* spine of helix; *insertion,* anterior border of helix; *innervation,* auriculotemporal and posterior auricular; *action,* tenses skin of auditory canal. <musculus helicis major> B

10 **helicis minor muscle** *origin,* anterior rim of helix; *insertion,* concha; *innervation,* temporal, posterior auricular. <musculus helicis minor> B

11 ***muscle of terminal notch*** an inconstant muscular slip continuing forward from the tragicus muscle to bridge the incisure of the cartilaginous meatus. <musculus incisurae terminalis> B

12 ***pyramidal muscle of auricle*** a prolongation of the fibers of the tragicus to the spine of the helix. <musculus pyramidalis auriculae> B

13 **tragicus muscle** a short, flattened vertical band on the lateral surface of the tragus, innervated by the auriculotemporal and posterior auricular nerves. <musculus tragicus> B

14 **oblique muscle of auricle** *origin,* cranial surface of concha; *insertion,* cranial surface of auricle above concha; *innervation,* temporal and posterior auricular (branches of facial). <musculus obliquus auriculae> C

15 **posterior auricular ligament** the auricular ligament that passes from the eminence of the concha to the mastoid part of the temporal bone. <ligamentum auriculare posterius> C

16 **superior auricular ligament** the auricular ligament that passes from the spine of the helix to the superior margin of the bony external acoustic meatus. <ligamentum auriculare superius> C

17 **transverse muscle of auricle** *origin,* cranial surface of auricle; *insertion,* circumference of auricle; *innervation,* posterior auricular; *action,* retracts helix. <musculus transversus auriculae> C

18 **annular stapedial ligament** a ring of fibrous tissue that attaches the base of the stapes to the oval window of the inner ear; called also *annular ligament of base of stapes.* <ligamentum anulare stapediale> D

19 **anterior ligament of malleus** a fibrous band that extends from the neck of the malleus just above the anterior process to the anterior wall of the tympanic cavity close to the petrotympanic fissure. Some of the fibers pass through the fissure to the spine of the sphenoid bone. <ligamentum mallei anterius> D

20 **anterior malleolar fold of mucous coat of tympanic cavity** a fold in the tunica mucosa of the tympanic cavity, reflected from the tympanic membrane over the anterior process and ligament of the malleus and part of the chorda tympani nerve. <plica mallearis anterior tunicae mucosae cavitatis tympanicae> D

21 **lateral ligament of malleus** a triangular fibrous band that passes from the posterior portion of the tympanic incisure to the head or neck of the malleus. <ligamentum mallei laterale> D

22 ***ligaments of auditory ossicles*** the ligaments of the auditory ossicles, comprising the anterior, lateral, and superior ligaments of the malleus, the posterior and superior ligaments of the incus, and the annular ligament of the stapes. <ligamenta ossiculorum auditus> D

23 **posterior ligament of incus** a fibrous band by which the tip of the short limb of the incus is fixed to the fossa of the incus. <ligamentum incudis posterius> D

24 **posterior malleolar fold of mucous coat of tympanic cavity** a fold of the tunica mucosa of the tympanic cavity, extending from the manubrium of malleus to the posterior wall of the cavity. <plica mallearis posterior tunicae mucosae cavitatis tympanicae> D

25 ***stapedial fold*** a mucosal fold that passes from the posterior wall of the tympanic cavity along the tympanic membrane and surrounds the stapes. <plica stapedialis> D

26 **stapedial membrane** a membrane filling the arch formed by the crura and base of the stapes. <membrana stapedialis> D

27 **superior ligament of incus** a fibrous band that passes from the body of the incus to the roof of the tympanic cavity just back of the superior ligament of the malleus. <ligamentum incudis superius> D

28 **superior ligament of malleus** a delicate fibrous strand passing from the roof of the tympanic cavity to the head of the malleus. <ligamentum mallei superius> D

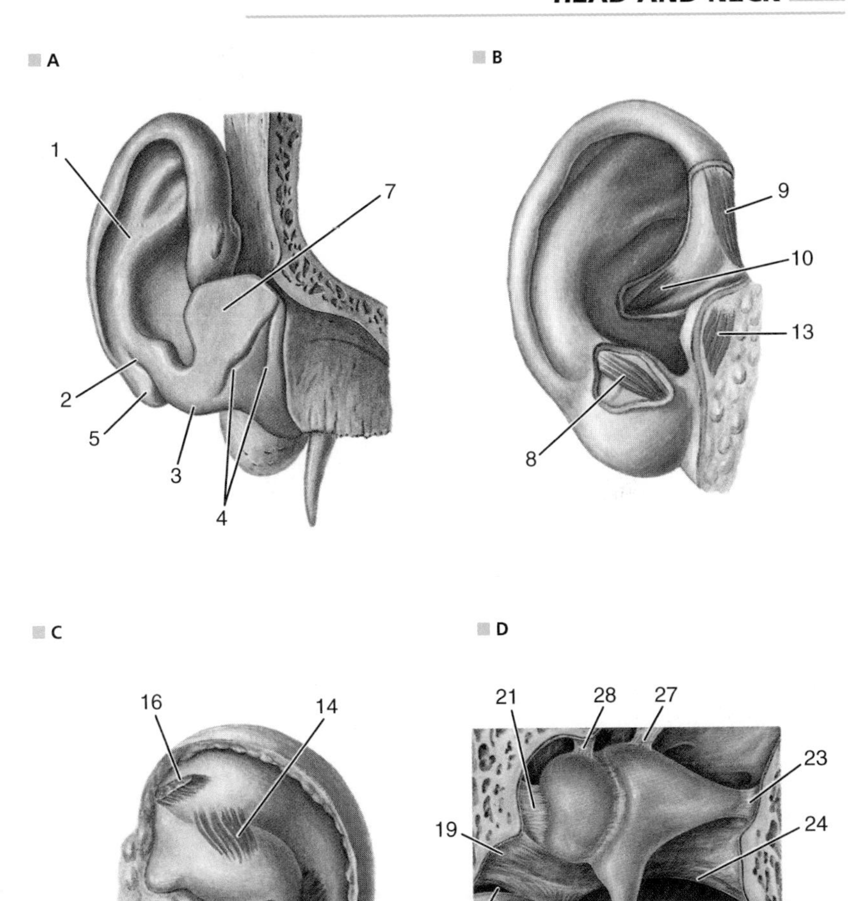
A
B
1
7
9
10
13
2
5
3
4
8
C
D
16
14
17
15
21
28
27
23
19
24
20
18
26

Temporal and Infratemporal Fossae

1 ***deep layer of temporal fascia*** the deep portion of the fascia investing the temporalis muscle. <lamina profunda fasciae temporalis> A

2 **superficial layer of temporal fascia** the superficial portion of the fascia investing the temporalis muscle. <lamina superficialis fasciae temporalis> A

3 **temporal fascia** a strong fibrous sheet covering the temporalis muscle, consisting of deep and superficial layers which attach inferiorly to the zygomatic arch. Called also *temporal aponeurosis.* <fascia temporalis> A

4 **articular disk of temporomandibular joint** a plate of fibrocartilage or fibrous tissue that divides the temporomandibular joint into two separate cavities; its circumference is connected to the articular capsule. Called also *meniscus of temporomandibular joint.* <discus articularis articulationis temporomandibularis> B

5 **articular surface of mandibular fossa** the articular surface found in the deep part of the mandibular fossa. <facies articularis fossae mandibularis> B

6 **lateral ligament of temporomandibular joint** a strong triangular fibrous band that is attached superiorly by its base to the zygomatic process of the temporal bone, passes down on the lateral side of the joint in contact with the capsule, and is inserted by its apex into the lateral and posterior surfaces of the neck of the condyloid process of the mandible. Called also *temporomandibular ligament.* <ligamentum laterale articulationis temporomandibularis> C

7 ***medial ligament of temporomandibular joint*** the medial ligament of the temporomandibular joint. <ligamentum mediale articulationis temporomandibularis> C

8 **deep part of masseter muscle** the part whose fibers arise from the medial surface of the zygomatic arch and the fascia over the temporalis muscle, and are directed in a vertical inferior direction. <pars profunda musculi masseteris> D

9 **masseter muscle** *origin,* SUPERFICIAL PART—zygomatic process of maxilla and inferior border of zygomatic arch, DEEP PART—inferior border and medial surface of zygomatic arch; *insertion,* SUPERFICIAL PART—angle and ramus of mandible, DEEP PART—superior half of ramus and lateral surface of coronoid process of mandible; *innervation,* masseteric, from mandibular division of trigeminal; *action,* raises mandible, closes jaws. <musculus masseter> D

10 ***masseteric fascia*** a layer of fascia covering the masseter muscle. <fascia masseterica> D

11 **superficial part of masseter muscle** the part of the muscle whose fibers arise from the anterior part of the zygomatic arch and are directed inferiorly and posteriorly. <pars superficialis musculi masseteris> D

12 **infratemporal fossa** the area on the side of the cranium limited superiorly by the infratemporal crest, posteriorly by the mandibular fossa, anteriorly by the infratemporal surface of the maxilla, and laterally by the zygomatic arch and part of the ramus of the mandible; called also *zygomatic fossa* and *infratemporal region.* <fossa infratemporalis> E

13 **temporal fossa** the area on the side of the cranium outlined posteriorly and superiorly by the temporal lines, anteriorly by the frontal and zygomatic bones, laterally by the zygomatic arch, and inferiorly by the infratemporal crest. <fossa temporalis> E

14 **temporalis muscle** *origin,* temporal fossa and fascia; *insertion,* coronoid process of mandible; *innervation,* mandibular; *action,* closes jaws. <musculus temporalis> F

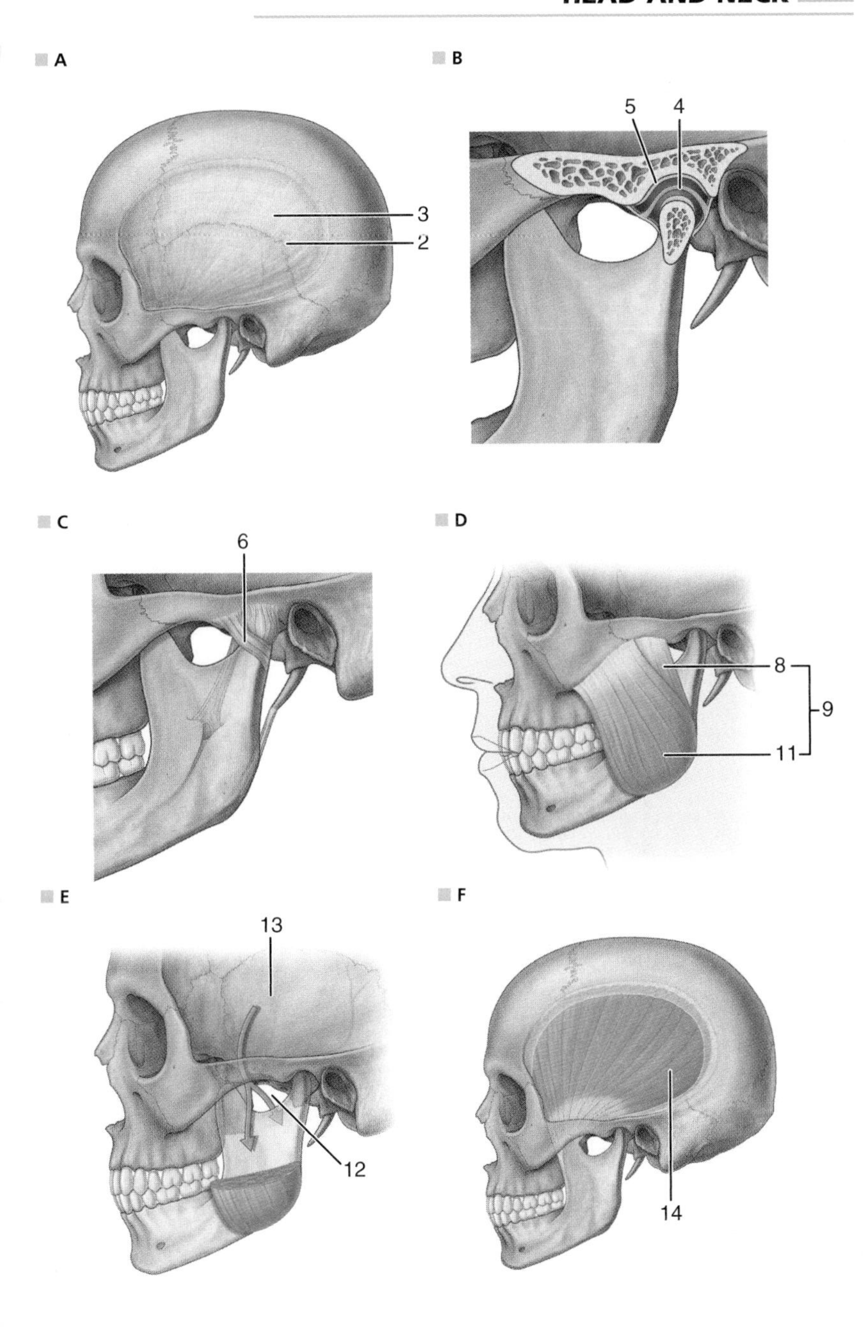
A
3
2
B
5
4
C
6
D
8
9
11
E
13
12
F
14

HEAD AND NECK

1 **lateral pterygoid muscle** (2 heads): *origin,* SUPERIOR HEAD—infratemporal surface of greater wing of sphenoid and infratemporal crest; INFERIOR HEAD—lateral surface of lateral pterygoid plate; *insertion,* neck of condyle of mandible, temporomandibular joint capsule; *innervation,* mandibular; *action,* protrudes mandible, opens jaws, moves mandible from side to side. <musculus pterygoideus lateralis> A

2 **medial pterygoid muscle** *origin,* medial surface of lateral pterygoid plate, tuberosity of maxilla; *insertion,* medial surface of ramus and angle of mandible; *innervation,* mandibular; *action,* closes jaws. <musculus pterygoideus medialis> A

3 **deep temporal nerves** usually two in number, anterior and posterior, with a third middle one often seen: *origin,* mandibular nerve; *distribution,* temporalis muscles; *modality,* motor. <nervi temporales profundi> B

4 **lateral pterygoid nerve** *origin,* mandibular nerve; *distribution,* lateral pterygoid muscle; *modality,* motor. <nervus pterygoideus lateralis> B

5 **masseteric nerve** *origin,* mandibular division of trigeminal nerve; *distribution,* masseter muscle and temporomandibular joint; *modality,* motor and general sensory. <nervus massetericus> B

6 **medial pterygoid nerve** *origin,* mandibular nerve; *branches,* nerve to tensor tympani, nerve to tensor veli palatini; *distribution,* medial pterygoid, tensor tympani, and tensor veli palatini muscles; *modality,* motor. <nervus pterygoideus medialis> B

7 **meningeal branch of mandibular nerve** a branch that arises from the trunk of the mandibular nerve, re-enters the cranium through the foramen spinosum, accompanies the middle meningeal artery to supply the dura mater, and also helps innervate the mucous membrane of the mastoid air cells. <ramus meningeus nervi mandibularis> B

8 **auriculotemporal nerve** *origin,* by two roots from the mandibular nerve; *branches,* anterior auricular nerve, nerve of external acoustic meatus, parotid branches, branch to tympanic membrane, and branches communicating with facial nerve; its terminal branches are superficial temporal to the scalp; *modality,* general sensory. <nervus auriculotemporalis> C

9 **inferior alveolar nerve** *origin,* mandibular nerve; *branches,* inferior dental plexus, mylohyoid and mental nerves; *modality,* motor and general sensory. <nervus alveolaris inferior> C

10 **lingual nerve** *origin,* mandibular nerve, descending to the tongue, first medial to the mandible and then under cover of the mucous membrane of the mouth; *branches,* sublingual nerve, lingual branches, branches to isthmus of fauces, and branches communicating with the hypoglossal nerve and chorda tympani; *modality,* general sensory. <nervus lingualis> C

11 **mandibular nerve** one of three terminal divisions of the trigeminal nerve, passing through the foramen ovale to the infratemporal fossa. *Origin,* trigeminal ganglion; *branches,* meningeal ramus, masseteric, deep temporal, lateral and medial pterygoid, buccal, auriculotemporal, lingual, and inferior alveolar nerves; *distribution,* extensive distribution to muscles of mastication, skin of face, mucous membrane of mouth, and teeth; see individual branches; *modality,* general sensory and motor. <nervus mandibularis> C

12 ***ganglionic branches of mandibular nerve to otic ganglion*** branches of the mandibular nerve that communicate with the otic ganglion. <rami ganglionares nervi mandibularis ad ganglion oticum> D

13 **otic ganglion** a parasympathetic ganglion in the infratemporal fossa, medial to the mandibular nerve and just inferior to the foramen ovale: its preganglionic fibers are derived from the glossopharyngeal nerve via the lesser petrosal nerve, and its postganglionic fibers supply the parotid gland. Sensory and postganglionic sympathetic fibers pass through the ganglion. <ganglion oticum> D

14 **anterior deep temporal artery** *origin,* maxillary artery; *branches,* to zygomatic bone and greater wing of sphenoid bone; *distribution,* temporal muscle, and anastomoses with middle temporal artery. <arteria temporalis profunda anterior> E

15 ***anterior tympanic artery*** *origin,* maxillary artery; *branches,* none; *distribution,* tympanic cavity. <arteria tympanica anterior> E

16 **buccal artery** *origin,* maxillary artery; *branches,* none; *distribution,* buccinator muscle, mucous membrane of mouth. Called also *buccinator artery.* <arteria buccalis> E

17 ***deep auricular artery*** *origin,* maxillary artery; *branches,* none; *distribution,* skin of auditory canal, tympanic membrane, temporomandibular joint. <arteria auricularis profunda> E

18 **inferior alveolar artery** *origin,* maxillary artery; *branches,* dental, peridental, mental, and mylohyoid branches; *distribution,* lower jaw, lower lip, and chin. Called also *inferior dental artery* and *mandibular artery.* <arteria alveolaris inferior> E

19 **masseteric artery** *origin,* maxillary artery; *branches,* none; *distribution,* masseter muscle. <arteria masseterica> E

20 **maxillary artery** external carotid artery; *branches,* deep auricular, anterior tympanic, inferior alveolar, middle meningeal, pterygomeningeal, masseteric, anterior and posterior deep temporal, buccal, posterior superior alveolar, infraorbital, descending palatine, and sphenopalatine arteries, and artery of pterygoid canal; *distribution,* both jaws, teeth, muscles of mastication, ear, meninges, nose, paranasal sinuses, palate. Called also *internal maxillary artery.* <arteria maxillaris> E

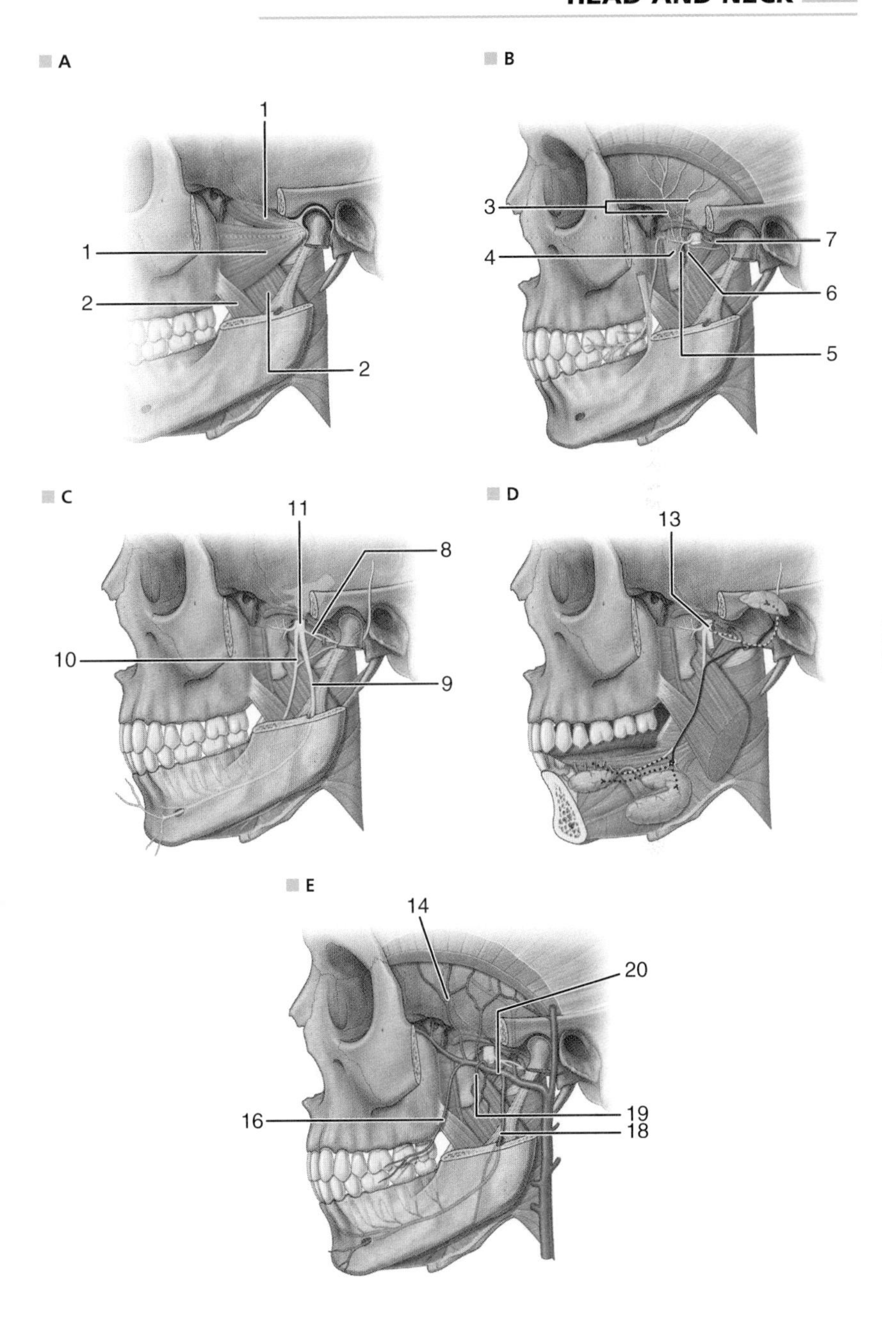
A
1
1
2
2
B
3
4
7
6
5
C
11
8
10
9
D
13
E
14
20
16
19
18

1 **middle temporal artery** *origin,* superficial temporal artery; *branches,* none; *distribution,* temporal region. <arteria temporalis media> A

2 **mylohyoid branch of inferior alveolar artery** a branch that descends with the mylohyoid nerve in the mylohyoid sulcus to supply the floor of the mouth. <ramus mylohyoideus arteriae alveolaris inferioris> A

3 **posterior deep temporal artery** *origin,* maxillary artery; *branches,* none; *distribution,* temporalis muscle, and anastomoses with middle temporal artery. <arteria temporalis profunda posterior> A

4 **pterygoid branches of maxillary artery** branches that supply the pterygoid muscles. <rami pterygoidei arteriae maxillaris> A

5 **sphenopalatine artery** *origin,* maxillary artery; *branches,* posterior lateral nasal artery and posterior septal branches; *distribution,* structures adjoining nasal cavity, the nasopharynx. Called also *nasopalatine artery.* <arteria sphenopalatina> A

6 **middle meningeal artery** *origin,* maxillary artery; *branches,* frontal, parietal, and lacrimal anastomotic, accessory meningeal, and petrosal rami, and superior tympanic artery; *distribution,* cranial bones, dura mater. <arteria meningea media> A

7 ***accessory branch of middle meningeal artery*** a branch arising from the middle meningeal artery, or directly from the maxillary artery, and entering the middle cranial fossa through the foramen ovale to supply the trigeminal ganglion, walls of the cavernous sinus, and neighboring dura mater. <ramus accessorius arteriae meningeae mediae> A

8 ***deep temporal veins*** veins that drain the deep portions of the temporal muscle and empty into the pterygoid plexus. <venae temporales profundae> B

9 **maxillary veins** veins from the pterygoid plexus, usually forming a single short trunk, passing back and uniting with the superficial temporal vein in the parotid gland to form the retromandibular vein. <venae maxillares> B

10 ***middle meningeal veins*** the venae comitantes of the middle meningeal artery, which end in the pterygoid venous plexus. <venae meningeae mediae> B

11 **pterygoid plexus** a network of veins corresponding to the second and third parts of the maxillary artery; situated on the lateral surface of the medial pterygoid muscle and on both surfaces of the lateral pterygoid muscle, and draining into the facial vein. <plexus pterygoideus> B

12 ***vein of pterygoid canal*** one of the veins that pass through the pterygoid canal and empty into the pterygoid plexus; called also *vidian vein.* <vena canalis pterygoidei> B

13 **artery of pterygoid canal** 1. *origin,* maxillary artery; *branches,* pterygoid; *distribution,* roof of pharynx, auditory tube. Called also *vidian artery.* 2. *origin,* internal carotid artery; *branches,* none; *distribution,* pterygoid canal, anastomosing with the artery of the pterygoid canal that branches from the maxillary artery. <arteria canalis pterygoidei> C

14 **maxillary nerve** one of the three terminal divisions of the trigeminal nerve, passing through the foramen rotundum of the sphenoid bone, and entering the pterygopalatine fossa. *Origin,* trigeminal ganglion; *branches,* meningeal branch, zygomatic, superior alveolar, and infraorbital nerves, ganglionic branches to pterygopalatine ganglion, and, indirectly, the branches of the pterygopalatine ganglion; *distribution,* extensive distribution to skin of face and scalp, mucous membrane of maxillary sinus and nasal cavity, and teeth; *modality,* general sensory. <nervus maxillaris> D

15 **descending palatine artery** *origin,* maxillary artery; *branches,* greater and lesser palatine arteries; *distribution,* soft palate, hard palate, tonsil. <arteria palatina descendens> C

Pterygopalatine Fossa

16 **pterygopalatine fossa** a small space between the anterior aspect of the root of the pterygoid process of the sphenoid bone and the posterior aspect of the maxilla. Called also *pterygomaxillary fossa.* <fossa pterygopalatina> E

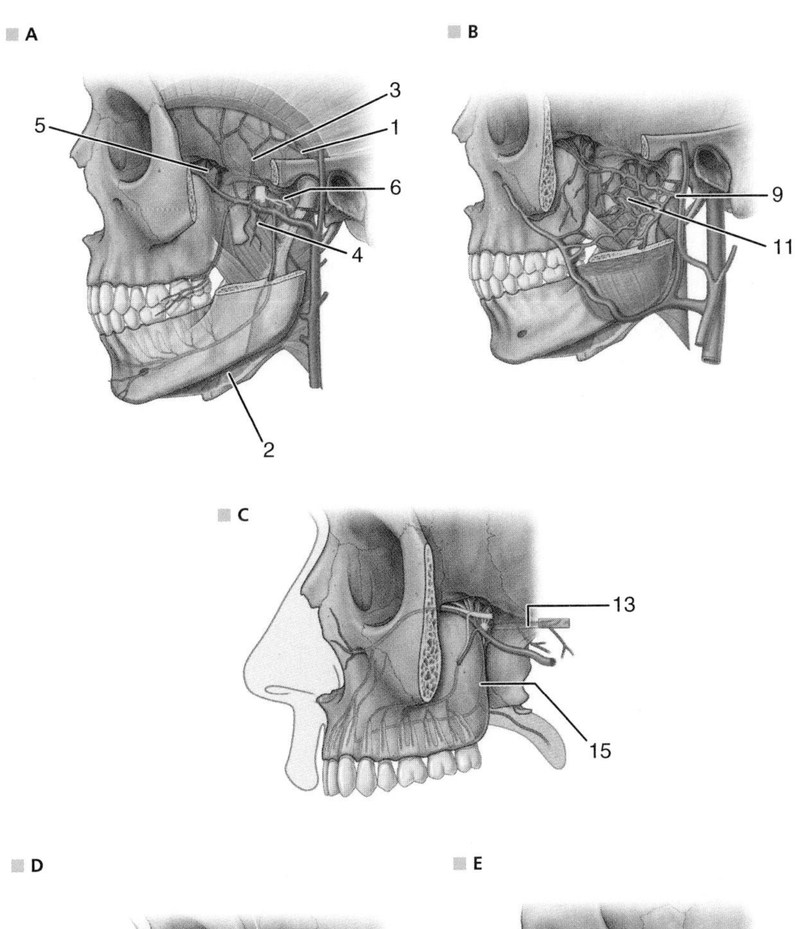
A
B
C
5
3
1
6
4
2
9
11
13
15

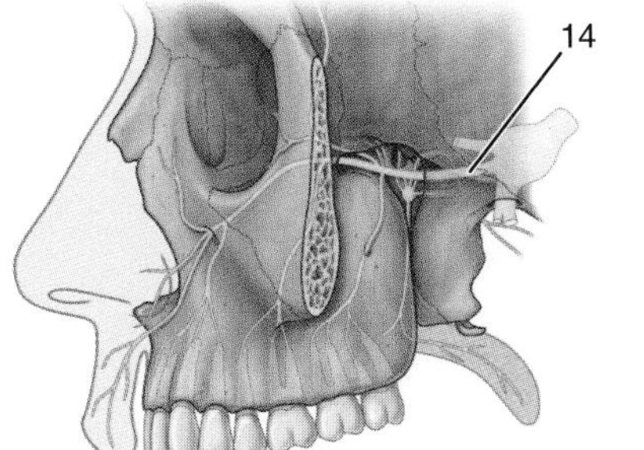
D
14

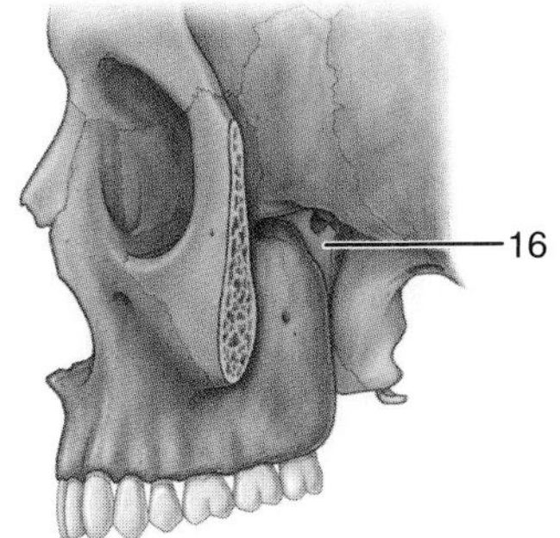
E
16

1 **lesser palatine canals** openings in the palatine bone that branch off the greater palatine canal to carry the lesser and middle palatine nerves and vessels to the roof of the mouth; they end at the lesser palatine foramina. Called also *accessory palatine canals.* <canales palatini minores> A

2 **greater palatine nerve** *origin,* pterygopalatine ganglion; *branches,* posterior inferior (lateral) nasal branches; *distribution,* emerges through the greater palatine foramen and supplies the palate; *modality,* parasympathetic, sympathetic, and general sensory. <nervus palatinus major> B

3 **lesser palatine nerves** *origin,* pterygopalatine ganglion; *distribution,* emerge through the lesser palatine foramen and supply the soft palate and tonsil; *modality,* parasympathetic, sympathetic, and general sensory. <nervi palatini minores> B

4 **middle superior alveolar branch of maxillary nerve** a branch from the infraorbital nerve that innervates the premolar teeth of the upper jaw by way of the superior dental plexus; *modality,* general sensory. <ramus alveolaris superior medius nervi maxillaris> B

5 **ganglionic branches of maxillary nerve to pterygopalatine ganglion** fibers connecting the maxillary nerve to the pterygopalatine ganglion. <rami ganglionares nervi maxillaris ad ganglion pterygopalatinum> C

6 **nerve of pterygoid canal** *origin,* union of deep and greater petrosal nerves; *distribution,* pterygopalatine ganglion and branches; *modality,* parasympathetic and sympathetic. <nervus canalis pterygoidei> C

7 **orbital branches of pterygopalatine ganglion** branches passing from the pterygopalatine ganglion through the inferior orbital fissure to supply the orbital periosteum and the ethmoidal and sphenoidal sinuses; *modality,* general sensory and parasympathetic. <rami orbitales nervi maxillaris> C

8 **pterygoid canal** a horizontal canal that passes forward through the base of the medial pterygoid plate of the sphenoid bone to open into the posterior wall of the pterygopalatine fossa just medial and inferior to the foramen rotundum; it transmits the pterygoid vessels and nerves. Called also *vidian canal.* <canalis pterygoideus> C

9 **pterygopalatine ganglion** a parasympathetic ganglion in the pterygopalatine fossa; its preganglionic fibers are derived from the facial nerve via the greater petrosal nerve and the nerve of the pterygopalatine canal. Its postganglionic fibers supply the lacrimal, nasal, and palatine glands; sensory and sympathetic fibers pass through the ganglion. Called also *Meckel's ganglion.* <ganglion pterygopalatinum> C

10 **sphenopalatine foramen** an opening on the medial wall of the pterygopalatine fossa, interconnecting this fossa with the nasal cavity, and transmitting the sphenopalatine artery and nasal nerves. Called also *pterygopalatine foramen.* <foramen sphenopalatinum> C

11 **deep petrosal nerve** *origin,* internal carotid plexus; *distribution,* joins greater petrosal nerve to form nerve of pterygoid canal, and supplies lacrimal, nasal, and palatine glands via pterygopalatine ganglion and its branches; *modality,* sympathetic. <nervus petrosus profundus> D

12 **greater palatine artery** *origin,* descending palatine; *branches,* none; *distribution,* hard palate. <arteria palatina major> E

13 **lesser palatine arteries** *origin,* descending palatine; *branches,* none; *distribution,* soft palate, tonsil. <arteriae palatinae minores> E

14 **pharyngeal branch of artery of pterygoid canal** a branch lying medial to the pterygopalatine ganglion. <ramus pharyngeus arteriae canalis pterygoidei> E

15 **inferior posterior nasal branches of greater palatine nerve** nerve branches of the maxillary nerve, usually branches of the greater palatine nerve, that supply the middle and inferior nasal meatus and inferior conchae; *modality,* general sensory. <rami nasales posteriores inferiores nervi palatini majoris> F

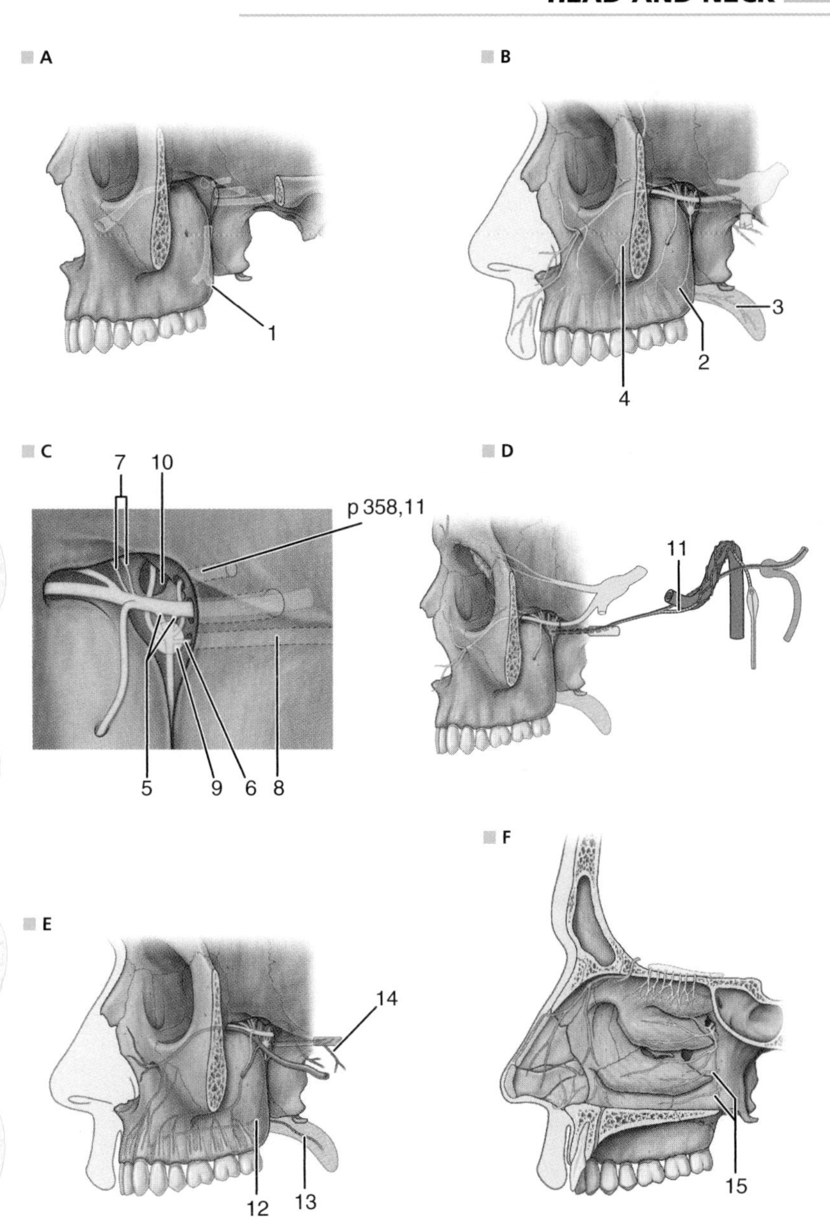
A
1
B
3
2
4
C
7
10
p 358,11
5
9
6
8
D
11
E
14
12
13
F
15

Pharynx

1 **pharynx** the musculomembranous passage between the mouth and posterior nares and the larynx and esophagus. The part above the level of the soft palate is the *nasopharynx.* The lower portion consists of two sections: the *oropharynx* and the *laryngopharynx.* Called also *throat.* A

2 **epiglottis** the lidlike cartilaginous structure overhanging the entrance to the larynx and serving to prevent food from entering the larynx and trachea while swallowing. A

3 **laryngopharynx** the portion of the pharynx that lies below the upper edge of the epiglottis and opens into the larynx and esophagus. <pars laryngea pharyngis> A

4 ***lateral pharyngeal space*** the part of the peripharyngeal space that is lateral to the pharynx. <spatium lateropharyngeum> A

5 ***lymphoid nodules of pharyngeal tonsils*** small collections of lymphoid tissue associated with the pharyngeal tonsils. <noduli lymphoidei tonsillae pharyngealis> A

6 ***mucosa of pharynx*** the mucous membrane (tunica mucosa) lining the pharynx. <tunica mucosa pharyngis> A

7 **nasopharynx** the part of the pharynx that lies above the level of the soft palate. <pars nasalis pharyngis> A

8 **oropharynx** the division of the pharynx lying between the soft palate and the upper edge of the epiglottis. <pars oralis pharyngis> A

9 **palatine tonsil** either of two small, almond-shaped masses located between the palatoglossal and palatopharyngeal arches, one on either side of the oropharynx, composed mainly of lymphoid tissue, covered with mucous membrane, and containing various crypts and many lymph follicles. Called also *tonsil* and *faucial tonsil.* <tonsilla palatina> A

10 ***peripharyngeal space*** the space around the pharynx, which is filled with areolar tissue; it is subdivided into the lateral pharyngeal and the retropharyngeal spaces. <spatium peripharyngeum> A

11 ***pharyngeal bursa*** an inconstant blind sac located above the pharyngeal tonsil in the midline of the posterior wall of the nasopharynx; it represents persistence of an embryonic communication between the anterior tip of the notochord and the roof of the pharynx. <bursa pharyngealis> A

12 **pharyngeal cavity** the space enclosed by the walls of the pharynx. <cavitas pharyngis> A

13 ***pharyngeal glands*** mucous glands beneath the tunica mucosa of the pharynx. <glandulae pharyngeales> A

14 **pharyngeal opening of auditory tube** the opening at the inferior end of either auditory tube, located on each of the lateral walls of the pharynx, posterior and inferior to the posterior end of the inferior nasal concha. <ostium pharyngeum tubae auditivae> A

15 **pharyngeal recess** a wide, slitlike lateral extension in the wall of the nasopharynx, superior and posterior to the pharyngeal orifice of the auditory tube; called also *Rosenmüller's fossa.* <recessus pharyngeus> A

16 **pharyngeal tonsil** the diffuse lymphoid tissue and follicles in the roof and posterior wall of the nasopharynx; called also *adenoid tonsil.* <tonsilla pharyngea> A

17 ***pharyngobasilar fascia*** a strong fibrous membrane in the wall of the pharynx, lined internally with mucous membrane and incompletely covered on its outer surface by the overlapping constrictor muscles of the pharynx. It blends with the periosteum at the base of the skull. <fascia pharyngobasilaris> A

18 **salpingopalatine fold** the mucosal fold passing caudally from the auditory tube to the lateral pharyngeal wall. <plica salpingopalatina> A

19 **salpingopharyngeal fold** a mucosal fold passing caudally from the posterior lip of the pharyngeal opening of the auditory tube to the lateral pharyngeal wall. <plica salpingopharyngea> A

20 ***submucosa of pharynx*** the submucous layer of the pharynx. <tela submucosa pharyngis> A

21 ***tonsillar branches of glossopharyngeal nerve*** branches that supply the mucosa over the palatine tonsil and the adjacent portion of the soft palate; *modality,* general sensory. <rami tonsillares nervi glossopharyngei> A

22 ***tonsillar branches of lesser palatine nerves*** branches that innervate the palatine tonsils. <rami tonsillares nervorum palatinorum minorum> A

23 ***tonsillar capsule*** a fibrous capsule covering the lateral surface of the palatine tonsils and separating them from the underlying connective tissue. <capsula tonsillaris> A

24 ***tonsillar crypts of palatine tonsil*** crypts within a palatine tonsil, representing the blind ends of the tonsillar pits. <cryptae tonsillares tonsillae palatinae> A

25 ***tonsillar crypts of pharyngeal tonsil*** crypts found within a pharyngeal tonsil, representing the blind ends of the tonsillar pits. <cryptae tonsillares tonsillae pharyngeae> A

26 ***tonsillar crypts of tubal tonsil*** small invaginations extending into a tubal tonsil. <cryptae tonsillares tonsillae tubariae> A

27 ***tonsillar fossa*** the depression between the palatoglossal and palatopharyngeal arches in which the palatine tonsil is located; called also *tonsillar sinus.* <fossa tonsillaris> A

28 ***tonsillar pits of palatine tonsil*** the mouths of the tonsillar crypts of the palatine tonsils. <fossulae tonsillares tonsillae palatinae> A

29 ***tonsillar pits of pharyngeal tonsil*** the mouths of the tonsillar crypts of the pharyngeal tonsils. <fossulae tonsillares tonsillae pharyngeae> A

A

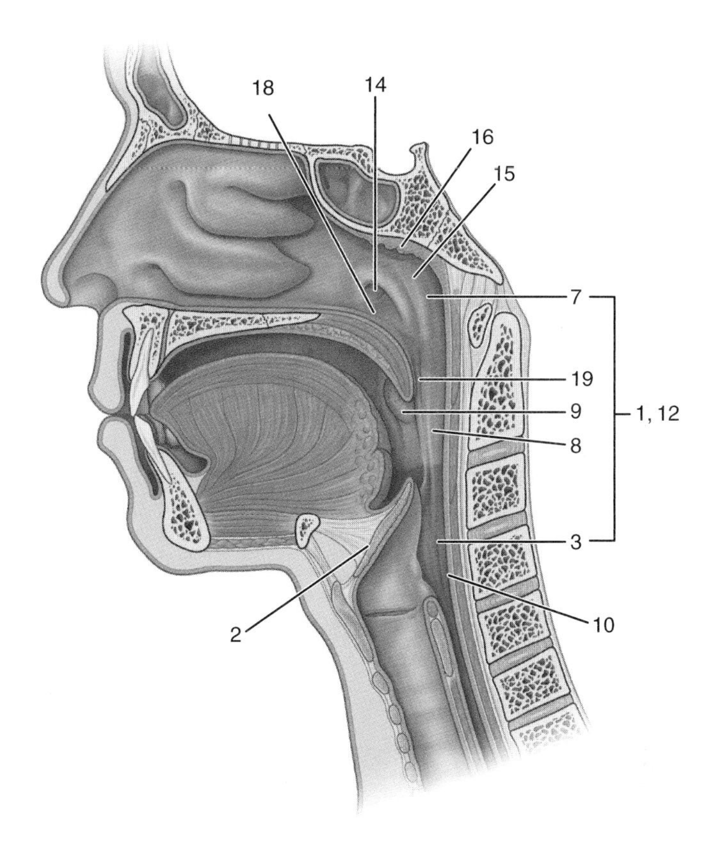

1 **torus levatorius** the mucosal fold covering the levator veli palatini muscle in the lateral wall of the nasal part of the pharynx. A

2 **torus tubarius** the projecting posterior lip of the pharyngeal opening of the auditory tube. A

3 **triangular fold** a fold of mucous membrane extending backward from the palatoglossal arch and covering the anteroinferior part of the palatine tonsil. <plica triangularis> A

4 ***tubal tonsil*** a collection of lymphoid tissue associated with the pharyngeal opening of the auditory tube. <tonsilla tubaria> A

5 **vault of pharynx** the archlike roof of the nasopharynx; called also *nasopharyngeal roof.* <fornix pharyngis> A

6 **pharyngeal nerve** a nerve running from the posterior part of the pterygopalatine ganglion, through the palatovaginal canal with the pharyngeal branch of the maxillary artery, to the mucous membrane of the nasal part of the pharynx posterior to the auditory tube. <nervus pharyngeus> B

7 ***glossopharyngeal part of superior constrictor muscle of pharynx*** the part of the superior constrictor muscle arising from the side of the root of the tongue. Called also *glossopharyngeus muscle.* <pars glossopharyngea musculi constrictoris pharyngis superioris> C

8 **inferior constrictor muscle of pharynx** *origin,* undersurfaces of cricoid and thyroid cartilages; *insertion,* median raphe of posterior wall of pharynx; *innervation,* glossopharyngeal, pharyngeal plexus, branches of superior laryngeal and recurrent laryngeal; *action,* constricts pharynx. <musculus constrictor pharyngis inferior> C

9 **middle constrictor muscle of pharynx** *origin,* cornua of hyoid and stylohyoid ligament; *insertion,* median raphe of posterior wall of pharynx; *innervation,* pharyngeal plexus of vagus and glossopharyngeal; *action,* constricts pharynx. <musculus constrictor pharyngis medius> C

10 **mylopharyngeal part of superior constrictor muscle of pharynx** the part of the superior constrictor muscle arising from the mylohyoid ridge of the mandible; called also *mylopharyngeal muscle.* <pars mylopharyngea musculi constrictoris pharyngis superioris> C

11 **pterygomandibular raphe** a tendinous line between the buccinator and the superior constrictor muscles of pharynx, from which the middle portions of both muscles originate. <raphe pterygomandibularis> C

12 **pterygopharyngeal part of superior constrictor muscle of pharynx** the part of the superior constrictor muscle arising from the caudal part and hamulus of the medial pterygoid plate; called also *pterygopharyngeus muscle.* <pars pterygopharyngea musculi constrictoris pharyngis superioris> C

13 **superior constrictor muscle of pharynx** *origin,* medial pterygoid plate, pterygomandibular raphe, mylohyoid ridge of mandible, and mucous membrane of floor of mouth; *insertion,* median raphe of posterior wall of pharynx; *innervation,* pharyngeal plexus of vagus; *action,* constricts pharynx. <musculus constrictor pharyngis superior> C

14 **thyropharyngeal part of inferior constrictor muscle of pharynx** the part of the inferior constrictor muscle arising from the thyroid cartilage; called also *thyropharyngeus muscle.* <pars thyropharyngea musculi constrictoris inferioris pharyngis> C

15 **pharyngeal raphe** a more or less distinct band of connective tissue extending downward from the base of the skull along the posterior wall of the pharynx in the median plane, and giving attachment to the constrictor muscles of the pharynx. <raphe pharyngis> D

A

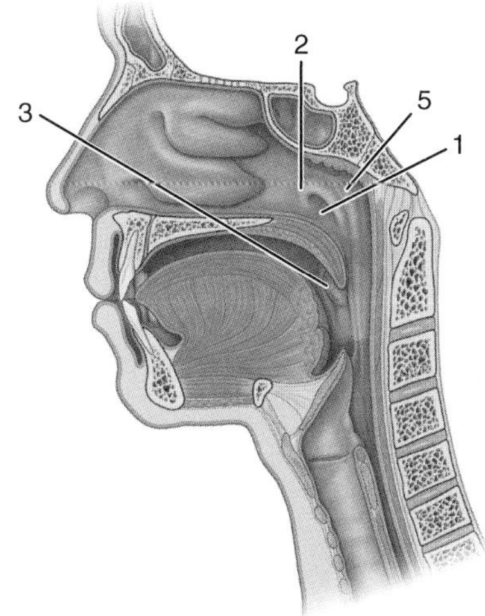

B

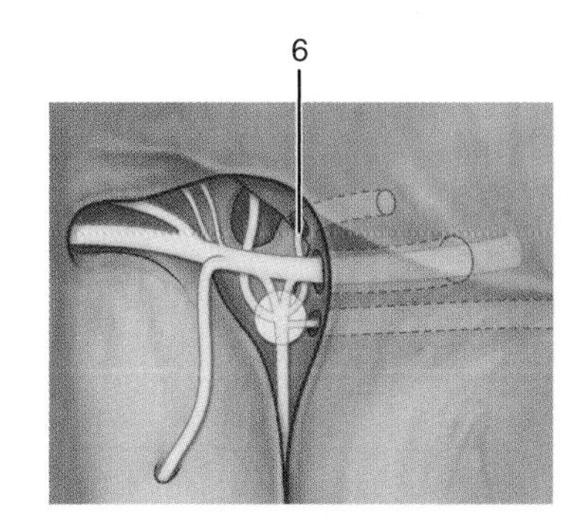

D

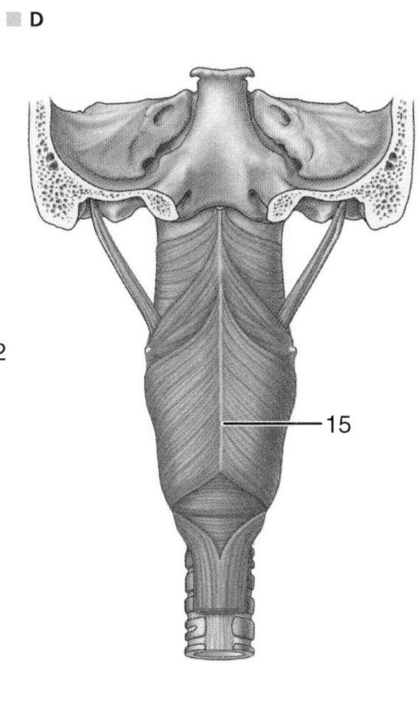

C

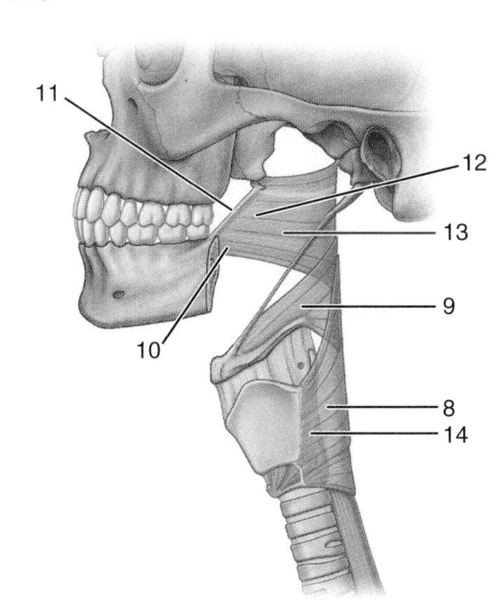

1 ***pharyngeal muscles*** the muscular coat of the pharynx, consisting of the three constrictor muscles and the stylopharyngeus, salpingopharyngeus, and palatopharyngeus muscles. <musculi pharyngis> A

2 **salpingopharyngeus muscle** *origin,* auditory tube near its orifice; *insertion,* posterior part of palatopharyngeus; *innervation,* pharyngeal plexus of vagus; *action,* elevates upper lateral wall of pharynx. <musculus salpingopharyngeus> A

3 **stylopharyngeus muscle** one of the intrinsic muscles of the larynx; *origin,* styloid process; *insertion,* thyroid cartilage and pharyngeal constrictors; *innervation,* pharyngeal plexus, glossopharyngeal; *action,* raises and dilates pharynx. <musculus stylopharyngeus> A

4 ***supratonsillar fossa*** the triangular space between the palatoglossal and palatopharyngeal arches superior to the tonsil. <fossa supratonsillaris> A

5 ***cricoesophageal tendon*** the tendon giving origin to the longitudinal fibers of the esophagus that come from the upper part of the lamina of the cricoid cartilage. <tendo cricooesophageus> B

6 **pharyngoesophageal constriction** the narrowing where the pharynx ends and the cervical esophagus begins, the site of the pharyngoesophageal sphincter. Called also *pharyngoesophageal junction.* <constrictio pharyngooesophagealis> B

7 **piriform recess [fossa]** a pear-shaped fossa in the wall of the laryngeal pharynx lateral to the arytenoid cartilage and medial to the lamina of the thyroid cartilage. <recessus piriformis> B

8 **tonsillar branch of facial artery** a vessel ascending from the facial artery on the pharynx to supply the tonsil and the root of the tongue. <ramus tonsillaris arteriae facialis> C

9 **pharyngeal branch of vagus nerve** any of several branches that innervate the pharyngeal muscles and mucosa and have motor and general sensory modalities. <ramus pharyngeus nervi vagi> D

10 **pharyngeal branches of glossopharyngeal nerve** branches innervating the mucous membrane of the oropharynx, with a general sensory modality. <rami pharyngei nervi glossopharyngei> D

11 **pharyngeal plexus of vagus nerve** a plexus formed chiefly by fibers from branches of the vagus nerves, but also containing fibers from the glossopharyngeal nerves and sympathetic trunks, and supplying motor, general sensory, and sympathetic innervation to the muscles and mucosa of the pharynx and soft palate, except for the tensor veli palatini muscle. <plexus pharyngeus nervi vagi> D

12 **epiglottic cartilage** the plate of cartilage that constitutes the central part of the epiglottis. <cartilago epiglottica> E

13 **epiglottic petiole** the pointed lower end of the epiglottic cartilage, which is attached to the back of the thyroid cartilage. <petiolus epiglottidis> E

14 **epiglottic tubercle** a posterior projection on the inferior part of the posterior surface of the epiglottic cartilage. <tuberculum epiglotticum> E

15 **epiglottic vallecula** a depression between the lateral and median glossoepiglottic folds on each side. <vallecula epiglottica> F

16 **lateral glossoepiglottic fold** either of two folds of mucous membrane extending, one on either side, between the base of the tongue and the epiglottis. <plica glossoepiglottica lateralis> F

17 **median glossoepiglottic fold** a single fold of mucous membrane between the two lateral glossoepiglottic folds, connecting the base of the tongue and the epiglottis. <plica glossoepiglottica mediana> F

18 **fauces** the passage from the mouth to the pharynx, including both the lumen and its boundaries. G

A

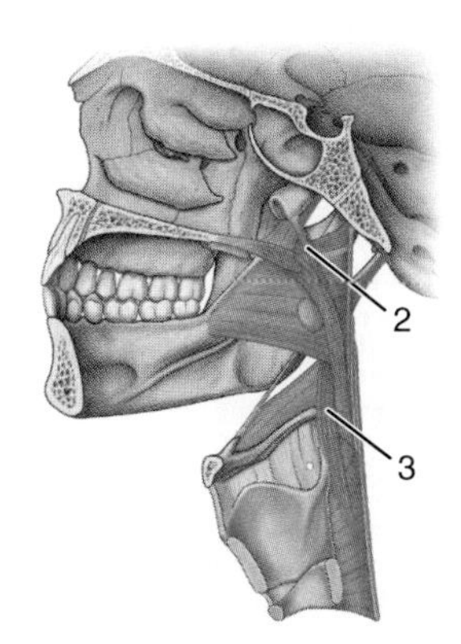

B

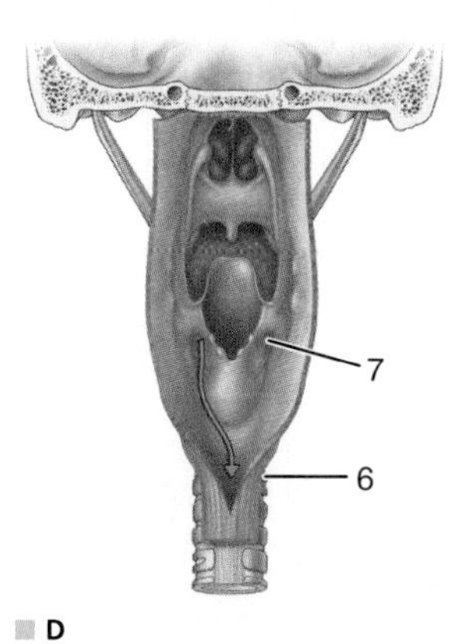

C

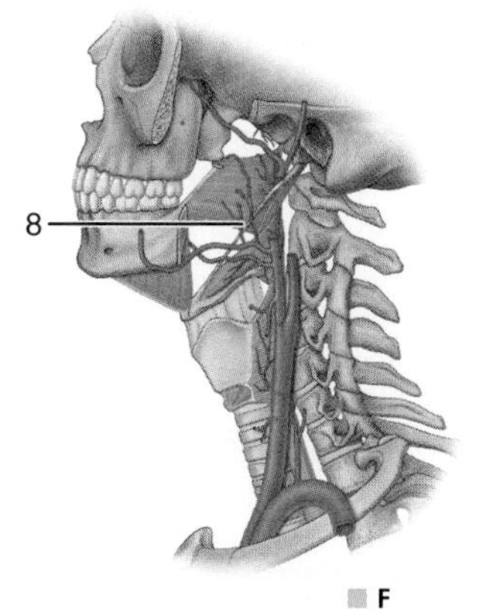

D

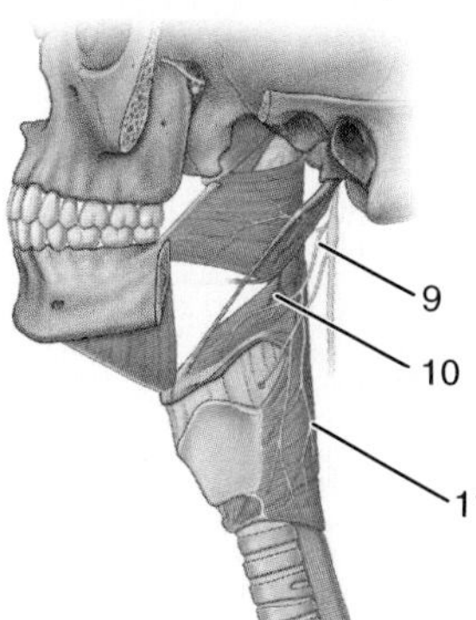

E

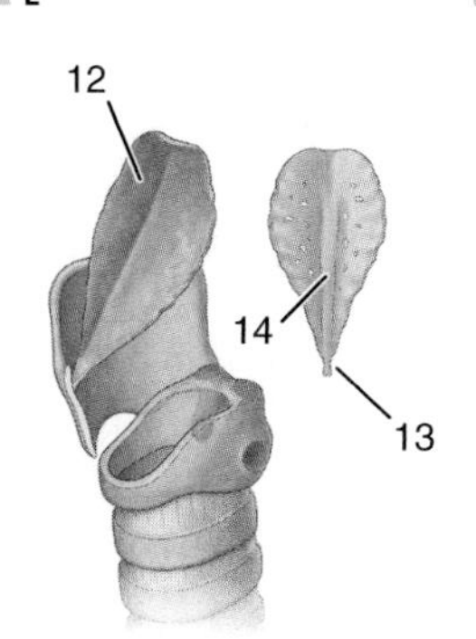

F

17 15 16

G

18

Larynx

1 **cricopharyngeal ligament** a ligament extending from the cricoid lamina to the midline of the pharynx. <ligamentum cricopharyngeum> A

2 **external branch of superior laryngeal nerve** the smaller of the two branches into which the superior laryngeal nerve divides, descending under cover of the sternothyroid muscle and innervating the cricothyroid and the inferior constrictor of the pharynx; *modality,* motor. <ramus externus nervi laryngei superioris> B

3 **thyroid cartilage** the largest cartilage of the larynx, with two broad, posteriorly diverging laminae and two pairs of horns, superior and inferior, that extend from the posterior borders of the laminae. <cartilago thyroidea> C

4 ***thyroid foramen*** an inconstantly present opening in the upper part of the lamina of the thyroid cartilage, resulting from incomplete union of the fourth and fifth branchial cartilages. <foramen thyroideum> C

5 **right/left lamina of thyroid cartilage** either of the broad plates that form the right and left sides of the cartilage, converging anteriorly to meet at the midline. <lamina cartilaginis thyroideae dextra/sinistra> C

6 **superior horn of thyroid cartilage** the superior extension of the posterior border of the thyroid cartilage. <cornu superius cartilaginis thyroideae> C

7 **superior thyroid notch** a deep notch in the upper portion of the anterior border of the thyroid cartilage. <incisura thyroidea superior> C

8 **superior thyroid tubercle** a more or less distinct tubercle at the superior extremity of the oblique line of the thyroid cartilage. <tuberculum thyroideum superius> C

9 **thyroepiglottic ligament** a fibrous band that attaches the epiglottic petiole to the thyroid cartilage just below the superior notch. <ligamentum thyroepiglotticum> D

10 **articular surface of arytenoid cartilage** the surface that articulates with the cricoid cartilage. <facies articularis cartilaginis arytenoidea> E

11 **cricoarytenoid joint** the synovial joint between the upper border of the cricoid cartilage and the base of the arytenoid cartilage. <articulatio cricoarytenoidea> E

12 ***cricoarytenoid ligament*** the ligament extending from the lamina of the cricoid cartilage to the medial surface of the base and muscular process of the arytenoid cartilage. <ligamentum cricoarytenoideum> E

13 **cricoid cartilage** a ringlike cartilage forming the lower and back part of the larynx. <cartilago cricoidea> E

14 **cuneiform cartilage** either of a pair of cartilages, one on the posterior border of each aryepiglottic fold; called also *Wrisberg's cartilage.* <cartilago cuneiformis> E

15 **lamina of cricoid cartilage** the broad posterior part of the cricoid cartilage. <lamina cartilaginis cricoideae> E

16 ***laryngeal cartilages and articulations*** cartilages of the larynx, including the cricoid, thyroid, and epiglottic, and two each of the arytenoid, corniculate, and cuneiform, together with the joints between them. <cartilagines et articulationes laryngis> E

17 **medial surface of arytenoid cartilage** the surface that faces medially toward the opposite arytenoid cartilage. <facies medialis cartilaginis arytenoideae> E

18 **muscular process of arytenoid cartilage** the lateral and posterior lower angular projection of the arytenoid cartilage to which the cricoarytenoid muscles are attached. <processus muscularis cartilaginis arytenoideae> E

19 **oblong fovea of arytenoid cartilage** a depression on the anterolateral surface of the arytenoid cartilage, separated from the triangular pit above by the arcuate crest; called also *oblong pit of arytenoid cartilage.* <fovea oblonga cartilaginis arytenoideae> E

20 **posterior surface of arytenoid cartilage** the concave dorsal surface, to which various laryngeal muscles are attached. <facies posterior cartilaginis arytenoideae> E

21 **triangular fovea of arytenoid cartilage** a depression on the anterolateral surface of the arytenoid cartilage, separated from the oblong pit below by the arcuate crest. <fovea triangularis cartilaginis arytenoideae> E

22 **vocal process** the process of the arytenoid cartilage to which the vocal ligament is attached. <processus vocalis> E

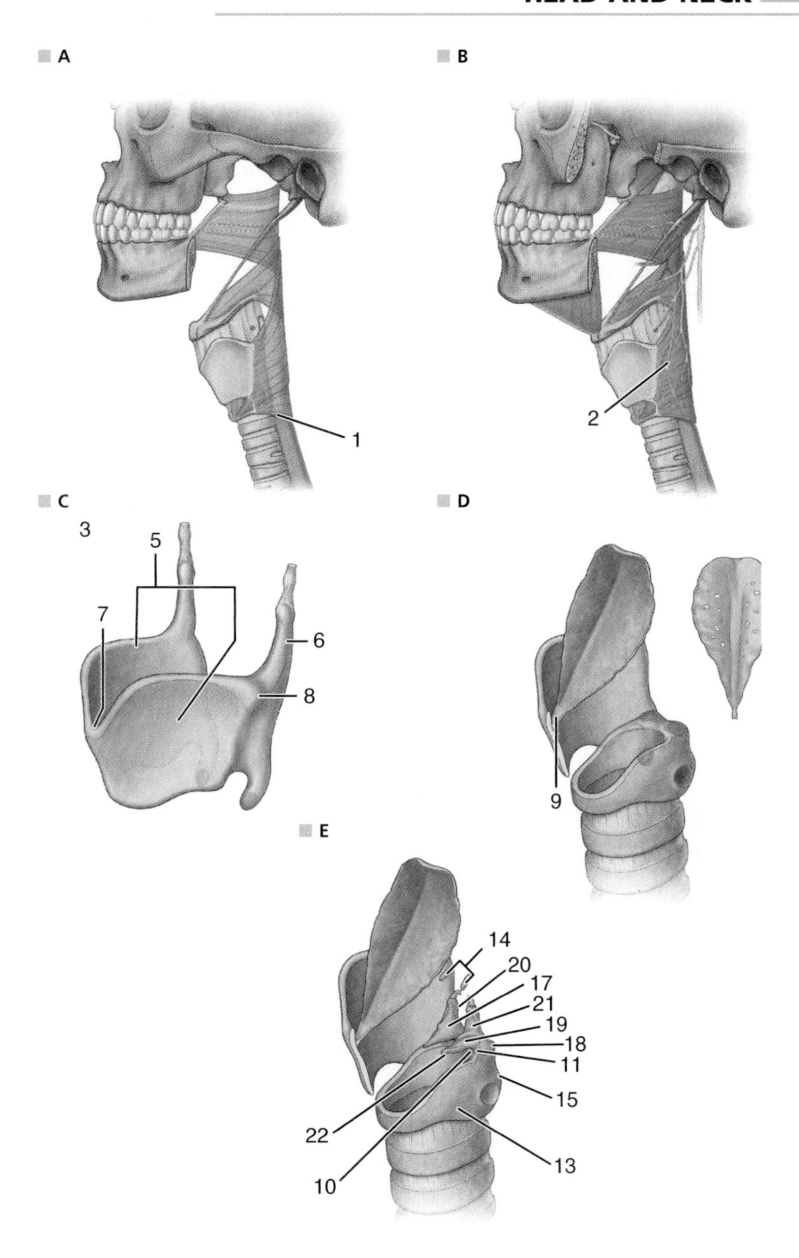
A
1
B
2
C
3
5
7
6
8
D
9
E
14
20
17
21
19
18
11
15
22
13
10

1 **ceratocricoid ligament** any of the three (anterior, lateral, or posterior) fibrous bands that serve to attach the capsule of the cricothyroid joint on either side. <ligamentum ceratocricoideum> A

2 **arch of cricoid cartilage** the slender anterior portion of the cricoid cartilage. <arcus cartilaginis cricoideae> B

3 **arytenoid articular surface of cricoid cartilage** the surface that articulates with the arytenoid cartilage. <facies articularis arytenoidea cartilaginis cricoideae> B

4 **inferior horn of thyroid cartilage** the inferior extension of the posterior border of the thyroid cartilage. <cornu inferius cartilaginis thyroideae> C

5 **inferior thyroid notch** a notch at the lower part of the anterior border of the thyroid cartilage. <incisura thyroidea inferior> A, C

6 **inferior thyroid tubercle** a more or less distinct tubercle at the inferior end of the oblique line of the thyroid cartilage. <tuberculum thyroideum inferius> C

7 **anterolateral surface of arytenoid cartilage** the external surface of the cartilage, which bears the triangular pit, the oblong pit, and the arcuate crest. <facies anterolateralis cartilaginis arytenoideae> D

8 **apex of arytenoid cartilage** the upper part of the arytenoid cartilage, which bends posteriorly and medially and connects with the corniculate cartilage. <apex cartilaginis arytenoideae> D

9 **arcuate crest of arytenoid cartilage** a ridge on the external surface of the arytenoid cartilage between the triangular pit and the oblong pit. <crista arcuata cartilaginis arytenoideae> D

10 **arytenoid cartilage** one of the paired, pitcher-shaped cartilages of the back of the larynx at the upper border of the cricoid cartilage; called also *triquetral cartilage.* <cartilago arytenoidea> D

11 **base of arytenoid cartilage** the triangular inferior part of the arytenoid cartilage, which bears the articular surface. <basis cartilaginis arytenoideae> D

12 ***colliculus of arytenoid cartilage*** a small eminence on the anterior margin and anterolateral surface of the arytenoid cartilage. <colliculus cartilaginis arytenoideae> D

13 **corniculate cartilage** a small nodule of cartilage at the apex of each arytenoid cartilage; called also Santorini's cartilage. <cartilago corniculata> D

14 **lateral thyrohyoid ligament** a round elastic cord that forms the posterior margin of the thyrohyoid membrane; it extends from the tip of the superior horn of the thyroid cartilage upward to the tip of the greater horn of the hyoid bone. <ligamentum thyrohyoideum laterale> E

15 **conus elasticus** the paired lateral portion of the fibroelastic laryngeal membrane, which extends upward in parallel thickenings from the cricoid cartilage to the vocal ligaments; called also cricovocal membrane and lateral cricothyroid ligament. D

16 **aryepiglottic fold** a fold of mucous membrane extending on each side between the lateral border of the epiglottis and the summit of the arytenoid cartilage. <plica aryepiglottica> F

17 **corniculate tubercle** a rounded eminence near the posterior end of the aryepiglottic fold, posterior to the cuneiform tubercle, corresponding to the corniculate cartilage. Called also *tubercle of Santorini.* <tuberculum corniculatum> F

18 **cuneiform tubercle** a rounded eminence in the posterior portion of the aryepiglottic fold, anterior to the corniculate tubercle, corresponding to the cuneiform cartilage. <tuberculum cuneiforme> F

19 ***mucosa of trachea*** the mucous membrane (tunica mucosa) lining the trachea. <tunica mucosa tracheae> G

20 **trachea** the cartilaginous and membranous tube descending from the larynx and branching into the right and left main bronchi. It is kept patent by a series of about twenty transverse horseshoe-shaped cartilages. Called also *windpipe.* G

21 **lesser supraclavicular fossa** the region of the neck in the depression posterior to the clavicle, about the interval between the two tendons of the sternocleidomastoid muscle. <fossa supraclavicularis minor> H

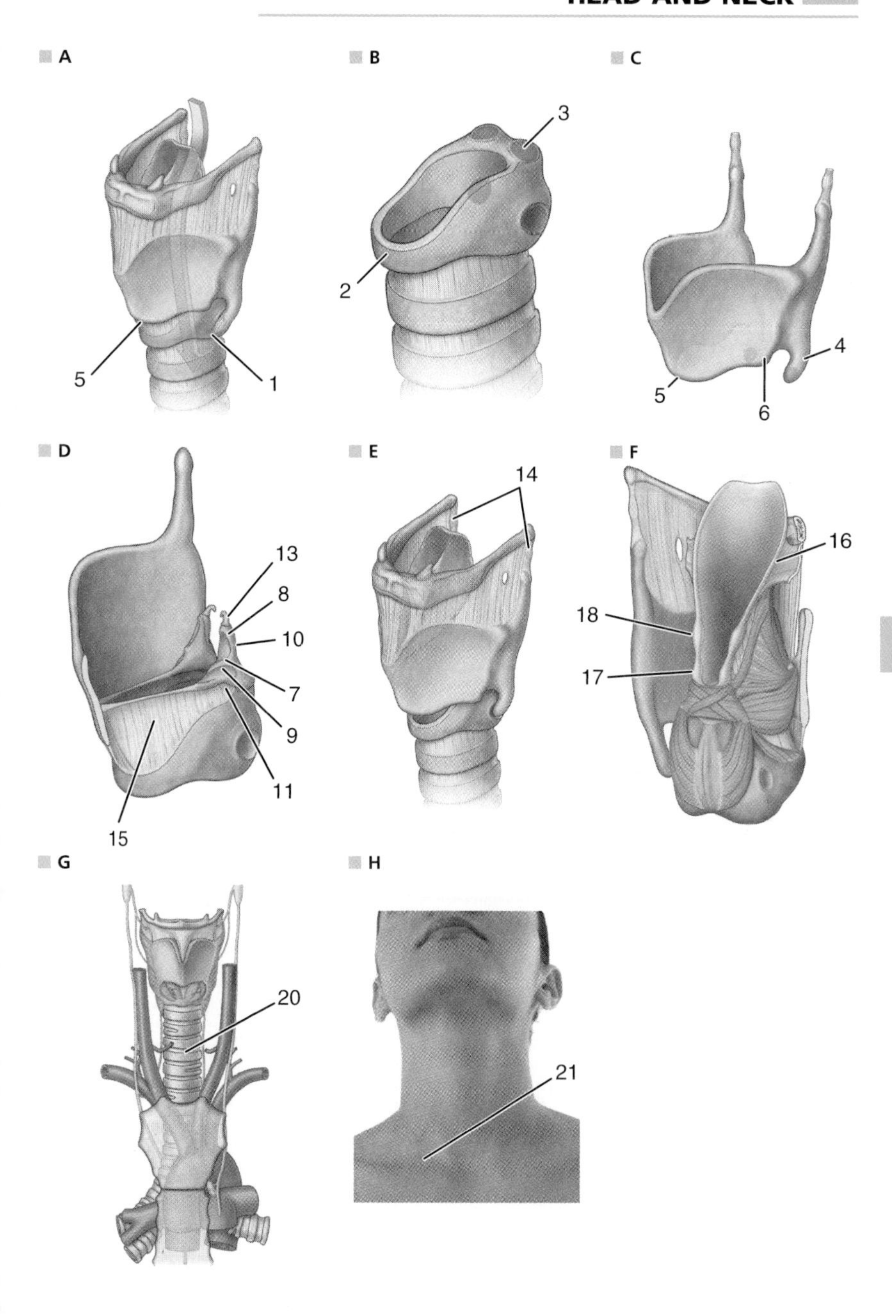
A
5
1
B
3
2
C
4
5
6
D
13
8
10
7
9
11
15
E
14
F
16
18
17
G
20
H
21

1 ***ceratocricoid muscle*** a name given a muscular fasciculus arising from the cricoid cartilage and inserted on the inferior horn of the thyroid cartilage considered one of the intrinsic muscles of the larynx. <musculus ceratocricoideus> A

2 **cricothyroid joint** the joint between the lateral aspect of the cricoid cartilage and the inferior horn of the thyroid cartilage. <articulatio cricothyroidea> A

3 **cricotracheal ligament** a narrow fibrous ring that connects the lower margin of the cricoid cartilage with the upper tracheal cartilage; it is continuous posteriorly with the membranous wall of the trachea. <ligamentum cricotracheale> A

4 ***laryngeal glands*** the mucous glands in the mucosa of the larynx. <glandulae laryngeales> A

5 ***larynx*** the musculocartilaginous structure, lined with mucous membrane, connected to the superior part of the trachea and to the pharynx inferior to the tongue and the hyoid bone; the essential sphincter guarding the entrance into the trachea and functioning secondarily as the organ of voice. It is formed by nine cartilages connected by ligaments and eight muscles. A

6 **median thyrohyoid ligament** the central, thicker portion of the thyrohyoid membrane; its broader upper part is attached to the body of the hyoid bone and its narrow lower end to the superior incisure of the thyroid cartilage. <ligamentum thyrohyoideum medianum> A

7 ***mucosa of larynx*** the mucous membrane lining the larynx. <tunica mucosa laryngis> A

8 **oblique line of thyroid cartilage** a line on the external surface of the lamina of the thyroid cartilage, extending between the two thyroid tubercles. <linea obliqua cartilaginis thyroideae> A

9 **thyrohyoid membrane** a broad fibroelastic sheet attached above to the upper margin of the posterior surface of the hyoid bone and below to the upper border of the thyroid cartilage. <membrana thyrohyoidea> A

10 ***thyroid articular surface of cricoid cartilage*** the surface that articulates with the thyroid cartilage. <facies articularis thyroidea cartilaginis cricoideae> A

11 **triticeal cartilage** a small cartilage in the thyrohyoid ligament. <cartilago triticea> A

12 **median cricothyroid ligament** the median or anterior part of the inferior, larger part of the fibroelastic laryngeal membrane, occurring as a flat band of white tissue continuous medially with the conus elasticus (cricovocal membrane) and cranially with the vocal fold and vocal ligament, and connecting the cricoid and thyroid cartilages; called also *anterior cricothyroid ligament*. <ligamentum cricothyroideum medianum> B

13 **interarytenoid fold** a median fold formed by mucous membrane anterior to the transverse arytenoid muscle as it protrudes into the larynx as the muscle approximates the arytenoid cartilages. <plica interarytenoidea> C

14 **interarytenoid notch** the posterior portion of the laryngeal outlet between the two arytenoid cartilages. <incisura interarytenoidea> C

15 **laryngeal inlet** the aperture by which the pharynx communicates with the larynx. <aditus laryngis> C

16 **fibroelastic membrane of larynx** the fibroelastic layer beneath the mucous coat of the larynx, comprising the quadrangular membrane and the conus elasticus. <membrana fibroelastica laryngis> C, D

17 **infraglottic cavity** the most inferior part of the laryngeal cavity, extending from the rima glottidis above to the cavity of the trachea below. Called also *subglottis*. <cavitas infraglottica> D

18 **laryngeal cavity** the space enclosed by the walls of the larynx. <cavitas laryngis> D

19 **laryngeal saccule** a diverticulum extending upward from the front of the laryngeal ventricle, between the vestibular fold medially and the thyroarytenoid muscle and thyroid cartilage laterally. <sacculus laryngis> D

20 **laryngeal ventricle** a lateral evagination of mucous membrane between the vocal and vestibular folds, reaching nearly to the angle of the thyroid cartilage. <ventriculus laryngis> D

21 **laryngeal vestibule** the portion of the laryngeal cavity above the vestibular folds. <vestibulum laryngis> D

22 ***pre-epiglottic fat body*** a mass of fatty tissue separating the lower anterior surface of the epiglottis from the thyroid cartilage and the thyrohyoid membrane. <corpus adiposum pre-epiglotticum> D

23 **quadrangular membrane** the upper part of the fibroelastic membrane of the larynx. <membrana quadrangularis> D

24 ***sesamoid cartilage of vocal ligament*** a small cartilage occasionally found within the vocal ligaments. <cartilago sesamoidea ligamenti vocalis> D

25 **vestibular fold** a fold of mucous membrane covering muscle in the larynx, separating the ventricle from the vestibule; called also *false vocal cord* or *fold*. <plica vestibularis> D

26 **vestibular ligament** the membrane that extends from the thyroid cartilage in front to the anterolateral surface of the arytenoid cartilage behind; it lies within the vestibular fold, above the vocal ligament. <ligamentum vestibulare> D

27 **vocal fold** a fold of mucous membrane covering the vocalis muscle in the larynx, forming the inferior boundary of the ventricle; called also *true vocal cord*. <plica vocalis> D

28 **vocal ligament** the elastic tissue membrane that extends from the thyroid cartilage in front to the vocal process of the arytenoid cartilage behind; it lies within the vocal fold, below the vestibular ligament. <ligamentum vocale> D

A

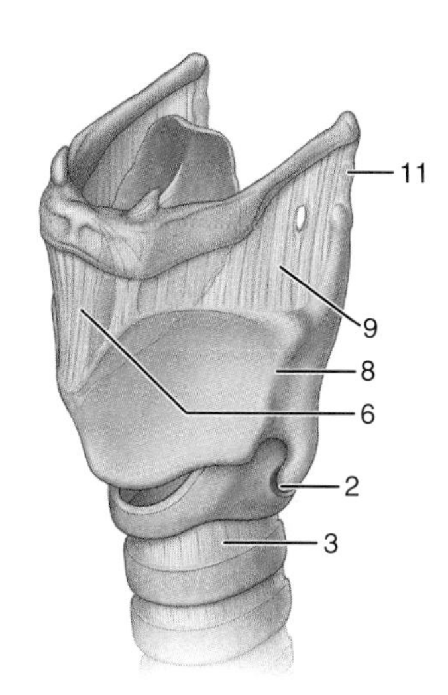

B

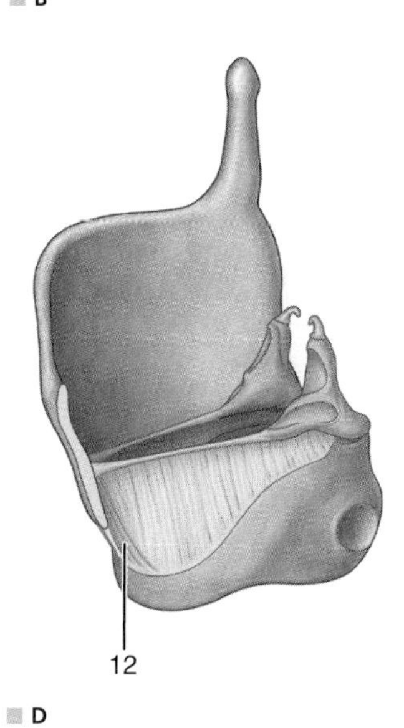

C

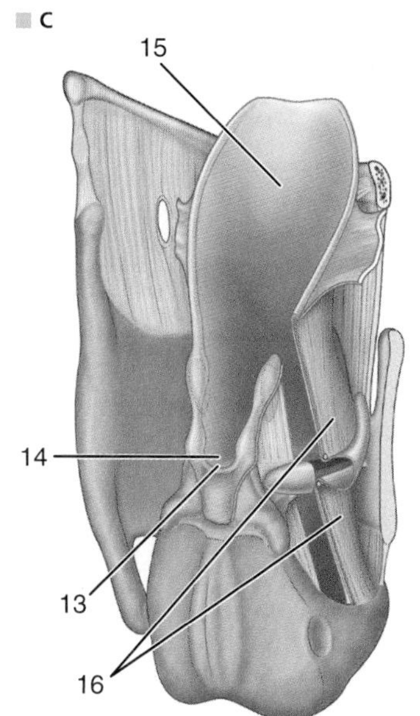

D

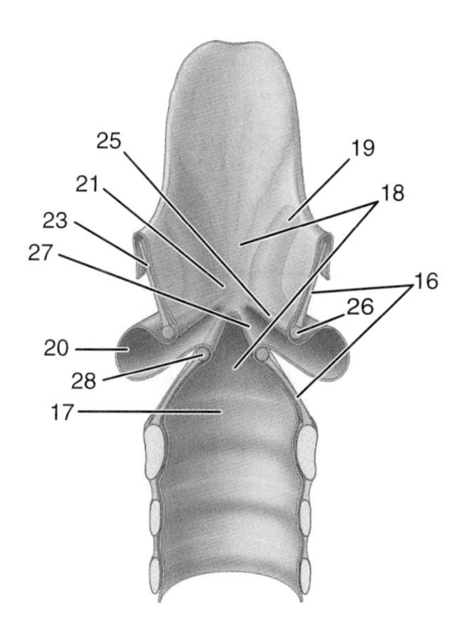

1 ***glottis*** the vocal apparatus of the larynx, consisting of the true vocal cords and the opening between them *(rima glottidis).* A

2 **intercartilaginous part of rima glottidis** the part of the rima glottidis between the arytenoid cartilages. <pars intercartilaginea rimae glottidis> A

3 **intermembranaceous part of rima glottidis** the part of the rima glottidis between the vocal folds. <pars intermembranacea rimae glottidis> A

4 **rima glottidis** the elongated opening between the true vocal cords and between the arytenoid cartilages. It consists of an intercartilaginous part and an intermembranous part. A

5 **rima vestibuli** the space between the right and left vestibular folds of the larynx. Called also *fissure of vestibule.* A

6 **oblique part of cricothyroid muscle** the fibers of the cricothyroid muscle that are inserted into the inferior horn, caudal margin, and inner surface of the thyroid cartilage. <pars obliqua musculi cricothyroidei> B

7 **straight part of cricothyroid muscle** the fibers of the cricothyroid muscle that are inserted into the straight margin of the thyroid cartilage. <pars recta musculi cricothyroidei> B

8 **vocalis muscle** one of the intrinsic muscles of the larynx; *origin,* angle between laminae of thyroid cartilage; *insertion,* vocal process of arytenoid cartilage; *innervation,* recurrent laryngeal; *action,* shortens and relaxes vocal folds. <musculus vocalis> C

9 ***laryngeal muscles*** the intrinsic and extrinsic muscles of the larynx, including the oblique and transverse arytenoid, ceratocricoid, lateral and posterior cricoarytenoid, cricothyroid, thyroarytenoid, and vocalis muscles. <musculi laryngis> D

10 **lateral cricoarytenoid muscle** one of the intrinsic muscles of the larynx; *origin,* lateral surface of cricoid cartilage; *insertion,* muscular process of arytenoid cartilage; *innervation,* recurrent laryngeal; *action,* approximates vocal folds. <musculus cricoarytenoideus lateralis> D

11 **oblique arytenoid muscle** one of the intrinsic muscles of the larynx; *origin,* posterior aspect of muscular process of arytenoid cartilage; *insertion,* apex of opposite arytenoid cartilage; *innervation,* recurrent laryngeal; *action,* closes inlet of larynx. <musculus arytenoideus obliquus> D

12 **posterior cricoarytenoid muscle** one of the intrinsic muscles of the larynx; *origin,* back of cricoid cartilage; *insertion,* muscular process of arytenoid cartilage; *innervation,* recurrent laryngeal; *action,* separates vocal folds. <musculus cricoarytenoideus posterior> D

13 **thyroarytenoid muscle** one of the intrinsic muscles of the larynx; *origin,* lamina of thyroid cartilage; *insertion,* muscular process of arytenoid cartilage; *innervation,* recurrent laryngeal; *action,* relaxes, shortens vocal folds. <musculus thyroarytenoideus> D

14 **thyroepiglottic part of thyroarytenoid muscle** fibers of the thyroarytenoid muscle that continue to the margin of the epiglottis; it closes the inlet to the larynx. Called also *thyroepiglottic muscle.* <pars thyroepiglottica musculi thyroarytenoidei> D

15 **transverse arytenoid muscle** one of the intrinsic muscles of the larynx; *origin,* posterior aspect of muscular process of arytenoid cartilage; *insertion,* posterior aspect of muscular process of opposite arytenoid cartilage; *innervation,* recurrent laryngeal; *action,* approximates arytenoid cartilages. <musculus arytenoideus transversus> D

16 **laryngeal prominence** a subcutaneous prominence on the front of the neck produced by the thyroid cartilage of the larynx; called also *Adam's apple.* <prominentia laryngea> E

17 ***subcutaneous bursa of laryngeal prominence*** a bursa anterior to the laryngeal prominence of the thyroid cartilage, under the skin. <bursa subcutanea prominentiae laryngeae> E

Nasal Cavity

18 ***mucosa of nasal cavity*** the mucous membrane (tunica mucosa) lining the nasal cavity. <tunica mucosa nasi> F

19 **nasal cavity** the portion of the passages of the respiratory system extending from the nares to the pharynx. It is divided into left and right halves by the nasal septum; its floor is the hard palate, which separates it from the oral cavity; and its lateral walls contain the nasal conchae and nasal meatus. <cavitas nasi> F

20 ***nasal glands*** numerous large mucous and serous glands in the respiratory part of the nasal cavity. <glandulae nasales> F

21 ***olfactory glands*** small mucous glands in the olfactory mucosa. <glandulae olfactoriae> F

22 ***vibrissae*** the hairs growing in the vestibular region of the nasal cavity. F

23 ***vomeronasal organ*** a short rudimentary canal just above the vomeronasal cartilage, opening in the side of the nasal septum and passing from there blindly upward and backward; called also *Jacobson's organ.* <organum vomeronasale> F

A

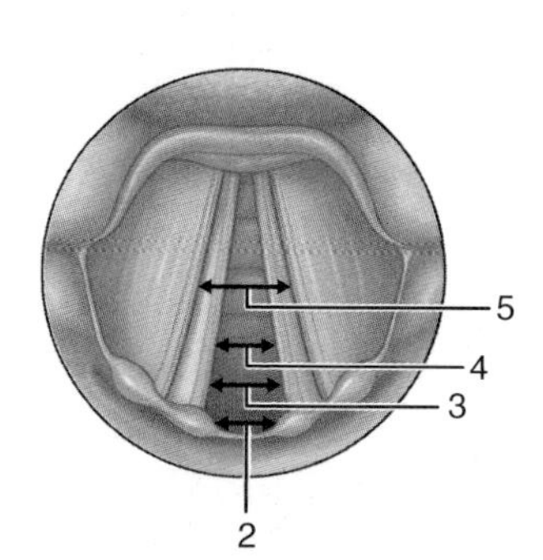

B

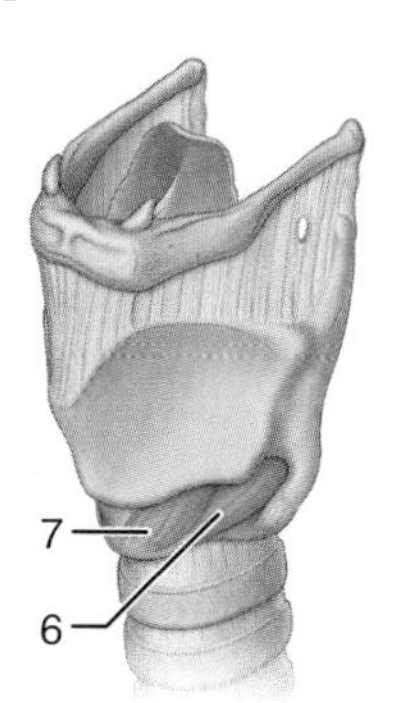

C

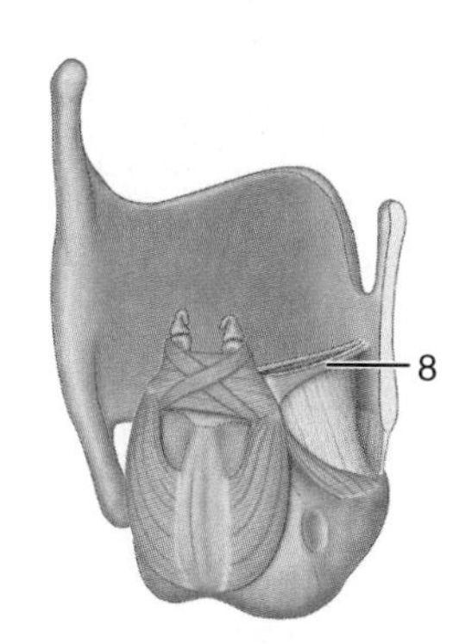

D

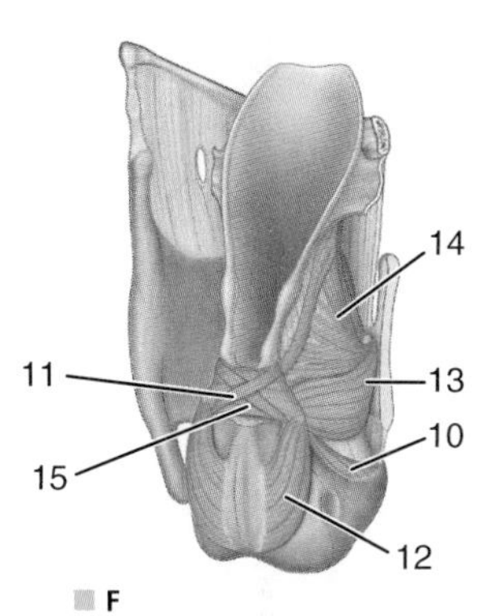

E

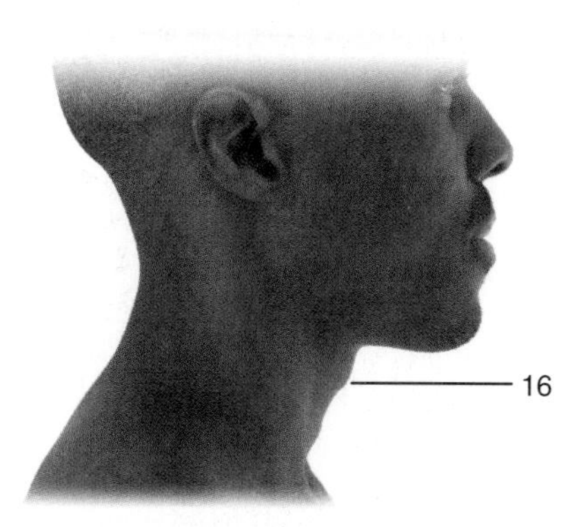

F

19

1 ***accessory nasal cartilages*** one or more small cartilages on either side of the nose between the greater alar and lateral nasal cartilages; called also *sesamoid cartilages of nose.* <cartilagines nasi accessoriae> A

2 **lateral crus of major alar cartilage** the part of the major alar cartilage that curves laterally around the naris and helps maintain its contour. <crus laterale cartilaginis alaris majoris> A

3 **lateral process of cartilage of nasal septum** a lateral expansion of the septal cartilage on either side of the nose, fused with the lateral nasal cartilage. <processus lateralis cartilaginis septi nasi> A

4 **major alar cartilage** either of two thin, curved cartilages, one on either side at the apex of the nose, each of which possesses a lateral and a medial crus. <cartilago alaris major> A

5 **medial crus of major alar cartilage** the part of the major alar cartilage, loosely attached to its fellow of the opposite side, and helping to form the mobile septum of the nose. <crus mediale cartilaginis alaris majoris> A

6 **minor alar cartilages** various small cartilages located in the fibrous tissue of the alae of nose posterior to a major alar cartilage; called also *accessory cartilages of nose* and *sesamoid cartilages.* <cartilagines alares minores> A

7 ***anterior ethmoidal cells*** ethmoidal air cells that open into the middle nasal meatus; they are often grouped with adjacent middle and posterior ethmoidal cells and called simply *ethmoidal cells* or *sinuses.* <cellulae ethmoidales anteriores> B

8 **ethmoidal cells** collective name for a type of paranasal sinus occurring in groups within the ethmoid bone and communicating with the ethmoidal infundibulum and bulla and the superior and highest meatuses. They are often subdivided into *anterior, middle,* and *posterior ethmoidal cells,* named according to the location of their openings into the nasal meatus. Called also *ethmoidal air cells.* <cellulae ethmoidales> B

9 **frontal sinus** one of the paired irregular shaped paranasal sinuses located in the frontal bone deep to the superciliary arch, separated from its fellow of the opposite side by a bony septum; it communicates by way of the nasofrontal duct with the middle meatus of the nasal cavity on the same side. <sinus frontalis> B

10 **maxillary sinus** one of the paired paranasal sinuses, located in the body of the maxilla on either side and communicating with the middle meatus of the nasal cavity on the same side. Called also *antrum of Highmore.* <sinus maxillaris> B

11 ***middle ethmoidal cells*** ethmoidal air cells that open into the middle nasal meatus; they are often grouped with adjacent anterior and posterior ethmoidal sinuses and called simply *ethmoidal cells.* <cellulae ethmoidales mediae> B

12 ***paranasal sinuses*** the mucosa-lined air cavities in the cranial bones which communicate with the nasal cavity, including the ethmoidal, frontal, maxillary, and sphenoidal sinuses. <sinus paranasales> B

13 ***posterior ethmoidal cells*** ethmoidal air cells that open into the superior nasal meatus; they are often grouped with adjacent middle and anterior ethmoidal sinuses and called simply *ethmoidal cells.* <cellulae ethmoidales posteriores> B

14 ***septum of frontal sinuses*** a thin lamina of bone in the lower part of the frontal bone, lying more or less in the median plane, that separates the frontal sinuses. <septum sinuum frontalium> B

15 **bony part of nasal septum** the bony part of the nasal septum, composed posterosuperiorly of the perpendicular plate of the ethmoid bone and posteroinferiorly of the vomer. Called also *bony* or *osseous nasal septum.* <pars ossea septi nasi> C

16 ***cartilage of nasal septum*** the hyaline cartilage forming the framework of the cartilaginous part of the nasal septum, adjacent to and partly fused with the lateral nasal cartilages; called also *septal cartilage of nose* and *quadrilateral cartilage.* <cartilago septi nasi> C

17 **cartilaginous part of nasal septum** the plate of cartilage forming the anterior part of the nasal septum. <pars cartilaginea septi nasi> C

18 **cuneiform part of vomer** the wedge-shaped anterior part of the vomer. <pars cuneiformis vomeris> C

19 **membranous nasal septum** the anterior inferior part of the nasal septum, beneath the cartilaginous part; it is composed of skin and subcutaneous tissues. <pars membranacea septi nasi> C

20 **mobile part of nasal septum** the part of the nasal septum at the apex of the nose, formed by skin, subcutaneous tissue, the greater alar cartilages, the membranous septum, and the columella. <pars mobilis septi nasi> C

21 **nasal septum** the partition separating the two nasal cavities in the midplane, composed of cartilaginous, membranous, and bony parts. <septum nasi> C

22 ***posterior process of cartilage of nasal septum*** a narrow flat strip of cartilage that extends backward and upward along the groove on the upper margin of the vomer and below the perpendicular plate of the ethmoid bone, from the septal cartilage nearly to the sphenoid bone. Called also *sphenoidal process of cartilage of nasal septum.* <processus posterior cartilaginis septi nasi> C

23 ***vomeronasal cartilage*** either of the two narrow, longitudinal strips of cartilage, one lying on either side of the anterior portion of the lower margin of the septal cartilage; called also *Jacobson's cartilage.* <cartilago vomeronasalis> C

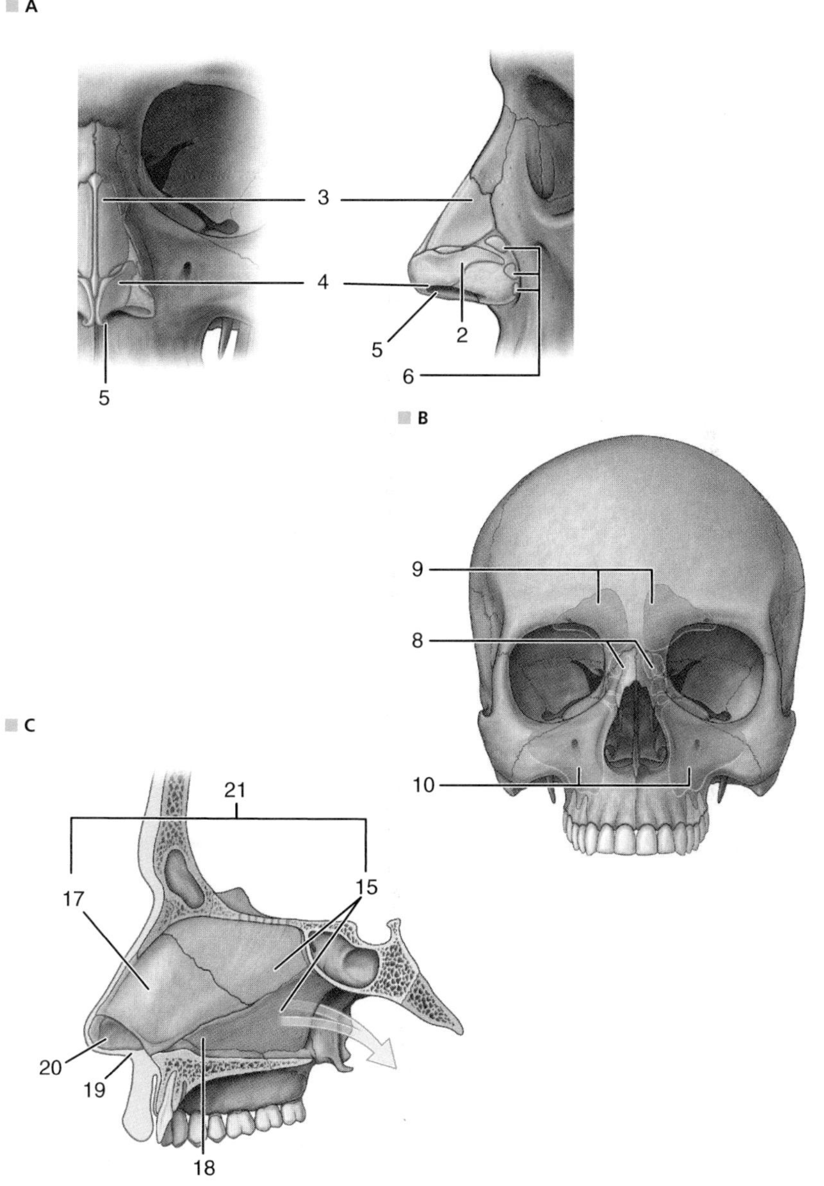
A
3
4
5
2
5
6
B
9
8
10
C
21
17
15
20
19
18

1 **lacrimoconchal suture** the line of junction between the lacrimal bone and the inferior nasal concha. <sutura lacrimoconchalis> A

2 **conchal crest of body of maxilla** an oblique ridge on the nasal surface of the body of the maxilla, just anterior to the lacrimal sulcus, which articulates with the inferior nasal concha; called also *inferior turbinal crest of maxilla.* <crista conchalis corporis maxillae> A

3 **conchal crest of palatine bone** a sharp transverse ridge, near the posterior edge of the palatine bone, which articulates with the inferior concha; called also *inferior turbinal crest of palatine bone.* <crista conchalis ossis palatini> A

4 **ethmoidal process of inferior nasal concha** a bony projection above and behind the maxillary process of the inferior nasal concha. <processus ethmoidalis conchae nasalis inferioris> A

5 **inferior nasal concha** a thin bony plate with curved margins, articulating with the ethmoid, maxilla, and lacrimal and palatine bones, and forming the lower part of the lateral wall of the nasal cavity, and the mucous membrane covering the plate; called also *inferior turbinate.* <concha nasi inferior> A

6 **lacrimal process of inferior nasal concha** a process of the inferior nasal concha that articulates with the lacrimal bone. <processus lacrimalis conchae nasalis inferioris> A

7 ***maxillary process of inferior nasal concha*** a bony process descending from the ethmoidal process of the inferior nasal concha. <processus maxillaris conchae nasalis inferioris> A

8 **middle nasal concha** the lower of two bony plates projecting from the inner wall of the ethmoid labyrinth and separating the superior from the middle meatus of the nose, and the mucous membrane covering the plate; called also *middle turbinate.* <concha nasi media> A

9 **sphenoethmoidal recess** the most superior and posterior part of the nasal cavity, above the superior nasal concha, into which the sphenoidal sinus opens. <recessus sphenoethmoidalis> B

10 **agger nasi** a ridgelike elevation midway between the anterior extremity of the middle nasal concha and the inner surface of the dorsum of the nose. B

11 **atrium of middle meatus** a depression in front of the middle nasal meatus, between the agger nasi and the middle nasal concha. <atrium meatus medii> B

12 ***highest nasal concha*** a thin bony plate occasionally found projecting from the inner wall of the ethmoid labyrinth above the bony superior nasal concha, and the mucous membrane covering the plate; called also *supreme nasal concha,* and *supreme* or *highest turbinate.* <concha nasi suprema> B

13 **limen nasi** the ridge at the junction of the lateral nasal cartilage and the lateral crus of the greater alar cartilage, marking the boundary between the vestibule of the nose and the nasal cavity proper. Called also *nasal valve.* B

14 **nasal vestibule** the anterior part of the nasal cavity, just superior to the nares and limited posteriorly by the limen nasi. It is lined with stratified squamous epithelium and contains hairs (vibrissae) and sebaceous glands. <vestibulum nasi> B

15 **nasopharyngeal meatus** the part of the nasal cavity posterior to the concha. <meatus nasopharyngeus> B

16 ***olfactory groove*** a shallow groove on the wall of the nasal cavity, passing upward from the level of the anterior end of the middle concha just above the agger nasi to the lamina cribrosa. <sulcus olfactorius nasi> B

17 **olfactory region** the superior part of the nasal cavity, the mucosa of which contains most of the receptors for the sense of smell. <pars olfactoria cavitatis nasi> B

18 ***respiratory region*** the part of the nasal cavity inferior to the olfactory region. <pars respiratoria cavitatis nasi> B

19 **superior nasal meatus** the narrow cavity below the superior nasal concha, with which the posterior ethmoidal cells communicate. <meatus nasi superior> B

20 **ethmoid bulla of nasal cavity** the large ethmoidal air cell lodged in the ethmoid bulla of ethmoid bone. <bulla ethmoidalis cavi nasi> C

21 ***ethmoidal infundibulum of ethmoid bone*** a variable sinuous passage extending upward from the middle nasal meatus through the ethmoidal labyrinth, communicating with the anterior ethmoidal cells and often with the frontal sinus. <infundibulum ethmoidale ossis ethmoidalis> C

22 **ethmoidal infundibulum of nasal cavity** a passage connecting the cavity of the nose with the anterior ethmoidal cells and the frontal sinus. <infundibulum ethmoidale cavitatis nasi> C

23 **inferior nasal meatus** the space beneath the inferior nasal concha, into which the nasolacrimal duct opens. <meatus nasi inferior> C

24 **maxillary hiatus** a small round or oval opening connecting the maxillary sinus and the middle nasal meatus; it is often indented so that the opening appears tubular. Called also *maxillary ostium.* <hiatus maxillaris> C

25 **middle nasal meatus** the space beneath the middle nasal concha, with which the anterior ethmoidal cells and frontal and maxillary sinuses communicate. <meatus nasi medius> C

26 **opening of frontal sinus** the external opening of the frontal sinus into the nasal cavity; its structure is variable, but it usually drains into the middle meatus. Called also *frontal ostium.* <apertura sinus frontalis> C

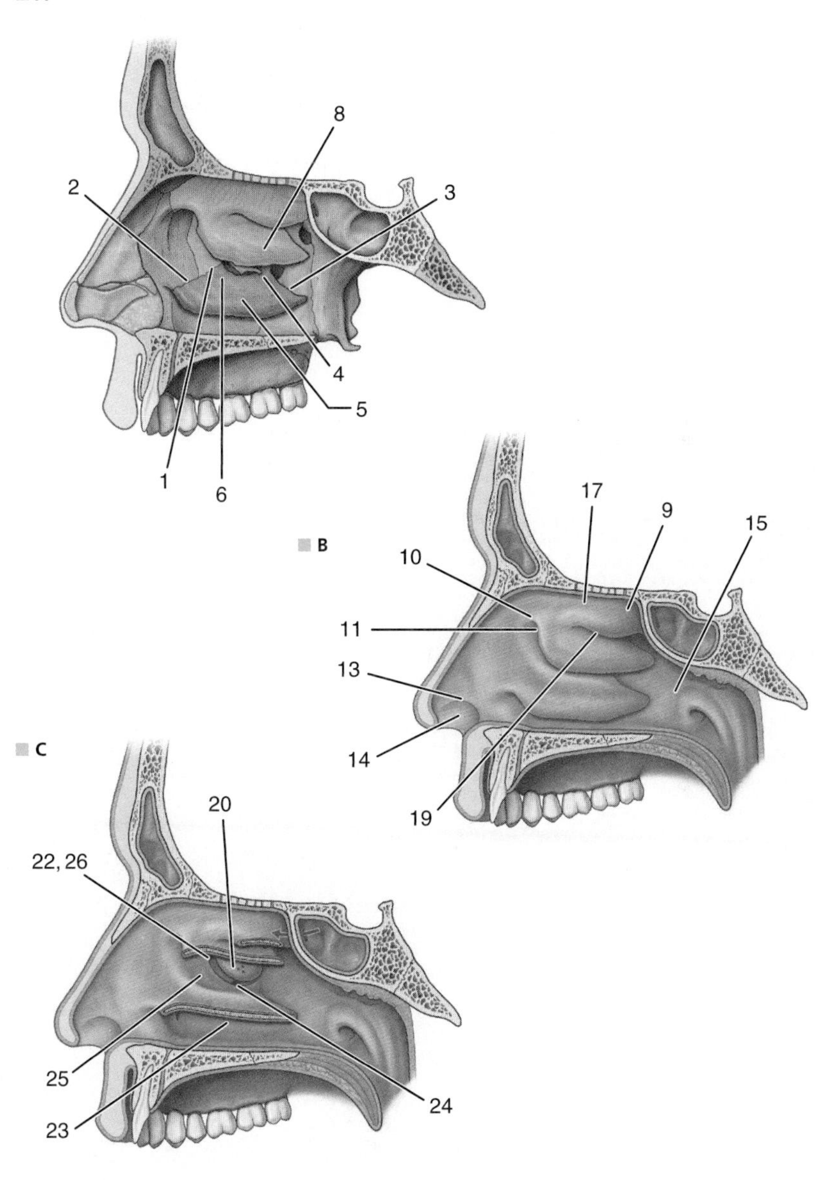
A
8
2
3
4
5
1
6
B
17
9
15
10
11
13
14
19
C
20
22, 26
25
23
24

1 **opening of sphenoid sinus** a round opening just above the superior nasal concha, connecting the sphenoid sinus and the nasal cavity. Called also *sphenoid ostium, sphenoid sinus ostium,* and *ostium of sphenoid sinus.* <apertura sinus sphenoidalis> A

2 **semilunar hiatus** the deep semilunar groove anterior and inferior to the bulla of the ethmoid bone; the anterior ethmoidal (air) cells, the maxillary sinus, and sometimes the frontonasal duct drain through it via the ethmoid infundibulum. <hiatus semilunaris> A

3 ***septum of sphenoidal sinuses*** a usually asymmetric, thin lamina of bone in the body of the sphenoid bone, lying more or less in the median plane and separating the sphenoidal sinuses. <septum sinuum sphenoidalium> A

4 **sphenoidal sinus** one of the paired paranasal sinuses, located in the anterior part of the body of the sphenoid bone and communicating with the superior meatus of the nasal cavity on the same side; it is separated from its fellow on the opposite side by a septum. <sinus sphenoidalis> A

5 **ala of vomer** one of the two lateral expansions on the superior border of the vomer, coming into contact with the sphenoidal process of the palatine bone and the vaginal process of the medial pterygoid plate. <ala vomeris> B

6 **choana** one of the pair of openings between the nasal cavity and the nasopharynx; called also *posterior nasal aperture.* B

7 **anterior lateral nasal branches of anterior ethmoidal artery** twigs of the anterior ethmoidal artery that descend into the nasal cavity with the anterior ethmoidal nerve and supply the lateral nasal wall. <rami nasales anteriores laterales arteriae ethmoidalis anterioris> C

8 **posterior lateral nasal arteries** *origin,* sphenopalatine artery; *branches,* none; *distribution,* frontal, maxillary, ethmoidal, and sphenoidal sinuses. <arteriae nasales posteriores laterales> C

9 **anterior septal branches of anterior ethmoidal artery** twigs of the anterior ethmoidal artery that descend into the nasal cavity with the anterior ethmoidal nerve and supply the nasal septum. <rami septales anteriores arteriae ethmoidalis anterioris> D

10 **nasal septal branch of superior labial artery** a branch that ramifies on the lower and front part of the nasal septum. <ramus septi nasi arteriae labialis superioris> D

11 **posterior septal branches of sphenopalatine artery** branches that anastomose with the ethmoidal arteries. <rami septales posteriores arteriae sphenopalatinae> D

12 **cavernous plexus of concha** any of the numerous venous plexuses in the thick mucous membrane of the nasal conchae. <plexus cavernosus conchae> E

13 **ethmoidal groove** a groove that extends the entire length of the posteromedial surface of the nasal bone and lodges the external nasal branch of the anterior ethmoid nerve. <sulcus ethmoidalis ossis nasalis> F

14 **internal nasal branches of infraorbital nerve** branches that innervate the mobile septum of the nose; *modality,* general sensory. <rami nasales interni nervi infraorbitalis> F

15 **lateral superior posterior nasal branches of maxillary nerve** branches that supply the superior and middle nasal conchae and the posterior ethmoidal sinuses; *modality,* general sensory. <rami nasales posteriores superiores laterales nervi maxillaris> F

A

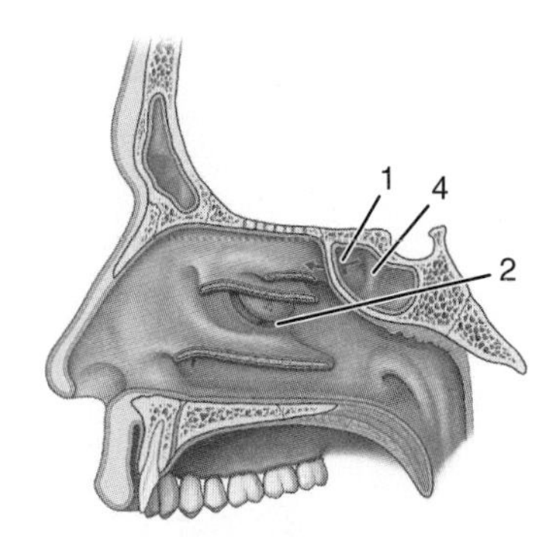

B

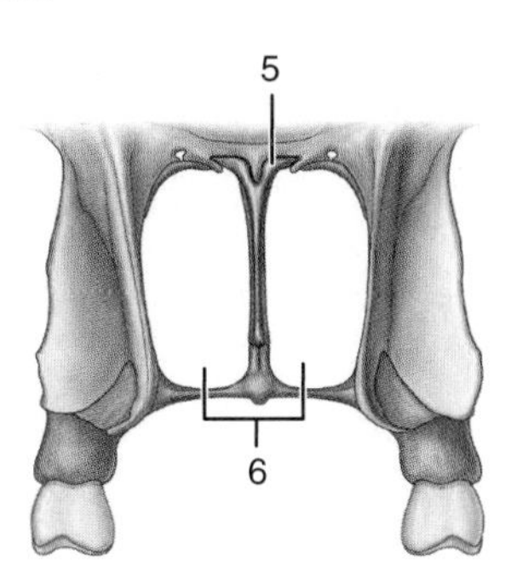

C

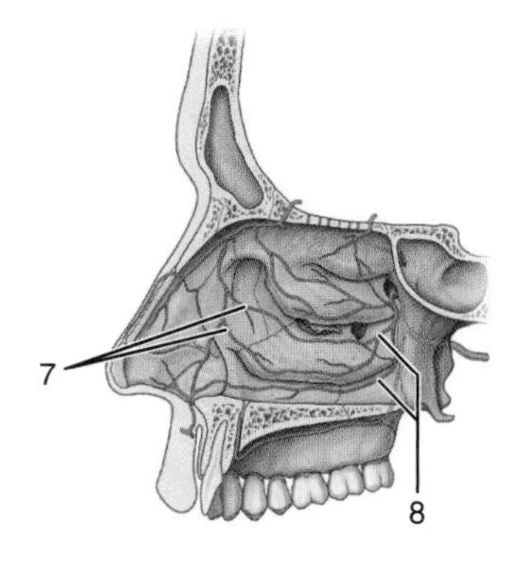

D

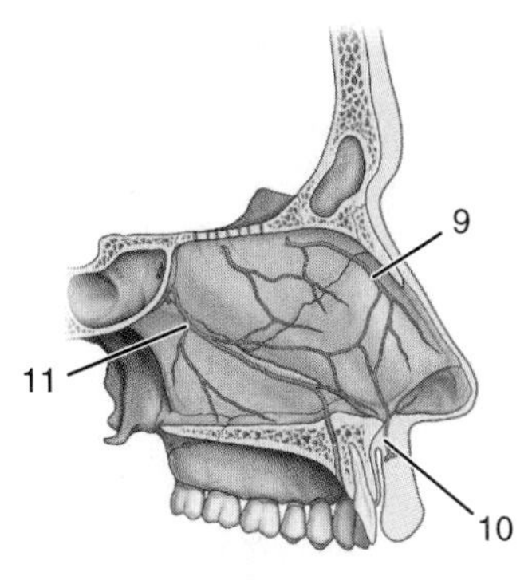

E

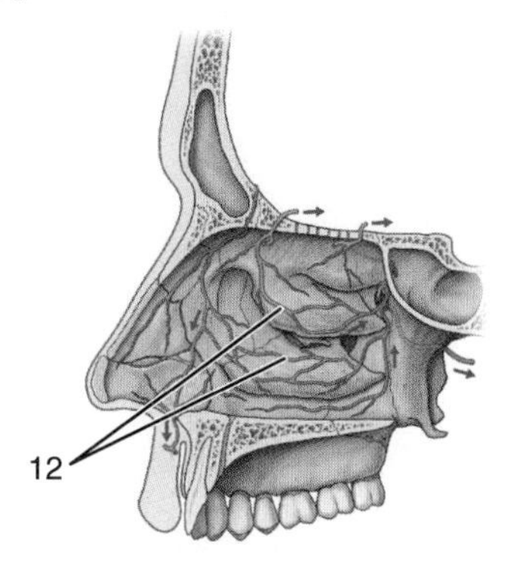

F

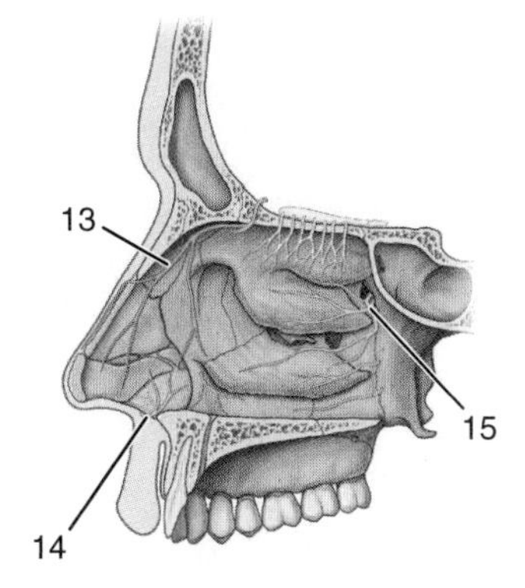

1 **internal nasal branches of anterior ethmoidal nerve** the medial and lateral branches that innervate the nasal septum and the mucous membrane of the lateral wall of the nasal cavity; *modality,* general sensory. <rami nasales interni nervi ethmoidalis anterioris> A

2 ***lateral internal nasal branches of anterior ethmoidal nerve*** branches arising from the internal nasal branches of the anterior ethmoidal nerve and innervating the mucosa of the lateral wall of the nasal cavity; *modality,* general sensory. <rami nasales interni laterales nervi ethmoidalis anterioris> A

3 **medial internal nasal branches of anterior ethmoidal nerve** branches arising from the internal nasal branches of the anterior ethmoidal nerve and supplying the nasal septum; *modality,* general sensory. <rami nasales interni mediales nervi ethmoidalis anterioris> A

4 ***medial superior posterior nasal branches of maxillary nerve*** nerve branches, usually branches of the nasopalatine nerve, that supply the nasal septum; *modality,* general sensory. <rami nasales posteriores superiores mediales nervi maxillaris> A

5 **nasopalatine nerve** *origin,* pterygopalatine ganglion; *distribution,* mucosa and glands of most of nasal septum and anterior part of hard palate; *modality,* parasympathetic and general sensory. <nervus nasopalatinus> A

6 **ala of nose** the flaring cartilaginous expansion forming the outer side of each naris. <ala nasi> B

7 **nares** the external orifices of the nose; called also *nostrils.* B

Oral Cavity

8 ***buccal glands*** the serous and mucous glands on the inner surface of the cheeks. <glandulae buccales> C

9 ***molar glands*** the glands on the external aspect of the buccinator muscle, their ducts piercing it to open on the internal aspect of the cheek; called also *retromolar glands.* <glandulae molares> C

10 ***mucous membrane of mouth*** the mucous membrane lining the oral cavity. <tunica mucosa oris> C

11 **oral cavity** the cavity of the mouth and associated structures, including the cheek, palate, oral mucosa, the glands whose ducts open into the cavity, the teeth, and the tongue. <cavitas oris> C

12 ***oral cavity proper*** the part of the oral cavity internal to the teeth. <cavitas oris propria> C

13 ***oral vestibule*** the portion of the oral cavity bounded on one side by the teeth and gingivae, or the residual alveolar ridges, and on the other side by the lips *(labial vestibule)* and cheeks *(buccal vestibule)*; called also *cavitas oris externa, cavum oris externum,* and *external oral cavity.* <vestibulum oris> C

14 **incisive canals** the small canals opening into the incisive fossa of the hard palate, and transmitting small vessels and nerves from the floor of the nose into the front part of the roof of the mouth. <canales incisivi> D

15 ***incisive duct*** a passage sometimes found in the incisive canal that interconnects the nasal and oral cavities during embryonic development; it occasionally fails to close. Called also *incisor duct.* <ductus incisivus> D

16 **geniohyoid muscle** *origin,* inferior mental spine; *insertion,* body of hyoid bone; *innervation,* a branch of first cervical nerve through hypoglossal; *action,* elevates and draws hyoid forward, or depresses mandible when hyoid is fixed by its depressors. <musculus geniohyoideus> E

17 **mylohyoid muscle** *origin,* mylohyoid line of mandible; *insertion,* body of hyoid bone and median raphe; *innervation,* mylohyoid branch of inferior alveolar; *action,* elevates hyoid bone, supports floor of mouth. <musculus mylohyoideus> E

18 **body of tongue** the larger anterior part of the tongue, in the floor of the mouth. <corpus linguae> F

19 **genioglossus muscle** *origin,* superior mental spine; *insertion,* hyoid bone and inferior surface of tongue; *innervation,* hypoglossal; *action,* protrudes and depresses tongue. <musculus genioglossus> F

20 **lingual tonsil** an aggregation of lymph follicles on the floor of the oropharyngeal passageway, at the root of the tongue. <tonsilla lingualis> F

21 ***lymphoid nodules of lingual tonsil*** lymphoid nodules on the root of the tongue, associated with the lingual tonsil. <noduli lymphoidei tonsillae lingualis> F

22 ***muscles of tongue*** the extrinsic and intrinsic muscles that move the tongue; called also *lingual muscles.* <musculi linguae> F

23 **root of tongue** the portion of the tongue posterior to the terminal sulcus, being attached inferiorly to the hyoid bone, and directed posteriorly and superiorly. <radix linguae> F

24 ***tonsillar crypts of lingual tonsil*** deep, irregular invaginations from the surface of the lingual tonsils. <cryptae tonsillares tonsillae lingualis> F

A

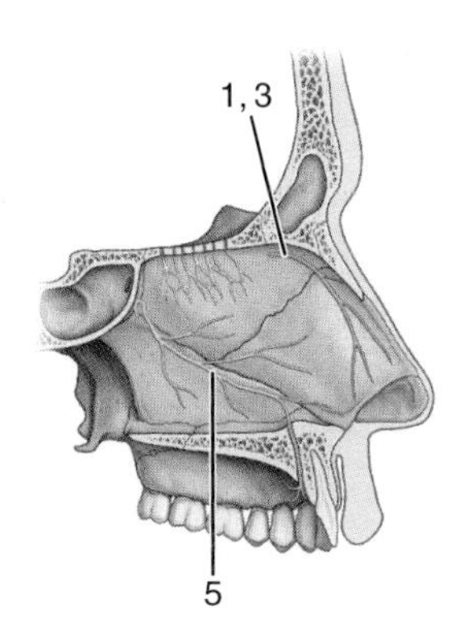

B

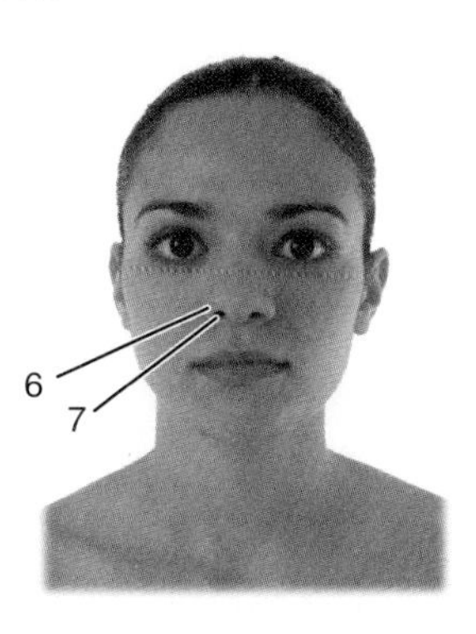

C

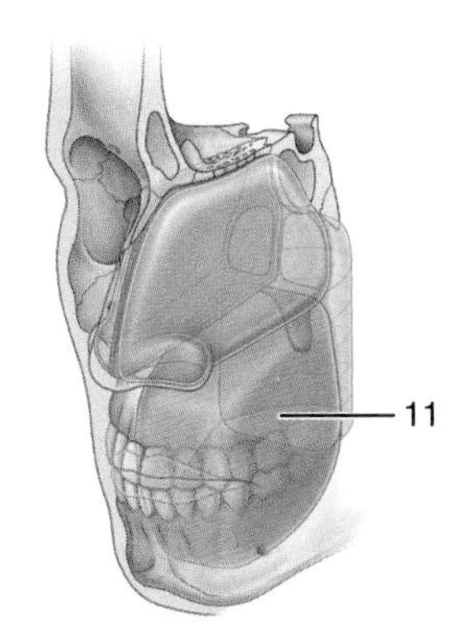

D

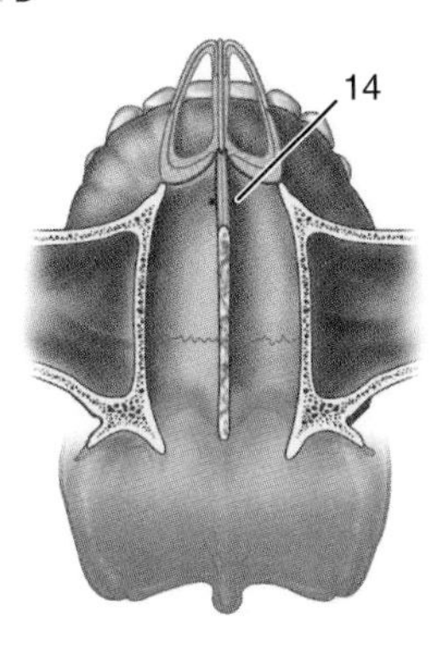

E

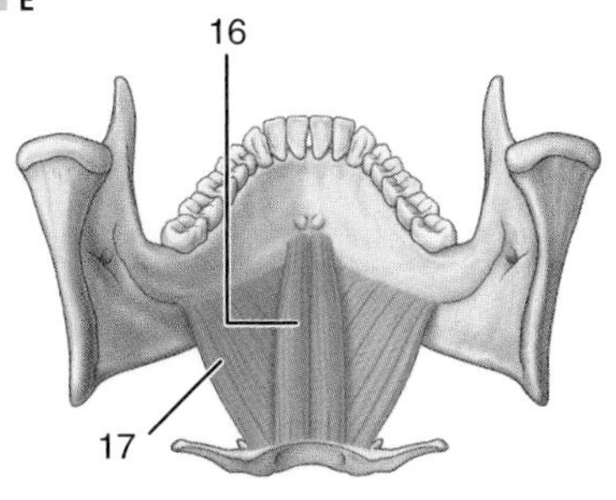

F

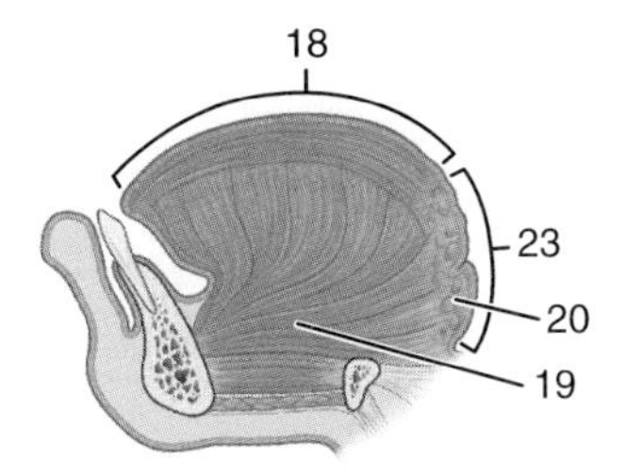

1 **anterior part of dorsum of tongue** the part of the dorsum of the tongue anterior to the terminal sulcus. <pars anterior dorsi linguae> A

2 **apex of tongue** the most distal portion of the tongue; called also *tip of tongue.* <apex linguae> A

3 **dorsum of tongue** the upper or posterosuperior surface of the tongue. <dorsum linguae> A

4 **filiform papillae** threadlike elevations that cover most of the tongue surface. <papillae filiformes> A

5 **foliate papillae** parallel mucosal folds on the margins of the tongue at the junction of its body and root. <papillae foliatae> A

6 **foramen cecum of tongue** a depression on the dorsum of the tongue at the end of the median sulcus, representing the remains of the superior end of the thyroglossal duct of the embryo. <foramen caecum linguae> A

7 **fungiform papillae** knoblike projections on the tongue, scattered singly among the filiform papillae. <papillae fungiformes> A

8 ***lingual aponeurosis*** the connective tissue framework of the tongue, supporting and giving attachment to the intrinsic and extrinsic muscles; composed of the connective tissue layer of the tunica mucosa, the lingual septum, and the posterior transverse expansion of the septum which attaches to the hyoid bone. <aponeurosis linguae> A

9 ***lingual glands*** the mucous and serous glands on the surface of the tongue. <glandulae linguales> A

10 ***lingual papillae*** the filiform, fungiform, vallate, foliate, and conical papillae of the tongue. <papillae linguales> A

11 **margin of tongue** the lateral border of the body of the tongue. <margo linguae> A

12 **median sulcus of tongue** a shallow groove on the dorsal surface of the tongue in the midline. <sulcus medianus linguae> A

13 ***mucous membrane of tongue*** the mucous membrane covering the tongue. <tunica mucosa linguae> A

14 **posterior part of dorsum of tongue** the part of the dorsum of the tongue posterior to the terminal sulcus. <pars posterior dorsi linguae> A

15 ***taste bud*** one of the minute, barrel-shaped terminal organs of the gustatory nerve, situated around the bases of the vallate, fungiform, and foliate papillae of the tongue. It contains several types of cells, including basal cells, taste cells, supporting cells, and some that are both supporting and taste cells. <caliculus gustatorius> A

16 ***taste pore*** the small opening of a taste bud onto the surface of the tongue. Called also *gustatory pore.* <porus gustatorius> A

17 **terminal sulcus of tongue** a more or less distinct groove on the tongue, extending from the foramen cecum anteriorly and lateralward to the margin of the tongue on either side, and dividing the dorsum of the tongue from the root. It is marked by a row of vallate papillae. <sulcus terminalis linguae> A

18 ***thyroglossal duct*** a duct in the embryo extending between the primordial thyroid and the posterior part of the tongue, which opens as the foramen caecum; the distal part usually differentiates to form the pyramidal lobe of the thyroid and the remainder becomes obliterated, but occasionally persists into adult life, giving rise to cysts, fistulas, or sinuses. <ductus thyroglossalis> A

19 ***tongue*** the movable, muscular organ on the floor of the mouth, subserving the special sense of taste and aiding in mastication, deglutition, and the articulation of sound; called also *glossa.* <lingua> A

20 **vallate papillae** the largest papillae of the tongue, 8 to 12 in number, arranged in the form of a V anterior to the terminal sulcus of the tongue. <papillae vallatae> A

21 **inferior longitudinal muscle of tongue** *origin,* inferior surface of tongue at base; *insertion,* tip of tongue; *innervation,* hypoglossal; *action,* changes shape of tongue in mastication and deglutition. <musculus longitudinalis inferior linguae> B

22 **lingual septum** the median vertical fibrous part of the tongue. <septum linguae> B

23 **superior longitudinal muscle of tongue** *origin,* submucosa and septum of tongue; *insertion,* margins of tongue; *innervation,* hypoglossal; *action,* changes shape of tongue in mastication and deglutition. <musculus longitudinalis superior linguae> B

24 **transverse muscle of tongue** *origin,* median septum of tongue; *insertion,* dorsum and margins of tongue; *innervation,* hypoglossal; *action,* changes shape of tongue in mastication and deglutition. <musculus transversus linguae> B

25 **vertical muscle of tongue** *origin,* dorsal fascia of tongue; *insertion,* sides and base of tongue; *innervation,* hypoglossal; *action,* changes shape of tongue in mastication and deglutition. <musculus verticalis linguae> B

26 **styloglossus muscle** *origin,* styloid process; *insertion,* margin of tongue; *innervation,* hypoglossal; *action,* raises and retracts tongue. <musculus styloglossus> B

A

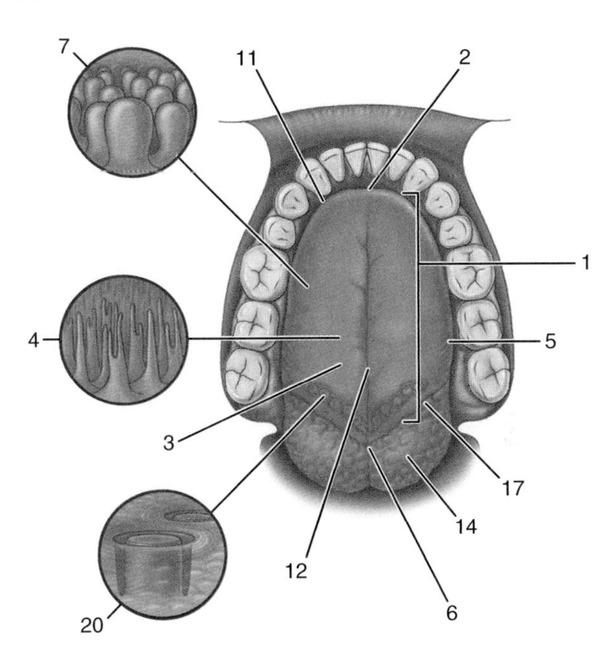

B

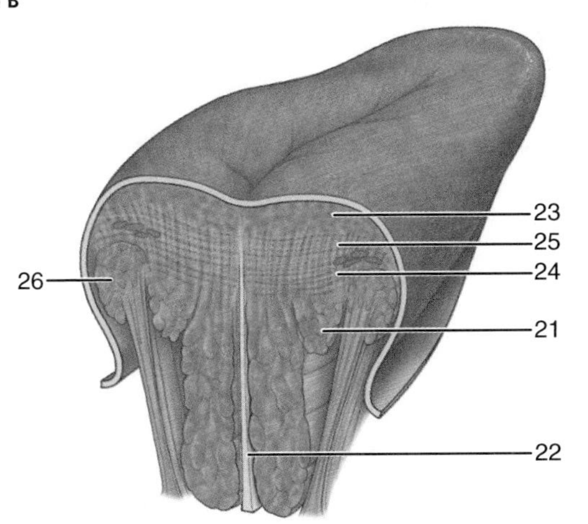

1 **oropharyngeal isthmus** the contricted aperture between the cavity of the mouth and the pharynx. <isthmus faucium> A

2 **palatoglossal arch** the anterior of the two folds of mucous membrane on either side of the oropharynx, connected with the soft palate and enclosing the palatoglossus muscle; called also *glossopalatine* or *anterior palatine arch* and *anterior column* or *pillar of fauces.* <arcus palatoglossus> A

3 **palatopharyngeal arch** the posterior of the two folds of mucous membrane on each side of the oropharynx, connected with the soft palate and enclosing the palatopharyngeus muscle; called also *pharyngopalatine arch, posterior palatine arch,* and *posterior column* or *pillar of fauces.* <arcus palatopharyngeus> A

4 **semilunar fold of fauces** a curved fold interconnecting the palatoglossal and palatopharyngeal arches and forming the upper boundary of the supratonsillar fossa. <plica semilunaris faucium> A

5 ***branches to isthmus of fauces*** branches from the lingual nerve to the isthmus of the fauces; *modality,* general sensory. <rami isthmi faucium nervi lingualis> B

6 **deep lingual artery** *origin,* lingual artery; *branches,* none; *distribution,* tongue. Called also *ranine artery.* <arteria profunda linguae> B

7 ***dorsal lingual branches of lingual artery*** branches of the lingual artery arising beneath the hyoglossus muscle and supplying the tonsil and the back of the tongue. <rami dorsales linguae arteriae lingualis> B

8 **dorsal lingual veins** veins that unite with a small vena comitans of the lingual artery and join the main lingual trunk. <venae dorsales linguae> B

9 **ganglionic branches of lingual nerve to submandibular ganglion** branches which interconnect the lingual nerve and the submandibular ganglion, and by which the ganglion is suspended from the nerve; they carry preganglionic fibers that derive from the chorda tympani and synapse in the submandibular ganglion, and postganglionic fibers. Called also *motor* or *sensory root of submandibular ganglion.* <rami ganglionares nervi lingualis ad ganglion submandibulare> B

10 **hyoglossus muscle** *origin,* body and greater cornu of hyoid bone; *insertion,* side of tongue; *innervation,* hypoglossal; *action,* depresses and retracts tongue. <musculus hyoglossus> B

11 **lingual branches of glossopharyngeal nerve** branches of the glossopharyngeal nerve that innervate the posterior third of the tongue; *modality,* general and special sensory. <rami linguales nervi glossopharyngei> B

12 **lingual branches of hypoglossal nerve** branches of the hypoglossal nerve that innervate the intrinsic and extrinsic muscles of the tongue; *modality,* motor. <rami linguales nervi hypoglossi> B

13 **lingual branches of lingual nerve** branches that innervate the anterior two-thirds of the tongue, adjacent areas of the mouth, and the gums; *modality,* general and special sensory. <rami linguales nervi lingualis> B

14 **lingual nerve** *origin,* mandibular nerve, descending to the tongue, first medial to the mandible and then under cover of the mucous membrane of the mouth; *branches,* sublingual nerve, lingual branches, branches to isthmus of fauces, and branches communicating with the hypoglossal nerve and chorda tympani; *modality,* general sensory. <nervus lingualis> B

15 ***sublingual artery*** *origin,* lingual artery; *branches,* none; *distribution,* sublingual gland. <arteria sublingualis> B

16 ***sublingual ganglion*** a ganglion of nerve cells sometimes found on the fibers passing distally from the submandibular ganglion to the lingual nerve. <ganglion sublinguale> B

17 ***sublingual nerve*** *origin,* lingual nerve; *distribution,* sublingual gland and overlying mucous membrane; *modality,* parasympathetic and general sensory. <nervus sublingualis> B

18 **submandibular ganglion** a parasympathetic ganglion located superior to the deep part of the submandibular gland, on the lateral surface of the hyoglossus muscle; its preganglionic fibers are derived from the facial nerve by way of the chorda tympani and lingual nerve, and its postganglionic fibers supply the submandibular and sublingual glands; sensory and postganglionic sympathetic fibers pass through the ganglion. <ganglion submandibulare> B

19 **papilla of parotid duct** the small papilla marking the orifice of the parotid duct in the mucous membrane of the cheek. <papilla ductus parotidei> C

20 **major salivary glands** the larger exocrine glands of the oral cavity, which together with the minor salivary glands secrete saliva; the group includes the sublingual, submandibular, and parotid glands. <glandulae salivariae majores> C

21 ***minor salivary glands*** the smaller exocrine glands of the oral cavity, which together with the major salivary glands secrete saliva; the group includes the labial, buccal, molar, palatine, and lingual glands, and the anterior lingual gland. <glandulae salivariae minores> C

22 ***major sublingual duct*** the duct that drains the sublingual gland and opens alongside the submandibular duct on the sublingual caruncle; called also *Bartholin's duct.* <ductus sublingualis major> C

23 **minor sublingual ducts** the ducts that drain the sublingual gland and open along the crest of the sublingual fold; called also *ducts of Rivinus.* <ductus sublinguales minores> C

A

C

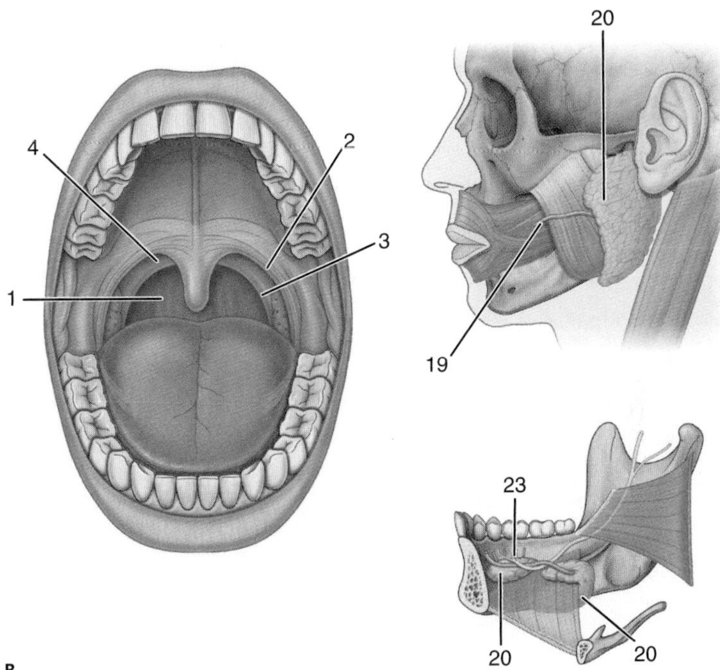

B

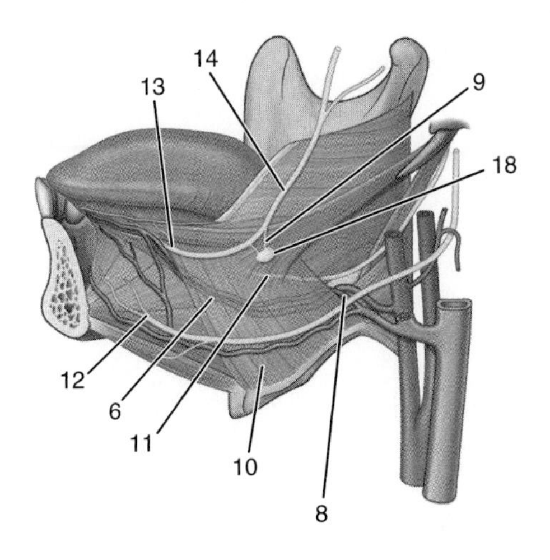

1 **sublingual gland** the smallest of the three salivary glands, occurring in pairs, predominantly mucous in type, and draining into the oral cavity through 10 to 30 sublingual ducts; called also *Rivinus gland.* <glandula sublingualis> A

2 **submandibular duct** the duct that drains the submandibular gland and opens at the sublingual caruncle; called also *Wharton's duct.* <ductus submandibularis> A

3 **submandibular gland** one of the three chief, paired salivary glands, predominantly serous, lying partly above and partly below the posterior half of the base of the mandible. <glandula submandibularis> A

4 **fimbriated fold** the lobulated fold running posteriorly and laterally from the anterior extremity of the frenulum of the tongue. <plica fimbriata> B

5 **frenulum of tongue** the vertical fold of mucous membrane inferior to the tongue, attaching it to the floor of the mouth. <frenulum linguae> B

6 **gingival papilla** a cone-shaped pad of the interdental gingiva filling the space between two contiguous teeth up to the contact area, as viewed from the labial, buccal, or lingual aspect; called also *interdental papilla* and *interproximal papilla.* <papilla gingivalis> B

7 **inferior surface of tongue** the under surface of the body of the tongue. <facies inferior linguae> B

8 **deep lingual vein** a vein that drains blood from the deep aspect of the tongue and joins the sublingual vein to form the vena comitans of the hypoglossal nerve. <vena profunda linguae> B

9 **sublingual caruncle** an eminence on each side of the frenulum of the tongue, at the apex of which are the openings of the major sublingual duct and the submandibular duct. <caruncula sublingualis> C

10 **sublingual fold** the elevation on the floor of the mouth under the tongue, covering part of the sublingual gland and containing its excretory ducts. <plica sublingualis> C

11 **hard palate** the anterior part of the palate, characterized by an osseous framework, covered superiorly by mucous membrane of the nasal cavity and, on its oral surface, by mucoperiosteum. <palatum durum> D

12 **incisive papilla** a rounded projection at the anterior end of the palatine raphe. <papilla incisiva> D

13 **palate** the partition separating the nasal and oral cavities, consisting anteriorly of a hard bony part and posteriorly of a soft fleshy part. <palatum> D

14 ***palatine glands*** the mucous glands on the soft palate and the posteromedial part of the hard palate. <glandulae palatinae> D

15 **palatine raphe** a narrow whitish streak at the midline of the roof of the mouth (both the hard and soft palates), extending from the incisive papilla to the tip of the uvula; it may present as a ridge in front and as a groove posteriorly. <raphe palati> D

16 **palatine uvula** the small, fleshy mass hanging from the soft palate above the root of the tongue, composed of the levator and tensor palatini muscles, musculus uvulae, connective tissue, and mucous membrane. <uvula palatina> D

17 **soft palate** the fleshy part of the roof of the mouth, extending from the posterior edge of the hard palate; from its free inferior border is a projection of variable length, the uvula. <palatum molle> D

18 **transverse palatine folds** four to six transverse ridges on the anterior part of the hard palate. Called also *palatine folds* and *palatine rugae.* <plicae palatinae transversae> D

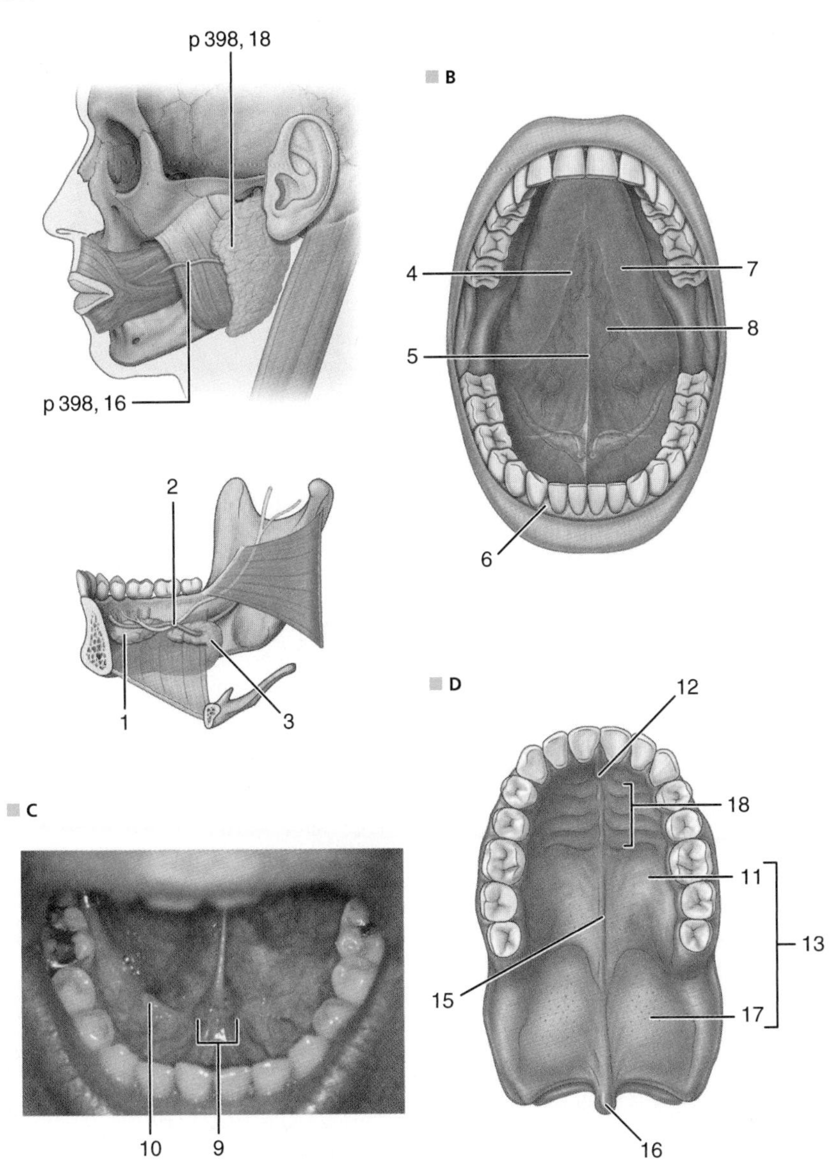
A
p 398, 18
p 398, 16
2
1
3
B
4
7
8
5
6
C
10
9
D
12
18
11
13
15
17
16

1 ***bursa of tensor veli palatini muscle*** a bursa between the hamular process of the sphenoid bone and the tendon of the tensor veli palatini. <bursa musculi tensoris veli palatini> A

2 ***nerve to tensor veli palatini*** *origin,* medial pterygoid nerve; *distribution,* tensor veli palatini muscle; *modality,* motor. <nervus musculi tensoris veli palatini> A

3 **palatine aponeurosis** a fibrous sheet in the anterior part of the soft palate, derived mainly from the tendons of the two tensor muscles, giving attachment to the musculus uvulae and to the palatopharyngeus and levator veli palatini muscles. <aponeurosis palatina> A

4 **tensor veli palatini muscle** *origin,* scaphoid fossa at base of medial pterygoid plate, wall of auditory tube, spine of sphenoid; *insertion,* aponeurosis of soft palate, horizontal part of palatine bone; *innervation,* mandibular; *action,* tenses soft palate, opens auditory tube. <musculus tensor veli palatini> A

5 **levator veli palatini muscle** *origin,* apex of petrous portion of temporal bone and cartilaginous part of auditory tube; *insertion,* aponeurosis of soft palate; *innervation,* pharyngeal plexus of vagus; *action,* raises and draws back soft palate. <musculus levator veli palatini> B

6 ***muscles of soft palate and fauces*** the intrinsic and extrinsic muscles that act upon the soft palate and the adjacent pharyngeal wall. <musculi palati mollis et faucium> B

7 **musculus uvulae** *origin,* posterior nasal spine of palatine bone and aponeurosis of soft palate; *insertion,* uvula; *innervation,* pharyngeal plexus of vagus; *action,* raises uvula. B

8 **palatoglossus muscle** *origin,* undersurface of soft palate; *insertion,* side of tongue; *innervation,* pharyngeal plexus of vagus; *action,* elevates tongue, constricts fauces. <musculus palatoglossus> B

9 **palatopharyngeus muscle** one of the intrinsic muscles of the larynx; *origin,* soft palate; *insertion,* aponeurosis of pharynx, posterior border of thyroid cartilage; *innervation,* pharyngeal plexus of vagus; *action,* aids in deglutition. <musculus palatopharyngeus> B

10 **bony palate** the bony part of the anterior two-thirds of the roof of the mouth, formed by the palatine processes of the maxillae and the horizontal plates of the palatine bones. <palatum osseum> C

11 ***incisive foramina*** the openings in the incisive fossa of the hard palate that transmit the nasopalatine nerves. Called also *foramina of Stensen.* <foramina incisiva> C

12 **incisive fossa** a slight depression on the anterior surface of the maxilla above the incisor teeth. <fossa incisiva> C

13 ***incisive suture*** an indistinct suture sometimes seen extending laterally from the incisive fossa to the space between the canine tooth and the lateral incisor, indicating the line of fusion between the premaxilla and the maxilla. <sutura incisiva> C

14 **lingual vein** the deep vein that follows the distribution of the lingual artery and empties into the internal jugular vein. <vena lingualis> D

15 ***sublingual vein*** a vein that follows the sublingual artery and opens into the lingual vein. <vena sublingualis> D

16 ***vena comitans of hypoglossal nerve*** a vessel, formed by union of the deep lingual vein and the sublingual vein, that accompanies the hypoglossal nerve; it empties into the facial, lingual, or internal jugular vein. <vena comitans nervi hypoglossi> D

17 **apex of cusp** the apex of the cusp of a tooth. <apex cuspidis dentis> E

18 **canine fossa** a wide depression on the external surface of the maxilla superolateral to the canine tooth socket; the levator anguli oris muscle arises from it. Called also *maxillary fossa.* <fossa canina> E

19 **canine tooth** the tooth immediately lateral to the lateral, or second, incisor; it has a long conical crown and the longest, most powerful root of all the teeth. Called also *canine.* <dens caninus> E

20 **diastema** a space between two adjacent teeth in the same dental arch. E

21 ***incisal margin*** the crest of the biting edge of an incisor tooth. <margo incisalis> E

22 **incisive bone** the portion of the maxilla that bears the incisor teeth. In humans, the embryonic bone called the premaxilla fuses with the maxilla proper to form the adult bone. Called also *premaxilla.* <os incisivum> E

23 **incisor tooth** either of the two most frontal teeth in each jaw, one on either side of the midline; it has a long root and is adapted for cutting. Called also *incisor.* <dens incisivus> E

24 **lingual surface** the surface of a tooth that faces inward toward the tongue (oral cavity), and opposite the vestibular (or facial) surface. Called also *oral surface.* <facies lingualis dentis> F

25 **maxillary dental arch** the portion of the dental arch formed by the teeth of the maxilla. Called also *superior dental arch.* <arcus dentalis maxillaris> F

26 **molar tooth** the most posterior teeth on either side in each jaw, totaling 8 in the deciduous dentition (2 on each side, upper and lower), and usually 12 in the permanent dentition (3 on each side, upper and lower). They are the grinding teeth, having large crowns with broad chewing surfaces. The upper molars characteristically have 4 major cusps and three roots. The lower first molars characteristically have 5 cusps, and the remaining lower molars 4 cusps. Normally all lower molars have two roots. The third molars ("wisdom teeth") are often malformed, but when developed normally their crown and root form corresponds in general with neighboring molars in the same jaw. Called also *molar.* <dens molaris> F

27 **premolar tooth** either of the two permanent teeth between the canine teeth and the molars. <dens premolaris> F

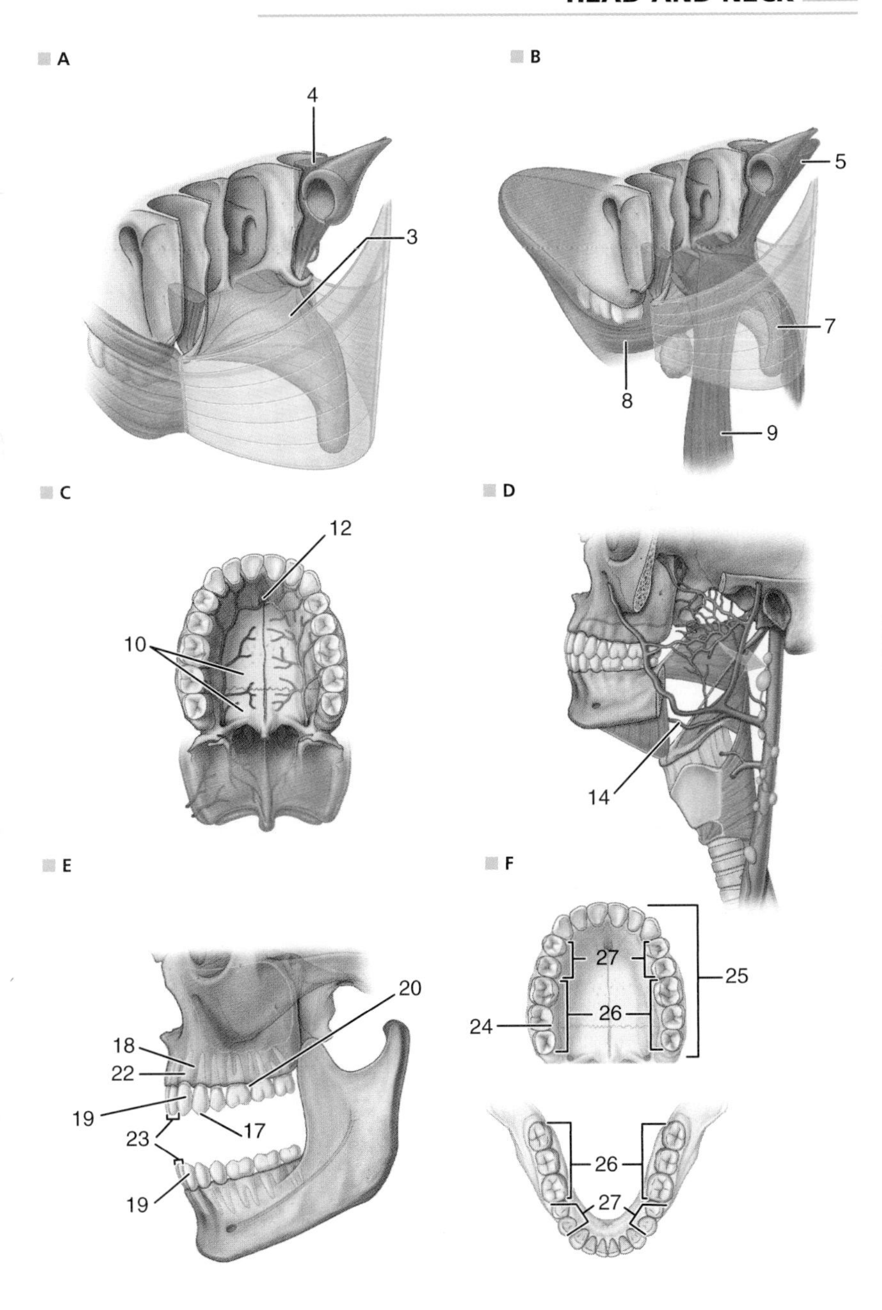
A
4
3
B
5
7
8
9
C
12
10
D
14
E
20
18
22
19
17
23
19
F
27
25
26
24
26
27

1 **third molar tooth** the last tooth on each side of each jaw. <dens molaris tertius> A

2 **buccal surface** the vestibular surface of the molars and premolars that faces the cheek. <facies buccalis dentis> B

3 **permanent teeth** the 32 teeth of the second dentition, which begin to appear in humans at about 6 years of age, starting with the first molars. <dentes permanentes> B

4 **deciduous teeth** the 20 teeth of the first dentition, which are shed and replaced by the permanent teeth. The first incisors appear at about the age of 6½ months, and all deciduous teeth generally erupt by 2½ years of age. <dentes decidui> C

5 **clinical crown** that portion of the tooth above the clinical root, i.e., the portion exposed beyond the gingiva, and thus visible in the oral cavity. <corona clinica> D

6 **clinical root** that portion of a tooth below the clinical crown, being attached to the gingiva or alveolus. <radix clinica> D

7 **crown of tooth** the upper part of the tooth, which joins the lower part, the root, at the neck at the cementoenamel junction, and terminates as the grinding surface of molar or premolar teeth or the cutting edge of incisors. Called also *anatomical* or *dental crown.* <corona dentis> D

8 ***dental tubercle*** a small elevation of indiscriminate size on some portion of the crown of a tooth, produced by extra formation of enamel. <tuberculum dentis> D

9 **apical foramen** a minute aperture usually at or near the apex of a root of a tooth but on occasion located on a side of a root, which gives passage to the vascular, lymphatic, and neural structures supplying the pulp; the main foramen sometimes branches near the apex to form two or more apical ramifications. Called also *pulpal* or *root foramen.* <foramen apicis dentis> E

10 ***approximal surface*** the area where the mesial and distal surfaces of the teeth touch each other; called also *centric stop* and *contact area* or *surface.* <facies approximalis dentis> E

11 **attached gingiva** the part of the gingiva that is firm and resilient and is bound to the underlying cementum and the alveolar bone, thus being immovable. <periodontium protectionis> E

12 **cementum** the bonelike rigid connective tissue covering the root of a tooth from the cementoenamel junction to the apex and lining the apex of the root canal; it also serves as an attachment structure for the periodontal ligament, thus assisting in tooth support. E

13 ***cingulum dentis*** the lingual lobe of an anterior tooth, making up the bulk of the cervical third of its lingual surface. E

14 ***contact area*** the area of the mesial or distal surface of a tooth that touches the adjoining tooth. <area contingens dentis> E

15 **coronal pulp** the portion of the dental pulp in the crown portion of the pulp cavity. <pulpa coronalis> E

16 **cusp of tooth** an elevation or mound on the crown of a tooth making up part of the occlusal surface; they are named for the tooth surface they are adjacent to, such as *buccal, lingual, and palatal cusps.* <cuspis dentis> E

17 **dental enamel** a hard, thin, translucent layer of calcified substance that envelops and protects the dentin of the crown of the tooth; it is the hardest substance in the body and is almost entirely composed of calcium salts. Called also *enamel.* <enamelum> E

18 ***dental papilla*** a small mass of condensed mesenchymal tissue in the enamel organ, which differentiates into the dentin and dental pulp. E

19 **dental pulp** the richly vascularized and innervated connective tissue of mesodermal origin contained in the central cavity of a tooth and delimited by the dentin, and having formative, nutritive, sensory, and protective functions. The portion within the tooth chamber proper is the *coronal pulp,* and that within the root is the *radicular pulp.* Called also *endodontium* and *tooth pulp.* <pulpa dentis> E

20 **dentin** the hard portion of the tooth surrounding the pulp, covered by enamel on the crown and cementum on the root, which is harder and denser than bone but softer than enamel. Sometimes spelled *dentine.* <dentinum> E

21 ***distal surface*** the proximal or contact surface of a tooth that is farthest from the midline of the dental arch. <facies distalis dentis> E

22 **free gingiva** the unattached portion of the gingiva, forming the wall of the gingival crevice. Called also *unattached gingiva.* <periodontium insertionis> E

23 **gingiva** that part of the oral mucosa overlying the crowns of unerupted teeth and encircling the necks of those that have erupted, serving as the supporting structure for subjacent tissues. It is formed by pale pink tissue immovably attached to the bone and the teeth, which joins the alveolar mucosa at the mucogingival junction. E

24 **gingival margin** the crest of the free gingiva that surrounds the teeth in a collarlike fashion, separated from the adjacent attached gingiva by the free gingival groove; it forms the wall of the gingival sulcus. Called also *free gum margin, gum margin,* and *marginal gingiva.* <margo gingivalis> E

25 **gingival sulcus** a shallow V-shaped space around the tooth, bounded by the tooth surface on one side and the epithelium lining the free margin of the gingiva on the other. <sulcus gingivalis> E

A

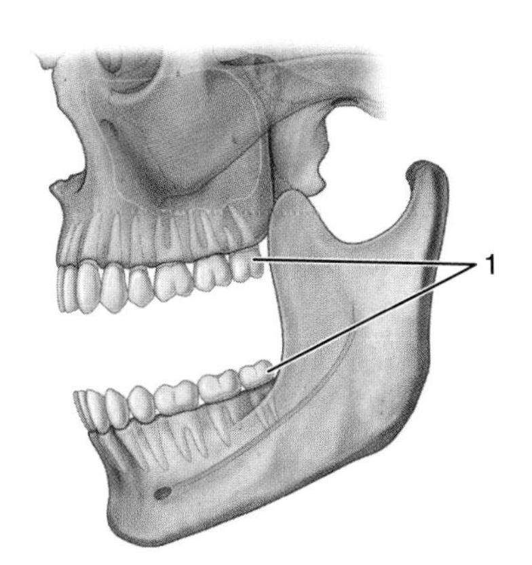

B

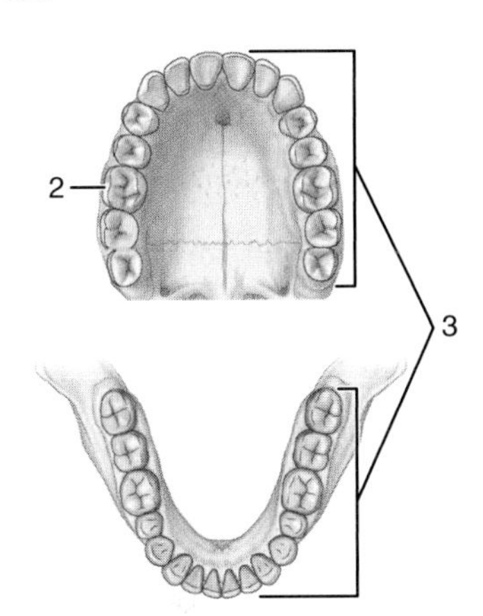

C

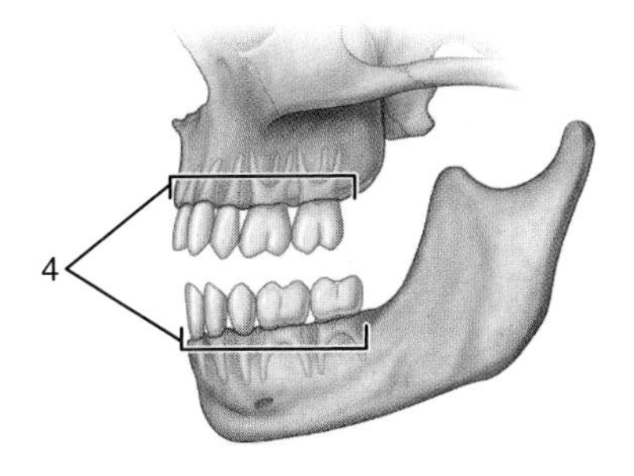

E

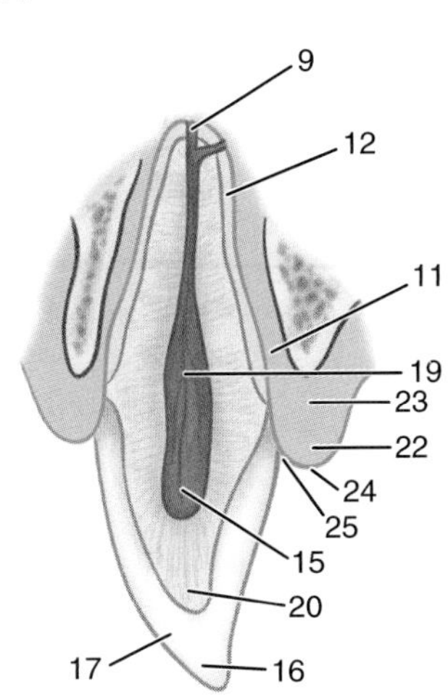

D

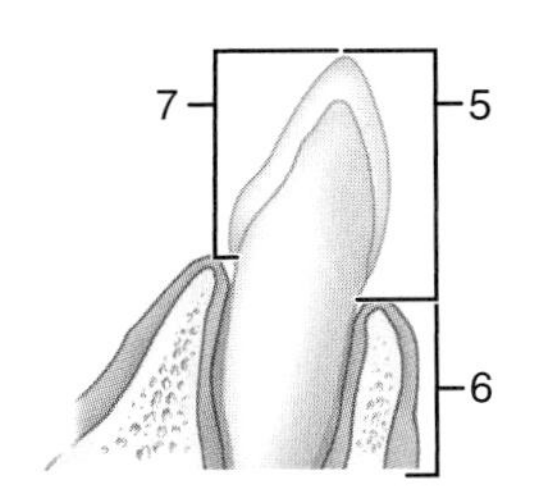

1 ***marginal ridge*** one of the elevated convex crests that form the mesial and distal borders of the occlusal surfaces of posterior teeth and of the lingual surfaces of anterior teeth. <crista marginalis> A

2 ***mesial surface*** the contact or proximal surface of a tooth that is closest to the midline of the dental arch. <facies mesialis dentis> A

3 **neck of tooth** the slightly constricted region of union of the crown and the root or roots of a tooth. <cervix dentis> A

4 ***occlusal curve*** the curve of a dentition on which the occlusal surfaces lie. <curvea occlusalis> A

5 ***occlusal surface*** the surface of the teeth coming in contact with those of the opposite jaw during the act of occlusion. In natural teeth, restricted to the anatomic tooth surfaces of the posterior teeth limited mesially and distally by the marginal ridges and buccally and lingually by the buccal and lingual boundaries of the cusp eminences. By extension, the term is used to designate the incisal surface of the anterior teeth. Called also *masticatory surface.* <facies occlusalis dentis> A

6 **palatal surface** the lingual surface of a maxillary tooth. <facies palatinalis dentis> A

7 **pulp cavity** the natural cavity in the central portion of a tooth occupied by the dental pulp, which is divided into the pulp chamber and the root canal. <cavitas dentis> A

8 **pulp chamber** the portion of the dental (pulp) cavity located in the tooth crown, occupied by the dental pulp. <cavitas coronae> A

9 **radicular pulp** the portion of the dental pulp in the root canal of a tooth. <pulpa radicularis> A

10 **root apex** the terminal end of the root of a tooth. <apex radicis dentis> A

11 **root canal** the portion of the dental pulp cavity in the root of a tooth, extending from the pulp chamber to the apical foramen; more than one canal may be present in a single root, two commonly being present in the mesial root of the mandibular first molar. Called also *pulp canal.* <canalis radicis dentis> A

12 **root of tooth** the portion of a tooth which is covered by cementum, proximal to the neck of the tooth and ordinarily embedded in the dental alveolus; called also *anatomical root.* <radix dentis> A

13 ***tooth*** one of the small bonelike structures of the jaws. <dens> A

14 ***transverse ridge*** an elevated crest coursing transversely across the occlusal surface of a mandibular premolar to link the apices of the buccal and lingual cusps. It comprises the buccal and lingual cusps and may be an uninterrupted prominence or may be sharply divided at its approximate midpoint by a groove. <crista transversalis> A

15 ***triangular ridge*** a ridge that descends from the tips of the cusps of molars and premolars toward the central part of the occlusal surface; so named because the slopes of each side of the ridge resemble two sides of a triangle. <crista triangularis> A

16 **vestibular surface** the surface of a tooth that is directed outward toward the vestibule of the mouth, including the buccal and labial surfaces, and opposite the lingual (or oral) surface. Called also *f. facialis dentis* and *facial surface.* <facies vestibularis dentis> A

17 **dentoalveolar syndesmosis** one of the fibrous joints by which a tooth is held in its socket. Called also *gomphosis.* <syndesmosis dentoalveolaris> B

18 **periodontal ligament** collagen fibers extending from the cementum to the alveolar bone and anchoring the teeth in their sockets. <desmodontium> B

19 ***periodontium*** the tissues that invest or help to invest and support the teeth, including the periodontal ligament, gingivae, cementum, and alveolar and supporting bone. B

20 **anterior superior alveolar arteries** *origin,* infraorbital artery; *branches,* dental and peridental branches; *distribution,* incisor and canine regions of upper jaw, maxillary sinus. Called also *anterior dental arteries.* <arteriae alveolares superiores anteriores> C

21 **dental branches of anterior superior alveolar arteries** branches that supply the incisor and canine teeth. <rami dentales arteriarum alveolarium superiorum anteriorum> C

22 **dental branches of inferior alveolar artery** branches arising from the inferior alveolar artery in the mandibular canal and supplying the inferior teeth. <rami dentales arteriae alveolaris inferioris> C

23 **dental branches of posterior superior alveolar artery** branches that supply the molar and premolar teeth. <rami dentales arteriae alveolaris superioris posterioris> C

24 **peridental branches of inferior alveolar artery** branches arising from the inferior alveolar artery in the mandibular canal and supplying the roots and pulp of the teeth. <rami peridentales arteriae alveolaris inferioris> C

25 **peridental branches of posterior superior alveolar artery** branches arising from the posterior superior alveolar artery and supplying the maxillary gingivae. <rami peridentales arteriae alveolaris superioris posterioris> C

26 **posterior superior alveolar artery** *origin,* maxillary artery; *branches,* dental and peridental branches; *distribution,* molar and premolar regions of upper jaw, maxillary sinus. Called also *posterior dental artery.* <arteria alveolaris superior posterior> C

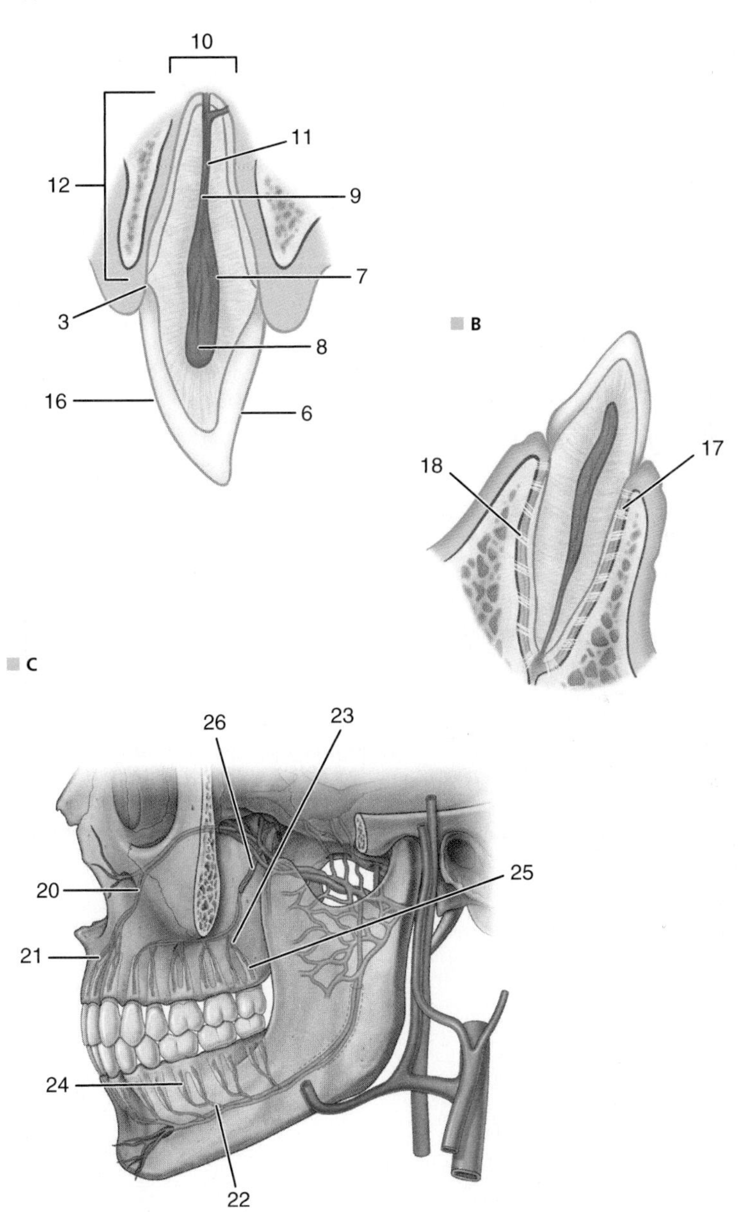
A
10
11
12
9
7
3
8
16
6
B
18
17
C
26
23
25
20
21
24
22

1 **anterior superior alveolar branches of maxillary nerve** branches from the infraorbital nerve that innervate the incisor and canine teeth of the upper jaw, help form the superior dental plexus, and give terminal twigs to the floor of the nose; *modality,* general sensory. <rami alveolares superiores anteriores nervi maxillaris> A

2 **gingival branches of mental nerve** branches that innervate the gums; *modality,* general sensory. <rami gingivales nervi mentalis> A

3 **inferior dental branches of inferior dental plexus** branches that innervate the lower teeth; *modality,* general sensory. <rami dentales inferiores plexus dentalis inferioris> A

4 **inferior dental plexus** a plexus of nerve fibers from the inferior alveolar nerve, situated around the roots of the lower teeth. <plexus dentalis inferior> A

5 ***inferior gingival branches of inferior dental plexus*** branches originating from the inferior dental plexus and innervating the gingivae of the lower jaw; *modality,* general sensory. <rami gingivales inferiores plexus dentalis inferioris> A

6 **posterior superior alveolar branches of maxillary nerve** branches that innervate the maxillary sinus, cheek, gums, and molar and premolar teeth of the upper jaw; they form part of the superior dental plexus; *modality,* general sensory. <rami alveolares superiores posteriores nervi maxillaris> A

7 ***superior alveolar nerves*** a term denoting collectively the alveolar branches arising from the maxillary and infraorbital nerves, innervating the teeth and gums of the upper jaw and the maxillary sinus, and forming the superior dental plexus. <nervi alveolares superiores> A

8 **superior dental branches of superior dental plexus** branches that innervate the teeth of the upper jaw; *modality,* general sensory. <rami dentales superiores plexus dentalis superioris> A

9 **superior dental plexus** a plexus of fibers from the superior alveolar nerves, situated around the roots of the upper teeth. <plexus dentalis superior> A

10 ***superior gingival branches of superior dental plexus*** branches arising from the superior dental plexus and innervating the gingivae of the upper jaw; *modality,* general sensory. <rami gingivales superiores plexus dentalis superioris> A

11 **oral fissure** the longitudinal opening between the lips. <rima oris> B

12 **oral region** the region of the face about the mouth. <regio oralis> B

NECK

General

13 **anterior cervical triangle** the region of the neck from the anterior border of the sternocleidomastoid muscle to the midline, subdivided into the submandibular, carotid, muscular, and submental triangles. C

14 **carotid triangle** the triangular region bounded by the posterior belly of the digastric muscle and the stylohyoid, the sternocleidomastoid muscle, and the superior belly of the omohyoid; called also *superior carotid triangle.* <trigonum caroticum> C

15 **muscular triangle** the part of the anterior cervical triangle medial to the omohyoid muscle; called also *omotracheal triangle.* <trigonum musculare> C

16 **occipital triangle** the area bounded by the sternocleidomastoid muscle anteriorly, the trapezius muscle posteriorly, and the omohyoid muscle inferiorly. C

17 **omoclavicular triangle** a deep region of the neck, corresponding to the greater supraclavicular fossa on the surface, in which the brachial plexus may be palpated, and by downward pressure the subclavian artery can be compressed against the first rib; called also *subclavian triangle.* <trigonum omoclaviculare> C

18 **posterior cervical triangle** the region of the neck posterior to the posterior border of the sternocleidomastoid muscle to the anterior border of the trapezius. Called also *nuchal region* and *posterior region of neck.* C

19 **submandibular triangle** the part of the anterior triangle bounded by the mandible, the stylohyoid muscle and posterior belly of the digastric muscle, and the anterior belly of the digastric muscle. <trigonum submandibulare> C

20 **submental triangle** the part of the anterior triangle bounded on either side by the anterior belly of the digastric muscle and below by the hyoid bone. <trigonum submentale> C

21 **neck** the part of the body connecting the head and trunk. <collum> D

Bones

22 ***cervical rib*** a supernumerary rib arising from a cervical vertebra, usually the seventh. <costa cervicalis>

23 **body of hyoid bone** the central portion of the hyoid bone to which the large and small horns are attached; called also *basihyal* and *basihyoid.* <corpus ossis hyoidei> E

24 **greater horn of hyoid bone** a bony projection passing posteriorly and superiorly from either side of the body of the hyoid bone. <cornu majus ossis hyoidei> E

25 **hyoid bone** a horseshoe-shaped bone situated at the base of the tongue, just superior to the thyroid cartilage. <os hyoideum> E

26 **lesser horn of hyoid bone** a small conical eminence projecting superiorly on either side of the hyoid bone at the angle of junction between the body and the greater horn. <cornu minus ossis hyoidei> E

27 ***retrohyoid bursa*** a bursa sometimes present behind the hyoid bone at the attachment of the sternohyoid muscle. <bursa retrohyoidea> E

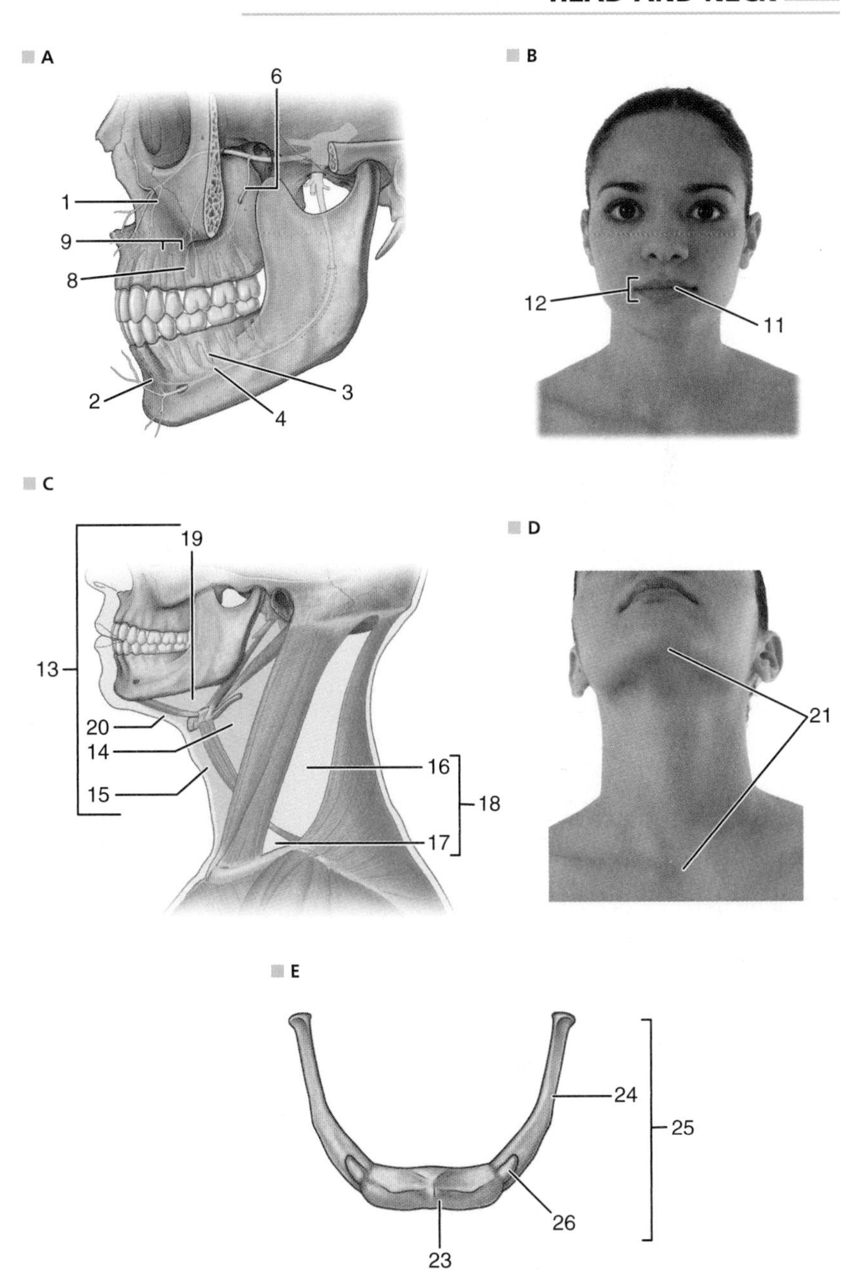
A
6
1
9
8
2
4
3
B
12
11
C
13
19
20
14
15
16
17
18
D
21
E
24
25
26
23

Fascial Layers

1 **buccopharyngeal fascia** a fibrous membrane forming the external covering of the constrictor muscles of the pharynx, and passing forward superiorly to the surface of the buccinator muscle. <fascia buccopharyngea> A

2 **carotid sheath of cervical fascia** the portion of the cervical fascia that encloses the carotid vessels and vagus nerve. <vagina carotica fasciae cervicalis> A

3 ***cervical fascia*** the fascia of the neck, consisting of a thin superficial layer, three deep layers *(superficial, pretracheal,* and *prevertebral),* and the *carotid sheath.* <fascia cervicalis> A

4 ***nuchal fascia*** the fascia on the muscles in the posterior region of the neck. <fascia nuchae> A

5 **pretracheal layer of cervical fascia** the middle layer of deep cervical fascia, which is anterior to the trachea and surrounds the thyroid gland. Called also *pretracheal fascia.* <lamina pretrachealis fasciae cervicalis> A

6 **prevertebral layer of cervical fascia** the layer of deep cervical fascia that is anterior to the vertebrae and posterior to the trachea and esophagus; it invests the scalene and levator scapulae muscles and is continuous laterally with the suprapleural membrane. Called also *prevertebral fascia.* <lamina prevertebralis fasciae cervicalis> A

7 **superficial layer of cervical fascia** the most superficial of the deep layers of cervical fascia, surrounding the neck superficial to the pretracheal layer; it invests the trapezius and sternocleidomastoid muscles and is attached posteriorly to the vertebrae. Called also *investing layer of cervical fascia.* <lamina superficialis fasciae cervicalis> A

8 **retropharyngeal space** the part of the peripharyngeal space that lies just behind the prevertebral layer of the deep cervical fascia, extending from the base of the skull to the level of the second thoracic vertebra. It is subdivided into the *retrovisceral space* and the *prevertebral space.* Called also *retroesophageal space.* <spatium retropharyngeum> B

Anterior Triangle

9 ***accessory thyroid glands*** small exclaves of the thyroid gland, found along the course of the thyroglossal duct and elsewhere; types (named for their locations) include *intrathoracic thyroid, lingual thyroid, suprahyoid thyroid,* and *retrosternal thyroid.* <glandulae thyroideae accessoriae> C

10 ***fibrous capsule of thyroid gland*** a connective tissue coat intimately adherent to the underlying gland. <capsula fibrosa glandulae thyroideae> C

11 **isthmus of thyroid gland** the band of tissue connecting the lobes of the thyroid gland. <isthmus glandulae thyroideae> C

12 **lobe of thyroid gland** either of the lobes (right or left) of the thyroid gland, located adjacent to either side of the trachea, cricoid cartilage, and thyroid cartilage. <lobus glandulae thyroideae> C

13 **lobules of thyroid gland** irregular areas on the surface of the thyroid gland produced by entrance into the gland of fibrous trabeculae from the sheath. <lobuli glandulae thyroideae> C

14 ***pyramidal lobe of thyroid gland*** an occasional third lobe that extends upward from the isthmus of the gland across the thyroid cartilage to the hyoid bone; it is the residuum of the thyroid stalk of the fetus. <lobus pyramidalis glandulae thyroideae> C

15 ***stroma of thyroid gland*** the tissue that forms the framework of the thyroid gland. <stroma glandulae thyroideae> C

16 **thyroid gland** an endocrine gland normally situated in the lower part of the front of the neck, consisting of two lobes, one on either side of the trachea and joined in front by a narrow isthmus. Called also *thyroid.* <glandula thyroidea> C

17 **inferior parathyroid gland** see *parathyroid glands.* <glandula parathyroidea inferior> D

18 **membranous wall of trachea** the posterior part of the wall of the trachea where the cartilaginous rings are deficient. <paries membranaceus tracheae> D

19 **parathyroid glands** small bodies apposed to the posterior surface of the thyroid gland, developed from the endoderm of the branchial clefts, occurring in a variable number of pairs, commonly two *(inferior* and *superior).* D

20 **superior parathyroid gland** see *parathyroid glands.* <glandula parathyroidea superior> D

21 **cervical part of esophagus** the part of the esophagus located in the cervical region, posterior to the trachea and the recurrent laryngeal nerves, anterior to the longus colli muscle and vertebral column, and medial to the lobes of the thyroid gland and the common carotid arteries. Called also *cervical esophagus.* <pars cervicalis oesophagi> E

22 **cervical part of trachea** the part of the trachea located in the cervical region, related anteriorly to the jugular venous arch, sternohyoid and sternothyroid muscles, isthmus of thyroid gland, inferior thyroid veins, thymus, thyroid ima artery, posteriorly to the esophagus and recurrent laryngeal nerves, and laterally to the lobes of the thyroid gland and common carotid arteries. <pars cervicalis tracheae> E

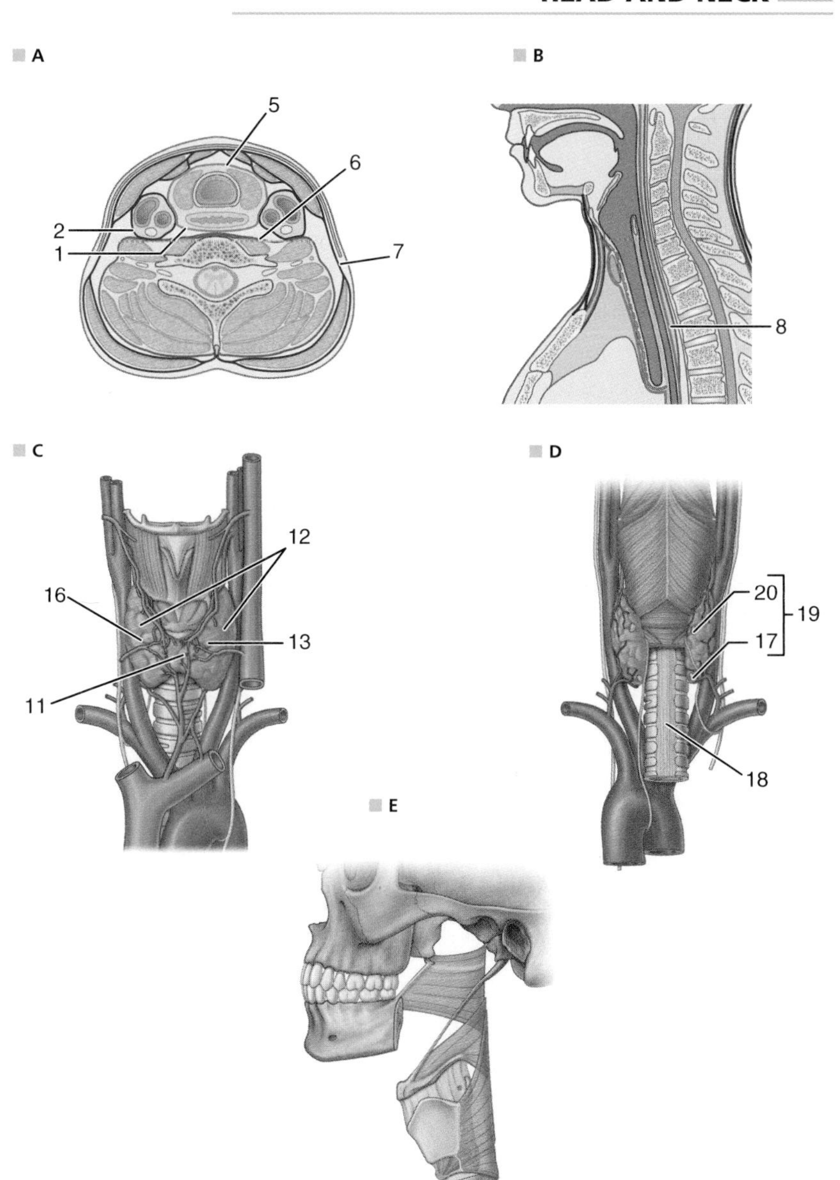
A
5
6
2
1
7
B
8
C
12
16
13
11
D
20
19
17
18
E
21
22

Muscles of Anterior Triangle of Neck

1 **sternocleidomastoid muscle** (2 heads): *origin,* STERNAL HEAD—manubrium of sternum, CLAVICULAR HEAD—superior surface of medial third of clavicle; *insertion,* mastoid process and superior nuchal line of occipital bone; *innervation,* accessory nerve and cervical plexus; *action,* flexes vertebral column, rotates head upward and to opposite side. <musculus sternocleidomastoideus> A

2 **anterior belly of digastric muscle** the shorter belly of the digastric muscle, arising from the digastric fossa on the mandible and extending posteriorly to join the posterior belly through an intermediate tendon attached to the hyoid bone. <venter anterior musculi digastrici> B

3 **digastric muscle** *origin,* ANTERIOR BELLY—digastric fossa on deep surface of inferior border of mandible near symphysis, POSTERIOR BELLY—mastoid notch of temporal bone; *insertion,* intermediate tendon on hyoid bone; *innervation,* ANTERIOR BELLY—mylohyoid, POSTERIOR BELLY—digastric branch of facial; *action,* elevates hyoid bone, lowers jaw. <musculus digastricus> B

4 **inferior belly of omohyoid muscle** a narrow band that attaches to the superior margin of the scapula. <venter inferior musculi omohyoidei> B

5 ***infrahyoid bursa*** a bursa sometimes present below the hyoid bone at the attachment of the sternohyoid muscle. <bursa infrahyoidea> B

6 ***infrahyoid muscles*** the muscles that anchor the hyoid bone to the sternum, clavicle, and scapula, including the sternohyoid, omohyoid, sternothyroid, and thyrohyoid muscles. <musculi infrahyoidei> B

7 ***levator glandulae thyroideae muscle*** an inconstant muscle originating on the isthmus or pyramid of the thyroid gland and inserting on the body of the hyoid bone. <musculus levator glandulae thyroideae> B

8 **omohyoid muscle** comprises two bellies (superior and inferior) connected by a central tendon that is bound to the clavicle by a fibrous expansion of the cervical fascia; *origin,* superior border of scapula; *insertion,* lateral border of hyoid bone; *innervation,* upper cervical through ansa cervicalis; *action,* depresses hyoid bone. <musculus omohyoideus> B

9 **posterior belly of digastric muscle** the longer belly of the digastric muscle, arising from the mastoid notch of the temporal bone and extending anteriorly to join the anterior belly through an intermediate tendon attached to the hyoid bone. <venter posterior musculi digastrici> B

10 **sternohyoid muscle** *origin,* manubrium of sternum, posterior sternoclavicular ligament, clavicle; *insertion,* body of hyoid bone; *innervation,* upper ansa cervicalis; *action,* depresses hyoid bone and larynx. <musculus sternohyoideus> B

11 **sternothyroid muscle** *origin,* manubrium of sternum; *insertion,* lamina of thyroid cartilage; *innervation,* ansa cervicalis; *action,* depresses thyroid cartilage. <musculus sternothyroideus> B

12 **stylohyoid muscle** *origin,* styloid process; *insertion,* body of hyoid bone; *innervation,* facial; *action,* draws hyoid and tongue superiorly and posteriorly. <musculus stylohyoideus> B

13 **superior belly of omohyoid muscle** a part that ascends and attaches to the hyoid bone. <venter superior musculi omohyoidei> B

14 ***suprahyoid muscles*** the muscles that attach the hyoid bone to the skull, including the digastric, stylohyoid, mylohyoid, and geniohyoid muscles. <musculi suprahyoidei> B

15 **thyrohyoid muscle** *origin,* lamina of thyroid cartilage; *insertion,* greater horn of hyoid bone; *innervation,* first cervical; *action,* raises and changes form of larynx. <musculus thyrohyoideus> B

16 **longus capitis muscle** *origin,* transverse processes of third to sixth cervical vertebrae; *insertion,* basilar part of occipital bone; *innervation,* branches from first, second, and third cervical; *action,* flexes head. <musculus longus capitis> C

17 **longus colli muscle** long muscle of neck: *origin,* SUPERIOR OBLIQUE PORTION—transverse processes of third to fifth cervical vertebrae; INFERIOR OBLIQUE PORTION—bodies of first to third thoracic vertebrae; VERTICAL PORTION—bodies of three upper thoracic and three lower cervical vertebrae; *insertion,* SUPERIOR OBLIQUE PORTION—tubercle of anterior arch of atlas; INFERIOR OBLIQUE PORTION—transverse processes of fifth and sixth cervical vertebrae; VERTICAL PORTION—bodies of second to fourth cervical vertebrae; *innervation,* anterior cervical; *action,* flexes and supports cervical vertebrae. <musculus longus colli> C

18 **rectus capitis anterior muscle** *origin,* lateral mass of atlas; *insertion,* basilar part of occipital bone; *innervation,* first and second cervical; *action,* flexes, supports head. <musculus rectus capitis anterior> C

19 **rectus capitis lateralis muscle** *origin,* upper surface of transverse process of atlas; *insertion,* jugular process of occipital bone; *innervation,* first and second cervical; *action,* flexes, supports head. <musculus rectus capitis lateralis> C

20 ***smallest scalene muscle*** a band occasionally found between the anterior and middle scalene muscles; *origin* transverse process of seventh cervical vertebra; *insertion* first rib, suprapleural membrane; *innervation* seventh cervical; *action* raises first rib, flexes and rotates cervical vertebrae, supports suprapleural membrane. <musculus scalenus minimus> C

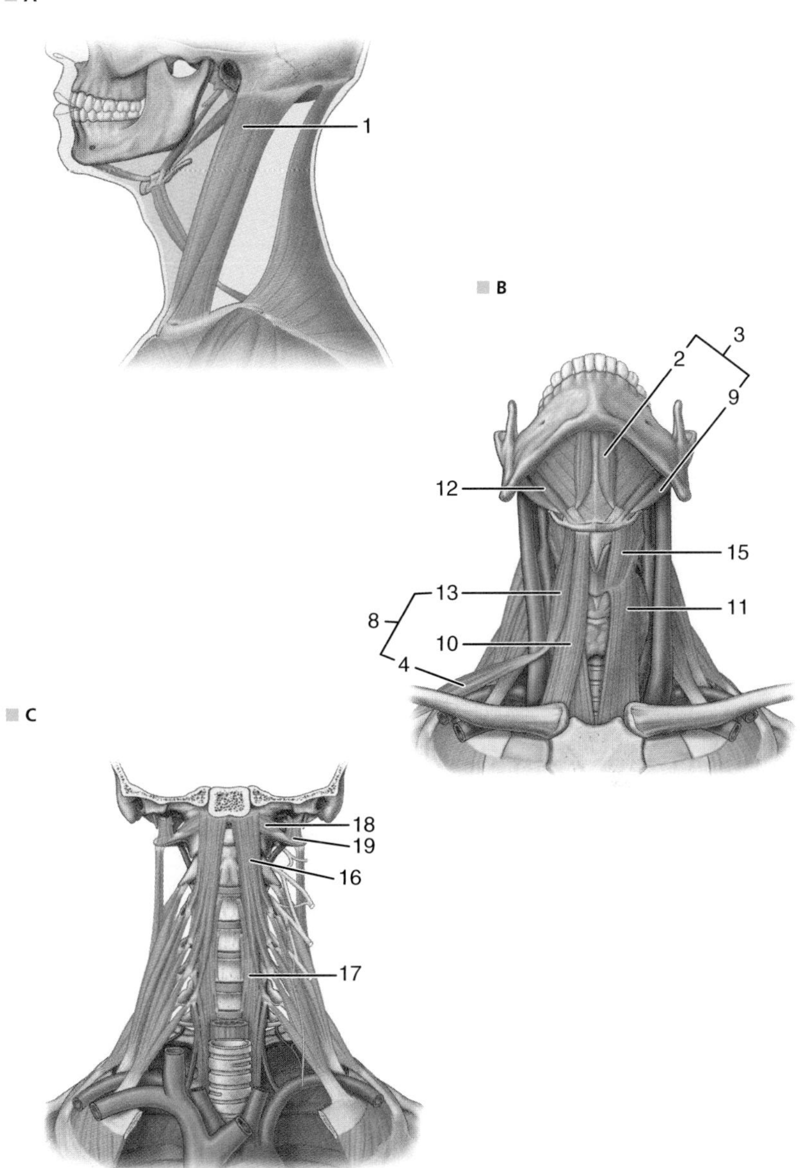
A
1
B
3
2
9
12
15
13
11
8
10
4
C
18
19
16
17

1 **buccopharyngeal part of superior constrictor muscle of pharynx** the part of the superior constrictor muscle arising from the pterygomandibular raphe; called also *buccopharyngeal muscle.* <pars buccopharyngea musculi constrictoris pharyngis superioris> A

2 **ceratopharyngeal part of middle constrictor muscle of pharynx** the part of the middle constrictor muscle arising from the greater horn of the hyoid bone; called also *ceratopharyngeal muscle.* <pars ceratopharyngea musculi constrictoris pharyngis medii> A

3 **chondropharyngeal part of middle constrictor muscle of pharynx** the part of the middle constrictor muscle arising from the lesser horn of the hyoid bone; called also *chondropharyngeal muscle.* <pars chondropharyngea musculi constrictoris pharyngis medii> A

4 **cricopharyngeal part of inferior constrictor muscle of pharynx** the part of the inferior constrictor muscle arising from the cricoid cartilage; called also *cricopharyngeal muscle.* <pars cricopharyngea musculi constrictoris pharyngis inferioris> A

5 ***trachealis muscle*** a transverse layer of smooth fibers in the dorsal portion of the trachea; *insertion,* tracheal cartilages; *innervation,* autonomic fibers; *action,* lessens caliber of trachea. <musculus trachealis> A

6 **cricothyroid muscle** one of the intrinsic muscles of the larynx; *origin,* front and side of cricoid cartilage; *insertion,* lamina of thyroid cartilage; *innervation,* external branch of superior laryngeal; *action,* tenses vocal folds. <musculus cricothyroideus> B

7 **aryepiglottic part of oblique arytenoid muscle** an inconstant fascicle of the oblique arytenoid muscle, originating from the apex of the arytenoid cartilage and inserting on the lateral margin of the epiglottis. Called also *aryepiglottic muscle.* <pars aryepiglottica musculi arytenoidei obliqui> C

8 **chondroglossus muscle** *origin,* medial side and base of lesser horn of hyoid bone; *insertion,* substance of tongue; *innervation,* hypoglossal; *action,* depresses, retracts tongue. <musculus chondroglossus> D

Nerves of Anterior Triangle of Neck

9 **internal carotid plexus** a nerve plexus on the internal carotid artery, formed by the internal carotid nerve, which supplies sympathetic fibers to the branches of the internal carotid artery, to the tympanic plexus, to the nerves in the cavernous sinus, and, directly or indirectly, to the cranial parasympathetic ganglia through which they pass. <plexus caroticus internus> E

10 **branch of glossopharyngeal nerve to carotid sinus** a branch that supplies the pressoreceptors and chemoreceptors of the carotid sinus and carotid body with visceral afferent fibers. <ramus sinus carotici nervi glossopharyngei> F

11 **glossopharyngeal nerve** ninth cranial nerve; *origin,* several rootlets from lateral side of upper part of medulla oblongata, between the olive and the inferior cerebellar peduncle; *branches,* tympanic nerve, pharyngeal, stylopharyngeal, carotid, tonsillar, and lingual rami, and rami communicating with the auricular and meningeal rami of the vagus nerve, with the chorda tympani, and with the auriculotemporal nerve; *distribution,* it has two enlargements (superior and inferior ganglia) and supplies the tongue, pharynx, and parotid gland-see individual branches; *modality,* motor, parasympathetic, and general, special, and visceral sensory. <nervus glossopharyngeus> F

12 ***stylopharyngeal branch of glossopharyngeal nerve*** a branch that supplies the stylopharyngeus muscle; *modality,* motor. <ramus musculi stylopharyngei nervi glossopharyngei> F

13 ***fold of superior laryngeal nerve*** a fold of mucous membrane in the larynx, overlying the superior laryngeal nerve. <plica nervi laryngei superior> G

14 **superior laryngeal nerve** *origin,* inferior ganglion of vagus nerve; *branches,* external, internal, and communicating branches; *distribution,* inferior constrictor of pharynx and cricothyroid muscles, and mucosa of epiglottis, base of tongue, and larynx; *modality,* motor, general sensory, visceral afferent, and parasympathetic. <nervus laryngeus superior> G

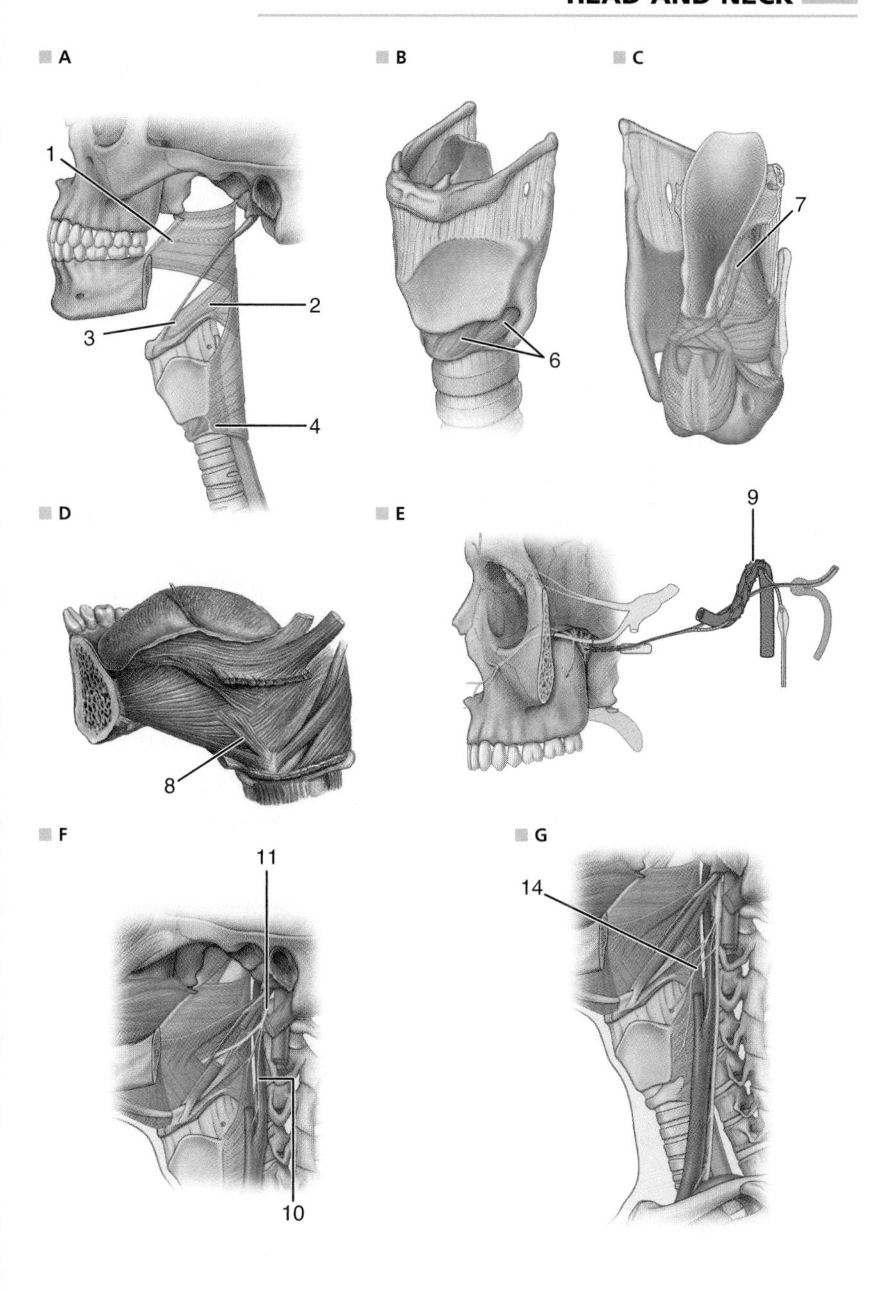
A
1
2
3
4
B
6
C
7
D
8
E
9
F
11
10
G
14

1 **ansa cervicalis** a nerve loop in the neck that supplies the infrahyoid muscles and that presents an anterior (superior) root, which connects with the hypoglossal nerve (and actually consists of fibers of the second or first cervical nerve), and an inferior root, which connects with the second and third cervical nerves. Called also *loop of hypoglossal nerve.* A

2 **inferior root of ansa cervicalis** a strand of filaments connecting the ansa cervicalis with branches of the second and third cervical nerves; called also *r. posterior ansae cervicalis* and *posterior root of ansa cervicalis.* <radix inferior ansae cervicalis> A

3 **superior root of ansa cervicalis** fibers of the first or second cervical nerve, descending in company with the hypoglossal nerve, connecting it with the ansa cervicalis and helping supply the infrahyoid muscles. Called also *anterior root of ansa cervicalis.* <radix superior ansae cervicalis> A

4 ***thyrohyoid branch of ansa cervicalis*** a branch from the superior root of the ansa cervicalis, innervating the thyrohyoid muscle; *modality,* motor. <ramus thyrohyoideus ansae cervicalis> A

5 ***pharyngeal branches of recurrent laryngeal nerve*** small nerve branches that innervate the inferior constrictor muscle of the pharynx. <rami pharyngei nervi laryngei recurrentis> C

6 **recurrent laryngeal nerve** *origin,* vagus nerve (chiefly the cranial part of the accessory nerve): *branches,* inferior laryngeal nerve and tracheal, esophageal, pharyngeal, and inferior cardiac branches; *modality,* parasympathetic, visceral afferent, and motor. <nervus laryngeus recurrens> C

7 ***tracheal branches of recurrent laryngeal nerve*** branches distributed to the tracheal mucosa; *modality,* general sensory. <rami tracheales nervi laryngei recurrentis> A

8 **anterior branch of great auricular nerve** a branch distributed to the skin of the face over the parotid gland; *modality,* general sensory. <ramus anterior nervi auricularis magni> B

9 **inferior branches of transverse nerve of neck** the more inferior of the branches that arise from the transverse cervical nerve near the anterior border of the sternocleidomastoid muscle, innervating skin and subcutaneous tissue in the anterior cervical region; *modality,* general sensory. <rami inferiores nervi transversi colli> B

10 ***cervical plexus*** a nerve plexus formed by the anterior rami of the upper four cervical nerves; arranged as an irregular series of loops, it gives off superficial branches (lesser occipital, greater auricular, transverse cervical, and supraclavicular nerves), and deep branches (phrenic, accessory phrenic, ansa cervicalis, and muscular nerves). <plexus cervicalis> E

11 **ansa subclavia** nerve filaments that pass anterior and posterior to the subclavian artery to form a loop interconnecting the middle and inferior cervical ganglia; called also *Vieussens' annulus.* D

12 **anterior rami of cervical nerves** eight nerve branches of which the upper four form the cervical plexus and the lower four form most of the brachial plexus. <rami anteriores nervorum cervicalium> D, F

13 ***common carotid plexus*** a nerve plexus on the common carotid artery, formed by branches of the internal and external carotid plexuses and the cervical sympathetic ganglia. <plexus caroticus communis> F

14 **external carotid nerves** *origin,* superior cervical ganglion; *distribution,* cranial blood vessels and glands via the external carotid plexus; *modality,* sympathetic. <nervi carotici externi> F

15 **external carotid plexus** a nerve plexus located around the external carotid artery, formed by the external carotid nerves from the superior cervical ganglion, and supplying sympathetic fibers which accompany the branches of the external carotid artery. <plexus caroticus externus> F

16 **inferior cervical cardiac nerve** *origin,* cervicothoracic ganglion; *distribution,* heart via cardiac plexus; *modality,* sympathetic (accelerator) and visceral afferent (chiefly pain). <nervus cardiacus cervicalis inferior> F

17 **inferior cervical ganglion** an inconstant ganglion formed by fusion of the lower two cervical ganglia in instances where the first thoracic ganglion remains separate. <ganglion cervicale inferioris> F

18 **internal carotid nerve** *origin,* superior cervical ganglion; *distribution,* cranial blood vessels and glands via internal carotid plexus; *modality,* sympathetic. <nervus caroticus internus> F

19 ***jugular nerve*** a branch of the superior cervical ganglion that communicates with the vagus and glossopharyngeal nerves. <nervus jugularis> F

20 ***laryngopharyngeal branches of superior cervical ganglion*** branches from the superior cervical ganglion to the larynx and walls of the pharynx; *modality,* sympathetic. <rami laryngopharyngei ganglii cervicalis superioris> F

21 **middle cervical cardiac nerve** *origin,* middle cervical ganglion; *distribution,* heart; *modality,* sympathetic (accelerator) and visceral afferent (chiefly pain). <nervus cardiacus cervicalis medius> F

22 **middle cervical ganglion** a variable ganglion, often fused with the vertebral ganglion, on the sympathetic trunk at about the level of the cricoid cartilage; its postganglionic fibers are distributed mainly to the heart, cervical region, and upper limb. <ganglion cervicale medium> F

23 **superior cervical cardiac nerve** *origin,* superior cervical ganglion; *distribution,* heart; *modality,* sympathetic (accelerator). <nervus cardiacus cervicalis superior> F

A

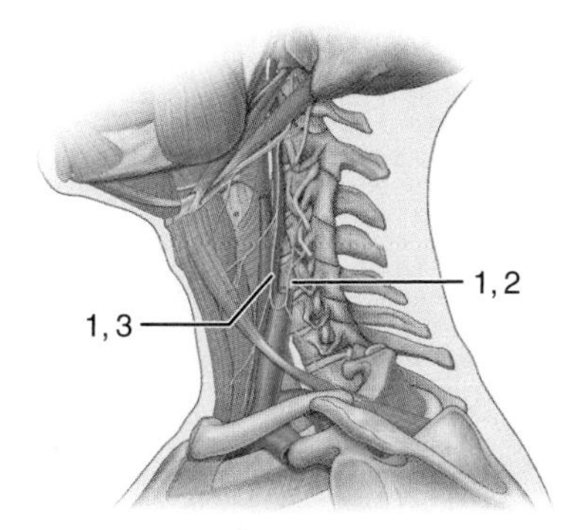
1, 2
1, 3

B

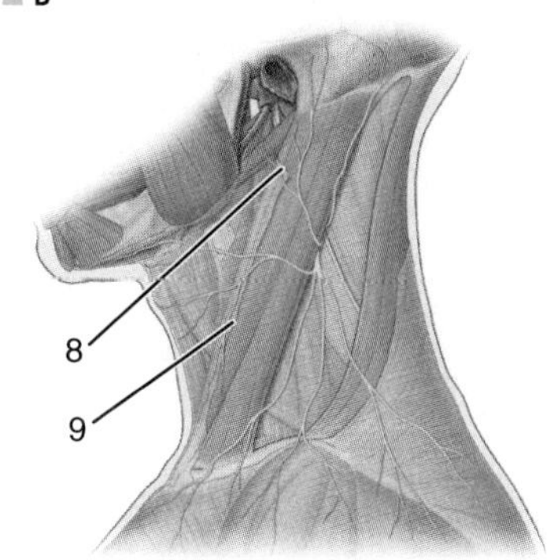
8
9

C

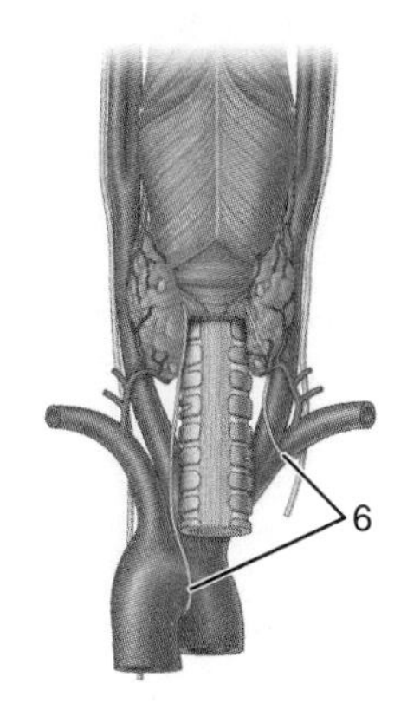
6

D

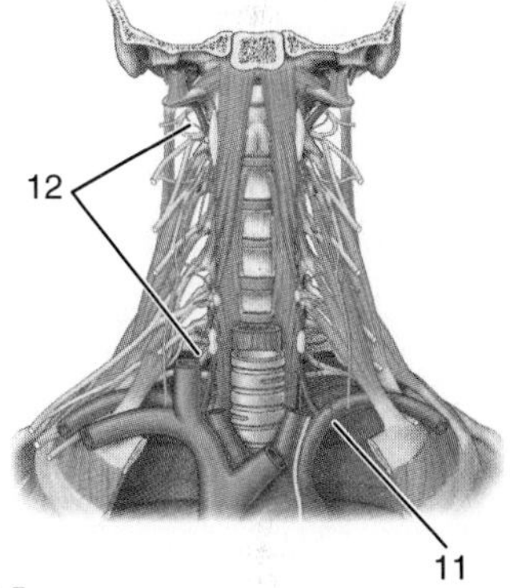
12
11

E

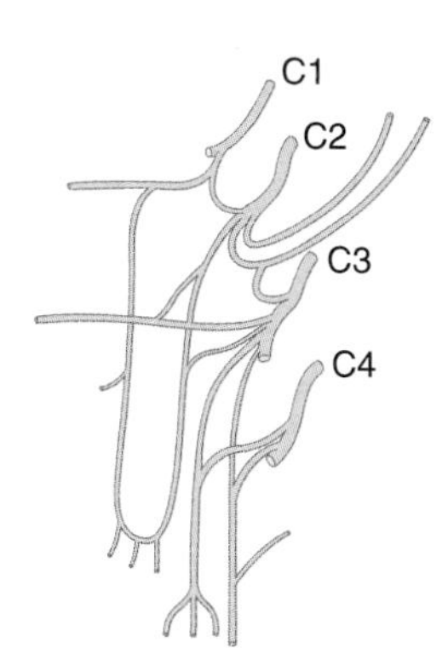
C1
C2
C3
C4

F

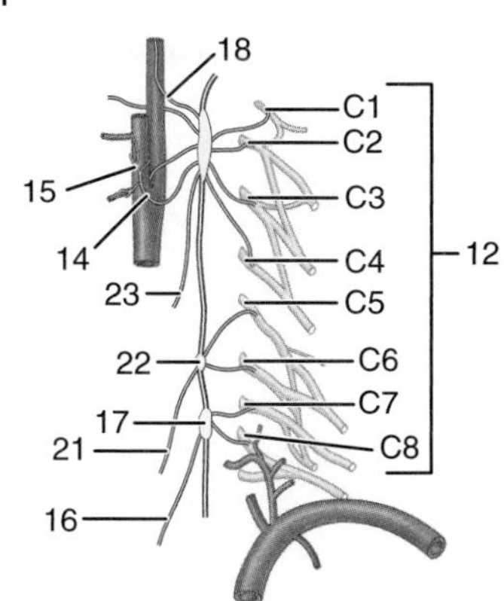
18
C1
C2
15
C3
14
C4
12
23
C5
22
C6
C7
17
21
C8
16

1 **superior cervical ganglion** the uppermost ganglion on the sympathetic trunk, lying behind the internal carotid artery and in front of the second and third cervical vertebrae; it gives rise to postganglionic fibers to the heart via cervical cardiac nerves, to the pharyngeal plexus and thence to the larynx and pharynx, and to the head via the external and internal carotid plexuses. <ganglion cervicale superius> A

2 ***vertebral ganglion*** a small ganglion almost always present between the middle and inferior sympathetic ganglia, usually anterior to the vertebral artery; it contributes to the ansa subclavia and sends postganglionic fibers to the vertebral nerve and plexus and to the brachial plexus. <ganglion vertebrale> A

Vessels of Anterior Triangle of Neck

3 **anterior jugular vein** a vein that arises under the chin, passes down the neck, and opens into the external jugular or the subclavian vein or into the jugular venous arch. <vena jugularis anterior> B

4 **external jugular vein** the vein that begins in the parotid gland behind the angle of the jaw by union of the retromandibular and the posterior auricular vein, passes down the neck, and opens into the subclavian, the internal jugular, or the brachiocephalic vein. <vena jugularis externa> B

5 **jugular venous arch** a transverse connecting trunk between the anterior jugular veins of either side. <arcus venosus jugularis> B

6 **retromandibular vein** the vein that is formed in the upper part of the parotid gland behind the neck of the mandible by union of the maxillary and superficial temporal veins; it passes downward through the gland, communicates with the facial vein, and emerging from the gland joins with the posterior auricular vein to form the external jugular vein. <vena retromandibularis> B

7 ***stylomastoid vein*** a vein following the stylomastoid artery and emptying into the retromandibular vein. <vena stylomastoidea> B

8 **anterior branch of the superior thyroid artery** a branch principally supplying the anterior surface of the thyroid gland, and anastomosing with the artery of the opposite side. <ramus glandularis anterior arteriae thyroideae superioris> C

9 **ascending pharyngeal artery** *origin,* external carotid; *branches,* posterior meningeal, pharyngeal, and inferior tympanic; *distribution,* pharynx, soft palate, ear, meninges. <arteria pharyngea ascendens> C

10 ***auricular branch of occipital artery*** an inconstant branch of the occipital artery that helps supply the medial aspect of the pinna. <ramus auricularis arteriae occipitalis> C

11 ***caroticotympanic arteries*** branches of the petrous part of the internal carotid artery that supply the tympanic cavity. <arteriae caroticotympanicae> C

12 **carotid bifurcation** the site where the common carotid artery divides into the external carotid artery and internal carotid artery, usually marked by a dilatation, the carotid sinus. <bifurcatio carotidis> C

13 ***carotid body*** a small neurovascular structure lying in the bifurcation of the right and left carotid arteries that functions as an arterial chemoreceptor. <glomus caroticum> C

14 **carotid sinus** the dilated portion of the internal carotid artery, situated above the division of the common carotid artery into its two main branches, or sometimes on the terminal portion of the common carotid artery, containing in its wall pressoreceptors that are stimulated by changes in blood pressure. <sinus caroticus> C

15 ***carotid siphon*** the innermost section of the petrosal part of the internal carotid artery just before the artery enters the cranial cavity. <siphon caroticum> C

16 **cervical part of internal carotid artery** an unbranched part located in the carotid triangle of the neck. <pars cervicalis arteriae carotidis internae> C

17 **common carotid artery** *origin,* brachiocephalic trunk (right), aortic arch (left); it divides the external and internal carotid arteries. <arteria carotis communis> C

18 ***cricothyroid branch of superior thyroid artery*** a vessel running medially over the cricothyroid muscle, toward the cricothyroid ligament, and anastomosing with its fellow of the opposite side. Called also *cricothyroid artery.* <ramus cricothyroideus arteriae thyroideae superioris> C

19 **external carotid artery** *origin,* common carotid; *branches,* superior thyroid, ascending pharyngeal, lingual, facial, sternocleidomastoid, occipital, posterior auricular, superficial temporal, maxillary; *distribution,* neck, face, skull. <arteria carotis externa> C

20 **inferior bulb of jugular vein** a dilatation of the internal jugular vein just before it joins the brachiocephalic vein. <bulbus inferior venae jugularis> C

21 ***inferior tympanic artery*** *origin,* ascending pharyngeal; *branches,* none; *distribution,* tympanic cavity. <arteria tympanica inferior> C

22 ***infrahyoid branch of superior thyroid artery*** a vessel running along the inferior border of the hyoid bone, supplying the infrahyoid region, and anastomosing with its fellow of the opposite side. <ramus infrahyoideus arteriae thyroideae superioris> C

A

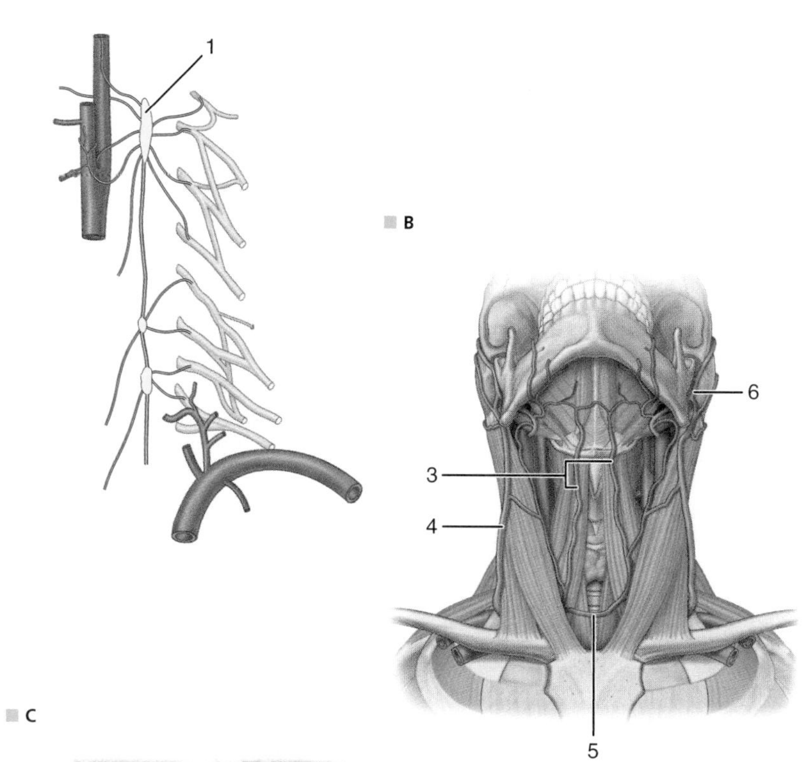

C

16
9
19
14
12
8
17
20

1 **internal carotid artery** divided into four parts: *cervical, petrous, cavernous,* and *cerebral; origin,* common carotid; *branches,* numerous, including (petrous part) caroticotympanic arteries; (cavernous part) tentorial basal, tentorial marginal, meningeal, and cavernous branches, and inferior hypophysial artery; (cerebral part) ophthalmic, superior hypophysial, posterior communicating, anterior choroidal, anterior cerebral, and middle cerebral arteries; *distribution,* middle ear, brain, pituitary gland, orbit, choroid plexus. <arteria carotis interna> A

2 ***internal carotid venous plexus*** a venous plexus around the petrosal portion of the internal carotid artery, through which the cavernous sinus communicates with the internal jugular vein. <plexus venosus caroticus internus> A

3 **internal jugular vein** the vein that begins as the superior bulb in the jugular fossa, draining much of the head and neck; it descends with first the internal carotid and then the common carotid artery in the neck, and joins with the subclavian vein to form the brachiocephalic vein. <vena jugularis interna> A

4 **lateral branch of superior thyroid artery** a branch distributed to the lateral surface of the thyroid gland. <ramus glandularis lateralis arteriae thyroideae superioris> A

5 **lingual artery** *origin,* external carotid; *branches,* suprahyoid, sublingual, dorsal lingual, deep lingual; *distribution,* tongue, sublingual gland, tonsil, epiglottis. <arteria lingualis> A

6 ***linguofacial trunk*** the common trunk by which the facial and lingual arteries often arise from the external carotid artery. <truncus linguofacialis> A

7 **occipital artery** *origin,* external carotid; *branches,* auricular, meningeal, mastoid, descending, occipital, and sternocleidomastoid rami; *distribution,* muscles of neck and scalp, meninges, mastoid cells. <arteria occipitalis> A

8 ***pharyngeal branches of ascending pharyngeal artery*** irregular vessels supplying the pharynx. <rami pharyngeales arteriae pharyngeae ascendentis> A

9 ***pharyngeal plexus*** a venous plexus posterolateral to the pharynx, formed by the pharyngeal veins, communicating with the pterygoid venous plexus, and draining into the internal jugular vein. <plexus pharyngeus> A

10 ***pharyngeal veins*** veins that drain the pharyngeal plexus and empty into the internal jugular vein. <venae pharyngeae> A

11 ***sternocleidomastoid branch of superior thyroid artery*** a branch that arises from the superior thyroid artery, but sometimes directly from the external carotid artery, passing across the carotid sheath to supply the middle portion of the sternocleidomastoid muscle. <ramus sternocleidomastoideus arteriae thyroideae superioris> A

12 ***sternocleidomastoid branches of occipital artery*** branches of the occipital artery, usually an upper and a lower, that supply the sternocleidomastoid and adjacent muscles. <rami sternocleidomastoidei arteriae occipitalis> A

13 ***sternocleidomastoid vein*** a vein that follows the course of the homonymous artery and opens into the internal jugular vein. <vena sternocleidomastoidea> A

14 ***superior bulb of jugular vein*** a dilatation at the beginning of the internal jugular vein. <bulbus superior venae jugularis> A

15 **superior laryngeal artery** *origin,* superior thyroid artery; *branches,* none; *distribution,* larynx. <arteria laryngea superior> A

16 **superior thyroid artery** *origin,* external carotid artery; *branches,* hyoid, sternocleidomastoid, superior laryngeal, cricothyroid, muscular, and anterior, posterior, and lateral glandular branches; *distribution,* thyroid gland and adjacent structures. <arteria thyroidea superior> A

17 ***suprahyoid branch of lingual artery*** a branch passing along the upper border of the hyoid bone, supplying suprahyoid muscles and anastomosing with its fellow of the other side. <ramus suprahyoideus arteriae lingualis> A

18 **inferior thyroid vein** either of two veins, left and right, that drain the thyroid plexus into the left and right brachiocephalic veins; occasionally they may unite into a common trunk to empty, usually, into the left brachiocephalic vein. <vena thyroidea inferioris> B

19 **middle thyroid veins** veins that drain blood from the thyroid gland into the internal jugular vein. <venae thyroideae mediae> B

20 **superior thyroid vein** a vein arising from the upper part of the thyroid gland on either side, opening into the internal jugular vein, occasionally in common with the facial vein. <vena thyroidea superior> B

21 ***thyroid ima artery*** *origin,* arch of aorta, brachiocephalic trunk, or right common carotid, internal mammary, subclavian, or inferior thyroid arteries; *branches,* none; *distribution,* thyroid gland. <arteria thyroidea ima> B

22 ***esophageal branches of inferior thyroid artery*** branches that supply the esophagus. <rami oesophageales arteriae thyroideae inferioris> C

23 **glandular branches of inferior thyroid artery** branches distributed to the inferior surface of the thyroid gland. <rami glandulares arteriae thyroideae inferioris> C

24 **inferior thyroid artery** *origin,* thyrocervical trunk; *branches,* pharyngeal, esophageal, and tracheal branches, inferior laryngeal and ascending cervical arteries; *distribution,* thyroid gland and adjacent structures. <arteria thyroidea inferior> C

25 ***muscular branches of vertebral artery*** branches of the transverse part of the vertebral artery that supply the deep muscles of the neck and anastomose with the descending branch of the occipital artery and the deep cervical artery. <rami musculares arteriae vertebralis> C

A

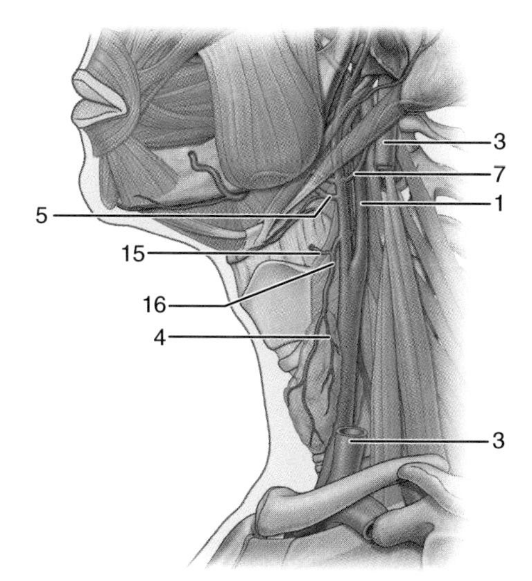

B

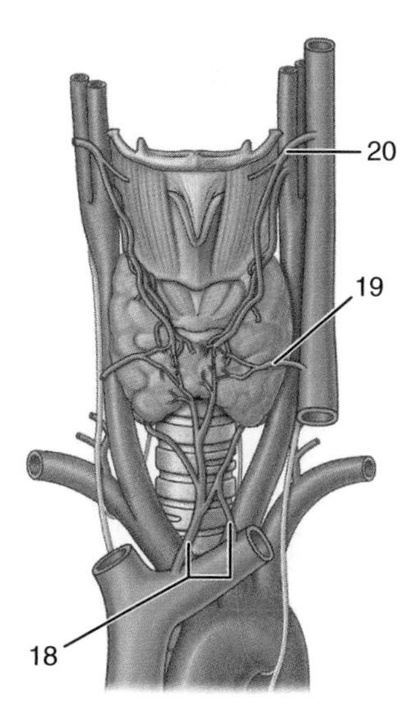

C

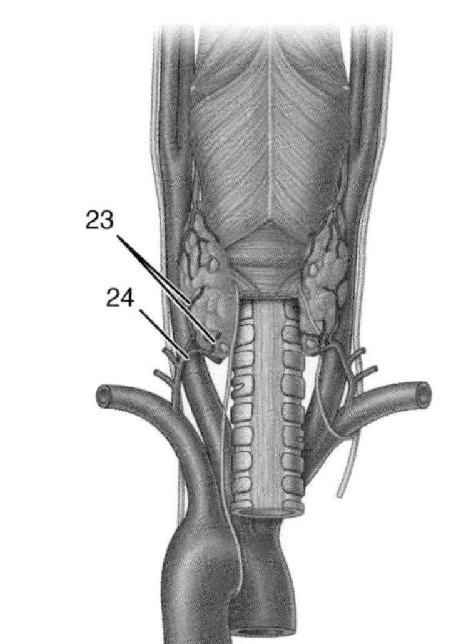

1 ***pharyngeal branches of inferior thyroid artery*** vessels that supply the pharynx. <rami pharyngeales arteriae thyroideae inferioris> A

2 **posterior glandular branch of the superior thyroid artery** a branch distributed mainly to the medial and lateral surfaces of the thyroid gland; it anastomoses with the inferior thyroid artery. <ramus glandularis posterior arteriae thyroideae superioris> A

3 ***tracheal branches of inferior thyroid artery*** vessels supplying the trachea. <rami tracheales arteriae thyroideae inferioris> A

4 **ascending cervical artery** *origin,* inferior thyroid artery or directly from thyrocervical trunk; *branches,* spinal branches; *distribution,* muscles of neck, vertebrae, vertebral canal. <arteria cervicalis ascendens> B

5 **costocervical trunk** an artery that arises from the back of the subclavian artery, arches backward, and at the neck of the first rib divides into the deep cervical and highest intercostal arteries, thus supplying blood to the structures of the first two intercostal spaces, the vertebral column, the muscles of the back, and the deep neck muscles. <truncus costocervicalis> B

6 **deep cervical artery** *origin,* costocervical trunk; *branches,* none; *distribution,* deep neck muscles. <arteria cervicalis profunda> B

7 ***deep cervical vein*** a vein that arises from a plexus in the suboccipital triangle, follows the deep cervical artery down the neck, and empties into the vertebral or the brachiocephalic vein. <vena cervicalis profunda> B

8 **prevertebral part of vertebral artery** the first part of the artery, from its branching off from the subclavian artery to the point at which it enters the tranverse process of the sixth cervical vertebra. <pars prevertebralis arteriae vertebralis> B

9 ***spinal branches of ascending cervical artery*** branches that help supply the vertebral canal. <rami spinales arteriae cervicalis ascendentis> B

10 **thyrocervical trunk** a short artery that arises from the convex side of the subclavian artery just medial to the anterior scalene muscle and at once divides into the inferior thyroid, transverse cervical, and suprascapular arteries, supplying thyroid, neck, and scapular regions. <truncus thyrocervicalis> B

11 **vertebral artery** divided into four parts: the *first* or *prevertebral part,* the *second* or *cervical part,* the *third* or *atlantal part,* and the *fourth* or *intracranial part; origin,* subclavian artery; *branches,* (cervical part) spinal and muscular branches; (intracranial part) anterior spinal artery and posterior inferior cerebellar artery and its branches, meningeal branches, and lateral and medial medullary branches; *distribution,* muscles of neck, vertebrae, spinal cord, cerebellum, interior of cerebrum. <arteria vertebralis> B

12 **inferior laryngeal artery** *origin,* inferior thyroid artery; *branches,* none; *distribution,* larynx, trachea, esophagus. <arteria laryngea inferior> C

13 ***inferior laryngeal vein*** a vein draining blood from the larynx into the inferior thyroid vein. <vena laryngea inferior> D

14 **superior laryngeal vein** a vein that drains blood from the larynx into the superior thyroid vein. <vena laryngea superior> D

15 **ascending palatine artery** *origin,* facial artery; *branches,* none; *distribution,* soft palate, wall of pharynx, tonsil, auditory tube. <arteria palatina ascendens> E

 A

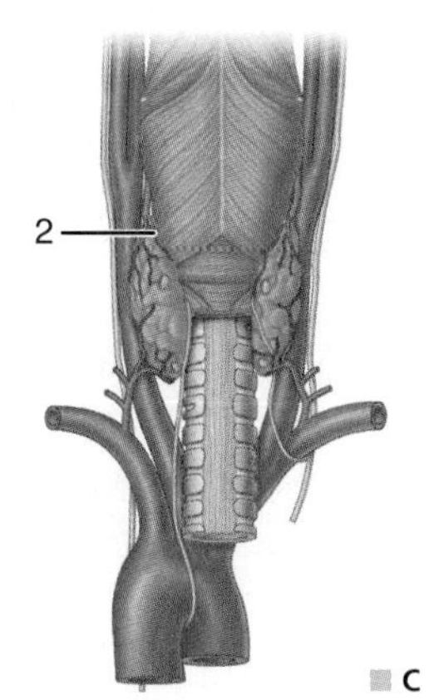

 B

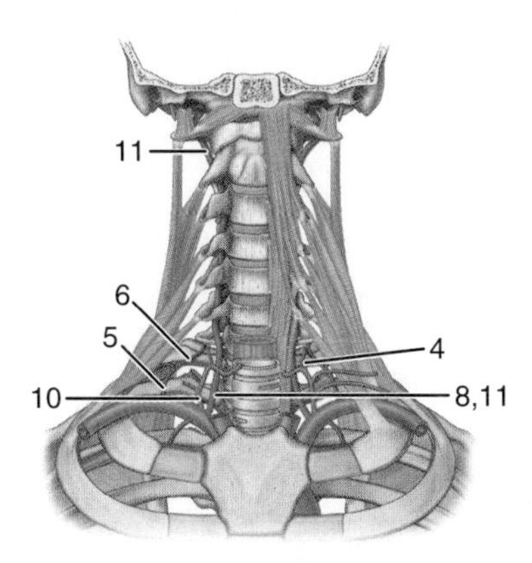

C

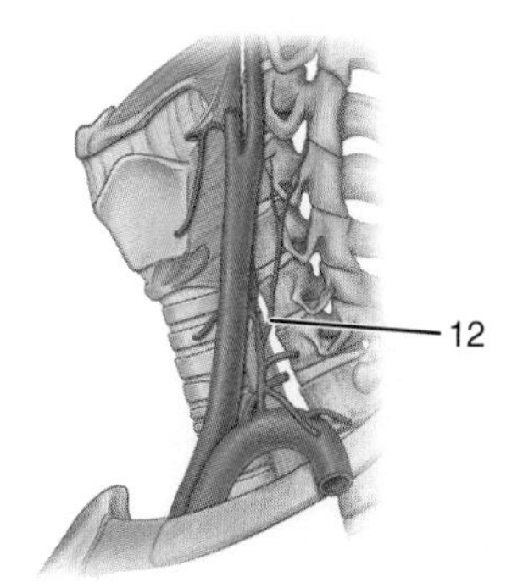

D

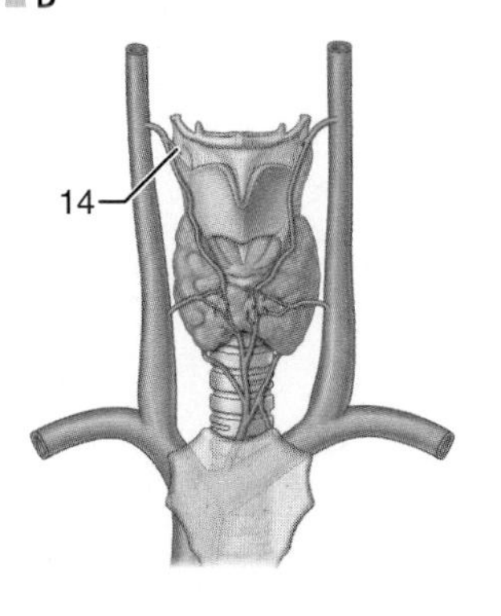

E

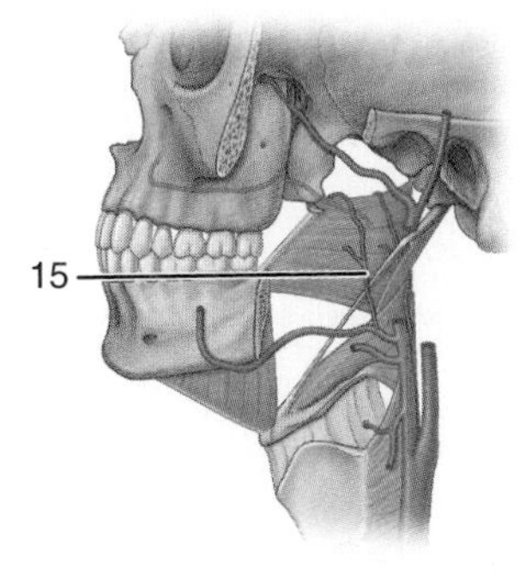

Posterior Triangle

1 **anterior scalene muscle** *origin,* transverse processes of third to sixth cervical vertebrae; *insertion,* scalene tubercle of first rib; *innervation,* fourth to sixth cervical; *action,* raises first rib, flexes cervical vertebrae forward and laterally and rotates to opposite side. <musculus scalenus anterior> A

2 **middle scalene muscle** *origin,* transverse processes of second to seventh cervical vertebrae and often atlas; *insertion,* upper surface of first rib; *innervation,* third to eighth cervical; *action,* raises first rib, flexes cervical vertebrae forward and laterally and rotates to opposite side. <musculus scalenus medius> A

3 **posterior scalene muscle** *origin,* posterior tubercles of transverse processes of fourth to sixth cervical vertebrae; *insertion,* second rib; *innervation,* sixth to eighth cervical; *action,* raises second rib, flexes cervical vertebrae laterally. <musculus scalenus posterior> A

Nerves of Posterior Triangle of Neck

4 **supraclavicular part of brachial plexus** the part of the brachial plexus lying in the cervical region above the level of the clavicle, in which arise the dorsal scapular, long thoracic, and suprascapular nerves, and the nerve to the subclavius muscle. <pars supraclavicularis plexus brachialis> B

5 **great auricular nerve** *origin,* cervical plexus-C2-C3; *branches,* anterior and posterior branches; *distribution,* skin over parotid gland and mastoid process, and both surfaces of auricle; *modality,* general sensory. <nervus auricularis magnus> C

6 **lesser occipital nerve** *origin,* superficial cervical plexus (C2-C3); *distribution,* ascends behind the auricle and supplies some of the skin on the side of the head and on the cranial surface of the auricle; *modality,* general sensory. <nervus occipitalis minor> C

7 ***muscular branches of accessory nerve*** the branches that supply the sternocleidomastoid and trapezius muscles. <rami musculares> C

8 ***posterior branch of great auricular nerve*** a branch, formed by division of the great auricular nerve, that innervates the skin over the mastoid process and the back of the external ear; *modality,* general sensory. <ramus posterior nervi auricularis magni> C

9 **superior branches of transverse cervical nerve** the upper of the branches that arise from the transverse cervical nerve near the anterior border of the sternocleidomastoid muscle, innervating skin and subcutaneous tissue in the anterior cervical region; *modality,* general sensory. <rami superiores nervi transversi colli> C

10 **transverse cervical nerve** *origin,* cervical plexus (C2-C3); *branches,* superior and inferior rami; *distribution,* skin on side and front of neck; *modality,* general sensory. Called also *transverse nerve of neck.* <nervus transversus colli> C

Vessels of Posterior Triangle of Neck

11 **transverse cervical artery** *origin,* subclavian artery; *branches,* deep and superficial branches; *distribution,* root of neck, muscles of scapula. <arteria transversa cervicis> D

12 ***transverse cervical veins*** veins that follow the transverse cervical artery and open into the subclavian vein. <venae transversae cervicis> D

Neck Lymphatics

13 **cervical part of thoracic duct** the portion of the thoracic duct in the neck. <pars cervicalis ductus thoracici> E, F

14 **jugular trunk** either of the two lymphatic trunks (right and left) that drain the deep cervical lymph nodes; on the right side they drain into the right lymphatic duct or subclavian vein, and on the left side into the thoracic duct or subclavian vein. <truncus jugularis> E, F

15 **subclavian trunk** either of two lymphatic trunks (right and left) that drain the axillary lymph nodes; the one on the right drains into the right lymphatic duct or subclavian vein, and the one on the left drains into the thoracic duct or the subclavian vein. <truncus subclavius> E, F

A

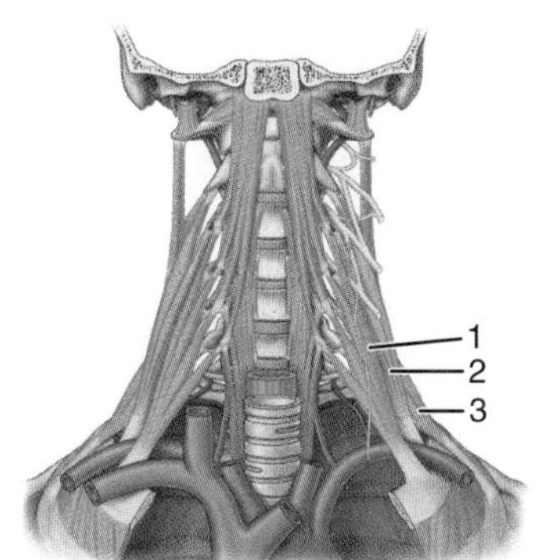

B

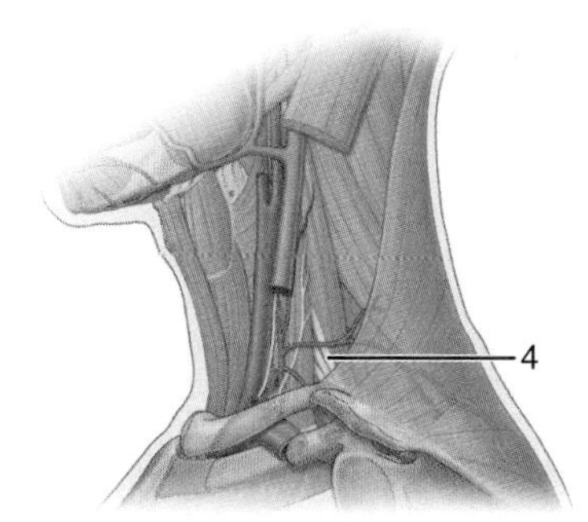

C

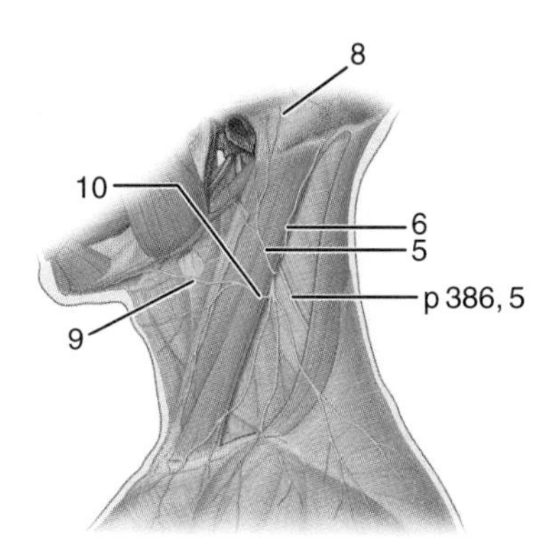

D

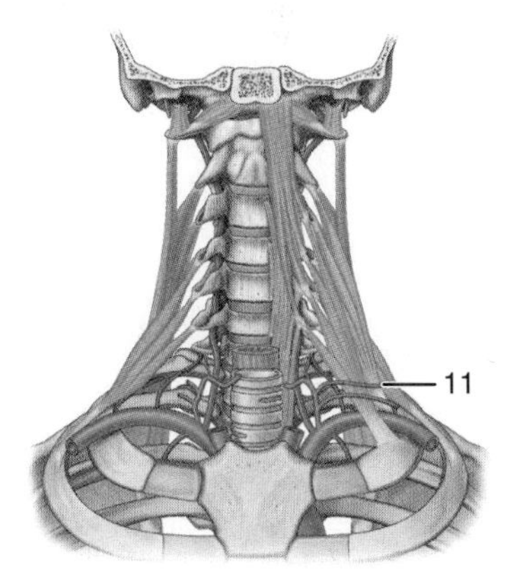

E

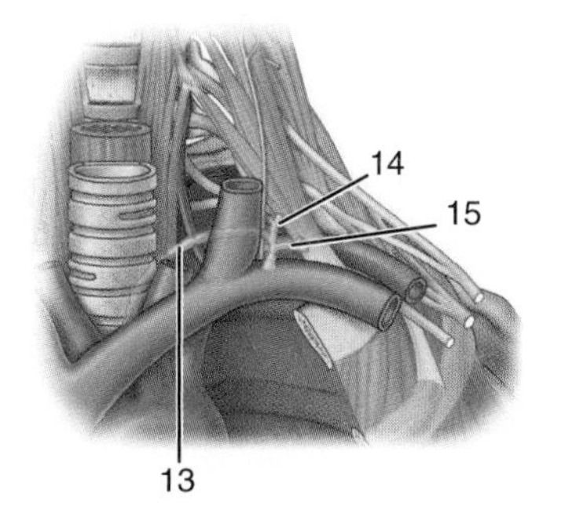

F

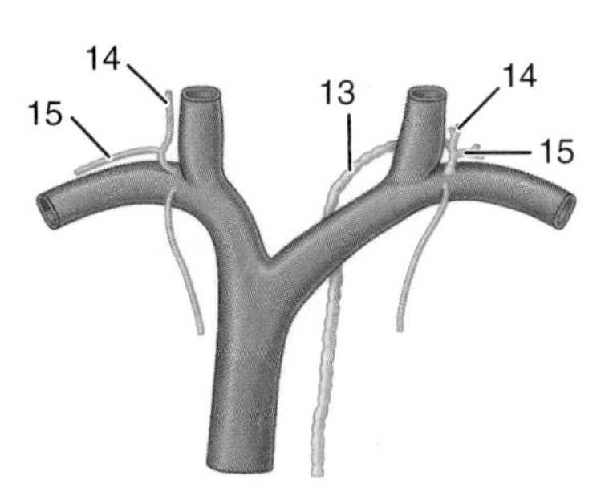

1 ***accessory nodes*** a chain of lymph nodes of the inferior deep lateral cervical group that follow the spinal accessory nerve and receive lymph from the occipital, postauricular, and suprascapular nodes and from the scalp, neck, and shoulder. <nodi lymphoidei accessorii> A

2 ***anterior cervical lymph nodes*** a group of lymph nodes anterior to the larynx and trachea, consisting of superficial vessels on the anterior jugular vein and deep vessels on the middle cricothyroid ligament as well as anterior to the trachea. <nodi lymphoidei cervicales anteriores> A

3 ***deep anterior cervical lymph nodes*** a group of numerous large lymph nodes that form a chain along the internal jugular vein, extending from the base of the skull to the root of the neck, situated near the pharynx, esophagus, and trachea; they receive lymph from both superficial and deep structures. <nodi lymphoidei cervicales anteriores profundi> A

4 **deep cervical lymph nodes** a group of lymph nodes adjacent to the carotid sheath, partly deep to the sternocleidomastoid muscle and extending into the subclavian triangle. They receive lymph from the back of the scalp and neck, the tongue, the superficial pectoral region, and part of the arm and drain into the jugular trunk. <nodi lymphoidei cervicales laterales profundi inferior inferiores> A

5 ***deep parotid lymph nodes*** lymph nodes on the lateral wall of the pharynx lying deep to or embedded in the deep substance of the parotid gland, through which lymph drains from the external acoustic meatus, auditory tube, tympanic cavity, soft palate, and posterior nasal cavity. <nodi lymphoidei parotidei profundi> A

6 ***infra-auricular deep parotid lymph nodes*** deep parotid lymph nodes situated below the ear. <nodi lymphoidei parotidei profundi infraauriculares> A

7 ***infrahyoid lymph nodes*** lymph nodes lying beneath the deep cervical fascia anterior to the thyrohyoid membrane that receive lymph from the anterior cervical nodes and epiglottic region and drain into the deep cervical nodes. <nodi lymphoidei infrahyoidei> A

8 ***intraglandular deep parotid lymph nodes*** deep parotid lymph nodes situated within the substance of the parotid gland. <nodi lymphoidei parotidei profundi intraglandulares> A

9 **jugulodigastric lymph node** one of the deep lateral cervical lymph nodes lying on the internal jugular vein at the level of the greater horn of the hyoid bone, i.e., just below the posterior belly of the digastric muscle. <nodus lymphoideus jugulodigastricus> A

10 **jugulo-omohyoid lymph node** one of the deep lateral cervical lymph nodes lying on the internal jugular vein just above the tendon of the omohyoid muscle. <nodus lymphoideus juguloomohyoideus> A

11 ***lingual lymph nodes*** deep cervical lymph nodes receiving afferent vessels from the tongue. <nodi lymphoidei linguales> A

12 ***malar lymph node*** one of a variable number of facial lymph nodes situated in the region of the zygomaticus minor muscle. <nodus lymphoideus malaris> A

13 ***mandibular lymph node*** one of a variable number of facial lymph nodes situated near the angle of the mandible, into which lymph from some of the superficial tissues of the head and neck is drained. <nodus lymphoideus mandibularis> A

14 **mastoid lymph nodes** lymph nodes, two or three on each side, that are superficial to the mastoid attachment of the sternocleidomastoid muscle and deep to the posterior auricular muscle; they drain the nasal fossae and paranasal sinuses, hard and soft palate, middle ear, and nasopharynx and oropharynx. Called also *retroauricular lymph nodes.* <nodi lymphoidei mastoidei> A

15 **occipital lymph nodes** several small nodes near the occipital insertion of the semispinalis capitis muscle. <nodi lymphoidei occipitales> A

16 ***preauricular deep parotid lymph nodes*** deep parotid lymph nodes situated in front of the ear. <nodi lymphoidei parotidei profundi preauriculares> A

17 ***prelaryngeal lymph nodes*** deep anterior cervical lymph nodes situated in front of the larynx that help drain the thyroid gland. <nodi lymphoidei prelaryngei> A

18 ***pretracheal lymph nodess*** deep anterior cervical lymph nodes situated in front of the trachea near the inferior thyroid veins. <nodi lymphoidei pretracheales> A

19 ***retropharyngeal lymph nodes*** deep lateral cervical lymph nodes, one median and two lateral groups, situated behind the upper part of the pharynx, especially concerned with drainage of the nasal fossae, paranasal sinuses, hard and soft palates, middle ear, nasopharynx, and oropharynx. <nodi lymphoidei retropharyngeales> A

20 **submandibular lymph nodes** the three to six nodes alongside the submandibular gland, through which lymph drains from the adjacent skin and mucous membrane. <nodi lymphoidei submandibulares> A

21 **submental lymph nodes** nodes under the chin into which the lymph from some of the superficial tissues of the head and neck is drained. <nodi lymphoidei submentales> A

22 ***superficial cervical lymph nodes*** lymph nodes along the external jugular vein; they receive afferent vessels from the auricle and parotid region. <nodi lymphoidei cervicales superficiales> A

A

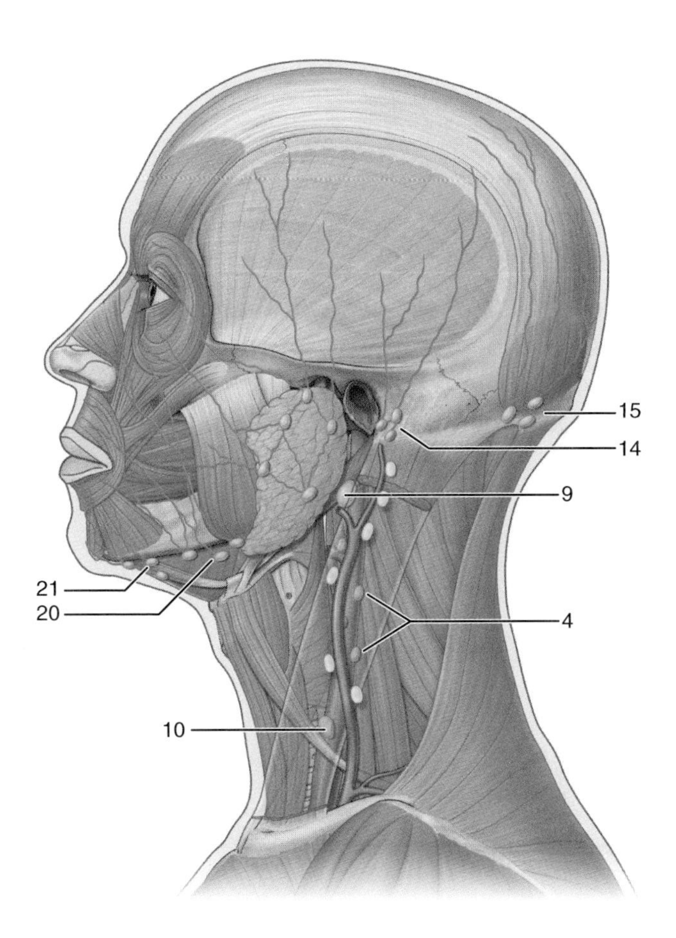

1 ***superficial lateral cervical lymph nodes*** lymph nodes situated along the external jugular vein that send efferent vessels to the deep lateral cervical lymph nodes. <nodi lymphoidei cervicales laterales superficiales> A

2 **superficial parotid lymph nodes** lymph nodes lying in the subcutaneous tissue of the parotid gland directly in front of the tragus. <nodi lymphoidei parotidei superficiales> A

3 ***superior deep cervical lymph nodes*** a group of lymph nodes adjacent to the carotid sheath deep to the sternocleidomastoid muscle; they receive lymph from a number of structures of the head and neck and drain into the inferior deep cervical nodes or the jugular trunk. <nodi lymphoidei cervicales laterales profundi superiores> A

4 ***supraclavicular lymph nodes*** the deep lateral cervical lymph nodes situated inferior to the omohyoid muscle, extending into the omoclavicular portion of the posterior triangle of the neck. <nodi lymphoidei supraclaviculares> A

5 ***thyroid lymph nodes*** deep anterior cervical lymph nodes situated around the thyroid gland. <nodi lymphoidei thyroidei> A

A

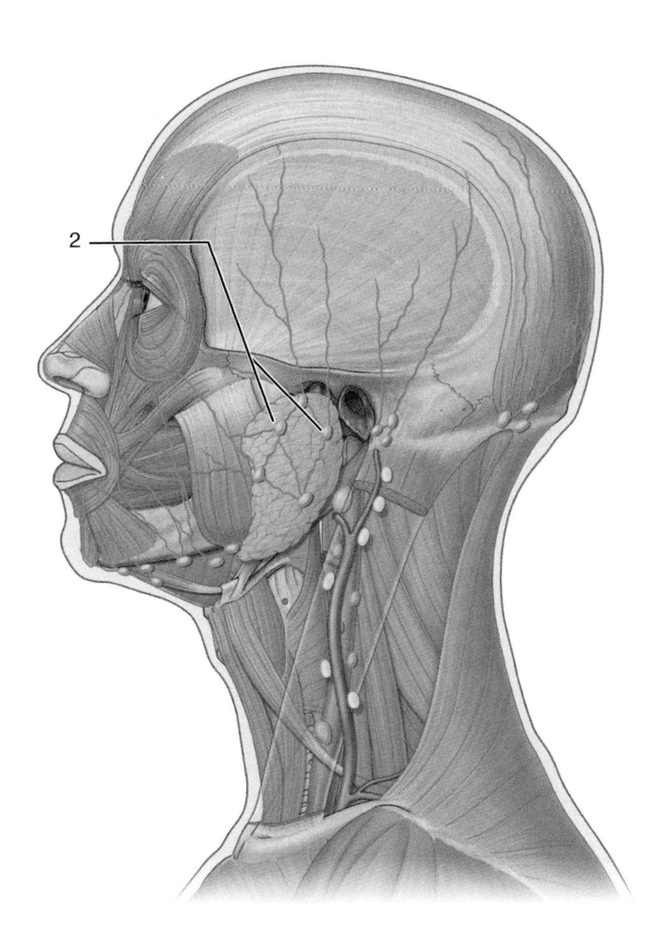

INDEX

E

I

Q